PHYSIOLOGY
AND
BIOCHEMISTRY
OF
EXTREMOPHILES

PHYSIOLOGY AND BIOCHEMISTRY OF EXTREMOPHILES

EDITED BY
CHARLES GERDAY
AND
NICOLAS GLANSDORFF

ASM PRESS

Washington, D.C.

Copyright © 2007   ASM Press
American Society for Microbiology
1752 N St., N.W.
Washington, DC 20036-2904

Library of Congress Cataloging-in-Publication Data

Physiology and biochemistry of extremophiles / edited by C. Gerday and N. Glansdorff.
  p. ; cm.
  Includes bibliographical references and index.
  ISBN-10: 1-55581-422-0 (hardcover: alk. paper)
  ISBN-13: 978-1-55581-422-9 (hardcover: alk. paper)
  Extreme environments—Microbiology.   I. Gerday, Charles.   II. Glansdorff, Nicolas.
  III. American Society for Microbiology.
  [DNLM: 1. Archaea—physiology. 2. Adaptation, Physiological. 3. Bacterial
  Proteins—physiology. 4. Biodiversity. 5. Environmental Microbiology.   QW 52 P5786 2007]

QR100.9.P59 2007
579'.1758—dc22

2006102166

All rights reserved
Printed in the United States of America

10 9 8 7 6 5 4 3 2 1

Address editorial correspondence to ASM Press, 1752 N St., N.W., Washington, DC 20036-2904, U.S.A.

Send orders to: ASM Press, P.O. Box 605, Herndon, VA 20172, U.S.A.
Phone: 800-546-2416; 703-661-1593
Fax: 703-661-1501
E-mail: books@asmusa.org
Online: http://estore.asm.org

Cover figure: The inset illustration on the front cover is the effluent channel of Twin Butte Vista Spring (Yellowstone National Park, Wyoming, USA), a typical alkaline siliceous spring (background) with a cyanobacterial mat in the foreground (courtesy of Debra Naylor).

CONTENTS

Contributors • vii
Foreword • *Koki Horikoshi* xi
Preface • xv

I. Introduction

1. Extremophiles and the Origin of Life • 3
Sara Islas, Ana María Velasco, Arturo Becerra, Luis Delaye, and Antonio Lazcano

II. Thermophiles

2. Thermal Environments and Biodiversity • 13
Elizabeth A. Burgess, Isaac D. Wagner, and Juergen Wiegel

3. Functional Genomics in Thermophilic Microorganisms • 30
Frank T. Robb and Deborah T. Newby

4. How Nucleic Acids Cope with High Temperature • 39
Henri Grosjean and Tairo Oshima

5. How Thermophiles Cope with Thermolabile Metabolites • 57
Jan Massant

6. Temperature-Dependent Molecular Adaptation Features in Proteins • 75
Sandeep Kumar, Sunil Arya, and Ruth Nussinov

7. The Physiological Role, Biosynthesis, and Mode of Action of Compatible Solutes from (Hyper)Thermophiles • 86
Helena Santos, Pedro Lamosa, Tiago Q. Faria, Nuno Borges, and Clélia Neves

8. Membrane Adaptations of (Hyper)Thermophiles to High Temperatures • 104
Arnold J. M. Driessen and Sonja V. Albers

III. Psychrophiles

9. Ecology and Biodiversity of Cold-Adapted Microorganisms • 119
Don A. Cowan, Ana Casanueva, and William Stafford

10. Life in Ice Formations at Very Cold Temperatures • 133
Jody W. Deming

11. Lake Vostok and Subglacial Lakes of Antarctica: Do They Host Life? • 145
Guido di Prisco

12. Psychrophiles: Membrane Adaptations • 155
Nicholas J. Russell

13. Cold-Adapted Enzymes • 165
Tony Collins, Salvino D'Amico, Jean-Claude Marx, Georges Feller, and Charles Gerday

14. The Cold-Shock Response • 180
Masayori Inouye and Sangita Phadtare

15. Perception and Transduction of Low Temperature in Bacteria • 194
S. Shivaji, M. D. Kiran, and S. Chintalapati

16. An Interplay between Metabolic and Physicochemical Constraints: Lessons from the Psychrophilic Prokaryote Genomes • 208
Antoine Danchin

IV. Halophiles

17. Biodiversity in Highly Saline Environments • 223
Aharon Oren

18. Response to Osmotic Stress in a Haloarchaeal Genome: a Role for General Stress Proteins and Global Regulatory Mechanisms • 232
Guadalupe Juez, David Fenosa, Aitor Gonzaga, Elena Soria, and Francisco J. M. Mojica

19. Molecular Adaptation to High Salt • 240
Frederic Vellieux, Dominique Madern, Giuseppe Zaccai, and Christine Ebel

V. Acidophiles

20. Physiology and Ecology of Acidophilic Microorganisms • 257
D. Barrie Johnson

21. Acidophiles: Mechanisms To Tolerate Metal and Acid Toxicity • 271
Sylvia Franke and Christopher Rensing

22. Genomics of Acidophiles • 279
A. Angelov and W. Liebl

VI. Alkaliphiles

23. Environmental and Taxonomic Biodiversities of Gram-Positive Alkaliphiles • 295
Isao Yumoto

24. Bioenergetic Adaptations That Support Alkaliphily • 311
Terry Ann Krulwich, David B. Hicks, Talia Swartz, and Masahiro Ito

VII. Piezophiles

25. Microbial Adaptation to High Pressure • 333
Douglas H. Bartlett, Federico M. Lauro, and Emiley A. Eloe

VIII. Exobiology

26. Astrobiology and the Search for Life in the Universe • 351
Giles M. Marion and Dirk Schulze-Makuch

IX. Biotechnology

27. Extremophiles, a Unique Resource of Biocatalysts for Industrial Biotechnology • 361
Garabed Antranikian and Ksenia Egorova

X. Lessons from Extremophiles

28. Lessons from Extremophiles: Early Evolution and Border Conditions of Life • 409
Ying Xu and Nicolas Glansdorff

Index • 423

CONTRIBUTORS

Sonja V. Albers
Molecular Microbiology, Groningen Biomolecular Sciences and Biotechnology Institute, University of Groningen, Haren, The Netherlands

Garabed Antranikian
Institute of Technical Microbiology, Hamburg University of Technology, Kasernenstrasse 12, D-21073 Hamburg, Germany

Sunil Arya
Department of Biological Sciences and Bio-Engineering, Indian Institute of Technology—Kanpur, Kanpur, Uttar Pradesh 208016, India

Arturo Becerra
Facultad de Ciencias, UNAM, Apdo. Postal 70-407, Cd. Universitaria, 04510 México D.F., Mexico

Nuno Borges
Instituto de Tecnologia Química e Biológica, Universidade Nova de Lisboa, Rua da Quinta Grande 6, Apartado 127, 2780-156 Oeiras, Portugal

Elizabeth A. Burgess
Savannah River Ecology Laboratory, Drawer E, Aiken, South Carolina 29803

Tony Collins
Laboratory of Biochemistry, Institute of Biochemistry B6, University of Liege, Sart-Tilman, B-4000 Liege, Belgium

Salvino D'Amico
Laboratory of Biochemistry, Institute of Biochemistry B6, University of Liege, Sart-Tilman, B-4000 Liege, Belgium

Antoine Danchin
Genetics of Bacterial Genomes URA 2171 CNRS, Institut Pasteur, 28, rue du Docteur Roux, 75724 Paris Cedex 15, France

Luis Delaye
Facultad de Ciencias, UNAM, Apdo. Postal 70-407, Cd. Universitaria, 04510 México D.F., Mexico

Guido di Prisco
Institute of Protein Biochemistry (IBP), National Research Council (CNR), Via Pietro Castellino 111, I-80131 Naples, Italy

Arnold J. M. Driessen
Molecular Microbiology, Groningen Biomolecular Sciences and Biotechnology, Institute and Materials Science Center Plus, University of Groningen, Haren, The Netherlands

Ksenia Egorova
Institute of Technical Microbiology, Hamburg University of Technology, Kasernenstrasse 12, D-21073 Hamburg, Germany

Tiago Q. Faria
Instituto de Tecnologia Química e Biológica, Universidade Nova de Lisboa, Rua da Quinta Grande 6, Apartado 127, 2780-156 Oeiras, Portugal

Georges Feller
Laboratory of Biochemistry, Institute of Biochemistry B6, University of Liege, Sart-Tilman, B-4000 Liege, Belgium

David Fenosa
División de Microbiología, Campus de San Juan, Universidad Miguel Hernández, San Juan de Alicante, 03550 Alicante, Spain

Sylvia Franke
Department of Soil, Water, and Environmental Science, University of Arizona, Shantz Building 38, Room 429, Tucson, Arizona 85721

Charles Gerday
Laboratory of Biochemistry, Institute of Biochemistry B6, University of Liege, Sart-Tilman, B-4000 Liege, Belgium

Nicolas Glansdorff
J.M. Wiame Research Institute for Microbiology, Vrije Universiteit Brussel, B-1070 Brussels, Belgium

Aitor Gonzaga
División de Microbiología, Campus de San Juan, Universidad Miguel Hernández, San Juan de Alicante, 03550 Alicante, Spain

Henri Grosjean
CNRS, Laboratoire d'Enzymologie et Biochimie Structurales, 1 Avenue de la Terrasse, Gif-sur-Yvette, F-91198, France

David B. Hicks
Department of Pharmacology and Biological Chemistry, Mount Sinai School of Medicine, New York, New York 10029

Koki Horikoshi
Japan Agency for Marine-Earth Science and Technology, Yokosuka, Kanagawa 237-0061, Japan

Masayori Inouye
Department of Biochemistry, Robert Wood Johnson Medical School, 675 Hoes Lane, Piscataway, New Jersey 08854

Sara Islas
Facultad de Ciencias, UNAM, Apdo. Postal 70-407, Cd. Universitaria, 04510 México D.F., Mexico

Masahiro Ito
Graduate School of Life Sciences, Toyo University, Oura-gun, Gunma 374-0193, Japan

D. Barrie Johnson
School of Biological Sciences, University of Wales, Bangor LL57 2UW, United Kingdom

Guadalupe Juez
División de Microbiología, Campus de San Juan, Universidad Miguel Hernández, San Juan de Alicante, 03550 Alicante, Spain

Terry Ann Krulwich
Department of Pharmacology and Biological Chemistry, Box 1603, Mount Sinai School of Medicine, 1 Gustave L. Levy Place, New York, New York 10029

Sandeep Kumar
Department of Biology, Zanvyl Krieger School of Arts and Sciences, Johns Hopkins University, Baltimore, Maryland 21218-2685

Pedro Lamosa
Instituto de Tecnologia Química e Biológica, Universidade Nova de Lisboa, Rua da Quinta Grande 6, Apartado 127, 2780-156 Oeiras, Portugal

Antonio Lazcano
Facultad de Ciencias, UNAM, Apdo. Postal 70-407, Cd. Universitaria, 04510 México D.F., Mexico

Giles M. Marion
Division of Earth and Ecosystem Sciences, Desert Research Institute, 2215 Raggio Parkway, Reno, Nevada 89512

Jean-Claude Marx
Laboratory of Biochemistry, Institute of Biochemistry B6, University of Liege, Sart-Tilman, B-4000 Liege, Belgium

Jan Massant
Erfelijkheidsleer en Microbiologie (MICR), Vrije Universiteit Brussel, Pleinlaan 2, B-1050 Brussels, Belgium

Francisco J. M. Mojica
División de Microbiología, Campus de San Vicente, Universidad de Alicante, San Vicente del Raspeig, 03080 Alicante, Spain

Clélia Neves
Instituto de Tecnologia Química e Biológica, Universidade Nova de Lisboa, Rua da Quinta Grande 6, Apartado 127, 2780-156 Oeiras, Portugal

Deborah T. Newby
Idaho National Laboratory, PO Box 1625, Idaho Falls, Idaho 83415

Ruth Nussinov
SAIC-Intramural Research Program, Center for Cancer Research Nanobiology Program, National Cancer Institute-Frederick, Frederick, Maryland 21702-1201

Tairo Oshima
Institute of Environmental Microbiology, Kyowa Kako Co., Ltd., 2-15-5 Tadao, Machida, Tokyo 194-0035, Japan

Sangita Phadtare
Department of Biochemistry, Robert Wood Johnson Medical School, 675 Hoes Lane, Piscataway, New Jersey 08854

Christopher Rensing
Department of Soil, Water, and Environmental Science, University of Arizona, Shantz Building 38, Room 429, Tucson, Arizona 85721

Frank T. Robb
Center of Marine Biotechnology, 701 E. Pratt Street, Baltimore, Maryland 21202

Helena Santos
Instituto de Tecnologia Química e Biológica, Universidade Nova de Lisboa, Rua da Quinta Grande 6, Apartado 127, 2780-156 Oeiras, Portugal

Dirk Schulze-Makuch
Department of Geology, Washington State University, Pullman, Washington 99164

Elena Soria
División de Microbiología, Campus de San Vicente, Universidad de Alicante, San Vicente del Raspeig, 03080 Alicante, Spain

Talia Swartz
Department of Pharmacology and Biological Chemistry, Mount Sinai School of Medicine, New York, New York 10029

Ana María Velasco
Facultad de Ciencias, UNAM, Apdo. Postal 70-407, Cd. Universitaria, 04510 México D.F., Mexico

Isaac D. Wagner
Department of Microbiology, The University of Georgia, Athens, Georgia 30602-2605

Jürgen Wiegel
Department of Microbiology, The University of Georgia, Athens, Georgia 30602-2605

Ying Xu
Marine Sciences Research Center, Stony Brook University, Stony Brook, New York 11794

Isao Yumoto
National Institute of Advanced Industrial Science and Technology (AIST), Research Institute of Genome-Based Biofactory, Sapporo 062-8517, Japan

FOREWORD

EARLY HISTORY OF EXTREMOPHILES

In June 1965, Thomas Brock, a microbiologist, discovered in the thermal vents of Yellowstone National Park a new form of bacteria, *Thermus aquaticus*, that can survive at near-boiling temperatures. At that time, the upper temperature for life was thought to be 73°C, but he found that one particular spring, Octopus Spring, had large amounts of pink, filamentous bacteria at temperatures of 82 to 88°C. Here were organisms living at temperatures above "the upper temperature for life." He isolated and collected many microbes from this geothermal area. It is worth mentioning that *T. aquaticus* strain YT-1 was the first to be used as source of Taq polymerase. Later on, his group showed that *T. aquaticus* was widespread in hot-water environments (Brock, 1997). His findings paved the way for a new thermophilic microbiology: taxonomy, physiology, enzymology, molecular biology, genetics, etc. By now many microbiologists have focused on how life adapted to hot environments.

In 1968, the author of this foreword was visiting Florence, Italy, looking at the Renaissance buildings, which are so different from Japanese architectural works of art. About 500 years ago, no Japanese could have imagined this Renaissance culture. Then suddenly a voice whispered in his ears, "There might be a whole new world of microorganisms in yet-unexplored environments. Could there be an entirely unknown domain of microorganisms at alkaline pH?" Upon his return to Japan, he prepared an alkaline medium containing 1% sodium carbonate, distributed small amounts of soil collected from various areas around the Institute of Physical and Chemical Research (RIKEN), Wako, Japan, into 30 test tubes, and incubated them overnight at 37°C. To his surprise, various microorganisms flourished in all the test tubes. This was for him the first encounter with such bacteria. Since then, he has isolated a great number of similar microorganisms and purified many of their enzymes (Horikoshi, 2006). He named these microorganisms that grow well in alkaline environments "alkaliphiles" and conducted systematic microbial physiological studies of them. It was a great surprise to see that these microorganisms, many of which are completely different from those previously reported, were widely distributed throughout the globe (even at the deepest point of the Pacific Ocean, at a depth of about 10,900 m, in the Mariana Trench) and that they produced new substances (Takami, 1999). Over the past three decades, studies on these microorganisms have focused on the enzymology, physiology, ecology, taxonomy, molecular biology, and genetics of alkaliphilic microorganisms, opening a new chapter in microbiology. Now, a major question arises: "Why do alkaliphiles require alkaline environments?" The cell surface of alkaliphiles can keep the intracellular pH values of about 7 to 8 in alkaline environments of pH 10 to 13. How pH homeostasis is maintained is one of the most fascinating aspects of alkaliphiles physiology. In order to understand this simple but difficult phenomenon, many microbiologists carried out several basic experiments. Whole genome sequences of three alkaliphilic *Bacillus* strains have been completed. Still, crucial gene(s) responsible for alkaliphily remains to be identified.

Thermophiles and alkaliphiles are now just two examples of "extremophiles," microorganisms capable of living under extreme conditions. Other extreme habitats where interesting bacteria live are those with low pH, high salt, high pressure, and low temperature. Extremophiles have become "hot" basic research topics for the biotechnology industry.

DEFINITION OF EXTREMOPHILES

R. D. MacElroy (1974) first coined the term "extremophiles" in a paper entitled "Some comments on the evolution of extremophiles." An extremophile is an organism which thrives under conditions that from a human perspective are clearly "extreme." Extremophiles often require such conditions for growth. Though the definition of "extreme" is anthropocentric, from the point of view of the organism its environment is, of course, completely normal. From the mid-1970s, the number of these novel microorganisms has been increasing. Studies on extremophiles have progressed to the extent that there are now regular international "extremophile" symposia, as well as dedicated scientific journals: *Extremophiles* and *Archaea*.

WHAT ARE EXTREMOPHILES?

Most extremophiles are microorganisms. The presently known upper temperature limit is 121°C for archaea, 95°C for bacteria, and 62°C for single-celled eukaryotes in contrast to multicellular eukaryotes which are unable to grow at above 50°C. Many species can survive extreme conditions in a dormant state but are not capable of growing or reproducing indefinitely under such conditions.

Terms used to describe extremophiles include the following:

- *Acidophilic*: An organism that requires an acid growth medium with an optimal growth pH of 3 or below.
- *Alkaliphilic*: An organism that requires an alkaline growth medium and that displays optimal growth at pH values usually above 9.
- *Endolithic*: An organism that lives within or penetrates deeply into stony substances.
- *Halophilic*: An organism requiring at least 0.2 M salt for growth.
- *Hyperthermophilic*: An organism having an optimal growth temperature of 80°C or higher.
- *Hypolithic*: An organism that colonizes the underside of stony substances in arid deserts.
- *Metallotolerant*: An organism capable of tolerating high levels of heavy metals, such as copper, cadmium, arsenic, and zinc.
- *Oligotrophic*: An organism capable of growth in nutritionally limited environments.
- *Piezophilic*: An organism that lives optimally at high hydrostatic pressure.
- *Psychrophilic*: An organism having an optimal growth temperature of 15°C, or lower, and a maximum temperature of 20°C.
- *Radioresistant*: An organism resistant to high levels of ionizing radiation.
- *Thermophilic*: An organism that can thrive at temperatures between 60 and 80°C.
- *Toxitolerant*: An organism able to withstand high levels of damaging agents: for example, living in water saturated with benzene or toluene or in the water-core of a nuclear reactor.
- *Xerophilic*: An organism capable of growth at low water activity; for example, extreme halophile or endolith.

Some extremophiles may fall under multiple categories, for example living inside hot rocks deep under the Earth's surface.

APPLICATIONS OF EXTREMOPHILES

From a commercial perspective, enzymes from extremophiles—extremozymes—have an important impact. As an example, alkaline proteases, derived from alkaliphilic species, constitute an important group of enzymes that find applications primarily as protein-degrading additives in detergents. These enzymes can be subjected to harsh environments—including elevated temperature, high pH, surfactants, bleach chemicals, and chelating agents—which limit applications of many other enzymes because of their low activity or stability. Enzyme production for detergents is a huge market—constituting approximately 30% of the total enzymes produced worldwide (Horikoshi, 1999, 2006).

The use of extremophiles in industrial processes has opened a new era in biotechnology. Each group has unique features that can be exploited to provide biomolecules for the industry. A wide variety of applications is illustrated in Table 1.

EPILOGUE

It is generally accepted that less than 5% of all existing microorganisms have been obtained in pure culture and tested for their biosynthetic potential. The remaining 95% either have not been discovered or are impossible to grow reproducibly in culture. Fortunately, the biotechnological potential of extremophiles has spurred an exponential increase in the isolation and characterization of new organisms.

Table 1. Some biotechnological products and applications of extremophiles

Extremophiles	Products	Applications
Acidophiles	Sulfur oxidizer	Waste treatment and degumming organic acid and solvents Recovery of metals and desulfurication
Alkaliphiles	Proteases, cellulases, lipases, amylases, and pullulanase Elastase and keratinases Xylanases Cyclodextrins	Detergents Hide dehairing Paper bleaching Foodstuffs, chemicals, cosmetics, and pharmaceuticals
Halophiles	Lipids Compatible solutes γ-Linoleic acid and β-carotene	Liposomes for drug delivery and cosmetics Protein, DNA, and cell protectants Health foods, dietary supplements, and food color
Psychrophiles	Alkaline phosphatase Proteases, lipases, cellulases, and amylases Polyunsaturated fatty acids	Molecular biology Detergents Food additives and dietary supplements
Thermophiles	DNA polymerases	DNA amplication by PCR
Hyperthermophiles	Proteases, lipases, and pullulanase Amylases Xylanases	Detergents Baking and brewing Paper bleaching

As result of these investigations, physiology, molecular biology, and genetics of extremophiles have been developed quickly.

Numerous biologists are extensively studying extremophiles by new genomic approaches, such as metagenomics, that prove very useful methods to understand these organisms. Beside this, however, we should investigate how to isolate and cultivate the so-called non-cultivable microorganisms as living creatures, even if it involves somewhat boring and time-consuming experiments. Combining in silico approaches with "wet" biology should bring about major progress in the physiology of extremophiles and answer many questions that presently puzzle biologists.

REFERENCES

Brock, T. D. 1997. The value of basic research: discovery of *Thermus aquaticus* and other extreme thermophiles. *Genetics* **146**: 1207–1210.

Horikoshi, K. 1999. Alkaliphiles: some applications of their products for biotechnology. *Microbiol. Mol. Biol. Rev.* **63**:735–750.

Horikoshi, K. (ed.). 2006. *Alkaliphile—Genetic Properties and Application of Enzymes*. Springer, Heidelberg, Germany.

MacElroy, R. D. 1974. Some comments on the evolution of extremophiles. *Biosystems* **6**:74–75.

Takami, H. 1999. Isolation and characterization of microorganisms from deep-sea mud, p. 3–26. *In* K. Horikoshi and K. Tsujii (ed.), *Extremophiles in Deep-Sea Environments*. Springer-Verlag, New York, NY.

Koki Horikoshi
Kanagawa, Japan

PREFACE

The topics covered in this book are related to organisms, called extremophiles, that from an anthropocentric point of view are considered to be odd and mysterious. Indeed, these organisms not only survive but happily thrive in environments characterized by physical and chemical parameters that render them completely inhospitable to human beings. For this reason, however, and already for more than a century, they have attracted the curiosity of researchers who have slowly but progressively invested more and more energy to unravel the characteristics and adaptation strategies of these extreme organisms. The most thoroughly investigated have been those living in hot-spring waters, because these environments are not only spectacular but also easily accessible and distributed in numerous sites all over our planet. Furthermore, it was previously believed that the primeval cell, about 4 billion years ago, was a thermophile, the only one able to survive in conditions of very high temperature supposed to prevail at that time. Because elevated temperatures are very harmful to many living organisms, those microorganisms that are capable of surmounting this adverse condition became the target of most of the early work on extremophiles, since they were assumed to be closely related to this primeval cell. For reasons that are detailed in this book, however, this concept is nowadays largely questioned; until very recently, though, it had almost become a dogma. All organisms living in permanently hot environments with temperatures close to or even higher than the boiling point of water are microorganisms (in this case, prokaryotes exclusively) and hence are relatively easily investigated. Moreover, it was discovered that low pH is often associated with high-temperature conditions because some of these organisms rely on sulfur and produce sulfuric acid. Hence, this connection readily led to a new field of interest: adaptation to low external pHs. The discovery of high-pH habitats such as soda lakes completed this field of investigation related to organisms capable of living at pHs close to 0 or approaching 14.

Microorganisms were then also discovered in extremely salty environments such as the Dead Sea, which displays an overall salt concentration of 340 g/l, equivalent to nearly 6 M NaCl, more than 10 times the concentration of common seawater. Also surprising was the discovery that permanently cold environments such as the Antarctic continent were hosts to an abundant and diversified microbial fauna capable of displaying metabolic fluxes at the low temperature of their environment comparable to those of their temperate homologs. In the meantime the little-known deep sea was being investigated, cold, dark, relatively poor in oxygen and in which organisms capable of sustaining, pressures as high as 1,000 atm were discovered. When adding to this list those organisms capable of surviving in highly toxic heavy-metal environments or of resisting harmful radiations, one has an extraordinarily diversified world of organisms characterized by the development of adaptation strategies coping with one of the parameters described above and sometimes more than one, such as heat and acidity or low temperature and high pressure. The aim of this book is to shed some light on these strategies and, in particular, to describe their molecular aspects.

We thank all the authors, who also are leading scientists in their field, for having agreed to participate in this endeavor. The role of a scientific editor is to gather the best possible scientists for each of the selected topics, even though these persons are also usually very busy and highly solicited researchers submerged by numerous commitments. However since they were conscious that they have an important educative role most of the contacted contributors readily accepted to participate in writing a chapter for the book. We are especially grateful to those authors who, despite illness, personal problems, or at first sight insurmountable tasks, gathered enough spirit to force themselves to limits and produce excellent manuscripts. We also thank ASM Press for having invested in us the necessary trust in our capacity to successfully achieve the editing of this book.

C. Gerday and N. Glansdorff
August 2006

I. INTRODUCTION

Chapter 1

Extremophiles and the Origin of Life

Sara Islas, Ana María Velasco, Arturo Becerra, Luis Delaye, and Antonio Lazcano

INTRODUCTION

During the past 2 decades, the description of a diverse assortment of prokaryotic species that thrive under extreme environments that used to be considered inhospitable has broadened our understanding of the range of conditions under which life can persist. The discovery of a number of archaeal and bacterial species that live under extreme environmental conditions, which include high temperature, ionizing radiation, pressure, UV light, or salinity; low or high levels of pH; or very low levels of nutrients, light, or water, has raised a number of issues that range from suggestions that extremophiles may be considered models of primordial organisms (Yayanos et al., 1981; Wiegel and Adams, 1998; Di Giulio, 2000, 2003), to the possibility that their lifestyles may provide insights into extraterrestrial habitats where life could develop (Kasting et al., 1993; Cavicchioli, 2002; Cleaves and Chalmers, 2004).

The possibility that the basic traits of the genetic code originated under acidic (Di Giulio, 2005a) or high-pressure (Di Giulio, 2005b) environmental conditions has been raised. With the exception of heat-loving prokaryotes, however, the phylogenetic distribution of other extremophiles in molecular cladograms does not provide clues to their possible antiquity. The fact that the deepest, shorter branches of rRNA-based molecular phylogenies are occupied by hyperthermophiles has led to the idea of a hot origin of life (Pace, 1991; Stetter, 1994). Although this is not a new idea (Harvey, 1924; Fox and Dose, 1977), the study of hydrothermal vents and their complex microbiotas (Corliss et al., 1981; Holm, 1992), as well as the demonstration that pyrite is a strong reductant (Wächtershäuser, 1988, 1990), has revitalized this idea.

At first glance, in fact, both the molecular and the paleontological fossil records appear to support such possibility of a hyperthermophilic origin of life. Life on Earth arose early. Large-scale analyses suggest that the surface of the primitive Earth was extremely hot soon after its formation. The planet is generally thought to have remained molten for some time after its formation 4.6×10^9 years ago (Wetherill, 1990), but mineralogical evidence of a 4.4×10^9-year-old hydrosphere implies that its surface rapidly cooled down (Wilde et al., 2001). However, there is theoretical and empirical evidence that the planet underwent late accretion impacts (Byerly et al., 2002; Schoenberg et al., 2002) that may have boiled off the oceans as late as 3.8×10^9 years ago (Sleep et al., 1989).

The proposals of a high-temperature origin of life face major problems, including the chemical decomposition of presumed essential biochemical compounds such as amino acids, nucleobases, RNA, and other thermolabile molecules, whose half-lives for decomposition at temperatures between 250°C and 350°C are at the most a few minutes (White, 1984; Miller and Bada, 1988). Furthermore, given the huge gap existing in current descriptions of the evolutionary transition between the prebiotic synthesis of biochemical compounds and the last common ancestor (LCA) of all extant living beings, it is probably naïve to attempt to describe the origin of life and the nature of the first living systems from molecular phylogenies. As reviewed here, although the primitive Earth may have been an extreme environment, the extrapolation of molecular phylogenies into prebiotic times is misleading (Miller and Lazcano, 1995; Bada and Lazcano, 2002; Islas et al., 2003), and a more likely alternative is the possibility that (hyper)thermophilic microbial lifestyles are in fact the outcome of secondary adaptations that developed during early stages of cell evolution.

S. Islas, A. M. Velasco, A. Becerra, L. Delaye, and A. Lazcano • Facultad de Ciencias, UNAM, Apdo. Postal 70-407, Cd. Universitaria, México D.F., Mexico.

THE PRIMITIVE EARTH AS AN EXTREME ENVIRONMENT

There is no geological evidence of the environmental conditions on the early Earth at the time of the origin of life, nor are any molecular or physical remnants preserved that provide information about the evolutionary processes that preceded the appearance of the first cellular organisms found in the early fossil record. Direct information is lacking not only on the composition of the terrestrial atmosphere during the period of the origin of life but also on the temperature, ocean pH values, and other general and local environmental conditions that may have been important for the emergence of living systems.

The available evidence suggest, in fact, that soon after its formation, Earth must have been a dry, hot planet, perhaps not even endowed with an atmosphere (Kasting, 1993a). During the Hadean period, volatile components that were trapped inside the accreting planet were degassed from the interior of the juvenile Earth to form a secondary atmosphere. Owing to the almost simultaneous formation of Earth's core with accretion, the metallic iron was removed from the upper mantle, which would have allowed the volcanic gases to remain relatively reduced and produce a very early atmosphere that contained species such as CH_4, NH_3, and H_2 (Bada and Lazcano, in press). Because the temperature at the surface was high enough to prevent any water from condensing, the atmosphere would have consisted mainly of superheated steam along with the other gases (Kasting, 1993b). Owing to large impact events, such as the one that created the Moon, it is possible that secondary atmospheres were lost several times and would have been regenerated by further outgassing from the interior and as well as resupplied from later comet-like impactors (Oró, 1961).

It is reasonable to assume that the atmosphere that developed on Earth over the period 4.4–3.8 Gya was probably dominated by steam until the surface temperatures dropped to ~100°C (depending on the pressure), at which point water condensed out to form early oceans (Wilde et al., 2001). Reduced chemical species, such as methane and ammonia, which were mainly supplied by volcanic outgassing, are very sensitive to UV radiation that penetrated through the atmosphere because of the lack of a protective ozone layer. These molecules were probably destroyed by photodissociation, although there might have been a steady-state equilibrium between these two processes that allowed a significant amount of these reduced species to be present in the atmosphere (Tian et al., 2005). However, currently it is generally accepted that the early atmosphere was dominated by neutral chemical species such as CO_2, CO, and N_2. A similar atmosphere is present on Venus today, although it is much denser than the atmosphere of the early Earth. In summary, the current models for the early terrestrial atmosphere suggest that it consisted of a weakly reducing mixture of CO_2, N_2, CO, and H_2O, with lesser amounts of H_2, SO_2, and H_2S. Reduced gases such as CH_4 and NH_3 are considered to be nearly absent or present only in localized regions near volcanoes or hydrothermal vents (Bada and Lazcano, 2006).

A TIMESCALE FOR THE ORIGIN AND EARLY EVOLUTION OF LIFE

It is not possible to assign a precise chronology to the appearance of life. Identification of the oldest paleontological traces of life remains a contentious issue (van Zullen et al., 2002). The early Archean geological record is scarce, and most of the sediments that have been preserved from those times have been metamorphosed to a considerable extent. There is evidence, however, that life emerged on Earth very early in its history. Although microstructures present in the 3.5×10^9-year-old Apex cherts of the Australian Warrawoona formation have been interpreted as cyanobacterial remnants (Schopf, 1993), alternative interpretations suggest that these structures are the outcome of abiotic hydrothermal processes (Brasier et al., 2002; García-Ruiz et al., 2003).

However, there is evidence of biological sulfate-reducing activity from the Dresser Formation (Shen et al., 2001), and isotopic analyses of 3.4×10^9-year-old South African cherts suggest that they were inhabited by anaerobic photosynthetic prokaryotes in a marine environment (Tice and Lowe, 2004). Recent analysis of methane-bearing inclusions from the 3.5×10^9 million-year-old hydrothermal precipitates from Pilbara, Western Australia have shown that they are depleted in ^{13}C values, which may indicate the onset of microbial methanogenesis (Ueno et al., 2006). These results strongly support the idea that the early Archean Earth was teeming with prokaryotes that included anoxygenic phototrophs, sulfate reducers, and methanogenic archaea (Canfield, 2006). This implies that the origin of life must have taken place as soon as the conditions were suitable to permit its survival and suggests that the critical factor may have been the presence of liquid water, which became possible as soon as the planet's surface finally cooled below the boiling point of water.

MOLECULAR CLADISTICS AND THE ORIGIN AND EARLY EVOLUTION OF LIFE

It is unlikely that data on how life originated will be provided by the geological record. There is neither direct evidence of the environmental conditions on Earth at the time of the origin of life nor any fossil register of the evolutionary processes that preceded the appearance of the first cells. Direct information is lacking not only on the composition of the terrestrial atmosphere during the period of the origin of life but also on the temperature, ocean pH values, and other general and local environmental conditions that may have been important for the emergence of living systems (Bada and Lazcano, 2006).

Moreover, the attributes of the first living organisms are unknown. They were probably simpler than any cell now alive and may have lacked not only protein-based catalysis but perhaps even the familiar genetic macromolecules, with their ribose-phosphate backbones. It is possible that the only property they shared with extant organisms was the structural complementarity between monomeric subunits of replicative informational polymers, e.g., the joining together of residues in a growing chain whose sequence is directed by preformed polymers. Such ancestral polymers may have not even involved nucleotides. Accordingly, the most basic questions pertaining to the origin of life relate to much simpler replicating entities predating by a long series of evolutionary events the oldest recognizable heat-loving prokaryotes represented in molecular phylogenies.

The variations of traits common to extant species can be easily explained as the outcome of divergent processes from an ancestral life form that existed prior to the separation of the three major biological domains, i.e., the LCA or cenancestor. No paleontological remains will bear testimony to its existence, as the search for a fossil of the cenancestor is bound to prove fruitless. From a cladistic viewpoint, the LCA is merely an inferred inventory of features shared among extant organisms, all of which are located at the tip of the branches of molecular phylogenies.

Reticulate phylogenies greatly complicate the reconstruction of cenancestral traits, driven in part by the impact of lateral gene acquisition, as revealed by the discrepancies of different gene phylogenies with the canonical rRNA tree, and in part by the surprising complexity of the universal ancestor, as suggested by direct backtrack characterizations of the oldest node of universal cladograms. Inventories of LCA genes include sequences that originated in different precenancestral epochs (Delaye and Lazcano, 2000; Anantharaman et al., 2002). The origin of the mutant sequences ancestral to those found in all extant species and the divergence of the *Bacteria*, *Archaea*, and *Eucarya* were not synchronous events, i.e., the separation of the primary domains took place later, perhaps even much later, than the appearance of the genetic components of their LCA (Delaye et al., 2002,

2005). The cenancestor is thus one of the last evolutionary outcomes of a series of ancestral events, including lateral gene transfer, gene losses, and paralogous duplications that took place before the separation of *Bacteria*, *Archaea*, and *Eucarya* (Lazcano et al., 1992; Glansdorff, 2000; Castresana, 2001; Delaye et al., 2002, 2005).

Although hyperthermophiles may be displaced from their basal position if molecular markers other than elongation factors or ATPase subunits are employed (Forterre et al., 1993; Klenk et al., 1994), or if alternative phylogeny-building methodologies are used (Brochier and Philippe, 2002), it can be argued that rRNA-based phylogenies provide one of the best-preserved historical records of cell evolution (Woese, 2002). However, the recognition that the deepest branches in rooted universal phylogenies are occupied by hyperthermophiles does not provide by itself conclusive proof of a heat-loving LCA, or much less of a hot origin of life. Analysis of the correlation of the optimal growth temperature of prokaryotes and the G+C nucleotide content of 40 rRNA sequences through a complex Markov model has led Galtier et al. (1999) to conclude that the universal ancestor was a mesophile. This possibility has been contested by Di Giulio (2000), who has argued for a thermophilic or hyperthermophilic LCA. However, because the time factor is absent from the methodology developed by Galtier et al. (1999), the inferred low G+C content of the cenancestral rRNA does not necessarily belong to the cenancestor itself but may correspond to a mesophilic predecessor that may have been located along the trunk of the universal tree.

The rooting of universal cladistic trees determines the directionality of evolutionary change and allows the recognition of ancestral from derived characters. Determination of the rooting point of a tree normally imparts polarity to most or all characters (Scotland, 1992). It is, however, important to distinguish between ancient and primitive organisms. Organisms located near the root of universal rRNA-based trees are cladistically ancient, but they are not endowed with primitive molecular genetic apparatus, nor do they appear to be more rudimentary in their metabolic abilities than their aerobic counterparts (Islas et al., 2003). Primitive living systems would initially refer to pre-RNA worlds, in which life may have been based on polymers using backbones other than ribose phosphate and possibly bases different from adenine, uracil, guanine, and cytosine (Levy and Miller, 1998), followed by a stage in which life was based on RNA both as the genetic material and as catalysts (Joyce, 2002; Dworkin et al., 2002).

Molecular cladistics may provide clues to some very early stages of biological evolution, but it is difficult to see how the applicability of this approach can be extended beyond a threshold that corresponds to a period of cellular evolution in which protein biosynthesis was already in operation, i.e., an RNA/protein world. Older stages are not yet amenable to molecular phylogenetic analysis. A cladistic approach to the origin of life itself is not feasible because all possible intermediates that may have once existed have long since vanished, and the temptation to do otherwise is best resisted.

CHEMICAL EVOLUTION AND EXTREME ENVIRONMENTS

The hypothesis that the first organisms were anaerobic heterotrophs is based on the assumption that abiotic organic compounds were a necessary precursor to the appearance of life. The first successful synthesis of organic compounds under plausible primordial conditions was accomplished by the action of electric discharges acting for a week over a mixture of CH_4, NH_3, H_2, and H_2O and led to a complex mixture of monomers that included racemic mixtures of several proteinic amino acids, in addition to hydroxy acids, urea, and other molecules (Miller, 1953). Prebiotic synthesis of amino acids is largely followed by a Strecker synthesis, which involves the aqueous-phase reactions of highly reactive intermediates. Detailed studies of the equilibrium and rate constants of these reactions demonstrated that both amino acids and hydroxy acids can be synthesized at high dilutions of HCN and aldehydes in a simulated primitive ocean. The reaction rates depend on temperature, pH, and HCN, NH_3, and aldehyde concentrations and are rapid on a geological timescale; the half-lives for the hydrolysis of the intermediate products in the reactions, amino- and hydroxynitriles, are <1000 years at 0°C, and there are no known slow steps (Miller and Lazcano, 2002).

The remarkable ease with which adenine can be synthesized by the aqueous polymerization of ammonium cyanide demonstrated the significance of HCN and its derivatives in prebiotic chemistry (Oró, 1960). The prebiotic importance of HCN has been further substantiated by the discovery that the hydrolytic products of its polymers include amino acids, purines, and orotic acid, which is a biosynthetic precursor of uracil (Ferris et al., 1978). The reaction of cyanoacetylene or cyanoacetaldehyde (a hydrolytic derivative of HCN) with urea (Ferris et al., 1968) leads to high yields of cytosine and uracil, especially under simulated evaporating pond conditions, which increase the urea concentration (Robertson and Miller, 1995).

The easiness of formation under reducing conditions ($CH_4 + N_2$, $NH_3 + H_2O$, or $CO_2 + H_2 + N_2$) in one-pot reactions of amino acids, purines, and pyrimidines strongly suggests that these molecules were present in the prebiotic broth. Experimental evidence suggests that urea, alcohols, sugars formed by the nonenzymatic condensation of formaldehyde, a wide variety of aliphatic and aromatic hydrocarbons, urea, carboxylic acids, and branched and straight fatty acids, including some which are membrane-forming compounds, were also components of the primitive soup. The remarkable coincidence between the molecular constituents of living organisms and those synthesized in prebiotic experiments is too striking to be fortuitous, and the robustness of this type of chemistry is supported by the occurrence of most of these biochemical compounds in the 4.5×10^9-year-old Murchison carbonaceous meteorite, which also yields evidence of liquid water in its parent body (Ehrenfreund et al., 2002).

Can the environmental conditions under which hyperthermophiles live provide insights into prebiotic processes? The low pH values under which a number of archaeal species thrive, for instance, are known to inhibit several prebiotic processes, including the formose condensation that leads to sugars, the HCN polymerization, and the Strecker synthesis of amino acids, which depends on the nucleophilicity of ammonia and cyanide anion, both of which have pKa of ~9.2. It is true, of course, that high temperatures allow chemical reactions to go faster, and the primitive enzymes, once they appeared, could have been less efficient (Harvey, 1924). However, high-temperature regimes would lead to the following:

1. reduced concentrations of volatile intermediates, such as HCN, H_2CO, and NH_3;
2. lower steady-state concentrations of prebiotic precursors such as HCN, which at temperatures a little >100°C undergoes hydrolysis to form formamide and formic acid and, in the presence of ammonia, NH_4HCO_3;
3. instability of reactive chemical intermediates such as amino nitriles ($RCHO(NH_2)CN$), which play a central role in the Strecker synthesis of amino acids;
4. loss of organic compounds by thermal decomposition and diminished stability of genetic polymers.

Extremophilic genomes are protected against thermal decomposition by many enzyme-dependent mechanisms (Grogan, 1998), but these would have not been available during prebiotic times or at the time of the origin of life. In fact, the existence of an RNA world with ribose appears to be incompatible with a (hyper)thermophilic environment. Survival of nucleic acids is limited by the hydrolysis of phosphodiester bonds (Lindahl, 1993), and the stability of Watson–Crick helices (or their pre-RNA equivalents) is strongly diminished by high temperatures. For an RNA-based biosphere, the reduced thermal stability on the geological timescale of ribose and other sugars is the worst problem (Larralde et al., 1995), but the situation is equally bad for pyrimidines, purines, and some amino acids. As reviewed elsewhere (Miller and Lazcano, 1995), the half-life of ribose at 100°C and pH 7 is only 73 min, and other sugars (2-deoxyribose, ribose-5-phosphate, and ribose 2,4-biphosphate) have comparable half-lives (Larralde et al., 1995). The half-life for hydrolytic deamination of cytosine at 100°C lies between 19 and 21 days (Shapiro and Klein, 1966; Garrett and Tsau, 1972; Levy and Miller, 1998), although at 100°C the half-life of uracil is ~12 years (Levy and Miller, 1998). At 100°C, the thermal stability of purines is also reduced—between 204 and 365 days for adenine (Frick et al., 1987; Shapiro, 1995; Levy and Miller, 1998)—with comparable values for guanine (Levy and Miller, 1998).

An alternative to the problem of low half-lives of biochemical monomers at temperatures of 100°C or more is to assume an autotrophic origin of life. The most elaborate chemoautotrophic origin-of-life scheme has been proposed by Wächtershäuser (1988). According to his hypothesis, life began with the appearance of an autocatalytic two-dimensional chemolithotrophic metabolic system based on the formation of the highly insoluble mineral pyrite. The synthesis in the activated form of organic compounds such as amino acid derivatives, thioesters, and keto acids is assumed to have taken place on the surface of FeS and FeS_2 in environments that resemble those of deep-sea hydrothermal vents. Replication followed the appearance of nonorganismal iron sulfide-based two-dimensional life, in which chemoautotrophic carbon fixation took place by a reductive citric acid cycle, or reverse Krebs cycle, of the type originally described for the photosynthetic green sulfur bacterium *Chlorobium limicola*. Molecular phylogenetic trees show that this mode of carbon fixation and its modifications (such as the reductive acetyl coenzyme A [acetyl-CoA] or the reductive coenzyme A [malonyl-CoA] pathways) are found in anaerobic archaea and the most deeply divergent eubacteria, which has been interpreted as evidence of its primitive character (Maden, 1995).

The reaction $FeS + H_2S = FeS_2 + H_2$ is indeed a very favorable one. It has an irreversible, highly exergonic character with a standard free energy change $\Delta G° = -9.23$ kcal/mol, which corresponds to a reduction potential $E° = -620$ mV. Thus, the FeS/H_2S combination is a strong reducing agent and has been shown to provide an efficient source of electrons for the reduction of organic compounds under mild conditions. Although pyrite-mediated CO_2 reduction to amino acids, purines, and pyrimidines is yet to be achieved, the FeS/H_2S combination is a strong reducing agent that has been shown to reduce nitrate and acetylene as well as to induce peptide bonds that result from the activation of amino acids with carbon monoxide and (Ni, Fe)S (Maden, 1995; Huber and Wächtershäuser, 1998). Acetic acid and pyruvic acid have been synthesized from CO under simulated hydrothermal conditions in the presence of sulfide minerals (Huber and Wächtershäuser, 1997; Cody et al., 2000). However, the empirical support for Wächtershäuser's central tenets is meager. The compounds synthesized using pyrite would also be exposed to decomposition at high temperatures. More importantly, life does not consist solely of metabolic cycles, and none of these experiments proves that enzymes and nucleic acids are the evolutionary outcome of multistep autocatalytic metabolic cycles surface bound to FeS/FeS_2 or some other mineral. In fact, experiments using the FeS/H_2S combination are also compatible with a more general, modified model of the primitive soup in which pyrite formation is recognized as an important source of electrons for the reduction of organic compounds (Bada and Lazcano, 2002, 2006).

CONCLUSIONS

As the initially molten young Earth cooled down, global temperatures of 100°C must have been reached but could not have persisted for more than 20 million years (Sleep et al., 2001). Deep-sea hydrothermal vents and other local high-temperature milieu have existed throughout the history of the planet and have played a major role in shaping the primitive environments (Holm, 1992; Wiegel and Adams, 1998). However, the rapidity of thermal decomposition of amino acids, nucleobases, and genetic polymers, which is very short on the geological timescale, is a strong argument against a hot origin of life in hydrothermal vents.

Because high salt concentrations protect DNA and RNA against heat-induced damage (Marguet and Forterre, 1994; Tehei et al., 2002), it could be hypothesized that this and other nonbiological mechanisms such as adsorption to mineral surfaces and formation of clay–nucleic acid complexes (Franchi et al., 1999) played a significant role in the preservation of organic compounds and genetic polymers in the primitive environments. However, such mechanisms would be inefficient at temperatures >100°C. Because adsorption involves the formation of weak noncovalent bonds, mineral-based concentration would have been most effective at low temperatures (Sowerby et al., 2001); at high temperatures, any adsorbed monomers would drift away into the surrounding aqueous environment and become hydrolyzed. As shown by the Cu^{2+}-montmorillonite enhancing effect of the decomposition of adenine to hypoxanthine (Strasak and Sersen, 1991), association of organic compounds with some minerals may in fact reduce their half-lives.

The remarkable coincidence between the monomeric constituents of living organisms and those synthesized in laboratory simulations of the prebiotic environment is too striking to be fortuitous, but at the same time the hiatus between the primitive soup and the RNA world, i.e., the evolutionary stage prior to the development of proteins and DNA genomes during which early life forms largely based on ribozymes may have existed, is discouragingly enormous. Accordingly, if hyperthermophily is not truly primordial, then heat-loving lifestyles may be relics of a secondary adaptation that evolved after the origin of life and before or soon after the separation of the major lineages (Miller and Lazcano, 1995; Forterre, 1996). As argued here, the so-called root of universal trees does not correspond to the first living system but is the tip of a trunk of still undetermined length in which the history of a long (but not necessarily slow) series of archaic evolutionary events, such as an explosion of gene families and multiple events of lateral gene transfer, is still preserved. Is it possible that traces of the emergence of hyperthermophily persist in the molecular records of earliest biological evolution somewhere along the trunk of rRNA-based phylogenic trees? If hyperthermophiles were not the first organisms, then their basal position in molecular trees could be explained as follows:

1. a relic from early Archean high-temperature regimes that may have resulted from a severe impact regime (Sleep et al., 1989; Gogarten-Boekels et al., 1994);

2. adaptation of *Bacteria* to extreme environments by lateral transfer of reverse gyrase (Forterre et al., 2000) and other thermoadaptative traits from heat-loving Archaea;

3. outcompetition of older mesophiles by hyperthermophiles originally adapted to stress-inducing conditions other than high temperatures (Miller and Lazcano, 1995).

Although there have been considerable advances in the understanding of chemical processes that may have taken place before the emergence of the first living systems, life's beginnings are still shrouded in mystery. As argued here, the diversity of environmental conditions under which prokaryotes can thrive should be understood as evidence of their adaptability and not as evidence that the origin of life took place under extreme conditions. As Sulston and Ferry (2002) wrote, "the beginnings of [biological] evolution and the origins of life are one and the same. Once something is replicating with variations, it will bit by bit explore the possibilities of its environment."

REFERENCES

Anantharaman, V., E. V. Koonin, and L. Aravind. 2002. Comparative genomics and evolution of proteins involved in RNA metabolism. *Nucleic Acids Res.* **30**:1427–1464.

Bada, J. L. and A. Lazcano. The origin of life. *In* M. Ruse (ed.), *The Harvard Companion to Evolution*, in press. Harvard University Press, Cambridge, MA.

Bada, J. L., and A. Lazcano. 2002. Some like it hot, but not biomolecules. *Science* **296**:1982–1983.

Brasier, M., O. R. Green, A. P. Jephcoat, A. K. Kleppe, M. J. van Kranendonk, J. F. Lindsay, A. Steele, and N. V. Grassineau. 2002. Questioning the evidence for Earth's earliest fossils. *Nature* **416**:76–79.

Brochier, C., and H. Philippe. 2002. A non-hyperthermophilic ancestor for Bacteria. *Nature* **417**:244.

Byerly, G. R., D. R. Lowe, J. L. Wooden and X. Xie. 2002. An Archean impact layer from the Pilbara and Kaapvaal clatons. *Science* **297**:1325–1327.

Canfield, D. E. 2006. Gas with an ancient history. *Nature* **440**:426–427.

Castresana, J. 2001. Comparative genomics and bioenergetics. *Biochim. Biophys. Acta* **1506**:147–162.

Cavicchioli, R. 2002. Extremophiles and the search for extraterrestrial life. *Astrobiology* **2**:281–292.

Cleaves, H. J., and J. H. Chalmers. 2004. Extremophiles may be irrelevant to the origin of life. *Astrobiology* **4**:1–9.

Cody, G. D., N. Z. Boctor, T. R. Filley, R. M. Haze, J. H. Scott, A. Sharma, and H. S. Yoder, Jr. 2000. Primordial carbonylated iron-sulfur compounds and the synthesis of pyruvate. *Science* **289**:1337–1340.

Corliss, J. B., J. A. Baross, and S. E. Hoffman. 1981. An hypothesis concerning the relationship between submarine hot springs and the origin of life on Earth. *Oceanologica Acta* **4**(Suppl.):59–69.

Delaye, L., A. Becerra, and A. Lazcano. 2002. The nature of the last common ancestor, p. 34–47. *In* L. Ribas de Pouplana (ed.), *The Genetic Code and the Origin of Life*. Landes Bioscience, Georgetown, TX.

Delaye, L., A. Becerra, and A. Lazcano. 2005. The last common ancestor: what's in a name? *Origins Life Evol. Biosph.* **35**: 537–554.

Delaye, L., and A. Lazcano. 2000. RNA-binding peptides as molecular fossils, p. 285–288. *In* J. Chela-Flores, G. Lemerchand, and J. Oró (ed.), *Origins from the Big-Bang to Biology: Proceedings of the First Ibero-American School of Astrobiology*. Kluwer Academic Publishers, Dordrecht, The Netherlands.

Di Giulio, M. 2000. The universal ancestor lived in a thermophilic or hyperthermophilic environment. *J. Theor. Biol.* **203**:203–213.

Di Giulio, M. 2003. The universal ancestor and the ancestor of bacteria were hyperthermophiles. *J. Mol. Evol.* **57:**721–730.

Di Giulio, M. 2005a. A comparison of proteins from *Pyrococcus furiosus* and *Pyrococcus abyssi*: barophily in the physicochemical properties of amino acids and in the genetic code. *Gene* **346:**1–6.

Di Giulio, M. 2005b. The ocean abysses witnessed the origin of the genetic code. *Gene* **346:**7–12.

Dworkin, J. P., A. Lazcano, and S. L. Miller. 2002. The roads to and from the RNA world. *J. Theor. Biol.* **222:**127–134.

Ehrenfreund, P., W. Irvine, L. Becker, J. Blank, J. Brucato, L. Colangeli, S. Derenne, D. Despois, A. Dutrey, H. Fraaije, A. Lazcano, T. Owen, and F. Robert. 2002. Astrophysical and astrochemical insights into the origin of life. *Reports Prog. Phys.* **65:**1427–1487.

Ferris, J. P., P. D. Joshi, E. H. Edelson, and J. G. Lawless. 1978. HCN: a plausible source of purines, pyrimidines, and amino acids on the primitive Earth. *J. Mol. Evol.* **11:**293–311.

Ferris, J. P., R. A. Sanchez, and L. E. Orgel. 1968. Studies in prebiotic synthesis. III. Synthesis of pyrimidines from cyanoacetylene and cyanate. *J. Mol. Biol.* **33:**693–704.

Forterre, P. 1996. A hot topic: the origin of hyperthermophiles. *Cell* **85:**789–792.

Forterre, P., N. Benachenhou-Lahfa, F. Confalonieri, M. Duguet, C. Elie, and B. Labedan. 1993. The nature of the last universal ancestor and the root of the tree of life. *BioSystems* **28:**15–32.

Forterre, P., C. Bouthier de la Tour, H. Philippe, and M. Duguet. 2000. Reverse gyrase from hyperthermophiles: probable transfer of a thermoadaptation trait from Archaea to Bacteria. *Trends Genet.* **16:**152–154.

Fox, S. W., and K. Dose. 1977. *Molecular Evolution and the Origin of Life*. Dekker, New York, NY.

Franchi, M., E. L. Morassi Bonzi, P. L. Orioli, C. Vettori, and E. Gallori. 1999. Clay–nucleic acid complexes: characteristics and implications for the preservation of genetic material in primeval habitats. *Origins Life Evol. Biosph.* **29:**297–315.

Frick, L., J. P. Mac Neela, and R. Wolfenden. 1987. Transition state stabilization by deaminases: rates of nonenzymatic hydrolysis of adenosine and cytidine. *Bioorg. Chem.* **15:**100–108.

Galtier, N., N. Tourasse, and M. Gouy. 1999. A nonhyperthermophilic common ancestor to extant life forms. *Science* **283:**220–221.

García-Ruiz, J. M., S. T. Hyde, A. M. Carnerup, A. G. Christy, M. J. Van Kranendonk, and N. J. Welham. 2003. Self-assembled silica-carbonate structures and detection of ancient microfossils. *Science* **302:**1194–1197.

Garrett, E. R., and J. Tsau. 1972. Solvolyses of cytosine and cytidine. *J. Pharm. Sci.* **61:**1052–1061.

Glansdorff, N. 2000. About the last common ancestor, the universal life-tree and lateral gene transfer: a reappraisal. *Mol. Microbiol.* **38:**177–185.

Gogarten-Boekels, M., E. Hilario, and J. P. Gogarten. 1994. The effects of heavy meteorite bombardment on the early evolution of life—a new look at the molecular record. *Origins Life Evol. Biosph.* **25:**78–83.

Grogan, D. W. 1998. Hyperthermophiles and the problem of DNA instability. *Mol. Microbiol.* **28:**1043–1049.

Harvey, R. B. 1924. Enzymes of thermal algae. *Science* **60:**481–482.

Holm, N. G. (ed.). 1992. *Marine Hydrothermal Systems and the Origin of Life*. Kluwer Academic Publishers, Dordrecht, The Netherlands.

Huber, C., and G. Wächtershäuser. 1997. Activated acetic acid by carbon fixation on (Fe, Ni)S under primordial conditions. *Science* **276:**245–247.

Huber, C., and G. Wächtershäuser. 1998. Peptides by activation of amino acids with CO on (Ni, Fe)S surfaces and implications for the origin of life. *Science* **281:**670–672.

Islas, S., A. M. Velasco, A. Becerra, L. Delaye, and A. Lazcano. 2003. Hyperthermophily and the origin and earliest evolution of life. *Int. Microbiol.* **6:**87–94.

Joyce, G. F. 2002. The antiquity of RNA-based evolution. *Nature* **418:**214–221.

Kasting, J. F. 1993a. Earth's early atmosphere. *Science* **259:**920–926.

Kasting, J. F. 1993b. Early evolution of the atmosphere and ocean, p. 149–176. *In* J. M. Greenberg, C. X. Mendoza-Gomez, and J. Pirronello (ed.), *The Chemistry of Life's Origin*. Kluwer Academic Publishers, Dordrecht, The Netherlands.

Kasting, J. F., D. P. Whitmore, and R. T. Reynolds. 1993. Habitable zones around main sequence stars. *Icarus* **101:**108–128.

Klenk, H. P., P. Palm, and W. Zillig. 1994. DNA-dependent RNA polymerases as phylogenetic marker molecules. *Syst. Appl. Microbiol.* **16:**638–647.

Larralde, R., M. P. Robertson, and S. L. Miller. 1995. Rates of decomposition of ribose and other sugars: implications for chemical evolution. *Proc. Natl. Acad. Sci. USA* **92:**8158–8160.

Lazcano, A., G. E. Fox, and J. Oró. 1992. Life before DNA: the origin and early evolution of early Archean cells, p. 237–295. *In* R. P. Mortlock (ed.), *The Evolution of Metabolic Function*. CRC Press, Boca Raton, FL.

Levy, M., and S. L. Miller. 1998. The stability of the RNA bases: implications for the origin of life. *Proc. Natl. Acad. Sci. USA* **95:**7933–7938.

Lindahl, T. 1993. Instability and decay of the primary structure of DNA. *Nature* **362:**709–715.

Maden, B. E. H. 1995. No soup for starters? Autotrophy and origins of metabolism. *Trends Biochem. Sci.* **20:**337–341.

Marguet, E., and P. Forterre. 1994. DNA stability at temperatures typical for hyperthermophiles. *Nucleic Acids Res.* **22:**1681–1686.

Miller, S. L. 1953. A production of amino acids under possible primitive Earth conditions. *Science* **117:**528.

Miller, S. L., and J. L. Bada. 1988. Submarine hot springs and the origin of life. *Nature* **334:**609–611.

Miller, S. L., and A. Lazcano. 1995. The origin of life—did it occur at high temperatures? *J. Mol. Evol.* **41:**689–692.

Miller, S. L., and A. Lazcano. 2002. Formation of the building blocks of life, p. 78–112. *In* J. W. Schopf (ed.), *Life's Origin: The Beginnings of Biological Evolution*. California University Press, Berkeley, CA.

Oró, J. 1960. Synthesis of adenine from ammonium cyanide. *Biochem. Biophys. Res. Commun.* **2:**407–412.

Oró, J. 1961. Comets and the formation of biochemical compounds on the primitive earth. *Nature* **190:**442–443.

Pace, N. R. 1991. Origin of life—facing up the physical setting. *Cell* **65:**531–533.

Robertson, M. P., and S. L. Miller. 1995. An efficient prebiotic synthesis of cytosine and uracil. *Nature* **375:**772–774.

Schoenberg, R., B. S. Kamber, K. D. Collerson, and S. Moorbath. 2002. Tungsten isotope evidence from 3.8-Gyr metamorphosed sediments for early meteorite bombardment of the Earth. *Nature* **418:**403–405.

Schopf, J. W. 1993. Microfossils of the early Archaean Apex chert: new evidence for the antiquity of life. *Science* **260:**640–646.

Scotland, R. W. 1992. Character coding, p. 14–43. *In* P. L. Florey, C. J. Humphries, I. L. Kitching, R. W. Scotland, D. J. Siebert, and D. M. Williams (ed.), *Cladistics: A Practical Course in Systematics*. Clarendon Press, Oxford, United Kingdom.

Shapiro, R. 1995. The prebiotic role of adenine: a critical analysis. *Origins Life Evol. Biosph.* **25:**83–98.

Shapiro, R., and R. S. Klein. 1966. The deamination of cytidine and cytosine by acidic buffer solutions: mutagenic implications. *Biochemistry* **5:**2358–2362.

Shen, Y., R. Buick, and D. E. Canfield. 2001. Isotopic evidence for microbial sulphate reduction in the early Archaean era. *Nature* 410:77–81.

Sleep, N. H., K. J. Zahnle, J. F. Kastings, and H. J. Morowitz. 1989. Annihilation of ecosystems by large asteroid impacts on the early Earth. *Nature* 342:139–142.

Sleep, N. H., K. J. Zahnle, and P. S. Neuhoff. 2001. Initiation of clement surface conditions on the earliest Earth. *Proc. Natl. Acad. Sci. USA* 98:3666–3672.

Sowerby, S. J., C.-M. Mörth, and N. G. Holm. 2001. Effect of temperature on the adsorption of adenine. *Astrobiology* 1:481–488.

Stetter, K. O. 1994. The lesson of archaebacteria, p. 114–122. *In* S. Bengtson (ed.), *Early Life on Earth: Nobel Symposium No. 84*. Columbia University Press, New York, NY.

Strasak, M., and F. Sersen. 1991. An unusual reaction of adenine and adenosine on montmorillonite: a new way of prebiotic synthesis of some purine nucleotides? *Naturwissenschaften* 78:121–122.

Sulston, J., and G. Ferry. 2000. *The Common Thread*. Corgi Books, London, United Kingdom.

Tehei, M., B. Franzetti, M.-C. Maurel, J. Vergne, Hountondji, and G. Zaccai. 2002. The search for traces of life: the protective effect of salt on biological macromolecules. *Extremophiles* 6:427–430.

Tian, F., O. Toon, A. Pavlov, and H. De Sterck. 2005. A hydrogen-rich early Earth atmosphere. *Science* 308:1014–1015.

Tice, M. M., and D. R. Lowe. 2004. Photosynthetic microbial mats in the 3,416-Myr-old ocean. *Nature* 431:549–552.

Ueno, Y., K. Yamada, N. Yoshida, S. Maruyama, and Y. Isozaki. 2006. Evidence from fluid inclusions for microbial methanogenesis in the early Archean era. *Nature* 440:516–519.

van Zullen, M., A. Lepland, and G. Arrhenius. 2002. Reassessing the evidence for the earliest traces of life. *Nature* 418:627–630.

Wächtershäuser, G. 1988. Before enzymes and templates: theory of surface metabolism. *Microbiol. Rev.* 52:452–484.

Wächtershäuser, G. 1990. The case for the chemoautotrophic origins of life in an iron-sulfur world. *Origins Life Evol. Biosph.* 20:173–182.

Wetherill, G. W. 1990. Formation of the Earth. *Annu. Rev. Earth Planet. Sci.* 18:205–256.

White, R. H. 1984. Hydrolytic stability of biomolecules at high temperatures and its implication for life at 250°C. *Nature* 310:430–432.

Wiegel, J., and W. W. M. Adams (ed.) 1998. *Thermophiles: The Keys to Molecular Evolution and the Origin of Life?* Taylor & Francis, London, United Kingdom.

Wilde, S., A. J. W. Valley, W. H. Peck, and C. M. Graham. 2001. Evidence from detrital zircons for the existence of continental crust and oceans on the Earth 4.4 Gyr ago. *Nature* 409:175–178.

Woese, C. R. 2002. On the evolution of cells. *Proc. Natl. Acad. Sci. USA* 99:8742–8747.

Yayanos, A. A., A. S. Dietz, and R. van Boxtel. 1981. Obligate barophilic bacterium from the Mariana Trench. *Proc. Natl. Acad. Sci. USA* 78:5212–5215.

II. THERMOPHILES

Chapter 2

Thermal Environments and Biodiversity

ELIZABETH A. BURGESS, ISAAC D. WAGNER, AND JUERGEN WIEGEL

INTRODUCTION

Biodiversity is of interest as it relates to stability and productivity in ecosystems and sustainability of ecosystem functions and is measured generally as species or genetic diversity. When E. O. Wilson, who coined the term biodiversity, wrote *The Diversity of Life*, there were 4,800 species described in the "kingdom" *Monera* (Wilson, 1992). Presently, there are ~7,600 validly published prokaryotic (domain *Bacteria/Archaea*) species (http://www.bacterio.cict.fr/number.html#total, March 2006) and >210,000 aligned 16S rRNA sequences in the Ribosomal Database Project (RDP) database (http://rdp.cme.msu.edu/misc/news.jsp#mar0206, March 2006), and every month ~30–50 more are added. Wilson (1992) estimated the number of known species of all microorganisms to be 1.4 million. Among prokaryotes alone, as many as 1×10^9 species have been hypothesized to exist (Dykhuizen, 1998).

This vast diversity is due, in part, to prokaryotes' ability to exploit extreme environments, such as environments with elevated temperatures, as high as 121°C, referred to as thermal, thermobiotic, or hot environments. Each discovery of environments at high temperatures has opened a new reservoir of biodiversity. Prokaryotes take center stage at high temperatures, although novel eukaryotic microorganisms have also been discovered exploiting the warm edges of thermal environments (Desbruyeres and Laubier, 1980; Jones, 1981). Considering the wealth of data about and the amazing diversity of thermophilic prokaryotes, most of the discussion to follow will focus on *Bacteria* and *Archaea*.

Measuring the biodiversity of prokaryotes can be a daunting task in any environment. Depending on the questions to be answered and the definition of biodiversity, nearly any biological unit of interest may be used. Thus, following a brief introduction to life at high temperatures, this chapter will summarize some of the thermal environments on Earth and describe the taxonomic, genetic, metabolic, and ecological diversity of these environments. This chapter is written with the notion of illustrating that thermal environments are a fascinating source of local and global biodiversity of interest from many perspectives.

HIGH TEMPERATURES AND LIFE

Temperature, as an environmental factor, constrains all living microorganisms. In contrast to the upper temperature boundaries, the lower temperature boundaries for growth among microorganisms are not well defined (N. J. Russell, personal communication; Russell, 1990). Whether bacteria in the permafrost are "hibernating" for thousands to millions of years or just grow extremely slowly with doubling times of hundreds or even thousands of years is not yet unequivocally established. So far, no well-established growth at temperatures below −20°C (Rivkina et al., 2000) has been demonstrated for any member of the *Bacteria*, *Archaea*, or fungi (yeast, −17°C).

Broadly, microorganisms that "love" heat are known as thermophiles. A word of caution is warranted regarding the use of the term thermophilic. The term means different temperature ranges for different groups of microorganisms. For example, *Candida thermophile* is described as a thermophilic yeast with a maximum growth temperature (T_{max}) of 51°C. The optimal growth temperature (T_{opt}) for this microorganism is 30–35°C (Shin et al., 2001). Among *Bacteria*, this would be a thermotolerant species.

Recent observations of Pompeii worms (Wiegel, 1990, 1992) (*Alvinella pompejana*) persisting in

---

E. A. Burgess • Savannah River Ecology Laboratory, Drawer E, Aiken, SC 29803.   **I. D. Wagner** • Department of Microbiology, The University of Georgia, Athens, GA 30602-2605.   **J. Wiegel** • Department of Microbiology, 215 Biological Sciences, The University of Georgia, Athens, GA 30602-2605.

temperatures exceeding 80°C at the tips of their tails (Cary et al., 1998) have generated new debate about the temperature tolerances of eukaryotes (Chevaldonne et al., 2000). Eukaryotes generally do not survive temperatures >60°C (Tansey and Brock, 1972). The amoeba *Echinamoeba thermarum*, found in hot springs in many places on the globe, grows optimally at 50°C and thus is one of the few truly thermophilic eukaryotes (Baumgartner et al., 2003).

Bacteria and *Archaea* can be classified according to their optimal growth temperature as follows: mesophiles (T_{opt} 20–45°C), thermophiles (T_{opt} 45–80°C), and hyperthermophiles (T_{opt} >80°C) (Stetter, 1996; Cavicchioli and Thomas, 2000). Under the classification scheme of Wiegel (1990, 1998a), using both the minimal and the maximal growth temperatures, thermophiles are further subdivided into thermotolerant, which grow optimally at mesophilic temperatures but have maximum growth temperatures >50°C, thermophiles (T_{opt} 50–70°C), and extreme thermophiles (T_{opt} >70°C). Prokaryotes with T_{opt} >65°C and able to grow over a 35–40°C temperature span, e.g., able to grow <40°C and as high as 75°C, are considered temperature-tolerant extreme thermophiles (Wiegel, 1990, 1992). Similarly, there are hyperthermophiles, thermophiles, and mesophiles showing broad temperature spans for growth, all characterized by biphasic temperature–growth curves (Wiegel, 1990, 1998b). The record for the widest temperature span for growth, >50°C, is held by *Methanothermobacter* (basonym *Methanobacterium*) *thermautotrophicus*, able to grow from 22°C to 75°C (J. Wiegel, unpublished data). For reasons of simplification, in this chapter, the term thermophile will be used generally to include all microorganisms with T_{opt} >50°C. Hyperthermophiles, as defined by Stetter (1996), are distinguished from thermophiles only when necessary for clarity.

One might suppose that life exists in nearly any place where all the necessary requirements are met. Liquid water, essential for living microorganisms, is found at 300°C and higher under intense pressure, e.g., at deep-sea vents (Kelley et al., 2002). However, the thermostability of particular cellular constituents such as ATP, amino acids, and peptides indicates that 150°C may be the upper temperature limit for life (White, 1984; Wiegel and Ljungdahl, 1986; Wiegel and Adams, 1998). For a long time, the record for highest T_{max} was held by *Pyrolobus fumarii*, isolated from a deep-sea thermal black smoker vent chimney. *P. fumarii* has a T_{max} of 113°C and a T_{opt} of 106°C and is unable to grow <90°C (Blochl et al., 1997). Strain 121, a Fe(III)-reducing *Archaea* recently isolated from a hydrothermal vent along the Juan de Fuca Ridge, is reported to have a doubling time of 24 h at 121°C and remains viable after exposure to temperatures as high as 130°C (Kashefi and Lovley, 2003); however, this record has yet to be verified with more detailed analyses. In addition, some thermophiles, such as those belonging to the phylogenetic branch of gram-type positive (Wiegel, 1981) bacteria (i.e., *Firmicutes*) (Wiegel et al., 1981), form highly heat-resistant spores (Onyenwoke et al., 2004). The current record for the most heat-resistant spore is held by *Moorella* (basonym *Clostridium*) *thermoacetica* strain JW/DB-4. When grown autotrophically and sporulating at 60°C, this bacterium forms spores with a decimal reduction time (time of exposure to reduce viable spore counts by 90%) of nearly 2 h at 121°C. A subpopulation of spores apparently requires ~1 h at 100°C to become fully activated before germinating (Byrer et al., 2000).

Thermal environments have been the source of a wide variety of biotechnological advances. *Thermus aquaticus*, an aerobic, thermophilic bacterium, was isolated from Yellowstone National Park, Wyoming, in the late 1960s (Brock and Freeze, 1969), and the microorganism's DNA polymerase, coined *Taq*, has become an essential component of molecular biology. Additional polymerases and many other thermostable enzymes from thermophiles are now on the market, and their use in applications will increase in the future. Many chemical industrial processes employing high temperatures have benefited by using different groups of thermophiles for various applications (Wiegel and Ljungdahl, 1986; Lowe et al., 1993), such as anaerobic fermentative microorganisms for waste treatment and fuel production or sulfur-metabolizing organisms to remove sulfur compounds from crude oil.

The anaerobic, hyperthermophilic microorganisms branched early in the universal phylogenetic tree (Stetter, 1996). Considering that these microorganisms persist today in habitats that may have been present, and at times predominant, throughout Earth's geologic history, some assume that they represent linea closely descendent from the first living microorganisms on the planet (Baross, 1998; Schwartzman, 1998; Wachtershauser, 1998). The argument has been made that, considering an absence of geologic evidence from very early Earth and the effects of temperature on organic molecules, life did not originate in hyperthermal environments (Forterre, 1998; Miller and Lazcano, 1998). Still, in the evolution of eukaryotic life, there is evidence for thermophilic ancestors (Lake, 1988; Lake et al., 1998). The biphasic temperature–growth curves of many thermophiles growing at elevated temperatures and the existence of cryptic thermophiles are considered as additional arguments for the start of life in the range of 60–90°C and that hyperthermophiles as well as mesophiles and psychrophiles are adaptations to changed environments (Wiegel, 1998).

Looking for evidence of early life on Earth is similar to looking for the evidence that life, and the necessary water, once existed on Mars (Westall, 2005). As our perception of habitable environments for life has expanded to include hyperthermobiotic environments and combinations of acidic, alkaline, saline, or piezoelectric conditions in environments at elevated temperatures, our technological ability to investigate these habitats and identify evidence of life has also advanced, both on Earth and on other planets (Des Marais and Walter, 1999). For example, thermal intraterrestrial habitats have been recently identified on Earth, which may also persist on planets with assumed uninhabitable surface environments (Gold, 1992).

The island-like nature of thermal environments has made them popular models to test biogeographical hypotheses. Using similar strains of the thermophilic archaeon *Sulfolobus* originating from hot springs in Yellowstone National Park and Italy, Zillig et al. (1980) formulated the hypothesis that "geographical barriers between habitats of the same type do not exist for microorganisms." This hypothesis also corresponds to the oft-quoted hypothesis that "everything is everywhere and the environment selects" (Baas-Becking, 1934 citing Beijerinck, 1913). However, Whitaker et al. (2003) attribute genetic divergence detected by multilocus sequence analysis of strains of *Sulfolobus solfataricus* from five sites to geographic isolation. Papke and Ward (2004) reasoned that biogeography requires physical separation and gave several examples of mechanisms that could mediate such separation among prokaryotes. Experiments with thermophilic cyanobacterial mat samples from several locations supported these arguments (Papke et al., 2003). Multilocus enzyme electrophoretic analysis of *Rhodothermus marinus* isolates from hot springs in Iceland attributed variances to both genetic drift and differences among the four locations, or local selection (Petursdottir et al., 2000). Different species of thermophiles appear to have different modes of population distribution. Examples of cosmopolitan microorganisms from thermal environments include *Methanothermobacter thermautotrophicus*, *Thermanaerobacter* (basonym *Clostridium*) *thermohydrosulfuricus*, *Thermoanaerobacterium thermosaccharolyticum*, and *Geobacillus stearothermophilus* (Wiegel and Ljungdahl, 1986). Although spore formation is certain to be an advantage for high population dispersal, *M. thermautotrophicus* is nonsporulating. Alternatively, examples of species regarded as endemic include members of the aerobic, hyperthermophilic, archaeal genus *Aeropyrum*, which have yet to be isolated from anywhere other than the coastal geothermal fields of southwest Japan (Nakagawa et al., 2004a), and the genus *Methanothermus*, thus far found only in Iceland (Lauerer et al., 1986). Among macroscopic organisms, island biogeography is being reexamined (Lomolino, 2000), and the universality of new paradigms should be enhanced by recent research with thermophiles.

The interaction correlation between biogeography and biogeochemistry in thermal environments is also worthy of note. As an example, when examining the thus-far described anaerobic alkalithermophiles, three combinations can be defined: (i) relaxed biogeography and biogeochemistry, (ii) relaxed biogeography and restricted biogeochemistry, and (iii) restricted biogeography and relaxed biogeochemistry. For example, *Thermoanaerobacterium thermosaccharolyticum* and *Thermobrachium celere* have a relaxed biogeography and biogeochemistry. They have been isolated from a variety of environments from several locations including thermobiotic, mesobiotic, slightly alkaline and acidic environments. *Clostridium paradoxum* and *Clostridium thermoalcaliphilum*, isolated from sewage sludge on four different continents, but only from sewage sludge, thus have a relaxed biogeography but apparently restricted biogeochemical requirements. *Anaerobranca horikoshii* ($pH^{60°C}_{min}$ 6.9; $pH^{60°C}_{opt}$ 8.5; $pH^{60°C}_{max}$ 10.3) is an example of restricted biogeography and relaxed biogeochemistry. *A. horikoshii* has only been isolated from a specific area behind the Old Faithful ranger station in Yellowstone National Park, but from several pools in that area, representing a spectrum of pH values from acidic (pH ~5) to alkaline (pH ~8.5). Although relatively easy to isolate, strains of *A. horikoshii* have not been obtained from other areas of Yellowstone National Park or other countries, nor has its sequence been found in environmental 16S rRNA gene libraries (Engle et al., 1995).

Finally, biogeology/biogeochemistry is especially of interest in thermal environments, where mineralization is active and the role of prokaryotes in mineralization is being examined. An increasing number of studies in thermal environments are linking geochemical conditions and phenomena with specific microbial species and metabolisms (Nubel et al., 2002; Donahoe-Christiansen et al., 2004). Electron microscopy and denaturing gradient gel electrophoresis (DGGE) were well coupled with geochemical analyses in the investigation of arsenite-oxidizing communities at Yellowstone National Park (Jackson et al., 2001; Langner et al., 2001). Prokaryotes can couple mineral formation with energy generation, e.g., the formation of pyrite by *Thiomonas thermosulfatus*, strain 51 (Popa and Kinkle, 2004). Thermophilic metal ion reducers contribute to the deposition of precious metals by dissimilatory reductive precipitation (Kashefi et al., 2001 and references therein). Fe(III) reducers (e.g., Wiegel et al., 2003) frequently can reduce a variety of heavy metal ions and are assumed to have been involved in the deposition of low-temperature banded iron formations. The presence of different types of microorganisms may influence the morphology of siliceous sinters laid down along the edges of hot springs (Frankel and Bazylinski, 2003 and references therein; Konhauser et al., 2004). The formation of such sinters may serve to protect the sinter-forming microorganisms from ultraviolet radiation (Phoenix et al., 2001). Additionally, understanding the role of microorganisms in mineralization in contemporary thermal environments will enhance our understanding of ore deposition, biogeochemistry, and paleogeology (e.g., Karpov and Naboko, 1990; Russel et al., 1998).

TYPES OF THERMAL ENVIRONMENTS

Considering the constraint of high temperatures as an environmental requirement, thermophiles are still able to exploit a diverse array of habitats on Earth. The environment at the surface of a solfatara may have a pH of <1 (Stetter, 1989), and microorganisms at the carbonate-rich Lost City hydrothermal field (Kelley et al., 2005) or sun-heated salt lakes (Mesbah and Wiegel, 2005) may experience pH as high as 11. Water is readily available in circumneutral, freshwater hot springs, but there are thermal environments having low water potentials; e.g., in intraterrestrial environments because of high surface-area-to-water ratios (Pedersen, 2000) or in solar-heated soils and sediments because of evaporation and high salinity (Mesbah and Wiegel, 2005). Thermal environments may be grouped into several types, some of which are described below. Within each type, a variety of factors generate additional diversity.

Terrestrial, Nonanthropogenic Environments

Hot springs, geysers, solfatares (mud or paint pots), and mud or paint pots are found concentrated in volcanically active regions throughout the world, including Iceland, Western North America, New Zealand, Japan, Eastern Russia, and the rest of the so-called Pacific Ring of Fire (Waring, 1965). The Yellowstone National Park area of North America contains the greatest concentration of geothermal features on Earth and was the site of some of the earliest studies of life at high temperatures, dating back to 1897 (as referenced within Reysenbach and Shock, 2002). Depending on local water chemistry and hydrology, such environments can represent a spectrum of pH and geochemical conditions that can vary greatly across small distances (<20 m in the Uzon Caldera, Kamchatka, Russian Far East) (personal observation). Generally, solfatares are sulfur rich and acidic, and isolates from such environments are acidophilic and frequently sulfur metabolizers (Stetter, 1989). Alternatively, waters are more neutral to alkaline if richer in chloride salts or carbonate (Hedenquist, 1991; Zhao et al., 2005). Temperatures in these environments may range from freezing to boiling, even within a single pool or spring (Combie and Runnion, 1996). Such environments are also frequently enriched in elements such as arsenic (As), antimony (Sb), and mercury (Hg), which are toxic to humans, and gold (Au), silver (Ag), and copper (Cu), which are valuable commercially (Karpov and Naboko, 1990).

Solar-heated environments may occur anywhere on Earth receiving solar energy inputs. Such environments are likely inhabited by mesophilic, thermotolerant, and thermophilic microorganisms because solar energy can heat some soils to 60°C and shallow waters to 50°C at certain times of the day or year, as already pointed out by Brock (1970). Thermal environments on Earth's surface also experience evaporation, and thus many environments have elevated salinity and, therefore, halophilic inhabitants. For example, *Thermohalobacter berrensis*, a thermophilic and halophilic bacterium, was isolated from a solar saltern in France (Cayol et al., 2000). In the author's J. Wiegel's laboratory, halophilic (up to 25% NaCl 4.5 M Sodium ion as NaCl/Na_2Co_3), thermophilic (up to 75°C), and alkaliphilic (up to $pH^{25°C}$ 10.5) triple extremophiles, coined haloalkalithermophiles, have been isolated from dry salts from salt flats in Nevada and from sediments of athalassohaline lakes in Wadi An Natrun, Egypt (Mesbah and Wiegel, in press). Note that especially for acidic and alkaline conditions the pH at 25°C can differ from the pH at 60°C by >1 pH unit, and a superscript indicating the temperature at which pH was determined (e.g., $pH^{25°C}$) is recommended (Wiegel, 1998a).

Marine Environments

Marine thermal environments may occur at beaches, e.g., Hot Water Beach (Whitianga, New Zealand), Pozzuol: (Italy), or Savusavu (Fiji Island) (Waring, 1965); under <8 m of water, e.g., vents off the coast of Mílos Island, Greece (Sievert et al., 2000a); or under an abysmal 2,500 m of water, e.g., at deep-sea hydrothermal vents first discovered in 1977 near the Galápagos Islands (Corliss et al., 1979). Venting water can exceed 300°C, but in deep-sea vents it cools quickly upon mixing with cold, deep-sea water, and habitat types range from those preferred by hyperthermophiles to temperatures habitable by spychrophiles (Cary et al., 1998; Kelley et al., 2002). Black smoker chimneys, associated with volcanic spychrophiles activity, plate spreading zones generally are fueled by high concentrations of sulfides (Kelley et al., 2002). Serpentinite-hosted systems, like the Lost City hydrothermal field, are enriched in hydrogen and methane as energy sources (Kelley et al., 2005).

Organisms living at deep-sea hydrothermal vent areas generally cope with additional environmental extremes. For example, they must be at least piezotolerant and may be piezophilic (barophilic). *Thermococcus barophilus*, obtained from the Snake Pit region of the Mid-Atlantic Ridge, requires elevated pressure for growth at or above 95°C (Marteinsson et al., 1999), and *Pyrococcus* strain ES4 shows an extension of T_{max} under increased pressure (Pledger et al., 1994; Summit et al., 1998). At hydrothermal vents, the level of natural radioactivity can be 100 times greater than that at Earth's surface because of increased occurrence of elements such as ^{210}Pb, ^{210}Po, and ^{222}Rn (Cherry et al., 1992, cited in Jolivet et al., 2003). The archaea *Thermococcus gammatolerans*, isolated from a hydrothermal site in Guaymas Basin (Jolivet et al., 2003), *Thermococcus marinus* from the Snake Pit hydrothermal site on the Mid-Atlantic Ridge, and *Thermococcus radiotolerans* from a hydrothermal site in the Guaymas Basin (Jolivet et al., 2004) were isolated from enrichment culture subjected to γ-irradiation on the basis of the rationale that microorganisms from deep-sea hydrothermal vent areas would be adapted to elevated levels of radiation. Additionally, all organisms existing in marine environments also have some tolerance for moderate (around 3%) salinity.

Subsurface Environments

Subsurface thermal environments include petroleum reservoirs and geothermally heated lakes and aquifers. Activity in subsurface environments varies with the availability of nutrients, water, and energy based on depth, surrounding matrix, and source materials. Owing to heterogeneity in the composition of Earth's crust, a variety of sources for energy are readily available (Gold, 1992). Lethal temperatures may not occur until as much as 10,000 m below the surface (Pedersen, 2000), though in some areas, e.g., Uzon Caldera, temperatures well above 100°C can occur at depths of only a few meters (personal observation). A depth record for culturable life has been established at 5,278 m (Szewzyk et al., 1994).

Elevated temperatures found within petroleum reservoirs can be up to 130°C (Grassia et al., 1996). The geochemical conditions in reservoirs are variable because of age, source material, and surrounding geology, and prokaryote communities therein vary as well (Orphan et al., 2003) and include several thermophilic Fe(III) reducers (Slobodkin et al., 1999). Takahata et al. (2000) have proposed that microorganisms in these environments may face oligotrophic conditions. Microorganisms in petroleum reservoirs may be endemic or introduced during drilling or water injection (L'Haridon et al., 1995; Pedersen, 2000). The same is true of ultradeep gold mines in South Africa, where service water containing mesophilic bacteria and fungi can contaminate communities of thermophilic sulfate-reducing bacteria indigenous to the hot, subsurface rocks (Onstott et al., 2003).

Subsurface geothermal aquifers such as the well-known and expansive Great Artesian Basin of Australia are nonvolcanically heated but experience temperatures up to nearly 103°C (Kimura et al., 2005). Nonvolcanic geothermal environments are temporally stable as compared with volcanically heated environments and have low flow rates and long recharge times (e.g., 1,000 years). Thus, the Great Artesian Basin contains microbial communities of ancient composition (Kimura et al., 2005). Microorganisms isolated from these habitats include *Desulfotomaculum australicum* (Love et al., 1993), *Fervidobacterium gondwanense* (Andrews and Patel, 1996), and *Caloramator indicus*, isolated from an artesian aquifer in Surat District, Gujarat, India (Chrisostomos et al., 1996).

Anthropogenic Environments

Anthropogenic habitats include household compost piles and water heaters and industrial process environments and thermal effluent from power plants (Brock, 1970; Stetter, 1989). One of the earliest well known anaerobic thermophiles, *Thermoanaerobacter* (basonym *Clostridium*) *thermohydrosulfuricus*, was isolated from an Austrian sugar factory (Klaushofer and Parkkinen, 1965; Lee et al., 1993). Other thermophiles have been isolated from thermally polluted effluent from a carpet factory (Carreto et al., 1996), the smoldering slag heap of a uranium mine (Fuchs et al., 1996), and mushroom compost (Korn-Wendisch et al., 1995). Strains of *T. aquaticus* have been isolated from various anthropogenic thermal environments including hot tap water and greenhouse soil (Brock and Freeze, 1969).

Temporary Environments and Mesobiotic Environments with Thermal Microniches

Thermophiles can be isolated from various environments, such as animal droppings, manure piles, or compost, temporarily heated by biodegradation of organic material. One example is the acetogenic bacterium *Moorella thermoacetica* (Fontaine et al., 1942). Interesting temporary thermal environments also include sun-heated soils and sediments at the edges of lakes and puddles at the ocean, which can have temperatures up to 50°C but are frequently around 35–45°C (J. Wiegel unpublished data). Whereas most of the thermophiles isolated from these environments are *Firmicutes*, i.e., endospore-forming species, there are exceptions. One example is the archaeon *Methanothermobacter* (basonym *Methanobacterium*) *thermautotrophicus*, for which no resting cell forms are known. This species (or alikes) can be easily isolated from sun-heated black sediments of lakes and mesobiotic sewage plants, but it also has been isolated from sun-heated wood stumps in Georgia, United States (J. Wiegel and L. G. Ljungdahl, unpublished data), and mesobiotic environments such as cold stream sediments in Germany (Wiegel et al., 1981) or sediments of Lake Mendota, Wisconsin, for which temperatures have never been measured <16°C.

The chapter authors believe that many environments that are classified as mesobiotic from their bulk temperature measurements contain temporary thermal microniches, created by localized biodegradation of organic material. An example for this phenomenon could be *Methanothermobacter thermautotrophicus* strain ΔH (isolated from sewage sludge) as well as strain JW500 (isolated from river sediment in Georgia, United States), which have been shown to grow in complex media (simulating a more natural environment) at temperatures as low as 22°C (Wiegel, unpublished). In mineral media, T_{min} is given as 35°C (Zeikus and Wolfe, 1972). However, this methanogen still metabolizes and produces methane at temperatures as low as 16°C (but not measurable within 3 months at 12°C). Metabolic activity and growth at even lower temperatures could be possible (but probably with doubling times of years and thus difficult to measure in laboratory settings). *M. thermautotrophicus* can grow over a temperature span of 55°C, the record of the widest (measured) temperature span over which a microorganism can grow, and thus, in classification of Wiegel, is a temperature-tolerant thermophile (Wiegel, 1990). *C. thermoalcaliphilum*, for which no spore formation could be demonstrated, and although having a lower temperature optimum and smaller temperature span for growth than the methanogen, has been isolated from similar low-temperature sewer sludge (Li et al., 1994).

A second exemplary case is the recent isolation of the alkalithermophilic *T. celere* from several meadows and floodplains (Engle et al., 1996). Interestingly, from the many obtained strains for this species, those isolated from mesobiotic environments have doubling times below 20 min and as low as 10 min, whereas isolates from hot springs have doubling times >30 min (Wiegel, unpublished). A possible explanation for this difference in doubling times is that thermophilic microorganisms in mesobiotic environments must be able to maximize the use of temporarily occurring high-temperature growth conditions and must be able to start growing quickly and grow at high growth rates. Such properties have been found, for example, for the thermoalkaliphile *Clostridium paradoxum*, isolated from mesobiotic anaerobic digestor and aerobic oxidation basin of the Athens and Atlanta, GA, Municipal Sewage Plant; it has a doubling time of 16 min; and is able to start growing exponentially from a spore solution within 30 min from the time of inoculating spores into prewarmed, prereduced media (Y. Li and J. Wiegel, unpublished data). In contrast, thermophiles living in stable thermal environments such as hot spring pools and sediments, living all the time at their growth temperature and more or less at constant,—although possibly low-substrate concentrations, do not have that selection pressure for very rapid growth as long as their residence time in the pool is longer than their doubling time.

CULTURAL DIVERSITY

Most of the microorganisms from nearly all environments they inhabit are presently uncultured (Hugenholtz, 2002). Considering the extreme conditions in which most thermophiles thrive, some require special handling or novel approaches for their enrichment, culturing, and isolation (Mesbah and Wiegel, in press; Wiegel, 1986).

An overview of the validly published and well-characterized (i. e, those with reported optimal growth conditions) thermophilic and hyperthermophilic microorganisms in isolation can be given in graphs using the cardinal data pH_{opt} versus T_{opt} (Figs 1 and 2, http://www.bacterio.net, March 2006). Figures 1 and 2 further indicate the extreme optima beyond which no microorganisms have been isolated, because either they do not exist or applied isolation conditions were insufficient for their growth. As already pointed out by Stetter in 1989, thermophiles that grow optimally at low pH tend to be aerobic, especially among *Bacteria* (Fig. 2). At the highest known growth temperatures, most of the microorganisms appear to be neutronphilic and *Archaea* (Fig. 1). Such trends among microorganisms in culture are likely the result of multiple interacting factors of the environments and the microorganisms as well as the objectives and limitations of the enrichments and isolations. However,

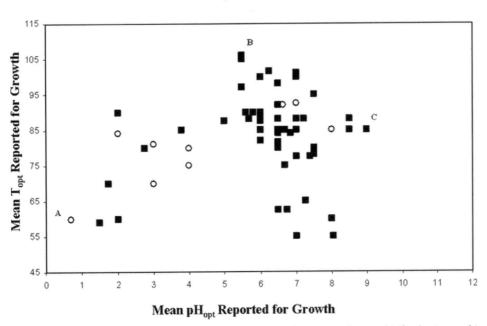

Figure 1. Thermophilic archaea: aerobic/microaerophilic/facultative aerobic archaea (○) and anaerobic/facultative aerobic archaea (■). A, *P. oshimae* and *P. torridus*: optimal growth at pH 0.7, 60°C (Schleper et al., 1995). B, *P. fumarii*: optimal growth at 106°C, pH 5.5 (Blochl et al., 1997). C, *Thermococcus acidaminovorans*: optimal growth at pH 9, 85°C (Dirmeier et al., 1998).

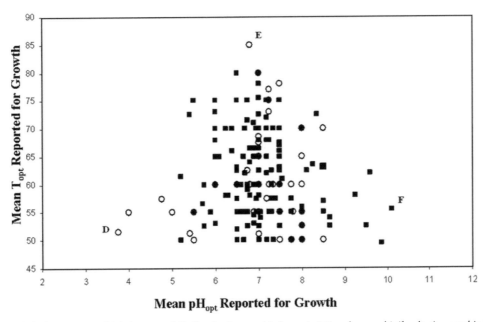

Figure 2. Thermophilic bacteria: aerobic/microaerophilic/facultative aerobic bacteria (○) and anaerobic/facultative aerobic bacteria (■). D, *Alicyclobacillus hesperidum*: optimal growth at pH 3.5–4.0, 50–53°C (Albuquerque et al., 2000). E, *Aquifex pyrophilus*: optimal growth at 85°C, pH 6.8 (Huber et al., 1992). F, *C. paradoxum*: optimal growth at $pH^{25°C}$ 10.1, 55–56°C (Li et al., 1993).

enrichments for specific types of microorganisms can yield commercially useful resources (Wiegel and Ljungdahl, 1986) or reveal information about how various environmental factors alter the distribution of different microorganisms (Grassia et al., 1996).

In 1996, Stetter identified 54 species of hyperthermophilic prokaryotes in 25 genera and 11 orders, and as of March 2006, among the well-characterized and validly published microorganisms, ~64 species of hyperthermophilic prokaryotes and >350 species of thermophiles have been isolated. Well represented among the ~30 genera of *Archaea* are *Thermococcus*, *Sulfolobus*, and *Methanococcus*. Other genera such as *Thermococcus*, *Sulfurisphaera*, *Methanotorris*, and *Acidilobus* at present contain few isolated species. Most of the hyperthermophiles are *Archaea*, though the bacterial genus *Aquifex* is represented by a hyperthermophilic strain and members of the bacterial genus *Thermotoga* have $T_{opt} \geq 80°C$. Of the *Bacteria* depicted in Fig. 2, there are >110 different genera represented. Except for the genera *Geobacillus*, *Thermoanaerobacter*, *Desulfotomaculum*, and *Clostridium*, each genus is represented by <10 isolates. This does not take into consideration recent debates on the taxonomy of these genera. For example, there are many species that do not belong to the genus *Clostridium* sensu stricto (Rainey et al., 2006).

PHYLOGENETIC AND GENETIC DIVERSITY

Upon an extensive analysis of bacterial evolution, Woese (1987) presented the small subunit rRNA gene as an espalier on which to train our universal phylogenetic tree. Among prokaryotes, within the 16S rRNA phylogeny, thermophily is polyphyletic. As mentioned above, hyperthermophiles, in general, branch early in the universal phylogenetic tree (Stetter, 1996). However, among thermophilic bacteria, there are mesophilic and thermophilic members in some of the same phylogenetic groups and genera (e.g., *Firmicutes*), perhaps as a result of adaptation to mesobiotic temperatures from thermophilic ancestors or via lateral gene transfer from thermophilic taxa (Wiegel, 1998b). Amplification of 16S rRNA genes directly from environmental DNA has revealed an astonishing amount of diversity among prokaryotes; e.g., Barns et al. (1994) and Hugenholtz et al. (1998) identified novel lineages of thermophilic *Archaea* and *Bacteria* in Obsidian Pool (OP) in Yellowstone National Park, via amplification of 16S rRNA genes in environmental DNA extracted from sediment samples. Sequences similar to the OP sequences have since been identified in many locations, including the Great Artesian Basin, Australia (Kimura et al., 2005), and thermal pools in Kamchatka, Russian Far East (E. A. Burgess, unpublished data). Environmental 16S rRNA sequence similarities to characterized species do not always directly correspond to the same in situ niche or physiology but have led to targeted culturing of novel microorganisms first identified in clone libraries (Huber et al., 1995; Ghosh et al., 2003).

In general, PCR biases, also present in the real-time PCR approach, limit the ability to determine quantitatively the relative abundance among all prokaryote populations and, consequently, the diversity in thermal environments via environmental 16S rRNA gene analyses (Wintzingerode et al., 1997; Baker and Cowan, 2004). Variations in experimental design and implementation among studies also inhibit direct comparisons. However, a brief, qualitative summary of richness as number of environmental 16S rRNA restriction enzyme phylotypes in a selection of environments is presented (Table 1). In general, these studies were coupled with some sequencing. Differences among dominant phylogenetic groups were observed. For example, from an artesian well in the Great Artesian Basin, Australia, nine *Archaeal* groups were dominated by methanogen-like sequences from *Euryarchaeota* (35 of 59 clones) (Kimura et al., 2005). From a hot spring in Thailand, 17 of 25 clones grouped with noncultivated *Crenarchaeota* sequences (Kanokratana et al., 2004). Sequences from deep-sea hydrothermal vents obtained by enrichment of in situ microorganisms within an innovative growth chamber (first deployed at the Mid-Atlantic Ridge and since then at other deep-sea hydrothermal vent locations) led to the identification of novel lineages among *Archaea* and *Bacteria* (Reysenbach et al., 2000).

Although many novel and diverse 16S rRNA sequences have been described from various thermal environments, some thermal environment communities may contain only a few phylogenetic types. Reysenbach and Shock (1994) identified only three major phylogenetic groups out of 35 clones analyzed from pink filaments collected at Octopus Spring, Yellowstone National Park. Environmental DNA from siliceous sinters in Yellowstone National Park contained predominantly 16S rRNA sequences related to *Thermocrinis ruber*, and the dominating abundance of *T. ruber* in the sinter community was verified by fluorescent in situ hybridization (FISH) (Blank et al., 2002). Interestingly, the populations were represented by diverse morphotypes when examined with electron microscopy (Blank et al., 2002). Thus, the resolution of 16S rRNA to represent genetic diversity in prokaryote communities appears to be limited. The diversity of other ribosomal operon sequences, such as internal transcribed spacer (ITS) regions and 23S rRNA genes, has been useful to differentiate

Table 1. Comparison of richness of restriction enzyme phylotypes (types) from clone libraries of environmental 16S rRNA genes from a small selection of thermal environments

Environment sampled	Specificity of primers used	No. of clones analyzed	No. of types identified	Reference
Hot spring, Yellowstone National Park	2 universal sets and bacterial set	>300	95 total from all three libraries	Hugenholtz et al., 1998
Hydrothermal vent, Mid-Atlantic	Archaeal set and bacterial set	85	38 (11 archaea, 27 bacteria)	Reysenbach et al., 2000
Two hydrothermal vents, Juan de Fuca	Archaeal set and bacterial set	141	64 and 59 total from each site	Alain et al., 2002
Petroleum reservoirs, California	Universal set and archaeal set	68	9 archaea	Orphan et al., 2003
Hot spring, Thailand	Archaeal set and bacterial set	200	24 archaea, 36 bacteria	Kanokratana et al., 2004
Artesian well, Australia	Archaeal set and Bacterial set	59 in each library	9 archaea, 7 bacteria	Kimura et al., 2005

among closely related strains of thermophilic prokaryotes (Moreno et al., 2002; Papke et al., 2003; I. D. Wagner, unpublished data).

Besides using ribosomal RNA genes for describing biodiversity, the sequences of "housekeeping" genes have proven to be very useful, especially when sequences from many strains are available (Santos and Ochman, 2004). However, not all genes are informative, as revealed by the presence/absence of sporulation-specific genes in the genomes of non- and asporulating *Firmicutes* (Onyenwoke et al., 2004). The selection of useful genes also depends on the physiological groups of interest. For example, novel lineages of dissimilatory sulfite reductase (*dsr*) genes have been identified in thermal environments and associated with high rates of sulfate reduction (Fishbain et al., 2003). Analysis of *dsr* has been used to compare prokaryote communities and sulfate-reducing bacterial lineages in thermal environments from underground mines in Japan to multiple hydrothermal vents (Nakagawa et al., 2002, 2004b) and ultradeep gold mines in Africa to basalt aquifers in Washington state (Baker et al., 2003), revealing that closely related *dsr* sequences are often found in distant locations.

Genomic analyses are the new avenue of investigation of genetic diversity. The information from genomic sequences (http://www.genomesonline.org) and the resulting proteomic information need to be linked to the geochemical processes occurring in the environment (Reysenbach and Shock, 2002) or the geographical distribution of prokaryote diversity (Whitaker et al., 2003). Furthermore, metagenomics has become a popular tool and, for example, has been used to identify a thermostable polymerase related to that of *Thermoplasma acidophilum* from a deep-sea vent sample (Moussard et al., 2006). The metagenomic approach is predicted to become much more frequently applied in a large variety of environments as sequencing becomes more affordable. Eventually, there will be a wealth of information available on the physiological and biogeochemical potential of prokaryote communities in the studied environments. Mining this metagenomic data will lead to in silico models of how the interwoven interactions between microorganisms, their metabolic activities, and the geochemistry of their environments occur in situ, and such models should be, at least in part, experimentally verifiable.

METABOLIC DIVERSITY

Although nearly all types of microbial metabolism are observed in thermal environments, chemolithotrophy (obtaining energy from inorganic electron-donating and -accepting reactions) is a cornerstone of hyperthermophilic communities. In general, chemolithotrophy has been observed both in autotrophic (obtaining carbon from inorganic sources such as CO_2) and in heterotrophic organisms (obtaining carbon from organic sources such as sugars or volatile fatty acids). Chemolithoautotrophy is considered the source of primary production in sunless environments and environments too hot for photoautotrophic production. For example, Blank et al. (2002) suggest that because chemolithoautotrophs, in particular bacteria of the order *Aquificales*, inhabit hot springs in Japan, Iceland, Kamchatka, and Yellowstone National Park, these types of microorganisms may be the primary producers in these ecosystems.

Many thermal environments are anaerobic or of reducing nature, either because of the remoteness of the environment from the atmosphere, the low solubility of oxygen in water at elevated temperature, as well as hypersalinity in some cases, or because of the inputs of gasses such as H_2 and reducing H_2S (Brock, 1970; Stetter, 1996). Thus, there is abundant chemical energy for chemolithotrophs, and anaerobic respiration predominates among respiring microorganisms. Amend and Shock (2001) calculated the energetics of many metabolic reactions in environments with elevated temperatures. Notably, anaerobic thermophilic *Bacteria* are generally unable to grow at acidic pH, and most acidophilic, thermophilic *Archaea* are obligate aerobes (Wiegel, 1998). There are neutrophilic, facultatively aerobic *Archaea*, such as *Pyrobaculum oguniense* (Sako et al., 2001), as well as examples of anaerobic acidophilic *Archaea*, such as *Stygiolobus azoricus* (Segerer et al., 1991). The most acidophilic *Archaea* are *Picrophilus oshimae* and *Picrophilus torridus*, growing optimally at pH ~0.7 (Schleper et al., 1995). The most acidophilic anaerobic bacteria presently are *Thermoanaerobacterium aotearoensis*, $pH^{60°C}_{opt}$ 5.2 and $pH^{60°C}$ range 3.8–6.8 (Liu et al., 1996), and *Lebetimonas acidiphila*, pH_{opt} 5.2 and pH range 4.2–7.0 (Takai et al., 2005), though for *L. acidiphila* the temperature of pH measurement has not been reported.

Most of the hyperthermophilic microorganisms are *Archaea*, and most of these perform chemolithotrophic metabolism for energy, including methanogenesis, sulfate reduction, sulfur oxidation, sulfur reduction, nitrate reduction, and hydrogen oxidation (Stetter, 1989; Slobodkin et al., 1999). Kletzin et al. (2004) provide a review of elemental sulfur metabolism in *Archaea*. Examples of physiological types among the *Archaea* are *Archaeoglobus*, *Thermodiscus*, *Thermoproteus*, *Acidianus*, and *Desulfurococcus* reducing sulfur or sulfate, *Sulfolobus* species generally oxidizing H_2S or elemental sulfur, *Methanothermus*, *Methanococcus*, and *Methanopyrus* being methanogenic, and *Pyrobaculum* and *Pyrolobus* reducing nitrate (Stetter, 1996).

Among *Bacteria* are anaerobic *Firmicutes* (gram-type positive, although among the thermophilic Firmicutes, many stain Gram reaction negative at all growth phases; Wiegel, 1981), such as the facultative chemolithoautotrophs *Moorella thermacetica*, producing acetate from CO_2/H_2 or from CO (100% gas phase) but also carrying out homoacetogenic fermentations from carbohydrates, and the anaerobe *Ammonifex degensii*, capable of forming ammonium from nitrate via chemolithoautotrophic growth (Huber et al., 1996). Other examples of chemolithoautotrophic thermophilic *Bacteria* are Fe(III) reducers such as *Thermolithobacter ferrireducens*, isolated from various hot springs, and *Thermolithobacter* (basonym *Carboxydothermus*) *carboxydivorans*, a hydrogenic CO utilizer (Sokolova et al., in press; Svetlichny et al., 1991; Wiegel et al., 2003). Recently, several novel, some obligately hydrogenic, CO utilizers have been described (Sokolova et al., 2005 and references therein). Photoheterotrophs such as *Chloroflexus aggregans*, *Chloroflexus aurantiacus*, *Heliobacterium modesticaldum*, and *Roseiflexus castenholzii* use light for generating energy but use organic carbon as their carbon source; i.e., they are unable to utilize CO_2 as would a photoautotroph (Pierson and Castenholz, 1974; Hanada et al., 1995, 2002; Kimble et al., 1995).

In some examples, in situ geochemistry of thermal environments may be shaping the dominant metabolisms or perhaps is shaped by the dominant metabolisms (Orphan et al., 2003). Microorganisms have been isolated which are obligate users of gasses escaping in volcanic regions, such as CO, as mentioned above. They are using a novel chemotrophic metabolism, i.e., using 100% CO as sole carbon and energy source via the reaction $CO + H_2O \rightarrow H_2 + CO_2$ (Svetlichny et al., 1991; Sokolova et al., 2005). Many thermal environments are enriched in elements that are toxic to humans, such as As and selenium (Se), and some microorganisms in these habitats use toxic, redox-active elements to gain energy, via either oxidation or reduction (Huber et al., 2000; Donahoe-Christiansen et al., 2004). Respirationally diverse Fe(III) reducers are frequently able to reduce many redox-active compounds in thermal environments (Kieft et al., 1999; Kashefi et al., 2001).

In situ radiolabel rate measurements and stable isotope analyses are yielding increasing amounts of information about the metabolism and ecology of microorganisms in thermal environments. Even with little knowledge of specific microorganisms responsible for activities in situ, metabolically active microorganisms can be quantified, the metabolic potential of the communities identified, and the distributions of different metabolic types determined (Burgess, unpublished; Harmsen et al., 1997; Slobodkin et al., 2001). Temperature limits for sulfate reduction have been expanded to nearly 100°C by radiotracer experiments in marine sediments (Jorgensen et al., 1992).

ECOLOGICAL DIVERSITY

There are diverse ecological interactions associated with the biology in geothermally and anthropogenically heated environments, and thermophilic prokaryotes have opened our eyes to many novel modes of life. The discovery of deep-sea hydrothermal vent communities demonstrated that life can exist at temperatures >100°C as well as at 2°C in the surrounding ocean, without inputs of solar energy, on the basis of associated microbial vent community (Corliss et al., 1979). Novel symbioses between eukaryotes and prokaryotes have been identified at deep-sea vents, such as the association between the tube worm *Riftia pachyptila* and chemosynthetic, sulfur-oxidizing bacteria (Cavanaugh et al., 1981) or the Pompeii worm, which may derive some of its thermotolerance from the eurythermal enzymes of a community of prokaryotes living on its back (Cottrell and Cary, 1999; Chevaldonne et al., 2000).

Microecology is a relatively young and rapidly expanding field, and thermal environments may be model systems for examining microecological concepts. Generally, the unit of ecological diversity is the species (Magurran, 1988). The species unit has become increasingly difficult to define among prokaryotes (Kampfer and Rossello-Mora, 2004; Gevers et al., 2005). Small subunit rRNA sequence diversity can be used as a proxy for species diversity (Stackebrandt et al., 2002) (Table 1). Moreover, an increasing number of statistically powerful tools have become available for comparison of 16S rRNA sequences (Singleton et al., 2001; Cole et al., 2005; Schloss and Handelsman, 2005). For example, the diversity of microorganisms within mats of cyanobacteria inhabiting hot springs has been examined with the intent to demonstrate the importance of a prokaryote species concept based on the role of the microorganisms in their environments (Ward, 1998; Ward et al., 1998). Classical ecological concepts such as energy partitioning and trophic transfer have been examined using models developed at Yellowstone National Park (Wiegert and Fraleigh, 1972). Viruses have been hypothesized as a top-down control on prokaryote communities (Torsvik et al., 2002). Perhaps the importance of this control may be measured in model thermal environment systems where viruses are present, novel, and quite diverse (Prangishvili and Garrett, 2004 and references therein).

Molecular tools have been largely responsible for many advances in our perception of the ecology in thermal environments, and the application of culture-independent methods has greatly enhanced our knowledge of the diversity and physiological potential of difficult-to-culture microorganisms that inhabit these environments. The effect of different geochemical parameters, including temperature, on the distribution and community structure of prokaryotes has been examined through genetics and the distribution of different metabolic types (Sievert et al., 1999, 2000b; Ramsing et al., 2000; Skirnisdottir et al., 2000; Norris et al., 2002; Orphan et al., 2003), and some of these data are demonstrating trends of prokaryote communities changing in predictable ways with changing environmental conditions. Methods such as FISH enable the examination of the structural distribution of microorganisms of known phylogenetic affiliations (Nubel et al., 2002). Lipids present within the membranes of prokaryotes can be diagnostic for various types of microorganisms and have provided insight into the distribution of microorganisms among different environments. For example, analysis of glycerol dialkyl glycerol tetraethers (GDGTs) from selected hot springs in Nevada revealed the presence of the archaeal lipid crenarchaeol, previously found only in low-temperature, marine environments. Additionally, DGGE band sequences of 16S rRNA genes from these springs were related to thermophilic *Crenarchaeota*, demonstrating that the presence of crenarchaeol is not exclusive to the cold-adapted, marine branch of the *Crenarchaeota* (Pearson et al., 2004). This conclusion has substantial implications for the theory of the origins of *Crenarchaeota* (Zhang et al., 2006).

CONCLUSIONS

As described above, thermal environments are an amazing reservoir of biodiversity. Especially among prokaryotes, some *Archaea* have optimum temperatures for growth in excess of 100°C. This vast biodiversity in turn has been a source of advancement in the fields of biotechnology, evolutionary biology, astrobiology, biogeography, biogeochemistry, and biogeology. Among the different types of environments with elevated temperatures, ranging from volcanic and geothermally heated to anthropogenic sources, to temporary thermal microniches in otherwise mesobiotic environments, there is a diverse array of unique habitats where often many extreme conditions converge, e.g., sun-heated haline and alkaline salt lakes. An important point that becomes evident when examining these environments and their biodiversity of archaea, bacteria, viruses, and even a few eukaryotic species is the vast diversity of approaches and methods that have been used in an attempt to measure it. A count of cultured thermophiles, >350 well-characterized and validly published, does not represent the much larger number of prokaryotes presently known only through environmental 16S rRNA sequences. Measuring the genetic diversity of 16S rRNA and functional genes, which has been an avenue for discovery of many enzymes for biotechnological applications and the isolation of novel microorganisms, provides only limited information about their in situ abundance and activity. Among the many modes of life, including metabolic pathways, found to be present in thermal environments are some that are unique to thermophiles and some that are employed by their mesophilic relatives. These diverse species, sequences, and their metabolisms in turn enhance the ecological diversity of the planet and play an important role in the evolution and stability of the planet's biota. In general, analysis of multiple approaches applied in single environments combined with that of similar approaches in different environments has enhanced the robustness of our understanding of the various high-temperature environments and the biodiversity they harbor. Considering how the initial discovery of life in shallow and deep-sea vents expanded our notion of global biodiversity, future approaches and discoveries, perhaps also in extraterrestrial thermal environments, will likely reveal additional information relevant to many fields of basic and applied science.

Acknowledgments. Owing to the wealth of research related to biodiversity of thermal environments and the pages allowed for this chapter, it was inevitable that we have been unable to include all the pertinent publications, and thus we apologize to all colleagues whose relevant work has not been cited. Details about the diverse adaptations of thermophiles to elevated temperatures are well covered in other chapters of this book. J. W. is especially indebted to Lars G. Ljungdahl (University of Georgia), who is celebrating his 80th birthday and—30 years ago—introduced him to the wonderful world of anaerobic thermophiles. The writing of this review and much of the cited work from the laboratory of J. Wiegel was supported through the NSF—Microbial Observatory grant NSF-MCB 0238407 and NSF-MIP 0348180.

REFERENCES

Alain, K., M. Olagnon, D. Desbruyeres, A. Page, G. Barbier, S. K. Juniper, J. Querellou, and M.-A. Cambon-Bonavita. 2002. Phylogenetic characterization of the bacterial assemblage associated with mucous secretions of the hydrothermal vent polychaete *Paralvinella palmiformis*. *FEMS Microbiol. Ecol.* **42:**463–476.

Albuquerque, L., F. A. Rainey, A. P. Chung, A. Sunna, M. F. Nobre, R. Grote, G. Antranikian, and M. S. da Costa. 2000. *Alicyclobacillus hesperidum* sp. nov. and a related genomic species from solfataric soils of Sao Miguel in the Azores. *Int. J. Syst. Evol. Microbiol.* **50:**451–457.

Amend, J. P., and E. L. Shock. 2001. Energetics of overall metabolic reactions of thermophilic and hyperthermophilic Archaea and bacteria. *FEMS Microbiol. Rev.* **25:**175–243.

Andrews, K. T., and B. K. Patel. 1996. *Fervidobacterium gondwanense* sp. nov., a new thermophilic anaerobic bacterium isolated from nonvolcanically heated geothermal waters of the Great Artesian Basin of Australia. *Int. J. Syst. Bacteriol.* **46:**265–269.

Baas-Becking, L. G. M. 1934. *Geobiologie of Inleiding Tot de Miliekunde*. Van Stockkum & Zoon, The Hague, Netherlands.

Baker, B. J., D. P. Moser, B. J. MacGregor, S. Fishbain, M. Wagner, N. K. Fry, B. Jackson, N. Speolstra, S. Loos, K. Takai, B. S. Lollar, J. Fredrickson, D. Balkwill, T. C. Onstott, C. F. Wimpee, and D. A. Stahl. 2003. Related assemblages of sulphate-reducing bacteria associated with ultradeep gold mines of South Africa and deep basalt aquifers of Washington State. *Environ. Microbiol.* **5:**267–277.

Baker, G. C., and D. A. Cowan. 2004. 16S rDNA primers and the unbiased assessment of thermophile diversity. *Biochem. Soc. Trans.* **32:**218–220.

Barns, S. M., R. E. Fundyga, M. W. Jefferies, and N. R. Pace. 1994. Remarkable archaeal diversity detected in a Yellowstone National Park hot spring environment. *Proc. Natl. Acad. Sci. USA* **91:**1609–1613.

Baross, J. A. 1998. Do the geological and geochemical records of the early Earth support the prediction from global phylogenetic models of a thermophilic cenancestor? p. 3–18. *In* J. Wiegel and M. W. W. Adams (ed.), *Thermophiles: The Keys to Molecular Evolution and the Origin of Life?* Taylor & Francis, Inc., Philadelphia, PA.

Baumgartner, M., A. Yapi, R. Grobner-Ferreira, and K. O. Stetter. 2003. Cultivation and properties of *Echinamoeba thermarum* n. sp., an extremely thermophilic amoeba thriving in hot springs. *Extremophiles* **7:**267–274.

Beijerinck, M. W. 1913. *De infusies en de ontdekking der backterien, Jaarboek van de Koninklijke Adakemie van Wetenschappen*. Muller, Amsterdam, Netherlands.

Blank, C. E., S. L. Cady, and N. R. Pace. 2002. Microbial composition of near-boiling silica-depositing thermal springs throughout Yellowstone National Park. *Appl. Environ. Microbiol.* **68:**5123–5135.

Blochl, E., R. Rachel, S. Burggraf, D. Hafenbradl, H. W. Jannasch, and K. O. Stetter. 1997. *Pyrolobus fumarii*, gen. and sp. nov., represents a novel group of archaea, extending the upper temperature for life to 113°C. *Extremophiles* **1:**14–21.

Brock, T. D. 1970. High temperature systems. *Annu. Rev. Ecol. Syst.* **1:**191–220.

Brock, T. D., and H. Freeze. 1969. *Thermus aquaticus* gen. n. and sp. n. a non-sporulating extreme thermophile. *J. Bacteriol.* **98:**289–297.

Byrer, D. E., F. A. Rainey, and J. Wiegel. 2000. Novel strains of *Moorella thermoacetica* from unusually heat-resistant spores. *Arch. Microbiol.* **174:**334–339.

Carreto, L., E. Moore, M. F. Nobre, R. Wait, R. W. Riley, R. J. Sharp, and M. S. da Costa. 1996. *Rubrobacter xylanophilus* sp. nov., a new thermophilic species isolated from a thermally polluted effluent. *Int. J. Syst. Bacteriol.* **46:**460–465.

Cary, S. C., T. Shank, and J. Stein. 1998. Worms bask in extreme temperatures. *Nature* **391:**545–546.

Cavanaugh, C. M., S. L. Gardiner, M. L. Jones, H. W. Jannasch, and J. B. Waterbury. 1981. Prokaryotic cells in the hydrothermal vent tube worm *Riftia pachyptila* Jones: possible chemoautotrophic symbionts. *Science* **213:**340–342.

Cavicchioli, R., and T. Thomas. 2000. Extremophiles, p. 317–337. *In* J. Lederberg (ed.), *Encyclopedia of Microbiology*, 2nd ed, vol. 2. Academic Press, San Diego, CA.

Cayol, J. L., S. Ducerf, B. K. Patel, J. L. Garcia, P. Thomas, and B. Ollivier. 2000. *Thermohalobacter berrensis* gen. nov., sp. nov., a thermophilic, strictly halophilic bacterium from a solar saltern. *Int J. Syst. Evol. Microbiol.* **50:**559–564.

Chevaldonne, P., C. R. Fisher, J. J. Childress, D. Desbruyeres, D. Jollivet, F. Zal, and A. Toulmond. 2000. Thermotolerance and the 'Pompeii worms'. *Mar. Ecol. Prog. Ser.* **208:**293–295.

Chrisostomos, S., B. K. C. Patel, P. P. Dwivedi, and S. E. Denman. 1996. *Caloramator indicus* sp. nov., a new thermophilic anaerobic bacterium isolated from the deep-seated nonvolcanically heated waters of an Indian artesian aquifer. *Int. J. Syst. Bacteriol.* **46:**497–501.

Cole, J. R., B. Chai, R. J. Farris, Q. Wang, S. A. Kulam, D. M. McGarrell, G. M. Garrity, and J. M. Tiedje. 2005. The Ribosomal Database Project (RDP-II): sequences and tools for high-throughput rRNA analysis. *Nucleic Acids Res.* **33:**D294–D296.

Combie, J., and K. Runnion. 1996. Looking for diversity of Yellowstone extremophiles. *J. Ind. Microbiol.* **17:**214–218.

Corliss, J. B., J. Dymond, L. I. Gordon, J. M. Edmond, R. P. von Herzen, R. D. Ballard, K. Green, D. Williams, R. D. Bainbridge,

K. Crane, and T. H. van Andel. 1979. Submarine thermal springs on the Galapagos Rift. *Science* **203**:1073–1083.

Cottrell, M. T., and S. C. Cary. 1999. Diversity of dissimilatory bisulfite reductase genes of bacteria associated with the deep-sea hydrothermal vent polychaete annelid *Alvinella pompejana*. *Appl. Environ. Microbiol.* **65**:1127–1132.

Des Marais, D. J., and M. R. Walter. 1999. Astrobiology: exploring the origins, evolution, and distribution of life in the universe. *Annu. Rev. Ecol. Syst.* **30**:397–420.

Desbruyeres, D., and L. Laubier. 1980. *Alvinella pompejana* gen. sp. nov., aberrant Ampharetidae from East Pacific Rise hydrothermal vents. *Oceanol. Acta* **3**:267–274.

Dirmeier, R., M. Keller, D. Hafenbradl, F. J. Braun, R. Rachel, S. Burggraf, and K. O. Stetter. 1998. *Thermococcus acidaminovorans* sp. nov., a new hyperthermophilic alkalophilic archaeon growing on amino acids. *Extremophiles* **2**:109–114.

Donahoe-Christiansen, J., S. D'Imperio, C. R. Jackson, W. P. Inskeep, and T. R. McDermott. 2004. Arsenite-oxidizing *Hydrogenobaculum* strain isolated from an acid-sulfate-chloride geothermal spring in Yellowstone National Park. *Appl. Environ. Microbiol.* **70**:1865–1868.

Dykhuizen, D. E. 1998. Santa Rosalia revisited: why are there so many species of bacteria? *Antonie Leeuwenhoek* **73**:25–33.

Engle, M., Y. Li, C. R. Woese, and J. Wiegel. 1995. Isolation and characterization of a novel alkalitolerant thermophile, *Anaerobranca horikoshii* gen. nov., sp. nov. *Int. J. Syst. Bacteriol.* **45**:454–461.

Engle, M., Y. Li, F. Rainey, S. DeBlois, V. Mai, A. Reichert, F. Mayer, P. Messner, and J. Wiegel. 1996. *Thermobrachium celere* gen. nov., sp. nov., a rapidly growing thermophilic, alkalitolerant, and proteolytic obligate anaerobe. *Int. J. Syst. Bacteriol.* **46**:1025–1033.

Fishbain, S., J. G. Dillon, H. L. Gough, and D. A. Stahl. 2003. Linkage of high rates of sulfate reduction in Yellowstone hot springs to unique sequence types in the dissimilatory sulfate respiration pathway. *Appl. Environ. Microbiol.* **69**:3663–3667.

Fontaine, F. E., W. H. Peterson, E. McCoy, M. J. Johnson, and G. J. Ritter. 1942. A new type of glucose fermentation by *Clostridium thermoaceticum* n. sp. *J. Bacteriol.* **43**:701–715.

Forterre, P. 1998. Were our ancestors actually hyperthermophiles? Viewpoint of a devil's advocate, p. 137–146. *In* J. Wiegel and M. W. W. Adams (ed.), *Thermophiles: The Keys to Molecular Evolution and the Origin of Life?* Taylor & Francis, London, United Kingdom.

Frankel, R. B. and D. A. Bazylinski. 2003. Biologically induced mineralization by bacteria. *Rev. Mineral. Geochem.* **54**:95–114.

Fuchs, T., H. Huber, K. Teiner, S. Burggraf, and K. O. Stetter. 1996. *Metallosphaera prunae*, sp. nov., a novel metal-mobilizing, thermoacidophilic Archaeum, isolated from a uranium mine in Germany. *Syst. Appl. Microbiol.* **18**:560–566.

Gevers, D., F. M. Cohan, J. G. Lawrence, B. G. Spratt, T. Coenye, E. J. Feil, E. Stackebrandt, Y. Van de Peer, P. Vandamme, F. L. Thompson, and J. Swigs. 2005. Re-evaluating prokaryotic species. *Nat. Rev. Microbiol.* **3**:733–739.

Ghosh, D., B. Bal, V. K. Kashyap, and S. Pal. 2003. Molecular phylogenetic exploration of bacterial diversity in a Bakreshwar (India) hot spring and culture of *Shewanella*-related thermophiles. *Appl. Environ. Microbiol.* **69**:4332–4336.

Gold, T. 1992. The deep, hot biosphere. *Proc. Natl. Acad. Sci. USA* **89**:6045–6049.

Grassia, G. S., K. M. McLean, P. Glenat, J. Bauld, and A. J. Sheehy. 1996. A systematic survey for thermophilic fermentative bacteria and archaea in high temperature petroleum reservoirs. *FEMS Microbiol. Ecol.* **21**:47–58.

Hanada, S., A. Hiraishi, K. Shimada, and K. Matsuura. 1995. *Chloroflexus aggregans* sp. nov., a filamentous phototrophic bacterium which forms dense cell aggregates by active gliding movement. *Int. J. Syst. Bacteriol.* **45**:676–681.

Hanada, S., S. Takaichi, K. Matsuura, and K. Nakamura. 2002. *Roseiflexus castenholzii* gen. nov., sp. nov., a thermophilic, filamentous, photosynthetic bacterium that lacks chlorosomes. *Int. J. Syst. Evol. Microbiol.* **52**:187–193.

Harmsen, H. J. M., D. Prieur, and C. Jeanthon. 1997. Distribution of microorganisms in deep-sea hydrothermal vent chimneys investigated by whole-cell hybridization and enrichment culture of thermophilic subpopulations. *Appl. Environ. Microbiol.* **63**:2876–2883.

Hedenquist, J. W. 1991. Boiling and dilution in the shallow portion of the Waiotapu geothermal system, New Zealand. *Geochim. Cosmochim. Acta* **55**:2753–2765.

Huber, R., S. Burggraf, T. Mayer, S. M. Barns, P. Rossnagei, and K. O. Stetter. 1995. Isolation of a hyperthermophilic archaeum predicted by *in situ* RNA analysis. *Nature* **376**:57–58.

Huber, R., P. Rossnagel, C. R. Woese, R. Rachel, T. A. Langworthy, and K. O. Stetter. 1996. Formation of ammonium from nitrate during chemolithoautotrophic growth of the extremely thermophilic bacterium *Ammonifex degensii* gen. nov. sp. nov. *Syst. Appl. Microbiol.* **19**:40–49.

Huber, R., M. Sacher, A. Vollman, H. Huber, and D. Rose. 2000. Respiration of arsenate and selenate by hyperthermophilic Archaea. *Syst. Appl. Microbiol.* **23**:305–314.

Huber, R., T. Wilharm, D. Huber, A. Trincone, S. Burggraf, H. Konig, R. Rachel, I. Rockinger, H. Fricke, and K. O. Stetter. 1992. *Aquifex pyrophilus* gen. nov., sp. nov., represents a novel group of marine hyperthermophilic hydrogen-oxidizing bacteria. *Syst. Appl. Microbiol.* **15**:340–351.

Hugenholtz, P. 2002. Exploring prokaryotic diversity in the genomic era. *Genome Biol.* **3**:reviews0003.1–reviews0003.8.

Hugenholtz, P., C. Pitulle, K. L. Hershberger, and N. R. Pace. 1998. Novel division level bacterial diversity in a Yellowstone hot spring. *J. Bacteriol.* **180**:366–376.

Jackson, C. R., H. W. Langner, J. Donahoe-Christiansen, W. P. Inskeep, and T. R. McDermott. 2001. Molecular analysis of microbial community structure in an arsenite-oxidizing acidic thermal spring. *Environ. Microbiol.* **3**:532–542.

Jolivet, E., E. Corre, S. L'Haridon, P. Forterre, and D. Prieur. 2004. *Thermococcus marinus* sp. nov. and *Thermococcus radiotolerans* sp. nov., two hyperthermophilic archaea from deep-sea hydrothermal vents that resist ionizing radiation. *Extremophiles* **8**:219–227.

Jolivet, E., S. L'Haridon, E. Corre, P. Forterre, and D. Prieur. 2003. *Thermococcus gammatolerans* sp. nov., a hyperthermophilic archaeon from a deep-sea hydrothermal vent that resists ionizing radiation. *Int. J. Syst. Evol. Microbiol.* **53**:847–851.

Jones, M. L. 1981. *Riftia pachyptila* Jones: observations on the Vestimentiferan Worm from the Galápagos Rift. *Science* **213**:333–336.

Jorgensen, B. B., M. F. Isaksen, and H. W. Jannasch. 1992. Bacterial sulfate reduction above 100°C in deep-sea hydrothermal vent sediments. *Science* **258**:1756–1757.

Kampfer, P., and R. Rossello-Mora. 2004. The species concept for prokaryotic microorganisms – an obstacle for describing diversity? *Poiesis Prax.* **4**:62–72.

Kanokratana, P., S. Chanapan, K. Pootanakit, and L. Eurwilaichitr. 2004. Diversity and abundance of Bacteria and Archaea in the Bor Khlueng Hot Spring in Thailand. *J. Basic Microbiol.* **44**:430–444.

Karpov, G. A., and S. I. Naboko. 1990. Metal contents of recent thermal waters, mineral precipitates and hydrothermal alteration in active geothermal fields, Kamchatka. *J. Geochem. Explor.* **36**:57–71.

Kashefi, K., and D. R. Lovley. 2003. Extending the upper temperature limit for life. *Science* 301:934.

Kashefi, K., J. M. Tor, K. P. Nevin, and D. R. Lovley. 2001. Reductive precipitation of gold by dissimilatory Fe(III)-reducing bacteria and archaea. *Appl. Environ. Microbiol.* 67:3275–3279.

Kelley, D. S., J. A. Baross, and J. R. Delaney. 2002. Volcanoes, fluids, and life at mid-ocean ridge spreading centers. *Annu. Rev. Earth Planet. Sci.* 30:385–491.

Kelley, D. S., J. A. Karson, G. L. Fruh-Green, D. R. Yoerger, T. M. Shank, D. A. Butterfield, J. M. Hayes, M. O. Schrenk, E. J. Olson, G. Proskurowski, M. Jakuba, A. Bradley, B. Larson, K. Ludwig, D. Glickson, K. Buckman, A. S. Bradley, W. J. Brazelton, K. Roe, M. J. Elend, A. Delacour, S. M. Bernasconi, M. D. Lilley, J. A. Baross, R. E. Summons, and S. P. Sylva. 2005. A serpentinite-hosted ecosystem: the Lost City of hydrothermal field. *Science* 307:1428–1434.

Kieft, T. L., J. K. Fredrickson, T. C. Onstott, Y. A. Gorby, H. M. Kostandarithes, T. J. Bailey, E. K. Kennedy, S. W. Li, A. E. Plymale, C. M. Spadoni, and M. S. Gray. 1999. Dissimilatory reduction of Fe(III) and other electron acceptors by a *Thermus* isolate. *Appl. Environ. Microbiol.* 65:1214–1221.

Kimble, L. K., L. Mandelco, C. R. Woese, and M. T. Madigan. 1995. *Heliobacterium modesticaldum* sp. nov., a thermophilic heliobacterium of hot springs and volcanic soils. *Arch. Microbiol.* 163:259–267.

Kimura, H., M. Sugihara, H. Yamamoto, B. K. Patel, K. Kato, and S. Hanada. 2005. Microbial community in a geothermal aquifer associated with the subsurface of the Great Artesian Basin, Australia. *Extremophiles* 9:407–414.

Klaushofer, H., and E. Parkkinen. 1965. Zur Frage der Bedeutung aerober und anaerober thermophiler Sporenbildner als Infekionsurasache in Rubenzucker-fabriken. I. *Clostridium thermohydrosulfuricum* eine neue Art eines saccharoseabbauenden, thermophilen, schwefelwasserstoffbilder Clostridiums. *Zeitschrift fur Zuckerindustrien Boehmen* 15:445–449.

Kletzin, A., T. Urich, F. Muller, T. M. Bandeiras, and C. M. Gomes. 2004. Dissimilatory oxidation and reduction of elemental sulfur in thermophilic Archaea. *J. Bioenerg. Biomembr.* 36:77–91.

Konhauser, K. O., B. Jones, V. R. Phoenix, G. Ferris, and R. W. Renaut. 2004. The microbial role in hot spring silification. *Ambio* 33:552–558.

Korn-Wendisch, F., F. Rainey, R. M. Kroppenstedt, A. Kempf, A. Majazza, H. J. Kutzner, and E. Stackebrandt. 1995. *Thermocrispum* gen. nov., a new genus of the order Actinomycetales, and description of *Thermocrispum municipale* sp. nov. and *Thermocrispum agreste* sp. nov. *Int. J. Syst. Bacteriol.* 45:67–77.

L'Haridon, S., A.-L. Reysenbach, P. Glenat, D. Prieur, and C. Jeanthon. 1995. Hot subterranean biosphere in a continental oil reservoir. *Nature* 377:223–224.

Lake, J. A. 1988. Origin of the eukaryotic nucleus determined by rate-invariant analysis of rRNA sequences. *Nature* 331:184–186.

Lake, J. A., R. Jain, J. E. Moore, and M. C. Rivera. 1998. Hyperthermophilic and mesophilic origins of the eukaryotic genome, p. 147–161. *In* J. Wiegel and M. W. W. Adams (ed.), *Thermophiles: The Keys to Molecular Evolution and the Origin of Life?* Taylor & Francis, Inc., Philadelphia, PA.

Langner, H. W., C. R. Jackson, T. R. McDermott, and W. P. Inskeep. 2001. Rapid oxidation of arsenite in a hot spring ecosystem, Yellowstone National Park. *Environ. Sci. Technol.* 35:3302–3309.

Lauerer, G., J. K. Kristjansson, T. A. Langworthy, H. Konig, and K. O. Stetter. 1986. *Methanothermus sociabilis* sp. nov., a second species within the *Methanothermaceae* growing at 97°C. *Syst. Appl. Microbiol.* 8:100–105.

Lee, Y. E., M. K. Jain, C. Lee, S. E. Lowe, and J. G. Zeikus. 1993. Taxonomic distinction of saccharolytic thermophilic anaerobes; description of *Thermoanaerobacterium xylanolyticum* gen. nov., sp. nov., and *Thermoanaerobacterium saccharolyticum* gen. nov., sp. nov.; reclassification of *Thermoanaerobium brockii*, *Clostridium thermosulfurogenes*, and *Clostridium thermohydrosulfuricum* E100-69 as *Thermoanaerobacter brockii* comb. nov., *Thermoanaerobacterium thermosulfurigenes* comb. nov., and *Thermoanaerobacter thermohydrosulfuricus* comb. nov., respectively; and transfer of *Clostridium thermohydrosulfuricum* 39E to *Thermoanaerobacter ethanolicus*. *Int. J. Syst. Bacteriol.* 43:41–51.

Li, Y., M. Engle, L. Mandelco, and J. Wiegel. 1994. *Clostridium thermoalcaliphilum* sp. nov., an anaerobic and thermotolerant facultative alkaliphile. *Int. J. Syst. Bacteriol.* 44:111–118.

Li, Y., L. Mandelco, and J. Wiegel. 1993. Isolation and characterization of a moderately thermophilic anaerobic alkaliphile, *Clostridium paradoxum* sp. nov. *Int. J. Syst. Bacteriol.* 43:450–460.

Liu, S.-Y., F. A. Rainey, H. W. Morgan, F. Mayer, and J. Wiegel. 1996. *Thermoanaerobacterium aotearoense*, sp. nov., a slightly acidophilic, anaerobic thermophile isolated from various hot springs in New Zealand and emendation of the genus *Thermoanaerobacterium*. *Int. J. Syst. Bacteriol.* 46:388–396.

Lomolino, M. V. 2000. A call for a new paradigm of island biogeography. *Global Ecol. Biogeogr.* 9:1–6.

Love, A. C., B. K. Patel, P. D. Nichols, and E. Stackebrandt. 1993. *Desulfotomaculum australicum*, sp. nov., a thermophilic sulfate-reducing bacterium isolated from the Great Artesian Basin in Australia. *Syst. Appl. Bacteriol.* 16:244–251.

Lowe, S. E., M. K. Jain, and J. G. Zeikus. 1993. Biology, ecology, and biotechnological applications of anaerobic bacteria adapted to environmental stresses in temperature, pH, salinity, or substrates. *Microbiol. Rev.* 57:451–509.

Magurran, A. E. 1988. *Ecological Diversity and Its Measurement*. Princeton University Press, Princeton, NJ.

Marteinsson, V. T., J. L. Birrien, A. L. Reysenbach, M. Vernet, D. Marie, A. Gambacorta, P. Messner, U. B. Sleytr, and D. Prieur. 1999. *Thermococcus barophilus* sp. nov., a new barophilic and hyperthermophilic archaeon isolated under high hydrostatic pressure from a deep-sea hydrothermal vent. *Int. J. Syst. Bacteriol.* 49:351–359.

Mesbah, N. M., and J. Wiegel. Isolation, cultivation and characterization of alkalithermophiles. *In Methods in Microbiology*. (Ed. A. Oren and F. A. Rainey) Academic Press / Elsevier. pp. 451–468.

Mesbah, N. M., and J. Wiegel. 2005. Halophilic thermophiles: a novel group of extremophiles, p. 91–118. *In* T. Satyanarayana and B. N. Johri (ed.), *Microbial Diversity: Current Perspectives and Potential Applications*. I.K. Interntational Pvt. Ltd., New Delhi, India.

Miller, S. L., and A. Lazcano. 1998. Facing up to chemical realities: life did not begin at the growth temperatures of hyperthermophiles, p. 127–133. *In* J. Wiegel and M. W. W. Adams (ed.), *Thermophiles: The Keys to Molecular Evolution and the Origin of Life?* Taylor & Francis, Inc., Philadelphia, PA.

Moreno, Y., M. A. Ferrrus, A. Vanoostende, M. Hernandez, R. Montes, and J. Hernandez. 2002. Comparison of 23S polymerase chain reaction–restriction fragment length polymorphism and amplified fragment length polymorphism techniques as typing systems for thermophilic campylobacters. *FEMS Microbiol. Lett.* 211:97–103.

Moussard, H., G. Henneke, D. Moreira, V. Jouffe, P. Lopez-Garcia, and C. Jeanthon. 2006. Thermophilic lifestyle for an uncultured Archaeon from hydrothermal vents: evidence from environmental genomics. *Appl. Environ. Microbiol.* 72:2268–2271.

Nakagawa, S., K. Takai, K. Horikoshi, and Y. Sako. 2004a. *Aeropyrum camini* sp. nov., a strictly aerobic, hyperthermophilic archaeon from a deep-sea hydrothermal vent chimney. *Int. J. Syst. Evol. Microbiol.* **54**:329–335.

Nakagawa, T., S. Hanada, A. Maruyama, K. Marumo, T. Urabe, and M. Fukuii. 2002. Distribution and diversity of thermophilic sulfate-reducing bacteria within a Cu–Pb–Zn mine (Toyoha, Japan). *FEMS Microbiol. Ecol.* **41**:199–209.

Nakagawa, T., S. Nakagawa, F. Inagaki, K. Takai, and K. Horikoshi. 2004b. Phylogenetic diversity of sulfate-reducing prokaryotes in active deep-sea hydrothermal vent chimney structures. *FEMS Microbiol. Lett.* **232**:145–152.

Norris, T., J. M. Wraith, R. W. Castenholz, and T. R. McDermott. 2002. Soil microbial community structure across a thermal gradient following a geothermal heating event. *Appl. Environ. Microbiol.* **68**:6300–6309.

Nubel, U., M. M. Bateson, V. Vandieken, A. Wieland, M. Kuhl, and D. M. Ward. 2002. Microscopic examination of distribution and phenotypic properties of phylogenetically diverse *Chloroflexaceae*-related bacteria in hot spring microbial mats. *Appl. Environ. Microbiol.* **68**:4593–4603.

Onstott, T. C., D. P. Moser, S. M. Pfiffner, J. K. Fredrickson, F. J. Brockman, T. J. Phelps, D. C. White, A. Peacock, D. Balkwill, R. Hoover, L. R. Krumholz, M. Borscik, T. L. Kieft, and R. Wilson. 2003. Indigenous and contaminant microbes in ultradeep mines. *Environ. Microbiol.* **5**:1168–1191.

Onyenwoke, R. U., J. A. Brill, K. Farahi, and J. Wiegel. 2004. Sporulation genes in members of the low G+C Gram-type-positive phylogenetic branch (*Firmicutes*). *Arch. Microbiol.* **182**:182–192.

Orphan, V. J., S. K. Goffredi, and E. F. DeLong. 2003. Geochemical influence on diversity and microbial processes in high temperature oil reservoirs. *Geomicrobiol. J.* **20**:295–311.

Papke, R. T., N. B. Ramsing, M. M. Bateson, and D. M. Ward. 2003. Geographical isolation in hot spring cyanobacteria. *Environ. Microbiol.* **5**:650–659.

Papke, R. T., and D. M. Ward. 2004. The importance of physical isolation to microbial diversification. *FEMS Microbiol. Ecol.* **48**:293–303.

Pearson, A., Z. Huang, A. E. Ingalls, C. S. Romanek, J. Wiegel, K. H. Freeman, R. H. Smittenberg, and C. L. Zhang. 2004. Nonmarine crenarchaeol in Nevada hot springs. *Appl. Environ. Microbiol.* **70**:5229–5237.

Pedersen, K. 2000. Exploration of deep intraterrestrial microbial life: current perspectives. *FEMS Microbiol. Lett.* **185**:9–16.

Petursdottir, S. K., G. O. Hreggvidsson, M. S. Da Costa, and J. K. Kristjansson. 2000. Genetic diversity analysis of *Rhodothermus* reflects geographical origin of the isolates. *Extremophiles* **4**:267–274.

Phoenix, V. R., K. O. Konhauser, D. G. Adams, and S. H. Bottrell. 2001. Role of biomineralization as an ultraviolet shield: implications for Archean life. *Geology* **29**:823–826.

Pierson, B. K., and R. W. Castenholz. 1974. A phototrophic gliding filamentous bacterium of hot springs, *Chloroflexus aurantiacus*, gen. and sp. nov. *Arch. Microbiol.* **100**:5–24.

Pledger, R. J., B. C. Crump, and J. A. Baross. 1994. A barophilic response by two hyperthermophilic, hydrothermal vent Archaea: an upward shift in the optimal temperature and acceleration of growth rate at supra-optimal temperatures by elevated pressure. *FEMS Microbiol. Ecol.* **14**:233–242.

Popa, R., and B. K. Kinkle. 2004. Isolation of *Thiomonas thermosulfatus* strain 51, a species capable of coupling biogenic pyritization with chemiosmotic energy transduction. *Geomicrobiol. J.* **21**:297–309.

Prangishvili, D., and R. A. Garrett. 2004. Exceptionally diverse morphotypes and genomes of crenarchaeal hyperthermophilic viruses. *Biochem. Soc. Trans.* **32**:204–208.

Rainey, F., R. Tanner, and J. Wiegel. 2006. Clostridiaceae. In *The Prokaryotes*. Springer Verlag, New York. Heidelberg., pp. 4:654–678.

Ramsing, N. B., M. J. Ferris, and D. M. Ward. 2000. Highly ordered vertical structure of *Synechococcus* populations within the one-millimeter-thick photic zone of a hot spring cyanobacterial mat. *Appl. Environ. Microbiol.* **66**:1038–1049.

Reysenbach, A. L., K. Longnecker, and J. Kirshtein. 2000. Novel bacterial and archaeal lineages from an in situ growth chamber deployed at a Mid-Atlantic Ridge hydrothermal vent. *Appl. Environ. Microbiol.* **66**:3798–3806.

Reysenbach, A. L., and E. Shock. 1994. Phylogenetic analysis of the hyperthermophile pink filament community in Octopus Spring, Yellowstone National Park. *Appl. Environ. Microbiol.* **60**:2113–2119.

Reysenbach, A., and E. L. Shock. 2002. Merging genomes with geochemistry in hydrothermal ecosystems. *Science* **296**:1077–1082.

Rivkina, E. M., E. I. Friedmann, C. P. McKay, and D. A. Gilichinsky. 2000. Metabolic activity of permafrost bacteria below the freezing point. *Appl. Environ. Microbiol.* **66**:3230–3233.

Russel, M. J., D. E. Daia, and A. J. Hall. 1998. The emergence of life from FeS bubbles at alkaline hot springs in an acid ocean, p. 77–1126. In J. Wiegel and M. W. W. Adams (ed.), *Thermophiles: The Keys to Molecular Evolution and the Origin of Life?* Taylor & Francis, London, United Kingdom.

Russell, N. J. 1990. Cold adaptation of microorganisms. *Philos. Trans. R. Soc. Lond. B.* **326**:595–608, discussion 608–611.

Sako, Y., T. Nunoura, and A. Uchida. 2001. *Pyrobaculum oguniense* sp. nov., a novel facultatively aerobic and hyperthermophilic archaeon growing at up to 97°C. *Int. J. Syst. Evol. Microbiol.* **51**:303–309.

Santos, S. R., and H. Ochman. 2004. Identification and phylogenetic sorting of bacterial lineages with universally conserved genes and proteins. *Environ. Microbiol.* **6**:754–759.

Schleper, C., G. Puehler, I. Holz, A. Gambacorta, D. Janekovic, U. Santarius, H. P. Klenk, and W. Zillig. 1995. *Picrophilus* gen. nov., fam. nov.: a novel aerobic, heterotrophic, thermoacidophilic genus and family comprising archaea capable of growth around pH 0. *J. Bacteriol.* **177**:7050–7059.

Schloss, P. D., and J. Handelsman. 2005. Introducing DOTUR, a computer program for defining operational taxonomic units and estimating species richness. *Appl. Environ. Microbiol.* **71**:1501–1506.

Schwartzman, D. W. 1998. Life was thermophilic for the first two-thirds of Earth history, p. 33–43. In J. Wiegel and M. W. W. Adams (ed.), *Thermophiles: The Keys to Molecular Evolution and the Origin of Life?* Taylor & Francis, Inc., Philadelphia, PA.

Segerer, A. H., A. Trincone, M. Gahrtz, and K. O. Stetter. 1991. *Stygiolobus azoricus* gen. nov., sp. nov. represents a novel genus of anaerobic, extremely thermoacidophilic archaebacteria of the order Sulfolobales. *Int. J. Syst. Bacteriol.* **41**:495–501.

Shin, K. S., Y. K. Shin, J. H. Yoon, and Y. H. Park. 2001. *Candida thermophila* sp. nov., a novel thermophilic yeast isolated from soil. *Int. J. Syst. Evol. Microbiol.* **51**:2167–2170.

Sievert, S. M., T. Brinkhoff, G. Muyzer, W. Ziebis, and J. Kuever. 1999. Spatial heterogeneity of bacterial populations along an environmental gradient at a shallow submarine hydrothermal vent near Milos Island (Greece). *Appl. Environ. Microbiol.* **65**:3834–3842.

Sievert, S. M., J. Kuever, and G. Muyzer. 2000a. Identification of 16S ribosomal DNA-defined bacterial populations at a shallow submarine hydrothermal vent near Milos Island (Greece). *Appl. Environ. Microbiol.* **66**:3102–3109.

Sievert, S. M., W. Ziebis, J. Kuever, and K. Sahm. 2000b. Relative abundance of Archaea and Bacteria along a thermal gradient of a shallow-water hydrothermal vent quantified by rRNA slot-blot hybridization. *Microbiology* **146:**1287–1293.

Singleton, D. R., M. A. Furlong, S. L. Rathbun, and W. B. Whitman. 2001. Quantitative comparisons of 16S rRNA gene sequence libraries from environmental samples. *Appl. Environ. Microbiol.* **67:**4374–4376.

Skirnisdottir, S., G. O. Hreggvidsson, S. Hjorleifsdottir, V. Marteinsson, S. Petursdottir, O. Holst, and J. K. Kristjansson. 2000. Influence of sulfide and temperature on species composition and community structure of hot spring microbial mats. *Appl. Environ. Microbiol.* **66:**2835–2841.

Slobodkin, A., B. Campbell, S. C. Cary, E. A. Bonch-Osmolovskaya, and C. Jeanthon. 2001. Evidence for the presence of thermophilic Fe(III)-reducing microorganisms in deep-sea hydrothermal vents at 13°N (East Pacific Rise). *FEMS Microbiol. Ecol.* **36:**235–243.

Slobodkin, A., D. G. Zavarzina, T. G. Sokolova, and E. A. Bonch-Osmolovskaya. 1999. Dissimilatory reduction of inorganic electron acceptors by thermophilic anaerobic prokaryotes. *Microbiology* **68:**522–542.

Sokolova, T. G., N. A. Kostrikina, N. A. Chernyh, T. V. Kolganova, T. P. Tourova, and E. A. Bonch-Osmolovskaya. 2005. *Thermincola carboxydiphila* gen. nov., sp. nov., a novel anaerobic, carboxydotrophic, hydrogenogenic bacterium from a hot spring of the Lake Baikal area. *Int. J. Syst. Evol. Microbiol.* **55:**2069–2073.

Sokolova, T. J Hanel, R. U. Onyenwoke, A. L. Reysenbach, A. Banta, R. Geyer, J. M. Gonzalez, W. B. Wjitman and J. Wiegel. 2007. Novel chemolithoautortropic, thermophilic, anaerobic bacteria Thermolithobacter ferrireducens, gen. nov., sp. nov. and thermolithobacter carboxydivorans sp. nov. Extremophiles **11:**145–157.

Stackebrandt, E., W. Frederiksen, G. M. Garrity, P. A. Grimont, P. Kampfer, M. C. Maiden, X. Nesme, R. Rossello-Mora, J. Swings, H. G. Truper, L. Vauterin, A. C. Ward, and W. B. Whitman. 2002. Report of the ad hoc committee for the re-evaluation of the species definition in bacteriology. *Int. J. Syst. Evol. Microbiol.* **52:**1043–1047.

Stetter, K. O. 1989. Extremely thermophilic chemolithoautotrophic Archaebacteria, p. 167–173. *In* H. G. Schlegel and B. Bowen (ed.), *Autotrophic Bacteria*. Springer-Verlag, New York, NY.

Stetter, K. O. 1996. Hyperthermophilic procaryotes. *FEMS Microbiol. Rev.* **18:**149–158.

Summit, M., B. Scott, K. Nielson, E. Mathur, and J. A. Baross. 1998. Pressure enhances thermal stability of DNA polymerase from three thermophilic organisms. *Extremophiles* **2:**339–345.

Svetlichny, V. A., T. G. Sokolova, M. Gerhardt, M. Ringpfeil, N. A. Kostrikina, and G. A. Zavarzin. 1991. *Carboxydothermus hydrogenoformans* gen. nov., sp. nov., a CO-utilizing thermophilic anaerobic bacterium from hydrothermal environments of Kunashir Island. *Syst. Appl. Microbiol.* **14:**254–260.

Szewzyk, U., R. Szewzyk, and T.-A. Stenstrom. 1994. Thermophilic, anaerobic bacteria isolated from a deep borehole in granite in Sweden. *Proc. Natl. Acad. Sci. USA* **91:**1810–1813.

Takahata, Y., M. Nishijima, T. Hoaki, and T. Maruyama. 2000. Distribution and physiological characteristics of hyperthermophiles in the Kubiki oil reservoir in Niigata, Japan. *Appl. Environ. Microbiol.* **66:**73–79.

Takai, K., H. Hirayama, T. Nakagawa, Y. Suzuki, K. H. Nealson, and K. Horikoshi. 2005. *Lebetimonas acidiphila* gen. nov., sp. nov., a novel thermophilic, acidophilic, hydrogen-oxidizing chemolithoautotroph within the 'Epsilonproteobacteria', isolated from a deep-sea hydrothermal fumarole in the Mariana Arc. *Int. J. Syst. Evol. Microbiol.* **55(Pt 1):**183–189.

Tansey, M. R., and T. D. Brock. 1972. The upper temperature limit for eukaryotic organisms. *Proc. Natl. Acad. Sci. USA* **69:**2426–2428.

Torsvik, V., L. Ovreas, and T. F. Thingstad. 2002. Prokaryotic diversity—magnitude, dynamics, and controlling factors. *Science* **296:**1064–1066.

Wachtershauser, G. 1998. The case for a hyperthermophilic, chemolithoautotrophic origin of life in an iron-sulfur world, p. 47–57. *In* J. Wiegel and M. W. W. Adams (ed.), *Thermophiles: The Keys to Molecular Evolution and the Origin of Life?* Taylor & Francis, Inc., Philadelphia, PA.

Ward, D. M. 1998. A natural species concept for prokaryotes. *Curr. Opin. Microbiol.* **1:**271–277.

Ward, D. M., M. J. Ferris, S. C. Nold, and M. M. Bateson. 1998. A natural view of microbial biodiversity within hot spring cyanobacterial mat communities. *Microbiol. Mol. Biol. Rev.* **62:**1353–1370.

Waring, G. A. 1965. Thermal springs of the United States and other countries of the world. A summary. Geological Survey Professional paper 492 (revised by R. R. Blankenship and R. Bentall), US Government Printing Office, Washington, D.C.

Westall, F. 2005. Early life on earth and analogies to Mars, p. 45–64. *In* T. Tokano (ed.), *Water on Mars and Life*, vol. 4. Springer-Verlag, Berlin, Germany.

Whitaker, R. J., D. W. Grogan, and J. W. Taylor. 2003. Geographic barriers isolate endemic populations or hyperthermophilic Archaea. *Science* **301:**976–978.

White, R. H. 1984. Hydrolytic stability of biomolecules at high temperatures. *Nature* **310:**430–432.

Wiegel, J. 1981. Distinction between the Gram reaction and the Gram type of bacteria. *Int. J. Syst. Bacteriol.* **31:**88.

Wiegel, J. 1986. Methods for isolation and study of thermophiles, p. 17–37. *In* T. D. Brock (ed.), *Thermophiles: General, Molecular, and Applied Microbiology*. John Wiley & Sons, Inc., Hoboken, NJ.

Wiegel, J. 1990. Temperature spans for growth: a hypothesis and discussion. *FEMS Microbiol. Rev.* **75:**155–170.

Wiegel, J. 1992. The obligately anaerobic thermophilic bacteria, p. 105–184. *In* J. K. Kristjansson (ed.), *Thermophilic Bacteria*. CRC Press LLC, Boca Raton, FL.

Wiegel, J. 1998a. Anaerobic alkalithermophiles, a novel group of extremophiles. *Extremophiles* **2:**257–267.

Wiegel, J. 1998b. Lateral gene exchange, an evolutionary mechanism for extending the upper or lower temperature limits for growth of microorganisms? A hypothesis, p. 177–185. *In* J. Wiegel and M. W. W. Adams (ed.), *Thermophiles: The Keys to Molecular Evolution and the Origin of Life?* Taylor & Francis, Inc., Philadelphia, PA.

Wiegel, J., and M. W. W. Adams (ed.). 1998. *Thermophiles: The Keys to Molecular Evolution and the Origin of Life?* Taylor & Francis, Inc., Philadelphia, PA.

Wiegel, J., M. Braun, and G. Gottschalk. 1981. *Clostridium thermoautotrophicum* specius novum, a thermophile producing acetate from molecular hydrogen and carbon dioxide. *Curr. Microbiol.* **5:**255–260.

Wiegel, J., J. Hanel, and K. Ayres. 2003. Chemolithoautotrophic thermophilic iron(III)-reducer, p. 235–251. *In* L. G. Ljungdahl, M. W. W. Adams, L. Barton, G. Ferry, and M. Johnson (ed.), *Biology and Physiology of Anaerobic Bacteria*. Springer-Verlag, New York, NY.

Wiegel, J., and L. G. Ljungdahl. 1986. The importance of thermophilic bacteria in biotechnology. *Crit. Rev. Biotechnol.* **3:**39–107.

Wiegert, R. G., and P. C. Fraleigh. 1972. Ecology of Yellowstone thermal effluent systems: net primary production and species diversity of a succesional blue-green algal mat. *Limnol. Oceanogr.* **17:**215–228.

Wilson, E. O. 1992. *The Diversity of Life*. Belknap Press of Harvard University Press, Cambridge, MA.

Wintzingerode, F. v., U. B. Gobel, and E. Stackebrandt. 1997. Determination of microbial diversity in environmental samples: pitfalls of PCR-based rRNA analysis. *FEMS Microbiol. Rev.* **21:**213–229.

Woese, C. R. 1987. Bacterial evolution. *Microbiol. Rev.* **51:**221–271.

Zeikus, J. G., and R. S. Wolfe. 1972. *Methanobacterium thermoautotrophicus* sp. n., an anaerobic, autotrophic, extreme thermophile. *J. Bacteriol.* **109:**707–713.

Zhang, C. L., A. Pearson, Y.-L. Li, G. Mills, and J. Wiegel. 2006. Thermophilic temperature optimum for crenarchaeol synthesis and its implications for archaeal evolution. *Appl. Environ. Microbiol.* **72:**4419–4422.

Zhao, W., C. S. Romanek, G. Mills, J. Wiegel, and C. L. Zhang. 2005. Geochemistry and microbiology of hot springs in Kamchatka, Russia. *Geol. J. China Univ.* **11:**217–223.

Zillig, W., K. O. Stetter, S. Wunderl, W. Schulz, H. Priess, and I. Scholz. 1980. The *Sulfolobus*-"*Caldariella*" group: taxonomy on the basis of the structure of DNA-dependent RNA polymerases. *Arch. Mikrobiol.* **125:**259–269.

Chapter 3

Functional Genomics in Thermophilic Microorganisms

FRANK T. ROBB AND DEBORAH T. NEWBY

INTRODUCTION: LIFE AT HIGH TEMPERATURE

Our concept of normality for living systems is centered on 37°C, pH 7, 50 mM NaCl, and atmospheric gases at normal pressure. Although humans are conceptually harnessed by the life-limiting conditions for higher eukaryotes, microbial life has adapted to extend well beyond the familiar mesophilic envelope and thrives in almost every earth environment! Thermophiles require elevated temperatures for growth. This distinguishes them from thermoduric microorganisms, such as endospore formers that have specialized adaptive strategies that allow them to survive but not grow at elevated temperatures. The temperature range 50°C and 75°C delimits thermophiles, whereas strains growing from 75°C to 90°C are extreme thermophiles, and strains that can grow >90°C are hyperthermophiles (Stetter, 1996). The hyperthermophiles that have been described to date are diverse, with 23 genera recognized as having hyperthermophilic members. Multiple genera, including *Pyrolobus*, *Pyrodictium*, *Pyrobaculum*, *Methanopyrus*, *Pyrococcus*, *Sulfolobus*, and *Archaeoglobus*, are exclusively hyperthermophilic. Although several bacterial species are also considered to be hyperthermophiles because they can grow at 90°C, isolates that grow optimally at ≥95°C are exclusively *Archaea*. These include a group of extremely thermophilic *Bacteria* with completed genome projects and include the hyperthermophiles *Thermotoga maritima* (Nelson et al., 1999) and *Aquifex aeolicus* (Deckert et al., 1998), the thermophile *Thermoanaerobacter tengcongensis* (Bao et al., 2002), and the acidophiles *Sulfolobus* spp. (She et al., 2001), *Picrophilus* spp. (Futterer et al., 2004) and *Thermoplasma* spp. (Ruepp et al., 2000) isolated from sulfurous volcanic environments and other acidic hot niches such as hot, wet coal dust. These hot habitats are the exclusive territory of prokaryotes, freed from competition from eukaryotes.

In this chapter, using the rapidly expanding set of whole-genome sequences now available, we examine the progress that has been made in understanding life at elevated temperatures. It is remarkable that the wide temperature range of microbial thermophiles is maintained by basic genomic organization and coding conventions that are similar to "normal" microbial cells. Consequently, the simple and powerful approach of whole-genome sequencing is providing the most informative method to explore the molecular basis of survival and growth at high temperature. In 1996, the second whole-genome sequence to be described was that of the deep-sea barophilic hyperthermophile *Methanococcus jannaschii* (Bult et al., 1996). Genome sequences of many thermophiles have been published and so-called functional genomic experiments have followed. Functional genomics is defined as "[a] field of molecular biology that is attempting to make use of the data produced by genome sequencing projects to describe genome function. Functional genomics uses high-throughput techniques like DNA microarrays, proteomics, metabolomics and mutation analysis to describe the function and interactions of genes" (http://www.wikipedia.org/wiki/Functional_genomics).

The published sequencing projects of thermophiles are described in Table 1. In many cases, more than one extreme condition is present; for example, many of the barophilic *Archaea* are hyperthermophiles that grow at temperatures near or above 100°C. Unfortunately, this list does not contain the high-temperature leader *Pyrolobus fumarii*, the best described hyperthermophile, isolated from a Mid-Atlantic vent at great depth (3650 m) and able to survive exposure to autoclave conditions for up to 1 h and able to grow at 113°C at a pressure of 25 MPa (Blochl et al., 1997). The optimal growth temperatures

---

F. T. Robb • Center of Marine Biotechnology, 701 E. Pratt Street, Baltimore, MD 21202. D. T. Newby • Idaho National Laboratory, PO Box 1625, Idaho Falls, ID 83415.

of the strains range from 55°C to 105°C, and their dominant modes of energy conservation cover a wide variety of metabolic strategies, ranging from heterotrophic fermentation, to anaerobic, autotrophic metabolism. Because it is impossible to apply many of the commonly used genetic tools and approaches to most hyperthermophiles and many extreme thermophiles because of their high growth temperatures and autotrophic energy conservation, postgenomic study methods are providing an alternative to genetic analysis. Comparative bioinformatic studies with genome sequences from groups of related organisms have also resulted in insights into mechanisms of genome plasticity and lateral gene transfer (LGT).

COMPARATIVE GENOMIC STUDIES ON ADAPTATIONS REQUIRED FOR THERMOPHILY AND STRUCTURAL GENOMICS

Table 1 references the genome sequences for three thermophilic methanogens, *Methanococcus jannaschii* (Bult et al., 1996), *Methanobacterium thermoautotrophicum* (Smith et al., 1997), and *Methanopyrus kandleri* (Slesarev et al., 2002), two *Sulfolobus* spp., *Sulfolobus tokodaii* (Kawarabayasi et al., 1999) and *Sulfolobus solfataricus* P1 (Brügger et al., in press), three *Pyrococcus* spp. (Ettema et al., 2001), and two *Thermoplasma* spp. (Kawashima et al., 2000). The high-temperature hyperthermophiles *M. kandleri*, a methanogen that grows at temperatures up to 110°C (Burggraf et al., 1991; Slesarev et al., 2002), and *Pyrobaculum aerophilum*, a microaerophilic, nitrate-reducing crenarchaeote that grows optimally at 100°C by nitrate reduction (Volkl et al., 1993), are examples of two autotrophic archaea with extraordinary resistance to high temperature. Moreover, an obligate aerobe that is able to grow at up to 100°C, *Aeropyrum pernix*, was sequenced, and this provides a comparison with the thermoacidophilic crenarchaeotes (Kawarabayasi et al., 1999). As a result, comparative genomic studies have become a key factor in understanding adaptive features enabling microbial survival and propagation in hot conditions. In many cases, this survival is due to protein structures that provide high intrinsic stability at high temperature. Molecular determinants of protein stability of hyperthermostable proteins have been vigorously debated and reviewed (Elcock, 1998; Haney et al., 1999). Several general strategies have emerged, including the conclusion that stable proteins have more compact structures with less internal voids, achieved by smaller loops and shorter N and C termini. In many cases, the proteins form larger oligomeric assemblies than are found in mesophiles, and the occurrence of ion pairs and ion-pair networks is more frequent in highly stable homologs of proteins. One of the major debates is the exact role that ion-pair interactions have in the stability of hyperstable proteins. In theory, electrostatic interactions promote robust internal bonding of protein structures due to their relatively long range (up to 4Å) compared with van der Waals forces. This increased range and their

Table 1. Sequenced thermophile genomes

Species	Genome size	Completion	T_{opt}	Energy conservation
M. jannaschii	1.66	1996 (Bult et al., 1996)	80	Methanogenesis
A. fulgidus VC16	2.18	1997 (Klenk et al., 1997)	83	Sulfate reduction
M. thermautotrophicus H	1.75	1997 (Smith et al., 1997)	65	Methanogenesis
P. horikoshii OT3	1.73	1998 (Kawarabayasi et al., 1999)	98	Heterotrophic/S_0 reduction
A. pernix K1	1.80	1999 (Kawarabayasi et al., 1999)	95	Heterotrophic
A. aeolicus[a]	1.80	1999 (Deckert et al., 1998)	85	H_2 oxidation
T. maritima MSB8[a]	1.86	1999 (Nelson et al., 1999)	80	Heterotrophic, S_0 reduction
P. abyssi	1.75	2000 (Cohen et al., 2003)	95	Heterotrophic/S_0 reduction
P. furiosus	1.91	2000 (Robb et al., 2001)	100	Heterotrophic/S_0 reduction
S. tokodaii 7	2.69	2000 (Kawarabayasi et al., 2001)	80	Heterotrophic, S oxidation
S. solfataricus P2	2.99	2001 (She et al., 2001)	85	Heterotrophic, S oxidation
T. volcanium GSS1	1.58	2001 (Kawashima et al., 1999)	60	Heterotrophic, S oxidation
Thermoplasma acidophilum	1.58	2001 (Ruepp et al., 2000)	63	Heterotrophic, S oxidation
M. kandleri AV19	1.69	2001 (Slesarev et al., 2002)	103	Methanogenesis
P. aerophilum IM2	2.22	2002 (Fitz-Gibbon et al., 2002)	100	Nitrate reduction
T. tengcongensis[a]	2.69	2002 (Bao et al., 2002)	75	Fermentation, S reduction
C. hydrogenoformans[a]	2.40	2006 (Wu et al., 2005)	78	CO oxidation, H_2 production

T_{opt}, optimal growth temperature.
[a] Bacteria; the remainder are Archaea.

insensitivity to the weakening of the hydrophobic effect near or above 100°C are important contributors to extreme thermal stability (Clark and Robb, 1999; Siegert et al., 2000; Karshikoff and Ladenstein, 2001; Sterner and Liebl, 2001). A recent comparative genomic study supports this conclusion. In an extensive comparative study with 12 genome sequences, including those of five thermophilic archaeons, one thermophilic bacterium, and six mesophilic bacteria, Das and Gerstein (2000) confirm that charged residues both within helices and in tertiary and quaternary structures are arranged preferentially to form ion pairs. The predominance of tertiary ion pairs and ion-pair networks has been suggested in earlier studies (Vetriani et al., 1998; Yip et al., 1998; Cambillau and Claverie, 2000). The completion of two genome sequences of *Thermoplasma* spp. has been very valuable as these strains have relatively low growth temperatures, ~60°C, and therefore, their proteomes provide a low-temperature baseline for comparison with high-temperature strains including the methanogens. The comparison of amino acid composition of the *Thermoplasma volcanium* proteome confirms the conclusions of two comparative studies (Kawashima et al., 2000; Ruepp et al., 2000; Angelov and Liebl, 2006). The general conclusion is that a shift occurs toward a higher abundance of charged amino acid residues such as Glu, Asp, Arg, and Lys with increasing growth temperatures, whereas polar residues such as Ser, Gln, Tyr, and Asn are reduced in frequency. While it is clear that charged residues can engage in ion-pair formation, it has also been argued that intracellular solubility of proteins in hyperthermophiles may depend on increased surface charge in high-salt conditions, accounting for the increase in overall charge of the proteome (Cambillau and Claverie, 2000).

Recent work has shown that moderate pressures ($\leq$100 MPa) below those normally needed for pressure-induced denaturation can dramatically stabilize proteins against thermoinactivation (Frankenberg et al., 2003; Tolgyesi et al., 2004). At very high temperatures, thermophilic enzymes exhibit significant stabilizing effects (Clark and Robb, 1999).

Access to the deduced proteome allows one to attempt to express a large proportion of the encoded proteins using recombinant expression techniques, giving rise to a relatively new field termed structural genomics. In practice, a structural genomic approach involving high-throughput expression, purification, and crystallization or nuclear magnetic resonance (NMR) spectroscopy is limited to those proteins that can be translated in an appropriate host, are correctly folded, and undergo multisubunit assembly and, in some cases, posttranslational modification. Membrane proteins of thermophiles in general are relatively inaccessible unless they can be expressed in a phylogenetically related host system. Despite these caveats, high-throughput expression and crystal structure determination projects for *T. maritima* (Lesley et al., 2002), *Methanococcus jannaschii* (Kim et al., 1998), and *Pyrococcus furiosus* (Scott et al., 2005) are underway. When a sufficient number of crystal structures of hyperstable proteins are resolved, the question as to whether their excess charges are buried or solvent exposed can be clarified by comparing the structures of mesophilic and thermostable proteins. For *T. maritima* and *P. furiosus*, many protein structures have already been resolved because these organisms are readily grown to high cell densities, allowing highly expressed native proteins to be purified from cell-free extracts by using conventional protocols. In this way, more than 100 native proteins from *P. furiosus* have been purified and characterized. Access to metalloproteins or glycolproteins with prosthetic groups already in place is the major advantage of this method. The success rates for correct expression of soluble proteins is rather low, about 19% for *P. furiosus* (Scott et al., 2005), suggesting that correct folding of hyperstable proteins may require exposure to conditions resembling the intracellular milieu, including high temperatures and novel solutes. Novel protocols and methods for expressing proteins may be needed before this postgenomic approach attains its full potential.

TRANSCRIPTIONAL ANALYSIS OF STRESS RESPONSES AND METABOLIC REGULATION

The lifestyles of thermophiles in geothermal or hydrothermal habitats expose them to thermal cycles, both above and below their optimal growth temperatures. Thus, conditions approaching a continuous alternation between both heat and cold shock may be normal. It is therefore surprising to find relatively few heat and cold shock genes in the genomes of thermophiles. The cold-shock response in *P. furiosus* has been recently thoroughly outlined using a whole-genome microarray, and exposure to a cold shock by rapid cooling from 95°C to 72°C. The *CipA* and *CipB* genes, which are related to cold-shock-responsive genes in bacteria, were also induced in *P. furiosus* (Weinberg et al., 2005). One of the cold adaptive proteins, superoxide reductase, in the hyperthermophile *P. furiosus* is active at temperatures well below the minimal growth temperature, 80°C (Weinberg et al., 2005). This is probably an adaptive response to combat oxidative damage if these oxygen sensitive cells are flushed out of the vent environment into cold, aerated seawater. Most hyperthermophiles grow to low terminal cell densities, thereby complicating the detection of heat-shock proteins. Inducing heat shock in hyperthermophiles may also require cultivation in vessels that are pressurized to prevent boiling.

An initial approach using whole-genome microarrays for transcriptional mapping is probably the most effective method to identify heat-shock proteins in hyperthermophiles. With multiple whole-genome analyses of hyperthermophiles, the first heat-shock regulons have been conveniently mapped by microarray analysis.

The first thermophilic methanogen to be analyzed comprehensively by microarray was *Methanocaldococcus jannaschii* (Boonyaratanakornkit et al., 2005). This study resulted in the discovery of a unique heat-shock-inducible prefoldin chaperone gene. The chaperone systems in hyperthermophilic Archaea have been proposed as a minimal set, forming pathways for protein folding containing the nascent polypeptide-associated complex (NAC), prefoldin (or small heat shock protein under heat-shock conditions), and the chaperonin HSP60 (Laksanalamai et al., 2004). The microarray study concluded that many genes found to be induced by heat shock in *P. furiosus*, such as the *small heat shock protein*, *AAA$^+$ ATPase*, and *htpX* (Shockley et al., 2003), were also part of the heat shock regulon in *Methanocaldococcus jannaschii*. However, the phr heat-shock regulator described for *P. furiosus* (Vierke et al., 2003) is absent in *M. jannaschii*. A heat-sensitive repressor of heat-shock responses in *Thermococcales*, phr has conserved binding sites, included in a 29-bp DNA sequence overlapping the transcription start site. Three sequences conserved in the binding sites of Phr—TTTA at −10, TGGTAA at the transcription start site, and AAAA at position +10—were required for Phr binding and are proposed as consensus regulatory sequences of *Pyrococcus* heat-shock promoters. Conservation of the phr was indeed found in other *Pyrococcus* heat-shock-regulated promoters (Laksanalamai et al., 2004). Interestingly, a comprehensive study of the heat-shock response in *Archaeoglobus fulgidus* revealed the presence of a putative regulator, AF XXXX, and conserved palindromic binding sites CTAAC N=5 GTTAG downstream of the promoters of the sHSP and AAA proteins but positioned upstream of the coding regions (Rohlin et al., 2005). There are two sHSP encoding genes in *A. fulgidus*, both of which are induced by heat stress and both contain the conserved binding site for the heat shock regulator HrcA.

P. furiosus was the first hyperthermophile to be isolated 20 years ago (Fiala and Stetter, 1986), and consequently, it has been the most actively studied, in terms of biochemistry and physiology. Several microarray studies have focused on understanding the metabolic capabilities of this organism. In an early study, activities of several key metabolic enzymes were evaluated when *P. furiosus* was grown on maltose and/or peptides, both with and without S^0 (Adams et al., 2001). This study revealed that *P. furiosus* is able to utilize both peptides and maltose as sources of C and that it is able to grow well in the absence of S^0 (Adams et al., 2001), metabolic characteristics that set it apart from most other S^0-reducing, heterotrophic hyperthermophiles. The observed differential enzyme activities provided the first evidence that C source and S^0 play a significant regulatory role in the metabolism of *P. furiosus* and generated questions regarding the interaction of S^0 with hydrogenases and the mechanism of S^0 reduction in *P. furiosus*. Capitalizing on the availability of the complete genome sequence of *P. furiosus*, more recent studies have employed microarray studies to further investigate the intriguing metabolism of this archaeon (Schut et al., 2001, 2003). The annotated genome of *P. furiosus* contains 2,065 open reading frames (ORFs), with ~50% designated as conserved hypothetical proteins with no similarity to characterized ORFs of other genomes (Robb et al., 2001). The ability to evaluate changes in the expression of ORFs of unknown function is a significant benefit of microarray technology. An array targeting a subset of ORFs believed to encode proteins involved in metabolic pathways, energy conservation, and metal metabolism was designed specifically to address the effects of S^0 on metabolism (Schut et al., 2003). Microarray results provided further evidence for S^0 regulation, including identification of two ORFs with 25-fold increased expression when cells were grown with S^0. The products of these ORFs—SipA and SipB (sulfur-induced proteins)—were hypothesized to be part of a novel S^0-reducing, membrane-associated, iron-sulfur cluster-containing complex (Schut et al., 2001). A complete-genome DNA microarray for *P. furiosus*, the first such microarray constructed for a hyperthermophile, evaluated the response of the transcriptome to changes in carbon source (peptide or maltose). Of the ORFs that showed >5-fold changes in expression between the growth conditions (8% of total ORFs), 65% appeared to be part of operons, indicating an unexpected degree of coordinate regulation in response to a change in C source.

The genome of the hydrogen-producing, carbon-monoxide-oxidizing thermophile, *Carboxydothermus hydrogenoformans*, was recently published (Wu et al., 2005), and a microarray study is currently in progress in order to determine the mechanism of accelerated hydrogen production when the culture is switched from heterotrophic to autotrophic growth (D. Newby, F. T. Robb, unpublished results, 2006). Hydrogen production continues at a lower rate in cells growing in heterotrophic mode, compared with cells grown with CO as sole C and energy source.

LGT, GENOME PLASTICITY, AND PHYLOGENY

Comparative genomics also provides insights into the mobility of chromosomal sections and LGT. Bacterial and archaeal thermophiles often share the same habitats, and there is abundant evidence from genomic analyses that LGT is common in the group. Heterotrophic thermophiles are also distinguished by frequent gene loss and gain. For example, the *T. maritima* genome has been estimated to have ~20% of genes that have primary homology to hyperthermophilic *Archaea*, principally *Pyrococcus* spp. (Nelson et al., 1999).

Through periodicity analysis, bacterial signatures that predominate in the *T. maritima* were distinguished from the archaeal signatures that mark the genes resembling those from *Archaea* (Worning et al., 2000). A recent study in which the sequenced strain *T. maritima* MSB8 was compared with nine closely related strains by subtractive sequencing confirmed the presence of hotspots of genome variability and suggested that active LGT was occurring (Mongodin et al., 2005). A significant microarray study suggesting that extensive "collective bargaining" occurs in hydrothermal vents was carried out by coculture of the heterotrophic *T. maritima* and the methanogen *Methanococcus jannaschii* (Johnson et al., 2006). The progress of a coculture from exponential into stationary phase was accompanied by the induction of 292 differentially expressed genes in *T. maritima*, ~10 times the number induced in a pure culture of *T. maritima* during the same growth phase transition. The crosstalk between these hyperthermophiles was revealed as a direct result of transcriptional profiling.

LGT and gene gains and losses are evident in the *Pyrococcus* genomes, which are significantly polymorphic with respect to whole pathways, including biosynthetic and degradative functions. The 1.95 Mbp *P. furiosus* genome is significantly larger than those of *Pyrococcus abyssi* (1.75 Mbp) and *Pyrococcus horikoshii* (1.73 Mbp). The *P. furiosus* and *P. abyssi* genomes encode trp, aro, arg, and ile/val operons (Maeder et al., 1999), unlike the genome of *P. horikoshii* (Kawarabayasi et al., 1998). The genomes of *P. abyssi* and *P. horikoshii* are similar in lacking biosynthetic pathways for histidine, cobalamin, a portion of the tricarboxylic acid (TCA) cycle, and the fermentation capacity (uptake and degradation) for starch, maltose, trehalose, and cellobiose. *P. horikoshii*, with the smallest genome of the pyrococci, is therefore auxotrophic for Val, Leu, and Ile and presumably the other aromatic amino acids, since it lacks the enzymes of the common aromatic pathway. *P. furiosus* and *P. abyssi* (Cohen et al., 2003) have the capacity to synthesize all nucleotides as well as vitamins B12 and B6, which is confirmed by their ability to grow in a defined medium in the absence of these nutrients. Ettema et al. (2001) have suggested that a mechanism of concerted, en bloc insertion and deletion is operating in the genomes of hyperthermophiles, using mechanisms similar to those described for bacteria, including transposition and gene conversion.

It is interesting to speculate that these three very similar genomes have been reduced in complexity during the descent of *Pyrococcus* spp. to more physically hostile environments in the deep-sea vents, or whether *P. furiosus*, having ascended from the depths, is acquiring new metabolic capacity in response to the availability of polysaccharides in its onshore niche. The available evidence supports the latter premise. The maltose region of *P. furiosus* is apparently a recently imported section of the genome that is not found in the other *Pyrococcus* spp. (DiRuggiero et al., 2000). The 19-kb section containing a maltose *EFG* operon similar in gene order and overall sequence to the *mal* operon of *Escherichia coli* is flanked by two of 23 insertion sequence (IS) elements in *P. furiosus* genome. By comparison with the *mal* region of a related archaeon, *Thermococcus litoralis*, it was discovered that the entire region was very similar in both strains. Very few mutations have occurred since the presumed divergence of the *P. furiosus* and *T. litoralis* versions of this region. Analysis of the terminal direct repeats of the transposon indicated that the region moved as a composite transposon (DiRuggiero et al., 2000). The distribution of IS elements in *Pyrococcus* spp. is extremely variable. *P. furiosus* has at least 29 full-length IS elements, which are very similar in distribution in *Pyrococcus woesei*; however, the genomes of *P. abyssi* and *P. horikoshii* are free from transposons. These findings lead to the proposal to rename *P. woesei* as a subtype of *P. furiosus* (Kanoksilapatham et al., 2004). A tandem IS insertion element consisting of two intact IS elements has been reported. Transfer is plausible because, although *T. litoralis* has a lower growth temperature range than *Pyrococcus* spp., its habitats overlap and *T. litoralis* was isolated from the same locale (the Italian shoreline) as *P. furiosus* (Fiala and Stetter, 1986).

These observations beg the question as to whether LGT is pervasive in all organisms sharing a niche. A recent analysis of the genome of *Picrophilus torridus*, a double extremophile with the ability to grow at pH 0 and an optimal growth temperature of 55°C, revealed new important information. The greatest proportion of proposed LGT events suggested by the genome were from acidophiles that were rather distantly related to *P. torridus*, which is in the same general phylogenetic area as *Thermoplasma* spp. Putative gene transfer events from *Sulfolobus* spp. and acid-tolerant bacteria to *P. torridus* were more prevalent than close homologs from *Thermoplasma* spp. (Futterer et al., 2004).

Genomic analysis has also cleared up a troubling anomaly within the phylogeny of *Archaea* based on 16S rRNA sequences, namely the placement of *M. kandleri*, a methanogen, as a basal root on the common archaeal branch (Burggraf et al., 1991). The remaining methanogens occupy one of the crowns of the *Euryarcheota*. *M. kandleri* is capable of growth under pressure at temperatures up to 110°C, making it the most thermophilic *Archaeon* to be sequenced to date. It also has the highest genomic G+C content of the methanogens (62%). Interestingly, phylogenetic analysis using alignments of concatenated sequences of ribosomal proteins and trees based on gene content strongly supported a monophyletic methanogen clade. It can be argued that this is a more reliable approach to organismal phylogeny, based on the surmise that genes encoding ribosomal proteins are relatively less likely to be transferred between unrelated strains. The *M. kandleri* placement in an extremely deeply branch on the 16S rRNA at the base of the *Archaeal* stem appears to be an artifact because of the extremely high G+C content of the rRNA (75%). Genomic sequencing resolves this anomaly. The genes for methanogens are also similar in amino acid sequence to those of other methanogens, and the conservation of gene order between the existing methanogen genomes and *M. kandleri* in these key loci concerned with energy conservation confirms that the methanogens are probably monophyletic. It follows that, at least in this case, extremely high G+C content is likely to be an adaptation allowing growth at up to 110°C. The analysis of the *T. tengcongensis* genome indicates that the %G+C of the genome is lower than the rRNA and tRNA G+C content. On the other hand, the growth temperature is positively correlated with the G+C content of stable RNAs, which rises to a maximum of 65 to 70% in hyperthermophiles. For example, the genomic G+C content of *P. abyssi* is 42%, whereas the rRNA is 68%. This is probably due to the structural constraints on stable RNA molecules, whose secondary structure must be maintained at high temperature. This disconnect between G+C contents can lead to artifactual rRNA-based phylogeny of hyperthermophiles.

The compact genome sequence of *M. kandleri* has the unusual feature of being essentially free of LGT events (Slesarev et al., 2002). The LGT events, or lack of them, in the *M. kandleri* genome provide support for the argument that multiple strains sharing a common habitat may be conducive to LGT. It is tempting to speculate that, physiologically, *M. kandleri* is a "lone ranger" at the top end of the temperature scale, with fewer opportunities for contact with live cells during active growth. Another consideration is the unusual physiology of *M. kandleri*, which has an extremely high concentration of phosphorylated compounds, including cyclic diphosphoglycerate, in its cytoplasm. One of the requirements for the eventual functioning of a protein that is imported into a new cell is that cytoplasmic composition permit the protein to function. The proteins of *M. kandleri* are highly unusual in that they have a requirement for molar concentrations of phosphate for optimal function in vitro. Enzyme stability and activity and subunit assembly were all adversely affected by low phosphate concentration, a condition referred to as lyotropic (Minuth et al., 1998). It is very likely that adaptive mutations required to enable functioning of proteins that were not lyotropic would be a significant barrier to incorporation of new genes by LGT.

The overall lack of conserved gene order in many thermophilic genomes begs the question as to the mechanisms that allow rapid rearrangement. The IS elements are an obvious source of nonreciprocal recombination and also, as repeated sequences, can be important in providing local regions of homology that promote insertion–deletion and inversion events by means of reciprocal recombination. IS elements are widely reported in thermophile genomes, and there is experimental evidence for their ongoing activity. For example, in the genome of *S. solfataricus*, the specific and frequent insertion of IS elements could be observed under positive selection for inactivation of *ura* genes by fluoroorotic acid (Martusewitsch et al., 2000). The occurrence of insertion elements of several types has been correlated with genome rearrangements during culture of thermophiles (R. Garrett, personal communication). The insertion of one of the *Pyrococcus* IS elements into the *napA* gene to disrupt it is presumed to be a recent event because the inactive *napA* gene did not accumulate mutations (Kanoksilapatham et al., 2004).

The known IS elements fall into two main types: autonomous IS elements and nonautonomous miniature inverted repeat element (MITE)-like elements. The latter are transmitted adventitiously using transposases from full-size IS elements and have appropriate direct repeats flanking variable-size "passenger" sequences. In addition, the integration of DNA viruses, plasmids, and DNA fragments from distant regions of the genomes can be transferred by an integrase mechanism *Sulfolobus* virus SSV1 (Brügger et al., in press) and also in the chromosomes of *A. pernix* and *P. horikoshii* where integron-like segments have been reported bordering inserted DNA elements such as the plasmid pXQ1.

CONCLUSION AND FUTURE TRENDS

The availability of deduced proteomes for thermophiles across the entire range of growth temperatures provides the opportunity for structural genomic studies. High-throughput recombinant gene expression studies are underway for *Methanocaldococcus jannaschii*, *T. maritima*, and *P. furiosus*. It could be argued that a structural genomic study of one of the "lower" thermophiles, such as *Thermus*, *Thermoplasma*, or *Thermoanaerobacter* spp., would provide the control experiment to determine the basis for protein thermostability by facilitating a comparative study together with the hyperthermophiles. In future, we can look forward to the availability of genome sequences of the "top" hyperthermophiles. The newly described archaeal Strain 121, isolated from a submarine vent, can grow slowly, at autoclave conditions (Kashefi and Lovley, 2003). Unfortunately, both *P. fumarii* and Strain 121 lack published genome sequences, which limits the scope of comparative studies.

The application of microarray-based studies, already underway using the *P. furiosus* genome information, will be important to examine global stress regulation. The recent discovery of a repressor that mediates the heat-shock response in *P. furiosus* and ongoing work with the bacterial-type repressor are promising rapid gains in information on gene regulation. In addition, although many of the details of the DNA replication fork are emerging from a study of the extreme thermophiles such as *S. solfataricus* or *M. thermautotrophicus*, the mechanism of initiation of DNA replication remains mechanism of initiation of DNA elusive. Presumably, the solution to this problem will also provide valuable information on the mechanisms of regulation of the cell cycle in the *Archaea*.

The mesophilic *Archaea* have been recently shown to encode the 22nd amino acid, pyrrolysine, using the UAG codon and a specialized UAG tRNA. Although not demonstrated for the thermophilic *Archaea* yet, it seems likely that a reexamination of existing genomes may reveal UAG codons with the appropriate context for pyrrolysine incorporation.

Thermophiles are advancing the ends of both basic and applied science. For instance, energy bioconversions in these cells are relevant to future production of nonfossil fuels because many of the hyperthermophiles are anaerobes that produce hydrogen or methane as gaseous end products. Further studies of the growth physiology and molecular biology of model organisms such as hyperthermophiles and halophiles will be necessary to determine their potential for the production of gas fuels and the potential application of their extremely thermostable enzymes in biotechnology.

Acknowledgments. We acknowledge support to F.T.R. from the National Science Foundation Assembling the Tree of Life and Microbial Observatories Programs, Grant MCB 0328337, a grant from the interinstitutional program in metabolic engineering, the National Institutes of Health and the Air Force Office of Scientific Research. This is Publication 06-144 from the Center of Marine Biotechnology.

REFERENCES

Adams, M. W., J. F. Holden, A. L. Menon, G. J. Schut, A. M. Grunden, C. Hour, A. M. Hutchins, J. F. Jenney, C. Ki, and K. Ma. 2001. Key role for sulfur in peptide metabolism and in regulation of three hydrogenases in the hyperthermophilic archaeon *Pyrococcus furiosus*. *J. Bacteriol.* **183:**716–724.

Angelov, A., and W. Liebl. 2006. Insights into extreme thermoacidophily based on genome analysis of *Picrophilus torridus* and other thermoacidophilic archaea. *J. Biotechnol.* **126:**3–10.

Bao, Q., Y. Tian, W. Li, Z. Xu, Z. Xuan, S. Hu, W. Dong, J. Yang, Y. Chen, Y. Xue, Y. Xu, X. Lai, L. Huang, X. Dong, Y. Ma, L. Ling, H. Tan, R. Chen, J. Wang, J. Yu, and H. Yang. 2002. A complete sequence of the *T. tengcongensis* genome. *Genome Res.* **12:**689–700.

Blochl, E., S. Burggraf, D. Hafenbradl, H. W. Jannasch, and K. O. Stetter. 1997. *Pyrolobus fumarii*, gen. and sp. nov., represents a novel group of archaea, extending the upper temperature limit for life to 113° C. *Extremophiles* **1:**14–21.

Boonyaratanakornkit, B. B., A. J. Simpson, T. A. Whitehead, C. M. Fraser, N. M. El-Sayed, and D. S. Clark. 2005. Transcriptional profiling of the hyperthermophilic methanarchaeon *Methanococcus jannaschii* in response to lethal heat and nonlethal cold shock. *Environ. Microbiol.* **7:**789–797.

Brügger, K., X. Peng, and R. A. Garrett. *Sulfolobus* genomes: mechanisms of rearrangement and change. In H.-P. Klenk and R. A. Garrett (ed.), *Archaeal Biology*, in press. Blackwell Publishing.

**Bult, C. J., G. J. Olsen, L. Zhou, R. D. Fleischmann, G. G. Sutton, J. A. Blake, L. M. FitzGerald, R. A. Clayton, J. D. Gocayne, A. R. Kerlavage, B. A. Dougherty, J. F. Tomb, M. D. Adams, C. L. Reich, R. Overbeek, E. F. Kirkness, K. G. Weinstock, J. M. Merrick, A. Glodek, J. L. Scott, N. S. Geoghagen, and J. C.

Venter. 1996. Complete genome sequence of the methanogenic archaeon, *Methanococcus jannaschii*. *Science* **273**:1058–1073.

Burggraf, S. K., P. Rouviere, and C. R. Woese. 1991. *Methanopyrus kandleri*: an archaeal methanogen unrelated to all other known methanogens. *Syst. Appl. Microbiol.* **14**:346–351.

Cambillau, C., and J. M. Claverie. 2000. Structural and genomic correlates of hyperthermostability. *J. Biol. Chem.* **275**:32383–32386.

Clark, D. S., and F. T. Robb. 1999. Adaptation of proteins from hyperthermophiles to high pressure and high temperature. *J. Mol. Microbiol. Biotechnol.* **1**:101–105.

Cohen, G. N., V. Barbe, D. Flament, M. Galperin, R. Heilig, O. Lecompte, O. Poch, D. Prieur, J. Querellou, and R. Ripp. 2003. An integrated analysis of the genome of the hyperthermophilic archaeon *Pyrococcus abyssi*. *Mol. Microbiol.* **47**:1495–1512.

Das, R., and M. Gerstein. 2000. The stability of thermophilic proteins: a study based on comprehensive genome comparison. *Funct. Integr. Genomics.* **1**:33–45.

Deckert, G., P. V. Warren, T. Gaasterland, W. G. Young, A. L. Lenox, D. E. Graham, R. Overbeek, M. A. Snead, M. Keller, R. Huber, S. J. Feldman, G. J. Olsen, and R. V. Swanson. 1998. The complete genome of the hyperthermophilic bacterium *Aquifex aeolicus*. *Nature.* **392**:353–358.

DiRuggiero, J., D. Dunn, D. L. Maeder, R. Holley-Shanks, J. Chatard, R. Horlacher, F. T. Robb, W. Boos, and R. B. Weiss. 2000. Evidence of recent lateral gene transfer among hyperthermophilic archaea. *Mol. Microbiol.* **38**:684–693.

Elcock, A. 1998. The stability of salt bridges at high temperatures: implications for hyperthermophilic proteins. *J. Mol. Biol.* **284**:489–502.

Ettema, T., J. v. d. Oost, and M. Huynen. 2001. Modularity in the gain and loss of genes: applications for function prediction. *Trends Genet.* **17**:485–487.

Fiala, G., and K. O. Stetter. 1986. *Pyrococcus furiosus* sp. nov. represents a novel genus of marine heterotrophic archaeabacteria growing optimally at 100°C. *Arch Microbiol.* **145**:56–61.

Fitz-Gibbon, S. T., H. Ladner, U. J. Kim, K. O. Stetter, M. L. Simon, and J. H. Miller. 2002. Genome sequence of the hyperthermophilic crenarchaeon *Pyrobaculum aerophilum*. *Proc. Natl. Acad. Sci. USA* **99**:984–989.

Frankenberg, R. J., M. Andersson, and D. S. Clark. 2003. Effect of temperature and pressure on the proteolytic specificity of the recombinant 20S proteasome from *Methanococcus jannaschii*. *Extremophiles* **7**:353–360.

Futterer, O., A. Angelov, H. Liesegang, G. Gottschalk, C. Schleper, B. Schepers, C. Dock, G. Antranikian, and W. Liebl. 2004. Genome sequence of *Picrophilus torridus* and its implications for life around pH 0. *Proc. Natl. Acad. Sci. USA* **101**:9091–9096.

Haney, P. J., G. L. Badger, G. L. Buldak, C. I. Reich, C. R. Woese, and G. J. Olsen. 1999. Thermal adaptation analyzed by comparison of protein sequences from mesophilic and extremely thermophilic *Methanococcus* species. *Proc. Natl. Acad. Sci. USA* **96**:3578–3583.

Johnson, M. R., S. B. Connors, C. I. Montero, C. J. Chou, K. R. Shockley, and R. M. Kelly. 2006. The *Thermotoga maritima* phenotype is impacted by syntrophic interaction with *Methanococcus jannaschii* in hyperthermophilic coculture. *Appl. Environ. Microbiol.* **72**:11–18.

Kanoksilapatham, W., D. L. Maeder, J. DiRuggiero, and F. T. Robb. 2004. A proposal to rename the hyperthermophile *Pyrococcus woesei* as *Pyrococcus furiosus* subsp. woesei. *Archaea* **1**:277–283.

Karshikoff, A., and R. Ladenstein. 2001. Ion pairs and the thermotolerance of proteins from hyperthermophiles: a "traffic rule" for hot roads. *Trends Biochem. Sci.* **26**:550–556.

Kashefi, K. L., and D. R. Lovley. 2003. Extending the upper temperature limit for life. *Science* **301**: 934.

Kawarabayasi, Y. S., Y. Hino, H. Horikawa, K. Jin-no, M. Takahashi, M. Sekine, S. Baba, A. Ankai, H. Kosugi, A. Hosoyama, S. Fukui, Y. Nagai, K. Nishijima, R. Otsuka, H. Nakazawa, M. Takamiya, Y. Kato, T. Yoshizawa, T. Tanaka, Y. Kudoh, J. Yamazaki, N. Kushida, A. Oguchi, K. Aoki, S. Masuda, M. Yanagii, M. Nishimura, A. Yamagishi, T. Oshima, and H. Kikuchi. 2001. Complete genome sequence of an aerobic thermoacidophilic crenarchaeon, *Sulfolobus tokodaii* strain7. *DNA Res.* **8**:123–140.

Kawarabayasi, Y., Y. Hino, H. Horikawa, S. Yamazaki, Y. Haikawa, K. Jin-no, M. Takahashi, M. Sekine, S. Baba, A. Ankai, H. Kosugi, A. Hosoyama, S. Fukui, Y. Nagai, K. Nishijima, H. Nakazawa, M. Takamiya, S. Masuda, T. Funahashi, T. Tanaka, Y. Kudoh, J. Yamazaki, N. Kushida, A. Oguchi, K. Aoki, K. Kubota, Y. Nakamura, N. Nomura, Y. Sako, and H. Kikuchi. 1999. Complete genome sequence of an aerobic hyper-thermophilic crenarchaeon, *Aeropyrum pernix* K1. *DNA Res.* **6**:83–101.

Kawarabayasi, Y. S., H. Horikawa, Y. Haikawa, Y. Hino, S. Yamamoto, M. Sekine, S. Baba, H. Kosugi, A. Hosoyama, Y. Nagai, M. Sakai, K. Ogura, R. Otsuka, H. Nakazawa, M. Takamiya, Y. Ohfuku, T. Funahashi, T. Tanaka, Y. Kudoh, J. Yamazaki, N. Kushida, A. Oguchi, K. Aoki, and H. Kikuchi. 1998. Complete sequence and gene organization of the genome of a hyper-thermophilic archaebacterium, *Pyrococcus horikoshii* OT3. *DNA Res.* **5**:147–155.

Kawashima, T., N. Amano, H. Koike, S.-L. Makino, S. Higuchi, Y. Kawashima-Ohya, K. Watanabe, M. Yamazaki, K. Kanehori, and T. Kawamoto. 2000. Archaeal adaptation to higher temperatures, revealed by genomic sequences of *Thermoplasma volcanium*. *Proc. Natl. Acad. Sci. USA* **97**:14257–14262.

Kawashima, T., Y. Yamamoto, H. Aramaki, T. Nunoshiba, T. Kawamoto, K. Watanabe, M. Yamazaki, K. Kanehori, N. Amano, Y. Ohya, K. Makino, and M. Suzuki. 1999. Determination of the complete genomic DNA sequence of *Thermoplasma volcanium* GSS1. *Proc. Jpn. Acad.* **75**:213–218.

Kim, K. K., H. Yokota, S. Santoso, D. Lerner, R. Kim, and S. H. Kim. 1998. Purification, crystallization, and preliminary X-ray crystallographic data analysis of small heat shock protein homolog from *Methanococcus jannaschii*, a hyperthermophile. *J. Struct. Biol.* **121**:76–80.

Klenk, P. H., A. R. Clayton, F. J. Tomb, O. White, E. K. Nelson, A. K. Ketchum, J. R. Dodson, M. Gwinn, K. E. Hickey, D. J. Peterson, L. D. Richardson, R. A. Kerlavage, E. D. Graham, C. N. Kyrpides, D. R. Fleischmann, J. Quackenbush, H. N. Lee, G. G. Sutton, S. Gill, F. E. Kirkness, A. B. Dougherty, K. McKenney, D. M. Adams, B. Loftus, S. Peterson, I. C. Reich, K. L. McNeil, H. J. Badger, A. Glodek, L. Zhou, R. Overbeek, D. J. Gocayne, F. J. Weidman, L. McDonald, T. Utterback, D. M. Cotton, T. Spriggs, P. Artiach, P. B. Kaine, M. S. Sykes, W. P. Sadow, P. K. D'Andrea, C. Bowman, C. Fujii, A. S. Garland, M. T. Mason, J. G. Olsen, M. C. Fraser, O. H. Smith, R. C. Woese, and C. J. Venter. 1997. The complete genome sequence of the hyperthermophilic, sulphate-reducing archaeon *Archaeoglobus fulgidus*. *Nature* **390**:364–370.

Laksanalamai, P., T. A. Whitehead, and F. T. Robb. 2004. Minimal protein-folding systems in hyperthermophilic archaea. *Nature Rev. Microbiol.* **2**:315–324.

Lesley, S. A., P. Kuhn, A. Godzik, A. M. Deacon, L. Mathews, A. Kreusch, G. Spraggon, H. E. Klock, D. McMullan, T. Shin, J. Vincent, A. Robb, L. S. Brinen, M. D. Miller, T. M. McPhillips, M. A. Miller, D. Scheibe, J. M. Canaves, C. Guda, L. Jaroszewski, T. L. Selby, M. A. Elsliger, J. Wooley, S. S. Taylor, K. O. Hodgson, L. A. Wilson, P. G. Schultz, and R. C. Stevens. 2002. Structural genomics of the *Thermotoga maritima* proteome implemented in a high-throughput structure determination pipeline. *Proc. Natl. Acad. Sci. USA* **399**:11664–11669.

Maeder, D. L., R. B. Weiss, D. M. Dunn, J. L. Cherry, J. M. Gonzalez, J. DiRuggiero, and F. T. Robb. 1999. Divergence of the hyperthermophilic archaea *Pyrococcus furiosus* and *P. horikoshii* inferred from complete genomic sequences. *Genetics* 152:1299–1305.

Martusewitsch, E., C. W. Sensen, and C. Schleper. 2000. High spontaneous mutation rate in the hyperthermophilic Archaeon *Sulfolobus solfataricus* is mediated by transposable elements. *J. Bacteriol.* 182:2574–2581.

Minuth, T., G. Frey, P. Lindner, R. Rachel, K. O. Stetter, and R. Jaenicke. 1998. Recombinant homo- and hetero-oligomers of an ultrastable chaperonin from the archaeon *Pyrodictium occultum* show chaperone activity in vitro. *Eur. J. Biochem.* 258:837–845.

Mongodin, E. F., I. R. Hance, R. T. DeBoy, S. R. Gill, S. Daugherty, R. Huber, C. M. Fraser, K. Stetter, and K. E. Nelson. 2005. Gene transfer and genome plasticity in *Thermotoga maritima*, a model hyperthermophilic species. *J. Bacteriol.* 187:4935–4944.

Nelson, K. E., S. R. Clayton, M. L. Gill, R. J. Gwinn, D. H. Dodson, E. K. Haft, J. D. Hickey, W. C. Peterson, K. A. Nelson, L. Ketchum, T. R. McDonald, J. A. Utterback, K. D. Malek, M. M. Linher, A. M. Garrett, M. D. Stewart, M. S. Cotton, C. A. Pratt, D. Phillips, D. Richardson, J. Heidelberg, G. G. Sutton, R. D. Fleischmann, J. A. Eisen, and C. M. Fraser. 1999. Evidence for lateral gene transfer between archaea and bacteria from genome sequence of *Thermotoga maritima*. *Nature* 399:323–329.

Robb, F. T., D. L. Maeder, J. R. Brown, J. DiRuggiero, M. D. Stump, R. K. Yeh, B. Weiss, and D. M. Dunn. 2001. Genomic sequence of hyperthermophile *Pyrococcus furiosus*: implications for physiology and enzymology. *Methods Enzymol.* 330:134–157.

Rohlin, L., J. D. Trent, K. Salmon, U. Kim, R. P. Gunsalus, and J. C. Liao. 2005. Heat shock response of *Archaeoglobus fulgidus*. *J. Bacteriol.* 187:6046–6057.

Ruepp, A., W. Graml, M. L. Santos-Martinez, K. K. Koretke, C. Volker, H. W. Mewes, D. Frishman, S. Stocker, A. N. Lupas, and W. Baumeister. 2000. The genome sequence of the thermoacidophilic scavenger *Thermoplasma acidophilum*. *Nature* 28:508–513.

Schut, G. J., S. D. Brehm, S. Datta, and M. W. Adams. 2003. Whole-genome DNA microarray analysis of a hyperthermophile and an archaeon: *Pyrococcus furiosus* grown on carbohydrates or peptides. *J. Bacteriol.* 185:3935–3947.

Schut, G. J., J. Zhou, and M. W. Adams. 2001. DNA microarray analysis of the hyperthermophilic archaeon *Pyrococcus furiosus*: evidence for a new type of sulfur-reducing enzyme complex. *J. Bacteriol.* 183:7027–7036.

Scott, R. A., N. J. Cosper, F. E. Jenney, and M. W. Adams. 2005. Bottlenecks and roadblocks in high-throughput XAS for structural genomics. *J. Synchrotron Rad.* 12:19–22.

She, Q., R. K. Singh, F. Confalonieri, Y. Zivanovic, G. Allard, M. J. Awayez, C. C. Chan-Weiher, I. G. Clausen, B. A. Curtis, A. De Moors, G. Erauso, C. Fletcher, P. M. Gordon, I. Heikamp-de Jong, A. C. Jeffries, C. J. Kozera, N. Medina, X. Peng, H. P. Thi-Ngoc, P. Redder, M. E. Schenk, C. Theriault, N. Tolstrup, R. L. Charlebois, W. F. Doolittle, M. Duguet, T. Gaasterland, R. A. Garrett, M. A. Ragan, C. W. Sensen, and J. Van der Oost. 2001. The complete genome of the crenarchaeon *Sulfolobus solfataricus* P2. *Proc. Natl. Acad. Sci. USA* 98:7835–7840.

Shockley, K., D. E. Ward, S. R. Chhabra, S. B. Conners, C. I. Montero, and R. M. Kelly. 2003. Heat shock response by the hyperthermophilic Archaeon *Pyrococcus furiosus*. *Appl. Environ. Microbiol.* 69:2365–2371.

Siegert, R., M. R. Leroux, C. Scheufler, F. U. Hartl, and I. Moarefi. 2000. Structure of the molecular chaperone prefoldin: unique interaction of multiple coiled coil tentacles with unfolded proteins. *Cell* 103:621–632.

Slesarev, A. I., K. S. Makarova, N. N. Polushin, O. V. Shcherbinina, V. V. Shakhova, G. I. Belova, L. Aravind, D. A. Natale, I. B. Rogozin, R. L. Tatusov, Y. I. Wolf, K. O. Stetter, A. G. Malykh, E. V. Koonin, and S. A. Kozyavkin. 2002. The complete genome of hyperthermophile *Methanopyrus kandleri* AV19 and monophyly of archaeal methanogens. *Proc. Natl. Acad. Sci. USA*. 99:4644–4649.

Smith, D. R., L. A. Doucette-Stamm, C. Deloughery, H.-M. Lee, J. Dubois, T. Aldredge, R. Bashirzadeh, D. Blakely, R. Cook, K. Gilbert, D. Harrison, L. Hoang, P. Keagle, W. Lumm, B. Pothier, D. Qiu, R. Spadafora, R. Vicare, Y. Wang, J. Wierzbowski, R. Gibson, N. Jiwani, A. Caruso, D. Bush, H. Safer, D. Patwell, S. Prabhakar, S. McDougall, G. Shimer, A. Goyal, S. Pietrovski, G. M. Church, C. J. Daniels, J. I. Mao, P. Rice, J. Nolling, and J. N. Reeve. 1997. Complete genome sequence of *Methanobacterium thermoautotrophicum* deltaH: functional analysis and comparative genomics. *J. Bacteriol.* 179:7135–7155.

Sterner, R., and W. Liebl. 2001. Thermophilic adaptation of proteins. *Crit. Rev. Biochem. Mol. Biol.* 36:39–106.

Stetter, K. O. 1996. Hyperthermophiles in the history of life. *Ciba Found. Symp.* 202:1–10; discussion 11–18.

Tolgyesi, E., C. S. Bode, L. Smelleri, D. R. Kim, K. K. Kim, K. Heremans, and J. Fidy. 2004. Pressure activation of the chaperone function of small heat shock proteins. *Cell. Mol. Biol.* 50:361–369.

Vetriani, C., D. Maeder, N. Tolliday, K. S. Yip, T. J. Stillman, K. L. Britton, D. W. Rice, H. H. Klump, and F. T. Robb. 1998. Protein thermostability above 100°C: a key role for ionic interactions. *Proc. Natl. Acad. Sci. USA* 95:12300–12305.

Vierke, G., A. Engelmann, C. Hebbeln, and M. Thomm. 2003. A novel archaeal transcriptional regulator of heat shock response. *J. Biol. Chem.* 278:18–26.

Volkl, P., R. Huber, E. Drobner, R. Rachel, S. Burggraf, A. Trincone, and K. O. Stetter. 1993. *Pyrobaculum aerophilum* sp. nov., a novel nitrate-reducing hyperthermophilic archaeum. *Appl. Environ. Microbiol.* 59:2918–2926.

Weinberg, M. V., G. Schut, S. Brehm, S. Datta, and M. W. Adams. 2005. Cold shock of a hyperthermophilic archaeon: *Pyrococcus furiosus* exhibits multiple responses to a suboptimal growth temperature with a key role for membrane-bound glycoproteins. *J. Bacteriol.* 187:336–348.

Worning, P., L. J. Jenson, K. E. Nelson, S. Brunak, and D. W. Ussery. 2000. Structural analysis of DNA sequence: evidence for lateral gene transfer in *Thermotoga maritima*. *Nucleic Acids Res.* 28:706–709.

Wu, M., Q. Ren, A. S. Durkin, S. C. Daugherty, L. M. Brinkac, R. J. Dodson, R. Madupu, S. A. Sullivan, J. F. Kolonay, W. C. Nelson, L. J. Tallon, K. M. Jones, L. E. Ulrich, J. M. Gonzalez, I. B. Zhulin, F. T. Robb, and J. A. Eisen. 2005. Life in hot carbon monoxide: the complete genome sequence of *Carboxydothermus hydrogenoformans* Z-2901. *PLoS Genetics* 1(5):e65.

Yip, K., K. L. Britton, T. Stillman, J. Lebbink, W. deVos, D. L. Maeder, F. T. Robb, C. Vetriani, and D. W. Rice. 1998. Insights into the molecular basis of thermal stability from the analysis of ion-pair networks in the glutamate dehydrogenase family. *Eur. J. Biochem.* 255:336–346.

Chapter 4

How Nucleic Acids Cope with High Temperature

HENRI GROSJEAN AND TAIRO OSHIMA

INTRODUCTION

Thermophilic organisms are a subgroup of extremophiles which are defined as having an optimum growth temperature above 45°C for moderate thermophiles, above 65°C for extreme thermophiles, and above 80°C for hyperthermophiles, some of which, such as *Pyrococcus*, *Pyrobaculum*, and *Methanopyrus*, are able to grow optimally at temperatures as high as 100 to 105°C (even higher in vegetative state or under high pressure). The hyperthermophilic genera are mostly *Archaea*, except for *Thermotoga* and *Aquifex* genus that belong to *Bacteria*. They thrive in very hot terrestrial habitats such as geysers, hot springs, and hot sediments of volcanic eruptions or near deep-sea hydrothermal vents and undersea volcanoes. The moderate thermophiles can be found almost everywhere, while most extreme thermophiles are found essentially in moderately hot environments (for details, see other chapters in this volume).

No eukaryotic cells have been found surviving above 60°C. Therefore, extreme thermophiles and hyperthermophiles are exclusively unicellular organisms lacking internal membranes, including nucleus and organelles. Because extreme thermophiles and hyperthermophiles cannot live at temperatures below 60°C, their macromolecules (DNA, RNA, and proteins/enzymes), which are composed of the same "building blocks" as those in mesophilic and psychrophilic organisms, are adapted to function only within a certain range of temperatures.

The existence of such "hot-loving" cells raises substantial questions related to the intrinsic thermoresistance of their various biomolecules, the existence of specific metabolic processes allowing cellular thermoprotection, and their evolutionary origin(s). It has been proposed that most, if not the majority, of the macromolecules in hyperthermophiles are somewhat sequestered (and thus protected) within some form of subcellular structures of which much remain to be discovered. Also, the various strategies leading to cell survival at high temperatures are necessarily metabolically interdependent. These intricate cellular strategies of thermoresistance/tolerance were developed during long-range evolution by microorganisms to adapt and finally become dependent on extreme living conditions as those existing in hot or very hot habitats (reviewed in Daniel and Cowan, 2000; Charlier and Droogmans, 2005; Forterre, 2006). In this chapter, we discuss these questions at the polynucleotide level—RNA, DNA, and their ribonucleoprotein derivatives (RNP/DNP). The same problems at the level of other biomolecules (proteins, small metabolites, and lipids), and aspects of transient thermotolerance of thermophilic organisms to sudden changes of growth temperature (thermal stress, heat shock, and transient thermoprotection), are treated independently in other chapters of this volume.

EFFECTS OF HEAT ON NUCLEIC ACIDS

Lessons From In Vitro Studies

When nucleic acids are heated in aqueous solution, two types of phenomena take place: denaturation of their architecture and chemical degradation of their building blocks (Fig. 1, upper part). In the case of double-stranded DNA, the helical structure is progressively lost, and at sufficiently elevated temperatures, the two strands ultimately separate (except for DNA plasmids or circularized DNA molecules). Concomitant to conformational denaturation, two main types of chemical degradations occur, mostly in the single-stranded portions of the denatured

---

H. Grosjean • Centre National de la Recherche Scientifique (CNRS), Laboratoire d'Enzymologie et Biochimie Structurales, 1 avenue de la Terrasse, F-91198 Gif-Sur-Yvette, France. T. Oshima • Institute of Environmental Microbiology, Kyowa Kako Co., Ltd, 2-15-5 Tadao, Machida, Tokyo 194-0035, Japan.

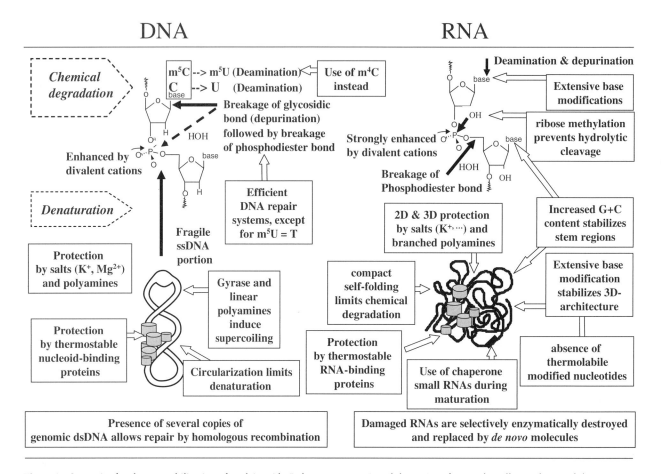

Figure 1. Strategies for thermostabilization of nucleic acids. In boxes are mentioned the various factors that allow a thermophilic organism to protect their nucleic acids against the deleterious effect of heat. A clear distinction between the giant extended macromolecule DNA and the more compact smaller RNA molecules has to be made. For details, see text.

DNA: chemical cleavages of the N-glycosidic bond between the sugar moiety and a base (especially with guanine and adenine, i.e., purines) (Lindahl and Nyberg, 1972; Lindahl and Karlstrom, 1973) and hydrolytic deaminations of cytosine and adenine to uracil and hypoxanthine (Lindahl and Nyberg, 1974; Karran and Lindahl, 1980; Frederico et al., 1990). Occasionally, base and sugar oxidation, base reduction, and/or alkylation occur due to radiation and/or presence in the biological fluids of active chemicals such as superoxide radicals, singlet oxygen, hydrogen peroxide, heavy metals or organic compounds, mutagens, and carcinogens (Lindahl, 1993).

Depurination and deamination of DNA are very slow at temperatures below 40°C, but both increase drastically as temperature increases. For example, in genomic DNA of a typical hyperthermophile (*Pyrococcus furiosus*, about 2 Mbp), it has been estimated that up to 1,000 purine bases are liberated after incubation for 1h in vitro at 70°C, i.e., about 0.025% of the bases, in fact about 0.1% of the total purines (Lindahl and Nyberg, 1972). Thus, at elevated temperatures, both depurination and cytidine/adenine deaminations in DNA, if not appropriately circumvented, can be a real problem in thermophiles.

DNA polynucleotide chain is naturally thermoresistant to spontaneous 3′ to 5′ phosphodiester bond hydrolysis. However, because the generation of apurinic/apyrimidic sites weakens the nearby phosphodiester bond, ssDNA fragments are generated upon heat treatment while nicks can appear in dsDNA (Lindahl and Andersson, 1972). This property has been exploited for ssDNA and RNA sequencing after in vitro reaction(s) with appropriate chemical reactants, leading to selective depurination of guanine or adenine of purified nucleic acid fragments (Maxam and Gilbert, 1977; Peattie, 1979).

RNA is more thermoresistant than DNA to depurination/depyrimidation, except for the naturally occurring modified purines N^7-methylguanosine (Wintermeyer and Zachau, 1970), wybutosine (Philippsen et al., 1968), and the pyrimidine dihydrouridine (House and Miller, 1996) that are highly thermolabile. Present at specific locations of the tRNA and rRNA molecules, these modified bases are the target sites for chemical thermodegradation. Also, among the four canonical nucleobases, cytosine is the less stable one at 100°C (Levy and Miller, 1998).

The main problem with RNA is the spontaneous hydrolysis of 3′ to 5′ phosphodiester bonds. This is due to the presence of a hydroxyl group in the 2′ position of every ribose (absent in deoxyribose) that promotes the intramolecular cleavage of the phosphodiester bond. At temperatures higher than 50°C and in the presence of Mg^{2+} ions, this phenomenon becomes dramatic, and upon extended incubation, RNA will ultimately be degraded to mononucleotides (Lindahl, 1967; Butzow and Eichhorn, 1975).

Despite the susceptibility of certain modified bases and of the ribonucleotide chain to thermal degradation, most naturally occurring tRNAs (especially those from hyperthermophilic organisms) appear fairly resistant to heat denaturation. For example, when tested in vitro in the presence of 10 mM Mg^{2+} ions, the melting temperatures (Tm) of the naturally occurring initiator tRNA^{Met} from *Pyrodictium occultum*—a hyperthermophilic archaeum that grows optimally at 105°C—were shown to be over 100°C (Ushida et al., 1996). Likewise, the melting temperatures of tRNA^{Ile} or tRNA^{Met} from *Thermus thermophilus*—an extreme thermophilic bacterium that grows optimally at 70°C—were shown to be over 86°C (Watanabe et al., 1980). Even tRNAs from mesophilic bacteria exhibit a melting temperature of at least 60 to 70°C, which is much higher than the optimal growth temperature of the corresponding microorganisms. Because of the difficulties to degrade chemically or enzymatically tRNAs from hyperthermophiles in vitro, very few sequences (only three) are available. The only tRNA sequences available come from PCR sequencing of tRNA genes (http://uni-bayreuth.de/departments/biochemie/trna); thus information is lacking concerning the modified nucleosides.

Nucleic Acids Are More Thermoresistant In Vivo than In Vitro

All the above estimates came from in vitro experiments with purified, not necessarily natural, RNA or DNA. In vivo, the half-lives of both RNA and DNA of thermophilic organisms are usually longer than that estimated in vitro, attesting to cellular strategies protecting the nucleic acids against the deleterious effects of heat. Even for mRNAs, their half-lives in the hyperthermophile *Sulfolobus solfataricus* range from minutes to hours when cells are grown at 80°C (as tested in a minimum salt medium supplemented only with glucose; generation time is 360 min) (Bini et al., 2002). These half-lives are of the same order of magnitude as those evaluated for the mesophilic bacterium *Methanococcus vannielii* (Hennigan and Reeve, 1994) and most eukaryotic cells such as *Saccharomyces cerevisiae* but obviously longer than those for a mesophilic bacterium such as *Escherichia coli*. The occurrence of relatively stable mRNA at 80°C in *Sulfolobus solfataricus* is surprising in light of what was said above. Therefore, either the intracellular composition of *Sulfolobus solfataricus* constrains the intrinsic instability of mRNA or protective processes (binding of proteins covering the RNA, transient association with special cellular substructure(s), and/or the existence of positively acting factors) are also in place to protect the fragile mRNA from hydrolytic degradation.

Notice that a single break in a compact folded RNA chain, such as tRNA or rRNA (but not mRNA), will not necessarily inactivate the molecule in vivo, since secondary and tertiary structures can prevent partial unfolding of the structural elements responsible for a particular function. Also, since RNA molecules always exist in multiple copies, and are continuously renewed during the cell life span, any chemical alteration occurring in only one or few of them has not much chance to be deleterious to the cell. Only when too many aberrant RNA molecules accumulate, one might expect impact on efficacy and accuracy of essential biosynthetic cellular metabolisms such as synthesis of informational DNA and mRNA.

Likewise, in the case of genomic DNA, the frequency of thermally induced backbone breaks and the rate of mutation in hyperthermophilic organisms appear by far less important in vivo as compared with what one could theoretically expect from the known intrinsic physical and chemical properties of the nucleic acids as measured in vitro (Peak et al., 1995; Grogan et al., 2001). Also, a single nick in the dsDNA may not affect dramatically the double-stranded structure around the nicked position, although it can have a profound consequence on supercoiling (see below). Nevertheless, any change in the chemical structure of the genomic DNA is potentially mutagenic, especially in coding regions of essential genes. If the frequency of these events exceeds the cellular capacity of repairing them accurately, the cell will not be able to survive. Here again, the internal chemical composition of the cell as well as efficient biochemical processes acting to protect and/or repair the damaged DNA molecules obviously allows hyperthermophiles to retain their genetic material intact.

HOW DO THERMOPHILES STABILIZE THEIR NUCLEIC ACIDS?

Despite the intrinsic potentiality of nucleic acids to degrade at elevated temperatures, many hyperthermophiles can survive at very high temperatures approaching or even surpassing the boiling point of water. The strategies developed during the evolution of these extremophiles are very diverse, and only collectively they allow efficient thermoprotection of both DNA and RNA. Figure 1 (bottom part) and the text below summarize the main features that contribute to the maintenance of the functional integrity of nucleic acids at high temperatures and/or appear to be correlated with thermophily.

Rule 1: Use of High G+C Content in RNA, but Not in DNA

As G:C base pairs in nucleic acids are more thermally stable than A:T base pairs, high G+C content may be a selective response to high temperature. From empirical data, an estimation of 5% increment in G+C content in the base-paired regions of tRNA brings about a 1.5°C rise in the melting temperature, with a theoretical upper limit of about 87°C for a poly(C)–poly(G) double helical structure under normal salt concentration (Oshima et al., 1976). As a matter of fact, an inspection of nucleotide sequences of tRNAs and tRNA genes from extreme thermophilic and hyperthermophilic organisms reveals that all stems of the cloverleaf are almost exclusively constituted of G:C base pairs (Yokoyama et al., 1987; Palmer et al., 1992; Marck and Grosjean, 2002). Also a linear correlation has been demonstrated between the optimal growth temperature and the G+C content in the secondary structure of rRNAs (Galtier and Lobry, 1997). Moreover, in tRNA and rRNA, the cytosine/uracil ratio normalized for guanine content increases with optimal growth temperature, suggesting that only most evolutionary conserved G:U pairs in mesophiles are converted to G:C in homolog RNAs of hyperthermophiles (Marck and Grosjean, 2002; Khachane et al., 2005). The adenine residues and more rarely uracil residues that remain in RNAs of hyperthermophiles are located in loops, non-structured regions, and key locations of stems known to be important as identity elements for recognition by proteins or enzymes (Varani and McClain, 2000). However, such an RNA stabilization strategy suffers from two drawbacks: first, a lesser variability among G+C-rich RNA populations (Xue et al., 2003) and second, the formation of non-functional two-dimensional and/or three-dimensional (3D) alternative structures (Urbonavicius et al., 2006). This last aspect of the problem is analogous to the A+U-rich RNAs of mitochondria (Helm, 2006).

In contrast, in single-stranded coding regions of mRNA, no correlation has been found between the G+C content and the optimal growth temperature of the cell (Galtier and Lobry, 1997; Hurst and Merchant, 2001; Lambros et al., 2003). A too high G+C content probably reduces the informational coding potentiality of mRNAs to an unacceptable lower vital limit. This situation may also favor the formation of highly stable stem loops or even pseudoknots along the mRNA molecule, and rRNA–mRNA base pairings, both of which can be a problem during translation of the mRNA. Instead, selection of certain dinucleotides at the first and second positions of codons in mRNAs (as well as of the corresponding genes in the chromosome) has been observed (Lynn et al., 2002). Such bias in codon usage do correlate with optimal growth temperature of the cell, with the bias being more pronounced in mRNAs of the hyperthermophiles (Kawashima et al., 2000; Nakashima et al., 2003; Paz et al., 2004). For example, codon GGC rather than CGC is used for tRNA-Ala, codon GGA rather than CGA or GCU is used for tRNA-Ser, and CCG rather than GCG is used for tRNA-Arg. The rationale is that the small helix forming between CC and GG is more stable than the ones forming between GC and CG or CG and GC (Borer et al., 1974; Bubienko et al., 1983), a situation that certainly exists also in codon–anticodon interaction during translation (Yokoyama et al., 1987; Marck and Grosjean, 2002). Moreover, the nucleotide at first wobble position of anticodon in abundant (major) tRNA from all extreme thermophilic and hyperthermophilic organisms is predominantly G (never A), thus allowing the formation of G:C pairs between the first letter of anticodon and the third letter of codon (Kagawa et al., 1984; Tong and Wong, 2004), a situation which should favor a more efficient translation process (Grosjean and Fiers, 1982).

Taken together, the above observations highlight the importance of ecological stress, as here the temperature, in the evolution of nucleic acid sequences and codon choice in mRNAs. The bias for G+C-rich stems in tRNA and rRNA, together with the preference of certain dinucleotides in coding sequences of mRNAs of hyperthermophiles, certainly reflects the need for optimizing the translation process (efficiency and accuracy) at high temperature.

Rule 2: Stabilization of Nucleic Acid Structures by Small Ligand Binding

From in vitro experiments with purified nucleic acids, it is known that Na^+, K^+, and Mg^{2+} protect dsDNA against chemical thermodegradation of the phophodiester bonds (Marguet and Forterre, 2001). High concentrations of monovalent cations reduce also the spontaneous chemical degradation of RNA (Hethke et al., 1999; Tehei et al., 2002), while Mg^{2+}

ions favor the hydrolytic degradation of ssRNA. Salts and specially Mg^{2+} ions act essentially as counterions to shield the highly negative phosphate backbone of nucleic acids and thereby aid nucleic acid folding (Londei et al., 1986; Serebrov et al., 2001). Hence, they stabilize the conformation of both DNA and RNA against thermodenaturation. In case of RNA, Mg^{2+} ions strongly bind to specific binding pockets in folded RNA (Brion and Westhof, 1997), some of which are being dependent on certain conserved modified nucleosides (Yue et al., 1994; Agris, 1996; Nobles et al., 2002).

Worth mentioning is that the intracellular K^+ concentration of certain hyperthermophilic euryarchaeota is much higher than that in mesophilic organisms. For example, in extreme thermophilic methanogens like *Methanothermus fervidus* (optimal growth temperature of 84°C) and Pyrococales such as *Pyrococcus furiosus* (optimal growth temperature of 100°C) for which a thorough analysis was performed, the intracellular K^+ concentration has been demonstrated to usually range between 700 and 900 mM, with a record of $1,060 \pm 30$ mM for *Methanothermus sociabilis* (Hensel and Konig, 1988). This is only five times less the normal intracellular K^+ concentration of halophilic archaeon such as *Halobacterium* (Brown, 1976). However, not all extreme thermophiles and hyperthermophiles (both bacteria and the Crenoarchaeota branch of *Archaea*) exhibit this feature.

Not only inorganic cations but also aliphatic, non-cyclic compounds containing two or more protonated amino nitrogen atoms, such as linear polyamines $NH_3^+[(CH_2)_{3-4}NH_2^+]_{x=1-4}(CH_2)_{3-4}NH_3^+$ or branched (ternary or quaternary) polyamines, can stabilize DNA and RNA. These compounds also neutralize the negative charges of phosphate, while long polyamines, such as thermine and spermine (tetramines) or caldopentamine and caldohexamine, can fix two or more distantly related phosphate groups in each strand of the nucleic acid. They are much more efficient than Mg^{2+} ions in stabilizing double helical structure of DNA, helical portions of RNAs, DNA–RNA hybrid, and more complex architectures of tRNA and rRNA (Basu and Marton, 1987; Terui et al., 2005; Xaplanteri et al., 2005). As a matter of fact, higher melting temperatures are observed when longer polyamines are added to tRNA or DNA in vitro, and in case of tRNAs, the highest melting temperature was recorded in the presence of a branched quaternary polyamine tetrakis(3-aminopropyl)ammonium (Terui et al., 2005). Probably, the "ball"-like structure of tetrakis(3-aminopropyl)ammonium, with all the positive charges on the surface, is more appropriate to interact with stem-and-loop structures of folded nucleic acids as RNAs.

Protection of single-stranded DNA from heat-induced depurination by polyamines has been demonstrated. Again, long polyamine chains are more efficient than short or branched polyamines, the order of protection efficiency being almost proportional to the number of nitrogen atoms in the cationic chains. This observation correlates with the fact that extreme thermophiles and hyperthermophiles often produce long bionic polyamines (Terui et al., 2005). However, once an abasic site appears in DNA, then polyamines, as any basic protein like histones, facilitate the hydrolysis of the nearest 3′ phosphodiester bond by inducing a beta-elimination reaction (Male et al., 1982).

From X-ray studies (Quigley et al., 1978; Jain et al., 1989) and molecular dynamics simulations (Bryson and Greenall, 2000), linear polyamines have been shown to interact with internal nucleic acid bases within the major groove of helical stems, in addition to the outside phosphate groups. Therefore, after binding to nucleic acids, polyamines can induce local structural distortions and help the nucleic acid either to condense into more compact and rigid structures (Basu and Marton, 1987; Vijayanathan et al., 2001) or to open locally to facilitate binding of proteins, including tRNA modification enzymes (Wildenauer et al., 1974). If the base sequence in DNA is appropriate, they also induce the conversion from the usual B-form to A- or Z-form, and if the DNA is circular, they promote negative supercoiling (Behe and Felsenfeld, 1981; Minyat et al., 1979; Hou et al., 2001). By strengthening preexisting interactions in the folding transition state, polyamines (as divalent cations) can also induce the formation of new cryptic polycationic binding sites. These changes can have important consequences for DNA–protein as well as RNA–protein interactions in the cell. Thus, naturally occurring polyamines play a major protective role in both thermodegradation and thermodenaturation of nucleic acids. However, the type of bionic polyamines, as well as their intracellular concentrations, varies much from one cell to another (Chen and Martynowicz, 1984; Oshima, 1989; Hamana and Matsuzaki, 1992).

Remarkably, in thermophilic *Eubacteria* and *Archaea*, many unusual polyamines are found (Oshima et al., 1987; Hamana et al., 1991, 1992, 1994). In extreme thermophilic bacteria *Thermotoga*, *Thermomicrobium*, and *Thermodesulfovibrio* and hyperthermophilic archaeons *Aeropyrum* and *Pyrodictium*, a large variety of long polyamines (mostly penta-amines and hexa-amines) are produced. In contrast, only branched polyamines, mostly tetrakis(3-aminopropyl)ammonium, are found in *Thermodesulfobacterium, Thermoleophilum, Aquifex,*

Hydrogenobacter (extreme thermophilic bacteria), *Methanococcus jannaschii* (extreme thermophilic archaeon), and *Pyrococcus* genus (hyperthermophilic archaea). Other extreme thermophilic bacteria, such as *Thermus* and *Rhodothermus*, produce, as major polyamines, both long and branched components, while *Geobacillus stearothermophilus* and *Bacillus acidocaldarius* (moderate thermophilic bacteria) and *Thermoplasma acidophilum* (moderate thermophilic archaeon) essentially produce the same short polyamines as the majority of mesophilic bacteria (e.g., spermine, spermidine, and putrescine). The extreme thermophilic archaea *Sulfolobus* (*acidocaldarius* and *tokodaii*) mainly produce homologs of these standard short polyamines norspermidine and bis(3-aminopropyl)1,3-diaminopropane (thermine).

The importance of these unusual naturally occurring cell-specific polyamines found in thermophiles and hyperthermophiles is suggested by the following observations (reviewed in Oshima, 1989): (i) the unusual long and branched polyamines are found only in extreme thermophiles and hyperthermophiles (bacteria and archaea), but not in mesophiles, moderate thermophiles (bacteria and archaea), or halophilic archaea (these contain high concentration of salts instead); (ii) cellular concentrations of long and branched polyamines are positively correlated with the growth temperature and increase only when the thermophiles are grown at the highest temperatures the cell can tolerate; (iii) knockout mutants of *Thermus thermophilus* lacking one of the genes coding for key enzymes of polyamine biosynthesis could not grow at high temperatures, and the growth was recovered only by the addition of some of polyamines in the culture medium; (iv) when performed at the temperature corresponding to the optimal physiological growth temperature of the cell, polyamines are required for in vitro protein biosynthesis, the highest activity being observed with the addition of tetrakis(3-aminopropyl)ammonium, precisely the branched polyamine that is found in many hyperthermophilic organisms.

Rule 3: Stabilization of Nucleic Acid Structures by Covalent Modification of Nucleosides

The majority of stable cellular RNAs, such as tRNA and rRNA molecules, contain a variety of modified nucleosides. They are all produced post-transcriptionally and catalyzed by a battery of RNA modification enzymes. To date, more than 100 different modified nucleosides have been identified in sequenced RNAs from various origins (for more information, see http://medlib.med.utah.edu/RNA mods/ and http://genesilico.pl/modomics/).

These modified nucleosides vary much in kind, degree of complexity, frequency of their occurrence, and origin of RNA. Some modified nucleosides are present in tRNA of most organisms, others are specific to particular RNA species or present in a limited number of evolutionary related organisms (*Bacteria*, *Eukarya*, or *Archaea*, Fig. 2). The common modified nucleosides are supposed to result from an early acquisition, while the species specific nucleosides probably arises from late acquisition during evolution (Levy and Miller, 1999; Cermakian and Cedergren, 1998).

An interesting peculiarity of tRNAs of archaeal hyperthermophiles is that they possess a remarkable number of doubly modified nucleosides, with covalent modification of the base in combination with the 2′-O-ribose methylation (ac^{4}Cm, m^{2}Gm, m$^2_2$Gm, m^{5}Cm, and m^{1}Im, Fig. 2; Edmonds et al., 1991; Kowalak et al., 1994; McCloskey et al., 2001). These hypermodified nucleosides exhibit exceptional conformational rigidity, and their presence in key positions of RNA molecules of hyperthermophiles may provide an important mechanism, not only for local but also for global thermal stabilization of the nucleic acid. ac^{4}Cm is the most rigid among the modified nucleosides studied so far (Kawai et al., 1989). Indeed, N^4-acetylation of cytosine and methylation of the sugar 2′-hydroxyl group independently stabilize the C3′-endo conformation of the ribose of cytidine, and both effects are additive, thereby favoring a local, very rigid A-type helical conformation of the RNA. Moreover, the electron-withdrawing property of the acetyl function at N^4 in ac^{4}C serves to make the cytidine N^4 proton more acidic, favoring a strong H-bond with guanine opposition of a Watson–Crick base pair. Similar conclusions were drawn for methylated bases in general (Engel and von Hippel, 1978; D'Andrea et al., 1983) and m^{2}Gm and m^{5}Cm in particular (Kawai et al., 1992). In addition to its potential stabilization role, the methyl group on 2′-hydroxyl of ribose allows to avoid hydrolysis of the corresponding phosphodiester bond in RNA. It also creates a more hydrophobic micro-environment which could help compaction of RNA and/or interaction with other molecules such as proteins.

Conversely, the thermolabile dihydrouridine (D) that obviously cannot form Watson–Crick base pair within a double helix (due to its non-aromatic character) allows more flexibility of the RNA backbone and thereby destabilizes locally the RNA molecules (Dallune et al., 1996). This modified nucleoside is totally absent in hyperthermophilic tRNAs, while it is abundant in psychrophilic bacteria (Dalluge et al., 1997). The systematic absence of another thermolabile nucleoside (m^{7}G) in hyperthermophiles is mentioned above.

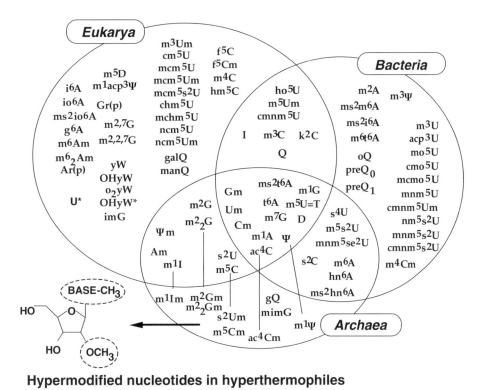

Hypermodified nucleotides in hyperthermophiles

Figure 2. Phylogenetic distribution of modified nucleosides in RNA from the three domains of life. Abbreviations of modified nucleosides are the conventional ones. For details, including the chemical structures, see in Limbach et al. (1995). Lines point out which ones among the hypermodified nucleosides in *Archaea* correspond to non-ribose methylated counterparts in *Eukarya* or *Bacteria*.

Evidence for thermodynamic stabilization of tRNA by posttranscriptional modifications was first obtained in vitro with *Escherichia coli* initiator tRNA$^{Met}_i$. By nuclear magnetic resonance spectrometry, Davanloo and collaborators (1979) have shown that methylation of uridine producing the conserved m^{5}U (thymidine) in position 54 of the T-loop resulted in an increase of the T_m by 6°C (local T_m) around position 54. The presence of a thiol group on m^{5}U54 (s^{2}T54) was further demonstrated to increase the average melting temperature(T_m) of tRNA^{Ile} of *Thermus thermophilus* by 3°C, as measured by UV spectrometry (Watanabe et al., 1976). Whereas in tRNA^{Trp} of *Bacillus subtilis*, the presence of the same thiol group on T54 increases the local T_m around the thiol group by 20°C (as measured by nuclear magnetic resonance spectrometry) (Gong et al., 2002). Such considerable induced thermal stability by just a simple methylation and/or thiolation of U54 results from the strategic location of this particular residue within the 3D architecture of the tRNA molecule. Indeed, not only the bulky highly polarizable 2-thiocarbonyl group favors the ribose puckering of s^{2}T to take preferentially the C3'-endo sugar conformation (as in A-form RNA) and thereby locally rigidifies the RNA molecule, but also it reinforces the reverse-Hoogsteen intra-loop pairing with A58 (m^{1}A58 in hyperthermophilic tRNAs), enhances the stacking interaction with a nearest-neighbor pseudouridine-55 (Davis, 1995), and allows site-specific coordination of Mg^{2+}, thus forming a rigid and characteristic intra-loop motif (Fig. 3) (see also in Romby et al., 1987; Becker et al., 1997; Nobles et al., 2002). In addition to this intra-loop s^{2}T effect, the strictly conserved A58 (or m^{1}A58 in hyperthermophilic tRNAs) forms an inter-strand stacking interaction with the conserved G18 (Gm18 in hyperthermophilic tRNAs) and G19 of the D-loop, this latter nucleoside being involved in a Watson–Crick tertiary pairing with C56 (Cm56 in archaea) of the T-loop. It results an intricate network of strongly interacting elements, including various modified nucleotides, as discussed in Yokoyama et al. (1987), Agris (1996), and Koshlap et al. (1999). These form the elbow region (3D core) of the characteristic L-shape functional tRNA molecule (T/D-loops interacting system, Fig. 3).

The importance of these modified nucleosides within the D/T-loops of tRNA in thermophily is also evident from the correlation between the optimal growth temperature of an organism and the level of several posttranscriptional modified nucleosides in RNAs (Agris et al., 1973; Kumagai et al., 1980; Noon et al., 2003). In particular, the degree of uridine thiolation (s^2T at position 54), 2′-O-methylguanosine (Gm at position 18), and N^1-methyladenosine (m^1A at position 58) in tRNA of hyperthermophilic bacterium *Thermus thermophilus* (optimum growth temperature 70°C) considerably increased when the temperature of the cell culture was raised from 50 to 80°C (Yokoyama et al., 1987; Shigi et al., 2006). The same situation exists along with elevation of the cultivation temperature of *Pyrococcus furiosus* (Kowalak et al., 1994). Likewise, the content of stabilizing 2′-O-methylribose in rRNA of *Sulfolobus solfataricus* increases by 20 to 25% with progressive increase of the cell culture temperature (Noon et al., 1998). In this latter case, methylation is performed by elaborated RNA methylation machinery involving several proteins and a battery of guide RNAs, in addition to the 2′-O-methyltransferase (Omer et al., 2000). It has been proposed that these guide RNAs also display an independent chaperone function during rRNA maturation by forming a short duplex (10 to 20 nucleotides in length) with the target RNA (Dennis et al., 2001). Last but not least, inactivation in *Thermus thermophilus* of the gene coding for N^1-methyladenosine (m^1A) in the T-loop of tRNA (position 58) results in a mutant strain with growth defect at 80°C (Droogmans et al., 2003), an observation that has been correlated with peculiarities of the corresponding tRNA:m^1A methyltransferase in *Archaea* (Roovers et al., 2004) and its importance for s^2T54 formation as we point out above (Shigi et al., 2006).

Thus, accumulation of selected modified nucleotides at few strategic positions within an RNA molecule appears as an efficient strategy to prevent thermodenaturation of whole RNA molecule in hyperthermophiles. On an evolution point of view, this requires the "invention" of many enzymes capable of catalyzing very specific base and ribose modifications. Obviously, different cells have developed independent posttranscriptional enzymatic machineries for covalent modification of their RNAs (tRNA, rRNA, and

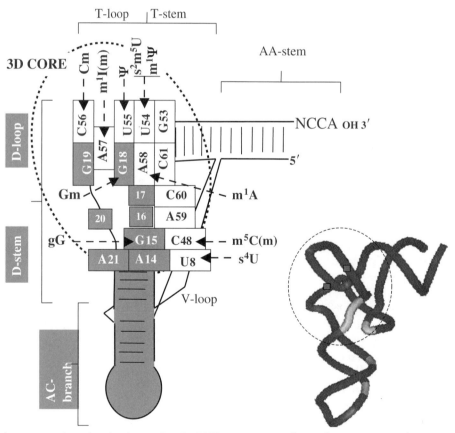

Figure 3. Schematic representation of tertiary interactions in tRNA structure. Numbers indicate conventional tRNA positions. Abbreviations of modified nucleosides are those of Fig. 2; see also in Sprinzl and Vassilenko (2005). Each nucleotide involved in stacking or base pairing with another nucleotide is represented by a rectangle. The rectangles representing nucleotides of the D-loop are in gray, while those representing nucleotides in the T-loop as well as nucleotides U8 and C48 located in between two stems are in white. Other parts of the tRNA molecule are represented by lines. Inside the dotted circle are elements that contributed to the 3D interaction, allowing an L-shaped spatial conformation to be formed from the 2D cloverleaf structure (see also Fig. 4). A wire representation of the 3D conformation is also indicated on the right side.

other RNAs not discussed here), some of which clearly depend on the thermal ecological niche of the microorganism. These enzymatic reactions are probably under genetic control, but to date, this remains an interesting hypothesis.

Genomic DNA of all types of organisms, including thermophiles, also contains minor modified bases. Their variety and their frequency of occurrence are considerably lower as compared with modified ribonucleosides in RNA. These DNA base modifications (almost essentially m^6A, m^5C, and m^4C) are catalyzed by specific enzymes (DNA methyltransferases) and arise at the polynucleotide level after their replication (Cheng, 1995). They have been shown to play a role in a variety of biological processes, including regulation of gene expression, DNA replication, and defense against restriction enzymes, but none of them (except one, see below) were demonstrated to be important in thermoresistance.

Methylation of C5 of pyrimidine ring of cytosine as in m^5C facilitates the spontaneous hydrolytic deamination of cytosine, especially at high temperatures (Cohen and Wolfenden, 1971; Wang et al., 1982; Shen et al., 1994; Zhang and Mathews, 1994). The corresponding deamination product becomes m^5U (deoxy-T), a normally occurring nucleoside in DNA. Unlike the uracil generated from non-modified cytosine, such deamination product of m^5C is not recognized by the ubiquitous U-DNA glycosylase repair system (Lindahl, 1979), and a high frequency of spontaneous mutation is expected if DNA contains too much m^5C. At temperatures above 80°C, this m^5C-to-m^5U(T) reaction can really become a problem for survival of the microorganism (Lutsenko and Bhagwat, 1999; Grogan, 2003).

To overcome this drawback, while maintaining a restriction/modification system operational, "Nature has invented" a completely different enzyme that catalyzes the direct transfer of a methyl group from S-adenosyl-methionine to exocyclic amine in position 6 of cytosine (m^4C), instead of in position C5 of pyrimidine ring as in m^5C. As a matter of fact, hyperthermophiles (*Bacteria* and *Archaea*) contains exclusively deamination-resistant m^4C residues (in addition to m^6A) and no detectable amount of m^5C (Ehrlich et al., 1985, 1986). This is again an interesting case of thermal adaptation of enzyme repertoire according to the ecological niche of the microorganism. This DNA:m^4C methyltransferase found only in *Archaea* and *Bacteria* has much in common with DNA:m^6A methyltransferase than with DNA:m^5C methyltransferase. Whether it originated in evolution by adaptation from a protein of an ancestral hyperthermophile is still an open question (Bujnicki and Radlinska, 1999).

Considering that during evolution DNA might have arisen from RNA, then DNA should be considered as an hypermodified nucleic acid in which: (i) every ribose now exists as a deoxyribose harboring an hydrogen atom in place of 2′-hydroxyl group, thus allowing better resistance to hydrolytic cleavage of the phosphodiester bonds, and (ii) all uracil bases methylated to 5-methyluracil (thymine), a modified base that was demonstrated, at least in tRNA to stabilize the nucleic acid. To achieve such extensive and quantitative hypermodifications of the primordial RNA into DNA, whatever the exact chemical composition of the primordial RNA was, the primitive cell had to evolve two major distinct types of enzymatic machineries. First one to reduce the 2′-hydroxyl group of ribonucleotides and the second one to methylate carbon-5 of the pyrimidine ring in every uracil residue to m^5U (thymine). To achieve such reactions, enzymes able to work at the mononucleotide rather than at the polynucleotide level had to evolve. Also novel specific RNA-dependent DNA polymerases able to discriminate between ribo- and deoxyribonucleotide building blocks and incorporating them with high efficiency into polymer had to be engineered. The only problem that arose later in evolution was with the site-specific enzymatic post-replicating formation of the easily deaminated m^5C in DNA, which in thermophilic organisms has been solved by engineering a novel DNA:m^4C methyltransferase instead of the DNA:m^5C methyltransferase as in all other mesophilic organisms examined so far. Different scenarios have been proposed for temporal appearance of these various enzymatic elements required for synthesis of DNA (Lazcano et al., 1988; Stubbe, 2000; Poole et al., 2000). With such a more stable nucleic acid as an RNA molecule (especially in long circularized dsDNA), faithful storage of genetic information in very long DNA polymers became possible, a property that RNA would not have been able to fulfill easily. Whether the progressive engineering of all these enzymes, including those we discussed above for posttranscriptional modifications of RNAs and of the DNA-repair/RNA-turnover machinery, occurred in a mesophilic or a thermophilic ancestral cell (LUCA) is discussed in another chapter of this volume.

Rule 4: Generation of Compact Tertiary Structures

Increasing the G+C content and the presence of sufficient amount of small cationic ligands allow enhancement of the stability of relatively long stretches of dsDNA and short stems in cellular dsRNA, as well as DNA–RNA complexes (during transcription) and RNA–RNA interaction (during translation). However, the same cannot be said for the spectacular thermoresistance of the complex architectures of large globular ssRNA like rRNA and giantess chromosomes.

In case of genomic dsDNA, nature has taken advantage of the stability of circular (topologically closed) double-stranded molecules, thus preventing the free rotation of the two strands of the helix, and at variance with linearized form of the same dsDNA (topologically open) molecule, it becomes naturally resistant to strand separation (Marguet and Forterre, 1994). Generation of supercoiled DNA (negative or positive supercoiling) provides an additional protection mechanism for genomic dsDNA (Forterre et al., 1992). This mechanism is facilitated by the presence of polyamines (see above) and by a family of DNA topoisomerases. In majority of microorganisms (mesophiles, moderate thermophiles, and few hyperthermophiles), a DNA gyrase produces negative supercoiling in the topologically closed genomic DNA as in bacterial plasmids. However, in majority of hyperthermophilic archaea and in few hyperthermophilic bacteria, a completely different type of DNA gyrase (called reverse gyrase) is found (Kikuchi and Asai, 1984; Bouthier de la Tour et al., 1990; Guipaud and Forterre, 2001). This reverse gyrase is a bifunctional enzyme resulting from a fusion between a gene coding for an ATP-dependent "helicase-like" protein and a type I DNA topoisomerase. When tested in vitro, it catalyzes the endoenergetic formation of positive superturns into dsDNA (Confalonieri et al., 1993; Rodriguez and Stock, 2002). This unique enzyme was probably evolved in a thermophilic organism and later transferred to a hyperthermophilic bacterium of the *Thermotogales* branch by lateral gene transfer (Forterre et al., 2000).

Because the reverse gyrase is a hallmark of hyperthermophilic organisms, it was assumed that positive supercoiling was the essence of its activity in vivo and thereby explained in large part the thermoresistance of genomic DNA at temperatures above 80°C (Forterre, 2002). However, recent data indicate that the problem is not so simple. Indeed, deletion of the gene coding for the reverse gyrase in the hyperthermophilic archaeon *Thermococcus kodakaraensis* KOD1 demonstrate that the enzyme is essential for cell growth only at extreme temperatures such as above 93°C, and not in the range of 80 to 90°C (Atomi et al., 2004). The need of active reverse gyrase at such extremely high temperatures might not be the generation of positive supercoiling of DNA as initially thought, but instead more subtle local protection of DNA against increasing number of damages (depurination, single and double-stranded breakages, and UV irradiation) due to harsh conditions of life at high temperatures (Napoli et al., 2004; Kampmann and Stock, 2004).

While a topologically close DNA, negatively or positively supercoiled, remains an important feature that conditions life at high temperature, the importance of the unique hyperthermophilic DNA-binding reverse gyrase in vivo might be more related to local rather than global structural needs, possibly in relation to DNA-repair processes, as well as for gene expression and/or regulation (including stress responses to either heat or cold shocks). The fact that the intracellular enzymatic activity of the reverse gyrase in *Sulfolobus* depends very much on the growth temperature (Lopez-Garcia and Forterre, 2000) together with the fact that *Thermotogales* family of bacteria, as well as the hyperthermophilic archaeon *Archaeoglobus fulgidus*, cited in Guipaud and Forterre (2001), possesses both DNA gyrase and reverse gyrase paid in favor of distinct roles for these two non-homologous enzymes.

In case of ssRNA, stability depends on several key structural elements, such as tertiary hydrogen bonds, pseudoknot fold, internal loop–loop interactions (kissing complexes), cross-strand purine stacking interactions, water- and/or metal-bridges, and the formation of local hydrophobic pockets (see above with ribose methylation), and this list is certainly not exhaustive. Recent studies on RNA structures at atomic resolution revealed additional recurrent folding motifs, such as kink-turn motif, GNRA tetraloop, A-minor motif, V-turn, bulged G motif, and bulge-helix-bulge motif. Only those of a simple molecule like tRNA are the best well understood to date (Westhof et al., 1988; Varani, 1995; Draper, 1996). Moreover, the crystal structure of rRNA within the context of the 50S and the 30S ribosomal subunits of thermophilic *Thermus thermophilus* has recently revealed the presence of several recurrent structural motifs which are essential for RNA folding and stability (Yusupov et al., 2001). All these various types of tertiary interactions and motifs probably act together in a synergistic way and allow the compaction of RNA conformation and/or modulate the flexibility/rigidity of the macromolecules according to their functional needs at a given temperature.

One such motif is the very stable loop–loop interacting system resembling the characteristic D/T interaction in tRNA as discussed above (Fig. 3). This architectural motif is found in rRNA of the ribosome (Lee et al., 2003) and the nucleic acid part of RNase P (Guo and Cech, 2002). The key component of this 3D structural motif is a U:A trans-reverse intra-loop Hoogsteen base pair within (in fact Psi:A in rRNA analogous to $s^2T:m^1A$ in tRNA) stacked on a canonical Watson–Crick base pair of one of the stem (usually

G:C). This motif is further stabilized by additional non-canonical H bonds, as well as by base stacking interactions with an other loop in trans of the RNA molecule (Draper, 1996; Conn and Draper, 1998; Nagaswamy and Fox, 2002).

An alternative way to obtain information on the molecular basis of RNA 3D stabilization is from in vitro selection studies (Wright and Joyce, 1997). The take-home-lesson with such an approach is that the stabilization of complex RNA architecture can be achieved through a small number of substitutions, which primarily increase the compactness of properly folded nucleic acids (Fang et al., 2001; Guo and Cech, 2002; Vergne et al., 2006). Evidently, removing water and divalent cations from the vicinity of most internal phosphodiester bonds by increasing local hydrophobicity is probably the main output of generating compact 3D architecture in both RNA and DNA of extreme thermophiles and hyperthermophiles. Different cellular strategies exist to achieve this goal for structural RNAs and genomic DNA respectively.

Rule 5: Generation of RNP Particles with Thermostable Proteins

Permanent or transient stabilization of specific portions of genomic DNA or RNA in vivo can be further dependent on a family of specific thermostable proteins. In case of DNA, chromatin proteins, such as basic histones, histone-like proteins, and a family of small DNA-binding proteins, Sul7d, Sac7, Sis7, Sso7, Alba, MC1, and HU, tightly bind to the nucleic acid (reviewed in Sandman and Reeve, 2005). It may be that reverse gyrase also acts as a chaperone to protect locally damaged DNA against thermal denaturation (Kampmann and Stock, 2004). These proteins prevent the free rotation of the two strands around each other and facilitate the compaction of DNA and/or condense linear dsDNA into regular, globular nucleosome. The dsDNA within these nucleoprotein complexes is also more resistant to heat denaturation (Lurz et al., 1986; Grayling et al., 1996; Guagliardi et al., 1997; Pereira and Reeve, 1998). Some of these proteins play a structural role only during specific steps of gene expression, transcription, replication, recombination, and/or even repair of DNA. They may modulate locally the topology of the genomic DNA molecule and/or simply prevent chemical damages to DNA by limiting access of the solvent to the nucleic acid (Lopez-Garcia and Forterre, 1999). However, due to the strong basic character of the histones and most DNA-binding proteins, they can catalyze chain scission at abasic sites by catalyzing a beta-elimination reaction, a drawback that might become a real problem at high temperatures if not appropriately circumvented.

Like for nucleosomes, interaction of the several ribosomal proteins (which are more numerous in the ribosomes of hyperthermophiles than those of mesophiles) (Kohrer et al., 1998; Conn et al., 1999; Gruber et al., 2003) with specific sites on 16S- and 23S-rRNA is essential to protect several exposed phosphodiester bonds from hydrolysis by water but mostly to form stable globular ribosomal RNP particles able to function at high temperatures. The particular strong binding of rRNA and ribosomal proteins (as compared with mesophilic systems) is testifying the difficulties to deplete all proteins from rRNA of hyperthermophilic ribosomes (Cammarano et al., 1983). Moreover, some of these proteins may have a chaperone function at some specific stage of the sequential rRNA maturation, thus solving in part the thermodynamic and kinetic folding problems and RNA–protein complex formation, which at high temperatures might not be a trivial problem (Herschlag, 1995; Weeks and Cech, 1996).

As far as tRNAs are concerned, transient association with abundant thermostable proteins like elongation factor and aminoacyl-tRNA ligases can also help to protect the nucleic acids at high temperatures. Other proteins able to bind stably or transiently to RNA (including mRNA) and specific to hyperthermophilic organisms have still to be discovered (see e.g., Morales et al., 1999). Also, depending on the temperature at which a given hyperthermophile is grown, the amount of certain proteins, or the activity of certain enzymes of the cell, may differ substantially. Above, we mentioned the cases of several RNA modification enzymes: tRNA and rRNA methyltransferases, tRNA thiolase, enzymes involved in the biosynthesis of long and branched polyamines, and reverse gyrase. Thanks to the development of DNA micro-array technology, coupled with quantitative PCR and sensitive detection of proteins by mass spectrometry, such phenomena will soon be more systematically investigated. These powerful biochemical and biophysical approaches may help to identify many other cases of proteins/enzymes involved in the complex process of thermoprotection and thermoadaption of nucleic acids (Lopez-Garcia and Forterre, 1999; Weinberg et al., 2005).

Rule 6: Controlling DNA Damages by Efficient DNA Repair

Despite the multiple deleterious effects of heat on genomic DNA, hyperthermophiles are able to preserve their genetic information from generation to generation (Grogan, 2004). As explained above, this is due to the existence of processes that allow the protection of nucleic acids not only from thermodenaturation and thermodegradation but also from a combinatory action of a battery of proteins and enzymes capable of repairing various types of DNA lesions. These are mostly DNA-glycosylases, suicide DNA-(de)-methyl-transferases, endonucleases, 5′ to 3′ exonucleases, DNA-ligase, and polymerases able to eliminate efficiently any abasic site, deaminated, alkylated, or oxidized base, and radiation-induced cross-link products that may appear accidentally in the DNA of the chromosome, especially at high temperatures. This adaptive base excision DNA-repair system is independent and differs from the independent inducible error-prone repair process. Detailed descriptions of these machineries, most of them being only recently identified in thermophiles, are out of the scope of the present review (for details see Grogan, 2006).

Because hyperthermophiles appear exceptionally resistant to radiation (DiRuggiero et al., 1997; Jolivet et al., 2003), a novel type of archaeal DNA-repair machinery specific to thermophiles is anticipated. At least, a systematic comparative analysis of conserved genes in genomes of hyperthermophiles supports this hypothesis (Eisen and Hanawalt, 1999; Makarova et al., 2002), but so far direct evidences are still lacking. The possibility exists that reverse gyrase, an almost ubiquitous enzyme in hyperthermophilic organisms, plays an essential role by combining its potential to maintain DNA topology and its participation in DNA-repair mechanism (Napoli et al., 2004; Kampmann and Stock, 2004).

Also, several copies of the chromosome are present in exponentially growing thermophilic archaea (Bernander and Poplawski, 1997). Together with the fact that homologous recombination is especially efficient in such microorganisms, as in bacteria (Grogan, 1996), exchanges of chromosomal material within and between hyperthermophilic archaea and bacteria living within the same ecological niche (lateral gene transfer) can occur (Forterre et al., 2000; Makarova and Koonin, 2003; Farahi et al., 2004). This efficient recombinational DNA-repair system, in complement with the arsenal of enzymes of the "classical" DNA-repair machinery as mentioned above, may be sufficient to preserve the genetic patrimony of prokaryotic cells that "love" high temperatures. It also allows more opportunities to borrow and further adapt interesting "inventions" from one type of organism to another via lateral gene transfer.

Rule 7: Eliminating RNA Damages by RNA Turnover

Despite the existence of enzymes capable of posttranscriptionally replacing a genetically encoded base by another in RNA, like tRNA:guanine G34 or G15 transglycosylases—also called queuine or archaeosine-insertases (Okada et al., 1979; Watanabe et al., 1997)—or able to delete/insert one or several nucleotides in an RNA molecule by a sophisticated RNA editing processes (see several chapters in Bass, 2000), existence of specific RNA-transglycosylases or editing machinery able to repair abnormal chemically damaged RNAs has not yet been described (see however Aas et al., 2003).

To cope with the thermosensibility of their RNAs (especially mRNAs), hyperthermophiles probably use their biological materials "quickly" enough and replace the damaged or degraded RNA before they poison the whole cellular machinery. Coupling gene transcription with translation of the nascent mRNA is obviously an advantage for the hyperthermophiles. Efficient elimination of abnormal, nonfunctional RNA molecules operated via specific degradative machineries (RNA surveillance-control quality and rapid RNA degradation—RTD—systems) is probably also vital for the thermophiles and hyperthermophiles. In *Bacteria* and *Eukarya*, a battery of enzymes and proteins organized in multicomponent complexes, collectively called degradosomes, have been identified (Vanzo et al., 1998; Anderson, 2005). The same type of machinery certainly exists in *Archaea* and especially in the hyperthermophiles, but systematic and thorough studies are still to be performed. As in the case of DNA-repair systems, the possibility exists that hyperthermophiles use more sophisticated or even novel type of degradosome machinery. Whatever the "bio-tactics" involved, life at a high temperature is a challenging enterprise, very endoenergetic (consume much ATP), and demanding for adaptation of many enzymes or even inventing new ones to fulfill a variety of difficult tasks.

CONCLUSION AND OUTLOOK

How nucleic acids of hyperthermophiles cope with high temperatures? This is a difficult question to answer because of the complexity of the problem. Indeed, the nucleic acid sequences and structures are so diverse that what apply to one type of macromolecules may not necessarily apply to another. Little is known about the intracellular chemical and the physical organization of the cell in which the DNA and RNA exist. Also, existence of different interdependent factors acting collectively is not easy to identify individually.

Stabilizing strategies of RNAs and DNAs may be classified into three major categories: (i) those which are intrinsic to the chemical structures of the nucleic acids; (ii) those which are dependent on extrinsic interactions with other biomolecules; and (iii) those which are dependent on a battery of enzymes able to detect and repair the DNA damage or to constantly renew functional RNA molecules. These adaptive "quality control" metabolic processes apply not only to DNA and RNA but also to proteins, lipids, and all the precursor biomolecules that are directly or indirectly involved in the processes of DNA and RNA synthesis and maturation. In thermophilic prokaryotes, these biosynthetic processes, of which much is still largely ignored, might be more elaborated and efficient than the homologous metabolic processes in mesophiles and psychrophiles. Complete understanding of how nucleic acids within the whole thermophilic organisms can cope with high temperature can be fully understood only if many synergistic factors, as those we discussed in this review, are considered. To illustrate such a complex interplay of many diversified factors, we summarize in Fig. 4 all what we have discussed in this review for only the family of tRNAs.

It is important to recall that majority of the identified biomolecules and metabolic processes that can account for the resistance of nucleic acids in thermophilic organisms come from in vitro studies. To better understand the importance of each individual parameter, better knowledge of the internal biochemical composition and, probably most important, the organization of these molecules within the cellular space are needed (Minton, 2001; Ellis, 2001; Ovadi and Saks, 2004). Genetic approach using mutant strains mutated in one or more biomolecules supposedly involved directly or indirectly in stabilization of nucleic acids should be more systematically used. Except in few cases related above, this genetic approach has been largely unexploited, because of insufficient knowledge of the genetic dynamics of the hyperthermophiles (Grogan, 2006). However, coupled with the recent available techniques of functional genomics and the potential predictive power of systematic in silico analyses of exponentially increasing numbers of completely sequenced genomes of

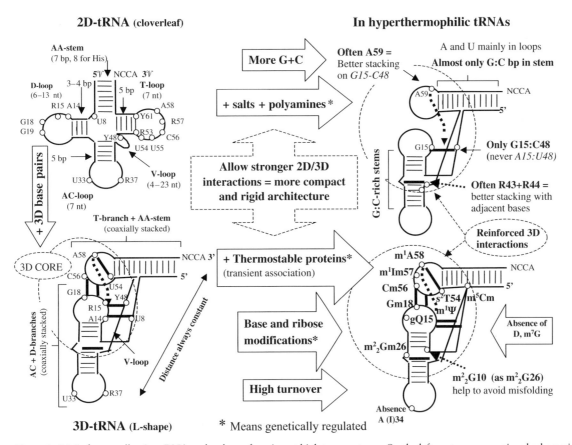

Figure 4. Main factors allowing tRNA molecules to function at high temperatures. On the left part are conventional schematic representations of 2D and 3D structures of tRNA. The remarkable features that are characteristic of a tRNA from a hyperthermophilic organism are indicated on the right side. In the boxes in the central part are indicated the various factors that allow a thermophilic organism to protect their nucleic acids against the deleterious effect of heat. For details, see text. Numbers indicate conventional tRNA positions; letters correspond to bases A, C, G, and U; R for purine, Y for pyrimidine, and N for any base.

thermophiles, it should allow the systematic identification of the molecular basis of thermophily, the discovery of novel aspects in *Archaea* (third domain of life), the establishment of a precise phylogenetic relationship of all existent forms of life, and possibly the development of useful derivatives for industrial applications (see also other chapters in this volume).

Acknowledgments. We thank Patrick Forterre (Pasteur Institute in Paris and University of Paris XI in Orsay) and Mark Helm (University of Heidelberg, Germany) for advice and sharing unpublished information. We also thank Basma El Yacoubi (University of Florida in Gainesville, USA) for improvements of the text. H. G. is Emeritus Researcher at University of Paris-South in Orsay, and his work was supported by a grant from the CNRS (Programme Interdépartemental de Géomicrobiologie des Environnements Extremes—Geomex 2002-03).

REFERENCES

Aas, P. A., M. Otterlei, P. O. Falnes, C. B. Vagbo, F. Skorpen, M. Akbari, O. Sundheim, M. Bjoras, G. Slupphaug, E. Seeberg, and H. E. Krokan. 2003. Human and bacterial oxidative demethylases repair alkylation damage in both RNA and DNA. *Nature* **421:**859–863.

Agris, P. F. 1996. The importance of being modified: roles of modified nucleosides and Mg^{2+} in RNA structure and function. *Prog. Nucleic Acid Res. Mol. Biol.* **53:**79–129.

Agris, P. F., H. Koh, and D. Soll. 1973. The effect of growth temperatures on the in vivo ribose methylation of *Bacillus stearothermophilus* transfer RNA. *Arch. Biochem. Biophys.* **154:**277–282.

Anderson, J. T. 2005. RNA turnover: unexpected consequences of being tailed. *Curr. Biol.* **15:**R635–R638.

Atomi, H., R. Matsumi, and T. Imanaka. 2004. Reverse gyrase is not a prerequisite for hyperthermophilic life. *J. Bacteriol.* **186:**4829–4833.

Bass, B. L. 2000. *RNA Editing*. Oxford University Press, Oxford, United Kingdom.

Basu, H. S., and L. J. Marton. 1987. The interaction of spermine and pentamines with DNA. *Biochem. J.* **244:**243–246.

Becker, H. F., Y. Motorin, M. Sissler, C. Florentz, and H. Grosjean. 1997. Major identity determinants for enzymatic formation of ribothymidine and pseudouridine in the T psi-loop of yeast tRNAs. *J. Mol. Biol.* **274:**505–518.

Behe, M., and G. Felsenfeld. 1981. Effects of methylation on a synthetic polynucleotide: the B–Z transition in poly(dG-m5dC). poly(dG-m5dC). *Proc. Natl. Acad. Sci. USA* **78:**1619–1623.

Bernander, R., and A. Poplawski. 1997. Cell cycle characteristics of thermophilic archaea. *J. Bacteriol.* **179:**4963–4969.

Bini, E., V. Dikshit, K. Dirksen, M. Drozda, and P. Blum. 2002. Stability of mRNA in the hyperthermophilic archaeon *Sulfolobus solfataricus*. *RNA* **8:**1129–1136.

Borer, P. N., B. Dengler, I. Tinoco, Jr., and O. C. Uhlenbeck. 1974. Stability of ribonucleic acid double-stranded helices. *J. Mol. Biol.* **86:**843–853.

Bouthier de la Tour, C., C. Portemer, M. Nadal, K. O. Stetter, P. Forterre, and M. Duguet. 1990. Reverse gyrase, a hallmark of the hyperthermophilic archaebacteria. *J. Bacteriol.* **172:**6803–6808.

Brion, P., and E. Westhof. 1997. Hierarchy and dynamics of RNA folding. *Annu. Rev. Biophys. Biomol. Struct.* **26:**113–137.

Brown, A. D. 1976. Microbial water stress. *Bacteriol. Rev.* **40:**803–846.

Bryson, K., and R. J. Greenall. 2000. Binding sites of the polyamines putrescine, cadaverine, spermidine and spermine on A- and B-DNA located by simulated annealing. *J. Biomol. Struct. Dyn.* **18:**393–412.

Bubienko, E., P. Cruz, J. F. Thomason, and P. N. Borer. 1983. Nearest-neighbor effects in the structure and function of nucleic acids. *Prog. Nucleic Acid Res. Mol. Biol.* **30:**41–90.

Bujnicki, J. M., and M. Radlinska. 1999. Molecular evolution of DNA-(cytosine-N4) methyltransferases: evidence for their polyphyletic origin. *Nucleic Acids Res.* **27:**4501–4509.

Butzow, J. J., and G. L. Eichhorn. 1975. Different susceptibility of DNA and RNA to cleavage by metal ions. *Nature* **254:**358–359.

Cammarano, P., F. Mazzei, P. Londei, A. Teichner, M. de Rosa, and A. Gambacorta. 1983. Secondary structure features of ribosomal RNA species within intact ribosomal subunits and efficiency of RNA-protein interactions in thermoacidophilic (*Caldariella acidophila*, *Bacillus acidocaldarius*) and mesophilic (*Escherichia coli*) bacteria. *Biochim. Biophys. Acta* **740:**300–312.

Cermakian, N., and R. Cedergren. 1998. Modified nucleotides always were: an evolutionary model, p. 535–541. *In* H. Grosjean, and R. Benne (ed.), *Modification and Editing of RNA*. ASM Press, Washington, DC.

Charlier, D., and L. Droogmans. 2005. Microbial life at high temperature, the challenges, the strategies. *Cell. Mol. Life Sci.* **62:**2974–2984.

Chen, K. Y., and H. Martynowicz. 1984. Lack of detectable polyamines in an extremely halophilic bacterium. *Biochem. Biophys. Res. Commun.* **124:**423–429.

Cheng, X. 1995. DNA modification by methyltransferases. *Curr. Opin. Struct. Biol.* **5:**4–10.

Cohen, R. M., and R. Wolfenden. 1971. The equilibrium of hydrolytic deamination of cytidine and N4-methylcytidine. *J. Biol. Chem.* **246:**7566–7568.

Confalonieri, F., C. Elie, M. Nadal, C. de La Tour, P. Forterre, and M. Duguet. 1993. Reverse gyrase: a helicase-like domain and a type I topoisomerase in the same polypeptide. *Proc. Natl. Acad. Sci. USA* **90:**4753–4757.

Conn, G. L., and D. E. Draper. 1998. RNA structure. *Curr. Opin. Struct. Biol.* **8:**278–285.

Conn, G. L., D. E. Draper, E. E. Lattman, and A. G. Gittis. 1999. Crystal structure of a conserved ribosomal protein-RNA complex. *Science* **284:**1171–1174.

D'Andrea, X., Alkema, X., Bell, X., Coddington, X., Haber, X., Hughes, X., and Neilson. 1983. Methylated bases stabilize short RNA duplex. *J. Am. Chem. Soc.* **105:**636–638.

Dalluge, J. J., T. Hamamoto, K. Horikoshi, R. Y. Morita, K. O. Stetter, and J. A. McCloskey. 1997. Posttranscriptional modification of tRNA in psychrophilic bacteria. *J. Bacteriol.* **179:**1918–1923.

Dalluge, J. J., T. Hashizume, A. E. Sopchik, J. A. McCloskey, and D. R. Davis. 1996. Conformational flexibility in RNA: the role of dihydrouridine. *Nucleic Acids Res.* **24:**1073–1079.

Daniel, R. M., and D. A. Cowan. 2000. Biomolecular stability and life at high temperatures. *Cell. Mol. Life Sci.* **57:**250–264.

Davanloo, P., M. Sprinzl, K. Watanabe, M. Albani, and H. Kersten. 1979. Role of ribothymidine in the thermal stability of transfer RNA as monitored by proton magnetic resonance. *Nucleic Acids Res.* **6:**1571–1581.

Davis, D. R. 1995. Stabilization of RNA stacking by pseudouridine. *Nucleic Acids Res.* **23:**5020–5026.

Dennis, P. P., A. Omer, and T. Lowe. 2001. A guided tour: small RNA function in Archaea. *Mol. Microbiol.* **40:**509–519.

DiRuggiero, J., N. Santangelo, Z. Nackerdien, J. Ravel, and F. T. Robb. 1997. Repair of extensive ionizing-radiation DNA damage at 95 degrees C in the hyperthermophilic archaeon *Pyrococcus furiosus*. *J. Bacteriol.* **179:**4643–4645.

Draper, D. E. 1996. Strategies for RNA folding. *Trends Biochem. Sci.* **21:**145–149.

Droogmans, L., M. Roovers, J. M. Bujnicki, C. Tricot, T. Hartsch, V. Stalon, and H. Grosjean. 2003. Cloning and characterization of tRNA (m¹A58) methyltransferase (TrmI) from *Thermus thermophilus* HB27, a protein required for cell growth at extreme temperatures. *Nucleic Acids Res.* **31:**2148–2156.

Edmonds, C. G., P. F. Crain, R. Gupta, T. Hashizume, C. H. Hocart, J. A. Kowalak, S. C. Pomerantz, K. O. Stetter, and J. A. McCloskey. 1991. Posttranscriptional modification of tRNA in thermophilic archaea (Archaebacteria). *J. Bacteriol.* **173:**3138–3148.

Ehrlich, M., M. A. Gama-Sosa, L. H. Carreira, L. G. Ljungdahl, K. C. Kuo, and C. W. Gehrke. 1985. DNA methylation in thermophilic bacteria: N4-methylcytosine, 5-methylcytosine, and N6-methyladenine. *Nucleic Acids Res.* **13:**1399–1412.

Ehrlich, M., K. F. Norris, R. Y. Wang, K. C. Kuo, and C. W. Gehrke. 1986. DNA cytosine methylation and heat-induced deamination. *Biosci. Rep.* **6:**387–393.

Eisen, J. A., and P. C. Hanawalt. 1999. A phylogenomic study of DNA repair genes, proteins, and processes. *Mutat. Res.* **435:**171–213.

Ellis, R. J. 2001. Macromolecular crowding: an important but neglected aspect of the intracellular environment. *Curr. Opin. Struct. Biol.* **11:**114–119.

Engel, J. D., and P. H. von Hippel. 1978. Effects of methylation on the stability of nucleic acid conformations. Studies at the polymer level. *J. Biol. Chem.* **253:**927–934.

Fang, X. W., B. L. Golden, K. Littrell, V. Shelton, P. Thiyagarajan, T. Pan, and T. R. Sosnick. 2001. The thermodynamic origin of the stability of a thermophilic ribozyme. *Proc. Natl. Acad. Sci. USA* **98:**4355–4360.

Farahi, K., G. D. Pusch, R. Overbeek, and W. B. Whitman. 2004. Detection of lateral gene transfer events in the prokaryotic tRNA synthetases by the ratios of evolutionary distances method. *J. Mol. Evol.* **58:**615–631.

Forterre, P. 2002. A hot story from comparative genomics: reverse gyrase is the only hyperthermophile-specific protein. *Trends Genet.* **18:**236–237.

Forterre, P. Strategies of hyperthermophiles in adaptation of nucleic acids at high temperatures. *In* C. Gerday, and N. Glansdorff (ed.), *Extremophiles, Encyclopedia of Life Support Systems*, in press. EOLSS Publishers, Oxford, United Kingdom.

Forterre, P., C. Bouthier De La Tour, H. Philippe, and M. Duguet. 2000. Reverse gyrase from hyperthermophiles: probable transfer of a thermoadaptation trait from archaea to bacteria. *Trends Genet.* **16:**152–154.

Forterre, P., F. Charbonnier, E. Marguet, F. Harper, and G. Henckes. 1992. Chromosome structure and DNA topology in extremely thermophilic archaebacteria. *Biochem. Soc. Symp.* **58:**99–112.

Frederico, L. A., T. A. Kunkel, and B. R. Shaw. 1990. A sensitive genetic assay for the detection of cytosine deamination: determination of rate constants and the activation energy. *Biochemistry* **29:**2532–2537.

Galtier, N., and J. R. Lobry. 1997. Relationships between genomic G+C content, RNA secondary structures, and optimal growth temperature in prokaryotes. *J. Mol. Evol.* **44:**632–636.

Gong, Q., Q. Guo, K. L. Tong, G. Zhu, J. T. Wong, and H. Xue. 2002. NMR analysis of bovine tRNA^{Trp}: conformation dependence of Mg^{2+} binding. *J. Biol. Chem.* **277:**20694–20701.

Grayling, R. A., K. Sandman, and J. N. Reeve. 1996. DNA stability and DNA binding proteins. *Adv. Protein Chem.* **48:**437–467.

Grogan, D. W. 2003. Cytosine methylation by the SuaI restriction-modification system: implications for genetic fidelity in a hyperthermophilic archaeon. *J. Bacteriol.* **185:**4657–4661.

Grogan, D. W. 1996. Exchange of genetic markers at extremely high temperatures in the archaeon *Sulfolobus acidocaldarius*. *J. Bacteriol.* **178:**3207–3211.

Grogan, D. W. 2007. Mechanisms of genome stability and evolution, p. 120–138. *In* R. Cavicchioli (ed.), *Archaea: Molecular and Cellular Biology.* American Society for Microbiology, Washington, D.C.

Grogan, D. W. 2004. Stability and repair of DNA in hyperthermophilic Archaea. *Curr. Issues Mol. Biol.* **6:**137–144.

Grogan, D. W., G. T. Carver, and J. W. Drake. 2001. Genetic fidelity under harsh conditions: analysis of spontaneous mutation in the thermoacidophilic archaeon *Sulfolobus acidocaldarius*. *Proc. Natl. Acad. Sci. USA* **98:**7928–7933.

Grosjean, H., and W. Fiers. 1982. Preferential codon usage in prokaryotic genes: the optimal codon–anticodon interaction energy and the selective codon usage in efficiently expressed genes. *Gene* **18:**199–209.

Gruber, T., C. Kohrer, B. Lung, D. Shcherbakov, and W. Piendl. 2003. Affinity of ribosomal protein S8 from mesophilic and (hyper)thermophilic archaea and bacteria for 16S rRNA correlates with the growth temperatures of the organisms. *FEBS Lett.* **549:**123–128.

Guagliardi, A., A. Napoli, M. Rossi, and M. Ciaramella. 1997. Annealing of complementary DNA strands above the melting point of the duplex promoted by an archaeal protein. *J. Mol. Biol.* **267:**841–848.

Guipaud, O., and P. Forterre. 2001. DNA gyrase from *Thermotoga maritima*. *Methods Enzymol.* **334:**162–171.

Guo, F., and T. R. Cech. 2002. Evolution of *Tetrahymena* ribozyme mutants with increased structural stability. *Nat. Struct. Biol.* **9:**855–861.

Hamana, K., H. Hamana, M. Niitsu, K. Samejima, T. Sakane, and A. Yokota. 1994. Occurrence of tertiary and quaternary branched polyamines in thermophilic archaebacteria. *Microbios* **79:**109–119.

Hamana, K., and S. Matsuzaki. 1992. Polyamines as a chemotaxonomic marker in bacterial systematics. *Crit. Rev. Microbiol.* **18:**261–283.

Hamana, K., M. Niitsu, S. Matsuzaki, K. Samejima, Y. Igarashi, and T. Kodama. 1992. Novel linear and branched polyamines in the extremely thermophilic eubacteria *Thermoleophilum*, *Bacillus* and *Hydrogenobacter*. *Biochem. J.* **284**(Pt 3):741–747.

Hamana, K., M. Niitsu, K. Samejima, and S. Matsuzaki. 1991. Polyamine distributions in thermophilic eubacteria belonging to *Thermus* and *Acidothermus*. *J. Biochem.* (Tokyo) **109:**444–449.

Helm, M. 2006. Post-transcriptional nucleotide modification and alternative folding of RNA. *Nucleic Acids Res.* **34:**721–733.

Hennigan, A. N., and J. N. Reeve. 1994. mRNAs in the methanogenic archaeon *Methanococcus vannielii*: numbers, half-lives and processing. *Mol. Microbiol.* **11:**655–670.

Hensel, R., and H. Konig. 1988. Thermoadaptation of methanogenic bacteria by intracellular ion concentration *FEMS Microbiol. Lett.* **49:**75–79.

Herschlag, D. 1995. RNA chaperones and the RNA folding problem. *J. Biol. Chem.* **270:**20871–20874.

Hethke, C., A. Bergerat, W. Hausner, P. Forterre, and M. Thomm. 1999. Cell-free transcription at 95 degrees: thermostability of transcriptional components and DNA topology requirements of *Pyrococcus* transcription. *Genetics* **152:**1325–1333.

Hou, M. H., S. B. Lin, J. M. Yuann, W. C. Lin, A. H. Wang, and L. Kan Ls. 2001. Effects of polyamines on the thermal stability and formation kinetics of DNA duplexes with abnormal structure. *Nucleic Acids Res.* 29:5121–5128.

House, C. H., and S. L. Miller. 1996. Hydrolysis of dihydrouridine and related compounds. *Biochemistry* 35:315–320.

Hurst, L. D., and A. R. Merchant. 2001. High guanine-cytosine content is not an adaptation to high temperature: a comparative analysis amongst prokaryotes. *Proc. Biol. Sci.* 268:493–497.

Jain, S., G. Zon, and M. Sundaralingam. 1989. Base only binding of spermine in the deep groove of the A-DNA octamer d(GTG-TACAC). *Biochemistry* 28:2360–2364.

Jolivet, E., F. Matsunaga, Y. Ishino, P. Forterre, D. Prieur, and H. Myllykallio. 2003. Physiological responses of the hyperthermophilic archaeon *Pyrococcus abyssi* to DNA damage caused by ionizing radiation. *J. Bacteriol.* 185:3958–3961.

Kagawa, Y., H. Nojima, N. Nukiwa, M. Ishizuka, T. Nakajima, T. Yasuhara, T. Tanaka, and T. Oshima. 1984. High guanine plus cytosine content in the third letter of codons of an extreme thermophile. DNA sequence of the isopropylmalate dehydrogenase of *Thermus thermophilus*. *J. Biol. Chem.* 259:2956–2960.

Kampmann, M., and D. Stock. 2004. Reverse gyrase has heat-protective DNA chaperone activity independent of supercoiling. *Nucleic Acids Res.* 32:3537–3545.

Karran, P., and T. Lindahl. 1980. Hypoxanthine in deoxyribonucleic acid: generation by heat-induced hydrolysis of adenine residues and release in free form by a deoxyribonucleic acid glycosylase from calf thymus. *Biochemistry* 19:6005–6011.

Kawai, G., T. Hashizume, T. Miyazawa, J. A. McCloskey, and S. Yokoyama. 1989. Conformational characteristics of 4-acetylcytidine found in tRNA. *Nucleic Acids Symp. Ser.* :61–62.

Kawai, G., Y. Yamamoto, T. Kamimura, T. Masegi, M. Sekine, T. Hata, T. Iimori, T. Watanabe, T. Miyazawa, and S. Yokoyama. 1992. Conformational rigidity of specific pyrimidine residues in tRNA arises from posttranscriptional modifications that enhance steric interaction between the base and the 2′-hydroxyl group. *Biochemistry* 31:1040–1046.

Kawashima, T., N. Amano, H. Koike, S. Makino, S. Higuchi, Y. Kawashima-Ohya, K. Watanabe, M. Yamazaki, K. Kanehori, T. Kawamoto, T. Nunoshiba, Y. Yamamoto, H. Aramaki, K. Makino, and M. Suzuki. 2000. Archaeal adaptation to higher temperatures revealed by genomic sequence of *Thermoplasma volcanium*. *Proc. Natl. Acad. Sci. USA* 97:14257–14262.

Khachane, A. N., K. N. Timmis, and V. A. dos Santos. 2005. Uracil content of 16S rRNA of thermophilic and psychrophilic prokaryotes correlates inversely with their optimal growth temperatures. *Nucleic Acids Res.* 33:4016–4022.

Kikuchi, A., and K. Asai. 1984. Reverse gyrase—a topoisomerase which introduces positive superhelical turns into DNA. *Nature* 309:677–681.

Kohrer, C., C. Mayer, O. Neumair, P. Grobner, and W. Piendl. 1998. Interaction of ribosomal L1 proteins from mesophilic and thermophilic Archaea and Bacteria with specific L1-binding sites on 23S rRNA and mRNA. *Eur. J. Biochem.* 256:97–105.

Koshlap, K. M., R. Guenther, E. Sochacka, A. Malkiewicz, and P. F. Agris. 1999. A distinctive RNA fold: the solution structure of an analogue of the yeast tRNA^{Phe} T-Psi-C domain. *Biochemistry* 38:8647–8656.

Kowalak, J. A., J. J. Dalluge, J. A. McCloskey, and K. O. Stetter. 1994. The role of posttranscriptional modification in stabilization of transfer RNA from hyperthermophiles. *Biochemistry* 33:7869–7876.

Kumagai, I., K. Watanabe, and T. Oshima. 1980. Thermally induced biosynthesis of 2′-O-methylguanosine in tRNA from an extreme thermophile, *Thermus thermophilus* HB27. *Proc. Natl. Acad. Sci. USA* 77:1922–1926.

Lambros, R. J., J. R. Mortimer, and D. R. Forsdyke. 2003. Optimum growth temperature and the base composition of open reading frames in prokaryotes. *Extremophiles* 7:443–450.

Lazcano, A., R. Guerrero, L. Margulis, and J. Oro. 1988. The evolutionary transition from RNA to DNA in early cells. *J. Mol. Evol.* 27:283–290.

Lee, J. C., J. J. Cannone, and R. R. Gutell. 2003. The lonepair triloop: a new motif in RNA structure. *J. Mol. Biol.* 325:65–83.

Levy, M., and S. L. Miller. 1999. The prebiotic synthesis of modified purines and their potential role in the RNA world. *J. Mol. Evol.* 48:631–637.

Levy, M., and S. L. Miller. 1998. The stability of the RNA bases: implications for the origin of life. *Proc. Natl. Acad. Sci. USA* 95:7933–7938.

Limbach, P. A., P. F. Crain, S. C. Pomerantz, and J. A. McCloskey. 1995. Structures of posttranscriptionally modified nucleosides from RNA. *Biochimie* 77:135–138.

Lindahl, T. 1979. DNA glycosylases, endonucleases for apurinic/apyrimidinic sites, and base excision-repair. *Prog. Nucleic Acid Res. Mol. Biol.* 22:135–192.

Lindahl, T. 1993. Instability and decay of the primary structure of DNA. *Nature* 362:709–715.

Lindahl, T. 1967. Irreversible heat inactivation of transfer ribonucleic acids. *J. Biol. Chem.* 242:1970–1973.

Lindahl, T., and A. Andersson. 1972. Rate of chain breakage at apurinic sites in double-stranded deoxyribonucleic acid. *Biochemistry* 11:3618–3623.

Lindahl, T., and O. Karlstrom. 1973. Heat-induced depyrimidination of deoxyribonucleic acid in neutral solution. *Biochemistry* 12:5151–5154.

Lindahl, T., and B. Nyberg. 1974. Heat-induced deamination of cytosine residues in deoxyribonucleic acid. *Biochemistry* 13:3405–3410.

Lindahl, T., and B. Nyberg. 1972. Rate of depurination of native deoxyribonucleic acid. *Biochemistry* 11:3610–3618.

Londei, P., J. Teixido, M. Acca, P. Cammarano, and R. Amils. 1986. Total reconstitution of active large ribosomal subunits of the thermoacidophilic archaebacterium *Sulfolobus solfataricus*. *Nucleic Acids Res.* 14:2269–2285.

Lopez-Garcia, P., and P. Forterre. 1999. Control of DNA topology during thermal stress in hyperthermophilic archaea: DNA topoisomerase levels, activities and induced thermotolerance during heat and cold shock in *Sulfolobus*. *Mol. Microbiol.* 33:766–777.

Lopez-Garcia, P., and P. Forterre. 2000. DNA topology and the thermal stress response, a tale from mesophiles and hyperthermophiles. *Bioessays* 22:738–746.

Lurz, R., M. Grote, J. Dijk, R. Reinhardt, and B. Dobrinski. 1986. Electron microscopic study of DNA complexes with proteins from the Archaebacterium *Sulfolobus acidocaldarius*. *EMBO J.* 5:3715–3721.

Lutsenko, E., and A. S. Bhagwat. 1999. Principal causes of hot spots for cytosine to thymine mutations at sites of cytosine methylation in growing cells. A model, its experimental support and implications. *Mutat. Res.* 437:11–20.

Lynn, D. J., G. A. Singer, and D. A. Hickey. 2002. Synonymous codon usage is subject to selection in thermophilic bacteria. *Nucleic Acids Res.* 30:4272–4277.

Makarova, K. S., L. Aravind, N. V. Grishin, I. B. Rogozin, and E. V. Koonin. 2002. A DNA repair system specific for thermophilic Archaea and bacteria predicted by genomic context analysis. *Nucleic Acids Res.* 30:482–496.

Makarova, K. S., and E. V. Koonin. 2003. Comparative genomics of Archaea: how much have we learned in six years, and what's next? *Genome Biol.* 4:115.

Male, R., V. M. Fosse, and K. Kleppe. 1982. Polyamine-induced hydrolysis of apurinic sites in DNA and nucleosomes. *Nucleic Acids Res.* 10:6305–6318.

Marck, C., and H. Grosjean. 2002. tRNomics: analysis of tRNA genes from 50 genomes of Eukarya, Archaea, and Bacteria reveals anticodon-sparing strategies and domain-specific features. *RNA* **8:**1189–1232.

Marguet, E., and P. Forterre. 1994. DNA stability at temperatures typical for hyperthermophiles. *Nucleic Acids Res.* **22:**1681–1686.

Marguet, E., and P. Forterre. 2001. Stability and manipulation of DNA at extreme temperatures. *Methods Enzymol.* **334:**205–215.

Maxam, A. M., and W. Gilbert. 1977. A new method for sequencing DNA. *Proc. Natl. Acad. Sci. USA* **74:**560–564.

McCloskey, J. A., D. E. Graham, S. Zhou, P. F. Crain, M. Ibba, J. Konisky, D. Soll, and G. J. Olsen. 2001. Post-transcriptional modification in archaeal tRNAs: identities and phylogenetic relations of nucleotides from mesophilic and hyperthermophilic Methanococcales. *Nucleic Acids Res.* **29:**4699–4706.

Minton, A. P. 2001. The influence of macromolecular crowding and macromolecular confinement on biochemical reactions in physiological media. *J. Biol. Chem.* **276:**10577–10580.

Minyat, E. E., V. I. Ivanov, A. M. Kritzyn, L. E. Minchenkova, and A. K. Schyolkina. 1979. Spermine and spermidine-induced B to A transition of DNA in solution. *J. Mol. Biol.* **128:**397–409.

Morales, A. J., M. A. Swairjo, and P. Schimmel. 1999. Structure-specific tRNA-binding protein from the extreme thermophile *Aquifex aeolicus*. *EMBO J.* **18:**3475–3483.

Nagaswamy, U., and G. E. Fox. 2002. Frequent occurrence of the T-loop RNA folding motif in ribosomal RNAs. *RNA* **8:**1112–1119.

Nakashima, H., S. Fukuchi, and K. Nishikawa. 2003. Compositional changes in RNA, DNA and proteins for bacterial adaptation to higher and lower temperatures. *J. Biochem.* (Tokyo) **133:**507–513.

Napoli, A., A. Valenti, V. Salerno, M. Nadal, F. Garnier, M. Rossi, and M. Ciaramella. 2004. Reverse gyrase recruitment to DNA after UV light irradiation in *Sulfolobus solfataricus*. *J. Biol. Chem.* **279:**33192–33198.

Nobles, K. N., C. S. Yarian, G. Liu, R. H. Guenther, and P. F. Agris. 2002. Highly conserved modified nucleosides influence Mg^{2+}-dependent tRNA folding. *Nucleic Acids Res.* **30:**4751–4760.

Noon, K. R., E. Bruenger, and J. A. McCloskey. 1998. Posttranscriptional modifications in 16S and 23S rRNAs of the archaeal hyperthermophile *Sulfolobus solfataricus*. *J. Bacteriol.* **180:**2883–2888.

Noon, K. R., R. Guymon, P. F. Crain, J. A. McCloskey, M. Thomm, J. Lim, and R. Cavicchioli. 2003. Influence of temperature on tRNA modification in archaea: *Methanococcoides burtonii* (optimum growth temperature Topt, 23 degrees C) and *Stetteria hydrogenophila* (Topt, 95 degrees C). *J. Bacteriol.* **185:**5483–5490.

Okada, N., S. Noguchi, H. Kasai, N. Shindo-Okada, T. Ohgi, T. Goto, and S. Nishimura. 1979. Novel mechanism of posttranscriptional modification of tRNA. Insertion of bases of Q precursors into tRNA by a specific tRNA transglycosylase reaction. *J. Biol. Chem.* **254:**3067–3073.

Omer, A. D., T. M. Lowe, A. G. Russell, H. Ebhardt, S. R. Eddy, and P. P. Dennis. 2000. Homologs of small nucleolar RNAs in Archaea. *Science* **288:**517–522.

Oshima, T. 1989. *Polyamines in Thermophiles. The Physiology of Polyamines*, p. 35–46, vol. 2. CRC Press, Boca Raton, FL.

Oshima, T., N. Hamasaki, M. Senshu, K. Kakinuma, and I. Kuwajima. 1987. A new naturally occurring polyamine containing a quaternary ammonium nitrogen. *J. Biol. Chem.* **262:**11979–11981.

Oshima, T., Y. Sakaki, N. Wakayama, K. Watanabe, and Z. Ohashi. 1976. Biochemical studies on an extreme thermophile *Thermus thermophilus*: thermal stabilities of cell constituents and a bacteriophage. *Experientia Suppl.* **26:**317–331.

Ovadi, J., and V. Saks. 2004. On the origin of intracellular compartmentation and organized metabolic systems. *Mol. Cell. Biochem.* **256-257:**5–12.

Palmer, J. R., T. Baltrus, J. N. Reeve, and C. J. Daniels. 1992. Transfer RNA genes from the hyperthermophilic Archaeon, *Methanopyrus kandleri*. *Biochim. Biophys. Acta* **1132:**315–318.

Paz, A., D. Mester, I. Baca, E. Nevo, and A. Korol. 2004. Adaptive role of increased frequency of polypurine tracts in mRNA sequences of thermophilic prokaryotes. *Proc. Natl. Acad. Sci. USA* **101:**2951–2956.

Peak, M. J., F. T. Robb, and J. G. Peak. 1995. Extreme resistance to thermally induced DNA backbone breaks in the hyperthermophilic archaeon *Pyrococcus furiosus*. *J. Bacteriol.* **177:**6316–6318.

Peattie, D. A. 1979. Direct chemical method for sequencing RNA. *Proc. Natl. Acad. Sci. USA* **76:**1760–1764.

Pereira, S. L., and J. N. Reeve. 1998. Histones and nucleosomes in Archaea and Eukarya: a comparative analysis. *Extremophiles* **2:**141–148.

Philippsen, P., R. Thiebe, W. Wintermeyer, and H. G. Zachau. 1968. Splitting of phenylalanine specific tRNA into half molecules by chemical means. *Biochem. Biophys. Res. Commun.* **33:**922–928.

Poole, A., D. Penny, and B. Sjoberg. 2000. Methyl-RNA: an evolutionary bridge between RNA and DNA? *Chem. Biol.* **7:**R207–R216.

Quigley, G. J., M. M. Teeter, and A. Rich. 1978. Structural analysis of spermine and magnesium ion binding to yeast phenylalanine transfer RNA. *Proc. Natl. Acad. Sci USA* **75:**64–68.

Rodriguez, A. C., and D. Stock. 2002. Crystal structure of reverse gyrase: insights into the positive supercoiling of DNA. *EMBO J.* **21:**418–426.

Romby, P., P. Carbon, E. Westhof, C. Ehresmann, J. P. Ebel, B. Ehresmann, and R. Giege. 1987. Importance of conserved residues for the conformation of the T-loop in tRNAs. *J. Biomol. Struct. Dyn.* **5:**669–687.

Roovers, M., J. Wouters, J. M. Bujnicki, C. Tricot, V. Stalon, H. Grosjean, and L. Droogmans. 2004. A primordial RNA modification enzyme: the case of tRNA (m^1A) methyltransferase. *Nucleic Acids Res.* **32:**465–476.

Sandman, K., and Reeve, J. N. 2005. Archaeal chromatin proteins: different structures but common function? *Curr. Opin. Microbiol.* **8:**656–661.

Serebrov, V., R. J. Clarke, H. J. Gross, and L. Kisselev. 2001. Mg^{2+}-induced tRNA folding. *Biochemistry* **40:**6688–6698.

Shen, J. C., W. M. Rideout III, and P. A. Jones. 1994. The rate of hydrolytic deamination of 5-methylcytosine in double-stranded DNA. *Nucleic Acids Res.* **22:**972–976.

Shigi, N., T. Suzuki, T. Terada, M. Shirouzu, S. Yokoyama, and K. Watanabe. 2006. Temperature-dependent biosynthesis of 2-thioribothymidine of *Thermus thermophilus* tRNA. *J. Biol. Chem.* **281:**2104–2113.

Sprinzl, M., and K. S. Vassilenko. 2005. Compilation of tRNA sequences and sequences of tRNA genes. *Nucleic Acids Res.* **33:**D139–D140.

Stubbe, J. 2000. Ribonucleotide reductases: the link between an RNA and a DNA world? *Curr. Opin. Struct. Biol.* **10:**731–736.

Tehei, M., B. Franzetti, M. C. Maurel, J. Vergne, C. Hountondji, and G. Zaccai. 2002. The search for traces of life: the protective effect of salt on biological macromolecules. *Extremophiles* **6:**427–430.

Terui, Y., M. Ohnuma, K. Hiraga, E. Kawashima, and T. Oshima. 2005. Stabilization of nucleic acids by unusual polyamines produced by an extreme thermophile, *Thermus thermophilus*. *Biochem. J.* **388:**427–433.

Tong, K. L., and J. T. Wong. 2004. Anticodon and wobble evolution. *Gene* **333:**169–177.

Urbonavicius, J., J. Armengaud, and H. Grosjean. Identity elements required for enzymatic formation of N(2),

N(2)-dimethylguanosine from N(2)-monomethylated derivative and its possible role in avoiding alternative conformations in archaeal tRNA. *J. Mol. Biol* XXX.

Ushida, C., T. Muramatsu, H. Mizushima, T. Ueda, K. Watanabe, K. O. Stetter, P. F. Crain, J. A. McCloskey, and Y. Kuchino. 1996. Structural feature of the initiator tRNA gene from *Pyrodictium occultum* and the thermal stability of its gene product, tRNA(imet). *Biochimie* **78:**847–855.

Vanzo, N. F., Y. S. Li, B. Py, E. Blum, C. F. Higgins, L. C. Raynal, H. M. Krisch, and A. J. Carpousis. 1998. Ribonuclease E organizes the protein interactions in the *Escherichia coli* RNA degradosome. *Genes Dev.* **12:**2770–2781.

Varani, G. 1995. Exceptionally stable nucleic acid hairpins. *Annu. Rev. Biophys. Biomol. Struct.* **24:**379–404.

Varani, G., and W. H. McClain. 2000. The G x U wobble base pair. A fundamental building block of RNA structure crucial to RNA function in diverse biological systems. *EMBO Rep.* **1:**18–23.

Vergne, J., J. A. H. Cognet, E. Szathmary, and M.-C. Maurel. In vitro selection of halo-thermophilic RNA reveals two families of resistant RNA. *Gene*, XXX.

Vijayanathan, V., T. Thomas, A. Shirahata, and T. J. Thomas. 2001. DNA condensation by polyamines: a laser light scattering study of structural effects. *Biochemistry* **40:**13644–13651.

Wang, R. Y., K. C. Kuo, C. W. Gehrke, L. H. Huang, and M. Ehrlich. 1982. Heat- and alkali-induced deamination of 5-methylcytosine and cytosine residues in DNA. *Biochim. Biophys. Acta* **697:**371–377.

Watanabe, K., T. Oshima, K. Iijima, Z. Yamaizumi, and S. Nishimura. 1980. Purification and thermal stability of several amino acid-specific tRNAs from an extreme thermophile, *Thermus thermophilus* HB8. *J. Biochem.* (Tokyo) **87:**1–13.

Watanabe, K., M. Shinma, T. Oshima, and S. Nishimura. 1976. Heat-induced stability of tRNA from an extreme thermophile, Thermus thermophilus. *Biochem. Biophys. Res. Commun.* **72:**1137–1144.

Watanabe, M., M. Matsuo, S. Tanaka, H. Akimoto, S. Asahi, S. Nishimura, J. R. Katze, T. Hashizume, P. F. Crain, J. A. McCloskey, and N. Okada. 1997. Biosynthesis of archaeosine, a novel derivative of 7-deazaguanosine specific to archaeal tRNA, proceeds via a pathway involving base replacement on the tRNA polynucleotide chain. *J. Biol. Chem.* **272:**20146–20151.

Weeks, K. M., and T. R. Cech. 1996. Assembly of a ribonucleoprotein catalyst by tertiary structure capture. *Science* **271:**345–348.

Weinberg, M. V., G. J. Schut, S. Brehm, S. Datta, and M. W. Adams. 2005. Cold shock of a hyperthermophilic archaeon: *Pyrococcus furiosus* exhibits multiple responses to a suboptimal growth temperature with a key role for membrane-bound glycoproteins. *J. Bacteriol.* **187:**336–348.

Westhof, E., P. Dumas, and D. Moras. 1988. Restrained refinement of two crystalline forms of yeast aspartic acid and phenylalanine transfer RNA crystals. *Acta Crystallogr. A* **44**(Pt 2)**:**112–123.

Wildenauer, D., H. J. Gross, and D. Riesner. 1974. Enzymatic methylations: III. Cadaverine-induced conformational changes of *E. coli* tRNA-fMet as evidenced by the availability of a specific adenosine and a specific cytidine residue for methylation. *Nucleic Acids Res.* **1:**1165–1182.

Wintermeyer, W., and H. G. Zachau. 1970. A specific chemical chain scission of tRNA at 7-methylguanosine. *FEBS Lett.* **11:**160–164.

Wright, M. C., and G. F. Joyce. 1997. Continuous in vitro evolution of catalytic function. *Science* **276:**614–617.

Xaplanteri, M. A., A. D. Petropoulos, G. P. Dinos, and D. L. Kalpaxis. 2005. Localization of spermine binding sites in 23S rRNA by photoaffinity labeling: parsing the spermine contribution to ribosomal 50S subunit functions. *Nucleic Acids Res.* **33:**2792–2805.

Xue, H., K. L. Tong, C. Marck, H. Grosjean, and J. T. Wong. 2003. Transfer RNA paralogs: evidence for genetic code-amino acid biosynthesis coevolution and an archaeal root of life. *Gene* **310:**59–66.

Yokoyama, S., K. Watanabe, and T. Miyazawa. 1987. Dynamic structures and functions of transfer ribonucleic acids from extreme thermophiles. *Adv. Biophys.* **23:**115–147.

Yue, D., A. Kintanar, and J. Horowitz. 1994. Nucleoside modifications stabilize Mg^{2+} binding in *Escherichia coli* tRNA-Val: an imino proton NMR investigation. *Biochemistry* **33:**8905–8911.

Yusupov, M. M., G. Z. Yusupova, A. Baucom, K. Lieberman, T. N. Earnest, J. H. Cate, and H. F. Noller. 2001. Crystal structure of the ribosome at 5.5 A resolution. *Science* **292:**883–896.

Zhang, X., and C. K. Mathews. 1994. Effect of DNA cytosine methylation upon deamination-induced mutagenesis in a natural target sequence in duplex DNA. *J. Biol. Chem.* **269:**7066–7069.

Chapter 5

How Thermophiles Cope with Thermolabile Metabolites

Jan Massant

INTRODUCTION

Thermophilic cells are well adapted to their habitats. Whereas the thermal profile of the major cell components such as proteins, nucleic acids, and lipids has been studied intensively, much less is known about the small molecules, metabolites, and coenzymes used by the thermophilic cell (Daniel and Cowan, 2000). Nonetheless, these small molecules are key elements of life, and they have critical roles at all levels of biological complexity (Schreiber, 2005). Most of the low-molecular-weight metabolites and coenzymes found in thermophiles and hyperthermophiles are the same as those found in mesophiles, yet many are thermolabile. The half-lives of these metabolites in aqueous media decrease rapidly at increasing temperatures. Only a few of them have been studied in more detail. Heat-sensitive metabolites involved in metabolic pathways in thermophiles and hyperthermophiles include phosphorylated intermediates in glucose metabolism, indole and phosphoribosyl anthranilate in tryptophan biosynthesis, phosphoribosylamine in purine biosynthesis, and carbamoyl phosphate in the biosynthesis of arginine and pyrimidines. Some of these intermediates decay with a very short half-life at elevated temperatures. Moreover, the hydrolysis of these thermolabile metabolites may give products that are toxic to the cell such as is the case with carbamoyl phosphate, which gives cyanate upon hydrolysis, able to carbamoylate any free amino group. An intriguing question is how thermophilic organisms, in particular hyperthermophilic ones, cope with these thermolabile metabolites. Here we take a closer look at glucose metabolism, tryptophan biosynthesis, de novo purine synthesis and biosynthesis of arginine and pyrimidines; metabolic pathways that involve thermally less stable intermediates and coenzymes and that have been studied in more detail in the past years.

GLUCOSE CATABOLISM

Glucose-degrading pathways involve several relatively thermally unstable metabolites such as the phosphorylated pathway intermediates, pyridine nucleotides NAD(H) and NADP(H) and adenylate nucleotide ATP (Figs 1 and 2). Among the most thermally unstable intermediates are glyceraldehyde-3-phosphate (GAP), with a half-life of 14.5 min at 60°C and 3.4 min at 80°C (Schramm et al., 2000), 1,3-bisphosphoglycerate (1,3-BPG), with a half-life of 1.6 min at 60°C (Dörr et al., 2003), and phosphoenolpyruvate (PEP), with a half-life of 6 min at 90°C (Schramm et al., 2000).

Glucose catabolism has been studied in more detail in the hyperthermophilic archaea *Pyrococcus furiosus*, *Thermoproteus tenax*, and *Sulfolobus solfataricus* and in the hyperthermophilic bacterium *Thermotoga maritima*. Hyperthermophilic archaea were found to degrade glucose to pyruvate via a modified Embden–Meyerhof (EM) pathway, a modified Entner–Doudoroff (ED) pathway, or a combination of both pathways (Selig et al., 1997; Ronimus and Morgan, 2003; Verhees et al., 2003; Ahmed et al., 2005). In contrast, *Thermotoga* uses the conventional forms of the EM and ED pathways for glucose degradation (Selig et al., 1997). The modifications of the EM pathway found in hyperthermophilic archaea are not yet completely understood. They were suggested to be related to life at high temperatures. In the modified EM pathway of *P. furiosus*, several novel enzymes were found to be involved, which include ADP-dependent glucokinase (GK), ADP-dependent phosphofructokinase (PFK), glyceraldehyde-3-phosphate ferredoxin oxidoreductase (GAPOR), PEP synthase, pyruvate ferredoxin oxidoreductase (POR), and ADP-forming acetyl coenzyme A (acetyl-CoA) synthetase (ACS).

Common to the modified EM pathways in hyperthermophilic archaea is the irreversible oxidation of

---

J. Massant • Erfelijkheidsleer en Microbiologie (MICR), Vrije Universiteit Brussel, Pleinlaan 2, B-1050 Brussels, Belgium.

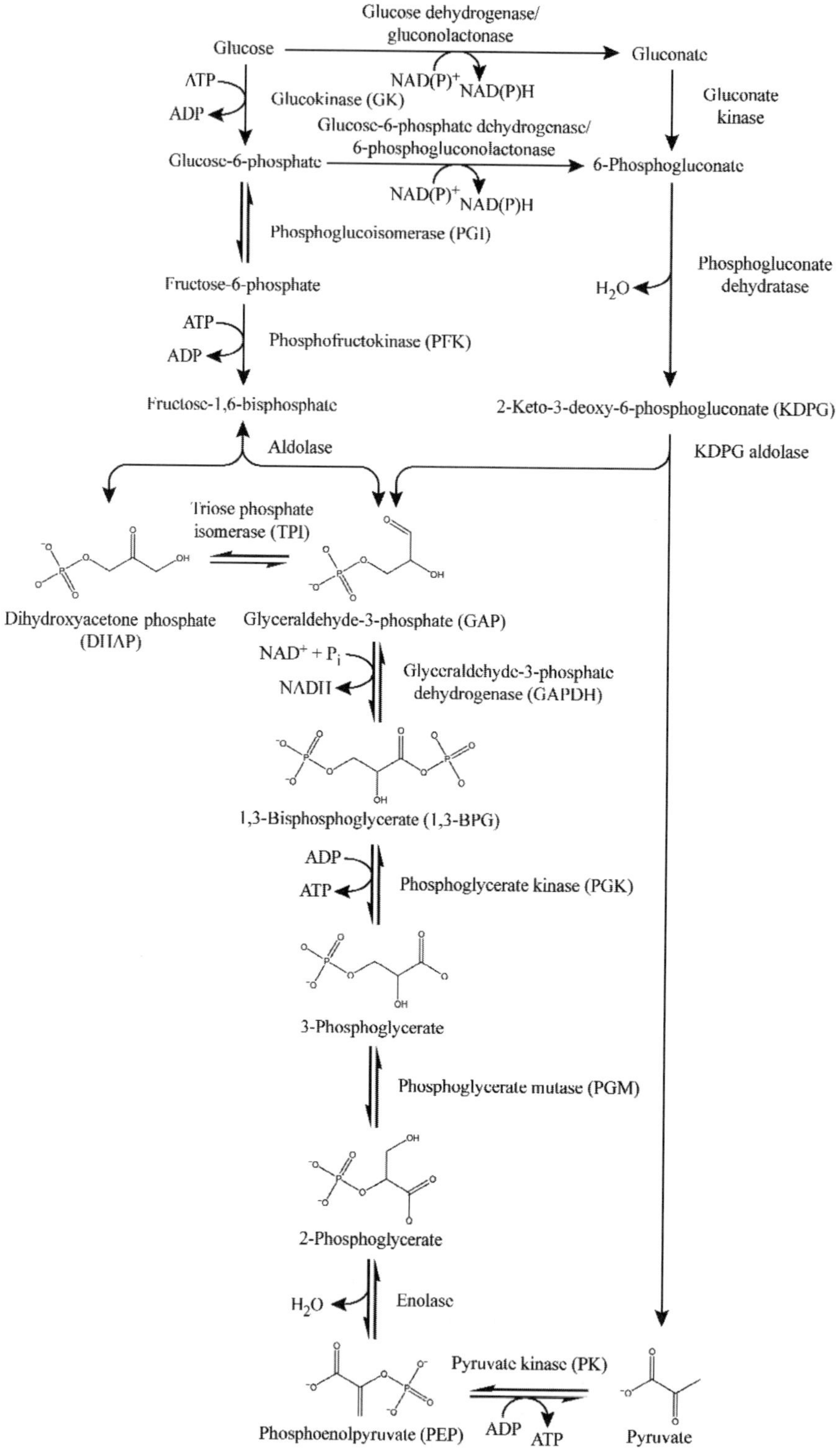

Figure 1. Embden–Meyerhof (EM) and Entner–Doudoroff (ED) glucose-degrading pathways.

GAP directly to 3-phosphoglycerate by a non-phosphorylating GAPOR found in *Pyrococcus*, *Thermococcus*, and *Desulfurococcus* strains or a non-phosphorylating NAD-dependent glyceraldehyde-3-phosphate dehydrogenase (GAPN) found in *T. tenax*. Thereby, the use of the heat-labile BPG is avoided, and pools of labile intermediates such as GAP and dihydroxyacetone phosphate (DHAP) upstream the GAPOR (or GAPN) reaction are decreased. The properties of pyruvate kinase (PK) could avoid the

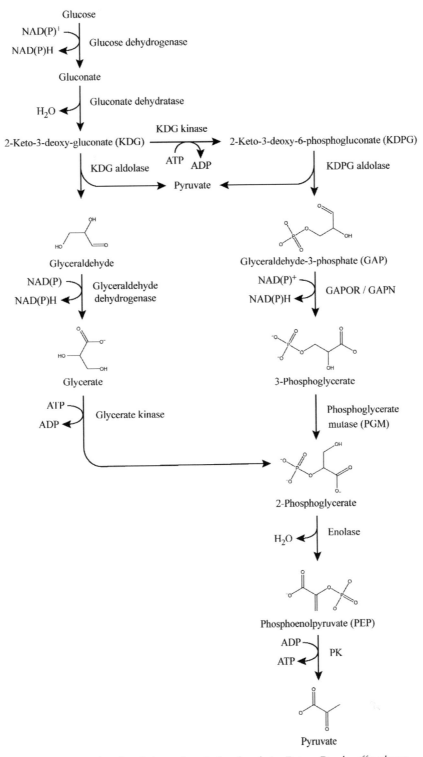

Figure 2. Non-phosphorylative and semi-phosphorylative Entner–Doudoroff pathway.

accumulation of PEP and other intermediates downstream the GAPOR reaction.

An ED pathway based on non-phosphorylated carbohydrates in hyperthermophilic archaea (Danson, 1988) was suggested to be used because of the greater stability of these intermediates compared with the phosphorylated forms (Siebers et al., 2004). However, evidence indicates the operation of both the non-phosphorylative and the semi-phosphorylative ED pathway in *T. tenax* and *S. solfataricus*. *T. tenax* uses a variant of the reversible EM pathway and two different modifications of the ED pathway (a non-phosphorylative and a semi-phosphorylative version) for carbohydrate metabolism (Ahmed et al., 2004; Siebers et al., 2004). Formation of heat-labile BPG is avoided by using the adapted EM and semi-phosphorylative ED pathways; the non-phosphorylative ED variant additionally avoids the formation of the two other relatively heat-labile intermediates (GAP and DHAP). The non-phosphorylative ED pathway might be appropriate for growth at the upper temperature range (Ahmed et al., 2004).

The catabolic EM pathway of *T. tenax* is characterized by significant deviations from the common glycolytic pathway: a non-phosphorylating GAPN avoids the formation of the labile 1,3-BGP and could compensate for the missing allosteric potential of PP_i-PFK and the reduced regulatory potential of PK, thus allowing the catabolism to take place without a bottleneck at the last step. The reduced allosteric regulation of PK precludes a considerable accumulation of intermediates such as PEP (Schramm et al., 2000). These modifications can be interpreted as thermoadaptative strategies to avoid thermally unstable intermediates or at least minimize their intracellular pools. The central carbohydrate metabolic pathways in the archaeal domain show versatility and flexibility. It is not yet clear what the adaptations to high temperatures are. The accumulation of free intermediates at high temperatures would result in an accelerated loss of these compounds and must therefore be avoided.

The formation of a glucose metabolon channeling the pathway intermediates might also provide a way of protecting the relatively thermally unstable intermediates. The presence of a glycolytic multienzyme complex in *Escherichia coli* was already proposed by Mowbray and Moses (1976). There is considerable evidence for the association of glycolytic enzymes and channeling of glycolytic intermediates (Orosz and Ovadi, 1987; Clegg and Jackson, 1990; Beeckmans et al., 1990; Gotz et al., 1999; Malaisse et al., 2004; Rakus et al., 2004; Maughan et al., 2005). Recently, channeling in glycolysis from fructose-1,6-bisphosphate all the way to CO_2 was demonstrated in *E. coli* (Shearer et al., 2005).

Triose phosphate isomerase (TPI) and phosphoglycerate kinase (PGK) activities are combined in a bifunctional fusion protein in *T. maritima*, but both reactions do not represent consecutive steps in the reaction pathway (Schurig et al., 1995). In *T. tenax*, the tetrameric TPI interacts specifically with glycerol-1-phosphate dehydrogenase (GLPDH) in gluconeogenesis (Walden et al., 2004a; Walden et al., 2004b). The interaction shifts the dimer/tetramer equilibrium to the tetrameric form and protects GLPDH against thermal inactivation, and in interaction with GLPDH, TPI has a higher specific activity. The specific interaction could reduce the build up of thermolabile intermediates in gluconeogenesis such as GAP and DHAP. The association of enzymes involved in glucose metabolism in hyperthermophiles and the channeling of intermediates deserves without doubt more attention and further investigation.

Compared to acetate and acetyl-CoA, acetyl phosphate is unstable. Formation of acetate directly from acetyl-CoA by (ADP forming) ACS rather than proceeding via acetyl phosphate (Schäfer et al., 1993) circumvents the use of acetyl phosphate as a metabolic intermediate in hyperthermophilic archaea. However, the same enzyme is also present in mesophilic archaea (Bräsen and Schönheit, 2004). Moreover, in the hyperthermophilic bacterium *T. maritima* conversion of acetyl-CoA into acetate passes via acetyl phosphate involving phosphate acetyltransferase and acetate kinase. The ADP-dependent ACS represents an enzyme typical of acetate-producing microorganisms rather than a characteristic of hyperthermophiles (Mai and Adams, 1996).

Important phosphorylated metabolites, also used in glucose metabolism, are ATP and ADP. The stability of adenylate nucleosides is in the order of PP_i/AMP>ADP>ATP (Daniel and Cowan, 2000). ATP is relatively unstable at 95°C, but its stability is greatly affected by pH and metal ions (Ramirez et al., 1980). Nonetheless, thermophilic energetics was shown to be consistent with adenylate metabolism in mesophilic organisms (Napolitano and Shain, 2005). In the thermophile *T. thermophilus*, ATP levels drop upon cold shock and heat shock at comparable levels (Napolitano and Shain, 2005).

Hyperthermophilic archaea were found to use the more stable compounds PP_i and ADP instead of ATP (Kengen et al., 1994). In *P. furiosus*, the modified EM pathway involves ADP-dependent kinases (Kengen et al., 1994). ADP-dependent Kinases, such as ADP-dependent GK of *P. furiosus* and *Thermococcus litoralis* and the bifunctional ADP-dependent GK/PKF of *Methanocaldococcus jannaschii*, required ADP, instead of ATP, as the phosphoryl group donor and they produce ADP. The presence of these ADP-dependent enzymes was discussed in terms of metabolic adaptation to high temperatures. However, because ATP-dependent hexokinases are also operative in hyperthermophiles (Dörr et al., 2003) and ADP-dependent enzymes in mesophiles, factors other than temperature might be responsible for this unusual nucleotide specificity. Furthermore, *P. furiosus* possesses in addition to its ADP-dependent sugar kinases an ATP-dependent galactokinase. Although ATP is less stable than ADP, with half-lives of 115 and 750 min at

90°C respectively, it is still more stable than several intermediates of the EM pathway. The use of ADP-dependent enzymes was suggested to be correlated with generally or transiently limiting ATP concentrations corresponding with a particular way of life (Dörr et al., 2003). The pyrophosphate-dependent PFK found in *T. tenax* uses PP$_i$ instead of ATP as phosphoryl donor for the PFK reaction (Siebers et al., 1998), but this is also the case in some eukaryotic parasites. Thus, the use of this enzyme does not seem to be a specific adaptation to a high-temperature life.

Hyperthermophiles were found to use non-haem iron proteins instead of NAD(P) (Daniel and Danson, 1995). At neutral pH, the half-life of NAD decreases rapidly at 100°C. NADP is even more unstable at these temperatures; it has a half-life of 2 min at 95°C (Lowry et al., 1961; Robb et al., 1992). Non-haem iron proteins are stable and functional at 100°C (Daniel and Danson, 1995; Adams, 1993). The redox reactions of glucose metabolism in *P. furiosus* are catalyzed by ferredoxin-linked oxidoreductases. A ferredoxin-dependent GAPOR replaces the usual GAPDH and phosphoglycerate kinase in the glycolytic pathway of *P. furiosus* (Mukund and Adams, 1995). Both these enzymes are present and function in gluconeogenesis when *P. furiosus* is grown on pyruvate (Schäfer and Schönheit, 1993). A phosphorylating NADP-dependent GAPN in *T. tenax* also fulfills anabolic purposes (Brunner et al., 2001). Pyruvate oxidation is also coupled to ferredoxin reduction in *P. furiosus* (Blamey and Adams, 1993) and other hyperthermophiles and involves POR. A POR, although with an enzyme mechanism different from that found in archaea, was also identified in *T. maritima* (Smith et al., 1994). Ferredoxin-linked oxidoreductases in other pathways were also identified in hyperthermophilic microorganisms (Daniel and Cowan, 2000). Because ferredoxin:oxidoreductases are found in all hyperthermophiles, their presence might be correlated with life at high temperatures. Despite their instability at high temperatures, NAD and NADP have been identified in cell-free extracts of *P. furiosus* (Pan et al., 2001). The overall cellular concentration of pyridine nucleotides is about half that of mesophiles, especially the NAD concentration is significantly lower. The identification of NAD(H) and NADP(H) in *P. furiosus* and the genes for the enzymes involved in their synthesis indicates that both pyridine-type nucleotides play important roles in the metabolism of this hyperthermophile (Pan et al., 2001). The presence of pyridine nucleotides in hyperthermophiles raises the question how adequate concentrations of these compounds can be maintained at high temperatures. Mechanisms by which hyperthermophiles stabilize these relatively thermally unstable nicotinamide nucleotides are not known. Protective strategies against thermal degradation could include the presence of specific proteins or organic compounds, a higher rate of biosynthesis, or the use of microsites with less destabilizing conditions (Pan et al., 2001). The thermal stability of many metabolites is highly dependent on physiological conditions. At pH 7, NADH has a half-life of a few minutes at 95°C, but at higher pH, the nucleotide is more stable (Walsh et al., 1983).

TRYPTOPHAN BIOSYNTHESIS

Tryptophan is synthesized from chorismate in seven enzyme-catalyzed steps (Fig. 3). The pathway was extensively studied in mesophiles *E. coli* and *Salmonella enterica* serovar Typhimurium (Pittard, 1996; Nichols, 1996). The entire *trp* operon from the hyperthermophilic bacterium *T. maritima* was cloned, and several of the enzymes were characterized in detail (Sterner et al., 1995). Tryptophan biosynthesis was also studied in detail in the thermoacidophilic archaeon *S. solfataricus* (Tutino et al., 1993). Some intermediates of tryptophan biosynthesis, such as phosphoribosyl anthranilate (PRAnt), are susceptible to thermal degradation and some, such as indole, are susceptible to loss. The comparison of the hyperthermophilic enzymes for tryptophan synthesis with the mesophilic homologs has suggested possible strategies by which hyperthermophiles cope with the thermolabile intermediates in tryptophan biosynthesis.

The most detailed characterized enzyme of tryptophan biosynthesis is without doubt tryptophan synthase (Anderson, 1999). In bacteria such as *E. coli* and serovar Typhimurium, tryptophan synthase is an $\alpha_2\beta_2$ tetrameric enzyme complex that catalyzes the final two steps in the biosynthesis of tryptophan and involves the conversion of indole 3-glycerol phosphate (IGP) and serine to tryptophan. Each α subunit (TrpA) catalyzes the cleavage of IGP to indole and GAP, whereas each β subunit (TrpB) catalyzes the condensation of indole with serine in a reaction mediated by pyridoxal phosphate. The three-dimensional crystal structure of the enzyme from serovar Typhimurium showed the existence of a hydrophobic tunnel 25 Å in length which connects the active sites of the α and β subunits of this enzyme to transfer indole (Hyde et al., 1988). Binding of the first substrate (IGP) induces a conformational change leading to the formation of a channel, while the second substrate (serine) activates the active site for the first substrate (Miles et al., 1999). Tryptophan synthase is one of the best-defined enzymes exhibiting metabolic channeling (Pan et al., 1997; Anderson, 1999).

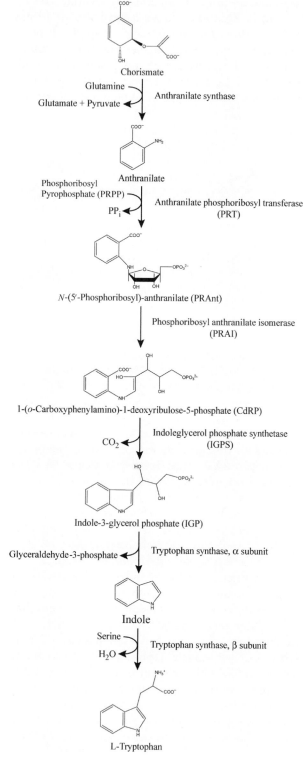

Figure 3. Tryptophan biosynthetic pathway.

The interaction between the α and β subunit from the *P. furiosus* enzyme was found to be extremely strong (Ogasahara et al., 2003). This strong interaction could reduce the leakage of indole from the tunnel connecting both the subunits. In *T. maritima*, a novel tryptophan synthase β subunit (TrpB2), a homolog of TrpB1, was found (Hettwer and Sterner, 2002). TrpB2 has high catalytic efficiency owing to a high affinity for indole, but in contrast to TrpB1, which forms with TrpA the tetrameric $\alpha_2\beta_2$ tryptophan synthase, it does not bind to TrpA. *TrpB2* genes are only found in hyperthermophiles (Hettwer and Sterner, 2002). The novel β subunit is proposed to act as an indole rescue protein that prevents the loss of this costly hydrophobic metabolite from the cell at high temperature. At the high physiological temperature of *T. maritima*, and other hyperthermophiles, a larger fraction of indole may leak from the channel connecting TrpA with TrpB1. Because of its high affinity for indole, TrpB2 would be ideally suited to scavenge the liberated indole and to prevent it from escaping through the cytoplasmic membrane.

An intermediate in the tryptophan biosynthetic pathway that is degraded rapidly at high temperatures is N-(5′-phoshoribosyl)-anthranilate (PRAnt) with a half-life of 39 s at 80°C. PRAnt is synthesized in the third step of tryptophan biosynthesis through the condensation of anthranilate and phosphoribosyl pyrophosphate (PRPP), an activated form of ribose phosphate, by an anthranilate phosphoribosyl transferase (PRT), and it is a substrate for phosphoribosyl anthranilate isomerase (PRAI). In *T. maritima*, spontaneous hydrolysis of the thermolabile PRAnt is outrunned by the very high catalytic efficiency of PRAI, resulting from a high affinity of the enzyme for its substrate (Sterner et al., 1996).

Besides being used for the synthesis of tryptophan, phosphoribosylamine (PRA) derivatives are also used in de novo purine synthesis and biosynthesis of histidine. As regards histidine biosynthesis in *T. maritima*, degradation of thermolabile N'-[(5′-phosphoribosyl)formimino]-5-aminoimidazole-4-carboxamide ribonucleotide (ProFAR) is also avoided by an unusual high catalytic efficiency of ProFAR isomerase (Henn-Sax et al., 2002). Hydrolysis of the labile PRA intermediates in tryptophan and histidine biosynthesis appears to be prevented by their rapid turnover. PRAI and ProFAR use a similar chemical reaction mechanism and are suggested to have evolved from a common ancestor (Henn-Sax et al., 2002).

With sedimentation equilibrium, Sterner et al. (1996) did not find evidence for a stable complex between PRAI and PRT. However, the interaction between both enzymes can be a transient one, and channeling of PRA nt was not investigated directly. The high affinity of the enzyme for its substrate could be compatible with channeling of the intermediate. In *E. coli*, PRAI is part of a bifunctional enzyme that also catalyzes the subsequent step in tryptophan biosynthesis (Creighton, 1970). PRAI of *Thermotoga* is a homodimeric enzyme (Hennig et al., 1997) and indole glycerol phosphate synthetase (IGPS) a monomer in solution (Merz et al., 1999; Knöchel et al., 2002). IGPS is present as a monomeric enzyme in most microorganisms, but it is occasionally part of a bi- or multifunctional enzyme, being fused to one or more of the other enzymes of the tryptophan biosynthetic pathway. IGPS of *T. maritima* also has a high affinity for its substrate 1-(o-carboxy-phenylamino)-1-deoxyribulose-5-phosphate (CdRP) and because of this also a high catalytic efficiency. However, CdRP is relatively stable and therefore does not demand rapid processing by IGPS. It was suggested that the similar catalytic efficiencies of PRAI and IGPS at 80°C might simply ensure unhindered processing of CdRP (Merz et al., 1999). The tendency for clustering of the *trp* genes has led in a few instances to fusions between adjacent genes. But these fusions have not necessarily resulted in a functional cooperativity or channeling of substrates between enzymes.

In comparison with PRAnt, PRPP is significantly more stable, but at high temperatures in the presence of divalent cations, PRPP is hydrolyzed to ribose-5-phosphate and pyrophosphate. PRPP is also required for de novo synthesis of purine, pyrimidine, and nicotinamide nucleotides and for histidine biosynthesis. At 80°C, the physiological temperature of *S. solfataricus*, 50% of PRPP is degraded after 2 min. Remarkably, PRT of *S. solfataricus* has a lower catalytic efficiency than its mesophilic homologs. The high affinity of PRT from *S. solfataricus* for anthranilate is interpreted to reduce the possible loss of the hydrophobic substrate through the cell wall at high temperatures (Ivens et al., 2001).

The first enzyme in tryptophan synthesis, anthranilate synthase, catalyzes the synthesis of anthranilate from chorismate and glutamine and is feedback inhibited by tryptophan (Knöchel et al., 1999; Morollo et al., 2001). It is a heterotetramer of anthranilate synthase (TrpE) and glutamine amidotransferase (TrpG) subunits. Ammonia produced by the TrpG subunit is channeled toward the active site of the TrpE subunit, protecting it from protonation because NH_4^+ is not a substrate for the TrpE subunit. Kinetic, mutational, and structural information suggests a model in which chorismate binding triggers a relative movement of the two domain tips of the TrpE subunit, activating the TrpG subunit and creating a channel for the passage of ammonia (Knöchel et al., 1999). The active sites of the TrpG and TrpE subunits face each other across the intersubunit interface but do not form a channel in the apoenzyme (Knöchel et al., 1999).

This kind of ammonia channeling is widespread in mesophiles among the family of glutamine amidotransferases (Zalkin, 1993). Glutamine amidotransferases are bifunctional enzyme complexes that catalyze two reactions at separate active sites. In the glutaminase reaction, the hydrolysis of glutamine yields ammonia, which in the subsequent synthase reaction is added to an acceptor substrate, specific to each glutamine amidotransferase. Structural and functional studies on several members of the glutamine amidotransferase family have identified ammonia tunnels that connect the glutaminase and synthase active sites of these enzyme complexes. Enzymes belonging to this large family include anthranilate synthase in tryptophan biosynthesis, glutamine phosphoribosyl pyrophosphate amidotransferase (GPAT) in de novo purine synthesis, carbamoyl phosphate synthetase (CPS) in arginine and pyrimidine biosynthesis, and imidazole glycerol phosphate synthase (ImGP synthase). The heterodimeric ImGP synthase links histidine and purine biosynthesis. The X-ray structure of the hyperthermophilic *T. maritima* enzyme reveals a putative tunnel for the transfer of ammonia over a distance of 25 Å in the complex (Douangamath et al., 2002). In yeast, ImGP synthase bears both the glutaminase and the cyclase catalytic activities on a single polypeptide chain (Chittur et al., 2000). The same enzymes of thermophilic organisms have to be adapted to high temperatures to prevent leakage of NH_3 from the channel.

DE NOVO PURINE NUCLEOTIDE BIOSYNTHESIS

The de novo purine nucleotide synthesis involves several thermally unstable metabolites (Fig. 4) (Zalkin and Nygaard, 1996): PRPP, ammonia, PRA, N-carboxyaminoimidazole ribonucleotide (NCAIR), folates (White, 1993, 1997; Ownby et al., 2005), and formyl phosphate.

Specifically, the first two reactions of the de novo purine synthetic pathway convert PRPP to glycinamide ribonucleotide (GAR) via the extremely labile PRA with a half-life of 5 s at 37°C, pH 7.5. In the sixth and seventh step of purine biosynthesis, aminoimidazole ribonucleotide (AIR) is converted to 4-carboxy 5-aminoimidazole ribonucleotide (CAIR) via the unstable intermediate NCAIR with a half-life of 54 s at 30°C, pH 7.8 (Mueller et al., 1994).

GPAT catalyzes the formation of PRA in the first reaction of de novo purine nucleotide synthesis in two steps and is the key regulatory enzyme in the pathway. Glutamine is hydrolyzed to glutamate and NH_3 at an N-terminal glutaminase site, and NH_3 is transferred to a distal PRPP site for synthesis of PRA.

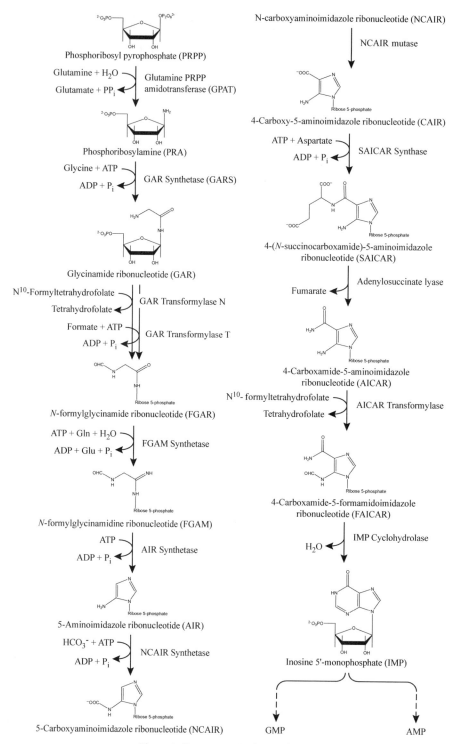

Figure 4. De novo purine biosynthesis.

Channeling of ammonia was reported for *E. coli* GPAT (Bera et al., 2000a) as well as for hyperthermophilic GPAT from *Aquifex aeolicus* (Bera et al., 2000b). A solvent inaccessible channel permits the transfer of the NH_3 intermediate between the two active sites. Results for the *E. coli* enzyme indicate that channel function and interdomain signaling are coupled (Bera et al., 2000a), preventing the wasteful hydrolysis of glutamine in the absence of PRA synthesis. Binding of PRPP to the PRT domain results in ordering of a PRT flexible loop, leading to the activation of the glutaminase domain and formation of a

20-Å hydrophobic channel for transfer of NH_3 (Krahn et al., 1997; Chen et al., 1999). Important from the point of view of adaptation to thermophily is that interdomain signaling (activation of glutaminase by PRPP) and channeling of NH_3 in the enzyme of *A. aeolicus* were found to be strongly temperature dependent (Bera et al., 2000a). Whereas channeling efficiency of ammonia in the hyperthermophilic GPAT was low at 37°C, efficiency increased to nearly 90% at 80°C.

PRA, the product of GPAT, is susceptible to hydrolysis to ribose-5-phosphate, with a half-life under physiological conditions of 5 s at 37°C and less than 1 s at 90°C (Rudolph and Stubbe, 1995; Bera et al., 2000b). The kinetics for the synthesis of GAR by GPAT and GAR synthetase (GARS), catalyzing the second step in purine biosynthesis, in *E. coli* (Rudolph and Stubbe, 1995) and *A. aeolicus* (Bera et al., 2000b) have suggested the direct transfer of PRA between the active sites of these two enzymes. However, gel chromatography, fluorescence spectroscopy, chemical cross-linking, and affinity gel chromatography were unsuccessful to provide evidence for a stable interaction between the two *E. coli* proteins, suggesting that the interaction between the two enzymes, indicated by the kinetic results, is a transient one (Rudolph and Stubbe, 1995). Interaction of hyperthermophilic GPAT and GARS was not investigated, but given the extreme instability of PRA at high temperatures, the existence of at least a transient complex in which the production of PRA could be tightly coupled to synthesis of GAR seems likely.

In most prokaryotes, formation of CAIR in the de novo purine biosynthetic pathway requires HCO_3^-, ATP, AIR, and the enzymes PurK and PurE. PurK catalyzes the ATP- and HCO_3^--dependent carboxylation of AIR to NCAIR. PurE catalyzes the unusual rearrangement of NCAIR to CAIR. The hyperthermophilic archaeon *Archaeoglobus fulgidus*, known to lack a PurK equivalent, only grows in CO_2-enriched environment. Similar to vertebrate organisms, CO_2, AIR, and PurE might be utilized by this hyperthermophile to make CAIR, avoiding the use of the thermolabile NCAIR (Meyer et al., 1999).

Based on circumstantial evidence, such as the clustering of non-sequential activities onto a single polypeptide chain (Aimi et al., 1990), the instable nature of some of the pathway intermediates (Rudolph and Stubbe, 1995; Mueller et al., 1994), and the copurification of the two transformylases (Smith et al., 1980), purine enzymes were proposed to associate and function in a multienzyme complex (Zalkin, 1993). However, clear evidence for such a purine complex was not yet obtained (Gooljarsingh et al., 2001).

CP METABOLISM IN (HYPER)THERMOPHILES

Carbamoyl phosphate (CP), the common precursor of the pyrimidine and arginine biosynthetic pathways (Fig. 5), is a very thermolabile intermediate (half-life of only 15 s at 80°C and less than 2 s at 100°C) and potentially toxic at high temperature (Legrain et al., 1995). Besides constituting a significant waste of energy for the cell, because CP synthesis requires ATP, the decomposition of CP in aqueous solutions at high temperatures leads to the accumulation of toxic amounts of cyanate which is a powerful and indiscriminate carbamoylating agent for free amino groups under such conditions. Since CP is both extremely labile at high temperatures and potentially toxic, it would appear to require metabolic protection *in vivo*. Therefore, organisms growing at high temperatures must have evolved strategies to protect CP from degradation.

Indications for metabolic channeling of CP in arginine and pyrimidine biosynthetic pathways were found for the extreme thermophilic bacteria *Thermus aquaticus* and *A. aeolicus* and the hyperthermophilic archaea *P. furiosus* and *Pyrococcus abyssi*. Isotopic competition experiments with *P. furiosus* cell-free extracts showed a marked preference of ornithine carbamoyltransferase (OTC) for CP synthesized by carbamate kinase (CK) rather than for CP added to the reaction mixture (Legrain et al., 1995). Moreover, the bisubstrate analogue N-δ-phosphonacetyl-L-ornithine (PALO) inhibits the formation of citrulline in the CK-OTC coupled reaction much less than in the uncoupled OTC reaction. These results suggested that CK and OTC form a channeling complex to protect CP from decomposition. In the extreme thermophilic bacterium, *T. aquaticus* ZO5, CP is also protected from thermodegradation by channeling toward the synthesis of citrulline and carbamoylaspartate as demonstrated by an isotopic competition assay (Van de Casteele et al., 1997). In the deep-sea hyperthermophilic archaeon *P. abyssi*, CP channeling in both arginine and pyrimidine biosynthetic pathways was also suggested by isotopic competition experiments and the kinetics of the coupled reactions between CK and both OTC and aspartate carbamoyltransferase (ATC) (Purcarea et al., 1999). Channeling was found leaky at ambient temperatures, but channeling efficiency appreciably increased at elevated temperatures. Recently, coupled reaction kinetics and transient time measurements have provided indications for channeling of CP in *A. aeolicus* (Purcarea et al., 2003). Channeling combined with the high affinity of ATC for its substrates ensures the efficient transfer of CP between the active sites of CPS and ATC.

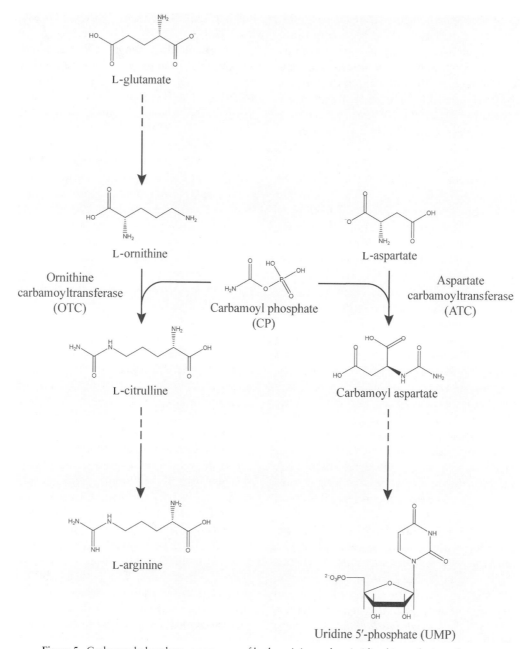

Figure 5. Carbamoyl phosphate, a precursor of both arginine and pyrimidine biosynthetic pathways.

Partial channeling of CP was suggested to occur in pyrimidine biosynthetic complexes from yeast (Belkaïd et al., 1988; Penverne et al., 1994) and mammals (Irvine et al., 1997) as well as between enzymes of the mammalian urea cycle (Cohen et al., 1992). In rat liver mitochondria, containing a high protein concentration and an extensive inner-membrane surface, CPS and OTC are abundant and thought to be associated with the inner-membrane surface (Powers-Lee et al., 1987; Cohen et al., 1992). These observations suggest that CP channeling is widespread and that, especially in thermophiles, it provides an efficient mechanism to protect this thermolabile metabolite from decomposition.

Intriguingly, the functional organization of the interacting enzymes and mechanism of channeling appears to differ in distinct organisms. Whereas in bacteria, such as *E. coli*, *Thermus*, and *Aquifex*, the first steps of pyrimidine biosynthesis are catalyzed by individual enzymes, in eukaryotes, such as mammals and yeast, a multifunctional protein is present. In mammals, a single 243-kDa multifunctional protein called CAD carries the CPS, ATC, and dihydroorotase (DHO) activities (Coleman et al.,

1977). The homologous multifunctional protein of yeast, encoded by the URA2 locus, possesses only the CPS and ATC activities; DHO is a separate enzyme. However, the yeast protein also contains a domain called pseudo-DHO (pDHO) that is not catalytically active, but its sequence is homologous to that of DHO of mammals and other organisms (Souciet et al., 1989). A linker region part of pDHO in the yeast protein and connecting DHO and ATC in CAD has been proposed to function as a spacer which ensures proximity of ATC and CPS active sites for the channeling of CP (Guy and Evans, 1994; Serre et al., 1998). The need for CP channeling thus may have provided the selective pressure for the evolution of the inactive pDHO domain in the yeast complex (Davidson et al., 1993). The efficient transfer of CP within the multienzyme polypeptide CAD is enhanced by "reciprocal" allosteric effects that protect CPS from inhibition by its product CP and increase the affinity of ATC for its substrates (Irvine et al., 1997; Serre et al., 1998). The in situ catalytic mechanism of ATC from yeast is altered by the fusion of the enzyme to CPS (Belkaïd et al., 1987).

Since multienzymatic polypeptides such as in yeast and mammals are not present in the hyperthermophilic organisms investigated, the enzymes from these thermophilic organisms must in vivo associate with a complex able to channel CP. Interestingly, Thermus ATC copurifies with DHO (Van de Casteele et al., 1994, 1997). Thermal stability of Thermus ATC is not an intrinsic property but comes from the association with DHO. The stability might further increase when CPS associates with the ATC–DHO complex. Perhaps, DHO is also required for association of CPS with ATC to form the CP-channeling complex. In A. aeolicus, DHO is activated upon association with ATC; without this interaction, DHO has no catalytic activity (Ahuja et al., 2004). Complex formation protects both enzymes against thermal degradation and might facilitate the transfer of the intermediate, carbamoylaspartate. Because indications for channeling of CP between CPS and ATC from A. aeolicus have been obtained with purified enzymes, DHO does not appear to be strictly necessary for channeling of CP in this organism. A mixture of A. aeolicus CPS, ATC, and DHO did not form a complex of all three enzymes.

Unlike in most prokaryotes, in the archaeon P. furiosus, CP is synthesized directly from ammonia, and the enzyme is enzymologically and structurally a CK (Durbecq et al., 1999; Uriarte et al., 1999; Ramón-Maiques et al., 2000). CKs from P. furiosus and Enterococcus faecalis were shown to be able to replace CPS in vivo in E. coli if ammonia and bicarbonate were present (Alcántara et al., 2000). The pyrococcal enzyme uses chemically made carbamate rather than ammonia and bicarbonate and catalyzes a reaction with the stoichiometry and equilibrium typical of a CK with one ATP molecule per CP molecule (Uriarte et al., 1999).

A carB gene, typical of the CPS present in E. coli and many other organisms (Charlier and Glansdorff, 2004), was found in the genome of P. furiosus but not in the genomes of the very related species P. abyssi and Pyrococcus horikoshii, making the situation more complex but also very interesting from the point of view of the evolution. The question rises what is the function of this putative CPS in P. furiosus. Could the enzyme play a physiological role at lower temperatures? Only the activity of the enzymes catalyzing the last three steps of the arginine biosynthetic pathway (OTC, argininosuccinate synthetase, and argininosuccinase) was detected in P. furiosus (Van de Casteele et al., 1990). However, P. furiosus, P. horikoshii, and P. abyssi have a gene cluster in their genome homologous to the argBCDE gene cluster found in other organisms and similar to that for lysine biosynthesis via the aminoadipic acetate pathway in T. thermophilus (Nishida et al., 1999). Their products may be active in ornithine as well as in lysine biosynthesis. Unlike P. furiosus, P. abyssi appears to lack the genes for the last steps in arginine biosynthesis in agreement with its arginine auxotrophy (Cohen et al., 2003). Comparison of the genomes of P. abyssi, P. furiosus, and P. horikoshii revealed a great genomic and metabolic plasticity of the three pyrococci (Lecompte et al., 2001).

A stable complex between CK and OTC or ATC from P. abyssi or P. furiosus could not be detected by size-exclusion chromatography (Purcarea et al., 1999; Massant, 2004). Neither elevated temperature, which promotes more efficient channeling, nor the presence of substrates, which could induce a change in enzyme conformation and thus promote complex formation, led to the formation of complexes detectable by gel filtration (Purcarea et al., 1999). However, several in vitro and in vivo methods, including coimmunoprecipitation, yeast two-hybrid analysis, gel filtration techniques, and isothermal titration calorimetry (ITC), have provided evidence for a biologically significant interaction between CK and OTC in arginine biosynthesis and between CK and ATC in pyrimidine biosynthesis (Massant et al., 2002; Massant and Glansdorff, 2004, 2005; Massant, 2004). Evidence for a direct physical interaction between CK and OTC and results indicating an interaction between CK and ATC support predictions made from kinetic analysis.

With affinity electrophoresis and coimmunoprecipitation experiments, indications were obtained for an interaction between CK and OTC of *P. furiosus* (Massant et al., 2002). This was the first evidence for a physical interaction between two hyperthermophilic enzymes involved in channeling of a thermolabile metabolite. Affinity electrophoresis experiments allowed the visualization of a direct interaction between both enzymes in conditions that mimic the crowded intracellular environment, and the resulting association between CK and OTC was shown to be specific. Further evidence for a biologically significant interaction between CK and OTC was obtained with coimmunoprecipitation in combination with cross-linking and addition of excess free enzyme. These results indicated that the interaction between both enzymes is relatively weak and unstable, at least in vitro and at ambient temperature. No influence of substrates on complex formation was observed. Using the yeast two-hybrid system and ITC, convergent and independent evidence for the association of CK and OTC was obtained (Massant and Glansdorff, 2005). The yeast two-hybrid screening allowed the detection in vivo of the interaction, thought to be weak or transient. Titration results indicated that the interaction is detectable in macromolecular crowding conditions. However, the kinetic and thermodynamic properties of the enzyme–enzyme interactions involved in the formation of the CP-channeling complex could not yet be determined. Results obtained with the Hummel–Dreyer method of equilibrium gel filtration have provided evidence for a direct interaction between CK and ATC and a first estimate of the binding constant between CK and ATC (Massant and Glansdorff, 2005). Because the interaction between CK and ATC appears to be very weak, it would be interesting to include DHO when investigating the interaction between CK and ATC further on. Yet it is important to note that, in contrast to the ATC of *Thermus* or the multifunctional protein from yeast or mammals, the pyrococcal ATC catalytic trimers are associated with regulatory dimers like the *E. coli* ATC. Allosteric regulation by nucleotide effectors and substrate cooperativity make the case for *P. furiosus* ATC more complex. These effects might influence the formation of the CK–ATC complex. The structures of the catalytic trimers of ATC and OTC are very similar; the structures of the holoenzymes, however, are not. Interest arises from the fact that the same enzyme (CK) can interact with two carbamoyltransferases of different quaternary structures.

The structure of *P. furiosus* OTC was determined at 2.7 Å (Villeret et al., 1998) but could recently be refined to 1.87 Å (Massant et al., 2003). While most mesophilic OTCs are trimeric, the hyperthermophilic protein is dodecameric, composed of four catalytic trimers disposed in a tetrahedral manner (Legrain et al., 1997). ATCs from *P. furiosus* and *P. abyssi* are also dodecamers but composed of two catalytic trimers and three regulatory dimers, similar to the *E. coli* enzyme (Purcarea et al., 1994; Massant, 2004; Lipscomb, 1994). The structure of the *P. abyssi* ATC catalytic trimer complexed with N-δ-phosphonacetyl-L-aspartate was determined at a resolution of 1.8 Å (Van Boxstael et al., 2003). The structure of the *P. furiosus* CK was solved at a resolution of 1.5 Å (Ramón-Maiques et al., 2000). An ADP molecule was found bound in the pyrococcal enzyme at the bottom of a large cavity, with the purine ring enclosed in a pocket specific to adenine. This bound nucleotide provided insight into substrate binding and catalytic mechanism. Crystal structures of the different enzymes involved in channeling of CP provide a basis for detailed mechanistic and structural information about the dynamic enzyme–enzyme interactions.

So far, co-crystallizations of CK and OTC or CK and ATC from *P. furiosus* have been unsuccessful. Based on the structures of CK and OTC, a mode of association that could allow channeling of CP was suggested by Ramón-Maiques et al. (2000). Corresponding to the relative sizes and shapes of both enzymes, the formation of the channeling complex was proposed to be nucleated by an OTC dodecamer associating with six CK dimers in such a way that the molecular twofold symmetry axes coincide. Docking simulations with the structures of the individual enzymes have been initiated with the MolFit program (Katchalski-Katzir et al., 1992; Ben-Zeev et al., 2003) in collaboration with Miriam Eisenstein from the Weizmann Institute, Israël (Massant, 2004). These docking simulations did not provide evidence for the kind of complex formation suggested by Ramón-Maiques et al. (2000). Neither could they provide a clear alternative for the structure of the channeling complex. Since docking simulations cannot take into account conformational changes or flexibility, they are only rough approximations. Electrostatic isopotential contours of the OTC catalytic trimer and the CK dimer as calculated with the Delphi program and the preliminary indications obtained by in silico docking suggest the possibility of formation of a CP-channeling complex guided by electrostatic forces. A role for electrostatics was also suggested by the coimmunoprecipitation and affinity electrophoresis experiments (Massant et al., 2002). Besides guiding the formation of the channeling complex, electrostatic forces might also play a role in guiding the negatively charged CP molecule between the active sites of CK and OTC. Electrostatic potentials and Brownian dynamics simulations have provided evidence for electrostatic channeling of oxaloacetate from malate dehydrogenase to citrate synthase in the pig mitochondria (Elcock and McCammon, 1996; Vélot et al., 1997) and of dihydrofolate between thymidilate synthase and dihydrofolate reductase (Elcock et al., 1996). Obviously, many questions about the formation and the structure of the CP-channeling complexes remain to be answered.

THERMAL ADAPTATION AND MACROMOLECULAR ORGANIZATION OF METABOLISM

Many of the low-molecular-weight metabolites and coenzymes used by hyperthermophiles in their biochemical pathways are thermally unstable. Some of them, such as PRA and CP, are extremely thermolabile. The thermal (in)stability of metabolites and coenzymes has been a somewhat neglected but essential aspect of the molecular physiology of hyperthermophiles. Hyperthermophiles appear to cope with the limitations posed by thermolabile metabolites and coenzymes by a range of mechanisms including rapid turnover or increased catalytic efficiency, local stabilization, substitution or bypassing, microenvironmental compartmentation, or metabolic channeling (Daniel and Cowan, 2000). However, detailed information about these mechanisms is still scarce.

An adequate supply of some thermo-sensitive compounds may be ensured by rapid re-synthesis. Such a strategy could however cause a loss of energy for the cell. Substitution or bypassing circumvents the thermal instability of a metabolite by using a more stable alternative compound or by use of an alternative pathway. The thermal stability of many metabolites (NADH, ATP) is highly dependent on physiological conditions. The use of microsites with less destabilizing conditions was suggested as a possible way to protect these molecules against thermal degradation (Daniel and Cowan, 2000). Thermophiles and hyperthermophiles accumulate compatible organic solutes (Martins and Santos, 1995). It was suggested that besides osmoprotection, these solutes might act as thermoprotectants (Borges et al., 2002; Santos and da Costa, 2002). The very high catalytic efficiency of certain enzymes has been suggested as a mechanism to overcome the thermal instability of their substrates (e.g., PRAI of *T. maritima*). As several examples show, metabolic channeling, the direct transfer of a metabolite between sequential enzymes in a metabolic pathway, could be an efficient strategy to protect chemically labile intermediates. The sine qua non for metabolic channeling is the existence of some sort of structural organization for the enzymes of the metabolic pathway.

It has become clear over the last decade that the cytoplasm is not a mere bag of enzymes and substrates but rather a highly organized and structured system (Ovádi and Saks, 2004). Compartmentation has since long been recognized as an important and probably necessary structural feature of the cell (Srere and Mosbach, 1974). But, while the idea of cytoplasmic microstructure is now well accepted (al-Habori, 1995; Hoppert and Mayer, 1999), the physiological implications are less clear. Compartmentation in living cells does not only lead to structural organization of the cell but creates also a unique physicochemical surrounding caused by different surface properties (e.g., surface charge) from phospholipid surfaces, proteins, and nucleic acid complexes. The cytoplasm of a living cell is crowded with solutes and macromolecules (Minton, 2001; Ellis, 2001). It is becoming more widely appreciated that under physiological conditions of crowding or confinement, the size- and shape-dependent reduction of volume available to every species of macromolecule results in major shifts in the rates and equilibria of a broad range of macromolecular reactions relative to those measured in dilute solution (Minton, 2000). Macromolecular crowding favors the formation of compact structures and macromolecular complexes and forces the formation of (soluble) homo- and hetero-oligomers (Minton, 2000). It is important to note that crowding does not make all macromolecules bind to one another but enhances the inherent tendency of macromolecules to associate specifically. Crowding in a cell appears for the most part selected for and specific since evolution seems to have conserved not only functional sites of protein molecules but also structural features that might determine the abilities of proteins to associate with one another (McConkey, 1982). The cell is a dynamic system involved in different physiological processes with a great number of interconnected reactions proceeding simultaneously. Cellular biochemistry appears to be largely run by a set of protein complexes, rather than by proteins that act individually and exist as isolated species (Alberts and Miake-Lye, 1992). Already some years ago, the enzymes of several metabolic pathways were suggested to be organized into structural and functional units (Srere, 1987). Several examples of the formation of enzyme–enzyme complexes, particularly among enzymes that are sequentially related in metabolic pathways, have been reported. In addition to stable multienzyme complexes and multifunctional enzymes that catalyze sequential reactions in a pathway, there are probably many specific interactions between soluble sequential enzymes of metabolic pathways. These associations are presumed to be able to channel metabolic intermediates.

In such a channeling complex, the intermediate can be transferred by different mechanisms: intermediates can be covalently linked to a swinging arm (Perham, 2000), they can be gated through a tunnel through the interior of the protein (Huang et al., 2001), or a favorable electrostatic field between the adjacent active sites can be used to constrain the intermediates within the channeling path along the surface of the protein (Elcock et al., 1996). Another mechanism of channeling could occur when the active sites of two enzymes are transiently brought into contact with each other, forming a cavity that permits direct transfer of the intermediate and sterically prevents its escape into the bulk phase (Spivey and Ovádi, 1998). Despite several examples, detailed mechanistic and structural information about transient enzyme-enzyme interactions is still scarce or lacking.

In this view, metabolic channeling of intermediates between physically associated enzymes that are sequential members of a metabolic pathway can be a major thermoprotective mechanism for thermolabile metabolites and therefore can play a critical role in the physiology of thermophiles. The intracellular structure and organization of metabolism might be an important factor in the adaptation of metabolic pathways to high temperatures. The clustering of functional related genes into operons might have been selected for as a mechanism facilitating the association of proteins stabilizing each other and/or channeling thermolabile intermediates (Glansdorff, 1999). For a better understanding of adaptation of metabolism to high temperatures, functional demands of proteins must be considered besides intrinsic stability (Walden et al., 2004b), and more data of a quantitative nature should be gathered on the actual stability of certain metabolites, especially in in vivo conditions.

Acknowledgments. Jan Massant is Postdoctoraal Onderzoeker of the FWO-Vlaanderen. I would like to thank Nicolas Glansdorff and Daniel Charlier for critical reading of the manuscript.

REFERENCES

Adams, M. W. W. 1993. Enzymes and proteins from organisms that grow near and above 100 degrees C. *Annu. Rev. Microbiol.* **47**:627–658.

Ahmed, H., B. Tjaden, R. Hensel, and B. Siebers. 2004. Embden–Meyerhof–Parnas and Entner–Doudoroff pathways in *Thermoproteus tenax*: metabolic parallelism or specific adaptation. *Biochem. Soc. Trans.* **32**:303–304.

Ahmed, H., T. J. G. Ettema, B. Tjaden, A. C. M. Geerling, J. van der Oost, and B. Siebers. 2005. The semi-phosphorylative Entner–Doudoroff pathway in hyperthermophilic archaea – a re-evaluation. *Biochem. J.* **390**:529–540.

Ahuja, A., Purcarea, C., Ebert, R., Sadecki, S., Guy, H. I., and D. R. Evans. 2004. *Aquifex aeolicus* dihydroorotase. Association with aspartate transcarbamoylase switches on catalytic activity. *J. Biol. Chem.* **279**:53136–53144.

Aimi, J., H. Oiu, J. Williams, H. Zalkin, and J. E. Dixon. 1990. De novo purine nucleotide biosynthesis: cloning of human and avian cDNAs encoding the trifunctional glycinamide ribonucleotide synthetase-aminoimidazole ribonucleotide synthetase-glycinamide ribonucleotide transformylase by functional complementation in *E. coli*. *Nucleic Acids Res.* **18**:6665–6672.

Alberts, B., and R. Miake-Lye. 1992. Unscrambling the puzzle of biological machines: the importance of the details. *Cell* **68**:415–420.

Alcántara, C., J. Cervera, and V. Rubio. 2000. Carbamate kinase can replace in vivo carbamoyl phosphate synthetase. Implications for the evolution of carbamoyl phosphate biosynthesis. *FEBS Lett.* **484**:261–264.

al-Habori, M. 1995. Microcompartmentation, metabolic channeling and carbohydrate metabolism. *Int. J. Biochem. Cell Biol.* **27**:123–132.

Anderson, K. S. 1999. Fundamental mechanisms of substrate channeling. *Methods Enzymol.* **308**:111–145.

Beeckmans, S., E. Van Driessche, and L. Kanarek. 1990. Clustering of sequential enzymes in the glycolytic pathway and the citric acid cycle. *J. Cell. Biochem.* **43**:297–306.

Belkaïd, M., B. Penverne, M. Denis, and G. Hervé. 1987. In situ behavior of the pyrimidine pathway enzymes in *Saccharomyces cerevisiae*. 2. Reaction mechanism of aspartate transcarbamylase dissociated from carbamylphosphate synthetase by genetic alteration. *Arch. Biochem. Biophys.* **254**:568–578.

Belkaïd, M., B. Penverne, and G. Herve. 1988. In site behavior of the pyrimidine pathway enzymes in saccharomyces cerevisiae, 3: Catalytic and regulatory properties of carbamoylphosphate synthetase: channeling of carbamoylphosphate to aspartate carbamoyl transferase. *Arch. Biochem. Biophys.* **262**:171–180.

Ben-Zeev, E., A. Berchanski, A. Heifetz, B. Shapira, and M. Eisenstein. 2003. Prediction of the unknown: inspiring experience with the CAPRI experiment. *Proteins* **52**:41–46.

Bera, A. K., J. L. Smith, and H. Zalkin. 2000a. Dual role for the glutamine phosphoribosylpyrophosphate amidotransferase ammonia channel. Interdomain signaling and intermediate channeling. *J. Biol. Chem.* **275**:7975–7979.

Bera, A. K., S. Chen, J. L. Smith, and H. Zalkin. 2000b. Temperature-dependent function of the glutamine phosphoribosylpyrophosphate amidotransferase ammonia channel and coupling with glycinamide ribonucleotide synthetase in a hyperthermophile. *J. Bacteriol.* **182**:3734–3739.

Blamey, J. M., and M. W. Adams. 1993. Purification and characterization of pyruvate ferredoxin oxidoreductase from the hyperthermophilic archaeon *Pyrococcus furiosus*. *Biochim. Biophys. Acta* **1161**:19–27.

Borges, N., A. Ramos, N. D. H. Raven, R. J. Sharp, and H. Santos. 2002. Comparative study of the thermostabilizing properties of mannosylglycerate and other compatible solutes on model enzymes. *Extremophiles* **6**:209–216.

Bräsen, C., and P. Schönheit. 2004. Unusual ADP-forming acetyl-coenzyme A synthetases from the mesophilic halophilic euryarchaeon *Haloarcula marismortui* and from the hyperthermophilic crenarchaeon *Pyrobaculum aerophilum*. *Arch. Microbiol.* **182**:277–287.

Brunner, N. A., B. Siebers, and R. Hensel. 2001. Role of two different glyceraldehydes-3-phosphate dehydrogenases in controlling the reversible Embden–Meyerhof–Parnas pathway in *Thermoproteus tenax*: regulation on protein and transcript level. *Extremophiles* **5**:101–109.

Charlier, D., and N. Glansdorff. 9 September 2004, posting date. Chapter 3.6.1.10. Biosynthesis of arginine and polyamines. *In* A. Böck et al. (ed.), EcoSal—*Escherichia coli and Salmonella: Cellular and Molecular Biology.* http://www.ecosal.org. ASM Press, Washington, DC.

Chen, S., J. W. Burgner, J. M. Krahn, J. L. Smith, and H. Zalkin. 1999. Tryptophan fluorescence monitors multiple conformational changes required for glutamine phosphoribosylpyrophosphate amidotransferase interdomain signaling and catalysis. *Biochemistry* **38:**11659–11669.

Chittur, S. V., Y. Chen, and V. J. Davisson. 2000. Expression and purification of imidazole glycerol phosphate synthase from *Saccharomyces cerevisiae. Protein Expr. Purif.* **18:**366–377.

Clegg, J. S., and S. A. Jackson. 1990. Glucose metabolism and the channeling of glycolytic intermediates in permeabilized L-929 cells. *Arch. Biochem. Biophys.* **278:**452–460.

Cohen, G. N., V. Barbe, D. Flament, M. Galperin, R. Heilig, O. Lecompte, O. Poch, D. Prieur, J. Quérellou, R. Ripp, J.-C. Thierry, J. Van der Oost, J. Weissenbach, Y. Zilvanovic, and P. Forterre. 2003. An integrated analysis of the genome of the hyperthermophilic archaeon *Pyrococcus abyssi. Mol. Microbiol.* **47:**1495–1512.

Cohen, N. S., C.-W. Cheung, E. Sijuwade, and L. Raijman. 1992. Kinetic properties of carbamoyl-phosphate synthase (ammonia) and ornithine carbamoyltransferase in permeabilized mitochondria. *Biochem. J.* **282:**173–180.

Coleman, P. F., D. P. Suttle, and G. R. Stark. 1977. Purification from hamster cells of the multifunctional protein that initiates *de novo* synthesis of pyrimidine nucleotides. *J. Biol. Chem.* **252:**6379–6385.

Creighton, T. E. 1970. N(5'-phosphoribosyl) anthranilate isomerase-indol-3-ylglycerol phosphate synthetase of tryptophan biosynthesis: relationship between the two activities of the enzyme from Escherichia coli: *Biochem. J.* **120,** 699–707.

Daniel, R. M., and D. A. Cowan. 2000. Biomolecular stability and life at high temperatures. *Cell. Mol. Life Sci.* **57:**250–264.

Daniel, R. M., and M. J. Danson. 1995. Did primitive microorganisms use nonhaem iron proteins in place of NAD/P? *J. Mol. Evol.* **40:**559–563.

Danson, M. J. 1988. Archaebacteria: the comparative enzymology of their central metabolic pathways. *Adv. Microb. Physiol.* **29:**165–231.

Davidson, J. N., K. C. Chen, R. S. Jamison, L. A. Musmanno, and C. B. Kern. 1993. The evolutionary history of the first three enzymes in pyrimidine biosynthesis. *Bioessays* **15:**157–164.

Dörr, C., M. Zaparty, B. Tjaden, H. Brinkman, and B. Siebers. 2003. The hexokinase of the hyperthermophile *Thermoproteus tenax:* ATP-dependent hexokinases and ADP-dependent glucokinases, two alternatives for glucose phosphorylation in Archaea. *J. Biol. Chem.* **278:**18744–18753.

Douangamath, A., M. Walker, S. Beismann-Driemeyer, M. C. Vega-Fernandez, R. Sterner, and M. Wilmanns. 2002. Structural evidence for ammonia tunneling across the (beta alpha)(8) barrel of the imidazole glycerol phosphate synthase bienzyme complex. *Structure* **10:**185–193.

Durbecq, V., C. Legrain, M. Roovers, A. Pierard, and N. Glansdorffa 1997. The carbomate kinase-like carbamoyl phosphate synthetase of the hyperthermophilic archaeor pyrococcus furiosus, a missing link in the evolution of carbamoyl phosphate biosynthesis. *Proc Natl Acad Sci* USA **94,** 12803–12808.

Elcock, A. H., and J. A. McCammon. 1996. Evidence for electrostatic channeling in a fusion protein of malate dehydrogenase and citrate synthase. *Biochemistry* **35:**12652–12658.

Elcock, A. H., M. J. Potter, D. A. Matthews, D. R. Knighton, and J. A. McCammon. 1996. Electrostatic channeling in the bifunctional enzyme dihydrofolate reductase-thymidilate synthase. *J. Mol. Biol.* **262:**370–374.

Ellis, R. J. 2001. Macromolecular crowding: obvious but underappreciated. *Trends Biochem. Sci.* **26:**597–604.

Glansdorff, N. 1999. On the origin of operons and their possible role in evolution toward thermophily. *J. Mol. Evol.* **49:**432–438.

Gooljarsingh, L. T., J. Ramcharan, S. Gilroy, and S. J. Benkovic. 2001. Localization of GAR transformylase in *Escherichia coli* and mammalian cells. *Proc. Natl. Acad. Sci. USA* **98:**6565–6570.

Gotz, R., E. Schluter, G. Shoham, and F. K. Zimmerman. 1999. A potential role of the cytoskeleton of *Saccharomyces cerevisiae* in a functional organization of glycolytic enzymes. *Yeast* **15:** 1619–1629.

Guy, H. I., and D. R. Evans. 1994. Cloning and expression of the mammalian multifunctional protein CAD in *Escherichia coli.* Characterization of the recombinant protein and a deletion mutant lacking the major interdomain linker. *J. Biol. Chem.* **269:**23808–23816.

Hennig, M., R. Sterner, K. Kirschner, and J. N. Jansonius. 1997. Crystal structure at 2.0 Å resolution of phosphoribosyl anthranilate isomerase from the hyperthermophile *Thermotoga maritima:* possible determinants of protein stability. *Biochemistry* **36:**6009–6016.

Henn-Sax, M., R. Thoma, S. Schmidt, M. Hennig, K. Kirschner, and R. Sterner. 2002. Two (βα)8-barrel enzymes of histidine and tryptophan biosynthesis have similar reaction mechanisms and common strategies for protecting their labile substrates. *Biochemistry* **41:**12032–12042.

Hettwer, S., and R. Sterner. 2002. A novel tryptophan synthase β-subunit from the hyperthermophile *Thermotoga maritima:* quaternary structure, steady-state kinetics, and putative physiological role. *J. Biol. Chem.* **277:**8194–8201.

Hoppert, M., and F. Mayer. 1999. Principles of macromolecular organization and cell function in Bacteria and Archaea. *Cell Biochem. Biophys.* **31:**247–284.

Huang, X., Holden, H. M., and F. M. Raushel. 2001. Channeling of substrates and intermediates in enzyme-catalyzed reactions. *Annu. Rev. Biochem.* **70:**149–180.

Hyde, C. C., S. A. Ahmed, E. A. Padlan, E. W. Miles, and D. R. Davies. 1988. Three-dimensional structure of the tryptophan synthase alpha 2 beta 2 multienzyme complex from *Salmonella typhimurium. J. Biol. Chem.* **263:**17857–17871.

Irvine, H. S., S. M. Shaw, A. Paton, and E. A. Carrey. 1997. A reciprocal allosteric mechanism for efficient transfer of labile intermediates between active sites in CAD, the mammalian pyrimidine-biosynthetic multienzyme polypeptide. *Eur. J. Biochem.* **247:**1063–1073.

Ivens, A., O. Mayans, H. Szadkowski, M. Wilmanns, and K. Kirschner. 2001. Purification, characterization and crystallization of thermostable anthranilate phosphoribosyltransferase from *Sulfolobus solfataricus. Eur. J. Biochem.* **268:**2246–2252.

Katchalski-Katzir, E., I. Shariv, M. Eisenstein, A. A. Friesem, C. Aflalo, and I. A. Vakser. 1992. Molecular surface recognition: Determination of geometric fit between proteins and their ligands by correlation techniques. *Proc. Natl. Acad. Sci. USA* **89:**2195–2199.

Kengen, S. W., F. A. De Bok, N. D. Van Loo, C. Dijkema, A. J. Stams, and W. M. De Vos. 1994. Evidence for the operation of a novel Embden–Meyerhof pathway that involves ADP-dependent kinases during sugar fermentation by *Pyrococcus furiosus. J. Biol. Chem.* **269:**17537–17541.

Knöchel, T., A. Ivens, G. Hester, A. Gonzalez, R. Bauerle, M. Wilmanns, K. Kirschner, and J. Jansonius. 1999. The crystal structure of anthranilate synthase from *Sulfolobus solfataricus:* functional implications. *Proc. Natl. Acad. Sci. USA* **96:**9479–9484.

Knöchel, T., A. Pappenberger, J. N. Jansonius, and K. Kirschner. 2002. The crystal structure of indoleglycerol-phosphate synthase from *Thermotoga maritima*. Kinetic stabilization by salt bridges. *J. Biol. Chem.* **277:**8626–8634.

Krahn, J. M., J. H. Kim, M. R. Burns, R. J. Parry, H. Zalkin, and J. L. Smith. 1997. Coupled formation of an amidotransferase interdomain ammonia channel and a phosphoribosyltransferase active site. *Biochemistry* **36:**11061–11068.

Larralde, R., M. P. Robertson, and S. L. Miller. 1995. Rates of decomposition of ribose and other sugars: implications for chemical evolution. *Proc. Natl. Acad. Sci. USA* **92:**8158–8160.

Lecompte, O., R. Ripp, V. Puzos-Barbe, S. Duprat, R. Heilig, J. Dietrich, J. C. Thierry, and O. Poch. 2001. Genome evolution at the genus level: comparison of three complete genomes of hyperthermophilic archaea. *Genome Res.* **11:**981–993.

Legrain, C., M. Demarez, N. Glansdorff, and A. Piérard. 1995. Ammonia-dependent synthesis and metabolic channelling of carbamoyl phosphate in the hyperthermophilic archaeon *Pyrococcus furiosus*. *Microbiology* **141:**1093–1099.

Legrain, C., V. Villeret, M. Roovers, D. Gigot, O. Dideberg, A. Piérard, and N. Glansdorff. 1997. Biochemical characterization of ornithine carbamoyltransferase from *Pyrococcus furiosus*. *Eur. J. Biochem.* **247:**1046–1055.

Lipscomb, W. N. 1994. Aspartate carbamoyltransferase from *Escherichia Coli*: activity and regulation. *Adv. Enzymol.* **68,** 67–151.

Lowry, O. H., J. V. Passonneau, and M. K. Rock. 1961. The stability of pyridine nucleotides. *J. Biol. Chem.* **236:**2756–2759.

Mai, X., and M. W. W. Adams. 1996. Purification and characterization of two reversible and ADP-dependent acetyl coenzyme A synthetases from the hyperthermophilic archaeon *Pyrococcus furiosus*. *J. Bacteriol.* **178:**5897–5903.

Malaisse, W. J., Y. Zhang, and A. Sener. 2004. Enzyme-to-enzyme channeling in the early steps of glycolysis in rat pancreatic islets. *Endocrine* **24:**105–109.

Martins, L. O., and H. Santos. 1995. Accumulation of mannosylglycerate and di-myo-inositol-phosphate by *Pyrococcus furiosus* in response to salinity and temperature. *Appl. Environ. Microbiol.* **61:**3299–3303.

Massant, J. 2004. Molecular physiology of hyperthermophiles: metabolic channeling of carbamoyl phosphate, a thermolabile and potentially toxic intermediate, p. 177. Ph.D. thesis. Vrije Universiteit Brussel, Brussels, Belgium.

Massant, J., and N. Glansdorff. 2004. Metabolic channelling of carbamoyl phosphate in the hyperthermophilic archaeon *Pyrococcus furiosus*: dynamic enzyme–enzyme interactions involved in the formation of the channelling complex. *Biochem. Soc. Trans.* **32:**306–309.

Massant, J., and N. Glansdorff. 2005. New experimental approaches to investigate interactions between *Pyrococcus furiosus* carbamate kinase and carbamoyltransferases, enzymes involved in the channeling of thermolabile carbamoyl phosphate. *Archaea*, **1,** 365–373.

Massant, J., J. Wouters, and N. Glansdorff. 2003. Refined structure of *Pyrococcus furiosus* ornithine carbamoyltransferase at 1.87 Å. *Acta Cryst.* **D59:**2140–2149.

Massant, J., P. Verstreken, V. Durbecq, A. Kholti, C. Legrain, S. Beeckmans, P. Cornelis, and N. Glansdorff. 2002. Metabolic channeling of carbamoyl phosphate, a thermolabile intermediate: evidence for physical interaction between carbamate kinase-like carbamoyl-phosphate synthetase and ornithine carbamoyltransferase from the hyperthermophile *Pyrococcus furiosus*. *J. Biol. Chem.* **277:**18517–18522.

Maughan, D. W., J. A. Henkin, and J. O. Vigoreaux. 2005. Concentrations of glycolytic enzymes and other cytosolic proteins in the diffusible fraction of a vertebrate muscle proteome. *Mol. Cell. Proteomics* **410:**1541–1549.

McConkey, E. H. 1982. Molecular evolution, intracellular organization, and the quinary structure of proteins. *Proc. Natl. Acad. Sci. USA* **79:**3236–3240.

Merz, A., T. Knöchel, J. N. Jansonius, and K. Kirschner. 1999. The hyperthermostable indoleglycerol phosphate synthase from *Thermotoga maritima* is destabilized by mutational disruption of two solvent-exposed salt bridges. *J. Mol. Biol.* **288:**753–763.

Meyer, E., T. J. Kappock, C. Osuji, and J. Stubbe. 1999. Evidence for the direct transfer of the carboxylate of N^5-4-carboxy-5-aminoimidazole ribonucleotide (N^5-CAIR) to generate 4-carboxy-5-aminoimidazole ribonucleotide catalyzed by *Escherichia coli* PurE, an N^5-CAIR mutase. *Biochemistry* **38:**3012–3018.

Miles, E. W., S. Rhee, and D. R. Davies. 1999. The molecular basis of substrate channeling. *J. Biol. Chem.* **274:**12193–12196.

Minton, A. P. 2000. Implications of macromolecular crowding for protein assembly. *Curr. Opin. Struct. Biol.* **10:**34–39.

Minton, A. P. 2001. The influence of macromolecular crowding and macromolecular confinement on biochemical reactions in physiological media. *J. Biol. Chem.* **276:**10577–10580.

Morollo, A. A., and M. J. Eck. 2001. Structure of the cooperative allosteric anthranilate synthase from *Salmonella typhimurium*. *Nat. Struct. Biol.* **8:**243–247.

Mowbray, J., and V. Moses. 1976. The tentative identification in *Escherichia coli* of a multienzyme complex with glycolytic activity. *Eur. J. Biochem.* **66:**25–36.

Mueller, E. J., E. Meyer, J. Rudolph, V. J. Davisson, and J. Stubbe. 1994. N^5-4-carboxy-5-aminoimidazole ribonucleotide: evidence for a new intermediate and two new enzymatic activities in the de novo purine biosynthetic pathway of *Escherichia coli*. *Biochemistry* **33:**2269–2278.

Mukund, S., and M. W. W. Adams. 1995. Glyceraldehyde-3-phosphate ferredoxin oxidoreductase, a novel tungsten-containing enzyme with a potential glycolytic role in the hyperthermophilic Archaeon *Pyrococcus furiosus*. *J. Biol. Chem.* **270:**8389–8392.

Napolitano, M. J., and D. H. Shain. 2005. Distinctions in adenylate metabolism among organisms inhabiting temperature extremes. *Extremophiles* **9:**93–98.

Nichols, B. P. 1996. Evolution of genes and enzymes of tryptophan biosynthesis, p. 2638–2648. *In* F. C. Neidhardt, R. Curtiss III, J. L. Ingraham, E. C. C. Lin, K. B. Low, B. Magasanik, W. S. Reznikoff, M. Riley, M. Schaechter, and H. E. Umbarger (ed.), *Escherichia coli and Salmonella: Cellular and Molecular Biology*, 2nd ed. ASM Press, Washington, DC.

Nishida, H., M. Nishiyama, N. Kobashi, T. Kosuge, T. Hoshino, and H. Yamane. 1999. A prokaryotic gene cluster involved in synthesis of lysine through the amino adipate pathway: a key to the evolution of amino acid biosynthesis. *Genome Res.* **9:**1175–1183.

Ogasahara, K., M. Ishida, and K. Yutani. 2003. Stimulated interaction between α and β subunits of tryptophan synthase from hyperthermophile enhances its thermal stability. *J. Biol. Chem.* **278:**8922–8928.

Orosz, F., and J. Ovadi. 1987. A simple approach to identify the mechanism of intermediate transfer: enzyme system related to triose phosphate metabolism. *Biochim. Biophys. Acta* **915:**53–59.

Ovádi, J., and V. Saks. 2004. On the origin of intracellular compartmentation and organized metabolic systems. *Mol. Cell. Biochem.* **256/257:**5–12.

Ownby, K., H. Xu, and R. H. White. 2005. A *Methanocaldococcus jannaschii* archaeal signature gene encodes for a 5-formaminoimidazole-4-carboxamide-1-β-D-ribofuranosyl 5′-monophosphate synthetase: a new enzyme in purine biosynthesis. *J. Biol. Chem.* **280:**10881–10887.

Pan, G., M. F. J. M. Verhagen, and M. W. W. Adams. 2001. Characterization of pyridine nucleotide coenzymes in the hyperthermophilic archaeon *Pyrococcus furiosus*. *Extremophiles* **5:** 393–398.

Pan, P., E. Woehl, and M. F. Dunn. 1997. Protein architecture, dynamics and allostery in tryptophan synthase channeling. *Trends Biochem. Sci.* 22:22–27.

Penverne, B., M. Belkaïd, and G. Hervé. 1994. In situ behavior of the pyrimidine pathway enzymes in *Saccharomyces cerevisiae*. 4. The channeling of carbamylphosphate to aspartate transcarbamylase and its partition in the pyrimidine and arginine pathways. *Arch. Biochem. Biophys.* 309:85–93.

Perham, R. N. 2000. Swinging arms and swinging domains in multifunctional enzymes: catalytic machines for multistep reactions. *Annu. Rev. Biochem.* 69:961–1004.

Pittard, A. J. 1996. Biosynthesis of aromatic amino acids, p. 458–484. *In* F. C. Neidhardt, R. Curtiss III, J. L. Ingraham, E. C. C. Lin, K. B. Low, B. Magasanik, W. S. Reznikoff, M. Riley, M. Schaechter, and H. E. Umbarger (ed.), *Escherichia coli and Salmonella: Cellular and Molecular Biology*, 2nd ed. ASM Press, Washington, DC.

Powers-Lee, S. G., R. A. Mastico and M. Bendayan. 1987. The interaction of rat liver carbamoyl phosphate synthetase and ornithine transcarbamoylase with inner mitochondrial membranes. *J. Biol. Chem.* 262:15683–15688.

Purcarea, C., A. Ahuja, T. Lu, L. Kovari, H. I. Guy, and D. R. Evans. 2003. *Aquifex aeolicus* aspartate transcarbamylase, an enzyme specialized for the efficient utilization of unstable carbamoyl phosphate at elevated temperature. *J. Biol. Chem.* 278:52924–52934.

Purcarea, C., G. Erauso, D. Prieur, and G. Hevvé. 1994. The catalytic and regulatory properties of aspartate carbamoyltransferase from pyrococcus abyss: a new deep-sea hyperthermophilic archaeobacterium. Microbiology 140, 1967–1975.

Purcarea, C., D. R. Evans, and G. Hervé. 1999. Channeling of carbamoyl phosphate to the pyrimidine and arginine biosynthetic pathways in the deep sea hyperthermophilic archaeon *Pyrococcus abyssi*. *J. Biol. Chem.* 274:6122–6129.

Rakus, D., M. Pasek, H. Krotkiewski, and A. Dzugaj. 2004. Interaction between muscle aldolase and muscle fructose 1,6-bisphophatase results in the substrate channeling. *Biochemistry* 43:14948–14957.

Ramirez, F., J. F. Marecek, and J. Szamosi. 1980. Magnesium and calcium ion effects on hydrolysis rates of adenosine-5′-triphosphate. *J. Org. Chem.* 45:4748–4752.

Ramón-Maiques, S., A. Marina, M. Uriarte, I. Fita, and V. Rubio. 2000. The 1.5 Å resolution crystal structure of the carbamate kinase-like carbamoyl phosphate synthetase from the hyperthermophilic archaeon *Pyrococcus furiosus*, bound to ADP, confirms that this thermostable enzyme is a carbamate kinase, and provides insight into substrate binding and stability in carbamate kinases. *J. Mol. Biol.* 299:463–476.

Robb, F. T., J. B. Park, and M. W. Adams. 1992. Characterization of an extremely thermostable glutamate dehydrogenase: a key enzyme in the primary metabolism of the hyperthermophilic archaebacterium *Pyrococcus furiosus*. *Biochim. Biophys. Acta* 1120:267–272.

Ronimus, R. S., and H. W. Morgan. 2003. Distribution and phylogenies of enzymes of the Embden–Meyerhof–Parnas pathway from archaea and hyperthermophilic bacteria support a gluconeogenic origin of metabolism. *Archaea* 1:199–221.

Rudolph, J., and J. Stubbe. 1995. Investigation of the mechanism of phosphoribosylamine transfer from glutamine phosphoribosylpyrophosphate amidotransferase to glycinamide ribonucleotide synthetase. *Biochemistry* 34:2241–2250.

Santos, H., and M. S. da Costa. 2002. Compatible solutes of organisms that live in hot saline environments. *Environ. Microbiol.* 4:501–509.

Schäfer, T., and P. Schönheit. 1993. Gluconeogenesis from pyruvate in the hyperthermophilic archaeon *Pyrococcus furiosus*: involvement of reactions of the Embden–Meyerhof pathway. *Arch. Microbiol.* 159:354–363.

Schäfer, T., M. Selig, and P. Schönheit. 1993. Acetyl-CoA synthetase (ADP forming) in archaea, a novel enzyme involved in acetate formation and ATP synthesis. *Arch. Microbiol.* 159: 72–83.

Schramm, A., B. Siebers, B. Tjaden, H. Brinkman, and R. Hensel. 2000. Pyruvate kinase of the hyperthermophilic crenarchaeote *Thermoproteus tenax*: physiological role and phylogenetic aspects. *J. Bacteriol.* 182:2001–2009.

Schreiber, S. L. 2005. Small molecules: the missing link in the central dogma. *Nat. Chem. Biol.* 1:64–66.

Schurig, H., N. Beaucamp, R. Ostendorp, R. Jaenicke, E. Adler, and J. R. Knowles. 1995. Phosphoglycerate kinase and triosephosphate isomerase from the hyperthermophilic bacterium *Thermotoga maritima* from a covalent bifunctional enzyme complex. *EMBO J.* 14:442–451.

Selig, M., K. B. Xavier, H. Santos, and P. Schönheit. 1997. Comparative analysis of Embden–Meyerhof and Entner–Doudoroff glycolytic pathways in hyperthermophilic archaea and the bacterium *Thermotoga*. *Arch. Microbiol.* 167:217–232.

Serre, V., H. Guy, X. Liu, B. Penverne, G. Hervé, and D. Evans. 1998. Allosteric regulation and substrate channeling in multifunctional pyrimidine biosynthetic complexes: analysis of isolated domains and yeast-mammalian chimeric proteins. *J. Mol. Biol.* 281:363–377.

Shearer, G., J. C. Lee, J. Koo, and D. H. Kohl. 2005. Quantitative estimation of channeling from early glycolytic intermediates to CO_2 in intact *Escherichia coli*. *FEBS J.* 272:3260–3269.

Siebers, B., B. Tjaden, K. Michalke, C. Dörr, H. Ahmed, M. Zaparty, P. Gordon, C. W. Sensen, A. Zibat, H.-P. Klenk, S. C. Schuster, and R. Hensel. 2004. Reconstruction of the central carbohydrate metabolism of *Thermoproteus tenax* by use of genomic and biochemical data. *J. Bacteriol.* 186:2179–2194.

Siebers, B., H.-P. Klenk, and R. Hensel. 1998. PPi-dependent phophofructokinase from *Thermoproteus tenax*, an archaeal descendant of an ancient line in phosphofructokinase evolution. *J. Bacteriol.* 180:2137–2143.

Smith, E. T., J. M. Blamey, and M. W. Adams. 1994. Pyruvate ferredoxin oxidoreductases of the hyperthermophilic archaeon, *Pyrococcus furiosus*, and the hyperthermophilic bacterium, *Thermotoga maritima*, have different catalytic mechanisms. *Biochemistry* 33:1008–1016.

Smith, G. K., W. T. Mueller, G. F. Wasserman, W. D. Taylor, and S. J. Benkovic. 1980. Characterization of the enzyme complex involving the folate-requiring enzymes of *de novo* purine biosynthesis. *Biochemistry* 19:4313–4321.

Souciet, J. L., M. Nagy, M. Le Gouar, F. Lacroute, and S. Potier. 1989. Organization of the yeast URA2 gene: identification of a defective dihydroorotase-like domain in the multifunctional carbamoylphosphate synthetase–aspartate transcarbamylase complex. *Gene* 79:59–70.

Spivey, H. O., and J. Ovádi. 1998. Substrate channeling. *Methods* 19:306–321.

Srere, P. A. 1987. Complexes of sequential metabolic enzymes. *Annu. Rev. Biochem.* 56:89–124.

Srere, P. A., and K. Mosbach. 1974. Metabolic compartmentation: symbiotic, organellar, multienzymic, and microenvironmental. *Annu. Rev. Microbiol.* 28:61–83.

Sterner, R., A. Dahm, B. Darimont, A. Ivens, W. Liebl, and K. Kirschner. 1995. (βα)8-barrel proteins of tryptophan biosynthesis in the hyperthermophile *Thermotoga maritima*. *EMBO J.* 14:4395–4402.

Sterner, R., G. R. Kleemann, H. Szadkowski, A. Lustig, M. Hennig, and K. Kirschner. 1996. Phosphoribosyl anthranilate isomerase from *Thermotoga maritima* is an extremely stable and active homodimer. *Protein Sci.* 5:2000–2008.

Tutino, M. L., G. Scarano, G. Marino, G. Sannia, and M. V. Cubellis. 1993. Tryptophan biosynthesis genes trpEGC in the thermoacidophilic archaebacterium *Sulfolobus solfataricus*. *J. Bacteriol.* **175:**299–302.

Uriarte, M., A. Marina, S. Ramón-Maiques, I. Fita, and V. Rubio. 1999. The carbamoyl-phosphate synthetase of *Pyrococcus furiosus* is enzymologically and structurally a carbamate kinase. *J. Biol. Chem.* **274:**16295–16303.

Van Boxstael, S., R. Cunin, S. Khan, and D. Maes. 2003. Aspartate transcarbamylase from the hyperthermophilic archaeon *Pyrococcus abyssi*: thermostability and 1.8 Å resolution crystal structure of the catalytic subunit complexed with the bisubstrate analogue N-phosphonacetyl-L-aspartate. *J. Mol. Biol.* **326:**203–216.

Van de Casteele, M., L. Desmarez, C. Legrain, P. G. Chen, K. Van Lierde, A. Piérard, and N. Glansdorff. 1994. Genes encoding thermophilic aspartate carbamoyltransferases of *Thermus aquaticus* ZO5 and *Thermotoga maritima* MSB8: modes of expression in *E. coli* and properties of their products. *Biocatalysis* **11:**165–179.

Van de Casteele, M., M. Desmarez, C. Legrain, N. Glansdorff, and A. Piérard. 1990. Pathways of arginine biosynthesis in extreme thermophilic archaeo- and eubacteria. *J. Gen. Microbiol.* **136:**1177–1183.

Van de Casteele, M., C. Legrain, L. Desmarez, P. G. Chen, A. Piérard, and N. Glansdorff. 1997. Molecular physiology of carbamoylation under extreme conditions: what can we learn from extreme thermophilic microorganisms? *Comp. Biochem. Physiol.* **118A:**463–473.

Vélot, C., M. B. Mixon, M. Teige, and P. A. Srere. 1997. Model of a quinary structure between Krebs TCA cycle enzymes: a model for the metabolon. *Biochemistry* **36:**14271–14276.

Verhees, C. H., S. W. M. Kengen, J. E. Tuininga, G. J. Schut, M. W. W. Adams, W. M. de Vos, and J. van der Oost. 2003. The unique features of glycolytic pathways in *Archaea*. *Biochem. J.* **375:**231–246.

Villerct, V., B. Clantin, C. Tricot, C. Legrain, M. Roovers, V. Stalon, N. Glamdorff, and J. Van Beeumen, 1998. The crystal structure of Pyrococcus furiosus ornithine carbamoyltransferase reveals a key role for the oligomerization in enzyme stability at high temperatures. *Proc. Natl. Acad. Sci. USA* **95,** 2801–2806.

Walden, H., G. L. Taylor, E. Lorentzen, E. Pohl, H. Lilie, A. Schramm, T. Knura, K. Stubbe, B. Tjaden, and R. Hensel. 2004a. Structure and function of a regulated archaeal triosephosphate isomerase adapted to high temperature. *J. Mol. Biol.* **342:**861–875.

Walden, H., G. Taylor, H. Lilie, T. Knura, and R. Hensel. 2004b. Triosephosphate isomerase of the hyperthermophile *Thermoproteus tenax*: thermostability is not everything. *Biochem. Soc. Trans.* **32:**305.

Walsh, K. A., R. M. Daniel, and H. W. Morgan. 1983. A soluble NADH dehydrogenase (NADH: ferricyanide oxidoreductase) from *Thermus aquaticus* strain T351. *Biochem. J.* **209:**427–433.

White, R. H. 1993. Structures of the modified folates in the thermophilic archaebacteria *Pyrococcus furiosus*. *Biochemistry* **32:**745–753.

White, R. H. 1997. Purine biosynthesis in the domain *Archaea* without folates or modified folates. *J. Bacteriol.* **179:**3374–3377.

Zalkin, H. 1993. Overview of multienzyme systems in biosynthetic pathways. *Biochem. Soc. Trans.* **21:**203–207.

Zalkin, H., and P. Nygaard. 1996. Biosynthesis of purine nucleotides, p. 561–579. *In* F. C. Neidhardt, R. Curtiss III, J. L. Ingraham, E. C. C. Lin, K. B. Low, B. Magasanik, W. S. Reznikoff, M. Riley, M. Schaechter, and H. E. Umbarger (ed.), *Escherichia coli and Salmonella: Cellular and Molecular Biology*, 2nd ed. ASM Press, Washington, DC.

Chapter 6

Temperature-Dependent Molecular Adaptation Features in Proteins

SANDEEP KUMAR, SUNIL ARYA, AND RUTH NUSSINOV

INTRODUCTION

Life on earth exists in several ecological niches under conditions which span extremes of temperature, pressure, salinity, and pH. In particular, microbial life is found in the most diverse conditions (see, e.g., Adams and Kelly, 1995). Temperature is a fundamental environmental factor with large variations at different geographical locations. Among the so-called "extremophiles," there are both the heat lovers (thermophiles/hyperthermophiles; archaea and bacteria found under conditions of high temperatures, such as deep-sea thermal vents) and cold lovers (psychrophiles such as microbes, protoctists, algae, and vertebrates like the Antarctic ice fish, *Chaenocephalus aceratus*). Many of these organisms also tolerate other extremes such as pH, salinity, radiation, pressure, and oxygen tension.

The high-temperature limit for the active microbial communities is well above 100°C. For example, *Pyrolobus fumarii* is one of the most thermophilic microorganism (Blochl et al., 1997). It grows in the walls of "black smoker" hydrothermal vent chimneys up to a temperature of 113°C. While thermophilic microorganisms can be found in intermittent heat conditions, for hyperthermophilic ones, the living habitats are hot springs and terrestrial as well as hydrothermal vents (Sterner and Liebl, 2001).

Cold environments on Earth are, in fact, much more common than the hot ones. These include cold, freezing conditions and ice. Psychrophiles grow optimally in such conditions. True psychrophiles have optimum growth temperatures of 15°C or less, although biological activity could be observed in the brine veins of sea-ice even at −20°C (Deming, 2002).

How these organisms live and thrive at extreme temperatures can teach us useful lessons, valuable not just for understanding physicochemical principles of life but also for several industrial applications. The purpose of this chapter is to compare and review the molecular adaptations shown by thermophilic and psychrophilic proteins. From a mesophilic point of view, the physicochemical challenges faced by the thermophiles and psychrophiles are not exactly opposite of each other. In order to function properly, thermophilic proteins have learnt how to "hold on" to their three-dimensional native conformations at high temperature while countering deleterious effects such as greater disorder, instability of the constituent amino acids, faster reaction rates, and greater mobility of the substrates (Kumar et al., 2000a; Sterner and Liebl, 2001; Kumar and Nussinov, 2001a). On the other hand, the challenge for the psychrophilic proteins is to overcome slower mobilities, low solubilities of substrates and apolar amino acids, and slower reaction rates and to ensure efficient catalysis in order to maintain metabolic flux (Russell, 2000; Feller and Gerday, 2003; Cavicchioli et al., 2002; Cavicchioli and Siddiqui, 2004).

At both high and low temperatures, the proteins need to be active to maintain the cellular machinery in functional state. The proteins appear to achieve this by modulating their conformational stability/flexibility. The overall fold appears to remain conserved among the homologous thermophilic, mesophilic, and psychrophilic proteins, and only rather minor adjustments are required for adaptation of the protein to high and low temperatures. Modulation of protein electrostatics appears to be one such adjustment. The thermophilic proteins optimize their electrostatic effects to preserve the integrity of active site regions

---

S. Kumar • Department of Biology, Zanvyl Krieger School of Arts and Sciences, Johns Hopkins University, Baltimore, MD 21218-2685. S. Arya • Department of Biological Sciences and Bio-Engineering, Indian Institute of Technology-Kanpur, Kanpur 208016, Uttar Pradesh, India. R. Nussinov • SAIC-Intramural Research Program, Center for Cancer Research Nanobiology Program, National Cancer Institute-Frederick, Frederick, MD 21702-1201. Department of Human Genetics, Sackler School of Medicine, Tel Aviv University, Tel Aviv 69978, Israel.

and oligomeric interfaces and to counter the deleterious effects of elevated temperatures. In contrast, the psychrophilic proteins use electrostatic effects to ensure proper protein solvation and flexibility, especially in the active site regions, via destabilization of the charged residues. The dual roles played by protein electrostatics could explain some of these adaptations in citrate synthase and a cold-adapted subtilase (Kumar and Nussinov, 2004a; Arnorsdottir et al., 2005). However, the modulation of the electrostatic effects, although most frequent, is not a universal route for temperature adaptation.

Owing to the large body of literature available, especially on thermophilic proteins, we found it impossible to present an exhaustive review. Instead, here we focus on the recent data, taking 2000 as the base year for most of the publications cited here. Moreover, we try to highlight those studies that reveal interesting observations for both the heat and cold adaptation of the proteins. We paid attention to analyses involving several proteins or genomes. Much of the review is focused on the role of protein electrostatics in the molecular adaptations shown by both the thermophilic and the psychrophilic proteins. Furthermore, we restrict ourselves to proteins that possess thermodynamically stable native three-dimensional structures in the living conditions of their source organisms, and all the adaptations to the extreme temperatures are intrinsic to their sequence and structural features. However, a significant number of biologically important proteins in organisms either are completely unfolded or contain unfolded/disordered regions at the relevant physiological conditions for functional reasons (Wright and Dyson, 1999, 2005; Dunker et al., 2001, 2002). Extremophilic organisms are no exception. For example, cold-shock protein from *Thermotoga maritima* (TmCsp) has thermodynamic stability of only 0.3 kcal/mol at 80°C, the living temperature of the organism (Kumar et al., 2001). Many of these proteins may stabilize upon binding to their partner protein(s), e.g., as in recognition and signal transduction (Gunasekaran et al., 2003). Small-molecule organic solutes called osmolytes could help stabilize the proteins in organisms exposed to environmental stresses (e.g., see Baskakov and Bolen, 1998; Sterner and Liebl, 2001).

MOLECULAR ADAPTATION BY THERMOPHILIC/HYPERTHERMOPHILIC PROTEINS

What sequence and structural adjustments the proteins make to cope with heat? A large number of factors have been proposed to contribute toward protein thermostability. These include greater hydrophobicity; greater occurrence of residues in α-helical conformation; deletion/shortening of surface loops; increased occurrence of amino acids with larger, branched, and charged functional groups; better packing in the protein core; smaller and fewer cavities; larger subunit interfaces and more extensive interaction across subunit interfaces facilitating stabilization via "fortified" oligomerization; higher oligomerization order; increased cation–pi interactions; and increased polar/charged interactions including hydrogen bonds, salt bridges, and their networks across the subunit interfaces and/or around the protein active sites. These factors were gleaned from the analyses based on a pair of homologous thermophilic and mesophilic protein structures or, at best, using the surveys of the available data on a modest number of protein families containing thermophilic and mesophilic proteins (Kumar and Nussinov, 2001a). It is encouraging that recent studies based on the analyses of large datasets taken from thermophilic organism genomes broadly confirm these trends (Bastolla et al., 2004; Friedman et al., 2004; Robinson-Rechavi and Godzik, 2005; Robinson-Rechavi et al., 2006).

Thermophilic proteins have specific amino acid composition requirements (Nakashima et al., 2003). In general, thermophilic proteins favor charged residues (Glu, Arg, and Lys) capable of providing increased formation of the ion pairs and their networks. Thermophilic proteins also have greater proline and reduced glycine content. Using the genomic data, Liang et al. (2005) found a number of amino acid coupling patterns which distinguish the thermophilic proteins from their mesophilic homologs. Using genomic data on eighteen mesophiles, four thermophiles, and six hyperthermophiles, De Farias and Bonato (2002) found that an increase in percent occurrence of Glu+Lys in hyperthermophilic proteins was consistently related to the decrease in Gln+His content.

It seems plausible that thermostable proteins are more rigid as compared with their mesophilic relatives, especially at room temperature. Increased rigidity correlates with the increased thermostability and is a conserved feature relevant to catalytic function (Fitzpatrick et al., 2001). However, Jaenicke (2000) suggests that the apparent rigidity of the thermophilic proteins at room temperature should not be overemphasized because these proteins may be optimally flexible at the living temperatures of their organisms. From the analysis of thermodynamic data, Kumar et al. (2001) also observed that thermophilic and mesophilic proteins had similar stabilities at the respective living temperatures of their source organisms. Butterwick et al. (2004) used nuclear magnetic resonance spin relaxation experiments to compare backbone dynamics of homologous ribonuclease HI from a thermophile *Thermus thermophilus* and a mesophile *Escherichia coli* at 310 K (37°C). The pico-second to nano-second timescale dynamics were similar for the two proteins. However, micro-second to milli-second timescale dynamics, measured by chemical exchange line broadenings, indicated increased activation barriers for residues associated with the function in *Thermus thermophilus* ribonuclease HI. Using a sequence-based technique to estimate local structural entropy in proteins, Chan et al. (2004) observed a direct linear relationship between the average structural entropy and protein thermostability.

Thermophilic proteins may have more apolar cores than their mesophilic cousins. A small surface/volume ratio and ionic bonds present on the surface confer compactness (Baker, 2004). There is a strong hydrophobic interaction at the dimer interface of maltosyltranferase, from *T. maritima*, which could be crucial to its stability (Baker, 2004). Earlier surveys of homologous thermophilic and mesophilic protein families did not find a consistent association between protein core packing/compactness and thermostability (Kumar et al., 2000a; Karshikoff and Ladenstein, 1998). However, recently, Pack and Yoo (2005) have observed that the thermophilic proteins contain greater frequencies of well-packed residues, especially in the protein core. A majority of proteins from *Thermotoga maritima* genome have higher contact order than their mesophilic homologs (Robinson-Rechavi and Godzik, 2005). Recently, Loladze and Makhatadze (2005) found hydrophobicity and α-helical propensity as important parameters at partially buried positions in ubiquitin to enhance its thermostability.

The α-helices in the thermophilic membrane proteins may be better packed (Schneider et al., 2002). Using non-redundant protein structural data from the Protein Data Bank (Berman et al., 2000), Kleiger et al. (2002) found helix interaction motifs AXXXA to be particularly abundant in the thermophilic protein structures.

Surface loops in the thermophilic proteins may be undesirable due to the increased mobility at high temperatures. Reduction in the protein flexibility via proline amino acid residue substitutions in the loop regions and/or deletion (shortening) of surface loops is considered to be potential routes to thermostability (Sterner and Liebl, 2001). Reduction in loop content for thermophilic proteins has been also indicated from the analyses of bacterial genomes (Chakravarty and Varadarajan, 2002; Thompson and Eisenberg, 1999). Another way for reducing protein flexibility and increasing protein stability is the incorporation of disulfide bonds. But, the disulfide bond containing proteins are often found in extracellular spaces and rarely in cell cytoplasm in vivo. Interestingly, however, in case of hyperthermophilic archaea *Pyrobaculum aerophilum* and *Aeropyrum pernix*, the analysis of genomic data indicates that their cytoplasmic proteins may be rich disulfide bonds. This observation points to different intracellular chemical compositions for these organisms (Mallick et al., 2002).

Protein oligomerization is another factor that has been often thought to help stabilize proteins at higher temperatures. Thermophilic proteins may have higher oligomerization order. Glansdorff (1999) argued that functionally related genes may have been forged into operons in ancient thermophilic archaeal and bacterial cells to facilitate the formation of multienzyme complexes to mutually stabilize thermolabile proteins and channel the substrates and metabolites which are also often thermolabile. This could have been a molecular strategy of evolution of the thermophilic life on the earth (Glansdorf, 1999). However, while operons and gene clusters are certainly found in modern archaeal and bacterial cells, the evidence for the implied higher oligomeric states for the thermophilic proteins appears inconsistent at present (Kumar et al., 2000a, 2000b; Dalhus et al., 2002; Ogasahara et al., 2003; Taka et al., 2005). For example, the enzyme phosphoribosyl anthranilate isomerase (PRAI) occurs as homodimer in *Thermotoga maritima* and *Thermus thermophilus* HB8 but occurs as a C-terminal domain in a monomeric bifunctional enzyme in *E. coli*. Dimer formation is thought to be responsible for the greater stability of TmPRAI, and the subunits of TmPRAI associate via hydrophobic interactions (Taka et al., 2005). Two interface loops, one in each monomer, consisting of hydrophobic residues are thought to be mainly responsible for the dimer formation by TmPRAI. The mesophilic EcPRAI domain does not contain the corresponding hydrophobic loop. However, Taka et al. (2005) also point that the hydrophobic loop sequence (50-LPPFV-53) is also found in other mesophilic PRAI sequences, thereby suggesting possible homodimer formation by those PRAIs. Another interesting example is that of the malate dehydrogenases (MDHs) from thermophilic and mesophilic phototropic bacteria. These MDHs are tetramers (dimers of dimers) rather than the usual homodimers as most MDHs are (Dalhus et al., 2002; Eijsink et al., 2004). Recently, Robinson-Rechavi et al. (2006) have studied all available crystal structures for the proteins from hyperthermophile *Thermotoga maritima* and their close homologs. This large study shows that oligomerization order is not a consistent factor in protein thermostability. In summary, a definitive answer to the question whether thermophilic proteins have higher oligomerization order needs further research.

Protein oligomerization could still contribute toward thermophilicity, if the subunit interfaces in the thermophilic proteins are strengthened as compared with their mesophilic homologs. This could be achieved by burial of greater non-polar surface area and formation of additional electrostatic interactions across the subunit interfaces. This fortification of the subunit interfaces, independent of oligomerization order, appears to be more consistent (Eijsink et al., 2004). It has been seen in a number of cases, including PRAI (Taka et al., 2005), tryptophan synthase (Ogasahara et al., 2003), MDH (Bjork et al., 2003), glutamate dehydrogenase (Lebbink et al., 2002; Kumar et al., 2000b and references therein), citrate synthase (Kumar and Nussinov, 2004a; Russell et al., 1998 and references therein), nitrogenase Fe (Sens and Peters, 2006), and the proteins in *Thermotoga maritima* (Robinson-Rechavi et al., 2006).

Among all the factors proposed to contribute toward protein thermostability, increase in charge–charge/polar interactions is the most consistent (Kumar et al., 2000a), although it does not account for the increased stability of the thermophilic proteins in all the cases. Cation–pi and weakly polar interactions can also contribute toward thermal stability of the proteins (Chakravarty and Varadarajan, 2002; Gromiha et al., 2002; Ibrahim and Pattabhi, 2004). Electrostatic interactions involving salt bridges and their networks have been particularly well studied. The thermophilic proteins often contain a greater number of salt bridges and their networks as compared with their mesophilic homologs. These additional interactions are most needed in the region(s) that may be important for structural and functional integrity of the protein, i.e., near the active sites and across the subunit interfaces. These interactions target protein structures at the important places by providing kinetic barriers to local unfolding and help resist conformational disintegration at the elevated temperatures (Kumar and Nussinov, 2001a; Kumar et al., 2000b; Nordberg et al., 2003). Using the structural bioinformatics analyses containing 127 thermophile–mesophile orthologs, Alsop et al. (2003) found a preferred increased in surface salt bridges (ion pairs) for the thermophilic proteins. Elimination of like charge repulsions and creation of opposite charge pairs on the protein surface can result in electrostatic optimization of the protein surface and thus confer thermostability on the mesophilic proteins (Torrez et al., 2003). Consistently, Fukuchi and Nishikawa (2001) have observed significant differences between the amino acid compositions of the surface residues of the thermophilic and mesophilic proteins. Robinson-Rechavi et al. (2006) also found significant increases in the density of salt bridges for the proteins of *Thermotoga maritima*.

How much do salt bridges contribute toward protein stability? It is difficult to obtain a precise estimate of the net free energy contribution by a salt bridge, both by experimental and by computational means, due to the limitations of the currently available methods (Bosshard et al., 2004). There are two experimental methods for estimating the contribution of a salt bridge toward protein stability, namely, pK_a change method and double-mutant cycle approach. The pK_a change method relies on the shifts in pK_a values of the ionizable groups upon protein folding due to the change in the electrostatic environments of the groups. The salt bridge in a protein is broken by the change in pH which results in protonation of acidic group or deprotonation of the basic group. This resulting change in the protein stability is then calculated using standard equations (see Bosshard et al., 2004; Marti and Bosshard, 2003 for details and example). In the double-mutant cycle approach, the charged residues forming the salt bridging pair in a protein are mutated to Ala both singly and simultaneously to yield two single mutants and a double mutant. The coupling energy between the salt bridging residues is then measured as the part of free energy change upon double mutation that cannot be accounted for by the sum of the free energy changes upon the single mutations (Bosshard et al., 2004; Makhatadze et al., 2003 for details). Both these experimental methods have several limitations (Bosshard et al., 2004) and cannot be applied to all salt bridges, especially those buried in the protein core.

Experimental methods cannot provide the net electrostatic free energy contribution of a salt bridge toward protein stability. This is because there is no way to switch "off" the atomic partial charges on the salt bridging residues in an experiment. However, this can be done computationally. In the Hendsch and Tidor (1994) method, the electrostatic free energy contribution by a salt bridge is estimated with respect to the hydrophobic isosteres of the charged residues. The hydrophobic isostere of a charged residue is the charged residue with all its atomic partial charges set to zero. The total electrostatic free energy of the salt bridge ($\Delta\Delta G_{tot}$) is given by the following equation:

$$\Delta\Delta G_{tot} = \Delta\Delta G_{dslv} + \Delta\Delta G_{brd} + \Delta\Delta G_{prt}$$

The desolvation energy ($\Delta\Delta G_{dslv}$) is usually an energy penalty paid by each charged residue in the folded state of the protein due to the loss of its hydration. The magnitude of this energy penalty depends upon the location of the charged residues in protein. The bridge energy ($\Delta\Delta G_{brd}$) is due to the favorable electrostatic interaction between the oppositely charged residues. The magnitude of this term varies with the geometrical orientation of the charged groups. The protein interaction energy ($\Delta\Delta G_{prt}$) measures the electrostatic interaction of the salt bridging residues with the charges in the rest of the protein. This term can be both favorable and unfavorable depending upon the context of the salt bridge in the protein. Depending upon the relative magnitudes of the three terms, the overall electrostatic free energy contribution ($\Delta\Delta G_{tot}$) of a salt bridge can be destabilizing, insignificant, or stabilizing (Hendsch and Tidor, 1994; Kumar and Nussinov, 1999). The computation of these terms requires electrostatic calculations. The calculations based on the solution of the Poisson–Boltzmann equation using DELPHI software are quite popular. Owing to the limits on available computational power, the calculations necessarily incorporate several simplifying assumptions and use of empirical parameter sets for atomic charges and radii. These parameters depend on the temperature and should be scaled appropriately for calculations involving thermophilic proteins (Elcock, 1998). The choice of empirical parameter sets and the accuracy of the protein structures used in the calculations affect the result of the calculations (Bosshard et al., 2004; Simonson et al., 2004). Simplifying assumptions such as use of a single dielectric constant for protein interior, neglect of internal water molecules in proteins, and treatment of solvent implicitly rather than in atomic details are among the limitations of the currently available electrostatic energy calculation protocols. The development of improved calculation protocols which could adequately account for all the complexities of non-homogenous distribution of the atomic charges and internal water molecules found in the protein structures remains an area of active research (e.g., Simonson et al., 2004).

Owing to the reasons cited above, the question whether salt bridges and their networks stabilize proteins has been debated in protein science literature (Hendsch and Tidor, 1994; Kumar and Nussinov, 1999; Dong and Zhou, 2002). However, the following is generally accepted: The contribution of a salt bridge formed by a pair of oppositely charged residues in a protein often varies greatly between being stabilizing and destabilizing depending upon a number of factors which include location of the charged residues in the protein, geometrical orientation of the side-chain charged groups with respect to one another and with respect to the other charges in the protein, and conformational flexibility of the protein. Moreover, the net contribution of the salt bridge is conformer population dependent and varies in different crystal forms and nuclear magnetic resonance structures of the protein as well as in different solution conditions (Kumar and Nussinov, 2001b).

Notwithstanding the above debate, electrostatic interactions, especially salt bridges and their networks, do appear to play important roles toward stabilizing the thermophilic proteins at elevated temperatures. Evidence in favor of this view has been accumulating in recent years. Much of this has been already reviewed (Kumar and Nussinov, 2001a; Sterner and Liebl, 2001; Karshikoff and Ladenstein, 2001). Below we present more recent results.

Elcock and coworkers have also found that the salt bridges are stabilizing toward the hyperthermophilic proteins (Elcock, 1998; Thomas and Elcock, 2004). Recently, they have performed explicit solvent molecular dynamics simulations to study how salt bridge interactions between two freely diffusing amino acids, lysine and glutamic acid, are affected by increase in temperature. The results show that salt bridge interactions are robust to the increase in temperature and are well suited for life under hot conditions (Thomas and Elcock, 2004). Using an off-lattice, three-color, 46 beads model toy protein, Wales and Dewsbury (2004) have simulated the effect of introducing a salt bridge on the energy landscape of the protein. Their results indicated that salt bridge formation could accelerate folding by making the global minima more accessible. Coupled with this, molecular dynamics simulations performed on C-terminal truncated staphylococcal nuclease (Snase) indicate that a salt bridge formed between Arg 105 and Glu 135 hinders protein unfolding because breaking the salt bridge presents a significant barrier to the transition of native state of Snase to the unfolded state (Gruia et al., 2003). Zhang et al. (2004) performed parallel molecular dynamics simulations on thermostable catechol 2,3-dioxygenase (TC23O). The salt bridge formed by residues Lys 188 and Glu 291 was found to contribute toward the stability of the folded protein. The dynamic transition temperature of a mutant lacking this salt bridge was reduced by 19°C. The salt bridge networks in the thermophilic proteins are found to be mostly stabilizing (Kumar et al., 2000b; Kumar and Nussinov, 2004a) and have been also experimentally shown to exhibit strong cooperativity in case of *Thermotoga maritima* glutamate dehydrogenase (Lebbink et al., 2002).

The increased and better optimized electrostatics for the thermophilic proteins has often been documented. Recently, Mozo-Villarias et al. (2003) have even developed a method based on the charge distribution in proteins, called dipole profile, to predict their thermal stability. Zhou (2002) has shown that the formation of a salt bridge or hydrogen bonding network around an ionized group in the folded protein results in reduced ΔC_p for thermophilic proteins. Recently, this notion has received experimental support from thermodynamic, mutational, and crystallographic studies of Lee et al. (2005) on homologous ribosomal proteins L30e from *Thermococcus celer* and yeast. The authors showed that mutation of Lys 9, Glu 90, and R92 to alanine destabilize *T. celer* L30e by disrupting favorable electrostatic interactions. These charges to neutral mutants also showed greater ΔC_p values than the wild-type protein. Another way to decrease ΔC_p in thermophilic proteins is to have smaller polypeptide chain lengths via deletion/reduction of surface loops. Recently, thermodynamic simulations performed for single domain proteins indicated that reduction in ΔC_p may be the primary route for thermostability, especially in case of hyperthermophilic proteins (Kumar et al., 2003; Kumar and Nussinov, 2004b).

Formation of hydrogen bonds, salt bridges, ion pairs, and their networks does not completely account for all the charged residues found in proteins. Nor do the charged residues restrict themselves to protein surface only. In fact, buried charges are commonly found in proteins, increasing often with the protein size (Kajander et al., 2000). Some of these could be unpaired. Unpaired buried charges could potentially destabilize the proteins due to the desolvation energy penalty that may remain uncompensated. Consistently, the thermophilic proteins have less buried charge (Kajander et al., 2000). Nevertheless, such charges do occur in the thermophilic proteins. Protein molecules often internalize water molecules. Such internal water molecules in proteins often play important roles in function, especially catalysis. At the elevated temperatures, it would be critical for the proteins to retain these internal water molecules to maintain proper function. The buried charged residues, especially those unpaired or un-networked, can coordinate these internal water molecules and restrain their movements and escape from the protein interior. In the process, these charged residues could provide additional kinetic barriers to protein denaturation at the high temperatures.

In the context of mesophilic proteins, removal of buried charges and repulsions rather than incorporation of specific additional salt bridges and networks can also lead to improved protein stability and adaptation to higher temperatures. One such example is of the cold-shock proteins. The cold-shock protein has emerged as an interesting model system for understanding the origin of protein thermostability due to the high sequence and structural similarities between the thermophilic and mesophilic homologs, particularly BcCSP (*Bacillus caldolyticus* cold-shock protein) and BsCSP (*Bacillus subtilis* cold-shock protein B) (Color Plate 1). Both the mesophilic and thermophilic cold-shock proteins have a native-like transition state (Perl et al., 1998), but they differ significantly in stability (Mueller et al., 2000; Perl et al., 2000). Furthermore, the differences appear to be due to only a dozen positions. These differences consist of eleven mutations (Leu2Gln, Glu3Arg, Ser11Asn, Phe15Tyr, Gln23Gly, Asp24Ser, Ser31Thr, Ala46Glu, Glu53Gln, Thr64Val, and Glu66Leu from BsCsp to BcCsp) and a deletion of the C-terminal residue (A67 in BsCSP is absent in the BcCSP sequence). Out of these, only two surface-exposed residue mutations, Glu3Arg and Glu66Leu, mostly contribute toward the greater thermostability of BcCSP (Perl et al., 2000). Recently, Dominy et al. (2004) examined the thermal stability differences between homologs of cold-shock proteins and reported that electrostatic interactions play an important role in determining the stability of these proteins at high temperatures. Removal of unfavorable electrostatic repulsions, rather than the formation of additional salt bridges, leads to greater thermostability in this case. That is, BcCSP appears to use destabilization of the unfolded state via charge–charge repulsions as a route to enhanced stability (Zhou and Dong, 2003). Garofoli et al. (2004) have performed molecular dynamics simulations of thermophilic BcCSP and mesophilic BsCSP and their mutants. Their results support the above hypothesis.

MOLECULAR ADAPTATION BY PSYCHROPHILIC PROTEINS

If one were to plot the optimal living temperatures of the source organisms along a thermometer, the psychrophilic and thermophilic proteins would lie on its opposite ends. However, the challenges faced by the psychrophilic proteins may not necessarily be the opposite of those faced by the thermophilic proteins. For example, temperature may not pose a direct threat to structural integrity of the psychrophilic proteins. This is because the living temperatures of the psychrophilic proteins, though low, are higher than the expected cold denaturation temperatures for the single domain proteins in water in the absence of denaturating agents (Kumar and Nussinov, 2003). However, the possibility of cold denaturation, in the case of multidomain proteins, cannot be fully ruled out (see e.g., Figure 2 in D'Amico et al., 2003a). Still rather than preserving the conformational integrity, the real challenge for the psychrophilic proteins is to maintain catalytic efficiency, activity, and, thus,

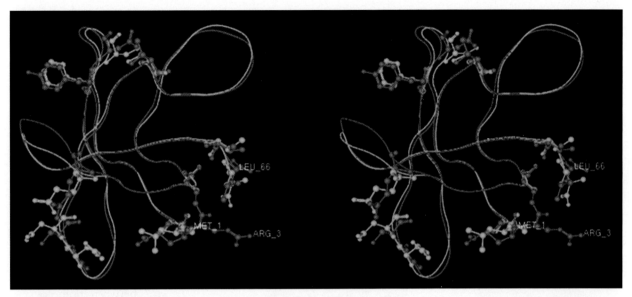

Color Plate 1 (Chapter 6). Stereo diagram depicting superimposition of highly homologous cold-shock proteins from *Bacillus caldolyticus* (red, PDB entry: 1C9O chain A, BcCSP) and *Bacillus subtilis* (green, PDB entry: 1CSP, BsCSP). The backbones of the two proteins are shown as thin lines and the nonconserved residues between the two protein structures are shown as balls and sticks (see text for details). For the sake of clarity, the conserved residues between the two proteins are not shown. The N- and C-termini are indicated by the first (Met 1) and the last residues (Leu 66) in BcCSP. Arg 3, which along with Leu 66 mostly confers thermostability on BcCSP (Perl et al., 2000), is also shown. Cold-shock proteins have emerged as a model system to understand molecular factors involved in protein thermostability. Modulation of protein electrostatics appears to be responsible for the greater stability of BcCSP. However, this is not achieved by incorporation of additional salt bridges or ion pairs in BcCSP. In this sense, the case of the cold-shock proteins appears exceptional.

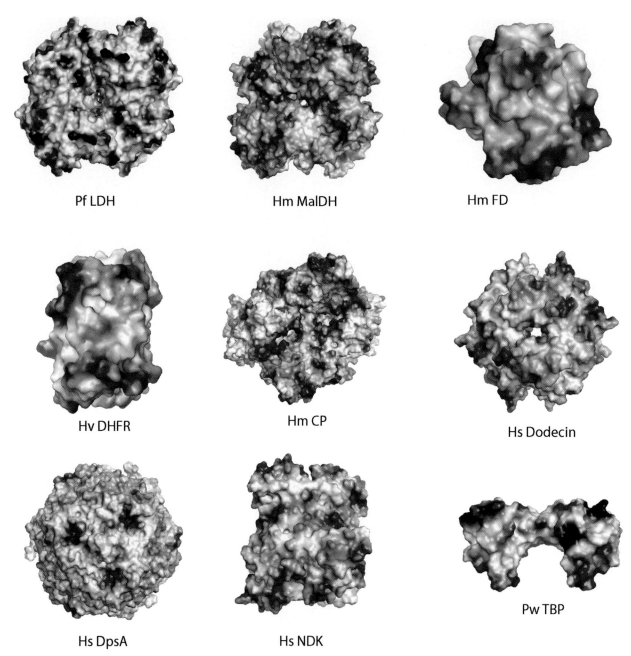

Color Plate 2 (Chapter 19). Crystallographic structures of halophilic proteins. PYMOL was used to represent the electrostatic potential at the surface of the proteins. The names of halophilic proteins are defined in Table 1. *Pf* LDH, the lactate dehydrogenase from the non-halophilic *Plasmodium falciparum*, is shown for comparison. *Pw* TBP is the TATA-binding box protein from the hyperthermophile *Pyrococcus woesei*.

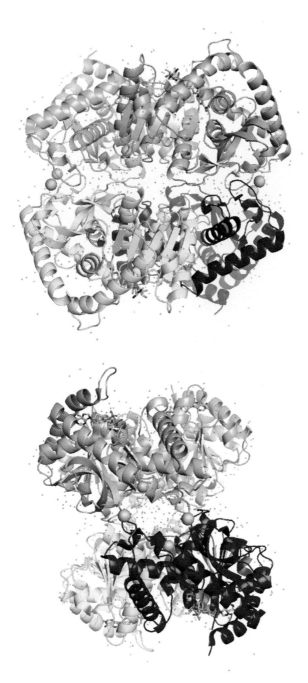

Color Plate 3 (Chapter 19). Crystallographic structures of R207S, R292S mutant of *Hm* MalDH. Two orthogonal views are presented, with the four polypeptide chains of the tetramer shown in ribbons of different colours; NADH is shown in a stick representation, chloride ions as grey balls, and water as small red dots.

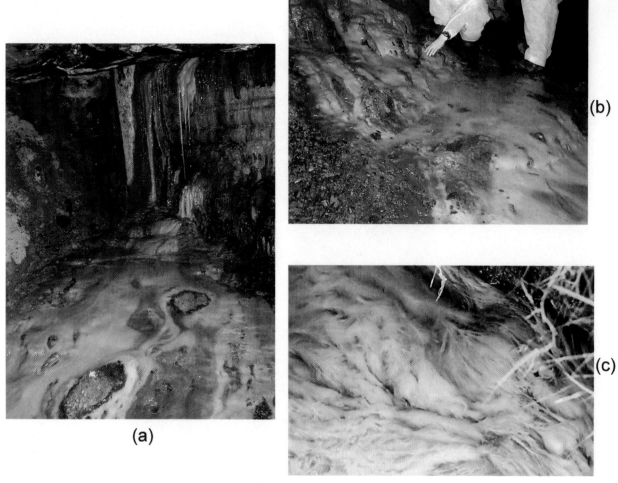

Color Plate 4 (Chapter 20). Macroscopic growths of acidophilic microbial communities at two mine sites in north Wales: (a) stalactite-like ("pipe") growths and (b) surface slimes within a disused pyrite mine and (c) streamer growths in acid waters draining an abandoned copper mine.

metabolic flux (Feller, 2003; Feller and Gerday, 2003; Claverie et al., 2003; Marx et al., 2004; Georlette et al., 2004). Psychrophilic proteins must overcome the expected lower solubilities of themselves as well as those of the substrates in water coupled with low reaction rates at cold temperatures.

As compared with the thermophilic and hyperthermophilic proteins, the amount of sequence and structural data available on the psychrophilic proteins is still small. Hence, fewer comparative studies on psychrophilic–mesophilic/psychrophilic–thermophilic protein pairs or psychrophilic–mesophilic–thermophilic protein families have been performed so far. Such studies have indicated a number of initial trends. For example, psychrophilic proteins often have fewer proline residues and disulfide bonds as compared with their thermophilic and mesophilic homologs. They also tend to contain glycine clusters, frequently, around the enzyme active sites (Russell, 2000; Feller and Gerday, 2003). Gianese et al. (2001) had analyzed of 21 psychrophilic enzymes along with their 427 mesophilic and thermophilic homologs. They found that Arg and Glu at exposed sites on α-helices may be replaced by Lys and Ala in psychrophilic proteins. However, sequence comparisons involving whole genomes for thermophilic, mesophilic, and psychrophilic organisms failed to identify any consistent trend (Nakashima et al., 2003). But Lys→Arg substitutions in alpha-amylase from *Pseudoalteromonas haloplanktis* facilitate its cold adaptation (Siddiqui et al., 2006). A comparison of the psychrophilic protein sequences available in SWISS-PROT database with their thermophilic and hyperthermophilic homologs showed a greater content for hydroxyl group containing residues Ser and Thr for the psychrophiles (Arya, 2005). A greater occurrence of these residues could potentially help in solvation of the psychrophilic proteins.

Psychrophilic proteins have high specific activities, yet their thermal stabilities are relatively low. Psychrophilic proteins are often more flexible, particularly, in the regions near the active sites (Russell et al., 1998; Kim et al., 1999; Russell, 2000; Feller, 2003; Marx et al., 2004). For example, Mavromatis et al. (2002) observed an increased number of glycine residues localized around the active site of alkaline phosphatase from Antarctic strain TAB5. This active site glycine cluster is thought to provide local mobility. Brandsdal et al. (1999) performed comparative molecular dynamic simulations on bovine and salmon trypsins. They did not find major differences in the global flexibilities of the two proteins. However, the active site of the cold-adapted salmon trypsin was found to be more flexible and relatively easily deformable. This may lower the activation barrier for ligand binding and catalysis. Gorfe et al. (2000) found the binding site electrostatic properties of the cold-adapted and mesophilic trypsins to be different especially in terms of the magnitude of the negative potential at the S1 specificity pocket. By introducing multiple mutations in psychrophilic alpha-amylase to mimic its mesophilic homolog the porcine alpha-amylase, D'Amico et al. (2003a) have revealed the central role played by weak interactions in modulating both the kinetic and thermodynamic parameters of enzyme. Using computer modeling, Tindbaek et al. (2004) "grafted" the highly flexible region in the binding site of subtilisin from Antarctic *Bacillus* TA39 into the mesophilic *Bacillus lentus* subtilisin, savinase. The hybrid protein showed increased activity at the low temperatures as well as increased local binding site and global flexibility.

Mesophilic proteins also often contain highly flexible, even intrinsically disordered, regions due to functional reasons (Dyson and Wright, 1999, 2005; Dunker et al., 2001, 2002). Such regions are often characterized by repetitiveness (low sequence complexity), weaker hydrophobic effect (lower occurrence of hydrophobic, aromatic, or bulky residues), and a higher incidence of polar and charged residues (see e.g., Romero et al., 2001). The intrinsic disorder of the mesophilic protein may be related to the conformational flexibility of the psychrophilic enzymes, and the amino acid compositional features in the two cases may be similar (Siddiqui et al., 2006) due to the common underlying physical reasons. The intrinsically disordered regions of the mesophilic proteins are often involved in recognition and cell signaling (Uversky et al., 2005). In such cases, one could postulate significant populations of "native-like" polypeptide conformations in solution which get selected in the presence of the binding partner(s) (Gunasekaran et al., 2003). For the psychrophilic enzymes, the conformational flexibility is for the catalysis, for the capture of the substrate molecules, and for the release of the product molecules. This also requires systemic flexibility of the enzyme active site where all the polypeptide conformations are native-like, have similar populations, and are located at the bottom of their energy funnels (see e.g., D'Amico et al., 2003b).

Bae and Phillips (2004) have solved and compared the crystal structures of three highly homologous (>65% sequence identity) adenylate kinases with equal protein length from psychrophile *Bacillus globisporus*, mesophile *Bacillus subtilis*, and thermophile *Bacillus stearothermophilus*. The three enzymes have different temperatures of optimal activity, reflecting the living temperatures of their source organisms. The three-dimensional structures of the three adenylate kinases are highly similar, with the root mean square values for the core and AMP-binding domains being less than 0.7 Å. The three adenylate kinases appear to adapt to the respective temperature conditions via subtle changes in the noncovalent intramolecular interactions. For the heat adaptation, these are electrostatic (increased ion pairs) and hydrophobic (greater burial of apolar surface area). The cold adaptation is essentially via small changes in hydrophobic interactions. The psychrophilic adenylate kinase has the least number of apolar atoms, but it has the greatest exposed apolar surface area, thereby destabilizing the folded state and making it more flexible. The psychrophilic adenylate kinase has the greatest average value for main chain atom B-factors and sparse crystal packing contacts, indicating its greater flexibility. Similar results were also obtained by Kim et al. (1999) who had compared homologous MDH from thermophilic, mesophilic, and psychrophilic organisms. They also noted the differences in the electrostatic potentials at the active site of the three MDHs which may be important for the efficient guidance of substrate into the active site.

That proteins could adapt to the diverse living temperatures of their source organisms by modulating the electrostatic effects came to light from studies on citrate synthase (Kumar and Nussinov, 2004a). In this study, the sequence, structural properties, and detailed electrostatic properties of citrate synthase from hyperthermophilic, mesophilic, and psychrophilic organisms were examined by computational means. First, the thermophilic and psychrophilic citrate synthases were found to be more similar to each other than to their mesophilic homolog, both sequence- and structure-wise. Moreover, both the thermophilic and psychrophilic citrate synthases showed increased occurrence of charged residues, salt bridges, and their networks. However, there were important differences in the location of the charged residues and salt bridges as well as in their networks between the thermophilic and mesophilic citrate synthases. The charged residues, salt bridges, and the salt bridge networks were mostly found in the dimer interface and active site regions of the thermophilic citrate synthase. In contrast, for the psychrophilic citrate synthase, these were distributed all over the protein structure and tended to be better surface exposed, ensuring better solvation of the protein molecule in water at low temperatures. The electrostatic free energy contribution toward protein stability ($\Delta\Delta G_{ele}$) by the charged residues showed greater variations in case of the psychrophilic citrate synthase than for the hyperthermophilic homolog. The citrate synthase active sites contain several charged residues. Those in the active site regions of the psychrophilic citrate synthase were found to be more destabilizing, ensuring greater flexibility. Recently, destabilization of the active site in the psychrophilic protein to ensure catalytic activity at the low temperatures was also suggested by experiments of D'Amico et al. (2003b) on homologous alpha-amylases from psychrophilic, mesophilic, and thermophilic organisms.

Acknowledgments. The authors thank Dr. Gerday for invitation to write this chapter. Sunil Arya thanks Department of Biotechnology, Ministry of Science and Technology for financial support. This project has been funded in whole or in part with federal funds from the National Cancer Institute, National Institutes of Health, under contract N01-CO-12400.

The content of this publication does not necessarily reflect the views or policies of the Department of Health and Human Services, nor does mention of trade names, commercial products, or organizations imply endorsement by the U.S. Government. This Research was supported [in part] by the Intramural Research Program of the NIH, National Cancer Institute, Center for Cancer Research.

REFERENCES

Adams, M. W. W., and R. M. Kelly. 1995. Enzymes from microorganisms in extreme environments. *Chem. Engng. News* **73:** 32–42.

Alsop, E., M. Silver, and D. R. Livesay. 2003. Optimized electrostatic surfaces parallel increased thermostability: a structural bioinformatics analysis. *Protein Eng.* **16:**871–874.

Arnorsdottir, J., M. M. Kristjansson, and R. Ficnor. 2005. Crystal structure of a subtilisin-like serine proteinase from a psychrotrophic Vibrio species reveals structural aspects of cold adaptation. *FEBS J.* **272:**832–845.

Arya, S. 2005. Phyletic and comparative statistical analysis of protein sequences as well as structures from the extremophilic bacteria. M.Tech. Thesis. Indian Institute of Technology, Kanpur, UP, India.

Bae, E., and G. N. Phillips, Jr. 2004. Structures and analysis of highly homologous psychrophilic, mesophilic, and thermophilic adenylate kinases. *J. Biol. Chem.* **279:**28202–28208.

Baker, P. J. 2004. From hyperthermophiles to psychrophiles: the structural basis of temperature stability of the amino acid dehydrogenases. *Biochem. Soc. Trans.* **32:**264–268.

Baskakov, I., and D. W. Bolen. 1998. Forcing thermodynamically unfolded proteins to fold. *J. Biol. Chem.* **273:**4831–4834.

Bastolla, U., A. Moya, E. Viguera, and R. C. van Ham. 2004. Genomic determinants of protein folding thermodynamics in prokaryotic organisms. *J. Mol. Biol.* **343:**1451–1466.

Berman, H. M., J. Westbrook, Z. Feng, G. Gilliland, T. N. Bhat, H. Weissig, I. N. Shindyalov, and P. E. Bourne. 2000. The Protein Data Bank. *Nucleic Acids Res.* **28:**235–242.

Bjork, A., D. Mantzilas, R. Sirevag, and V. G. Eijsink. 2003. Electrostatic interactions across the dimer–dimer interface contribute to the pH-dependent stability of a tetrameric malate dehydrogenase. *FEBS Lett.* **553:**423–426.

Blochl, E., R. Rachel, S. Burggraf, D. Hafenbradl, H. W. Jannasch, and K. O. Stetter. 1997. *Pyrolobus fumarii*, gen. and sp. nov., represents a novel group of archaea, extending the upper temperature limit for life to 113 degrees C. *Extremophiles* **1:** 14–21.

Brandsdal, B. O., E. S. Heimstad, I. Sylte, and A. O. Smalas. 1999. Comparative molecular dynamics of mesophilic and psychrophilic protein homologues studied by 1.2 ns simulations. *J. Biomol. Struct. Dyn.* **17:**493–506.

Bosshard, H. R., D. N. Marti, and I. Jelesarov. 2004. Protein stabilization by salt bridges: concepts, experimental approaches and clarification of some misunderstandings. *J. Molec. Recog.* 17:1–16.

Butterwick, J. A., J. P. Loria, N. S. Astrof, C. D. Kroenke, R. Cole, M. Rance, and A. G. Palmer III. 2004. Multiple time scale backbone dynamics of homologous thermophilic and mesophilic ribonuclease HI enzymes. *J. Mol. Biol.* 339:855–871.

Cavicchioli, R., and K. S. Siddiqui. 2004. Cold adapted enzymes, p. 615–638. *In* A. Pandey, C. Webb, C. R. Soccol, and C. Larroche (ed.), *Enzyme Technology*. AsiaTech Publishers Inc., New Delhi, India.

Cavicchioli, R., K. S. Siddiqui, D. Andrews, and K. R. Sowers. 2002. Low-temperature extremophiles and their applications. *Curr. Opin. Biotechnol.* 13:253–261.

Chakravarty, S., and R. Varadarajan. 2002. Elucidation of factors responsible for enhanced thermal stability of proteins: a structural genomics based study. *Biochemistry* 41:8152–8161.

Chan, C. H., H. K. Liang, N. W. Hsiao, M. T. Ko, P. C. Lyu, and J. K. Hwang. 2004. Relationship between local structural entropy and protein thermostability. *Proteins: Struct. Funct. Bioinfo.* 57:684–691.

Claverie, P., C. Viganob, J. M. Ruysschaertb, C. Gerday, and G. Feller. 2003. The precursor of a psychrophilic α-amylase: structural characterization and insights into cold adaptation. *Biochim. Biophys. Acta* 1649:119–122.

D'Amico, S., C. Gerday, and G. Feller. 2003a. Temperature adaptation of proteins: Engineering mesophilic-like activity and stability in a cold adapted alpha-amylase. *J. Mol. Biol.* 332:981–988.

D'Amico, S., J. C. Marx, C. Gerday, and G. Feller. 2003b. Activity–stability relationships in extremophilic enzymes. *J. Biol. Chem.* 278:7891–7896.

Dalhus, B., M. Saarinen, U. H. Sauer, P. Eklund, K. Johansson, A. Karlsson, S. Ramaswamy, A. Bjork, B., Synstad, K. Naterstad, R. Sirevag, and H. Eklund. 2002. Structural basis of thermophilic protein stability: Structures of thermophilic and mesophilic malate dehydrogenases. *J. Mol. Biol.* 318:707–721.

De Farias, S. T., and M. C. Bonato. 2002. Preferred codons and amino acid couples in hyperthermophiles. *Genome Biol.* 3:Preprint.

Deming, J. W. 2002. Psychrophiles and polar regions. *Curr. Opin. Micrbiol.* 5:301–309.

Dominy, B. N., H. Minoux, and C. L. Brooks III. 2004. An electrostatic basis for the stability of thermophilic proteins. *Proteins* 57:128–141.

Dong, F., and H. X. Zhou. 2002. Electrostatic contributions to T4 lysozyme stability: solvent-exposed charges versus semi-buried salt bridges. *Biophys. J.* 83:1341–1347.

Dunker, A. K., J. D. Lawson, C. J. Brown, R. M. Williams, P. Romero, J. S. Oh, C. J. Oldfield, A. M. Campen, C. M. Ratliff, K. W. Hipps, J. Ausio, M. S. Nissen, R. Reeves, C. Kang, C. R. Kissinger, R. W. Bailey, M. D. Griswold, W. Chiu, E. C. Garner, and Z. Obradovic. 2001. Intrinsically disordered protein. *J. Mol. Graph. Model.* 19:26–59.

Dunker, A. K., C. J. Brown, J. D. Lawson, L. M. Iakoucheva, and Z. Obradovic. 2002. Intrinsic disorder and protein function. *Biochemistry* 41:6573–6582.

Dyson, H. J., and P. E. Wright. 2005. Intrinsically unstructured proteins and their functions. *Nat. Rev. Mol. Cell Biol.* 6:197–208.

Eijsink, V. G. H., A. Bjork, S. Gaseidnes, R. Sirevag, B. Synstad, B. van den Burg, and G. Vriend. 2004. Rational engineering of enzyme stability. *J. Biotechnol.* 113:105–120.

Elcock, A. H. 1998. The stability of salt bridges at high temperatures: Implications for hyperthermophilic proteins. *J. Mol. Biol.* 284:489–502.

Feller, G. 2003. Molecular adaptations to cold in psychrophilic enzymes. *Cell. Mol. Life Sci.* 60:648–662.

Feller, G., and C. Gerday. 2003. Psychrophilic enzymes: Hot topics in cold adaptation. *Nat. Rev. Microbiol.* 1:200–208.

Fitzpatrick, T. B., P. Killer, R. M. Thomas, I. Jelesarov, N. Amrhein, and P. Macheroux. 2001. Chorismate synthase from the hyperthermophile *Thermotoga maritima* combines thermostability and increased rigidity with catalytic and spectral properties similar to mesophilic Counterparts. *J. Biol. Chem.* 276:18052–18059.

Friedman, R., J. W. Drake, and A. L. Hughes. 2004. Genome-wide patterns of nucleotide substitution reveal stringent functional constraints on the protein sequences of thermophiles. *Genetics* 167:1507–1512.

Fukuchi, S., and K. Nishikawa. 2001. Protein surface amino acid compositions distinctively differ between thermophilic and mesophilic Bacteria. *J. Mol. Biol.* 309:835–843.

Garofoli, S., M. Falconi, and A. Desideri. 2004. Thermostability of wild type and mutant cold shock proteins by molecular dynamics simulation. *J. Biomol. Struct. Dyn.* 21:771–779.

Georlette, D., V. Blaise, T. Collins, S. D'Amico, E. Gratia, A. Hoyoux, J. C. Marx, G. Sonan, G. Feller, and C. Gerday. 2004. Some like it cold: biocatalysis at low temperatures. *FEMS Microbiol. Rev.* 28:25–42.

Gianese, G., P. Argos, and S. Pascarella. 2001. Structural adaptation of enzymes to low temperatures. *Protein Eng.* 14:141–148.

Glansdorff, N. 1999. On the origin of operons and their possible role in evolution towards thermophily. *J. Mol. Evol.* 49:432–438.

Gorfe, A. A., B. O. Brandsdal, H. K. Leiros, R. Helland, and A. O. Smalas. 2000. Electrostatics of mesophilic and psychrophilic trypsin isoenzymes: qualitative evaluation of electrostatic differences at the substrate binding site. *Proteins: Struct. Funct. Bioinfo.* 40:207–217.

Gromiha, M. M., S. Thomas, and C. Santhosh. 2002. Role of cation–pi interactions to the stability of thermophilic proteins. *Prep. Biochem. Biotechnol.* 32:355–362.

Gruia, A. D., S. Fischer, and J. C. Smith. 2003. Molecular dynamics simulation reveals a surface salt bridge forming a kinetic trap in unfolding of truncated Staphylococcal nuclease. *Proteins: Struct. Funct. Bioinfo.* 50:507–515.

Gunasekaran, K., C. J. Tsai, S. Kumar, D. Zanuy, and R. Nussinov. 2003. Extended disordered proteins: targeting function with less scaffold. *Trends Biochem. Sci.* 28:81–85.

Hendsch, Z. S., and B. Tidor. 1994. Do salt bridges stabilize proteins? A continuum electrostatic analysis. *Protein Sci.* 3:211–226.

Ibrahim, B. S., and V. Pattabhi. 2004. Role of weak interactions in thermal stability of proteins. *Biochem. Biophys. Res. Commun.* 325:1082–1089.

Jaenicke, R. 2000. Do ultrastable proteins from hyperthermophiles have high or low conformational rigidity? *Proc. Natl. Acad. Sci. USA* 97:2962–2964.

Kajander, T., P. C. Kahn, S. H. Passila, D. C. Cohen, L. Lehtio, W. Adolfsen, J. Warwicker, U. Schell, and A. Goldman. 2000. Buried charged surface in proteins. *Structure* 8:1203–1214.

Karshikoff, A., and R. Ladenstein. 2001. Ion pairs and the thermotolerance of proteins from hyperthermophiles: a "traffic rule" for hot roads. *Trends Biochem. Sci.* 26:550–556.

Karshikoff, A., and R. Ladenstein. 1998. Proteins from thermophilic and mesophilic organisms essentially do not differ in packing. *Protein Eng.* 11:867-872.

Kim, S. Y., K. Y. Hwang, S. H. Kim, H. C. Sung, Y. S. Han, and Y. Cho. 1999. Structural basis for cold adaptation: sequence, biochemical properties, and crystal structure of malate dehydrogenase from a psychrophilic *Aquaspirillum arcticum*. *J. Biol. Chem.* 274:11761–11767.

Kleiger G., R. Grothe, P. Mallick, and D. Eisenberg. 2002. GXXXG and AXXXA: common alpha-helical interaction motifs in proteins, particularly in extremophiles. *Biochemistry* 41:5990–5997.

Kumar, S., and R. Nussinov. 2004a. Different roles of electrostatics in heat and in cold: adaptation by citrate synthase. *ChemBioChem.* **5**:280–290.

Kumar, S., and R. Nussinov. 2004b. Experiment-guided thermodynamic simulations on reversible two-state proteins: implications for protein thermostability. *Biophys. Chem.* **111**:235–246.

Kumar, S., C. J. Tsai, and R. Nussinov. 2003. Temperature range of thermodynamic stability for the native state of reversible two-state proteins. *Biochemistry* **42**:4864–4873.

Kumar, S., and R. Nussinov. 2001a. How do thermophilic proteins deal with heat? *Cell. Mol. Life Sci.* **58**:1216–1233.

Kumar, S., and R. Nussinov. 2001b. Fluctuations in ion pairs and their stabilities in proteins. *Proteins: Struct. Funct. Genet.* **43**:433–454.

Kumar, S., C. J. Tsai, and R. Nussinov. 2001. Thermodynamic differences among homologous thermophilic and mesophilic proteins. *Biochemistry* **40**:14152–14165.

Kumar, S., C. J. Tsai, and R. Nussinov. 2000a. Factors enhancing protein thermostability. *Protein Eng.* **13**:179–191.

Kumar, S., B. Ma, C. J. Tsai, and R. Nussinov. 2000b. Electrostatic strengths of salt bridges in thermophilic and mesophilic glutamate dehydrogenase monomers. *Proteins: Struct. Funct. Genet.* **38**:368–383.

Kumar, S., and R. Nussinov. 1999. Salt bridge stability in monomeric proteins. *J. Mol. Biol.* **293**:1241–1255.

Liang, H. K., C. M. Huang, M. T. Ko, and J. K. Hwang. 2005. Amino acid coupling patterns in thermophilic proteins. *Proteins: Struct. Funct. Bioinfo.* **59**:58–63.

Lebbink, J. H., V. Consalvi, R. Chiaraluce, K. D. Berndt, and R. Ladenstein. 2002. Structural and thermodynamic studies on a salt-bridge triad in the NADP-binding domain of glutamate dehydrogenase from *Thermotoga maritima*: Cooperativity and electrostatic contribution to stability. *Biochemistry* **41**:15524–15535.

Lee, C. F., M. D. Allen, M. Bycroft, and K. B. Wong. 2005. Electrostatic interactions contribute to reduced heat capacity change of unfolding in a thermophilic ribosomal protein L30e. *J. Mol. Biol.* **348**:419–431.

Loladze, V. V., and G. I. Makhatadze. 2005. Both helical propensity and side chain hydrophobicity at a partially exposed site in alpha-helix contribute to the thermodynamic stability of Ubiquitin. *Proteins: Struct. Funct. Bioinfo.* **58**:1–6.

Makhatadze, G. I., V. V. Loladze, D. N. Ermolenko, X. Chen, and S. T. Thomas. 2003. Contribution of surface salt bridges to protein stability: guidelines for protein engineering. *J. Mol. Biol.* **327**:1135–1148.

Mallick, P., D. R. Boutz, D. Eisenberg, and T. O. Yeates. 2002. Genomic evidence that the intracellular proteins of archaeal microbes contain disulfide bonds. *Proc. Natl. Acad. Sci. USA* **99**:9679–9684.

Marti, D. N., and H. R. Bosshard. 2003. Electrostatic interactions in leucine zippers: thermodynamic analysis of the contributions of Glu and His residues and the effect of mutating salt bridges. *J. Mol. Biol.* **330**:621–637.

Marx, J. C., V. Blaise, T. Collins, S. D'Amico, D. Delille, E. Gratia, A. Hoyoux, A. L. Huston, G. Sonan, G. Feller, and C. Gerday. 2004. A perspective on cold enzymes: current knowledge and frequently asked questions. *Cell. Mol. Biol.* **50**:643–655.

Mavromatis, K., I. Tsigos, M. Tzanodaskalaki, M. Kokkinidis, and V. Bouriotis. 2002. Exploring the role of a glycine cluster in cold adaptation of an alkaline phosphatase. *Eur. J. Biochem.* **269**:2330–2335.

Mozo-Villarias, A., J. Cedano, and E. Querol. 2003. A simple electrostatic criterion for predicting the thermal stability of proteins. *Protein Eng.* **16**:279–286.

Mueller, U., D. Perl, F. X. Schmid, and U. Heinemann. 2000. Thermal stability and atomic-resolution crystal structure of the *Bacillus caldolyticus* cold shock protein. *J. Mol. Biol.* **297**:975–988.

Nakashima, H., S. Fukuchi, and K. Nishikawa. 2003. Compositional changes in RNA, DNA and proteins for bacterial adaptation to higher and lower temperatures. *J. Biochem* (Tokyo) **133**:507–513.

Nordberg, K. E., S. J. Crennell, C. Higgins, S. Nawaz, L. Yeoh, D. W. Hough, and M. J. Danson. 2003. Citrate synthase from *Thermus aquaticus*: a thermostable bacterial enzyme with a five-membered inter-subunit ionic network. *Extremophiles* **7**:9–16.

Ogasahara, K., M. Ishida, and K. Yutani. 2003. Stimulated interaction between α and β subunits of tryptophan synthase from hyperthermophile enhances its stability. *J. Biol. Chem.* **278**:8922–8928.

Pack, S. P., and Y. J. Yoo. 2005. Packing-based difference of structural features between thermophilic and mesophilic proteins. *Int. J. Biol. Macromol.* **35**:169–174.

Perl, D., C. Welker, T. Schindler, K. Schroder, M. A. Marahiel, R. Jaenicke, and F. X. Schmid. 1998. Conservation of rapid two-state folding in mesophilic, thermophilic and hyperthermophilic cold shock proteins. *Nat. Struct. Biol.* **5**:229–235.

Perl, D., U. Mueller, U. Heinemann, and F. X. Schmid. 2000. Two exposed amino acid residues confer thermostability on a cold shock protein. *Nat. Struct. Biol.* **7**:380–383.

Robinson-Rechavi, M., and A. Godzik. 2005. Structural genomics of *Thermotoga maritima* proteins shows that contact order is a major determinant of protein thermostability. *Structure* **13**:857–860.

Robinson-Rechavi, M., A. Alibes, and A. Godzik. 2006. Contribution of electrostatic interactions, compactness and quaternary structure to protein thermostability: Lessons from structural genomics of *Thermotoga maritima*. *J. Mol. Biol.* **356**:547–557.

Romero, P., Z. Obradovic, X. Li, E. C. Garner, C. J. Brown, and A. K. Dunker. 2001. Sequence complexity of disordered protein. *Proteins: Struct. Funct. Genet.* **42**:38–48.

Russell, N. J. 2000. Toward a molecular understanding of cold activity of enzymes from psychrophiles. *Extremophiles* **4**:83–90.

Russell, R. J., U. Gerike, M. J. Danson, D. W. Hough, and G. L. Taylor. 1998. Structural adaptations of the cold-active citrate synthase from an Antarctic bacterium. *Structure* **6**:351–361.

Schneider, D., Y. Liu, M. Gerstein, D. M. Engelman. 2002. Thermostability of membrane protein helix–helix interaction elucidated by statistical analysis. *FEBS Lett.* **532**:231–236.

Sens, S., and J. W. Peters. 2006. The thermal adaptation of the nitrogenase Fe protein from thermophilic *Methanobacter thermoautotrophicus*. *Proteins: Struct. Funct. Bioinfo.* **62**:450–460.

Siddiqui, K. S., A. Poljak, M. Guilhaus, D. de Fancisci, P. M. G. Curmi, G. Feller, S. D'Amico, C. Gerday, V. N. Uversky, and R. Cavicchioli. 2006. The role of lysine versus arginine in enzyme cold-adaptation: Modifying lysine to homo-arginine stabilized the cold-adapted α-amylase from *Psuedoalteramonas haloplanktis*. *Proteins: Struct. Funct. Bioinf,* **64**:486–501.

Simonson, T., J. Carlsson, and D. A. Case. 2004. Proton binding to proteins: pk_a calculations with explicit and implicit solvent models. *J. Am. Chem. Soc.* **126**:4167–4180.

Sterner, R., and W. Liebl. 2001. Thermophilic adaptation of proteins. *Crit. Rev. Biochem. Mol. Biol.* **36**:39–106.

Taka, J., K. Ogasahara, J. Jeyakanthan, N. Kunishima, C. Kuroishi, M. Sugahara, S. Yokoyama, and K. Yutani. 2005. Stabilization due to dimer formation of phosphoribsyl anthranilate isomerase from *Thermus thermophilus* HB8: X-ray analysis and DSC experiments. *J. Biochem.* **137**:569–578.

Tindbaek, N., A. Svendsen, P. R. Oestergaard, and H. Draborg. 2004. Engineering a substrate specific cold adapted subtilisin. *Protein Eng. Des. Sel.* **17**:149–156.

Thomas, A. S., and A. H. Elcock. 2004. Molecular simulations suggest protein salt bridges are uniquely suited to life at high temperatures. *J. Am. Chem. Soc.* **126**:2208–2214.

Thompson, M. J., and D. Eisenberg. 1999. Transproteomic evidence of a loop-deletion mechanism for enhancing protein thermostability. *J. Mol. Biol.* **290**:595–604.

Torrez, M., M. Schultehenrich, and D. R. Livesay. 2003. Conferring thermostability to mesophilic proteins through optimized electrostatic surfaces. *Biophys. J.* **85**:2845–2853.

Uversky, V. N., C. J. Oldfield, and A. K. Dunker. 2005. Showing your ID: intrinsic disorder as an ID for recognition, regulation and cell signaling. *J. Mol. Recognit.* **18**:343–384.

Wales, D. J., and P. E. Dewsbury. 2004. Effect of salt bridges on the energy landscape of a model protein. *J. Chem. Phys.* **121**:10284–10290.

Wright, P. E., and H. J. Dyson. 1999. Intrinsically unstructured proteins: re-assessing the protein structure–function paradigm. *J. Mol. Biol.* **293**:321–331.

Zhang, J. H., L. L. Zhang, and L. X. Zhou. 2004. Thermostability of protein studied by molecular dynamics simulation. *J. Biomol. Struct. Dyn.* **21**:657–662.

Zhou, H. X. 2002. Towards the physical basis of thermophilic proteins: linking of enriched polar interactions and reduced heat capacity of unfolding. *Biophys. J.* **83**:3126–3133.

Zhou, H. X., and F. Dong. 2003. Electrostatic contributions to the stability of a thermophilic cold shock protein. *Biophys. J.* **84**:2216–2222.

Chapter 7

The Physiological Role, Biosynthesis, and Mode of Action of Compatible Solutes from (Hyper)Thermophiles

HELENA SANTOS, PEDRO LAMOSA, TIAGO Q. FARIA, NUNO BORGES, AND CLÉLIA NEVES

INTRODUCTION

Microorganisms can colonize environments with a wide range of salt concentrations, from freshwater and marine biotopes to hypersaline sites with NaCl concentrations up to saturation (Oren, 1999). Therefore, microbes must adapt their internal water activity to the external environment by accumulating osmotically active molecules, since a positive turgor is required for cell division. There are two main mechanisms by which microorganisms can cope with high salinity of the environment. One involves the influx of inorganic ions (the so-called salt-in-the-cytoplasm) and requires a drastic adaptation of the cellular structures to high ionic strength (Galinski, 1995). The most common and flexible strategy, however, involves the accumulation of small organic solutes, which are able to provide proper internal osmotic pressure without the requisite of special molecular adaptations (Santos and da Costa, 2002). This strategy is widespread in the biological world, being employed by organisms of varied lineages such as archaea, bacteria, yeast, filamentous fungi, and algae that rely exclusively on solute accumulation for osmoprotection.

To achieve osmotic balance in highly saline environments, these substances need to accumulate to high levels (up to the molar range of concentration) without disturbing cellular metabolism, and hence the term "compatible solutes" coined by Brown in the 1970s (Brown, 1976). Compatible solutes must be highly soluble and they usually belong to one of the following groups of compounds: amino acids, sugars, polyols, betaines, and ectoines (da Costa et al., 1998). The concept of compatible solute, initially restricted to osmoadaptation, has recently been extended to account for substances that protect cells against a variety of stress conditions like supraoptimal temperature, drying, or free radicals (da Costa et al., 1998). Accordingly, we will henceforth use the term compatible solute to designate small organic compounds that accumulate in response to an osmotic or heat stress and whose accumulation to high levels is compatible with cell metabolism.

Many thermophiles and hyperthermophiles (from now on designated (hyper)thermophiles) have been isolated from both fresh water and seawater sources. Compatible solute accumulation occurs not only in response to an increase in the external salinity but also in response to supraoptimal growth temperatures. The latter observation seems rather intriguing, especially if we consider that the external water activity remains practically unaltered when the temperature is raised. On the other hand, the superior protective effect of compatible solutes from (hyper)thermophiles upon cellular structures (namely proteins) is well documented (Hensel and König, 1988; Scholz et al., 1992; Ramos et al., 1997; Shima et al., 1998; Borges et al., 2002). From these two observations the inference of a link between compatible solute accumulation by (hyper)thermophiles and structural protection against heat damage appears inevitable. And if such a correlation is legitimate, a number of questions immediately arise. Do thermophiles accumulate the same solutes that are used by mesophiles for osmoadaptation? What is the nature of the eventual new solutes? Are they better suited for protection against heat damage? What is their mode of action? Which pathways are used for their synthesis? Do they accumulate to different extent in response to different stress factors? How is their accumulation regulated?

These are the questions we will address in this chapter. We may still not have the answers to many of them, but it is undeniable that knowledge in this area has advanced considerably in the last ten years, from a time when solutes from (hyper)thermophiles were largely unknown to the present day when many new solutes have been identified, several biosynthetic routes discovered, and the thermo-protecting properties

---

H. Santos, P. Lamosa, T. Q. Faria, N. Borges, and C. Neves. • Instituto de Tecnologia Química e Biológica, Universidade Nova de Lisboa, Rua da Quinta Grande 6, Apartado 127, 2780-156 Oeiras, Portugal.

of these solutes well illustrated, with applications envisaged in fields ranging from cosmetics to diagnosis and medicine.

WHAT COMPATIBLE SOLUTES OCCUR IN (HYPER)THERMOPHILES?

In general, compatible solutes accumulate to high levels in the cytoplasm. The relative abundance combined with the low molecular mass of these compounds greatly facilitates the task of their molecular identification by resorting to two powerful analytical techniques: nuclear magnetic resonance (NMR) and mass spectrometry (Santos et al., 2006). Today, the main bottleneck in the discovery of new compatible solutes resides in the extremely poor growth yields of many hyperthermophiles. Despite this difficulty, many organisms have been examined (Table 1, Fig. 1). Some solutes, like trehalose, α-glutamate, glycine betaines, or proline, are frequently found in nonthermophilic organisms; others like di-*myo*-inositol phosphates are restricted to (hyper)thermophiles of both archaea and bacteria domains; others still, like mannosylglycerate, are strongly associated with (hyper)thermophiles and appear only rarely in mesophiles. With respect to mannosylglycerate, however, recent evidence suggests that the distribution of this solute in mesophiles is probably far from being confined to the red algae, where it was initially identified (Empadinhas, 2004).

Compatible solutes exclusively or mainly found in (hyper)thermophiles will herein be called "hypersolutes" for the convenience of a short designation. In contrast to the solutes more commonly found in mesophiles, hypersolutes are generally negatively charged, and most fall into two categories: hexose derivatives with the hydroxyl group at carbon 1 usually blocked in an α configuration and polyol-phosphodiesters. The most representative compound in the first category is 2-α-O-mannosylglycerate (MG). Although MG was initially identified in red algae of the order *Ceramiales* (Bouveng et al., 1955), this is one of the most widespread solutes in (hyper)thermophiles, occurring in members of bacteria and archaea belonging to distant lineages (Fig. 2). Its structure is that of a mannose moiety blocked in the α-pyranoside configuration by a glycosidic linkage to the C_2 of glycerate. Among (hyper)thermophiles, MG was first found, along with its corresponding amide, mannosylglyceramide (MGA), in the thermophilic bacterium *Rhodothermus marinus* (Nunes et al., 1995; Silva et al., 1999). Since then, it has been reported to occur in the thermophilic bacteria *Thermus thermophilus* and *Rubrobacter xylanophilus*, in the crenarchaeotes *Aeropyrum pernix* and *Stetteria hydrogenophila*, in the euryarchaeotes *Archaeoglobus veneficus*, *Archaeo-globus profundus*, *Methanothermus fervidus*, and in the three genera of the order *Thermococcales*, *Thermococcus*, *Pyrococcus*, and *Palaeococcus* (see Table 1) (Martins and Santos, 1995; Martins et al., 1997; Lamosa et al., 1998; Gonçalves et al., 2003; Neves et al., 2005; our unpublished results). The level of MG increases primarily in response to osmotic stress, the only known exceptions to this behavior being found in bacteria of the genus *Rhodothermus* where it also increases at supraoptimal growth temperatures.

In contrast to the frequent occurrence of MG, its uncharged derivative, MGA, is extremely rare and was found in only a few strains of *Rhodothermus marinus*. Other variations upon the MG theme include mannosylglucosylglycerate (MGG) and glucosylglucosylglycerate (GGG), compounds that have only been found in *Petrotoga myotherma* and *Persephonella marina*, respectively. Another related compound is glucosylglycerate (GG), which is relatively common in halotolerant mesophilic bacteria and has been found in one thermophile, the bacterium *Persephonella marina* (our unpublished results).

Within the polyol-phosphodiester group, the most notable member is di-*myo*-inositol-1,1′-phosphate (DIP). This solute has never been found in organisms with optimal growth temperature below 50°C, and is accumulated by bacteria (*Thermotoga* and *Aquifex* spp.) as well as archaea (members of the genera *Pyrodictium*, *Pyrococcus*, *Thermococcus*, *Methanotorris*, *Aeropyrum*, and *Archaeoglobus*) in response to supraoptimal growth temperatures (Santos and da Costa, 2002). Examples of other polyol-phosphodiesters are diglycerol phosphate (DGP), only found in members of the genus *Archaeoglobus*, and glycerophosphoinositol, a structural chimera of DIP and DGP that was found only in two hyperthermophiles. A DIP derivative, di-mannosyl-di-*myo*-inositol phosphate as well as di-*myo*-inositol-1,3′-phosphate are present, along with DIP, in species of the genus *Thermotoga*, where their levels increase mainly in response to heat stress (Martins et al., 1996).

A few other solutes that do not fall under these two categories have been found in (hyper)thermophiles, that is the case of cyclic-2,3-bisphosphoglycerate (restricted to methanogens), 1,3,4,6-tetracarboxyhexane, galactosylhydroxylysine, and N[e]acetyl-β-lysine (see Table 1 and references therein).

Although (hyper)thermophiles use a variety of compatible solutes during thermoadaptation or osmoadaptation, there are some general trends in their response. As in other halophiles, accumulation by active transport is preferred over de novo synthesis, the latter being a more costly strategy (Oren, 1999).

Table 1. Distribution of compatible solutes of (hyper)thermophilic microorganisms

Organisms	Optimal growth temperature (°C)	cBPG	Tre	MG	DIP	DIP"	α-Glu	β-Glu	Asp	Other	Reference
Archaea											
Pyrolobus fumarii	106										Our unpublished results
Pyrodictium occultum	105				+		+				Martins et al., 1997
Pyrobaculum aerophilum	100		+		+						Martins et al., 1997
Pyrobaculum islandicum	100										Martins et al., 1997
Pyrococcus furiosus	100			↑(S)	↑(T)		+				Martins and Santos, 1995
Pyrococcus horikoshii	98		+	↑(S)	+		+				Empadinhas et al., 2001
Methanopyrus kandleri	98	+					+				Martins et al., 1997
Stetteria hydrogenophila	95		+	+	+						Our unpublished results
Aeropyrum pernix	90				+						Santos and da Costa, 2001
Methanotorris igneus	88				↑(T)		↑(S)	↑(S)			Ciulla et al., 1994; Robertson et al., 1990
Thermoproteus tenax	88		+								Martins et al., 1997
Thermococcus stetteri	87			↑(S)	↑(T); ↑(S)		+		↑(S)		Lamosa et al., 1998
Thermococcus celer	87			↑(S)	↑(T)		+		↑(T)		Lamosa et al., 1998
Thermococcus litoralis	85		↑(S)	↑(S)	↑(T)		+		↑(S)	GalHI	Lamosa et al., 1998
Thermococcus kodakaraensis	85				+		+		+		Our unpublished results
Methanocaldococcus jannaschii	85							↑(S)			Robertson et al., 1990
Palaeococcus ferrophilus	83			↑(T); ↑(S)			↑(T)				Neves et al., 2005
Methanothermus fervidus	83	+		+			+		+		Martins et al., 1997
Archaeoglobus fulgidus	83				↑(T)		+			DGP ↑(S)	Martins et al., 1997; Gonçalves et al., 2003
Archaeoglobus fulgidus (VC-16)	83				↑(T)		+			DGP ↑(S)	Gonçalves et al., 2003
Archaeoglobus profundus	83				+		+				Gonçalves et al., 2003
Acidianus ambivalens	80		+	+							Martins et al., 1997
Archaeoglobus veneficus	75			+	+		+			DGP	Gonçalves et al., 2003
Thermococcus zilligii	75										Lamosa et al., 1998
Sulfolobus sulfataricus	75		+								Martins et al., 1997
Metallosphaera sedula	75		+								Martins et al., 1997
Methanothermobacter thermautotrophicus	70	+					+			TCH	Gorkovenko et al., 1994
Methanothermococcus okinawensis	70						+		+		Our unpublished results

CHAPTER 7 • PHYSIOLOGICAL ROLE, BIOSYNTHESIS, AND MODE OF ACTION OF COMPATIBLE SOLUTES

Organism	T (°C)	Stress response	Solutes	Reference
Methanothermococcus thermolithotrophicus[a]	65	+ +	NAL	Robertson et al., 1990; Martin et al., 1999
Bacteria				
Thermotoga maritima	80	↑(S) ↑(T) +	DmDIP↑(T)	Martins et al., 1996
Thermotoga neapolitana	80	↑(S) ↑(T) + +	DmDIP↑(T)	Martins et al., 1996
Thermosipho africanus	75	+	Pro	Martins et al., 1996
Thermotoga thermarum	70	+		Martins et al., 1996
Marinitoga piezophila	70	+	Pro	Our unpublished results
Fervidobacterium islandicum	70	+		Martins et al., 1996
Persephonella marina	70	↑(S) +	GG, GGG	Our unpublished results
Thermus thermophilus	70	↑(S); ↑(T) +	GB	Nunes et al., 1995
Rhodothermus marinus	65	+ +	MGA↑(S)	Nunes et al., 1995; Silva et al., 1999
Petrotoga myotherma	55	+	MGG, Pro	Our unpublished results

[a] α-Glu, α-glutamate; Asp, aspartate; β-Glu, β-glutamate; cBPG, cyclic 2,3-bisphosphoglycerate; DGP, di-glycerol phosphate; DIP, di-*myo*-inositol-1,1′-phosphate; DIP″, di-*myo*-inositol-1,3′-phosphate; dmDIP, di-mannosyl-di-*myo*-inositol-1,1′-phosphate; GalHl, β-galactopyranosyl-5-hydroxylysine; GB, glycine betaine; GG, glucosylglycerate; GGG, glucosylglucosylglycerate; MG, α-mannosylglycerate; MGA, α-mannosylglyceramide; MGG, mannosylglucosylglycerate; NAL, N^α-acetyl-β-lysine; Pro, proline; TCH, 1,3,4,6-tetracarboxyhexane; Tre, trehalose. The plus sign indicates the presence of the solute in cases for which the response to environmental conditions was not reported. ↑(S) and ↑(T) indicate that the intracellular level of the solute increases in response to osmotic and heat stress, respectively.

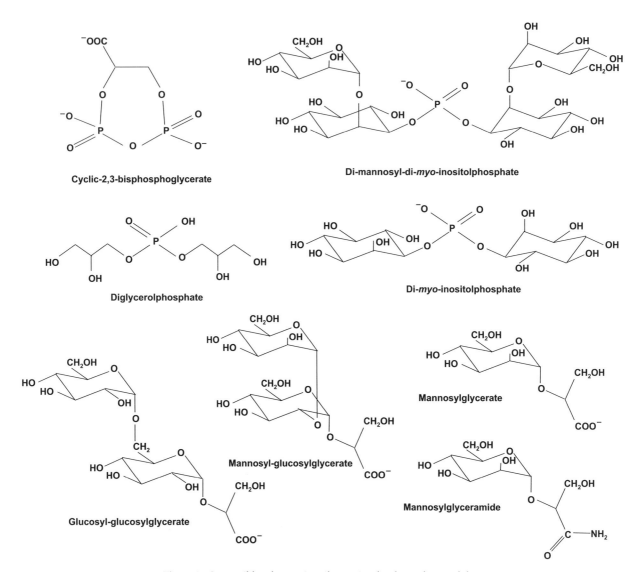

Figure 1. Compatible solutes primarily restricted to hyperthermophiles.

Therefore, the choice of compatible solutes and their relative amounts often depends on the growth medium. One of the best illustrations is provided by *Thermococcus litoralis*, which accumulates trehalose only if this solute is present in the growth medium; aspartate, although absent when the organism is grown on tryptone, becomes the major solute when grown on peptone, in which case it also accumulates several other amino acids and amino acid derivatives (Lamosa et al., 1998).

Another general trend refers to the differential pattern of solute accumulation in response to different stress factors. Often, MG, DGP, and amino acids accumulate preferentially in response to increased salinity while the level of DIP and DIP-derivatives responds to heat stress. These rules, however, present several exceptions due to the diversity of stress response patterns encountered even in the case of closely related species. A paradigmatic case is the comparison of compatible solute accumulation patterns among the members of the order *Thermococcales*. DIP occurs in all strains of *Thermococcus* and *Pyrococcus* examined to date (Empadinhas et al., 2001; Lamosa et al., 1998; Martins and Santos, 1995; Scholz et al., 1992) but is not present in *Palaeococcus ferrophilus*. In fact, this organism only accumulates aspartate (in response to increased salinity), glutamate (increased level at supraoptimal temperatures), and MG (which responds to both stresses) (Neves et al., 2005). In all organisms known to accumulate DIP, the level of this solute increases consistently in response to heat stress. In *Palaeococcus ferrophilus*, the only member of the *Thermococcales* that does not synthesize DIP, the specialized role of

CHAPTER 7 • PHYSIOLOGICAL ROLE, BIOSYNTHESIS, AND MODE OF ACTION OF COMPATIBLE SOLUTES 91

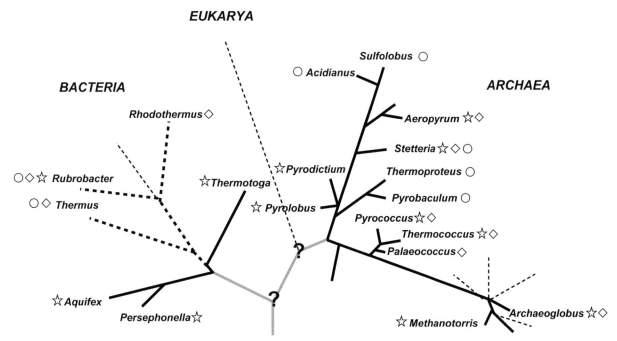

Figure 2. Distribution of trehalose (circles), mannosylglycerate (diamonds) and di-*myo*-inositol-1,1′-phosphate (stars) among (hyper)thermophiles. Tree of Life adapted from Blöchl et al., 1995. The question marks indicate unknown positions of the branching points between the three domains.

DIP in thermoadaptation seems to be replaced by glutamate and MG. Interestingly, a reverse situation is observed in the closely related *Thermococcus kodakaraensis*: MG is absent, and instead DIP, aspartate, and glutamate are accumulated under all conditions studied (our unpublished results).

The preferential involvement of MG and DIP in the osmotic and heat stress response, respectively, has been observed in all the members of the *Pyrococcus* and most *Thermococcus* species examined. It is particularly patent in *Pyrococcus furiosus* in which the MG level increases with increasing NaCl concentration of the medium, while DIP is the only solute accumulating under heat stress, its level increasing by over 10-fold for a temperature up-shift of 6°C from the optimal growth temperature (Martins and Santos, 1995).

BIOSYNTHESIS OF COMPATIBLE SOLUTES

Biosynthesis of Mannosylglycerate

Two distinct pathways for the synthesis of mannosylglycerate have been identified in *Rhodothermus marinus* (Martins et al., 1999). One of the pathways involves the direct condensation of GDP-mannose with D-glycerate to yield MG through the action of mannosylglycerate synthase (MGS) (Fig. 3). In the second pathway, mannosyl-3-phosphoglycerate synthase (MPGS) catalyzes the condensation of GDP-mannose with D-3-phosphoglycerate and produces mannosyl-3-phosphoglycerate, which is subsequently dephosphorylated by a specific phosphatase (mannosyl-3-phosphoglycerate phosphatase, MPGP).

The single-step pathway is rare and, thus far, has only been found in *Rhodothermus marinus* (Martins et al., 1999), and in red algae (our unpublished results). MGS from *Rhodothermus marinus* has been extensively studied: the native protein was purified and characterized, the encoding gene was identified, and the recombinant protein characterized. MGS activity is absolutely dependent on divalent metal ions, and is maximal at 85°C and pH 6.5 (Martins et al., 1999). At the amino acid sequence level, MGS shows similarity with glycosyltransferases from family 2, which comprises glycosyltransferases that invert the configuration of the anomeric carbon of the sugar donor during catalysis. However, MGS performs catalysis with retention of the anomeric configuration of the sugar donor (Martins et al., 1999). Therefore, a new class of glycosyltransferases, GT78, was created to include the MGS from *Rhodothermus marinus* (http://afmb.cnrs-mrs.fr/CAZY/). Recently, the three-dimensional structure of MGS was determined at and a proposal to explain the peculiar catalytic mechanism for glycosyl transfer with retention of the anomeric configuration was put forward (Flint et al., 2005). Interestingly, MGS shows broad

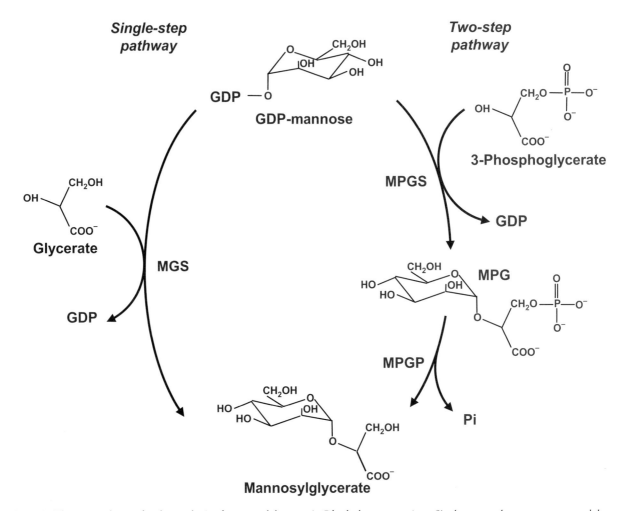

Figure 3. The two pathways for the synthesis of mannosylglycerate in *Rhodothermus marinus*. Single-step pathway uses mannosylglycerate synthase (MGS), while the two-step pathway involves the actions of mannosyl-3-phosphoglycerate synthase (MPGS) and mannosyl-3-phosphoglycerate phosphatase (MPGP).

substrate specificity with regard to sugar donor and also, to some extent, to the 3-carbon acceptor: GDP-mannose and D-glycerate are the preferred substrates, but GDP-glucose, GDP-fucose, UDP-mannose, and UDP-glucose can also be used as the sugar donor, and D-lactate and glycolate, as the 3-carbon acceptor. Moreover, the substrate specificity was clearly dependent on temperature (Flint et al., 2005).

In contrast to the rarity of the single-step pathway, the two-step pathway, which proceeds via a phosphorylated intermediate, seems to be a common route to produce MG. This pathway, initially identified in *Rhodothermus marinus* (Martins et al., 1999), has also been observed in *Pyrococcus horikoshii* (Empadinhas et al., 2001), *Thermus thermophilus* (Empadinhas et al., 2003), *Palaeococcus ferrophilus*, and *Thermococcus litoralis* (Neves et al., 2005). The MPGSs share many biochemical properties and kinetic parameters (Table 2). In particular, they show high substrate specificity for GDP-mannose and 3-phosphoglycerate (3-PGA) at the temperature examined (83°C), and exhibit comparable K_m values for these substrates. As expected, the optimal temperature for activity follows the optimal temperature for growth of the source organism.

Unexpectedly, the recombinant MPGS from *Rhodothermus marinus* has a low specific activity when compared with other homologous MPGSs characterized thus far (Empadinhas et al., 2001, 2003; Borges et al., 2004; Neves et al., 2005). An intriguing feature is the existence of about 30 extra amino acids in the C-terminus region. Truncation of this extension produced a protein with a specific activity of the same order of magnitude of the other MPGSs. Moreover, the activity of the complete MPGS was enhanced upon incubation with *Rhodothermus marinus* cell extracts, and protease inhibitors abolished activation. These results suggest the involvement of proteolytic activation; however, immunoblotting assays failed to detect the truncated form of MPGS in

Table 2. Biochemical properties of recombinant mannosyl-3-phosphoglycerate synthase from *Archaea* and *Bacteria*

Mannosyl-3-phosphoglycerate synthase	No. of amino acid residues	Molecular mass (kDa)	Optimal temperature (°C)	Optimal pH	Mg^{2+a}	Half-life (min)[b]	Substrate specificity[b]	K_m (mM) GDP-man	K_m (mM) 3-PGA	V_{max} (mmol/min mg)	Reference
Archaea											
Palaeococcus ferrophilus	393	44	90	~7.0	0	18	GDP-man; 3-PGA	ND	ND	331	Neves et al., 2005
Pyrococcus horikoshii	394	46	90–100	~7.6	46	16	GDP-man; 3-PGA	0.17	0.14	186	Empadinhas et al., 2001
Bacteria											
Rhodothermus marinus	427	49	80	~7.6	0	40	GDP-man; 3-PGA	0.5	0.6	~15	Borges et al., 2004
Thermus thermophilus	391	45	80–90	~7.0	0	189	GDP-man; 3-PGA	0.3	0.13	122	Empadinhas et al., 2003

[a] Values refer to the activity in the absence of Mg^{2+} as a percentage of the maximum activity obtained with 15 mM Mg^{2+}.
[b] Measured at the optimal temperature. GDP-man, guanosine-diphospho-D-mannose; 3-PGA, 3-phosphoglycerate; ND, not determined.

extracts derived from cells grown under different stress conditions. Therefore, the operation of this regulatory process remains uncertain (Borges et al., 2004). The MPGSs characterized to date produce mannosyl-3-phosphoglycerate with the same anomeric configuration of the substrate and accordingly have been classified as members of glycosyltransferases family GT55, which comprises GDP-mannose: α-mannosyltransferases that retain the anomeric configuration of the substrate (http://afmb.cnrs-mrs.fr/CAZY/).

Genes for mannosylglycerate synthesis

The two-step pathway appears to be much more widely disseminated in Nature than the single-step pathway. In fact, genes coding for the synthase of the single-step pathway are known only in *Rhodothermus marinus* (Martins et al., 1999), and in the red alga *Caloglossa leprieurii* (our unpublished results). In contrast, the genes encoding the synthase of the two-step pathway have been functionally identified in a number of organisms: *Rhodothermus marinus*, *Pyrococcus horikoshii*, *Thermus thermophilus*, *Palaeococcus ferrophilus*, and *Dehalococcoides ethenogenes*. In addition, a number of genes with considerable degree of identity appear scattered in the Tree of Life (Fig. 2 and 4). With the increasing number of gene sequences available it becomes apparent that these putative *mpgS* genes are not confined to (hyper)thermophiles, occurring in typical mesophiles such as the fungi *Magnaporthe grisea* and *Neurospora crassa* (Empadinhas, 2004) and in a number of uncultured archaea isolated from moderate environments. Apparently, MG synthesis is more widespread among mesophiles than presumed a few years ago when the amount of data was rather limited; however, it is worth pointing out that the occurrence of MG has not been shown in mesophiles, except for red algae, and in this case MG is not involved in osmoadaptation (Karsten et al., 1994).

The unrooted phylogenetic tree based on the alignment of the amino acid sequences of known or putative MPGSs is in general agreement with the phylogenetic organization of the organisms based on 16S rRNA sequences (Fig. 4). As expected, the bacterial MPGS from *Thermus thermophilus*, *Rhodothermus marinus*, and *Dehalococcoides ethenogenes* are grouped together in a separate cluster. The MPGS from *Aeropyrum pernix* forms a distinct branch together with homologous sequences present in fungi, not far from the two other crenarchaeote representatives.

Interestingly, there is a considerable degree of sequence conservation within MPGSs (at least 30% identity), even when they originate from distantly related phylogenetic groups. Among the archaea the notable prevalence of biosynthetic MG genes could be explained by vertical inheritance. For example, the representatives of the *Thermococcales* form a very tight cluster, in which the most divergent MPGS belongs to the genus *Palaeococcus*, which had an early separation from the common ancestor of *Thermococcus* and *Pyrococcus* (Neves et al., 2005). On the other hand, there are examples of closely related organisms that differ in respect to the ability to synthesize this

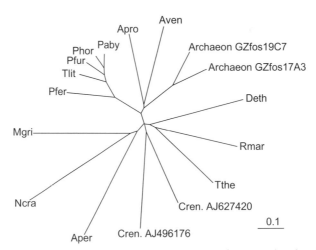

Figure 4. Unrooted phylogenetic tree based on known or putative sequences of mannosyl-3-phosphoglycerate synthase genes. The ClustalX and TreeView programs 5,7 were used for sequence alignment and to generate the phylogenetic tree. The significance of the branching order was evaluated by bootstrap analysis of 1000 computer-generated trees. Bar, 0.1 change per site. Abbreviations: Aper, *Aeropyrum pernix*; Apro, *Archaeoglobus profundus*; Aven, *Archaeoglobus venificus*; Deth, *Dehalococcoides ethenogenes*; Mgri, *Magnaporthe grisea*; Ncra, *Neurospora crassa*; Pfer, *Palaeococcus ferrophilus*; Paby, *Pyrococcus abyssi*; Pfur, *Pyrococcus furiosus*; Phor, *Pyrococcus horikoshii*; Rmar, *Rhodothermus marinus*; Tthe, *Thermus thermophilus*; Tlit, *Thermococcus litoralis*.

compatible solute, namely *Archaeoglobus fulgidus* in which the homologous genes for the synthesis of MG are absent in the genome of the type strain, but are present and functional in *Archaeoglobus fulgidus* 7324, *Archaeoglobus veneficus*, and *Archaeoglobus profundus* (Gonçalves et al., 2003). Apparently this was a common trait that was lost by the type strain. The evolution of MG biosynthesis is a fascinating topic but a meaningful discussion would demand more ample data sets and reliable tools for genome analysis.

In most organisms using the two-step pathway, the synthase and the phosphatase are encoded by two consecutive genes, designated *mpgS* and *mpgP*, respectively. In *Pyrococcus horikoshii* and all other members of the *Thermococcales* examined, *mpgS* and *mpgP* are organized in an operon-like structure which includes the genes encoding the enzymes that convert fructose-6P into GDP-mannose (M1P-GT/PMI and PMM) (Empadinhas et al., 2001; Neves et al., 2005) (Fig. 5). In *Thermus thermophilus*, *Aeropyrum pernix*, *Archaeoglobus* spp., and *Rhodothermus marinus*, the *mpgS* gene is located immediately upstream of the *mpgP*, but these organisms lack the genes for M1P-GT/PMI and PMM in adjacent location. An unusual gene fusion between *mpgS* and *mpgP* was found in the genome of the mesophilic bacterium *Dehalococcoides ethenogenes* (Empadinhas et al., 2004).

Regulation of mannosylglycerate synthesis

External salinity and growth temperature are physicochemical factors that clearly modulate the accumulation of MG (Martins and Santos, 1995; Santos and da Costa, 2002). So far, the regulation of the synthesis of MG, at the expression level, has only been studied with *Rhodothermus marinus*, the sole organism known to have two routes for the synthesis of MG (Martins et al., 1999). The reason for this pathway multiplicity was explained in a recent work (Borges et al., 2004). The single-step pathway is selectively expressed in response to heat stress, whereas the two-step pathway is predominantly active under osmotic stress. Therefore, the two biosynthetic pathways play specialized roles in the adaptation to heat and osmotic stress. The use of a single solute to respond to both temperature and salt stress is unusual, and it is still unclear whether the observed pathway duplicity for the synthesis of MG reflects an evolutionary advantage in the colonization of a wider range of hot saline environments.

Biosynthesis of Polyol-phosphodiesters

Biosynthesis of di-*myo*-inositol phosphate

Despite the fact that DIP was the first hypersolute to be discovered, the respective biosynthetic pathway has not been firmly established. DIP synthesis was investigated on *Methanotorris igneus* (Chen et al., 1998) and *Pyrococcus woesei* (Scholz et al., 1998) and, as a result, two different pathways have been proposed. Inositol-1-phosphate is synthesized from glucose-6-phosphate by inositol-1-phosphate synthase. The enzyme of the hyperthermophilic archaeon *Archaeoglobus fulgidus* has been characterized in detail (Chen et al., 2000). In *Methanotorris igneus*, part of the inositol-1-phosphate is dephosphorylated into inositol while the other part is activated to CDP-inositol, both molecules being then condensed to yield DIP by the action of a DIP synthase. A different pathway was proposed in *Pyrococcus woesei* (Scholz et al., 1998). In this case, two

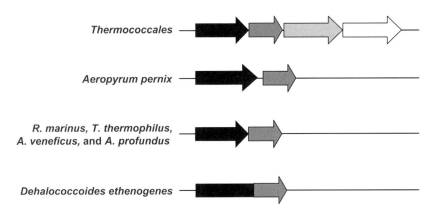

Figure 5. Genomic organization of mannosylglycerate biosynthesis via the two-step pathway. Black arrows indicate *mpgS* genes; dark grey arrows indicate the *mpgP* genes; light grey arrow represents the phosphomannose isomerase/mannose-1-phosphate guanylyltransferase; the white arrow indicates phosphomannomutase.

molecules of inositol-1-phosphate are condensed to yield DIP with the consumption of nucleoside triphosphate. Both proposed pathways were partially demonstrated in cell extracts, but the key enzymes (CTP-inositol cytidylyltransferase and the two DIP synthases) were not characterized, neither were the respective genes identified.

Biosynthesis of DGP

DGP biosynthesis was investigated by our team on *Archaeoglobus fulgidus*. The enzyme(s) responsible for the production of DGP is unknown, but the corresponding activities have been detected in cell extracts. Synthesis of DGP is strictly dependent on CDP-glycerol and glycerol-3-phosphate, in a process that resembles one of the above-mentioned pathways proposed for DIP biosynthesis.

THE PHYSIOLOGICAL RELEVANCE OF COMPATIBLE SOLUTE ACCUMULATION

The observation that some solutes are preferentially or strictly found in (hyper)thermophiles and/or their level increases primarily in response to supraoptimal growth temperatures led to the hypothesis that hypersolutes would be superior protectors of proteins and other cellular structures against heat damage (Santos and da Costa, 2001). However, their real importance as stabilizers in vivo has not been shown due to the lack of genetic tools for the efficient manipulation of (hyper)thermophiles. So far, the most elucidatory study involves the comparison of the osmotolerance of *Thermus* strains naturally lacking the ability to accumulate trehalose, MG, or both. Trehalose and MG appear to display a synergistic effect, allowing an increased osmotolerance of the organisms able to accumulate the two solutes (Alarico et al., 2005). Unfortunately, the thermotolerance of these strains was not reported.

The vast majority of enzymes from (hyper)thermophiles are heat stable (Hensel, 1993; Adams, 1993). It would be implausible that these enzymes could rely exclusively on extrinsic factors to withstand high temperatures. Nevertheless, some proteins display lesser stability than expected from the temperature range of the organism from which they originate, suggesting an important role of compatible solutes as stabilizers in vivo. Extensive in vitro data have been collected in recent years proving their superior ability as thermal stabilizers in vitro (Scholz et al., 1992; Ramakrishnan et al., 1997; Shima et al., 1998; Ramos et al., 1997; Borges et al., 2002; Lamosa et al., 2000, 2001; Faria et al., 2003, 2004), and it is tempting to speculate that these compatible solutes might have an auxiliary role in stabilizing biological structures in vivo as well. This view is further supported by a large number of examples of hypersolute accumulation in response to supraoptimal growth temperatures; however, the presence of compatible solutes is not a prerequisite for hyperthermophily. In fact, there are several hyperthermophiles, like *Thermotoga thermarum*, *Hydrogenobacter islandicum*, *Thermococcus zilligii*, and *Pyrobaculum islandicum*, which are isolated from nonsaline environments, that do not accumulate compatible solutes (Martins et al., 1996, 1997; Lamosa et al., 1998). Obviously, structural stabilization at elevated temperature involves more factors than just compatible solute accumulation. In fact, the two available examples of large-scale analysis of the heat stress response in hyperthermophilic archaea indicate that the general features are rather similar to those observed in mesophilic model microorganisms, involving up-regulation of post-transcriptional modifications, protein turnover, and chaperones (Rohlin et al., 2005; Schokley et al., 2003).

The hypothesis that exposure to a given stress confers cross-protection against other stresses is supported by a growing number of experimental observations. It is, therefore, reasonable to propose that the strategies optimized to cope with high salinity, namely those leading to the synthesis of osmolytes, could also be useful for protection against heat in halophilic (hyper)thermophiles. Still, the apparent trend of hypersolute specialization in osmotic or heat stress adaptation is an intriguing feature that remains to be elucidated.

WHY ARE NEGATIVELY CHARGED COMPATIBLE SOLUTES PREFERRED BY HYPERTHERMOPHILES?

As mentioned above, most of the solutes accumulated by (hyper)thermophiles are negatively charged in contrast with the noncharged or zwitterionic nature of solutes found in mesophiles. This observation together with the superior ability of these solutes to stabilize proteins, lead to the question whether the negative charge plays an important role in the stabilization mechanism.

In the case of solutes from hyperthermophiles, it was shown that the protecting effect was clearly dependent on the solute charge. The melting temperature of bovine ribonuclease A (RNase A) in the presence and absence of MG depends on the ionization state of the solute. A greater degree of stabilization is achieved at pH values above 5, when the solute is fully ionized. Moreover, as the solute's charge decreased, the melting temperature becomes identical to that observed in the absence of solutes (Fig. 6) (Faria et al., 2003).

Another illuminating example is provided by the comparison of the effect of a series of charged and uncharged solutes on the stability of two proteins that have opposite net charges at the working pH: staphylococcal nuclease and malate dehydrogenase. The negatively charged solutes (mannosylglycerate, DGP, di-*myo*-inositol phosphate) exert a remarkably higher protective effect on both enzymes than the uncharged solutes (trehalose, glycerol, ectoine, hydroxylectoine, and MGA) (Fig. 7) (our unpublished results). Glycerol, a canonical protein stabilizer, increased the melting temperature of staphylococcal nuclease by 3°C at 2.5 M concentration (not shown), whereas the increment achieved by 0.5 M MG was 8°C. Also, it is interesting to point out the poor performance of the salt KCl used here as a control for ionic strength. On the other hand, the effect exerted by the charged solutes depends on the particular protein/solute considered: for example, malate dehydrogenase is stabilized to a greater extent by DIP or MG, while DGP is a better stabilizer of staphylococcal nuclease, despite the identical net charge of these solutes.

At present, a satisfactory explanation for the preferential use of negatively charged solutes in organisms adapted to hot environments cannot be given because the mechanisms underlying solute stabilization by solutes are highly complex and poorly understood (see below). However, the superior ability of charged compounds, namely multi-carboxylic acids, to improve the thermal stability of proteins is well established (Borges et al., 2002; Kaushik and Bhat, 1999). Therefore, it is tempting to speculate that the observed correlation between charged solutes and hyperthermophiles could be associated with an evolutionary adaptation of these organisms to take advantage of the greater contribution of coulombic interactions at high temperatures.

UNDERSTANDING THE MOLECULAR BASIS OF PROTEIN STABILIZATION BY SOLUTES

Types of Molecular Interactions

To understand the mechanisms responsible for the kinetic and thermodynamic stabilization of proteins by solutes is an extremely difficult task, given the complexity and the large number of factors that may be involved, the relative contributions depending on the particular system. In principle, the stabilizing effect could be accomplished either via perturbation of the solvent or via more-direct protein/solute interactions. Excellent overviews on this topic are available (Ramos and Baldwin, 2002; Timasheff, 1993; Baldwin, 1996; Davis-Searles et al., 2001).

At a molecular level, the effect of solutes can be depicted by the following mechanisms: (i) solute exclusion from the protein vicinity affecting the organization of hydration water molecules; (ii) solute unspecific

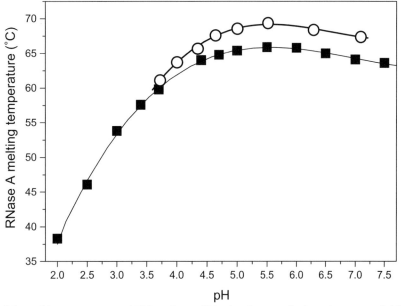

Figure 6. Dependence of the melting temperature of RNase A on pH in the absence of solutes (squares and thin line) and with 0.5 M mannosylglycerate (circles and thick line).

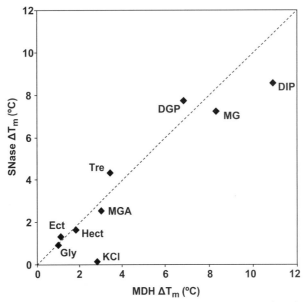

Figure 7. Effect of solutes on the melting temperature of staphylococcal nuclease (SNase) and pig heart malate dehydrogenase (MDH). Abbreviations: Tre, trehalose; MG, α-mannosylglycerate; MGA, α-mannosylglyceramide; DIP, di-*myo*-inositol-1,1'-phosphate; DGP, diglycerol phosphate; KCl, potassium chloride; Gly, glycerol; Ect, ectoine; Hect, hydroxyectoine.

binding to the protein by increasing the solute concentration in the protein hydration shell in comparison with the bulk concentration; (iii) specific binding of the solute to particular segments of the protein structure. In any of these cases there is a differential interaction of the solute with the native and unfolded protein forms. Thus, the equilibrium of protein unfolding is affected by the presence of solutes.

Regardless of the molecular mechanisms involved, if a solute acts as a stabilizer it must induce a decrease in the equilibrium constant of protein unfolding. Thermodynamically, from the Wyman relation, this means that the solute is either more strongly bound to the native than to the unfolded protein, or more excluded from the denatured state than from the native state (Timasheff, 1993). Even in the case of a binding solute, there must be exclusion upon unfolding in order to induce stabilization. In fact, when measuring binding of solute molecules (sugars and amino acids) to proteins, using dialysis equilibrium experiments, Timasheff and co-workers found negative binding, meaning that these solutes are excluded from the protein water interface. These results form the basis for the theory of preferential exclusion (Fig. 8) (Gekko and Timasheff, 1981; Arakawa and Timasheff, 1982a, 1982b, 1983). The exclusion of solutes from the protein vicinity can occur by different mechanisms. When the solute molecules are much bigger than water molecules, their penetration in the protein hydration shell is prevented, creating a water-enriched hydration layer due to steric exclusion of the solute. Upon unfolding, the exposed protein surface increases, and therefore exclusion will be higher on the unfolded protein, leading to stabilization (see Fig. 8). An example of this type of mechanism is the effect of polyethylene glycol on β-lactoglobulin (Arakawa and Timasheff, 1985a).

Another means of inducing solute exclusion from the protein vicinity is via an increase of the surface tension of water. The surface tension is a measure of the energy required to form an interface in a solution. As solutes perturb the cohesive forces between the water molecules, they affect the surface tension of the solution. A solute that increases the water surface tension will be driven out of the contact surfaces. Moreover, solute exclusion will be greater for the unfolded than for the native protein, since the protein-solvent surface increases upon unfolding. This is probably the most widespread mechanism and occurs in the presence of many sugars, amino acids, and other polyols that act as protein protectors (Lee and Timasheff, 1981; Arakawa and Timasheff, 1983; Arakawa and Timasheff, 1985b; Lin and Timasheff, 1996; Kaushik and Bhat, 1998). In summary, an increase of the water surface tension will lead to protein stabilization as long as other solute-protein interactions, such as solute binding, do not occur. For example, glycerol is a solute that causes a slight decrease of the water surface tension; it is excluded from nonpolar regions of the protein surface due to entropic effects but has an affinity for the polar

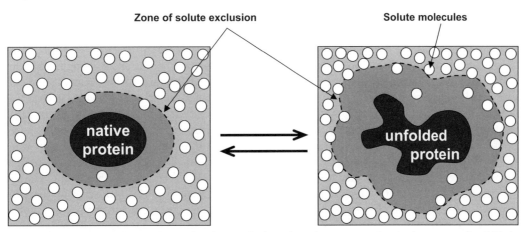

Figure 8. Schematic representation of the preferential exclusion of solutes from the protein surface in the native and unfolded states. Upon denaturation, the zone of exclusion increases and, although the same amount of solute molecules is depicted, these become locally more concentrated, i.e., their chemical potential has increased. As a consequence, to achieve thermodynamic equilibrium, the unfolding reaction is displaced to the native state, hence the solute acts as a stabilizer.

regions (Gekko and Timasheff, 1981). The overall result of glycerol action on the protein structure will depend on the relative contribution of each effect. However, upon unfolding, there is a large increase on the exposure of nonpolar residues, and glycerol will be more strongly excluded from the unfolded protein state. Thus, the presence of glycerol prevents protein unfolding (Gekko and Timasheff, 1981). Other example of solute-protein interactions dependent on the chemical nature of the protein surface is 2-methyl-2,4-pentanediol (MPD) (Pittz and Timasheff, 1978). MPD is repelled from electric charges but binds to nonpolar regions. In the native state, there is usually strong exclusion of MPD; however, binding overcomes exclusion in the unfolded state and, therefore, MPD is a strong precipitating agent and facilitates unfolding despite being excluded from the native protein state.

To understand the molecular basis of solute exclusion, Bolen and co-workers performed a systematic study of the energy involved in the transfer of the twenty amino acids and an analogue of the peptide backbone from pure water to solutions with various concentrations of sucrose, sarcosine, or urea (Liu and Bolen, 1995; Qu et al., 1998). The transfer free energies obtained are quite small, showing that the interactions of those solutes with the protein components are modest and that water is a strong competitor for interactions with the protein. The determination of the transfer free energy of the peptide backbone from water to solutions with stabilizing solutes was found to be largely positive (thermodynamically unfavorable) (Qu et al., 1998). This approach identifies the solvophobic effect of the backbone as the main driving force for stabilization and explains the relative lack of specificity of solute stabilizing properties; a certain degree of solute/protein specificity is due to the opposing contributions of the amino acid side chains (Liu and Bolen, 1995).

Studies involving organic charged molecules are very rare. Work on the protein stabilization induced by several carboxylic acids and citrate derivatives with different number of carboxylic groups and chain length demonstrated that protein thermal stabilization increases with the number of carboxylic acid groups and is not affected by chain length or the number and position of hydroxyl groups (Busby and Ingham, 1984; Kaushik and Bhat, 1999). More importantly, it was shown that the effect could not be explained solely by the increase on the water surface tension induced by the salts (Kaushik and Bhat, 1999). Therefore, interactions other than those mediated by the solvent should be taken into account. The multitude of possible interactions and the lack of precise mathematical models are the main reasons for the yet unaccomplished task of understanding and predicting protein stabilization effects by solutes.

Probing the Effect of Solutes on the Unfolding Pathway of Proteins

In an attempt to understand the effect of MG on the unfolding pathway of the model protein staphylococcal nuclease, picosecond time-resolved fluorescence spectroscopy was used (Faria et al., 2004). This is a very powerful technique that allows the identification and quantification of the different native and unfolded states of the protein. The profile of staphylococcal nuclease fluorescence decay times was evaluated along thermal unfolding in the absence and in the presence of 0.5 M MG (Fig. 9). The presence of MG did not alter the pattern of the fluorescence decay times, indicating that the native and denatured forms during thermal denaturation of the nuclease are identical with and without MG. Therefore, despite the notable increase on the protein melting temperature (about 8°C) induced by 0.5 M MG, no evidence was found for alterations of the nuclease unfolding pathway. The molecular mechanism for stabilization of staphylococcal nuclease by MG probably involves subtle changes with no detectable impact on the nature and population of protein states along denaturation, at least for those protein states that are accessible to detection by this technique.

Probing the Effect of Solutes on Protein Dynamics

The molecular basis for the protective effect of solutes upon proteins was first investigated in relation to possible changes in the three-dimensional protein structure, susceptible to explain the added stability. However, no measurable structural changes were detected by NMR. In the presence of 0.1 M DGP, a concentration capable of producing a fourfold increase in the half-life of *Desulfovibrio gigas* rubredoxin on thermal denaturation, no alteration of structure as probed by alterations of the proton chemical shift was detected (Lamosa et al., 2003). Similarly, no effect was reported for chymotrypsin inhibitor 2 and horse heart cytochrome *c* upon addition of 2 M glycine (Foord and Leatherbarrow, 1998). These results strengthened the view that stabilizing compatible solutes exert their action through changes in the solvent structure and/or by changing dynamic properties of the protein rather than causing substantial alteration in the protein structure itself.

The presence of MG induced an upshift in the temperature for optimum activity of RNase A, which suggests that this protein could acquire a less flexible structure in the presence of MG (Faria et al., 2003). Moreover, when investigating the changes in the dynamic behavior of wild-type and mutant rubredoxin as a consequence of solute addition, we found evidence for a compaction of the protein structure induced by the presence of DGP or MG (Lamosa et al., 2003; Pais et al., 2005). Since no structural alterations were detected, changes in the dynamic behavior of rubredoxin could be one of the keys to explaining the stabilization phenomenon in this protein (Lamosa et al., 2003). The dynamical parameters of the backbone amide groups of rubredoxin were determined through ^{15}N NMR relaxation measurements and show a rigidification expressed by a small but generalized increase in the order parameters upon solute addition. This overall rigidification became more pronounced when our analysis moved from fast to slow time scales. While ^{15}N NMR relaxation rates probe motions in the very fast time scale (10^{-10} to 10^{-5} s), amide exchange rates provide information regarding very slow events (up to minutes or hours).

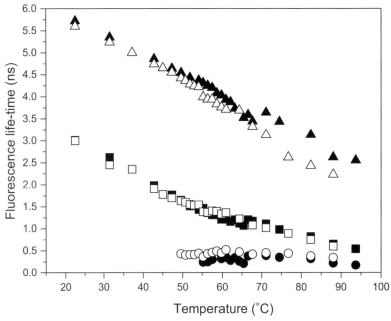

Figure 9. Fluorescence decay times of staphylococcal nuclease as a function of temperature in the presence of 0.5 M mannosylglycerate (solid symbols) and in the absence of solutes (open symbols). Native protein: triangles; denatured states of the protein: squares and circles.

In *Desulfovibrio gigas* rubredoxin, there is a remarkable decrease in the rates of NH exchange upon solute addition, meaning that the large-amplitude concerted motions are notably restricted by the addition of solutes. Overall, the results indicate that although the vibrational modes of individual atoms and small groups may be barely affected by the presence of solutes, it is clear that the concerted motion of larger protein segments becomes more restricted and this may be an important factor in protein stabilization (Lamosa et al., 2003).

POTENTIAL APPLICATIONS FOR SOLUTES FROM (HYPER)THERMOPHILES

Due to the enhanced ability to stabilize biological materials, the application of hypersolutes in industrial applications was soon envisioned, and several industrial patents on their uses have been filed (Santos et al., 1999, 2003, 2005).

Since a major drawback of enzyme usage in industrial processes is the enzyme's relative instability leading to degradation during long-term storage and repetitive use, there is a high interest in methodologies that increase enzyme stability. Therefore, high performance solutes able to stabilize enzymes at lower concentrations than conventional stabilizers are currently envisioned as additives in analytical/clinical test kits, development of heat-stable vaccines, biosensors, cosmetics, DNA amplification, and cell preservation. Moreover, the ability of these solutes to interfere with protein aggregation and fibril formation may provide clues for the design of chemical chaperones useful in the treatment and/or prevention of protein-misfolding associated diseases.

Acknowledgments. This work was funded by the European Commission Contract COOP-CT-2003-508644 and Fundação para a Ciência e a Tecnologia and FEDER, Portugal, POCTI/BIA-PRO/57263/2004 and POCTI/BIA-MIC/59310/2004. P. Lamosa, N. Borges, T. Q. Faria, and C. Neves acknowledge post-doc grants from FCT, Portugal (BPD/11511/2002, BPD/14841/2003, BPD/20352/2004 and BPD/7119/2001).

REFERENCES

Adams, M. 1993. Enzymes and proteins from organisms that grow near and above 100°C. *Annu. Rev. Microbiol.* 47:627–658.

Alarico, S., N. Empadinhas, C. Simões, Z. Silva, A. Henne, A. Mingote, H. Santos, and M. S. da Costa. 2005. Distribution of genes for synthesis of trehalose and mannosylglycerate in *Thermus* spp. and direct correlation of these genes with halotolerance. *Appl. Environ. Microbiol.* 71:2460–2466.

Arakawa, T., and S. N. Timasheff. 1982a. Stabilization of protein structure by sugars. *Biochemistry.* 21:6536–6544.

Arakawa, T., and S. N. Timasheff. 1982b. Preferential interactions of proteins with salts in concentrated solutions. *Biochemistry.* 21:6545–6552.

Arakawa, T., and S. N. Timasheff. 1983. Preferential interactions of proteins with solvent components in aqueous amino acid solutions. *Arch. Biochem. Biophys.* 224:169–177.

Arakawa, T., and S. N. Timasheff. 1985a. Mechanism of poly(ethylene glycol) interaction with proteins. *Biochemistry.* 24:6756–6762.

Arakawa, T., and S. N. Timasheff. 1985b. The stabilization of proteins by osmolytes. *Biophys. J.* 47:411–414.

Baldwin, R. L. 1996. How Hofmeister ion interactions affect protein stability. *Biophys. J.* 71:2056–2063.

Blöchl, E., S. Burggraf, G. Fiala, G. Lauerer, G. Huber, H. Huber, R. Rachel, A. Segerer, K. O. Stetter, and P. Völkl. 1995. Isolation, taxonomy and phylogeny of hyperthermophilic microorganisms. *World J. Microbiol. Biotechnol.* 11:9–16.

Borges, N., A. Ramos, N. D. H. Raven, R. J. Sharp, and H. Santos. 2002. Comparative study of the thermostabilizing properties of mannosylglycerate and other compatible solutes on model enzymes. *Extremophiles.* 6:209–216.

Borges, N., J. D. Marugg, N. Empadinhas, M. S. da Costa, and H. Santos. 2004. Specialized roles of the two pathways for the synthesis of mannosylglycerate in osmoadaptation and thermoadaptation of *Rhodothermus marinus*. *J. Biol. Chem.* 279:9892–9898.

Bouveng, H., B. Lindberg, and B. Wickberg. 1955. Low-molecular carbohydrates in algae. *Acta Chem. Scand.* 9:807–809.

Brown, A. D. 1976. Microbial water stress. *Bacteriol. Rev.* 40:803–846.

Busby, T. F., and K. C. Ingham. 1984. Thermal stabilization of antithrombin III by sugars and sugar derivatives and the effects of nonenzymatic glycosylation. *Biochim. Biophys. Acta.* 799:80–89.

Ciulla, R. A., S. Burggraf, K. O. Stetter, and M. F. Roberts. 1994. Occurrence and role of di-*myo*-inositol-1,1′-phosphate in *Methanococcus igneus*. *Appl. Env. Microbiol.* 60:3660–3664.

Chen, L., E. T. Spiliotis, and M. F. Roberts. 1998. Biosynthesis of di-*myo*-inositol-1,1′-phosphate, a novel osmolyte in hyperthermophilic archaea. *J. Bacteriol.* 180:3785–3792.

Chen, L., C. Zhou, H. Yang, and M. F. Roberts. 2000. Inositol-1-phosphate synthase from *Archaeoglobus fulgidus* is a class II aldolase. *Biochemistry.* 39:12415–12423.

da Costa, M. S., H. Santos, and E. A. Galinski. 1998. An overview of the role and diversity of compatible solutes in Bacteria and Archaea. *Adv. Biochem. Eng. Biotechnol.* 61:117–153.

Davis-Searles, P. R., A. J. Saunders, D. A. Erie, D. J. Winzor, and G. J. Pielak. 2001. Interpreting the effect of small uncharged solutes on protein-folding equilibria. *Annu. Rev. Biophys. Biomol. Struct.* 30:271–306.

Empadinhas, N., J. D. Marugg, N. Borges, H. Santos, and M. S. da Costa. 2001. Pathway for the synthesis of mannosylglycerate in the hyperthermophilic archaeon *Pyrococcus horikoshii*. Biochemical and genetic characterization of key enzymes. *J. Biol. Chem.* 276:43580–43588.

Empadinhas, N., L. Albuquerque, A. Henne, H. Santos, and M. S. da Costa. 2003. The bacterium *Thermus thermophilus*, like hyperthermophilic archaea, uses a two-step pathway for the synthesis of mannosylglycerate. *Appl. Environ. Microbiol.* 69:3272–3279.

Empadinhas, N., L. Albuquerque, J. Costa, S. H. Zinder, M. A. Santos, H. Santos, and M. S. da Costa. 2004. A gene from the mesophilic bacterium *Dehalococcoides ethenogenes* encodes a novel mannosylglycerate synthase. *J. Bacteriol.* 186: 4075–4084.

Empadinhas, N. 2004. Pathways for the synthesis of mannosylglycerate in prokaryotes: genes, enzymes and evolutionary implications. Ph. D. thesis, University of Coimbra, Portugal.

Faria, T. Q., S. Knapp, R. Ladenstein, A. L. Maçanita, and H. Santos. 2003. Protein stabilisation by compatible solutes: effect of mannosylglycerate on unfolding thermodynamics and activity of ribonuclease A. *ChemBioChem.* 4:734–741.

Faria, T. Q., J. C. Lima, M. Bastos, A. L. Macanita, and H. Santos. 2004. Protein stabilization by osmolytes from hyperthermophiles: effect of mannosylglycerate on the thermal unfolding of recombinant nuclease a from *Staphylococcus aureus* studied by picosecond time-resolved fluorescence and calorimetry. *J. Biol. Chem.* 279:48680–48691.

Flint, J., E. Taylor, M. Yang, D. N. Bolam, L. E. Tailford, C. Martinez-Fleites, E. J. Dodson, B. G. Davis, H. J. Gilbert, and G. J. Davies. 2005. Structural dissection and high-throughput screening of mannosylglycerate synthase. *Nat. Struct. Mol. Biol.* 12:608–614.

Foord, R. L., and R. J. Leatherbarrow. 1998. Effect of osmolytes on the exchange rates of backbone amide proton in proteins. *Biochemistry.* 37:2969–2978.

Galinski, E. A. 1995. Osmoadaptation in Bacteria. *Adv. Microb. Physiol.* 37:272–328.

Gekko, K., and S. N. Timasheff. 1981. Thermodynamic and kinetic examination of protein stabilization by glycerol. *Biochemistry.* 20:4677–4686.

Gonçalves, L. G., R. Huber, M. S. da Costa, and H. Santos. 2003. A variant of the hyperthermophile *Archaeoglobus fulgidus* adapted to grow at high salinity. *FEMS Microbiol. Lett.* 218:239–244.

Gorkovenko, A., M. F. Roberts, and R. H. White. 1994. Identification, biosynthesis, and function of 1,3,4,6-hexanetetracarboxylic acid in *Methanobacterium thermoautotrophicum* DeltaH. *Appl. Environ. Microbiol.* 60:1249–1253.

Hensel, R. and H. König. 1988. Thermoadaptation of methanogenic bacteria by intracellular ion concentration. *FEMS Microbiol. Lett.* 49:75–79.

Hensel, R. 1993. Proteins of extreme thermophiles. *New Comp. Biochem.* 26:209–221.

Karsten, U., K. D. Barrow, A. S. Mostaert, R. J. King, and J. A. West. 1994. ^{13}C- and ^{1}H-NMR studies on digeneaside in the red alga *Caloglossa leprieurii*. A re-evaluation of its osmotic significance. *Plant Physiol. Biochem.* 32:669–676.

Kaushik, J. K., and R. Bhat. 1998. Thermal stability of proteins in aqueous polyol solutions: role of the surface tension of water in the stabilizing effect of polyols. *J. Phys. Chem. B.* 102:7058–7066.

Kaushik, J. K. and R. Bhat. 1999. A mechanistic analysis of the increase in the thermal stability of proteins in aqueous carboxylic acid salt solutions. *Protein Sci.* 8:222–233.

Lamosa, P., L. O. Martins, M. S. da Costa, and H. Santos. 1998. Effects of temperature, salinity, and medium composition on compatible solute accumulation by *Thermococcus* spp. *Appl. Environ. Microbiol.* 64:3591–3598.

Lamosa, P., A. Burke, R. Peist, R. Huber, M. Y. Liu, G. Silva, C. Rodrigues-Pousada, J. LeGall, C. Maycock, and H. Santos. 2000. Thermostabilization of proteins by diglycerol phosphate, a new compatible solute from the hyperthermophile *Archaeoglobus fulgidus*. *Appl. Environ. Microbiol.* 66:1974–1979.

Lamosa, P., L. Brennan, H. Vis, D. L. Turner, and H. Santos. 2001. NMR structure of *Desulfovibrio gigas* rubredoxin: a model for studying protein stabilization by compatible solutes. *Extremophiles.* 5:303–311.

Lamosa, P., D. L. Turner, R. Ventura, C. Maycock, and H. Santos. 2003. Protein stabilization by compatible solutes: effect of diglycerol phosphate on the dynamics of *Desulfovibrio gigas* rubredoxin studied by NMR. *Eur. J. Biochem.* 270:4606–4614.

Lee, J. C., and S. N. Timasheff. 1981. The stabilization of proteins by sucrose. *J. Biol. Chem.* 256:7193–7201.

Lin, T. Y., and S. N. Timasheff. 1996. On the role of surface tension in the stabilization of globular proteins. *Protein Sci.* 5:372–381.

Liu, Y., and D. W. Bolen. 1995. The peptide backbone plays a dominant role in protein stabilization by naturally occurring osmolytes. *Biochemistry.* 34:12884–12891.

Martin, D. D., R. A. Ciulla, and M. F. Roberts. 1999. Osmoadaptation in *Archaea. Appl. Environ. Microbiol.* 65:1815–1825.

Martins, L. O., and H. Santos. 1995. Accumulation of mannosylglycerate and di-*myo*-inositol-phosphate by *Pyrococcus furiosus* in response to salinity and temperature. *Appl. Environ. Microbiol.* 61:3299–3303.

Martins, L. O., L. S. Carreto, M. S. da Costa, and H. Santos. 1996. New compatible solutes related to di-*myo*-inositol-phosphate in members of the order *Thermotogales*. *J. Bacteriol.* 178:5644–5651.

Martins, L. O., R. Huber, H. Huber, K. O. Stetter, M. S. da Costa, and H. Santos. 1997. Organic solutes in hyperthermophilic *Archaea. Appl. Environ. Microbiol.* 63:896–902.

Martins, L. O., N. Empadinhas, J. D. Marugg, C. Miguel, C. Ferreira, M. S. da Costa, and H. Santos. 1999. Biosynthesis of mannosylglycerate in the thermophilic bacterium *Rhodothermus marinus*. Biochemical and genetic characterization of a mannosylglycerate synthase. *J. Biol. Chem.* 274:35407–35414.

Neves, C., M. S. da Costa, and H. Santos. 2005. Compatible solutes of the hyperthermophile *Palaeococcus ferrophilus*: osmoadaptation and thermoadaptation in the order *Thermococcales*. *Appl. Environ. Microbiol.* 71:8091–8098.

Nunes, O. C., C. M. Manaia, M. S. da Costa, and H. Santos. 1995. Compatible solutes in the thermophilic bacteria *Rhodothermus marinus* and "*Thermus thermophilus*". *Appl. Environ. Microbiol.* 61:2351–2357.

Oren, A. 1999. Bioenergetic aspects of halophilism. *Microbiol. Mol. Biol. Rev.* 63:334–348.

Pais, T. M., P. Lamosa, W. dos Santos, J. LeGall, D. L. Turner, and H. Santos. 2005. Structural determinants of protein stabilization by solutes. The importance of the hairpin loop in rubredoxins. *FEBS J.* 272:999–1011.

Pittz, E. P., and S. N. Timasheff. 1978. Interaction of ribonuclease A with aqueous 2-methyl-2,4-pentanediol at pH 5.8. *Biochemistry* 17:615–623.

Qu, Y., C. L. Bolen, and D. W. Bolen. 1998. Osmolyte-driven contraction of a random coil protein. *Proc. Natl. Acad. Sci. USA.* 95:9268–9273.

Ramakrishnan, V., Q. Teng, and M. W. Adams. 1997. Characterization of UDP amino sugars as major phosphocompounds in the hyperthermophilic archaeon *Pyrococcus furiosus*. *J. Bacteriol.* 179:1505–1512.

Ramos, A., N. D. H. Raven, R. J. Sharp, S. Bartolucci, M. Rossi, R. Cannio, J. Lebbink, J. van der Oost, W. M. de Vos, and H. Santos. 1997. Stabilization of enzymes against thermal stress and freeze-drying by mannosylglycerate. *Appl. Environ. Microbiol.* 63:4020–4025.

Ramos, C. H. I., and R. L. Baldwin. 2002. Sulfate anion stabilization of native ribonuclease A both by anion binding and the Hofmeister effect. *Protein Sci.* 11:1771–1778.

Robertson, D. E., M. F. Roberts, N. Belay, K. O. Stetter, and D. R. Boone. 1990. Occurrence of β-glutamate, a novel osmolyte, in marine methanogenic bacteria. *Appl. Environ. Microbiol.* 56:1504–1508.

Rohlin, L., J. D. Trent, K. Salmon, U. Kim, R. P. Gunsalus, and J. C. Liao. 2005. Heat shock response in *Archaeoglobus fulgidus*. *J. Bacteriol.* 187:6046–6057.

Santos, H., P. Lamosa, A. Burke, and C. Maycock. 1999. Thermostabilisation, osmoprotection, and protection against desiccation of enzymes, and cell components by di-glycerol-phosphate. European patent no. 98670002.9.

Santos, H., and M. S. da Costa. 2001. Organic solutes from thermophiles and hyperthermophiles. *Methods Enzymol.* **334:** 302–315.

Santos, H., and M. S. da Costa. 2002. Compatible solutes of organisms that live in hot saline environments. *Environ. Microbiol.* **4:**501–509.

Santos, H., P. Lamosa, C. Jorge, and M. S. da Costa. 2003. Diglycosyl glyceryl compounds for the stabilisation or preservation of biomaterials. Submitted to the European Patent Office. (International publication number WO 2004/094631 A1).

Santos, H., P. Lamosa, N. D. Raven, L. G. Gonçalves, and M. V. Rodrigues. 2005. Glycerophosphoinositol as a stabilizer and/or preservative of biological materials. Submitted to the European Patent Office.

Santos, H., P. Lamosa, and N. Borges. 2006. Characterization of organic compatible solutes of thermophilic microorganisms, p. 171–198. *In* A. Oren and F. Rainey (ed.), *Methods in Microbiology: Extremophiles.* Elsevier, Amsterdam, The Netherlands.

Scholz, S., J. Sonnenbichler, W. Schäfer, and R. Hensel. 1992. Di-*myo*-inositol-1,1′-phosphate: a new inositol phosphate isolated from *Pyrococcus woesei. FEBS Lett.* **306:**239–242.

Scholz, S., S. Wolff, and R. Hensel. 1998. The biosynthesis pathway of di-*myo*-inositol-1,1′-phosphate in *Pyrococcus woesei. FEMS Microbiol. Lett.* **168:**37–42.

Shima, S., D. A. Hérault, A. Berkessel, and R. K. Thauer. 1998. Activation and thermostabilization effects of cyclic 2,3-diphosphoglycerate on enzymes from the hyperthermophilic *Methanopyrus kandleri. Arch. Microbiol.* **170:**469–472.

Schokley, K. R., D. E. Ward, S. R. Chhabra, S. B. Conners, C. I. Montero, and R. M. Kelly. 2003. Heat shock response by the hyperthermophilic archaeon *Pyrococcus furiosus. Appl. Environ. Microbiol.* **69:**2365–2371.

Silva, Z., N. Borges, L. O. Martins, R. Wait, M. S. da Costa, and H. Santos. 1999. Combined effect of the growth temperature and salinity of the medium on the accumulation of compatible solutes by *Rhodothermus marinus* and *Rhodothermus obamensis. Extremophiles.* **3:**163–172.

Timasheff, S. N. 1993. The control of protein stability and association by weak interactions with water: how do solvents affect these processes? *Annu. Rev. Biophys. Biomol. Struct.* **22:**67–97.

Chapter 8

Membrane Adaptations of (Hyper)Thermophiles to High Temperatures

ARNOLD J. M. DRIESSEN AND SONJA-VEERANA ALBERS

INTRODUCTION

By studying the basic properties of prokaryotes such as size, structure, and metabolic versatility, we now start to understand how these remarkable life forms are so adaptable to environments previously thought to be uninhabitable. It is well established that prokaryotes on Earth can utilize almost any redox couple that yields energy, taking advantage of this energy, while transforming these molecules during metabolism. The ability to grow at the expense of inorganic redox couples allows microbes to occupy niches not available to the more metabolically constrained eukaryotes. Furthermore, the simplicity of the prokaryotic cell structure renders them considerably more resistant to environmental variables (pH, salinity, and temperature) that are hostile or lethal to more complex organisms. This concerns environments with an exceptional low (pH <2.5) or high (pH >10) pH value, high salinity (saturated salt lakes), high pressure (deep sea), and extremely high (above 80°C up to 121°C) or low (around 0°C) temperatures. Insights into the molecular basis of extremophilicity can be used to explain the predominance of prokaryotes in extreme environments on Earth and to speculate perhaps how such life forms may develop on non-Earth environments. Remarkably, many of the organisms that grow in such extreme environments belong to the group of *Archaea*. Woese and coworkers (Woese et al., 1990; Woese, 2004) first recognized this group as a separate domain of life on Earth in addition to the previously known groups *Bacteria* and *Eukarya* (eukaryotes). *Archaea* are also prokaryotes, and in many aspects, they share a global ultrastructure which is very similar to *Bacteria*. On the other hand, their genomic composition differs from that of bacteria and eukaryotes. They can be classified phylogenetically as a separate phylum on the basis of their rRNA sequences, while genomic analysis indicates a mosaic organization wherein one-third of the genes found in *Archaea* do not have homologs in *Bacteria* and *Eukarya*.

Hyperthermophiles are organisms that thrive in extremely hot environments, i.e., hotter than around 80°C (Stetter, 1999). The optimal temperatures are between 80 and 113°C, while a recently discovered strain was found to double even during 24 h in an autoclave at 121°C (Cowen, 2004; Kashefi and Lovley, 2003). These organisms have intrigued mankind for many years, not only because they are a rich source of extremely stable enzymes that can be used in biotechnological applications (Antranikian et al., 2005) but also because lessons can be learned on the molecular basis of thermophilicity (Stetter, 1996). Obviously, thermophilicity originates from adaptations in enzyme and DNA structure (Stetter, 1999). However, in addition, thermophilicity demands special adaptations to the cell envelope and cellular membrane that forms the main barrier that separates the inside of the cell from the outside, the focus of this chapter. Typically, cytoplasmic membranes of bacteria (and eukaryotes) contain lipids that are mainly di-esters from glycerol and two fatty acyl chains. In contrast, archaeal membranes contain predominantly ether lipids in which two isoprenoid chains are ether linked to glycerol or another alcohol. This was first recognized by Kates and coworkers (Seghal et al., 1962) who found such lipids in the membrane of the extremely halophilic archaeon *Halobacterium cutirubrum*. In the meantime, a wide range of ether-type lipids, mostly phospholipids, glycolipids, and phosphoglycolipids, have been identified in various archaea (for a recent review see Koga and Morii, 2005). However, the presence of such ether lipids appears not to be associated with extremophilicity per se. The *Archaea* represent a diverse group of organisms that in addition to the hyperthermophiles and halophiles also include organisms that thrive in more moderate, non-hostile environments. The kingdom of

---

A. J. M. Driessen and S. V. Albers • Molecular Microbiology, Groningen Biomolecular Sciences and Biotechnology Institute, University of Groningen, Haren, The Netherlands.

the *Archaea* is subdivided into the subdomains Euryarchaeota and Crenarchaeota (Woese, 2004). The subdomain Euryarchaeota consists of methanogens, extreme halophiles, thermophiles, and extremely acidophilic thermophiles (Belly and Brock, 1972; Schleper et al., 1995b). Methanogens grow over the whole temperature spectrum where life is found: from cold (psychrophiles) (Saunders et al., 2003) via moderate (mesophiles) (Kandler and Hippe, 1977) to extremely hot environments (extreme thermophiles) (Burggraf et al., 1991). Crenarchaeota comprise one of the most thermophilic organisms known to date, such as *Pyrolobus fumarii* (Blochl et al., 1997), which is able to grow at temperatures up to 113°C, but also the intensively studied extreme hyperthermoacidophile *Sulfolobus acidocaldarius* (Brock et al., 1972) that thrives in hot acidic pools with a temperature of 85°C and a pH of 2.5. The only psychrophilic crenarchaeote discovered until now is *Cenarchaeum symbiosum* which symbiotically inhabits tissue of a temperate water sponge (Preston et al., 1996). This organism grows well at 10°C, which is more than 60°C lower than the growth temperature of all other Crenarchaeota found so far. Despite the enormous difference in extreme and moderate environments, all organisms known share the same biochemical basis for metabolism and proliferation. Here, we discuss our recent insights into the mechanisms of membrane adaptation of *Archaea* and *Bacteria* to high temperatures, with an emphasis on the structure and function of the lipids that constitute the membrane of hyperthermophiles.

FUNCTION OF THE CYTOPLASMIC MEMBRANE IN BIOENERGETICS

The cytoplasmic membrane plays an essential role in many metabolic processes, energy transduction, and signaling. It forms a barrier that separates the cell into two main compartments, the inside, the cytosol, and the outside, the external environment. This membrane consists of a layer of lipids in which proteins are embedded. Because of the permeability characteristics of the hydrophobic lipid membrane, the membrane is impermeable to most ions and polar solutes (nutrients); it controls the movement of solutes into and out of the cell. This requires specific mechanisms that involve proteinaceous transport systems and channels that catalyze the uptake and excretion of solutes and even the secretion of macromolecules such as proteins and oligosaccharides (Albers et al., 2001, 2006; Ren and Paulsen, 2005). In addition, the cytoplasmic membrane fulfils an essential function in the generation of metabolic energy by energy transduction. In this process, the energy of an electrochemical ion gradient across the membrane is transformed into other forms of energy or vice versa (Mitchell and Moyle, 1967; Mitchell, 1967, 1972). Primary energy sources are solar light and the oxidation of all kinds of organic molecules. Metabolic energy can also be obtained in the form of ATP and ADP by substrate level phosphorylation processes. All metabolic energy-generating processes are closely linked, and together they determine the energy status of the cell. The energy transducing systems are located in the cytoplasmic membrane. They utilize the various forms of energy (redox, chemical, or solar) to pump protons or sodium ions across the membrane into the external medium. This activity results in the generation of a transmembrane electrochemical gradient of protons or sodium ions. When protons are extruded, the resulting electrochemical gradient exerts a force on the protons to pull them back into the cell. This force is termed the proton motive force (PMF). The PMF consists of two components: the ΔpH, i.e., the transmembrane concentration gradient of protons, and the $\Delta\psi$, the transmembrane electrical potential, caused by the transport of electrical charge:

$$\text{PMF} = \Delta\psi - 2.303 \, (RT/F)\Delta\text{pH}$$

expressed in mV, in which R is the gas constant, T the absolute temperature (K), and F the Faraday constant. The effect of 1 unit pH difference is 54 mV at 0°C, 59 mV at 25°C, 70 mV at 80°C, and 77 mV at 115°C. In all living organisms, the resulting PMF is negative and typically in the order of −180 mV. The driving force on the protons is directed into the cell. However, the two components of the PMF may have a different sign depending on the environmental conditions (Fig. 1). In organisms that live around pH 7 (neutrophiles), the $\Delta\psi$ has a negative sign (inside negative), whereas the ΔpH is positive ($\text{pH}_{\text{in}}-\text{pH}_{\text{out}}$; inside alkaline versus outside acidic). In organisms that thrive in an extremely acidic environment (acidophiles), the ΔpH is also positive but extremely large as organisms in general try to maintain the intracellular pH near neutrality (Krulwich et al., 1978; Michels and Bakker, 1985; Stingl et al., 2002; van de Vossenberg et al., 1998). In the absence of substantial $\Delta\psi$, this large ΔpH would amount to a very high PMF which is difficult to maintain. In these cells, the large ΔpH is counterbalanced by an inversed $\Delta\psi$ (inside positive), reducing the absolute magnitude of the PMF to values that support growth (see also chapter on *Proton and Ion Transport, Intracellular pH Regulation*). On the other hand, in organisms that thrive in an extremely alkaline environment (alkaliphiles), the ΔpH has a reversed sign (inside acid versus outside alkaline), and a sufficiently large PMF is maintained by the presence of a very large $\Delta\psi$

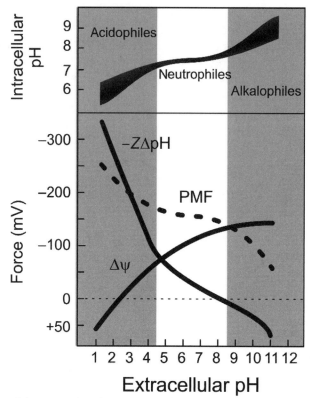

Figure 1. Schematic representation of the magnitude and composition of the proton motive force (PMF) and intracellular pH as a function of the extracellular pH for acido-, neutro-, and alkaliphilic bacteria. The compositions of the PMF, i.e., the transmembrane electrical potential ($\Delta\psi$) and pH gradient ($-Z\Delta pH$), are indicated separately. The scheme is a mosaic obtained from bioenergetic studies of various bacteria, but depending on the membrane proton permeability, the exact magnitude of the various components of the PMF may be different for individual species.

(inside negative) (Krulwich et al., 1998; Padan et al., 2005) (see also chapter on *Proton Transport and Bioenergetics*).

In analogy with the PMF, sodium ion pumps can generate a sodium motive force (SMF) (Albers et al., 2001). The PMF or SMF can be used to transduce their potential energy to metabolic energy requiring processes such as ATP synthesis from ADP and phosphate, transport of specific solutes across the membrane, flagellar rotation, and maintenance of the intracellular pH and turgor. Importantly, this type of energy transduction can only operate if the transmembrane gradient of protons or sodium ions can be maintained across the membrane. Therefore, the lipid composition of the cytoplasmic membrane is such that the ion and proton permeability is limited.

Biological membranes consist of a bi- or monolayer of lipid molecules which form a matrix for the hydrophobic membrane proteins. In nature, an enormous diversity of lipids is found. Lipids have a polar headgroup that sticks into the water phase and hydrophobic hydrocarbon chains that are oriented to the interior of the membrane (Fig. 2A). At the growth temperature of a given organism, the membranes are in a liquid-crystalline state (Melchior and Steim, 1976; Melchior, 2006) which implies substantial dynamics of lipid movement (Fig. 2B), ensuring optimal functioning of the membrane proteins. In particular, the middle of the membrane is liquid like. The structure of the membrane is mainly held together by noncovalent bonds such as Van der Waals bonds and electric interactions. The permeability of membranes to small solutes and ions is restricted due to the high energy that is required for the transfer of a polar solute or ion from the aqueous phase into the apolar interior of the membrane. The lipid layer forms a suitable matrix for proteins such as transport proteins that generate and maintain specific solute concentration gradients across the membrane. Because of the low endogenous permeability of the membrane, the energy expenses needed for maintaining such gradients are limited. Essential, the passive permeation of protons and ions means energy dissipation, which is mostly a loss of heat. However, the composition of the membranes is also of a highly dynamic nature, and even their lipid composition can vastly differ depending on the growth conditions. By changing the lipid composition, organisms can control various

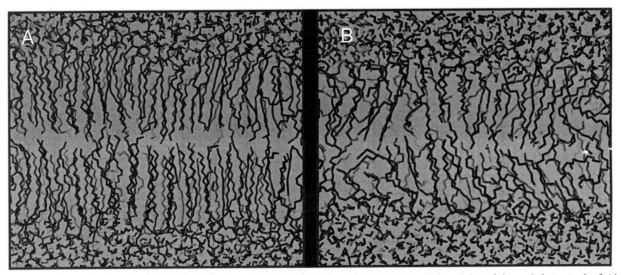

Figure 2. Model of a phospholipid membrane in the gel (A) and fluid (B) phase showing the high mobility of the acyl chains in the fluid membrane phase. The picture represents a slab image of 1-palmitoyl 2-oleoyl phosphatidyl choline bilayers that were obtained by molecular dynamics simulations as described by Heller et al. (1993). The picture was generated with PyMOL (http://pymol.sourceforge.net/).

physicochemical characteristics of the membrane, such as the membrane fluidity and ion permeability in response to environmental factors such as temperature, salinity, or pH. The rate at which protons leak inward is determined by the proton permeability of the membrane and the magnitude of the PMF across the membrane. A proper balance between proton permeability and the rate of outward proton pumping is needed to sustain the PMF. Therefore, any change in lipid composition requires that the PMF remains at a viable level.

CORE LIPIDS OF THE MEMBRANES OF (HYPER-)THERMOPHILES

Membranes of bacteria mainly contain phospholipids with a core structure consisting of a glycerol, a three-carbon alcohol, to which two fatty acid acyl chains are linked via ester bonds (Fig. 3A) (Raetz and Dowhan, 1990). Diacylglycerol functions as the core structure. To the C-3 hydroxyl group of the glycerol platform, a phosphate is linked to which another alcohol can be linked. Phospholipids derived from glycerol are termed phosphoglycerides. The simplest phosphoglyceride is diacylglycerol 3-phosphate which lacks the alcohol addition to the phosphate. All major phosphoglycerides are derived from diacylglycerol 3-phosphate by the formation of an ester bond between the phosphate group and the hydroxyl group of one of several alcohols such as the amino acid serine, ethanolamine, choline, glycerol, or inositol. The fatty acids used in bacteria usually contain an even number of carbon atoms, typically between 14 and 24, most commonly with 16- and 18-carbon fatty acids. The hydrocarbon chain may be saturated or it may contain one or more double bonds, which are usually in the *cis* configuration. The properties of the hydrocarbon region of the phospholipids are dependent on chain length and degree of saturation. Unsaturated fatty acids have lower melting points than saturated acids with the same acyl chain length. Chain length and the position of methyl branching also influence the melting point. These lipids are organized in a bilayer in which the carbon chains are directed toward the inner side of the membrane. In the liquid-crystalline state, the lipid acyl chains are highly dynamic and mobile (see Fig. 1A), and they do not resemble the straight carbon chains that are often depicted in cartoons. The various lipid shapes and the high mobility on one hand ensure proper interactions with the membrane proteins but on the other hand, result in a high occurrence of small membrane defects that occasionally allow the formation of transmembrane water channels along which protons may cross the membrane. Most thermophilic bacteria have a phospholipid composition similar to that of mesophiles, except that the acyl chain composition is vastly different to accommodate the higher growth temperatures as will be discussed later.

The archaeal membrane lipids differ in composition from those of bacteria in three important ways. First, the lipid acyl chains are joined to a glycerol backbone by ether rather than ester linkages. The ether linkage is more resistant to hydrolysis. Second, the acyl chains are branched rather than linear. They are built

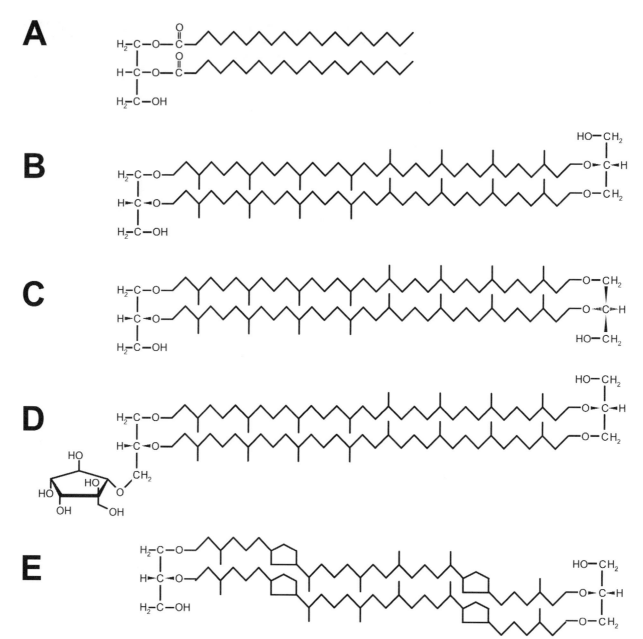

Figure 3. Core structures of phospholipids in bacteria and tetraether lipids in *Archaea*. (A) Diacylglycerol in bacteria; and the archaeal tetraether lipids; (B) caldarchaeol; (C) isocaldarchaeol; (D) calditoglycerocaldarchaeol; and (E) crenarchaeol.

up from repeats of a full saturated five-carbon fragment, isoprene unit, resulting in a phytanyl chain. These phytanyl chains contain methyl groups at every fourth carbon atom in the backbone. Most of the archaeal lipid acyl chains are fully saturated isoprenoids (De Rosa and Gambacorta, 1988; Kates, 1996; Koga et al., 1993; Koga and Morii, 2005; Yamauchi and Kinoshita, 1995) that are resistant to oxidation. The ability of archaeal lipids to resist hydrolysis and oxidation may help these organisms to withstand the extreme environmental conditions. Finally, the stereochemistry of the central glycerol is inverted as compared with the ester-based phospholipids. The isoprenoid chains of archaeal lipids may not only be linked to a glycerol, but also other polyols such as nonitol are used. The structures of archaeal membrane lipids have been extensively reviewed (see for instance Koga and Morii, 2005), and here we will only focus on the core lipid structures.

Halobacteria and most archaea growing under moderate conditions contain lipids which consist of a C_{20} diether lipid core (Kates et al., 1993; Kates, 1996; Koga and Morii, 2005). These lipids form bilayers in a similar way as the ester lipids. Membrane

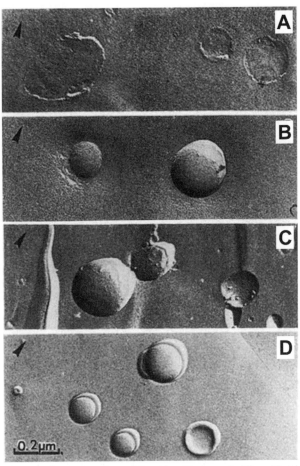

Figure 4. Freeze-fracture (A, C) and freeze-etch (B, D) replicas of *S. acidocaldarius* and of *E. coli* lipid vesicles. Freeze-fracture of *S. acidocaldarius* (A) and *E. coli* lipid (B). The tetraether lipid vesicles show no fracture face as they form monolayers that cannot be cut in the middle of the membrane. Freeze-etching of *S. acidocaldarius* (C) and *E. coli* lipid (D), showing the surface of the vesicles. The arrow indicates the direction of shadowing. Bar = 200 nm. Taken from Elferink et al. (1992) with permission.

spanning (bolaform amphiphilic) tetraether lipids are found in extreme thermophiles and acidophiles (De Rosa and Gambacorta, 1988). The core structure, also termed caldarchaeol, consists of an antiparallel arrangement of the two glycerol units (Fig. 3B). The parallel isomer is named isocaldarchaeol (Fig. 3C). These lipids have C_{40} isoprenoid acyl chains which span the entire membrane (Relini et al., 1996). In the hyperthermoacidophile *Sulfolobus solfataricus*, two kinds of tetraether core lipids are found, the diglycerocaldarchaeol (Fig. 3B) and glycerononitocaldarchaeol (Fig. 3D). In the latter core lipid, the two polyol backbones that are bound to the C_{40} isoprenoid diol are glycerol and nonitol (also called calditol). In nature and in fossil records, also other structures are found such as C_{30}, C_{31}, and C_{35} chains, with a limited number of methyl branches (Schouten et al., 2000). The mono- and dimethyl branched hydrocarbons might not have been derived from archaea but from thermophilic bacteria such as *Thermotoga*. Another modification which is of special interest is the cyclopentane ring containing lipid crenarchaeol as found in *Sulfolobus* spp. (Fig. 3E) (De Rosa and Gambacorta, 1988; Gulik et al., 1988, 1985). Freeze-fracturing of these membranes reveals that cleavage between two leaflets of the membrane does not occur, which means that the water-facing sides of the membrane are connected and cannot be separated (Beveridge et al., 1993; Choquet et al., 1992; Elferink et al., 1992) (Fig. 4). Tetraether lipids from *Thermoplasma acidophilum* and *Sulfolobus solfataricus* form monolayer black lipid membranes of a constant thickness of 2.5 to 3.0 nm (Elferink et al., 1992; Gliozzi et al., 1982; Stern et al., 1992) (Fig. 4). This monolayer type of organization gives the membrane a high degree of rigidity (Bartucci et al., 2005; Elferink et al., 1994; Fan et al., 1995; Gliozzi et al., 1983; Mirghani et al., 1990; Thompson et al., 1992).

Methanopyrus kandleri is a hyperthermophilic methanogen that grows optimal at 98°C (Burggraf et al., 1991). Membranes of this organism contain an unsaturated diether core lipid termed digeranylgeranylglycerol (unsaturated archaeol) (Hafenbradl et al., 1996, 1993). Via a phosphate, the glycerol moiety of this lipid can be found to be bound to choline, ethanolamine, inositol, glycerol, serine, or N-acetylglucosamine, giving rise to six different phospholipid species. Unsaturated archaeol- and unsaturated hydroxyarchaeol-based phospholipids have also been found in the psychrophilic methoarchaeon *Methanococcoides burtoni* isolated from Ace lake in Antarctica (Nichols et al., 2004). Moreover, recent studies on the lipids of methanogens revealed that they reflect the phylogenetic relationships of archaeal organisms and that the lipid composition can be used as a tool for taxonomic and ecological studies of *Archaea* (Schouten et al., 2000; Sinninghe Damste et al., 2002a, 2002b).

PHYSIOCHEMICAL PROPERTIES OF MEMBRANES OF (HYPER)THERMOPHILES

As mentioned before, the presence of ether linkages in archaeal lipids is not an exclusive adaptation to the extreme growth environments. However, a main feature of the ether-linked archaeal lipids is their phase transition temperature that is much lower than that of most fatty acyl ester lipids. For instance, the phase transition temperature of the total polar tetraether lipid extract from *Thermoplasma acidophilum* is between −20 and −15°C (Blocher et al., 1990), whereas bacterial phospholipids ester linked to typical 16- and 18-carbon saturated acyl chains show melting points in the range of 40 to 50°C. The melting point of fatty acyl ester lipid membranes is different depending on the chain length, number of double bonds, and the position of methyl branching. Typically, bacterial membranes consist of a mixture of lipids that ensure the membrane to be in a liquid-crystalline phase at the respective growth temperature. So, in psychrophilic bacteria, mostly single and double unsaturated acyl chains are found with a bias toward the shorter acyl chain lengths (≤18). On the other hand, in thermophiles, the lipids are most saturated and on average with long acyl chain lengths (≥16) (Russell, 1983).

Because of the low melting point of the archaeol- and caldarchaeol-based polar lipid membranes of archaea, these membranes are in a liquid-crystalline phase over a wide temperature range of 0 and 100°C that is physiologically relevant (De Rosa et al., 1986). One of the reasons for the higher stability and melting behavior of the phytanyl chain could be the reduced segmentary motion of tertiary carbon atoms (i.e., rotation of carbon atoms that are bound to three other C-atoms, resulting in kinks in the acyl chain). The segmental motion in the phytanyl chains is hindered due to the methyl side groups, which is particularly pronounced in the lamellar phase and prevents kink formation in the phytanyl chains. This restriction in hydrocarbon chain mobility may not only affect the melting behavior of these lipids, but it may also reduce the permeability of the archaeal membrane as will be discussed in the next section.

Another characteristic feature of the archaeal lipid membrane is their extremely low permeability to solutes and ions. Liposomes composed of tetraether lipids derived from *Sulfolobus acidocaldarius* are more stable than those of bacterial bilayer lipids and exhibit a very low proton permeability at moderate temperatures (Elferink et al., 1994; Gliozzi et al., 1982; van de Vossenberg et al., 1995). Even at extreme temperatures, the proton permeability of tetraether lipids is sufficiently low to allow the generation of a high PMF (see below). Such liposomes are essential impermeable for small solutes such as the fluorophore carboxyfluorescein, and the permeation characteristics are only marginally temperature dependent consistent with the lack of a lipid phase transition in the temperature range studied (Chang, 1994; Elferink et al., 1994; Mirghani et al., 1990; Relini et al., 1996). A study on synthetic membrane spanning lipids revealed that in particular, the highly branched isoprenoid chains are responsible for the lowered proton and carboxyfluorescein permeability (Yamauchi and Kinoshita, 1995; Yamauchi et al., 1994). For this reason, phospholipids with a phytanyl acyl chain are often used in black lipid membranes to allow the undisturbed study of channel proteins.

Ether links are far more resistant to oxidation and high temperatures than ester links. Consequently, liposomes prepared from archaeal tetraether lipids are more thermostable (Chang, 1994; Elferink et al., 1994). Therefore, ether linkages are considered to be much more suitable for a thermophilic way of life. However, the chemical stability of these lipids does not seem to be an absolute requirement. For instance, a significant fraction of the lipids of the hyperthermophilic archaeon *Methanopyrus kandleri* consist of the chemically labile allyl ether lipid (Hafenbradl et al., 1993). Such a bond is as labile as an ester bond. In contrast to ester links, ether links are not susceptible to degradation at alkaline pH (saponification) and enzymatic degradation by phospholipases (Choquet et al., 1994). The stability of liposomes of tetraether lipids is superior to cholesterol-stabilized liposomes prepared from saturated synthetic lipids that resemble bacterial lipids (Horikoshi, 1998). For this reason, liposomes composed of tetraether lipids are considered to be promising vehicles for drug delivery in medical applications (Conlan et al., 2001; Krishnan et al., 2003; Patel and Sprott, 1999; Patel et al., 2002).

MEMBRANE ADAPTATIONS TO HEAT STRESS

Bacteria respond to changes in ambient temperature through adaptations of the lipid composition of their cytoplasmic membranes (Gaughran, 1947). These changes are needed to keep the membrane in a liquid-crystalline state (Russell, 1983; Russell and Fukunaga, 1990) and to limit the proton permeation rates. At higher temperatures, this can be done by increasing the chain length of the lipid acyl chains, the ratio of iso/anteiso branching, and/or the degree of saturation of the acyl chain (Prado et al., 1988b, 1988a; Reizer et al., 1985; Svobodová and Svoboda, 1988). Because isoprenoid ether lipid membranes of *Archaea* are in a liquid-crystalline state at all physiologically relevant temperatures, archaea live at as low as 1°C and as high as 115°C with the same archaeol- and caldarchaeol-based polar lipid membranes. These lipids also ensure a low proton permeability of the membrane (see next section). Thus, they can meet the two requirements for a functional biological membrane, i.e., a liquid-crystalline phase and a low permeability. Many archaea have caldarchaeol-based polar lipids with smaller amounts of archaeol-based lipid, for instance the hyperthermophilic *Pyrococcus furiosus* (optimal growth temperature, 98°C) (Sprott et al., 1997), the moderately thermophilic *Methanothermobacter thermautotrophicus* (65°C) (Nishihara and Koga, 1987), the mesophilic *Methanobacterium formicicum* (37°C) (Koga et al., 1993), and the *Methanogenium cariaci* (23°C) (Koga et al., 1993). The unsaturated archaeol is not restricted to the psychrophilic *Methanococcides burtonii* that can grow at 2°C (Nichols et al., 2004) but is also found in the hyperthermophilic *Methanopyrus kandleri* (98°C) (Nishihara et al., 2002), although at lower concentration. Overall, it seems that the core lipid membrane composition in archaea appears much more invariant to the growth temperature than in bacteria. In some archaea, the lipid composition can be controlled in response to temperature changes. For instance, *Sulfolobus solfataricus* and *Thermoplasma* that contain a high percentage of tetraether lipids in their total lipids (above 90%) alter the degree of cyclization (number of cyclopentane groups) of the C_{40} isoprenoid in their tetraether lipids with increasing growth temperatures (De Rosa and Gambacorta, 1988; Langworthy, 1982). By increasing the degree of the cyclization of the C_{40} isoprenoid chains, the lipids can be packed more tightly, which results in a more restricted motion of the lipids and prevents that the membrane becomes too fluid. *Methanocaldococcus jannaschii* can accommodate its lipid composition upon an increase in the growth temperature by increasing the amount of the more thermostable tetraether lipids relative to diether lipids (Sprott et al., 1991). Also in this case, cyclization of the chains tends to decrease the motion of the lipids and therefore contributes to maintaining membrane fluidity and proton permeability at an acceptable level at the elevated growth temperature. Finally, the content of trans-unsaturation of the isoprenoid chains has been reported to decrease with higher growth temperature in *Methanococcoides burtonii* (Nichols et al., 2004), a phenomenon that is due to the temperature-dependent expression of a desaturase (Goodchild et al., 2004; Saunders et al., 2003) (see also chapter 12 "Psychrophiles: Membrane Adaptations," by Nicholas J. Russell). It is well possible that other changes in lipid composition are prevalent in hyperthermophilic archaea, such as alterations in polar headgroup, the abundance of minor lipid components, and others. Unfortunately, to this date, only few systematic studies have been performed, concerning the adaptations of membrane lipid composition in hyperthermophilic archaea.

MEMBRANE PERMEABILITY UNDER PHYSIOLOGICAL CONDITIONS

High temperatures impose a burden on the cellular metabolism and require a higher stability of enzymes and other macromolecules (Adams, 1993). Since the basis for membrane permeation is diffusion (mainly the diffusion of water in case of proton permeation), the ion permeability of the membrane increases with the temperature. When the coupling ions, i.e., protons or sodium ions, permeate too fast, cells are unable to establish a sufficiently high PMF or SMF to sustain growth. The permeability of the cytoplasmic membrane is thus, amongst others, a major factor in determining the maximum growth temperature. Liposomes have been prepared from lipids extracted from a variety of organisms that grow optimally at different temperatures. With each of these liposomes, the proton permeability increases with the temperature, with activation energies for proton permeation ($\Delta G'_{H+}$) ranging from 40 to 55 kJ/mol (Fig. 5). The membranes of these liposomes become highly permeable to protons at temperatures above the growth temperature of the microorganism from which the lipids were derived. On the other hand, the

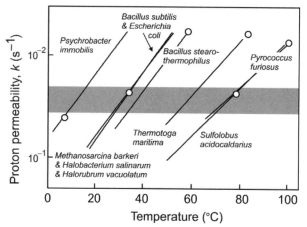

Figure 5. Schematic presentation of the proton permeabilities of membranes from *Archaea* and *Bacteria* that live at different temperatures. Data were obtained by measuring the proton permeabilities of liposomes made of the lipids of the respective organisms at different temperatures. At the respective growth temperatures, the proton permeabilities fall within a narrow window (gray bar). The bacteria *T. maritima* and *B. stearothermophilus* have a permeability that is higher than that in the other organisms at their respective growth temperature. From van de Vossenberg et al. (1995) and adapted, with permission.

sodium ion permeability of these same liposomes is several orders of magnitudes lower than the proton permeability. The basal sodium ion permeability was found to depend on the temperature with an activation energy $\Delta G'_{Na+}$ of about 47 kJ/mol, which is of the same order as observed for proton permeations. However, the sodium ion permeability was barely dependent on the composition of the membranes (van de Vossenberg et al., 1995). The most important finding of these studies is that the proton permeability of most bacterial and all archaeal membranes at the temperature of growth is maintained within a narrow window (proton permeability coefficient near 10^{-9} cm/s) (van de Vossenberg et al., 1995), suggesting a degree of homeostasis (Fig. 5). The homeostasis of proton permeability is termed "homeo-proton permeability adaptation." On the basis of these findings, it has been suggested that the proton permeability of the membrane can be modulated by adjusting the lipid composition of the membranes. This concept was validated in the bacterium *Bacillus subtilis* grown at and within the boundaries of its growth temperature range of 13 to 50°C (van de Vossenberg et al., 1999). Over this whole temperature range, the proton permeability of the *B. subtilis* membranes at the respective growth temperatures was relatively constant. Interestingly, the fluidity of the membrane did not remain constant but increased significantly with temperature. From the observations described above, it is evident that the proton permeability can be an important growth-limiting factor at the upper boundary of the growth temperature.

In contrast, the permeability of the membranes for sodium ions at different growth temperatures is not constant but increases exponentially with temperature in a similar way for all organisms studied. Since the sodium permeability is several orders of magnitude lower than the proton permeability, it will be possible to maintain a high SMF even at high temperatures. The lipid composition of the membrane thus has only a minor effect on the membrane permeability to sodium ions, while the rate of sodium ion permeation seems mainly to be influenced by the temperature.

Unlike in psychrophilic and mesophilic bacteria and archaea, in thermophilic bacteria studied thus far, the proton permeability of their membranes at the respective growth temperatures appears much higher than the proton permeability found at the growth temperature in the other organisms (van de Vossenberg et al., 1995). These thermophilic bacteria, such as *B. stearothermophilus* and *Thermotoga maritima*, appear to be unable to reduce the proton permeability of their membrane at the high temperatures at which they grow. Thermophilic bacteria thus have at their growth temperature membranes that are relatively leaky for protons. Some moderately thermophilic bacteria can compensate for the high proton leak by drastically increasing the respiration rate and therefore the rate of proton pumping (de Vrij et al., 1988). Other thermophilic bacteria shift their energy transduction mechanisms to the less permeable sodium ion as coupling ion. This is for instance observed in the bacterium *Caloramator fervidus* (Speelmans et al., 1993b, 1993a), an organism that can grow at a higher temperature than *B. stearothermophilus*, i.e., 70 versus 65°C (Esser and Souza, 1974; Patel et al., 1987). *C. fervidus* has a Na^+-translocating ATPase that excretes sodium ions at the expense of ATP (Lolkema et al., 1994). As a result, an SMF is generated that is the driving force for energy requiring processes such as solute transport. However, a major penalty of this way of life is an inability to maintain a constant intracellular pH. Because of the high proton permeability, growth of *C. fervidus* is confined to a narrow niche, i.e., an environment with a pH near neutrality. In the thermoalkaliphile *Anaerobranca gottschalkii* (*Thermoalkalibacter bogoriae*), the PMF is a result of a reversed ΔpH (inside acid) and a normal $\Delta\psi$ (inside negative) (Prowe et al., 1996). This reversed ΔpH is most likely generated by the passive influx and presumably high membrane permeability to protons in response to the $\Delta\psi$ generated by Na^+ pumping. In this way, these cells can maintain a near-to-neutral cytoplasmic pH despite a highly alkaline environment.

Thermoacidophiles experience a very high ΔpH across their membranes, which needs to be maintained at the high growth temperatures. For instance, this ΔpH can be up to 4 pH units in *Picrophilus oshimae*, an archaeon that grows at 65°C and pH 0.8 (Schleper et al., 1995a; van de Vossenberg et al., 1998). The very high ΔpH is compensated by an inverted $\Delta\psi$ (negative inside) to yield a normal PMF. Inversion of the $\Delta\psi$ is likely due to the active influx of positively charged ions such as K^+. Obviously, a large ΔpH can only be maintained when the proton permeability of the membrane is very low. Strikingly, the proton permeability of the membranes from thermoacidophilic archaea *P. oshimae* and *S. solfataricus* is at the elevated temperatures at which the organisms grow as low as that of membranes of mesophilic bacteria grown at the mesophilic growth temperatures. It therefore appears that the growth temperature has the most dramatic effect on the proton permeability of biological membranes and that the temperature-dependent variations in membrane composition are adaptations that restrict the membrane ion permeability to allow the generation of a PMF or SMF at a viable level during growth. To this end, however, only few studies that record the thermodynamic parameters in (hyper)thermophiles have been reported, and it would be of great interest to compare the levels of the PMF in organisms that have been grown in a wide temperature range.

PERSPECTIVE AND CONCLUDING REMARKS

Of all extreme conditions that extremophiles face, temperature seems to have a major effect on the physiochemical properties of membranes and influences the phase behavior and permeability characteristics of the membrane. The lipids and membrane proteins of (hyper-)thermophiles are well adapted to this environmental stress factor. In particular, the tetraether lipids of hyperthermophilic archaea are well equipped to maintain a viable PMF across their membrane under these extreme heat conditions. Also, major changes in lipid composition have been observed in some archaea in response to growth temperature, and these likely have an effect on the proton permeability. In addition, minor lipid components may also have profound effects on the overall membrane stability. The ability to grow at a particular temperature is not only determined by membrane composition but is the integrated result of the thermostability of all metabolic enzymes and cellular structures (Adams, 1993). In addition, many hyperthermophiles also face other extreme conditions such as acidic pH or high pressure. High pressure can promote the growth of some hyperthermophiles without altering its temperature growth range (Bernhardt et al., 1988; Canganella et al., 1997), but the exact molecular basis of this barophilic behavior is unknown (Horikoshi, 1998). Possibly, barophilicity also involves specific adaptations to the membrane lipid composition, but this has not yet been explored. Interestingly, tetraether liposomes are extremely pressure resistant in vitro (Choquet et al., 1994). The thermoresistance and tolerance of the membranes of hyperthermophiles is likely a result of an interplay between lipids and proteins. Indeed, as a result of protein–lipid interactions, the membrane composition has been shown to influence protein thermostability. For instance, reconstitution studies with terminal oxidase complexes isolated from different sources demonstrate that thermostability is determined both by the protein origin and by the lipid composition (Elferink et al., 1995). At this stage, this interplay between protein and lipid is only poorly understood and will remain a challenge for future work.

Acknowledgments. A. J. M. Driessen was supported by a the Van der Leeuw Programme of the Earth and Life Sciences Foundation (ALW), which is subsidized by the Dutch Science Organization (NWO). S. V. Albers was supported by a VENI- and VIDI-grant from NWO.

REFERENCES

Adams, M. W. 1993. Enzymes and proteins from organisms that grow near and above 100 degrees C. *Annu. Rev. Microbiol.* 47:627–658.

Albers, S. V., Z. Szabo, and A. J. M. Driessen. 2006. Protein secretion in the Archaea: multiple paths towards a unique cell surface. *Nat. Rev. Microbiol.* 4:537–547.

Albers, S. V., J. L. C. M. van de Vossenberg, A. J. M. Driessen, and W. N. Konings. 2001. Bioenergetics and solute uptake under extreme conditions. *Extremophiles* 5:285–294.

Antranikian, G., C. E. Vorgias, and C. Bertoldo. 2005. Extreme environments as a resource for microorganisms and novel biocatalysts. *Adv. Biochem. Eng. Biotechnol.* 96:219–262.

Bartucci, R., A. Gambacorta, A. Gliozzi, D. Marsh, and L. Sportelli. 2005. Bipolar tetraether lipids: chain flexibility and membrane polarity gradients from spin-label electron spin resonance. *Biochemistry* 44:15017–15023.

Belly, R. T., and T. D. Brock. 1972. Cellular stability of a thermophilic, acidophilic mycoplasma. *J. Gen. Microbiol.* 73:465–469.

Bernhardt, G., R. Jaenicke, H. D. Ludemann, H. Konig, and K. O. Stetter. 1988. High pressure enhances the growth rate of the thermophilic archaebacterium *Methanococcus thermolithotrophicus* without extending its temperature range. *Appl. Environ. Microbiol.* 54:1258–1261.

Beveridge, T. J., C. G. Choquet, G. B. Patel, and G. D. Sprott. 1993. Freeze-fracture planes of methanogen membranes correlate with the content of tetraether lipids. *J. Bacteriol.* 175:1191–1197.

Blocher, D., R. Gutermann, B. Henkel, and K. Ring. 1990. Physicochemical characterization of tetraether lipids from *Thermoplasma acidophilum*. V. Evidence for the existence of a metastable state in lipids with acyclic hydrocarbon chains. *Biochim. Biophys. Acta* 1024:54–60.

Blochl, E., R. Rachel, S. Burggraf, D. Hafenbradl, H. W. Jannasch, and K. O. Stetter. 1997. *Pyrolobus fumarii*, gen. and sp. nov., represents a novel group of archaea, extending the upper temperature limit for life to 113 degrees C. *Extremophiles* 1:14–21.

Brock, T. D., K. M. Brock, R. T. Belly, and R. L. Weiss. 1972. *Sulfolobus*: a new genus of sulfur-oxidizing bacteria living at low pH and high temperature. *Arch. Mikrobiol.* 84:54–68.

Burggraf, S., K. O. Stetter, P. Rouviere, and C. R. Woese. 1991. *Methanopyrus kandleri*: an archaeal methanogen unrelated to all other known methanogens. *Syst. Appl. Microbiol.* 14:346–351.

Canganella, F., J. M. Gonzalez, M. Yanagibayashi, C. Kato, and K. Horikoshi. 1997. Pressure and temperature effects on growth and viability of the hyperthermophilic archaeon *Thermococcus peptonophilus*. *Arch. Microbiol.* 168:1–7.

Chang, E. L. 1994. Unusual thermal stability of liposomes made from bipolar tetraether lipids. *Biochem. Biophys. Res. Commun.* 202:673–679.

Choquet, C. G., G. B. Patel, T. J. Beveridge, and G. D. Sprott. 1992. Formation of unilamellar liposomes from total polar lipid extracts of methanogens. *Appl. Environ. Microbiol.* 58:2894–2900.

Choquet, C. G., G. B. Patel, T. J. Beveridge, and G. D. Sprott. 1994. Stability of pressure-extruded liposomes made from archaeobacterial ether lipids. *Appl. Microbiol. Biotechnol.* 42:375–384.

Conlan, J. W., L. Krishnan, G. E. Willick, G. B. Patel, and G. D. Sprott. 2001. Immunization of mice with lipopeptide antigens encapsulated in novel liposomes prepared from the polar lipids of various Archaeobacteria elicits rapid and prolonged specific protective immunity against infection with the facultative intracellular pathogen, *Listeria monocytogenes*. *Vaccine* 19:3509–3517.

Cowen, D. A. 2004. The upper temperature of life—where do we draw the line? *Trends Microbiol.* 12:58–60.

De Rosa, M., and A. Gambacorta. 1988. The lipids of archaebacteria. *Prog. Lipid Res.* 27:153–175.

De Rosa, M., A. Gambacorta, and A. Gliozzi. 1986. Structure, biosynthesis, and physicochemical properties of archaebacterial lipids. *Microbiol. Rev.* 50:70–80.

de Vrij, W., R. A. Bulthuis, and W. N. Konings. 1988. Comparative study of energy-transducing properties of cytoplasmic membranes from mesophilic and thermophilic *Bacillus* species. *J. Bacteriol.* 170:2359–2366.

Elferink, M. G., T. Bosma, J. S. Lolkema, M. Gleiszner, A. J. M. Driessen, and W. N. Konings. 1995. Thermostability of respiratory terminal oxidases in the lipid environment. *Biochim. Biophys. Acta* 1230:31–37.

Elferink, M. G., J. G. de Wit, R. Demel, A. J. M. Driessen, and W. N. Konings. 1992. Functional reconstitution of membrane proteins in monolayer liposomes from bipolar lipids of *Sulfolobus acidocaldarius*. *J. Biol. Chem.* 267:1375–1381.

Elferink, M. G., J. G. de Wit, A. J. M. Driessen, and W. N. Konings. 1994. Stability and proton-permeability of liposomes composed of archaeal tetraether lipids. *Biochim. Biophys. Acta* 1193:247–254.

Esser, A. F., and K. A. Souza. 1974. Correlation between thermal death and membrane fluidity in *Bacillus stearothermophilus*. *Proc. Natl. Acad. Sci. USA* 71:4111–4115.

Fan, Q., A. Relini, D. Cassinadri, A. Gambacorta, and A. Gliozzi. 1995. Stability against temperature and external agents of vesicles composed of archael bolaform lipids and egg PC. *Biochim. Biophys. Acta* 1240:83–88.

Gaughran, E. R. L. 1947. The saturation of bacterial lipids as a function of temperature. *J. Bacteriol.* 53:506–509.

Gliozzi, A., R. Rolandi, M. De Rosa, and A. Gambacorta. 1982. Artificial black membranes from bipolar lipids of thermophilic Archaebacteria. *Biophys. J.* 37:563–566.

Gliozzi, A., R. Rolandi, M. De Rosa, and A. Gambacorta. 1983. Monolayer black membranes from bipolar lipids of archaebacteria and their temperature-induced structural changes. *J. Membr. Biol.* 75:45–56.

Goodchild, A., N. F. Saunders, H. Ertan, M. Raftery, M. Guilhaus, P. M. Curmi, and R. Cavicchioli. 2004. A proteomic determination of cold adaptation in the Antarctic archaeon, *Methanococcoides burtonii*. *Mol. Microbiol.* 53:309–321.

Gulik, A., V. Luzzati, M. De Rosa, and A. Gambacorta. 1985. Structure and polymorphism of bipolar isopranyl ether lipids from archaebacteria. *J. Mol. Biol.* 182:131–149.

Gulik, A., V. Luzzati, M. DeRosa, and A. Gambacorta. 1988. Tetraether lipid components from a thermoacidophilic archaebacterium. Chemical structure and physical polymorphism. *J. Mol. Biol.* 201:429–435.

Hafenbradl, D., M. Keller, and K. O. Stetter. 1996. Lipid analysis of *Methanopyrus kandleri*. *FEMS Microbiol. Lett.* 136:199–202.

Hafenbradl, D., M. Keller, R. Thiericke, and K. O. Stetter. 1993. A novel unsaturated archaeal ether core lipid from the hyperthermophile *Methanopyrus kandleri*. *Syst. Appl. Microbiol.* 16:165–169.

Heller, M., M. Schaeffer, and K. Schulten. 1993. Molecular dynamics simulation of a bilayer of 200 lipids in the gel and in the liquid-crystal phases. *J. Phys. Chem.* 97:8343–8360.

Horikoshi, K. 1998. Barophiles: deep-sea microorganisms adapted to an extreme environment. *Curr. Opin. Microbiol.* 1:291–295.

Kandler, O., and H. Hippe. 1977. Lack of peptidoglycan in the cell walls of *Methanosarcina barkeri*. *Arch. Microbiol.* 113:57–60.

Kashefi, K., and D. R. Lovley. 2003. Extending the upper temperature limit for life. *Science* 301:934.

Kates, M. 1996. Structural analysis of phospholipids and glycolipids in extremely halophilic archaebacteria, p. 113–128. *In J. Microbiol. Meth.* 25.

Kates, M., N. Moldoveanu, and L. C. Stewart. 1993. On the revised structure of the major phospholipid of *Halobacterium salinarium*. *Biochim. Biophys. Acta* 1169:46–53.

Koga, Y., and H. Morii. 2005. Recent advances in structural research on ether lipids from archaea including comparative and physiological aspects. *Biosci. Biotechnol. Biochem.* 69:2019–2034.

Koga, Y., M. Nishihara, H. Morii, and M. Kagawa-Matsushita. 1993. Ether polar lipids of methanogenic bacteria: structures, comparative aspects, and biosyntheses. *Microbiol. Rev.* 57:164–182.

Krishnan, L., S. Sad, G. B. Patel, and G. D. Sprott. 2003. Archaeosomes induce enhanced cytotoxic T lymphocyte responses to entrapped soluble protein in the absence of interleukin 12 and protect against tumor challenge. *Cancer Res.* 63:2526–2534.

Krulwich, T. A., L. F. Davidson, S. J. Filip, Jr., R. S. Zuckerman, and A. A. Guffanti. 1978. The protonmotive force and β-galactoside transport in *Bacillus acidocaldarius*. *J. Biol. Chem.* 253:4599–4603.

Krulwich, T. A., M. Ito, R. Gilmour, D. B. Hicks, and A. A. Guffanti. 1998. Energetics of alkaliphilic *Bacillus* species: physiology and molecules. *Adv. Microb. Physiol.* 40:401-438.

Langworthy, T. A. 1982. Lipids of *Thermoplasma*. *Methods Enzymol.* 88:396-406.

Lolkema, J. S., G. Speelmans, and W. N. Konings. 1994. Na(+)-coupled versus H(+)-coupled energy transduction in bacteria. *Biochim. Biophys. Acta* 1187:211-215.

Melchior, D. L. 2006. Lipid phase transitions and regulation of membrane fluidity in prokaryotes. *Curr. Top. Membr. Transp.* 17:263-316.

Melchior, D. L., and J. M. Steim. 1976. Thermotropic transitions in biomembranes. *Annu. Rev. Biophys. Bioeng.* 5:205-238.

Michels, M., and E. P. Bakker. 1985. Generation of a large, protonophore-sensitive proton motive force and pH difference in the acidophilic bacteria *Thermoplasma acidophilum* and *Bacillus acidocaldarius*. *J. Bacteriol.* 161:231-237.

Mirghani, Z., D. Bertoia, A. Gliozzi, M. De Rosa, and A. Gambacorta. 1990. Monopolar-bipolar lipid interactions in model membrane systems. *Chem. Phys. Lipids* 55:85-96.

Mitchell, P. 1967. Translocations through natural membranes. *Adv. Enzymol. Relat. Areas Mol. Biol.* 29:33-87.

Mitchell, P. 1972. Chemiosmotic coupling in energy transduction: a logical development of biochemical knowledge. *J. Bioenerg.* 3:5-24.

Mitchell, P., and J. Moyle. 1967. Chemiosmotic hypothesis of oxidative phosphorylation. *Nature* 213:137-139.

Nichols, D. S., M. R. Miller, N. W. Davies, A. Goodchild, M. Raftery, and R. Cavicchioli. 2004. Cold adaptation in the Antarctic Archaeon *Methanococcoides burtonii* involves membrane lipid unsaturation. *J. Bacteriol.* 186:8508-8515.

Nishihara, M., and Y. Koga. 1987. Extraction and composition of polar lipids from the archaebacterium, *Methanobacterium thermoautotrophicum*: effective extraction of tetraether lipids by an acidified solvent. *J. Biochem.* (Tokyo) 101:997-1005.

Nishihara, M., H. Morii, K. Matsuno, M. Ohga, K. O. Stetter, and Y. Koga. 2002. Structural analysis by reductive cleavage with LiAlH4 of an allyl ether choline-phospholipid, archaetidylcholine, from the hyperthermophilic methanoarchaeon *Methanopyrus kandleri*. *Archaea* 1:123-131.

Padan, E., E. Bibi, M. Ito, and T. A. Krulwich. 2005. Alkaline pH homeostasis in bacteria: new insights. *Biochim. Biophys. Acta* 1717:67-88.

Patel, B. K. C., C. Monk, H. Littleworth, H. W. Morgan, and R. M. Daniel. 1987. *Clostridium fervidus* sp. nov., a new chemoorganotrophic acetogenic thermophile. *Int. J. Syst. Bacteriol.* 37:123-126.

Patel, G. B., A. Omri, L. Deschatelets, and G. D. Sprott. 2002. Safety of archaeosome adjuvants evaluated in a mouse model. *J. Liposome Res.* 12:353-372.

Patel, G. B., and G. D. Sprott. 1999. Archaeobacterial ether lipid liposomes (archaeosomes) as novel vaccine and drug delivery systems. *Crit. Rev. Biotechnol.* 19:317-357.

Prado, A., M. S. da Costa, J. Laynez, and V. M. Madeira. 1988a. Physical properties of membrane lipids isolated from a thermophilic eubacterium (*Thermus* sp.). *Adv. Exp. Med. Biol.* 238:47-58.

Prado, A., M. S. da Costa, and V. M. Madeira. 1988b. Effect of growth temperature on the lipid composition of two strains of *Thermus* sp. *J. Gen. Microbiol.* 134:1653-1660.

Preston, C. M., K. Y. Wu, T. F. Molinski, and E. F. DeLong. 1996. A psychrophilic crenarchaeon inhabits a marine sponge: *Cenarchaeum symbiosum* gen. nov., sp. nov. *Proc. Natl. Acad. Sci. USA* 93:6241-6246.

Prowe, S. G., J. L. C. M. van de Vossenberg, A. J. M. Driessen, G. Antranikian, and W. N. Konings. 1996. Sodium-coupled energy transduction in the newly isolated thermoalkaliphilic strain LBS3. *J. Bacteriol.* 178:4099-4104.

Raetz, C. R., and W. Dowhan. 1990. Biosynthesis and function of phospholipids in *Escherichia coli*. *J. Biol. Chem.* 265:1235-1238.

Reizer, J., N. Grossowicz, and Y. Barenholz. 1985. The effect of growth temperature on the thermotropic behavior of the membranes of a thermophilic *Bacillus*. Composition-structure-function relationships. *Biochim. Biophys. Acta* 815:268-280.

Relini, A., D. Cassinadri, Q. Fan, A. Gulik, Z. Mirghani, R. M. De, and A. Gliozzi. 1996. Effect of physical constraints on the mechanisms of membrane fusion: bolaform lipid vesicles as model systems. *Biophys. J.* 71:1789-1795.

Ren, Q., and I. T. Paulsen. 2005. Comparative analyses of fundamental differences in membrane transport capabilities in prokaryotes and eukaryotes. *PLoS Comput. Biol.* 1:e27.

Russell, N. J. 1983. Adaptation to temperature in bacterial membranes. *Biochem. Soc. Trans.* 11:333-335.

Russell, N. J., and N. Fukunaga. 1990. A comparison of thermal adaptation of membrane lipids in psychrophilic and thermophilic bacteria. *FEMS Microbiol. Rev.* 75:171-182.

Saunders, N. F., T. Thomas, P. M. Curmi, J. S. Mattick, E. Kuczek, R. Slade, J. Davis, P. D. Franzmann, D. Boone, K. Rusterholtz, R. Feldman, C. Gates, S. Bench, K. Sowers, K. Kadner, A. Aerts, P. Dehal, C. Detter, T. Glavina, S. Lucas, P. Richardson, F. Larimer, L. Hauser, M. Land, and R. Cavicchioli. 2003. Mechanisms of thermal adaptation revealed from the genomes of the Antarctic Archaea *Methanogenium frigidum* and *Methanococcoides burtonii*. *Genome Res.* 13:1580-1588.

Schleper, C., G. Puehler, I. Holz, A. Gambacorta, D. Janekovic, U. Santarius, H. P. Klenk, and W. Zillig. 1995a. *Picrophilus* gen. nov., fam. nov.: a novel aerobic, heterotrophic, thermoacidophilic genus and family comprising archaea capable of growth around pH 0. *J. Bacteriol.* 177:7050-7059.

Schleper, C., G. Puhler, B. Kuhlmorgen, and W. Zillig. 1995b. Life at extremely low pH. *Nature* 375:741-742.

Schouten, S., E. C. Hopmans, R. D. Pancost, and J. S. Sinninghe Damste. 2000. Widespread occurrence of structurally diverse tetraether membrane lipids: evidence for the ubiquitous presence of low-temperature relatives of hyperthermophiles. *Proc. Natl. Acad. Sci. USA* 97:14421-14426.

Seghal, S. N., M. Kates, and N. E. Gibbons. 1962. Lipids of *Halobacterium cutirubrum*. *Can. J. Biochem. Physiol.* 40:69-81.

Sinninghe Damste, J. S., W. I. Rijpstra, E. C. Hopmans, F. G. Prahl, S. G. Wakeham, and S. Schouten. 2002a. Distribution of membrane lipids of planktonic Crenarchaeota in the Arabian Sea. *Appl. Environ. Microbiol.* 68:2997-3002.

Sinninghe Damste, J. S., S. Schouten, E. C. Hopmans, A. C. van Duin, and J. A. Geenevasen. 2002b. Crenarchaeol: the characteristic core glycerol dibiphytanyl glycerol tetraether membrane lipid of cosmopolitan pelagic crenarchaeota. *J. Lipid Res.* 43:1641-1651.

Speelmans, G., B. Poolman, T. Abee, and W. N. Konings. 1993a. Energy transduction in the thermophilic anaerobic bacterium *Clostridium fervidus* is exclusively coupled to sodium ions. *Proc. Natl. Acad. Sci. USA* 90:7975-7979.

Speelmans, G., B. Poolman, and W. N. Konings. 1993b. Amino acid transport in the thermophilic anaerobe *Clostridium fervidus* is driven by an electrochemical sodium gradient. *J. Bacteriol.* 175:2060-2066.

Sprott, G. D., B. J. Agnew, and G. B. Patel. 1997. Structural features of ether lipids in the archaeobacterial thermophiles *Pyrococcus furiosus*, *Methanopyrus kandleri*, *Methanothermus fervidus*, and *Sulfolobus acidocaldarius*. *Can. J. Microbiol.* 43:467-476.

Sprott, G. D., M. Meloche, and J. C. Richards. 1991. Proportions of diether, macrocyclic diether, and tetraether lipids in

Methanococcus jannaschii grown at different temperatures. *J. Bacteriol.* **173:**3907–3910.

Stern, J., H. J. Freisleben, S. Janku, and K. Ring. 1992. Black lipid membranes of tetraether lipids from *Thermoplasma acidophilum*. *Biochim. Biophys. Acta* **1128:**227–236.

Stetter, K. O. 1996. Hyperthermophiles in the history of life. *Ciba Found. Symp.* **202:**1–10.

Stetter, K. O. 1999. Extremophiles and their adaptation to hot environments. *FEBS Lett.* **452:**22–25.

Stingl, K., E. M. Uhlemann, R. Schmid, K. Altendorf, and E. P. Bakker. 2002. Energetics of *Helicobacter pylori* and its implications for the mechanism of urease-dependent acid tolerance at pH 1. *J. Bacteriol.* **184:**3053–3060.

Svobodová, J., and P. Svoboda. 1988. Membrane fluidity in *Bacillus subtilis*. Physical change and biological adaptation. *Folia Microbiol* (Praha) **33:**161–169.

Thompson, D. H., K. F. Wong, R. Humphry-Baker, J. J. Wheeler, J.-M. Kim, and S. B. Rananavare. 1992. Tetraether bolaform amphiphiles as models of archaebacterial membrane lipids: Raman spectroscopy, 31 P NMR, X-ray scattering, and electron microscopy. *J. Am. Chem. Soc.* **114:**9035–9042.

van de Vossenberg, J. L. C. M., A. J. M. Driessen, M. S. da Costa, and W. N. Konings. 1999. Homeostasis of the membrane proton permeability in *Bacillus subtilis* grown at different temperatures. *Biochim. Biophys. Acta* **1419:**97–104.

van de Vossenberg, J. L. C. M., A. J. M. Driessen, W. Zillig, and W. N. Konings. 1998. Bioenergetics and cytoplasmic membrane stability of the extremely acidophilic, thermophilic archaeon *Picrophilus oshimae*. *Extremophiles* **2:**67–74.

van de Vossenberg, J. L. C. M., T. Ubbink-Kok, M. G. Elferink, A. J. M. Driessen, and W. N. Konings. 1995. Ion permeability of the cytoplasmic membrane limits the maximum growth temperature of bacteria and archaea. *Mol. Microbiol.* **18:**925–932.

Woese, C. R. 2004. The archaeal concept and the world it lives in: A retrospective. *Photosyn. Res.* **80:**361–372.

Woese, C. R., O. Kandler, and M. L. Wheelis. 1990. Towards a natural system of organisms: proposal for the domains Archaea, Bacteria, and Eucarya. *Proc. Natl. Acad. Sci. USA* **87:**4576–4579.

Yamauchi, K., and M. Kinoshita. 1995. Highly stable lipid membranes from archaebacterial extremophiles. *Prog. Polym. Sci.* **18:**763–804.

Yamauchi, K., Y. Yoshida, T. Moriya, K. Togawa, and M. Kinoshita. 1994. Archaebacterial lipid models: formation of stable vesicles from single isoprenoid chain-amphiphiles. *Biochim. Biophys. Acta* **1193:**41–47.

III. PSYCHROPHILES

Chapter 9

Ecology and Biodiversity of Cold-Adapted Microorganisms

DON A. COWAN, ANA CASANUEVA, AND WILLIAM STAFFORD

INTRODUCTION

Habitats for cold-adapted microorganisms represent a large proportion of Earth's area. Much of the oceans, which cover some 70% of Earth's surface, are at a temperature of 2.5 to 5°C. Polar regions, including Antarctica and those portions of North America and Europe that lie within the Arctic circle, constitute some 20% of the world's land surface area. The montane regions of Europe (the Alps), Asia (the Himalayas), the Americas (the Rocky Mountains and the South American Alps), and others constitute a further 5% surface area. The number of man-made habitats (refrigeration and freezer systems) contributes only a small proportion but has substantial economic relevance. The combination of low temperatures and low liquid water availability contributes to making these regions extremely inhospitable to all forms of life. For example, mean annual Antarctic air temperatures are well below 0°C and temperatures as low as −80°C have been recorded in winter months. However, during the short polar summer seasons, temperatures exceed 0°C, causing snowmelt and water flow via small streams to lakes and oceans. The seasonal process of encasement of microbial communities into ice, exposure to severe winter conditions, and release again in summertime seeds the global ocean with cold-adapted microorganisms.

Cold-adapted organisms are generally classed in two overlapping groups: psychrophiles and psychrotrophs (or psychrotolerants). This phenotypic distinction (Table 1) has been devised from comparisons of the growth properties of cold-adapted organisms (Russell, 1998; Morita, 1975). Alternative terminologies have been proposed. The terms stenopsychrophile (true psychrophile) and eurypsychrophile (psychrotolerant or psychrotroph) have been suggested for organisms with restricted and broad growth temperature ranges, respectively (Feller and Gerday, 2003). For convenience, in this chapter we will use the term psychrophile in a generic sense to refer to all organisms growing in cold environments.

The ability of microorganisms to survive and grow in cold environments is the result of a range of unique molecular and physiological adaptations. Cold-adapted enzymes tend to have an increased structural flexibility, resulting in reduced activation energies and a consequent increase in catalytic efficiency. Cell membranes contain unique lipid constituents to maintain fluidity and enable the transport of substrates. The ability to rapidly synthesize cryoprotectants such as exopolysaccharides and cold-shock proteins is also a critical factor in microbial survival under low temperature (Finegold, 1986; Russell, 1997; D'Amico et al., 2002; Feller, 2003; Golovlev, 2003; Georlette et al., 2004).

The diversity of microorganisms inhabiting some cold environments has been extensively investigated (see, e.g., Vishniac, 1993). However, most of such studies are culture dependent. It is now recognized that only 1 to 10% of microbial species present in any habitat can be detected by culture-dependent techniques (Head et al., 1998). Many microorganisms are fastidious, coculture dependent, or in a viable but nonculturable state (Amann et al., 1995; Holmes et al., 2000; McDougald et al., 1998; Suzuki and Giovannoni, 1996; Waterbury et al., 1979). Culture-independent techniques such as small subunit

Table 1. Cardinal temperatures for cold-adapted organisms

Class	T_{min}	T_{opt}	T_{max}
Psychrophiles	<0	15	20
Psychrotrophs	>0	>20	>30

T_{min}, T_{opt}, and T_{max} are the minimum, optimal, and maximum growth temperatures, in °C, respectively.

D. A. Cowan, A. Casanueva, and W. Stafford • Department of Biotechnology, University of the Western Cape, Bellville 7535, Cape Town, South Africa.

rRNA gene sequence analysis and in situ hybridization (Amann et al., 1995, 2001) have proved to be invaluable in phylogenetic studies of soil microbiota, often revealing taxa not represented by any cultivated organisms (Ward, 2002). With the application of modern molecular techniques, we have the capacity to gain enormous insight into the diversity and distribution of cold-adapted microorganisms. However, it is well understood that such methods provide no useful information on the functional roles of the organisms detected.

In this chapter, we aim to review the diversity of cold-adapted microorganisms known to exist in each of the major cold environments. We acknowledge that an organism isolated from or a phylotype detected in a cold environment cannot, a priori, be assumed to be a psychrophile or psychrotroph.

MARINE ENVIRONMENTS

Deep Marine Environments

The deep oceans and the sediments on the deep seafloor have an average temperature of ~3°C. Other factors potentially limiting microbial growth and diversity include low nutrient availability, a complete absence of light, and very high pressures (equivalent to +1 atm per 10 m below sea level: Morita, 1986). Despite these conditions, numerous microorganisms (Table 2) have been identified in and/or isolated from deep marine systems (Takami et al., 2002) and even up to 800 m below the deep seafloor (Parkes et al., 2000). However, relatively little is known about deep marine microbial communities and their role in these ecosystems.

Bacteria are found throughout the water column but are generally more numerous at depths <1,000 m (Fuhrman and Davis, 1997), presumably reflecting C/N availability. The γ-*Proteobacteria* are ubiquitous and numerous, suggesting that they play an important role in marine prokaryotic communities, including those in deep marine sediments (López-García et al., 2001b). Deep marine systems have proved to be a rich source of novel microbial taxa. While *Rhodococcus* strains isolated from depths up to 10,897 m had high similarity to terrestrial rhodococci (Colquhoun et al., 1998), deep marine sediments have yielded numerous novel actinobacterial taxa (Pathom-Aree et al., 2006a; 2006b). Large bacterial

Table 2. Microbial genera commonly isolated from psychrophilic habitats

Habitat	Genera	Reference
Deep marine	*Marinobacter*	Edwards et al., 2003; Pathom-Aree et al., 2006a, 2006b; Lauro et al., 2004
	Halomonas	
	Dermacoccus	
	Kocuria	
	Micromonospora	
	Streptomyces	
	Williamsia	
	Tsukamurella	
	Clostridium	
Polar marine	*Alteromonas*	Mergaert et al., 2001; Junge et al., 2002
	Colwellia	
	Glaciecola	
	Pseudoalteromonas	
	Shewanella	
	Polaribacter	
Terrestrial Antarctic	*Pseudomonas*	Reddy et al., 2004; Bozal et al., 2003; Gupta et al., 2004; Yi and Chun, 2006
	Psychrobacter	
	Arthrobacter	
	Flavobacterium	
Permafrost	*Psychrobacter*	Bakermans et al., 2003
	Arthrobacter	
	Frigoribacterium	
	Subtercola	
	Microbacterium	
	Rhodococcus	
	Bacillus	

populations associated with uncultured SAR11 and SAR406 clusters have been identified in the deep Atlantic and Pacific Oceans (Giovannoni et al., 2005). The former is thought to be responsible for a previously unaccounted global carbon fixation and phototrophy in the oceans because SAR11 bacteria contain novel bacteriorhodopsins.

Studies on marine Archaeal diversity have been performed in numerous locations at various depths, many of which may not constitute psychrophilic environments (Murray et al., 1998; Massana et al., 1997; DeLong et al., 1999; Fuhrman and Davis, 1997). However, most of these studies suggest that Archaeal diversity is low in surface waters and becomes greater at depths >100 m, accounting for 20 to 30% of prokaryotes in deep waters. It is reasonable to assume that much of this diversity is comprised of psychrotrophic or psychrophilic taxa. Archaeal diversity is currently clustered into four phylogenetic groups: marine group I, which consists of *Crenarchaeota* and is represented by the sponge symbiont *Cenarchaeum symbiosum*; marine groups II and III, which consist of *Euryarchaeota* and are distantly related to *Thermoplasmales*; and the recently described marine group IV, which includes a novel set of deep branching phylotypes (López-García et al., 2001b).

Representatives of marine group I increase in number with increasing depth, reaching close to 40% of the total microbial community at depths >1,000 m (Karner et al., 2001), where their numbers may equal or exceed that of *Bacteria*. Members of marine group II are more abundant in relatively shallow waters (Massana et al., 1997) and may have little psychrophilic representation. Marine groups II and III have been identified in 3,000-m deep Antarctic polar front waters (López-García et al., 2001a) and are probably virtually all psychrophilic. Marine group IV is widely distributed, but numbers also increase with increasing depth (López-García et al., 2001b). Deep marine *Archaea* have been shown to actively take up amino acids, suggesting a heterotrophic physiology (Ouverney and Fuhrman, 2000).

Few studies on fungal diversity in deep cold marine environments have been reported. Yeast populations decrease with increasing depth and distance from land (Nagahama et al., 2001), suggesting a terrestrial (and largely mesophilic) origin. However, filamentous fungi have been isolated from abyssal seawater (Takami et al., 1997), and yeasts such as *Rhodotorula* and *Sporobolomyces* have been isolated from the deep ocean sediments.

Deep marine psychrophiles may play an important role in global nutrient turnover. Dissimilatory sulfate reduction is one of the most important bacterial reactions in anoxic marine sediments, and it is thought to account for approximately half of the total organic carbon remineralization (Knoblauch et al., 1999). Psychrophilic sulfate reducers were found to have higher specific metabolic rates than their mesophilic counterparts. A range of environmentally important turnover processes, including organic carbon oxidation, methane production and consumption, and reduction of sulfate, nitrate, and manganese, have been demonstrated experimentally in various shallow psychrophilic marine sediments (D'Hondt et al., 2004). In situ temperatures of between 1 and 20°C suggest that this environment is essentially psychrophilic.

Antarctic Marine Environments

The southern oceans surrounding the Antarctic continent have an average depth of 4,000 to 5,000 m, with only limited areas of shallow water (Bano et al., 2004). There is little freshwater inflow and therefore a limited input of terrestrial-derived micronutrients. A large portion of the southern ocean is covered with the Antarctic ice pack, with a seasonal temperature range of −2 to +10°C (Davidson, 1998). Large seasonal fluctuations in bacterial biomass reflect ambient temperatures and light availability (Murray et al., 1998) and are driven by phototrophic populations. Large populations of heterotrophic flavobacteria are also found in surface seawater: population numbers correlate positively with seawater chlorophyll *a* and nutrient concentrations and are thought to represent a direct trophic link to primary productivity (Abell and Bowman, 2005).

Archaea are abundant in Antarctic marine surface waters, comprising up to 34% of the prokaryotic biomass (DeLong et al., 1994). Archaeal abundance increases in late winter and early spring, in line with increasing water temperatures and light availability (Murray et al., 1998). Archaeal phylogenetic diversity is dominated by group I crenarchaeotes and group II euryarchaeotes (Murray et al., 1998; Bano et al., 2004). Observed increases in Archaeal abundance with increasing depth (Bano et al., 2004), where light levels, oxygen concentrations, and temperatures may be lower, suggest that the factors driving Archaeal productivity are complex.

Sea Ice

Sea ice is one of the coldest environments on Earth, with temperatures ranging from 0 to −35°C. As seawater freezes, salts are expelled from the ice crystal matrix and concentrated in a network of brine-filled channels and pores (Mock and Thomas, 2005). The coldest ice forms so far examined (sea ice at −20°C) contain metabolically active bacteria within the brine channel liquid phase. Labeling experiments using [^{3}H]-thymidine and ^{14}C-leucine have shown evidence of metabolic activity even at −17°C (Carpenter et al., 2000; Junge et al., 2001). The large surface areas of the brine channel walls are colonized by algae and bacteria, with an estimated 6.4% of the brine channel surface covered by microorganisms (Mock and Thomas, 2005). Together, these observations suggest that sea ice does not merely represent a repository of inactive encapsulated microbial cells (Bowman et al., 1997), but may constitute a unique extreme habitat for psychrophilic microorganisms.

Sea-ice samples typically contain large populations of viruses, bacteria, and eukaryotes but lower numbers of *Archaea* (Brown and Bowman, 2001; Mock and Thomas, 2005). Eukaryotes include autotrophic and heterotrophic nanoplankton such as dinoflagellates, flagellates, and ciliates (Brown and Bowman, 2001; Bowman et al., 1997). The *Bacteria* are mostly cold-adapted halotolerant heterotrophs (Mock and Thomas, 2005) with members of the γ-*Proteobacteria* such as *Roseobacter* dominating the microbial population (Brinkmeyer et al., 2003).

Glacial Ice

The relatively low salt content of (terrestrial) glacial ice does not facilitate the formation of saline liquid inclusions as found in sea ice. As a consequence, although viable microorganisms can be readily isolated from glacial ice samples (Gounot, 1976; Shivaji et al., 2005), such terrestrial ice bodies may, in general, represent passive systems for the accumulation and stabilization of incident microbial cells, rather than substrates for active microbial communities.

One obvious exception to this position is found in *cryoconite holes*. Solid bodies (stones, soil, etc.) deposited on glacial surfaces are subject to solar warming, which gradually melts the underlying ice, forming a liquid inclusion in the surrounding solid ice (Gerdel and Drouet, 1960; Christner et al., 2003). Cryoconite holes contain very active and complex microscopic and macroscopic communities, and a wide range of bacteria, algae, diatoms, fungi, and rotifers have been identified (Wharton et al., 1985). Bacteria isolated from cryoconite holes are members of the β-*Proteobacteria*, *Cytophagales*, and *Actinobacteria*. Culture-independent methods have identified phylotypes assigned to cyanobacterial, γ-proteobacterial, acidobacterial, *Cytophagales*, planctomycete, α-proteobacterial, *Gemmimonas*, verrucomicrobial, and actinobacterial taxa. Eukaryote diversity included algae, tardigrades, rotifers, fungi, and ciliates, emphasizing the complex trophic structure contained within the cryoconite hole habitats.

SOIL ENVIRONMENTS

Polar Soils

Exposed soils represent a very low proportion of the total land mass in polar regions. For example, >98% of the total land area of the Antarctic continent is permanently covered by ice (http://www.cia.gov/cia/publications/factbook/geos/ay.html#Geo). The Dry Valleys of Eastern Antarctica are the most extreme example of polar soils and are arguably the coldest and driest deserts on Earth. Soil water contents of 0.5 to 2% wt/wt and seasonal temperatures of −30 to +15°C are typically recorded (Cowan and Ah Tow, 2004). However, surface ground temperatures average around +15°C during periods of direct sunlight, with huge and rapid fluctuations—temperature changes from −15 to +27.5°C within a 3-h period have been recorded (Wynn-Williams, 1990; Cameron, 1974). These freeze–thaw cycles are potentially lethal to soil microorganisms (Vishniac, 1993). The desiccating conditions and extreme temperatures, together with osmotic stress due to high salt concentrations and a low content of organic matter (0.064 ± 0.035% total organic carbon) (Matsumoto et al., 1983), impose major limitations on microbial growth and survival.

Culture-dependent studies have shown that Antarctic Dry Valley mineral soils contain low levels of microorganisms (e.g., 10^2 to 10^4 g^{-1}: Cameron, 1969; Cameron et al., 1970). However, in situ ATP analysis data have indicated that levels of microbial biomass are possibly 3 to 4 orders of magnitude higher than previously reported (Cowan et al., 2002). Most culturable bacteria in Dry Valley mineral soils (Table 2) are found toward the surface, and most microorganisms isolated from this habitat are aerobic heterotrophs (Cameron et al., 1970; Cameron, 1971). Culture-based studies identified a restricted number of largely aerobic cosmopolitan taxa (Vishniac, 1993). Gram-negative bacteria are dominated by coryneforms (*Arthrobacter*, *Brevibacterium*, *Cellulomonas*, and *Corynebacterium*), and gram-positive bacteria included *Bacillus*, *Micrococcus*, *Nocardia*, *Streptomyces*, *Flavobacterium*, and the pseudomonads (Cameron et al., 1972). Cyanobacteria have been frequently identified in moist soils that are exposed to glacial meltwaters (de la Torre et al., 2003; Taton et al., 2003).

Phylogenetic analysis of DNA extracts from mineral soils collected from the slopes of the Miers

Valley (Ross Desert, Eastern Antarctica) has indicated that as many as 50% of the retrieved sequences were from uncultured bacteria (Smith et al., 2006). Most of the remaining sequences could be assigned to the *Actinobacteria* (27%), *Bacteroidetes* (11%), *Acidobacteria* (6%), and *Verrucomicrobia* (6%). In a separate phylogenetic study performed on mineral soil collected from heavily impacted gravels from McMurdo Station (Ross Island, Eastern Antarctica), phylotypes belonging to the *Actinobacteria* (12.5%), *Bacteroidetes* (25%), *Firmicutes* (12.5%), *Planctomycetes* (12.5%), and *Proteobacteria* (37.5%) were identified (E. M. Kuhn and D. A. Cowan, unpublished data). Both cosmopolitan and indigenous fungal, yeast, and protozoan species have been isolated from McMurdo Dry Valley mineral soils. Yeasts are relatively abundant in the surface horizons of certain moist Antarctic soils (Atlas et al., 1978). The predominant algae isolated from Antarctic soil were the oscillatorioids, *Microcoleus* sp., *Schizothrix* spp., *Anacystis* spp., and *Coccochloris* spp.

When both *Bacteria* and *Archaea* have been surveyed across a range of polar and other cold environments, the *Bacteria* have generally shown a vastly greater diversity than the *Archaea* (Brown and Bowman, 2001; Brambilla et al., 2001; López-García et al., 2001a).

Montane Environments

Montane soils are widely distributed across all continental systems. Such high-altitude soils are subjected to large temperature fluctuations and regular freeze–thaw cycles. Microbial biomass has been shown to increase in autumn and winter in Alpine and Arctic soils, most likely responding to seasonal increase in plant-derived (deciduous) organic substrates (Nemergut et al., 2005; Kobaba et al., 2004; Lipson et al., 2002; Schadt et al., 2003). There are clear changes in microbial community structure from winter to summer (Lipson et al., 2002). During the winter, microbial populations are metabolically active under snow cover, serving as a sink for CO_2 and N (Schadt et al., 2003). During the snowmelt, there is a release of CO_2 and N, which is thought to contribute to the growth of plants during the warmer summer season (Lipson et al., 2002). Analysis of the bacterial composition indicated an abundance of *Acidobacteria* in spring, while *Verrucomicrobia*, β-*Proteobacteria*, and *Bacteriodetes* increased in summer and winter (Lipson and Schmidt, 2004). Fungal diversity was dominated by the *Ascomycota*, although several new taxa have been identified (Schadt et al., 2003). Twelve novel psychrophilic *Microbotryomycetidae* strains have been recently isolated from montane soils (Bergauer et al., 2005).

Lithic Communities

Microorganisms that avoid climatic extremes through their association with porous and translucent rocks are termed endoliths (Nienow and Friedmann, 1993; Siebert et al., 1996; Wynn-Williams, 1990). The most important lithic characteristics are porosity (providing interstitial spaces for microbial colonization) and translucence (facilitating photosynthetic activity). Endolithic communities are wholly dependent on photoautotrophic energy capture, although growth may be limited by CO_2 availability because of low diffusion rates in the crystalline rock strata (Vishniac, 1993). Water supply is also limiting (Friedmann and McKay, 1985), although water availability from the upward diffusion of water vapor or transport of liquid water from a melting ice/permafrost is also possible. Lithic communities have been defined on the basis of localization. *Chasmoendolithic* organisms inhabit cracks in weathering rocks, *cryptoendolithic* organisms exist in the interstices of crystalline rock structures, while *hypolithic* microbial communities reside on the underside of translucent stones. Chasmoendolithic microbial communities are widespread in ice-free areas in polar and montane regions because freeze fracturing of rock strata is a common phenomenon. These biotopes are principally occupied by lichens and cyanobacteria (Nienow and Friedmann, 1993), but little is known of the diversity of the associated nonphotosynthetic prokaryote communities.

The cryptoendoliths have been divided into two community types: lichen-dominated and cyanobacteria-dominated. A recent comprehensive phylogenetic study of both lichen- and cyanobacteria-dominated communities revealed an extensive and varied bacterial population (de la Torre et al., 2003). The lichen-dominated community represents a symbiotic association between the fungus *Texosporium sancti-jacobi* (29%) and the green alga *Trebouxia jamesii* (22%), with other fungal and algal phylotypic signals representing a minor proportion of the total clones (<2%). Bacterial phylotypes comprised >15% of the clones sequenced, with members of the *Cytophagales* being predominant. Phylotypes belonging to the *Actinobacteria* (*Blastococcus* sp., *Microsphaera* sp., *Sporichthya* sp., and *Rhodococcus* sp.), the α-*Proteobacteria* (*Sphingomonas* sp., uncultured clones), the γ-*Proteobacteria* (*Acinetobacter* spp.), and the *Planctomycetales* were also identified. In the lichen-dominated community, cyanobacteria of the *Leptolyngbya–Phormidium–Plectonema* group

(Castenholz and Waterbury, 1989) constituted over 30% of clones sequenced. Heterotrophic bacterial phylotypes comprised a further 60%, falling into two major groups: the α-*Proteobacteria* (~34%, virtually all *Blastomonas* sp.) and the *Thermus–Deinococcus* group (26%) (de la Torre et al., 2003). The presence of the *Blastomonas*-like phylotypes suggests that the cyanobacteria are not the sole contributors to primary productivity in endolithic communities because *Blastomonas* spp. are aerobic anoxygenic phototrophs (Yurkov and Beatty, 1998; Nienow and Friedmann, 1993).

Fellfield Communities

Fellfield soils (moist high-silt-content soils and drier sand or gritty ash soils, with discontinuous cryptogamic vegetation) are largely restricted to the warmer, more northerly regions of the Antarctic continent (the Antarctic peninsula), to the high Arctic latitudes, and to offshore islands (such as the Antarctic Signy and Marion Islands). The physical topology of fellfields is governed by climatic factors (frost heave and particle sorting) (Chambers, 1967), desiccation, meltwater, and wind erosion (Wynn-Williams, 1990). The particle sorting action of the freeze–thaw process generates semiregular surface structures including polygons (Chambers, 1967). The concentrations of nutrients such as nitrogen, phosphorous, and organic carbon are high compared with desert mineral soils (Matsumoto et al., 1983; Smith and Tearle, 1985). The fellfield soils of the Antarctic Peninsula support a low diversity of vegetation comprised of mosses, liverworts, lichens, and algae. An extensive microbial population exists, with the cyanobacteria *Phormidium autumnale* and *Pseudanabaena catenata* dominating and subdominant populations of the diatom *Pinnularia borealis* (Davey and Clarke, 1991). Other species include *Nostoc* sp., *Achnanthes lapponica*, *Chlamydomonas chlorostellata*, *Planktospheaerella terrestris*, *Cylindrocystis brebissonii*, *Cosmarium undulatum*, and *Netrium* sp. The continental and maritime fellfield communities support substantial heterotrophic microbial populations (Davis, 1981; Wynn-Williams, 1989). A variety of bacterial (Miwa, 1975), yeast (Montes et al., 1999), and filamentous fungi (Selbmann et al., 2002) have been isolated.

Ornithogenic Environments

Ornithogenic soils (modified or generated by the presence of birds) offer localized and specialized environments that differ widely from other polar microbial habitats (Ugolini, 1972). Ornithogenic soils occur worldwide but are particularly significant in polar regions because of the density of breeding seabirds. These environments are unique in that they are not dependent on *in situ* photoautotrophy, have continuous exogenous nutrient supplementation and maintain very high nutrient levels. The nutrient status is consistent with high levels of microbial biomass, as estimated by direct microscopic counts (2×10^{10} cells g^{-1}: Bowman et al., 1996) and by ATP analysis (5×10^7 to 7×10^8 cells g^{-1}: Cowan et al., 2002). Microscopic examination of Antarctic ornithogenic soils has shown that up to 50% of the microbiota were gram-negative non-motile cocci (Ramsey and Stannard, 1986). Although only a low proportion of the cell types were culturable (Bowman et al., 1996), several new species of *Psychrobacter* have been isolated, all of which could grow on urate as a sole C source.

MAN-MADE PSYCHROPHILIC ENVIRONMENTS

The modern need for preservation of food and other substances has created a large number of manmade cold environments. These typically include refrigerators (4 to 6°C), open chiller display units (10 to 12°C), freezers (–20°C), and liquid nitrogen storage facilities (–196°C) (Morris, 2005). Other than the latter, all provide habitats for psychrophilic organisms. Many of the microorganisms capable of growing in these habitats (e.g., *Brochothrix thermosphacta*, *Pseudomonas* spp., and *Micrococcus* spp.) are implicated in food spoilage (Champagne et al., 1994; Borch et al., 1996; Russell, 2002) and therefore are of considerable economic significance. In addition, some microorganisms responsible for food poisoning, such as *Listeria monocytogenes*, *Yersinia enterocolitica*, and *Aeromonas hydrophila*, are psychrotolerant (Russell, 2002; Daskalov, 2005). A low pH, high salt content, and/or packaging in a modified atmosphere is required to effectively prevent the growth of these pathogens in refrigerated foods (Fredricksson-Ahomaa and Korkeala, 2003).

Owing to the presence of thermolabile enzymes in psychrotrophic organisms, mild heat treatment has been used to reduce their number in refrigerated foods (Russell, 2002). The addition of probiotics, such as *Lactobacillus casei* and *Lactococcus lactis*, can inhibit the growth of pathogens such as *A. hydrophila* (Daskalov, 2005); and Antarctic microorganisms have been shown to produce growth-inhibiting proteins active against *L. monocytogenes*, *Pseudomonas fragi*, and *B. thermosphacta*

(O'Brien et al., 2004). This finding potentially offers a new range of cold-adapted probiotics and antibiotics for enhanced preservation of refrigerated food.

GENOMICS OF PSYCHROPHILES

A number of psychrophilic and psychrotolerant microbial genomes have been and currently are being sequenced using high-throughput DNA sequencing of shotgun libraries. To date, only five genomes have been fully sequenced, and at least another dozen are in progress (Table 3). One of the more important applications of these sequence data will be to identify sequence determinants of psychrophilicity. However, there are intrinsic difficulties in interpreting amino acid changes in psychrophilic proteins in the context of thermal adaptation with a background of general evolution and genetic drift. Principal component analysis has been successfully used to statistically identify thermal trends in amino acid composition. For example, significantly high levels of noncharged polar amino acids (particularly glutamine and threonine) and low hydrophobic amino acid content (particularly leucine) were identified in the deduced proteins from the archaeal psychrophiles *Methanogenium frigidum* and *Methanococcoides burtonii* DSM6242 (Saunders et al., 2003). Detailed analysis of >1,000 protein-homology models showed that proteins from archaeal psychrophiles tended to have more glutamine, threonine, and hydrophobic residues in the solvent-accessible area. A similar study noted a trend toward increased polar residues (particularly serine), substitution of aspartate for glutamate, and a general decrease in charged residues on the surface of psychrophilic proteins from *Colwellia psychrerythraea* 34H, giving further support to the theory that increased flexibility and reduced thermostability contribute to enzyme cold adaptation (Methé et al., 2005).

BIOTECHNOLOGICAL APPLICATIONS OF PSYCHROPHILES

Biotechnologically Important Properties

At present, the biotechnological applications of psychrophilic organisms and their products are limited in scope (Cavicchioli et al., 2002). However, the evolution of these organisms to function optimally at low temperatures has endowed their structural and functional systems with unique properties, most notably the high catalytic activity of psychrophilic enzymes at low temperatures. With a general trend in the global biotechnology industries toward more energy-efficient and "greener" systems, it seems reasonable to predict that low-temperature biocatalysis will attract increasing attention.

Psychrophilic enzymes have similar specific activities to their mesophilic and thermophilic equivalents at their respective optimal temperatures (Cavicchioli et al., 2002) but are rapidly inactivated at elevated temperatures. These two characteristics, high catalytic activity at low and moderate temperatures and easy inactivation by moderate increases in temperature, are of potential use in several areas of biotechnology. Both the food and chemical biotransformation industries could benefit from the inherent controllability of psychrophilic enzymes.

The biosynthesis of nonpolar fatty acid esters, peptides, and oligosaccharide derivatives is often difficult because of the low solubility of substrates and products in aqueous media (Gerday et al., 2000). It has been suggested that psychrophilic enzymes may be effective biocatalysts in microaqueous solvents because of their inherent flexibility, which could counteract the structural rigidification that occurs in low-water-activity organic solvent systems (Gerday et al., 2000; Cavicchioli et al., 2002). However, this potential must also be considered in light of the low conformational stability of psychrophilic enzymes.

Bioremediation and Natural Cycle Processes

In cold environments, psychrophiles are major contributors to the carbon, nitrogen, phosphorous, and sulfur cycles (Russell, 1998). There is thus considerable scope for these organisms to use low-molecular-weight environmental pollutants as substrates. Low-temperature biodegradation of polyols from aircraft deicing fluids has been demonstrated in contaminated soil (Klecka et al., 1993). Natural psychrophilic microbial populations have also been shown to effectively degrade waste diesel oil (Margesin and Schinner, 2001; Margesin et al., 2003).

Food Industry

A potential advantage of using cold-adapted enzymes is the ability to induce thermal inactivation at moderate temperatures while avoiding any modification of the product (Cavicchioli et al., 2002). For example, proteases are used to tenderize meat (Gerday et al., 2000). Mild heat inactivation of psychrophilic proteases would prevent excessive hydrolysis. Similar applications in the dairy industries have been considered (Demirjian et al., 2001; Gerday et al., 2000), and psychrophilic β-galactosidase has been used to

Table 3. Psychrophile genome sequencing projects[a]

Species	Origin	Location	Reference
M. frigidum	Antarctica, Ace Lake	Amersham Biosciences, Piscataway, NJ, U.S.A.	http://psychro.bioinformatics.au/genomes/index.php; Saunders et al., 2003
M. burtonii DSM6242	Antarctica; Ace Lake	Joint Genome Institute, Walnut Creek, CA, U.S.A.	http://psychro.bioinformatics.unsw.edu.au/genomes/index.php; Saunders et al., 2003
C. symbiosum	Symbiont of marine sponge	Joint Genome Institute, Walnut Creek, CA, U.S.A.	http://img.jgi.doe.gov/cgi-bin/pub/main.cgi?page=taxonDetail&taxon_oid=630050000
C. psychrerythraea 34H	Arctic marine sediments	The Institute for Genomic Research, Rockville, MD, U.S.A.	http://www.tigr.org/tdb/mdb/mdbinprogress.html
Pseudoalteromonas haloplanktis TAC125	Antarctic marine water	Genoscope, Evry cedex, France	http://www.genoscope.cns.fr/agc/mage/psychroscope; Medigue et al., 2005
Halorubrum lacusprofundi	Antarctic Deep Lake	University of Maryland, **Institute for Systems Biology Seattle**, U.S.A.	http://genomics.com/genomes/index.htm; Goo et al., 2004
Vibrio salmonicida	Atlantic salmon, Norway	University of Tromsø, Tromsø, Norway; Wellcome Trust Sanger Institute, Cambridge, U.K	http://www.sanger.ac.uk/Projects/
Photobacterium profundum SS9	Amphipod homogenate from 2.5 km deep in the Sulu Sea	CRIBI Biotechnology Centre, University of Padua, Padua, Italy	http://img.jgi.doe.gov/cgi-bin/pub/main.cgi
Shewanella violacea DSS12	Deep-sea mud (5.1 km) from Ryukyu Trench, Japan	AB, Keio University, NAIST; JAMSTEC, Japan	http://genomics.com/genomes/index.htm
Shewanella frigidimarina NCMB400	Sea Ice, Antarctica	Joint Genome Institute, Walnut Creek, CA, U.S.A.	http://genome.jgi-psf.org/draft_microbes/shefr/shefr.download.html
Psychrobacter sp. 273–4	**Siberian permafrost core (20,000–40,000 years old)**	Joint Genome Institute, Walnut Creek, CA, U.S.A.	http://genome.jgi-psf.org/finished_microbes/psy24/psy24.home.html
Shewanella benthica KT99	9,000 m, Tonga-Kermadec Trench	Moore Foundation, San Francisco, CA; **The Institute for Genomic Research, Rockville, MD**; Scripps Institute of Oceanography, La Jolla, CA, U.S.A.	https://research.venterinstitute.org/moore/
Psychromonas sp. CNPT3	5,800 m, Central North Pacific Ocean	Moore Foundation, San Francisco, CA; **The Institute for Genomic Research, Rockville, MD**; Scripps Institute of Oceanography, La Jolla, CA, U.S.A.	https://research.venterinstitute.org/moore/

Moritella sp.	3,600 m, Pacific Ocean	Moore Foundation, San Francisco, CA; **The Institue for Genomic Research, Rockville, MD**; Scripps Institute of Oceanography, La Jolla, CA, U.S.A.	https://research.venterinstitute.org/moore/
***Desulfotalea psychrophila* LSv54**	**Arctic marine sediments**	**MPI, Bremen, Germany**	**http://www.regx.de/blast/index.html; Rabus et al., 2004**
Exiguobacterium 255-15	Siberian permafrost	Joint Genome Institute, Walnut Creek, CA, U.S.A.	http://www.jgi.psf.org/draft_microbes/exigu.home.html
Flavobacterium psychrophilum	Salmon pathogen	NCCCWA, Kearneysville, WV, U.S.A.	http://genomics.com/genomes/index.htm
Psychroflexus torquis	Sea-ice algal assemblage, Prydz Bay, Antarctica	Moore Foundation, San Francisco, CA, U.S.A.; **The Institute for Genomic Research, Rockville, MD, U.S.A.**; University of Tasmania, Australia	https://research.venterinstitute.org/moore/
Polaribacter filamentous	Surface seawater, 350 km north of Deadhorse, Alaska	Integrated Genomics, Chicago, IL, U.S.A.	http://ergo.integratedgenomics.com/ERGO_supplement/genomes.html
Polaribacter irgensii	Nearshore marine waters of Antarctic Peninsula	Moore Foundation, San Francisco, CA; **The Institute for Genomic Research, Rockville, MD, U.S.A.**	https://research.venterinstitute.org/moore
Renibacterium salmoninarum	Diseased Chinook salmon	Genome Center, USA; Integrated Genomics, Chicago, IL, U.S.A.	http://www.nwfsc.noaa.gov/research/divisions/reutd/fhm/rs-genome/index.html
Leifsonia-related PHSC20-c1	Nearshore marine waters of Antarctic Peninsula	Moore Foundation, San Francisco, CA; Venter Institute, Rockville, MD; Desert Research Institute, Reno, NV U.S.A.	http://genomics.com/genomes/index.htm

[a] Completely sequenced genomes are shown in boldface.

decrease the lactose content of milk in cold storage (Fernandes et al., 2002).

In many sectors of the food industry, low-temperature processing is mandatory in order to protect the quality of the product. An example of a current high-volume enzyme application, where psychrophilic homologs might be viable alternatives, is the use of pectinases in the fruit juice industry to enhance juice extraction, reduce viscosity, and assist with clarification (Gerday et al., 2000).

Ice-nucleating proteins produced by certain bacteria stimulate the formation of ice crystals at high sub-zero temperatures (Kawahara, 2002). Possible uses of these proteins include ice-cream manufacturing and the preparation of synthetic snow. Alternatively, antifreeze proteins produced by psychrophiles could have applications in the food industry for products where low-temperature storage is critical but where ice formation would damage texture or structure.

Cleaning Additives

Proteases, lipases, α-amylases, and celluloses are all used as detergent additives. The cold-adapted counterparts of these enzymes could be used for cold washing, resulting in reduced energy consumption and reduction in wear and tear of garments (Gerday et al., 2000). Contact lens cleaning solutions contain proteinases such as subtilisin to remove the proteinaceous buildup on the lenses. Psychrophilic subtilisins have been isolated, and work is ongoing to explore the stability of these enzymes for industrial applications (Tindbaek et al., 2004).

Textile Industry

Cellulases are regularly used in the textile industry for biopolishing and stonewashing of cotton fibers (Miettinen-Oinonen et al., 2005). Cotton-fiber ends normally protrude from the main fibers from tissues, causing reduced smoothness of the fabric. The use of cold-adapted cellulases would decrease the temperature required for inactivation, thereby improving the resultant mechanical resistance of the fabric, which may be compromised by residual enzyme activity.

Molecular Biology

In molecular biology, there are numerous sequential protocols where enzymes used in a step must be deactivated before the subsequent step (Cavicchioli et al., 2002). Psychrophilic enzymes are typically completely inactivated at temperatures well below strand denaturation temperatures, and their use could minimize additional cleanup steps. The development of low-temperature protein expression systems could also be advantageous because expression of recombinant proteins at temperatures <30°C is often attempted as a means of preventing protein–protein aggregation and inclusion body formation (Carrio and Villaverde, 2002). With *Escherichia coli* and other mesophilic expression systems, this often results in reduced yields of recombinant protein. Expression in a psychrophilic host could potentially enhance the yields of correctly folded functional recombinant proteins, and various Antarctic bacteria have been investigated as hosts for recombinant gene expression (Cavicchioli et al., 2002).

CONCLUSIONS

Habitats for cold-adapted microorganisms are widespread on Earth. These habitats vary enormously in their properties, including the nature of their psychrophily. For example, marine psychrophilic habitats are relatively constant with respect to environmental variables, whereas some physical characteristics of terrestrial polar habitats may fluctuate substantially on an annual, diurnal, or even hourly basis. Some have existed for significant portion of the evolution of the planet, whereas others are the result of very recent industrialization processes.

A likely consequence of the variations in origin and physicochemical properties of psychrophilic habitats is that each will harbor substantially different microbial populations. Furthermore, it follows that members of these populations will have different physiological and biochemical characteristics that reflect their specific adaptations to these habitats. The general belief that obligate psychrophiles dominate cold marine systems, whereas facultative psychrophiles (psychrotrophs) are more common in terrestrial systems, is circumstantial evidence to support this view. The recent isolation of numerous novel lineages of actinomycetes from deep marine sediments (Pathom-Aree et al., 2006b) also supports this suggestion.

However, it is very early to confirm or refute these predictions. With the current consensus that culture-dependent studies of diversity provide a wholly unrealistic picture of true microbial diversity, accurate comparisons can only be made using data from culture-independent surveys. To the authors' knowledge, no such comprehensive comparative studies have yet been undertaken to date. Even the phylogenetic analyses of localized niches in wider psychrophilic environments (e.g., Brambilla et al., 2001; Christner et al., 2003; de la Torre et al., 2003; Smith et al., 2006) are restricted by two important factors. Firstly, the cost of DNA sequencing and analysis typically limits such studies to the identification of a few hundred unique phylotypes, typically representing only a small proportion of the total microbial diversity. Secondly, the current status of 16S rRNA gene data is such that a substantial proportion of queries to international databases (typically 20 to 30%) return little or no useful information. Several factors are likely to alleviate these limitations over the next decades: (i) the development of high-throughput sequencing with the concomitant reduction in price will enable more laboratories to obtain better coverage of the microbial diversity in any sample; (ii) new approaches to the rapid characterization of microbial species diversity are being developed; and (iii) increased emphasis on culturing "uncultured" microorganisms will slowly but steadily increase the depth of data supporting current sequence databases.

Psychrophilic environments are endowed with substantial economic value, given that many are also favored sites for current and future tourism (e.g., montane regions, the ice-free Antarctic margins). However, some of these sites are also highly sensitive to environmental impacts. The slow growth rates of psychrophilic microbial communities make them potentially very susceptible to physical disturbance, a factor which is at the forefront of concerns about the projected increase in tourist activity on the Antarctic continent. Climatic factors such as global warming could potentially have more dramatic and devastating impact on psychrophilic environments and their specialized microbial populations.

The wider economic potential of psychrophilic habitats includes the intrinsic microbial populations, as targets for biotechnological exploitation. Both the steady growth of new methods for metagenomic gene recovery and the continued expansion of the industrial enzyme market suggest that psychrophiles, even if relatively unexploited at present, will play an increasingly important role in the future of biotechnology.

REFERENCES

Abell, G. C., and J. P. Bowman. 2005. Ecological and biogeographic relationships of class Flavobacteria in the Southern Ocean. *FEMS Microbiol. Ecol.* 51:265–277.

Amann, R., B. M. Fuchs, and S. Behrens. 2001. The identification of microorganisms by fluorescence in situ hybridisation. *Curr. Opin. Biotechnol.* 12:231–246.

Amann, R. I., W. Ludwig, and K. H. Schleifer. 1995. Phylogenetic identification and in situ detection of individual microbial cells without cultivation. *Microbiol. Rev.* 59:143–169.

Atlas, R. M., M. E. Di Menna, and R. E. Cameron. 1978. Ecological investigations of yeasts in Antarctic soils. *Antarct. Res. Ser.* 30:27–34.

Bakermans, C., A. I. Tsapin, V. Souza-Egipsy, D. A. Gilichinsky, and K. H. Nealson. 2003. Reproduction and metabolism at −10°C of bacteria isolated from Siberian permafrost. *Environ. Microbiol.* 5:321–326.

Bano, N., S. Ruffin, B. Ransom, and J. T. Hollibaugh. 2004. Phylogenetic composition of Arctic Ocean archaeal assemblages and comparison with Antarctic assemblages. *Appl. Environ. Microbiol.* 70:781–789.

Bergauer, P., P. A. Fonteyne, N. Nolard, F. Schinner, and R. Margesin. 2005. Biodegradation of phenol and phenol-related compounds by psychrophilic and cold-tolerant alpine yeasts. *Chemosphere* 59:909–918.

Borch, E., M. L. Kant-Muermans, and Y. Blixt. 1996. Bacterial spoilage of meat and cured meat products. *Int. J. Food Microbiol.* 33:103–120.

Bowman, J. P., J. Cavanagh, J. J. Austin, and K. Sanderson. 1996. Novel *Psychrobacter* species from Antarctic ornithogenic soils. *Int. J. Syst. Bacteriol.* 46:841–848.

Bowman, J. P., S. A. McCammon, M. V. Brown, D. S. Nichols, and T. A. McMeekin. 1997. Diversity and association of psychrophilic bacteria in Antarctic sea ice. *Appl. Environ. Microbiol.* 63:3068–3078.

Bozal, N., M. J. Montes, E. Tudela, and J. Guinea. 2003. Characterization of several *Psychrobacter* strains isolated from Antarctic environments and description of *Psychrobacter luti* sp. nov. and *Psychrobacter* fozii sp. nov. *Int. J. Syst. Evol. Microbiol.* 53:1093–1100.

Brambilla, E., H. Hippe, A. Hagelstein, B. J. Tindall, and E. Stackebrandt. 2001. 16S rDNA diversity of cultured and uncultured prokaryotes of a mat sample from Lake Fryxell, McMurdo Dry Valleys, Antarctica. *Extremophiles* 5:23–33.

Brinkmeyer, R., K. Knittel, J. Jürgens, H. Weyland, R. Amann, and E. Helmke. 2003. Diversity and structure of bacterial communities in Arctic versus Antarctic pack ice. *Appl. Environ. Microbiol.* 69:6610–6619.

Brown, M. V., and J. P. Bowman. 2001. A molecular phylogenetic survey of sea-ice microbial communities (SIMCO). *FEMS Microbiol. Ecol.* 35:267–275.

Cameron, R. E. 1969. Cold desert characteristics and problems relevant to other arid lands, p. 167–205. *In* W. G. McGinnies and B. J. Goldman (ed.), *Arid Lands in Perspective*. American Association for the Advancement of Science, Washington, D.C.

Cameron, R. E., J. King, and C. N. David. 1970. Microbial ecology and microclimatology of soil sites in Dry Valleys of Southern Victoria Land, Antarctica, p. 702–716. *In* M. W. Holdgate (ed.), *Antarctic Ecology, Volume 1*. Academic Press, London, United Kingdom.

Cameron, R. E. 1971. Antarctic soil microbial and ecological investigations, p. 137–189. *In* L. O. Ouam and H. D. Porter (ed.), *Research in the Antarctic*. American Association for the Advancement of Science, Washington, D.C.

Cameron, R. E. 1974. Application of low latitude microbial ecology to high latitude deserts, p. 71–90. *In* T. L. Smiley and J. H. Zumberge (ed.), *Polar Deserts and Modern Man*. University of Arizona Press, Tucson, AZ.

Cameron, R. E., F. A. Morelli, and R. M. Johnson. 1972. Bacterial species in soil and air of the Antarctic continent. *Antarct. J. U.S.* 7:187–189.

Carpenter, E. J., S. Lin, and D. G. Capone. 2000. Bacterial activity in South Pole snow. *Appl. Environ. Microbiol.* 66:4514–4517.

Carrio, M. M., and A. Villaverde. 2002. Construction and deconstruction of bacterial inclusion bodies. *J. Biotechnol.* 96:3–12.

Castenholz, R. W., and J. B. Waterbury. 1989. Group I. Cyanobacteria, p. 1710–1728. *In* J. T. Staley, M. O. P. Bryant, N. Pfennig, and J. G. Holt (ed.), *Bergey's Manual of Systematic Bacteriology*. Williams & Wilkins, Baltimore, MD.

Cavicchioli, R., K. S. Siddiqui, D. Andrews, and K. R. Sowers. 2002. Low-temperature extremophiles and their applications. *Curr. Opin. Biotechnol.* 13:253–261.

Chambers, M. J. G. 1967. Investigations of patterned ground at Signey Island, South Orkney Islands. III. Miniature patterns, frost heaving and general conclusions. *Br. Antarct. Surv. Bull.* **12:**1–22.

Champagne, C. P., R. R. Laing, D. Roy, A. A. Mafu, and M. W. Griffiths. 1994. Psychrotrophs in dairy products: their effects and their control. *Crit. Rev. Food Sci. Nutr.* **34:**1–30.

Christner, B. C., B. H. Kvitko, and J. N. Reeve. 2003. Molecular identification of Bacteria and Eukarya inhabiting an Antarctic cryoconite hole. *Extremophiles* **7:**177–183.

Colquhoun, J. A., S. C. Heald, L. Li, J. Tamaoka, C. Kato, K. Horikoshi, and A. T. Bull. 1998. Taxonomy and biotransformation activities of some deep-sea actinomycetes. *Extremophiles* **2:**269–277.

Cowan, D. A., and L. Ah Tow. 2004. Endangered Antarctic microbial communities. *Annu. Rev. Microbiol.* **58:**649–690.

Cowan, D. A., N. J. Russell, A. Mamais, and D. M. Sheppard. 2002. Antarctic Dry Valley mineral soils contain unexpectedly high levels of microbial biomass. *Extremophiles* **6:**431–436.

D'Amico, S., P. Claverie, T. Collins, D. Georlette, E. Gratia, A. Hoyoux, M. A. Meuwis, G. Feller, and C. Gerday. 2002. Molecular basis of cold adaptation. *Philos. Trans. R. Soc. Lond. B* **357:**917–925.

D'Hondt, S., B. B. Jørgensen, D. J. Miller, A. Batzke, R. Blake, B. A. Cragg, H. Cypionka, G. R. Dickens, T. Ferdelman, K. U. Hinrichs, N. G. Holm, R. Mitterer, A. Spivack, G. Wang, B. Bekins, B. Engelen, K. Ford, G. Gettemy, S. D. Rutherford, H. Sass, C. G. Skilbeck, I. W. Aiello, G. Guerin, C. H. House, F. Inagaki, P. Meister, T. Naehr, S. Niitsuma, R. J. Parkes, A. Schippers, D. C. Smith, A. Teske, J. Wiegel, C. Naranjo Padilla, and J. L. Solis Acosta. 2004. Distributions of microbial activities in deep subseafloor sediments. *Science* **306:**2216–2221.

Daskalov, H. 2005. The importance of *Aeromonas hydrophila* in food safety. *Food Control.* **17:**474–483.

Davey, M. C., and K. J. Clarke. 1991. The spatial distribution of microalgae in Antarctic fellfield soils. *Antarct. Sci.* **3:**257–263.

Davidson, A. T. 1998. The impact of UVB radiation on marine plankton. *Mutat. Res.* **422:**119–129.

Davis, R. C. 1981. Structure and function of two Antarctic terrestrial moss communities. *Ecolog. Monogr.* **51:**125–143.

de la Torre, J. R., B. M. Goebel, E. I. Friedmann, and N. R. Pace. 2003. Microbial diversity of cryptoendolithic communities from the McMurdo Dry Valleys, Antarctica. *Appl. Environ. Microbiol.* **69:**3858–3867.

DeLong, E. F., K. Y. Wu, B. B. Prezelin, and R. V. Jovine. 1994. High abundance of Archaea in Antarctic marine picoplankton. *Nature* **371:**695–697.

DeLong, E. F., L. T. Taylor, T. L. Marsh, and C. M. Preston. 1999. Visualization and enumeration of marine planktonic archaea and bacteria by using polyribonucleotide probes and fluorescent in situ hybridization. *Appl. Environ. Microbiol.* **65:**5554–5563.

Demirjian, D. C., F. Morís-Varas, and C. S. Cassidy. 2001. Enzymes from extremophiles. *Curr. Opin. Chem. Biol.* **5:**144–151.

Edwards, K. J., D. R. Rogers, C. O. Wirsen, and T. M. McCollom. 2003. Isolation and characterization of novel psychrophilic, neutrophilic, Fe-oxidizing, chemolithoautotrophic α- and γ-*Proteobacteria* from the deep sea. *Appl. Environ. Microbiol.* **69:**2906–2913.

Feller, G. 2003. Molecular adaptations to cold in psychrophilic enzymes. *Cell. Mol. Sci.* **60:**648–662.

Feller, G., and C. Gerday. 2003. Psychrophilic enzymes: hot topics in cold adaptation. *Nat. Rev. Microbiol.* **1:**200–208.

Fernandes, S., B. Geueke, O. Delgado, J. Coleman, and R. Hatti-Kaul. 2002. Beta-galactosidase from a cold-adapted bacterium: purification, characterization and application for lactose hydrolysis. *Appl. Microbiol. Biotechnol.* **58:**313–321.

Finegold, L. 1986. Molecular aspects of adaptation to extreme cold environments. *Adv. Space Res.* **6:**257–264.

Fredriksson-Ahomaa, M., and H. Korkeala. 2003. Low occurrence of pathogenic *Yersinia enterocolitica* in clinical, food, and environmental samples: a methodological problem. *Clin. Microbiol. Rev.* **16:**220–229.

Friedmann, E. I., and C. P. McKay. 1985. Methods for the continuous monitoring of snow: application to the cryptoendolithic microbial community of Antarctica. *Antarct. J. U.S.* **20:**179–181.

Fuhrman, J. A., and A. A. Davis. 1997. Widespread archaea and novel Bacteria from the deep sea as shown by 16S rRNA gene sequences. *Ecol. Prog. Ser.* **150:**275–285.

Georlette, D., V. Blaise, T. Collins, S. D'Amico, E. Gratia, A. Hoyoux, J. C. Marx, G. Sonan, G. Feller, and C. Gerday. 2004. Some like it cold: biocatalysis at low temperatures. *FEMS Microbiol. Rev.* **28:**25–42.

Gerday, C., M. Aittaleb, M. Bentahir, J. P. Chessa, P. Claverie, T. Collins, S. D'Amico, J. Dumont, G. Garsoux, D. Georlette, A. Hoyoux, T. Lonhienne, M. A. Meuwis, and G. Feller. 2000. Cold-adapted enzymes: from fundamentals to biotechnology. *Trends Biotechnol.* **18:**103–107.

Gerdel, R. W., and F. Drouet. 1960. The cryoconite in the Thule area, Greenland. *Trans. Am. Microsc. Soc.* **79:**256–272.

Giovannoni, S. J., L. Bibbs, J. C. Cho, M. D. Stapels, R. Desiderio, K. L. Vergin, M. S. Rappe, S. Laney, L. J. Wilhelm, H. J. Tripp, E. J. Mathur, and D. F. Barofsky. 2005. Proteorhodopsin in the ubiquitous marine bacterium SAR11. *Nature* **438:**82–85.

Golovlev, E. L. 2003. Bacterial cold shock response at the level of DNA transcription, translation and chromosome dynamics. *Mikrobiologiia* **72:**5–13.

Goo, A. Y., J. Roach, G. Glusman, N. S. Baliga, K. Deutsch, M. Pan, S. Kennedy, S. DasSarma, W. V. Ng, and L. Hood. 2004. Low-pass sequencing for microbial comparative genomics. *BMC Genomics.* **5:**1–19.

Gounot, A. M. 1976. Effects of temperature on the growth of psychrophilic bacteria from glaciers. *Can. J. Microbiol.* **22:**839–846.

Gupta, P., G. S. Reddy, D. Delille, and S. Shivaji. 2004. *Arthrobacter gangotriensis* sp. nov. and *Arthrobacter kerguelensis* sp. nov. from Antarctica. *Int. J. Syst. Evol. Microbiol.* **54:**2375–2378.

Head, I. M., J. R. Saunders, and R. W. Pickup. 1998. Microbial evolution, diversity, and ecology: a decade of ribosomal RNA analysis of uncultivated microorganisms. *Microb. Ecol.* **35:**1–21.

Holmes, A. J., J. Bowyer, M. P. Holley, M. O'Donoghue, M. Montgomery, and M. R. Gillings. 2000. Diverse, yet-to-be-cultured members of the *Rubrobacter* subdivision of the Actinobacteria are widespread in Australian arid soils. *FEMS Microbiol. Ecol.* **33:**111–120.

Junge, K., C. Krembs, J. Deming, A. Stierle, and H. Eicken. 2001. A microscopic approach to investigate bacteria under in situ conditions in sea ice samples. *Ann. Glaciol.* **33:**304–310.

Junge, K., J. F. Imhoff, J. T. Staley, and J. W. Deming. 2002. Phylogenetic diversity of numerically important Arctic sea-ice bacteria cultured at subzero temperature. *Microb. Ecol.* **43:**315–328.

Karner, M. B., E. F. DeLong, and D. M. Karl. 2001. Archaeal dominance in the mesopelagic zone of the Pacific Ocean. *Nature* **409:**507–510.

Kawahara, H. 2002. The structures and functions of ice crystal-controlling proteins from bacteria. *J. Biosci. Bioeng.* **94:**492–496.

Klecka, G. M., C. L. Carpenter, and B. D. Landenberger. 1993. Biodegradation of aircraft deicing fluids in soil at low temperatures. *Ecotoxicol. Environ. Saf.* **25:**280–295.

Knoblauch, C., B. B. Jørgensen, and J. Harder. 1999. Community size and metabolic rates of psychrophilic sulfate-reducing bacteria in Arctic marine sediments. *Appl. Environ. Microbiol.* 65:4230–4233.

Lauro, F. M., G. Bertoloni, A. Obraztsova, C. Kato, B. M. Tebo, and D. H. Bartlett. 2004. Pressure effects on *Clostridium* strains isolated from a cold deep-sea environment. *Extremophiles* 8:169–173.

Lipson, D. A., and S. K. Schmidt, 2004. Seasonal changes in alpine soil microbial community in the Colorado Rocky Mountains. *Appl. Environ. Microbiol.* 70:2867–2879.

Lipson, D. A., C. W. Schadt, and S. K. Schmidt. 2002. Changes in soil microbial community structure and function in an alpine dry meadow following spring snow melt. *Microb. Ecol.* 43:307–314.

López-García, P., A. López-López, D. Moreira, and F. Rodríguez-Valera. 2001a. Diversity of free-living prokaryotes from a deep-sea site at the Antarctic Polar Front. *FEMS Microbiol. Ecol.* 36:193–202.

López-García, P., D. Moreira, A. López-López, and F. Rodríguez-Valera. 2001b. A novel haloarchaeal-related lineage is widely distributed in deep oceanic regions. *Environ. Microbiol.* 3: 72–78.

Margesin, R., and F. Schinner. 2001. Bioremediation (natural attenuation and biostimulation) of diesel-oil-contaminated soil in an alpine glacier skiing area. *Appl. Environ. Microbiol.* 67:3127–3133.

Margesin, R., D. Labbé, F. Schinner, C. W. Greer, and L. G. Whyte. 2003. Characterization of hydrocarbon-degrading microbial populations in contaminated and pristine alpine soils. *Appl. Environ. Microbiol.* 69:3085–3092.

Massana, R., A. E. Murray, C. M. Preston, and E. F. DeLong. 1997. Vertical distribution and phylogenetic characterization of marine planktonic Archaea in the Santa Barbara Channel. *Appl. Environ. Microbiol.* 63:50–56.

Matsumoto, G., K. Chikazawa, H. Murayama, T. Torii, H. Fukushima, and T. Hanya. 1983. Distribution and correlation of total organic carbon and mercury in Antarctic dry valley soils, sediments and organisms. *Geochem. J.* 17:241–246.

McDougald, D., S. A. Rice, D. Weichart, and S. Kjelleberg. 1998. Noncultureability: adaptation or debilitation. *FEMS Microbiol. Ecol.* 25:1.

Medigue, C., E. Krin, G. Pascal, V. Barbe, A. Bernsel, P. N. Bertin, F. Cheung, S. Cruveiller, S. D'Amico, A. Duilio, G. Fang, G. Feller, C. Ho, S. Mangenot, G. Marino, J. Nilsson, E. Parrilli, E. P. Rocha, Z. Rouy, A. Sekowska, M. L. Tutino, D. Vallenet, G. von Heijne, and A. Danchin. 2005. Coping with cold: the genome of the versatile marine Antarctica bacterium *Pseudoalteromonas haloplanktis* TAC125. *Genome Res.* 15:1325–1335.

Mergaert, J., A. Verhelst, M. C. Cnockaert, T. L. Tan, and J. Swings. 2001. Characterization of facultative oligotrophic bacteria from polar seas by analysis of their fatty acids and 16S rDNA sequences. *Syst. Appl. Microbiol.* 24:98–107.

Methé, B. A., K. E. Nelson, J. W. Deming, B. Momen, E. Melamud, X. Zhang, J. Moult, R. Madupu, W. C. Nelson, R. J. Dodson, L. M. Brinkac, S. C. Daugherty, A. S. Durkin, R. T. DeBoy, J. F. Kolonay, S. A. Sullivan, L. Zhou, T. M. Davidsen, M. Wu, A. L. Huston, M. Lewis, B. Weaver, J. F. Weidman, H. Khouri, T. R. Utterback, T. V. Feldblyum, and C. M. Fraser. 2005. The psychrophilic lifestyle as revealed by the genome sequence of *Colwellia psychrerythraea* 34H through genomic and proteomic analyses. *Proc. Natl. Acad. Sci. USA* 102:10913–10918.

Miettinen-Oinonen, A., M. Paloheimo, R. Lantto, and P. Suominen. 2005. Enhanced production of cellobiohydrolases in *Trichoderma reesei* and evaluation of the new preparations in biofinishing of cotton. *J. Biotechnol.* 116:305–317.

Miwa, T. 1975. Clostridia in the soil of Antarctica. *Jpn. J. Med. Sci. Biol.* 28:201–213.

Mock, T., and D. N. Thomas. 2005. Recent advances in sea-ice microbiology. *Environ. Microbiol.* 7:605–619.

Montes, M. J., C. Belloch, M. Galiana, M. D. Garcia, C. Andres, S. Ferrer, J. M. Torres-Rodriguez, and J. Guinea. 1999. Polyphasic taxonomy of a novel yeast isolated from Antarctic environment; description of *Cryptococcus victoriae* sp. nov. *Syst. Appl. Microbiol.* 22:97–105.

Morita, R. Y. 1975. Psychrophilic bacteria. *Bacteriol. Rev.* 39:144–167.

Morita, R. Y. 1986. Pressure as an extreme environment, p. 171–186. *In* R. A. Herbert and G. A. Codd (ed.), *Microbes in Extreme Environments*. Academic Press, London, United Kingdom.

Morris, G. J. 2005. The origin, ultrastructure, and microbiology of the sediment accumulating in liquid nitrogen storage vessels. *Cryobiology* 50:231–238.

Murray, A. E., C. M. Preston, R. Massana, L. T. Taylor, A. Blakis, K. Wu, and E. F. DeLong. 1998. Seasonal and spatial variability of bacterial and archaeal assemblages in the coastal waters near Anvers Island, Antarctica. *Appl. Environ. Microbiol.* 64:2585–2595.

Nagahama, T., M. Hamamoto, T. Nakase, H. Takami, and K. Horikoshi. 2001. Distribution and identification of red yeasts in deep-sea environments around the northwest Pacific Ocean. *Antonie Leeuwenhoek* 80:101–110.

Nemergut, D. R., E. K. Costello, A. F. Meyer, M. Y. Pescador, M. N. Weintraub, and S. K. Schmidt. 2005. Structure and function of alpine and arctic soil microbial communities. *Res. Microbiol.* 156:775–784.

Nienow, J. A., and E. I. Friedmann. 1993. Terrestrial lithophytic (rock) communities, p. 343–412. *In* E. I. Friedmann (ed.), *Antarctic Microbiology*. Wiley-Liss, New York, NY.

O'Brien, A., R. Sharp, N. J. Russell, and S. Roller. 2004. Antarctic bacteria inhibit growth of food-borne microorganisms at low temperatures. *FEMS Microbiol. Ecol.* 48:157–167.

Ouverney, C. C., and J. A. Fuhrman. 2000. Marine planktonic archaea take up amino acids. *Appl. Environ. Microbiol.* 66:4829–4833.

Parkes, R. J., B. A. Cragg, and P. Wellsbury. 2000. Recent studies on bacterial populations and processes on sub-seafloor sediments: a review. *Hydrogeol. J.* 8:11.

Pathom-Aree, W., Y. Nogi, I. C. Sutcliffe, A. C. Ward, K. Horikoshi, A. T. Bull, and M. Goodfellow. 2006a. *Williamsia marianensis* sp. nov., a novel actinomycete isolated from the Mariana Trench. *Int. J. Syst. Evol. Microbiol.* 56:1123–1126.

Pathom-Aree, W., J. E. Stach, A. C. Ward, K. Horikoshi, A. T. Bull, and M. Goodfellow. 2006b. Diversity of actinomycetes isolated from Challenger Deep sediment (10,898 m) from the Mariana Trench. *Extremophiles* 10:181–189.

Rabus, R., A. Ruepp, T. Frickey, T. Rattei, B. Fartmann, M. Stark, M. Bauer, A. Zibat, T. Lombardot, I. Becker, J. Amann, K. Gellner, H. Teeling, W. D. Leuschner, F. O. Glockner, A. N. Lupas, R. Amann, and H. P. Klenk. 2004. The genome of *Desulfotalea psychrophila*, a sulfate-reducing bacterium from permanently cold Arctic sediments. *Environ. Microbiol.* 6:887–902.

Ramsey, A. J., and R. E. Stannard. 1986. Numbers and viability of bacteria in ornithogenic soils of Antarctica. *Polar Biol.* 5:195–198.

Reddy, G. S., G. I. Matsumoto, P. Schumann, E. Stackebarndt, and S. Shivaji. 2004. Psychrophilic pseudomonads from Antarctica: *Pseudomonas antarctica* sp. nov., *Pseudomonas meridiana* sp. nov. and *Pseudomonas proteolytica* sp. nov. *Int. J. Syst. Evol. Microbiol.* 54:713–719.

Russell, N. J. 1997. Psychrophilic bacteria—molecular adaptations of membrane lipids. *Comp. Biochem. Physiol. A.* 118:489–493.

Russell, N. J. 1998. Molecular adaptations in psychrophilic bacteria: potential for biotechnological applications. *Adv. Biochem. Eng. Biotechnol.* **61:**1–21.

Russell, N. J. 2002. Bacterial membranes: the effects of chill storage and food processing. An overview. *Int. J. Food Microbiol.* **79:**27–34.

Saunders, N. F., T. Thomas, P. M. Curmi, J. S. Mattick, E. Kuczek, R. Slade, J. Davis, P. D. Franzmann, D. Boone, K. Rusterholtz, R. Feldman, C. Gates, S. Bench, K. Sowers, K. Kadner, A. Aerts, P. Dehal, C. Detter, T. Glavina, S. Lucas, P. Richardson, F. Larimer, L. Hauser, M. Land, and R. Cavicchioli. 2003. Mechanisms of thermal adaptation revealed from the genomes of the Antarctic Archaea *Methanogenium frigidum* and *Methanococcoides burtonii*. *Genome Res.* **13:**1580–1588.

Schadt, C. W., A. P. Martin, D. A. Lipson, and S. K. Schmidt. 2003. Seasonal dynamics of previously unknown fungal lineages in tundra soils. *Science* **301:**1359–1361.

Selbmann, L., S. Onofri, M. Fenice, F. Federici, and M. Petruccioli. 2002. Production and structural characterisation of the exopolysaccharide of the Antarctic fungus *Phoma herbarum* CCFEE 5080. *Res. Microbiol.* **153:**585–592.

Shivaji, S., P. Chaturvedi, G. S. Reddy, and K. Suresh. 2005. *Pedobacter himalayensis* sp. nov., from the Hamta glacier located in the Himalayan mountain ranges of India. *Int. J. Syst. Evol. Microbiol.* **55:**1083–1088.

Siebert, J., P. Hirsch, B. Hoffmann, C. G. Gliesche, K. Piessl, and M. Jendrach. 1996. Cryptoendolithic microorganisms from Antarctic sandstone of Linnaeus Terrace (Asgard Range): diversity, properties and interactions. *Biodiv. Conserv.* **5:**1337–1363.

Smith, H. G., and P. V. Tearle. 1985. Aspects of microbial and protozoan abundances in Signy Island fellfields. *Br. Antarct. Surv. Bull.* **68:**83–90.

Smith, J. J., L. A. Tow, W. Stafford, C. Cary, and D. A. Cowan. 2006. Bacterial diversity in three Antarctic cold desert mineral soils. *Microb. Ecol.* **51:**413–421. [Epub ahead of print]

Suzuki, M. T., and S. J. Giovannoni. 1996. Bias caused by template annealing in the amplification of mixtures of 16S rRNA genes by PCR. *Appl. Environ. Microbiol.* **62:**625–630.

Takami, H., Y. Takaki, and I. Uchiyama. 2002. Genome sequence of *Oceanobacillus iheyensis* isolated from the Iheya Ridge and its unexpected adaptive capabilities to extreme environments. *Nucleic Acids Res.* **30:**3927–3935.

Takami, H., A. Znone, F. Fuji, and K. Horikoshi. 1997. Microbial flora in the deepest sea mud of the Mariana Trench. *FEMS Microbiol. Lett.* **152:**279–285.

Taton, A., S. Grubisic, E. Brambilla, R. De Wit, and A. Wilmotte. 2003. Cyanobacterial diversity in natural and artificial microbial mats of Lake Fryxell McMurdo Dry Valleys, Antarctica: a morphological and molecular approach. *Appl. Environ. Microbiol.* **69:**5157–5169.

Tindbaek, N., A. Svendsen, P. R. Oestergaard, and H. Draborg. 2004. Engineering a substrate-specific cold-adapted subtilisin. *Protein Eng. Des. Sel.* **17:**149–156.

Ugolini, F. C. 1972. Ornithogenic soils of Antarctica *Antarct. Res. Ser.* **20:**181–193.

Vishniac, H. S. 1993. The microbiology of Antarctic soils, p. 297–341. *In* E. I. Friedmann (ed.), *Antarctic Microbiology*. Wiley-Liss, New York, NY.

Ward, B. B. 2002. How many species of prokaryotes are there? *Proc. Natl. Acad. Sci. USA* **99:**10234–10236.

Waterbury, J. B., S. W. Watson, R. R. L. Guillard, and L. E. Brand. 1979. Widespread occurrence of a unicellular, marine, planktonic cyanobacterium. *Nature* **277:**293–294.

Wharton, R. A., C. P. McKay, G. M. Simmons, and B. C. Parker. 1985. Cryoconite holes on glaciers. *Bioscience* **35:**499–503.

Wynn-Williams, D. D. 1989. TV image analysis of microbial communities in Antarctic fellfields. *Polarforschung* **58:**239–250.

Wynn-Williams, D. D. 1990. Ecological aspects of Antarctic microbiology. *Adv. Microbial Ecol.* **11:**71–146.

Yi, H., and J. Chun. 2006. *Flavobacterium weaverense* sp. nov. and *Flavobacterium segensis* sp. nov., novel psychrophiles isolated from the Antarctic. *Int. J. Syst. Evol. Microbiol.* **56:**1239–1244.

Yurkov, V. V., and J. T. Beatty. 1998. Aerobic anoxygenic phototrophic bacteria. *Microb. Mol. Biol. Rev.* **62:**695–724.

Physiology and Biochemistry of Extremophiles
Edited by C. Gerday and N. Glansdorff
© 2007 ASM Press, Washington, D.C.

Chapter 10

Life in Ice Formations at Very Cold Temperatures

JODY W. DEMING

INTRODUCTION

Much has been written about microbial life in ice in recent years (e.g., reviews by Thomas and Dieckmann, 2002; Lizotte, 2003; Priscu and Christner, 2004; Rivkina et al., 2004; Mock and Thomas, 2005; Deming and Eicken, 2007). The subject has risen in visibility as Earth's polar and high-altitude environments are radically affected by climate warming, and thus loss of ice (Osterkamp, 2003; Gregory et al., 2004; Alley et al., 2005; Stroeve et al., 2005), and as space exploration and new thrusts in astrobiology focus on the prevalence of water-ice in our solar system (Murray et al., 2005; Richardson and Mischna, 2005). In this century we may witness the progressive extinction of ice-dependent organisms on Earth, even as we may discover, embracing the "science of optimism" (Sullivan and Baross, 2007), the signatures of life—or life itself—in extraterrestrial ices.

The most vibrant and extensive of within-ice microbial ecosystems are those that develop in seasonal ice formed from seawater. During the sunlit season, massive blooms of sea-ice algae and bacteria can rival planktonic production achieved over a comparable period of time (Gosselin et al., 1997), supporting in turn an extensive array of grazers dependent on them, from viruses to crustaceans still within the ice to apex predators feeding from above or below the ice (Thomas and Dieckmann, 2002). Why does sea ice favor such vibrant ecosystems, while other forms of ice (e.g., glacial ice) are more often characterized as preserving environments? Fundamentally, the more generously an ice matrix is permeated, even flushed, by the liquid phase, the greater the opportunity for diffusional and transport processes that favor nutrient delivery and exchange and thus biological production. For a given ice volume and temperature, sea ice contains a greater fraction of interior liquid than freshwater ice (Junge et al., 2004a; Deming and Eicken, 2007) (Table 1). At relatively warm ice temperatures (above −5°C), the ice is well channelized (Fig. 1) and flushed by underlying seawater (Eicken, 2003). Even at very cold temperatures, the principal freeze-depressing impurities in seawater—its salts—ensure the presence of liquid inclusions or brine-filled pores within the sea-ice matrix. These briny pores have recently been shown to remain interconnected on a scale relevant to microbes, even during the coldest months of winter when sea-ice temperatures can drop to −20°C (Fig. 2) and below, potentially approaching the seawater eutectic of −55°C where the atmosphere is sufficiently cold and the ice free of insulating snow. This eutectic also happens to be the average planetary surface temperature of Mars (Catling and Kasting, 2007).

Our planet has been losing its sea ice in recent decades and most notably in the Arctic, where the loss rate of about 8% per decade over the last 25 years (Stroeve et al., 2005 and citations therein) shows no signs of slowing (Lindsay and Zhang, 2005). Because Arctic sea ice also experiences lower temperatures in winter than the seasonal sea ice that surrounds the Antarctic continent, it emerges as a uniquely severe yet permissive (in terms of available liquid) habitat for microbial life. What have we learned about adaptive microbial strategies for surviving in the coldest of these saline ice environments? What organisms, unique genes, or biological products may the planet lose as these Arctic habitats are driven, seemingly relentlessly, to extinction by climate change? Other forms of ice, of course, are also experiencing reductions due to atmospheric warming (Alley et al., 2005; Smith et al., 2005). A comparative analysis of some of these ice types—specifically glacial ice over Greenland and Antarctica, and the "permanently" frozen soils or permafrost of the high Arctic during winter—can enlighten information emerging from microbial studies of winter sea ice and its liquid phase.

J. W. Deming • School of Oceanography, University of Washington, Seattle, WA 98195.

Table 1. Comparative features of three types of ice formations subject to very cold temperature

Feature	Polar glacial ice	Arctic winter permafrost	Arctic winter sea ice
Temperature (°C) range (over depth range)	−55 to 0 (0–3 km)	−37[a] to −2 (0–60 m)[a]	−40 to −2 (0–2 m)
Age of ice (year)[b]	1 to >10^5	1 to >10^6	1 to 10
Mean fraction of liquid[c]	0.0001	0.01	0.08
Salinity of liquid phase (‰, wt/vol)	<0.1[d]	~240–38[e]	~240–38[e]
DOC content of liquid phase	μM[d]	mM[e]	mM[e]
Lithogenic content (%, wt/vol)	<0.1	~70	<0.1 to >10
Bacterial[f] content (no. of cells per unit indicated)	10^1–10^2 (ml of ice^{-1})	10^7–10^9 (g of soil^{-1})	10^3–10^8 (ml of brine^{-1})
Viral content (no. of VLP ml of brine^{-1})[g]	Present, not quantified	High, not quantified	10^6–10^8

[a] From Alt and Labine (1998). Note that permafrost can extend below a soil depth of 1,600 m in some parts of the Russian Arctic (Washburn, 1980).
[b] Note that the older the ice formation (more deeply buried or removed from atmospheric influence), the warmer its temperature.
[c] From Deming and Eicken (2007). Note that glacial ice can also contain a significant fraction of gas; Arctic sea ice, visible amounts of entrained sediments; and Arctic permafrost, meter-scale lenses of liquid brine called cryopegs (Gilichinsky et al., 2003, 2005) and wedges of solid ice (Shur et al., 2005).
[d] Salt (and DOC) composition differs from seawater; the liquid phase is reported to be acidic (Price, 2000) and its DOC content low (Priscu et al., 2006).
[e] Cryopegs (see footnote c above) have temperatures of −9 to −11°C and salinities of 150–200‰ (Gilichinsky et al., 2005); the salinity (and DOC) of liquid inclusions (in pores or as films on grain surfaces) in winter permafrost over a wider temperature range is not well known but expected to resemble that determined for winter sea-ice brines (Krembs et al., 2002; Junge et al., 2004b).
[f] Referring to the sum of *Bacteria* and *Archaea*, since the microscopic methods used to obtain these estimates do not distinguish between the two domains: data for clean polar ice from Christner et al. (2005), for Arctic winter sea ice from Helmke and Weyland (1995) and Junge et al. (2001, 2004b), for permafrost from Vorobyova et al. (1997).
[g] Among these ice types, viral-like particles (VLP) have been quantified only for Arctic winter sea ice (Wells and Deming, 2006); the presence of VLP has been reported in warm Lake Vostok ice (Priscu et al., 2006) and of tomato virus throughout Greenland glacial ice (Castello et al., 1999); VLP abundance in permafrost is expected to be high, given high bacterial abundance in permafrost and high viral content of unfrozen soils and sediments (Danovaro et al., 2001).

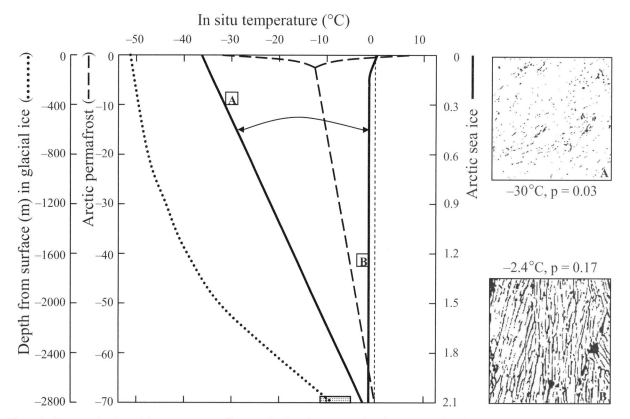

Figure 1. Comparative (generic) temperature profiles over depth in three types of ice formations: glacial ice at South Pole, Antarctica (dotted line), with basement temperature varying depending on method of assessment (hatched box; adapted from Price et al., 2002); Arctic permafrost, with the seasonal swing in the active surface layer (bold dashed lines; adapted from Hinkel et al., 2003; Osterkamp, 2003; Oelke and Zhang, 2004; Smith et al., 2005); and Arctic sea ice, with a seasonal swing that results in its melting, first at the surface as atmospheric temperatures warm in summer (bold lines; adapted from Krembs et al., 2002; Eicken, 2003). Upper right image (A) depicts pore microstructure (liquid phase in black) in a 20-mm wide section of sea ice at −30°C; lower right image (B) same, but at −2.4°C with greater pore connectivity (see Eicken et al., 1998; Deming and Huston, 2000).

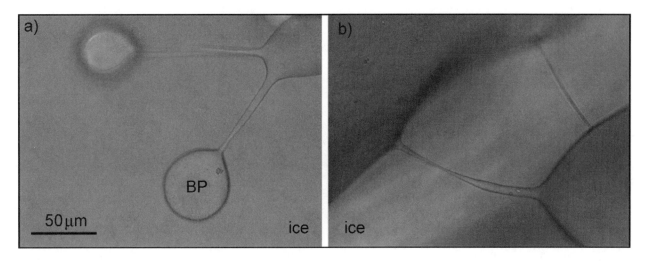

Figure 2. Microscopic images of thin sections of Arctic winter sea ice, taken by transmission light at −20°C, showing the size and interconnectivity of brine pores (BP) and some particulate matter within them; note scale bar of 50 μm (courtesy of H. Eicken; see Stierle and Eicken, 2002, for details).

For this chapter, I focus on ice temperatures of −10°C and below, referring to them as "very cold," and on organisms (and their viruses) from the domains of *Bacteria* and *Archaea*, referring to them generically as "microbes" or "bacteria." Bacteria have long been known to grow at warmer ice temperatures of −9°C and above (ZoBell, 1934; Baross and Morita, 1978). Some recent culture work still draws a line at −10°C for bacterial growth (Bakermans et al., 2003), but the line is a moving target (Christner, 2002; Breezee et al., 2004). In fact, colder natural ice formations have rarely been explored as viable microbial habitats until recently. With climate warming and ice loss continuing as predicted, examining what we do know about life in very cold ice appears timely.

COMPARISON OF VERY COLD GLACIAL ICE, PERMAFROST, AND SEA ICE

A comparative analysis of three types of ice formations that experience very cold temperatures—high altitude glacial ice overlying Greenland and Antarctica, Arctic winter permafrost, and Arctic winter sea ice (see Table 1 and Fig. 1)—benefits from the additional endmember factors that make each ice type unique (see Deming and Eicken, 2007, for a comparison that includes other frozen environments). Glacial ice is the cleanest of these ice formations, where "clean" refers to a relatively low mass of lithogenic inclusions, primarily dust particles deposited with the snow that, once consolidated, form glacial ice. Its microbial content is correspondingly low, from nearly undetectable (Bulat et al.,

2004) to a few hundred cells per milliter of melted ice, after accounting for contaminants introduced during ice-coring and sample handling (Rogers et al., 2004; Christner et al., 2005). Selected viruses have been detected in glacial ice (Castello et al., 1999; Priscu et al., 2006) but not quantified. The interior liquid phase of glacial ice is the most limited among ice formations; it exists primarily as nanometer-scale films between ice-crystal grains and on surfaces of entrained dust particles (Price, 2000). The liquid pores at triple-point junctures of glacial ice crystals as envisioned by Price (2000), however, are mere shadows of those that occur in winter sea ice (Eicken, 2003; Junge et al., 2004a, 2004b). More microbially relevant measurements of the dissolved inorganic and organic content of the liquid phase in glacial ice are needed, but current information suggests an acidic and oligotrophic milieu (Price, 2000; Priscu et al., 2006). The extent to which microbes deposited in glacial ice eventually segregate into this liquid phase, versus freezing (or remaining frozen) into the solid phase, perhaps via ice-nucleating proteins on their cell surfaces (Cohet and Widehem, 2000), is not known.

Whether or not the liquid phase of glacial ice is sufficient to support microbial activity in situ is a matter of recent interest and debate (Price and Sowers, 2004; Priscu et al., 2006; Sowers et al., 2006). Aiding the debate is the test-tube demonstration of DNA synthesis by cultured bacteria after suspension in distilled water with a labeled DNA precursor (^{3}H-thymidine), freezing, and incubation at −15°C (Christner, 2002). With few exceptions, however (e.g., Abyzov et al., 1998; Castello et al., 1999), biological research has not been focused on the cleanest and coldest of glacial ice but on the deepest, oldest, and "dirtiest" portions of the thick (~3 km) glacial ice masses overlying Greenland and Lake Vostok in Antarctica. Over Greenland, the basal layer has entrained sediments during glacial-ice movement over soil (Sheridan et al., 2003; Miteva et al., 2004; Priscu et al., 2006); at the base of the glacial ice overlying Lake Vostok, the lake-accreted ice, with its different form and history, has accrued a higher particle load than glacial ice (reviewed by Priscu et al., 2006). The dirtier the ice, the more the mineral surface area within it: the availability of liquid in surface films is maximized, as is the abundance of bacteria, since they entrain with sediment grains and other particles. At these depths, of course, the ice itself is also relatively warm (above –10°C; see Fig. 1). Taken together, the conditions within the deepest portions of glacial ice seem the most promising for detecting in situ microbes that may be metabolically active (thus attracting most biological study), even if such activity must still be strongly limited by the liquid phase within the ice matrix. The latter expectation and the age of the deepest glacial ice, on the order of hundreds of thousands of years (Christner et al., 2003), has led to the notion that glacial ice constitutes an "ice museum" of microbes from the distant past (Priscu et al., 2006).

If even relatively warm glacial ice is considered life-preserving, then what can we expect in the vast volume of ice that is both clean and very cold, experiencing temperatures down to –55°C near the surface? Glaciogeochemists have forayed into this icy territory in search of long-term climate records, since glacial ice has also entrapped greenhouse gases from its formation time. In the process, they have stumbled across indirect evidence in glacial ice (at an age of ~140,000 years) above Lake Vostok, which is consistent with in situ bacterial activity (Sowers, 2001). The evidence, derived from coincident maxima of N_2O, dust, and bacterial content in ice-core depth profiles and anomalous $d^{15}N$-to-$d^{18}O$ ratios in the trapped N_2O, suggests bacterial nitrification (conversion of NH_4 to N_2O) at temperatures as low as –40°C, even if at an extremely low rate (calculated carbon turnover time for a single cell, assuming all cells present are nitrifying, of 10^8 years; Price and Sowers, 2004). Subsequent pure culture work with a known bacterial nitrifier (and CO_2-fixer) has yielded empirical evidence for nitrification at temperatures of –8 to –10°C, an exciting new low for this process (Sowers et al., 2006), but one still in the "permissive" range and well short of the in situ glacial ice temperature. To the extent that chemolithotrophy relies upon inorganic gases rather than dissolved organic compounds (for heterotrophy), typically supplied by diffusion or advection and thus requiring a significant liquid phase, this work in glacial ice points to metabolisms worth targeting in any very cold ice formation.

Permanently frozen soil, or permafrost, represents an endmember to glacial ice in being primarily lithogenic by definition, with only a limited volume of either frozen or liquid water held within the soil matrix (see Table 1). Arctic permafrost can be interrupted at depth by meter-scale wedges of "solid" ice that cause the phenomenon of frost-heave (Shur et al., 2005) or by meter-scale lenses of liquid brine at –10 to –12°C, of special interest as geological remnants of seawater incursions known as cryopegs (Gilichinsky et al., 2003, 2005), but the comparison here is with the frozen environment that is predominantly mineral grains or soil. The mean liquid fraction in this permafrost is much greater than that found in glacial ice, due in part to the seasonal evaporative cycle that increases its ionic strength, but is still less than in sea ice. Because soils are rich in microbial biomass and diversity prior to freezing (unlike the snow leading to glacial ice), the deepest layers of permafrost, some of which can be several million years old (Rivkina et al., 2000), represent a microbial museum surpassing glacial ice. Only the active, upper (and younger) layer of permafrost, however, is subjected to seasonal variations in temperature (Shur et al., 2005) that include the very cold temperatures addressed by this chapter (see Fig. 1). Even so, this upper layer rarely sees temperatures approaching the eutectic of its liquid phase, given the insulating nature of snow and soil against the winter atmosphere. Of the ice formations being compared here, permafrost is the least well characterized on a scale relevant to the individual microbe (see Rivkina et al., 2000, and citations therein). Its mineral content, occupying the most of its volume (~70%), presents special challenges to such microscale analyses. Both the salt content and the concentration of dissolved organic matter in the liquid phase, however, are expected to be high, given the evaporative cycle and the organic-rich nature of the soils (Hinzman et al., 1998), even if the mean liquid fraction of permafrost falls below that of sea ice (see Table 1). At winter temperatures, microbes still apparently experience only nanometer thicknesses of liquid, according to calculations by Rivkina et al. (2000).

Of these ice formations that can experience very cold temperatures, Arctic winter sea ice represents the endmember for mean liquid fraction within the ice (see Table 1). The fraction of liquid inclusions is relatively large as a result of the non-linear process of ice-formation from seawater, which promotes the retention of pockets of seawater as the ice grows. The wonders and vagaries of sea-ice formation are described in detail elsewhere (Eicken, 2003), but the final winter result is an ice formation that is not as "solid" as glacial ice or permafrost (made uniquely solid by its mineral content) at a comparable temperature. Winter sea ice is considered nearly impermeable on a bulk scale at very cold temperatures (Eicken, 2003), but it retains an interconnected network (see Fig. 2) of interior liquid veins and pores of diverse sizes and shapes (Light et al., 2003) on a scale that generously accommodates microbes. Interior fluid flow on the microscale is considered possible, given the vertical thermal gradient through sea ice (Eicken, 2003; Deming and Eicken, 2007), and the brine itself is rich in dissolved organic carbon and nitrogen (DOC and DON, respectively) (Thomas et al., 1995; Krembs et al., 2002; Junge et al., 2004b). Application of newly developed microscopic techniques to artificial sea ice, grown from seawater containing a known number of bacteria pre-stained for detection, has indicated that 95% of bacteria partition into the liquid phase as the ice crystals form and grow (Junge et al., 2001). A few cells were observed in the solid phase, but the use of ice-nucleating proteins on the surface of the cell as a means to freeze into the ice does not appear to be a major bacterial strategy for inhabiting sea ice (as it may be for snow and glacial ice). Instead, microbes follow the water, as allowed by the natural freezing rate of sea ice, and end up in highly saline brine inclusions for the winter. Whereas bacteria in glacial ice may be confronted with a nanometer film of acidic liquid, bacteria in winter sea ice (and their counterparts in permafrost) face high salt concentrations (see Table 1) usually considered preservative, if not osmotically damaging.

Yet, natural populations of bacteria in Arctic winter sea ice have been shown to be metabolically active down to −20°C, at brine salinities to 210‰ (Junge et al., 2004b), and those in Arctic permafrost, to at least −10°C (salinity not reported; Rivkina et al., 2000). Pure culture work has confirmed these temperatures: a permafrost isolate grows at −10°C in a medium with 3% salt (Bakermans et al., 2003); an Arctic sea-ice isolate grows at −12°C in a seawater-glycerol medium (Breezee et al., 2004); and another Arctic isolate makes protein in artificial sea ice at −20°C, where the salinity of brine inclusions is 210‰ (Junge et al., 2006). What enables bacteria and their enzymes to function at very cold temperatures in brines of 21% salt? How have they solved the paradox of presumably prohibitive salt concentrations providing the very liquid phase required for cellular and biochemical reactions?

THE FREEZE-CONCENTRATION EFFECT IN VERY COLD SEA ICE

As atmospheric conditions sharpen the temperature gradient across sea ice during winter, with surface temperatures typically dropping to −20° or lower (see Fig. 1), more interior water moves into the solid phase and the liquid fraction of the ice matrix decreases accordingly. With shrinking pore space, the encased liquid becomes increasingly salty. Temperature thus determines the salinity of the brine, though not in linear fashion (Cox and Weeks, 1983), in part because the subzero precipitation points of individual sea salts differ (Assur, 1960). For example, Mirabilite ($NaSO_4 \cdot 10H_2O$) precipitates at −8.2°C, while Hydrohalite ($NaCl \cdot 2H_2O$) precipitates at −22.9°C (for details, see Eicken, 2003). Nevertheless, even at the coldest temperatures experienced by sea ice (cited as about −40°C by Eicken, 2003), a significant liquid fraction remains.

Not only do salts become concentrated as the temperature drops, dissolved organic and particulate materials, including bacteria, find themselves encased in spaces of diminishing volume. Relative to inorganic salts, the freeze-concentration process for organic elements in sea ice is poorly known (Giannelli et al., 2001; Thomas and Papadimitriou, 2003), especially for very cold sea ice. Information relevant to the spatial scale of microbial life in winter sea ice has become available recently with studies of Arctic sea ice (and simulations thereof) near Barrow, Alaska. In addition to observing microscopically that most bacteria partition into the liquid phase during sea-ice formation (Junge et al., 2001), we learned that gelatinous, extracellular polysaccharide substances, called EPS or simply exopolymers, were present throughout the liquid inclusions of winter sea ice (Krembs et al., 2002). The gelatinous nature of EPS presents some operational challenges in sampling and analysis compared to dissolved and particulate organic matter, which are commonly defined by filter pore-size separations (e.g., DOC < 0.7 μm < POC). Nevertheless, sampling across the temperature gradient inherent to winter ice cores revealed a freeze-concentration effect for EPS (>0.4 μm, sometimes called particulate EPS): the amount of EPS (and DOC), when scaled to the liquid volume within a 10-cm section of ice, was higher at colder temperatures (Fig. 3). The results of complementary experiments with artificial ice and additional considerations (Krembs et al., 2002) led us to conclude that EPS is also produced in situ in the ice over the temperature range of −5 to −25°C (the lowest tested). We proposed cryoprotection as the biological "incentive" for EPS production.

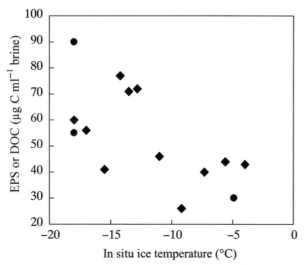

Figure 3. Concentration of EPS (◆) or DOC (•) scaled to the brine volume of ice sections cored from Arctic winter sea ice, where the sampled temperature gradient spanned –4 to –18°C (data from Krembs et al., 2002).

Cryoprotection encompasses more than the concept of freeze-depression, as the rest of this chapter considers.

Although the initial work on exopolymers in winter sea ice focused on algae as the dominant producers (Krembs et al., 2002 and citations therein), subsequent work has highlighted bacterial associations with EPS (Meiners et al., 2003). Detailed microscopic study of EPS particles recovered from (Antarctic) sea ice shortly after its formation in autumn revealed a predictable and positive association of bacteria with the EPS, deemed a colonization process (Meiners et al., 2004). Sea-ice bacteria grown in culture have been shown to produce copious amounts of EPS (Helmke and Weyland, 1995), especially when incubated at suboptimal growth or subzero temperatures (–2°C) (Mancuso Nichols et al., 2004; Nichols et al., 2005a, 2005b). Although tests at even colder temperatures or within an ice matrix have not yet been reported, the trend continues in that direction (J. G. Marx and J. W. Deming, unpublished data). In general, EPS production at subzero temperatures is considered a stress response to low temperature, but the precise benefits of this potentially costly production effort remain to be determined. They can be predicted, however.

PREDICTABLE BENEFITS OF EXOPOLYMERS IN VERY COLD ICE

First efforts to evaluate bacterial activity in Arctic winter sea ice revealed that the attached state favors activity at very cold temperatures, down to –20°C. Virtually all (>98%) cells scored as active by the 5-cyano-2, 3-ditolyl tetrazolium chloride (CTC) stain (Fig. 4), indicating oxygen-based respiration, or by the fluorescent in situ hybridization technique, indicating abundant protein-making machinery, were evaluated as attached to a surface (Junge et al., 2004b). The available surfaces ranged from entrained sediment grains and organic detrital particles to the walls of ice crystals framing a pore space (see Fig. 2). Members of the *Cytophaga-Flavobacterium-Bacteroides* (CFB) group of *Bacteria*, many of which are known for exopolymer production and an attached lifestyle, constituted a greater fraction of the total bacterial population in the colder ice sections, likely accounting for a sizeable fraction of the active cells at the coldest temperatures (see Fig. 4). In other fluid-bathed environments, microbes attach to a surface as a result of their own hydrophobicity, electrostatic surface charges, and/or the adhesive properties of exopolymers (Decho, 1990). When the bathing fluid is brine, charge differences are minimized and the availability of EPS emerges as the key factor.

The benefits of attaching to a surface within the brine pores of sea ice could be envisioned in an optimal foraging context (Vetter et al., 1998), whereby attachment provides a microbe with direct access to organic resources sorbed to the same surface, as occurs in other, especially oligotrophic, environments (Geesey and White, 1990). In a brine pore already

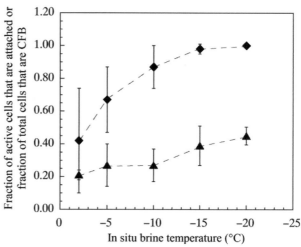

Figure 4. Fraction of actively respiring cells determined by CTC stain and evaluated as originally attached to surfaces in the brine pores of Arctic winter sea ice (◆), where the sampled temperature gradient spanned –2 to –20°C, and fraction of the total bacterial population that fluoresced in response to in situ hybridization probes for *Cytophaga-Flavobacterium-Bacteroides* (▲), a group generally known for EPS production and an attached lifestyle. Error bars are standard error of the mean for triplicate samples; at –20°C, the symbol for CTC-stained bacteria obscures the error bars (data from Junge et al., 2004b).

enriched with DOC, however, the more immediate benefit of embedding in a surface-sorbed matrix of EPS may be osmotic. A cellular coating of EPS is already known to provide a physical-chemical buffer against chemical challenges to the microbe (Mancuso Nichols et al., 2005); in this case, the hydrated state of gelatinous EPS may prevent the cell from experiencing the full brunt of the dissolved salt beyond its exopolymeric coating. In the attached state, the cell could benefit even further from the reduction in cell-surface exposure (Murray and Jumars, 2002). An implication of this salt-buffering concept is that the salinity of a brine inclusion may not be uniform within a pore; in providing for lower salt concentrations within its polymeric matrix, the presence of EPS may create a higher salt concentration in the remaining fluid, further depressing the freezing point and maintaining the volume of occupyable pore space, yet another benefit.

At the nanometer scale, the gelatinous nature of EPS has also been suggested to provide physical scaffolding within which extracellular enzymes may be held, not simply close to the cell (an obvious benefit; Vetter et al., 1998) but in a favorable catalytic state. The nearby localization of secreted enzymes for the benefit of the cell is well known (Decho, 1990; Mancuso Nichols et al., 2005), but how the EPS-enzyme interaction may function, especially at very cold temperatures in a briny fluid, is not. We began to consider the possibilities in Arctic winter sea ice, as a result of late-summer surveys (1997–1999) of Arctic environments in search of cold-active extracellular enzymes. Using fluorescently tagged substrate analogs to assay extracellular enzyme activity (EEA) in a variety of sample types as a function of temperature, we determined that multi-year sea ice (melted into sterile EEA-free seawater to enable the assays) supported the highest frequency of unusually cold-adapted EEA (Fig. 5), defined as activity with a $T_{opt} \leq 15°C$ (very low, relative to T_{opt} for virtually all enzymes purified and characterized in the laboratory) (Deming and Baross, 2002). Multi-year sea ice has weathered at least one winter season without melting; enzymes (and their bacterial sources) retained within its pore spaces, in spite of brine expulsion over the seasons, would have experienced very cold temperatures cyclically (inter-annually), imposing a selection.

Taking advantage of an overwintering Arctic expedition in 2003–2004 (Langlois et al., 2006), I sampled first-year winter sea ice (no multi-year ice was within reach) and measured EEA over the temperature range of –12 to 22°C and brine salinities of 22–220‰ (Fig. 6) (Deming, 2004). The different salt concentrations were achieved by melting sea ice into artificial (EEA-free) brine solutions at salt concentrations calculated to yield a final meltwater of the

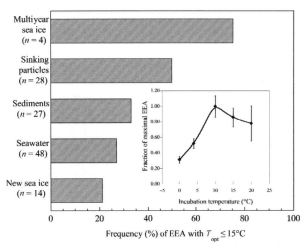

Figure 5. Frequency of detection of cold-adapted extracellular protease activity ($T_{opt} \leq 15°C$; A. L. Huston and J. W. D., unpublished), as measured by the protocols of Huston et al. (2000); inset depicts an individual experiment based on a sample of sinking particulate matter originally at –1°C (data from Huston et al., 2000).

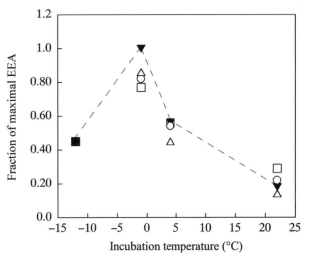

Figure 6. Temperature-dependent extracellular enzyme activity (EEA) measured by fluorescent substrate assay (for proteases, as in Huston et al., 2000) in first-year Arctic winter sea-ice brines from Franklin Bay, N.W.T. (from Deming, 2004). Ice sections originally at –20°C in situ were melted into 0.2-μm filtered, artificial brine solutions to achieve final meltwater salinities (‰) of 220 (squares), 210 (circles), 180 (triangles), or 22 (inverted triangles). The dotted line connects the solid symbols, which represent EEA at salinities nearest to those of the microbial habitat at each temperature.

desired salinity (a variant of the isohaline-isothermal melting approach described by Junge et al., 2004b). This experimental approach yielded very low but significant rates of enzyme activity and the unprecedented but repeated finding of an EEA T_{opt} of –1°C (see Fig. 5). (Only the highest salinity was tested at –12°C, since the samples would have frozen at lower salinities.) At –1°C, the coldest test temperature where salinity was varied, higher brine salinities reduced EEA, suggesting a need for enzyme protection against the high salinities of natural winter sea-ice brines. As argued earlier, EPS could provide this benefit.

We conducted some initial laboratory tests of the idea that EPS could protect enzymes in the cold, working with a purified extracellular protease, an aminopeptidase produced by *Colwellia psychrerythraea* strain 34H (Huston et al., 2004). This strain, an obligately psychrophilic and halophilic bacterium, also produces EPS and most abundantly when incubated at subzero temperatures (Huston et al., 2004; J. G. Marx and J. W. D., unpublished). The annotation of its whole genome sequence suggests that it may invest considerable energy in producing and exporting a wide variety of enzymes and exopolymers (Methé et al., 2005). Though strain 34H was originally isolated from Arctic sediments, other strains of *C. psychrerythraea* have been found in sea ice, from both poles (Bowman et al., 1997; Brinkmeyer et al., 2003). When the purified test enzyme from strain 34H was incubated at –1°C in a sea-salts solution in the absence of EPS, no residual activity (assayed at room temperature) remained after 48 h; when the enzyme/sea-salts solution was amended with a crude extract of (protease-free) EPS, not only did the enzyme retain activity for the duration of the experiment (48 h) but its activity was enhanced. Although more work along these lines needs to be done (and is under way), the added benefit of enzyme stabilization or protection against denaturing forces can be added to the list of potential benefits of EPS to ice-encased microbes.

Sources of extracellular enzymes within the brine inclusions of winter sea ice are not limited to the encased microbes. Enzymes may have been present in the initial seawater prior to freezing, and thus derived from an unknown myriad of possible organisms, only to partition and eventually concentrate in the brine inclusions during the freezing process and onset of winter. Rarely are any organisms other than bacteria detected in the coldest horizons of winter sea ice, due to space limitations and ice damage to eukaryotic structures as brine pores shrink during the freezing process (Krembs et al., 2002), but that same ice damage would release enzymes into the liquid phase of the ice. Theoretically (empirical studies are lacking), any of these enzymes could also benefit from the stabilizing effect of EPS.

Another more paradoxical source of extracellular enzymes in winter sea-ice brines is the viral population (see Table 1). Warm-season sea ice harbors some of the highest concentrations of virus-like particles observed anywhere in nature (Maranger et al., 1994), up to 10^9 ml of brine^{-1} (Gowing, 2003; Gowing et al., 2004), and active (infective) bacteriophage have been isolated from summer sea ice using cold-adapted hosts (Borriss et al., 2003). Recently, both abundant viruses, approaching 10^8 ml of brine^{-1}, and active bacteriophage have been reported for first-year winter sea ice (Wells and Deming, 2006). A diffusion-based model, assuming viral partitioning into the liquid phase along with bacteria and accounting for the brine concentrating factor inherent to winter sea ice, yielded contact rates between viruses and bacteria in winter sea-ice brines orders of magnitude higher than those in seawater (Wells and Deming, 2006). Increases in viral numbers and bacterial growth were also detected under simulated winter sea-ice brine conditions at –12°C and 160‰ (Fig. 7; Wells and Deming, 2006). Although not yet examined for sea-ice phage, many infective bacteriophage from other environments are known to carry active enzymes, particularly depolymerases, for the purpose of degrading the outer coatings of a host for eventual entry into its cytoplasm (Hughes et al., 1998). Just as the presence of EPS under subzero conditions can enhance the lifetime and activity of extracellular proteases, it may also protect viral enzymes, favoring their ability to infect ice-encased bacteria (Fig. 8). If the infection leads to cell lysis, the benefits accrue to neighboring cells in the form of freshly released, nitrogen-rich organic solutes. If the infection leads instead to lysogeny, the attack could prove beneficial if a gene conferring a competitive advantage transfers in the process.

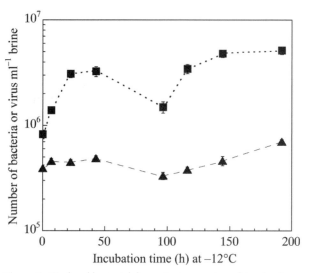

Figure 7. Viral and bacterial dynamics in a section of Arctic winter sea ice melted into an artificial (0.2-μm filtered) brine solution to achieve a final sample salinity of 160‰, mimicking the in situ brine salinity of the ice at –12°C (data from Wells and Deming, 2006). In many cases, symbols (▲ for bacteria, ■ for viruses) for the mean of the microscopic counts obscure the error bars (standard error of the mean).

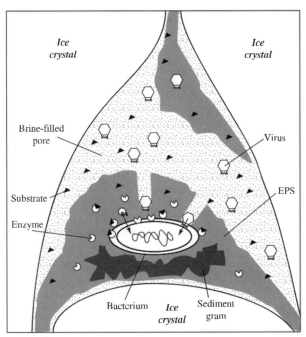

Figure 8. Schematic diagram of the potential roles of EPS in winter sea-ice brine pores (including freeze depressant, attachment facilitator, salt segregator, and physical buffer against osmotic damage) and as enzyme localizer and stabilizer (including virus-borne polysaccharide depolymerases; adapted from Hughes et al., 1998).

THE POTENTIAL FOR DISCOVERY

In considering the potential benefits of EPS to microbial inhabitants of the very cold and briny fluids of Arctic winter sea ice (see Fig. 8), numerous research questions emerge. Some are specific to sea ice but many pertain to any ice formation that experiences very cold temperatures. For example, how does a bacterium attach to the surface of an ice crystal when the system is wetted with brine (as in sea ice and permafrost) or with an acidic fluid (as in glacial ice)? Can complex biofilms develop on ice surfaces, as they do on other inanimate surfaces under milder conditions? What molecular or nanoscale architectural mechanisms allow EPS to stabilize an enzyme against denaturation in an unfrozen brine? Do some viruses package their depolymerases in ways that enhance activity in the unfrozen fluids of an ice formation or are they dependent on the presence of organism-provided EPS to infect a host? Is carrying their own enzymes a hallmark of cold-active viruses? Does lateral gene transfer occur in ice formations that experience very cold temperatures? If so, do the ice-encased organisms benefit from this process by gaining a survival trait that prevents their extinction or perhaps an improved ability to dominate once milder conditions prevail?

The knowledge base on microbial life in very cold ice formations has developed sufficiently in recent years that the pace of progress in answering these and other questions may now increase. Although the general research focus has been on heterotrophic *Bacteria*, especially in the case of winter sea ice, chemolithotrophic *Bacteria* and all types of *Archaea* await study from many perspectives. Permafrost has long been known to harbor methanogenic *Archaea*, as do soils prior to freezing. Although they clearly activate upon warming, the ability of methanogens to consume and produce gases (consuming CO_2 and H_2, yielding CH_4) at very cold temperatures is not clear. The potential for gas-consuming (NH_4^+ and CO_2) nitrifying chemolithotrophs, whether bacterial or archaeal, as identified in very cold glacial ice (Sowers, 2001; Sowers et al., 2006) suggests the conceptual model that gas-based microbes in general may fare well in very cold ice formations. Even in winter sea ice, where organic solutes abound and sufficient liquid exists for their diffusive and even advective transport, *Archaea* have recently been discovered. Just as the fraction of CFB increased with decreasing temperatures (see Fig. 5), so did the fraction of *Archaea* (data not shown; see Junge et al., 2004b). Although of unknown physiology in that study and present only in small numbers (1–3% of the total microbial community), a gas-based livelihood could explain their persistence in the face of a dominant population of heterotrophic bacteria. Still-more-recent work on first-year Arctic winter sea ice has confirmed the presence of methane-cycling *Archaea* and their persistence through conditions of –28°C and brine salinity of 230‰ (Collins and Deming, 2006). Do cold-adapted *Archaea* express EPS or are gas-based physiologies free of the need for this multifaceted benefaction? Do they escape or invite viral attack?

A key evolutionary question for microbial life in very cold ice formations on this planet or elsewhere involves lateral gene transfer, whether mediated by viruses or free DNA—do conditions in very cold ice formations actually promote genetic exchange between organisms and even between domains, as the high phage-host contact rate model mentioned earlier suggests? If gas-based physiologies are an advantage in a severely liquid-limited system, especially in an ancient ice formation, can we detect gene-exchange processes that may have favored their retention or expansion within such a stressed system? Genomics and proteomics will lead the way in answering these questions. The detection of methanogenic archaeal genes in *C. psychrerythraea* strain 34H (Methé et al., 2005) provides a tantalizing start. The availability of large volumes of very cold brine buried in ancient permafrost (Gilichinsky et al., 2003, 2005) holds great promise for direct environmental interrogations of both viral and microbial genomes.

Extremely cold ice formations—those with temperatures that fall below the eutectic of seawater (−55°C)—have rarely, if ever been, examined for the presence of living microbes. Such environments—surface glacial ice in winter, ice aggregates in the troposphere—are either rare or generally inaccessible on Earth. Elsewhere—on Mars and Europa—they are commonplace. We know about bacterial survival under deeply frozen conditions in a −80°C freezer or in −196°C liquid nitrogen storage tanks, but from the applied perspective of preserving cultures for research. Such work has taught that high solute concentrations can make a difference to the survival of a culture if coupled with vitrification (Dumont et al., 2006), the rapid freezing process by which water moves from the liquid to glass (solid) phase, bypassing ice-crystal formation and thus ice-crystal damage to cell membranes. In what state do cells exist in vitrified glass? Conventional wisdom assumes an entirely inactive state. New experiments with *C. psychrerythraea* strain 34H suggest otherwise (Junge et al., 2006). If the starting suspension of cells is in a sea-salt solution enriched with EPS, protein synthesis appears to continue even at subeutectic temperatures. Tracers to measure the synthesis must enter the cell at lower temperatures permissive of solute diffusion, but incorporation into protein appears possible via conformational changes of enzymes that do not require diffusion. The wonders of life in the extreme cold continue to beckon.

Acknowledgments. Preparation of this chapter benefited from exchanges with former and current students, Adrienne Huston, Karen Junge, Llyd Wells and Eric Collins, and with colleagues John Baross, Hajo Eicken, Christopher Krembs, and Joe Marx. Shelly Carpenter assisted with graphics.

NSF-OPP, NASA-ABI, and Washington State Sea Grant provided financial support.

REFERENCES

Abyzov, S. S., I. N. Mitskevich, and M. N. Poglazova. 1998. Microflora of the deep glacier horizons of central Antarctica. *Microbiology (Moscow)* 67:66–73.

Alley, R. B., P. U. Clark, P. Huybrechts, and I. Joughin. 2005. Ice-sheet and sea-level changes. *Science* 310:456–460.

Alt, B., and C. Labine. 1998. Meteorology and soil temperatures, Hot Weather Creek, Ellesmere Island, NWT, Canada. *In* International Permafrost Association, Data and Information Working Group, comp., *Circumpolar Active-Layer Permafrost System (CAPS)*, version 1.0. CD-ROM available from National Snow and Ice Data Center, nsidc@kryos.colorado.edu, Boulder: NSIDC, University of Colorado at Boulder.

Assur, A. 1960. Composition of sea ice and its tensile strength. *SIPRE Res. Rep.* 44:1–49.

Bakermans, C., A. I. Tsapin, V. Souza-Egipsy, D. A. Gilichinsky, and K. H. Nealson. 2003. Reproduction and metabolism at −10°C of bacteria isolated from Siberian permafrost. *Environ. Microbiol.* 5:321–326.

Baross, J. A., and R. Y. Morita. 1978. Microbial life at low temperatures: ecological aspects, p. 9–71. *In* D. J. Kushner (ed.), *Microbial Life in Extreme Environments.* Academic Press, New York, NY.

Borriss, M., E. Helmke, R. Hanschke, and T. Schweder. 2003. Isolation and characterization of marine psychrophilic phage-host systems from Arctic sea ice. *Extremophiles* 7:377–384.

Bowman, J. P., S. A. McCammon, M. V. Brown, D. S. Nichols, and T. A. McMeekin. 1997. Diversity and association of psychrophilic bacteria in Antarctic sea ice. *Appl. Environ. Microbiol.* 63:3068–3078.

Breezee, J., N. Cady, and J. T. Staley. 2004. Subfreezing growth of the sea ice bacterium *Psychromonas ingrahamii. Microb. Ecol.* 47:300–304.

Brinkmeyer, R., K. Knittel, J. Jürgens, H. Weyland, R. Amann, and E. Helmke. 2003. Diversity and structure of bacterial communities in Arctic versus Antarctic pack ice. *Appl. Environ. Microbiol.* 69:6610–6619.

Bulat, S. A., I. A. Alekhina, M. Blot, J.-R. Petit, M. de Angelis, D. Wagenbach, V. Y. Lipenkov, L. P. Vasilyeva, D. M. Wloch, D. Raynard, and V. V. Lukin. 2004. DNA signature of thermophilic bacteria from the aged accretion ice of Lake Vostok, Antarctica: implications for searching for life in extreme icy environments. *Intl. J. Astrobiol.* 3:1–12.

Castello, J. D., S. O. Rogers, W. T. Starmer, C. M. Catranis, L. Ma, G. D. Bachand, Y. Zhao, and J. E. Smith. 1999. Detection of tomato mosaic tobamovirus RNA in ancient glacial ice. *Polar Biol.* 22:207–212.

Catling, D., and J. F. Kasting. Planetary atmospheres and life. *In* W. T. Sullivan and J. A. Baross (ed.), *Planets and Life: The Emerging Science of Astrobiology*, Cambridge University Press, Cambridge, United Kingdom, in press.

Christner, B. C. 2002. Incorporation of DNA and protein precursors into macromolecules by bacteria at −15°C. *Appl. Environ. Microbiol.* 68:6435–6438.

Christner, B. C., E. Moseley-Thompson, L. G. Thompson, and J. N. Reeve. 2003. Bacterial recovery from ancient glacial ice. *Environ. Microbiol.* 5:433–436.

Christner, B. C., J. A. Mikucki, C. M. Foreman, J. Denson, and J. C. Priscu. 2005. Glacial ice cores: A model system for developing extraterrestrial decontamination protocols. *Icarus* 174:572–584.

Cohet, N., and P. Widehem. 2000. Ice crystallization by *Pseudomonas syringae. Appl. Microbiol. Biotechnol.* 54:153–161.

Collins, R. E., and J. W. Deming. 2006. Persistence of Archaea in sea ice, p. 214. *In Symposia Poster Presentations, Astrobiology* 6:174–221.

Cox, G. F. N., and W. F. Weeks. 1983. Equations for determining the gas and brine volume in sea-ice samples. *J. Glaciol.* 29:306–316.

Danovaro, R., A. Dell'Anno, A. Trucco, M. Serresi, and S. Vanucci. 2001. Determination of virus abundance in marine sediments. *Appl. Environ. Microbiol.* 67:1384–1387.

Decho, A. W. 1990. Microbial exopolymer secretions in ocean environments: Their role(s) in food webs and marine processes. *Oceanogr. Mar. Biol. Annu. Rev.* 28:73–153.

Deming, J. W. 2004. New directions in the study of bacteria inhabiting very cold sea-ice formations. *EOS Trans. AGU* 85(47), Fall Meeting Suppl., Abstract B23C-01.

Deming, J. W., and J. A. Baross. 2002. Search and discovery of microbial enzymes from thermally extreme environments in the ocean, p. 327–362. *In* R. P. Dick and R. G. Burns (ed.), *Enzymes in the Environment, Activity, Ecology and Applications.* Marcel Dekker Publishers, New York, NY.

Deming, J. W., and H. Eicken. 2007. Life in ice. *In* W. T. Sullivan and J. A. Baross (ed.), *Planets and Life: The Emerging Science of*

Astrobiology. Cambridge University Press, Cambridge, United Kingdom (in press).

Deming, J. W., and A. L. Huston. 2000. An oceanographic perspective on microbial life at low temperatures with implications for polar ecology, biotechnology and astrobiology, p. 149–160. *In* J. Seckbach (ed.), *Cellular Origins and Life in Extreme Habitats.* Kluwer Publishers, Dordrecht, The Netherlands.

Dumont, F., P. A. Marechal, and P. Gervais. 2006. Involvement of two specific causes of cell mortality in freeze-thaw cycles with freezing to −196°C. *Appl. Environ. Microbiol.* **72:**1330–1335.

Eicken, H. 2003. From the microscopic, to the macroscopic, to the regional scale: Growth, microstructure and properties of sea ice, p. 22–81. *In* D. N. Thomas and G. S. Dieckmann (ed.), *Sea Ice: An Introduction to Its Physics, Chemistry, Biology and Geology,* Blackwell Science, Oxford, United Kingdom.

Eicken, H., J. Weissenberger, I. Bussmann, J. Freitag, W. Schuster, F. Valero Delgado, K.-U. Evers, P. Jochmann, C. Krembs, R. Gradinger, F. Kindemann, F. Cottier, R. Hall, P. Wadhams, M. Reisemann, H. Kouse, J. Ikavalko, G. H. Leonard, H. Shen, S. F. Ackley, and L. H. Smedsrud. 1998. Ice tank studies of physical and biological sea-ice processes, p. 363–370. *In* H. T. Shen (ed.), *Ice in Surface Waters,* Balkema, Rotterdam, The Netherlands.

Geesey, G. G., and D. C. White. 1990. Determination of bacterial growth and activity at solid-liquid interfaces. *Annu. Rev. Microbial.* **44:**579–602.

Giannelli, V., D. N. Thomas, C. Haas, G. Kattner, H. A. Kennedy, and G. S. Dieckmann. 2001. Behaviour of dissolved organic matter and inorganic nutrients during experimental sea ice formation. *Ann. Glaciol.* **33:**317–321.

Gilichinsky, D., E. Rivkina, V. Shcherbakova, K. Laurinavichuis, and J. M. Tiedje. 2003. Supercooled water brines within permafrost—an unknown ecological niche for microorganisms: a model for Astrobiology. *Astrobiology* **3:**331–341.

Gilichinsky, D., E. Rivkina, C. Bakermans, V. Shcherbakova, L. Petrovskaya, S. Ozerskaya, N. Ivanushkina, G. Kochkina, K. Laurinavichuis, S. Pacheritsina, R. Fattakhova, and J. M. Tiedje. 2005. Biodiversity of cryopegs in permafrost. *FEMS Microb. Ecol.* **53:**117–128.

Gosselin, M., M. Levasseur, P. A. Wheeler, R. A. Horner, and B. C. Booth. 1997. New measurements of phytoplankton and ice algal production in the Arctic Ocean. *Deep-Sea Res. II* **44:**1623–1644.

Gowing, M. M. 2003. Large viruses and infected microeukaryotes in Ross Sea summer pack ice habitats. *Mar. Biol.* **142:**1029–1040.

Gowing, M. M., D. L. Garrison, A. H. Gibson, J. M. Krupp, M. O. Jeffries, and C. H. Fritsen. 2004. Bacterial and viral abundance in Ross Sea summer pack ice communities. *Mar. Biol.* **142:**1029–1040.

Gregory, J. M., P. Huybrechts, and S. C. B. Raper. 2004. Threatened loss of the Greenland ice-sheet. *Nature* **428:**616.

Helmke, E., and H. Weyland. 1995. Bacteria in the sea ice and underlying water on the eastern Weddell Sea in midwinter. *Mar. Ecol. Prog. Ser.* **117:**269–287.

Hinkel, K. M., F. E. Nelson, W. Parker, V. Romanovsky, O. Smith, W. Tucker, T. Vinson, and L. W. Brigham (U.S. Arctic Research Commission Permafrost Task Force). 2003. *Climate Change, Permafrost, and Impacts on Civil Infrastructure.* Special Report 01–03, U.S. Arctic Research Commission, Arlington, Virginia.

Hinzman, L. D., D. J. Goering, and D. L. Kane. 1998. A distributed thermal model for calculating soil temperature profiles and depth of thaw in permafrost regions. *J. Geophys. Res.* **103(D22):**28,975–28,991.

Hughes, K. A., I. W. Sutherland, and M. V. Jones. 1998. Biofilm susceptibility to bacteriophage attack: the role of phage-borne polysaccharide depolymerase. *Microbiology* **144:**3039–3047.

Huston, A. L., B. B. Krieger-Brockett, and J. W. Deming. 2000. Remarkably low temperature optima for extracellular enzyme activity from Arctic bacteria and sea ice. *Environ. Microbiol.* **2:**383–388.

Huston, A. L., B. Methé, and J. W. Deming. 2004. Purification, characterization and sequencing of an extracellular cold-active aminopeptidase produced by marine psychrophile *Colwellia psychrerythraea* strain 34H. *Appl. Environ. Microbiol.* **70:**3321–3328.

Junge, K., C. Krembs, J. Deming, A. Stierle, and H. Eicken. 2001. A microscopic approach to investigate bacteria under in-situ conditions in sea-ice samples. *Ann. Glaciol.* **33:**304–310.

Junge, K., J. W. Deming, and H. Eicken. 2004a. A microscopic approach to investigate bacteria under in situ conditions in Arctic lake ice: initial comparisons to sea ice, p. 381–388. *In* R. Norris and F. Stootman (ed.), *Bioastronomy 2002: Life among the Stars,* Astronomical Society of the Pacific, IAU Symposium Series, Vol. 213. International Astronomical Union, Paris, France.

Junge, K., H. Eicken, and J. W. Deming. 2004b. Bacterial activity at −2 to −20°C in Arctic wintertime sea ice. *Appl. Environ. Microbiol.* **70:**550–557.

Junge, K., H. Eicken, B. D. Swanson, and J. W. Deming. 2006. Bacterial incorporation of leucine into protein down to −20°C with evidence for potential activity in subeutectic saline ice formations. *Cryobiology* **52:**417–429.

Krembs, C., J. W. Deming, K. Junge, and H. Eicken. 2002. High concentrations of exopolymeric substances in wintertime sea ice: implications for the polar ocean carbon cycle and cryoprotection of diatoms. *Deep-Sea Res. I* **49:**2163–2181.

Langlois, A., C. J. Mundy, and D. G. Barber. 2007. On the winter evolution of snow thermophysical properties over landfast first-year sea ice. *Hydrological Process.* (in press).

Light, B., G. A. Maykut, and T. C. Grenfell. 2003. Effects of temperature on the microstructure of first-year Arctic sea ice. *J. Geophys. Res.* **108(C2),** 3051, doi:10.1029/2001JC000887.

Lindsay, R. W., and J. Zhang. 2005. The thinning of Arctic sea ice, 1988–2003: have we passed a tipping point? *J. Climate* **18:**4879–4894.

Lizotte, M. P. 2003. The microbiology of sea ice, p. 184–210. *In* D. N. Thomas and G. S. Dieckmann (ed.), *Sea Ice: an Introduction to Its Physics, Chemistry, Biology and Geology,* Blackwell Science, Oxford, United Kingdom.

Mancuso Nichols, C., J. Guezennec, and J. P. Bowman. 2005. Bacterial exopolysaccharides from extreme marine environments with special considerations of the Southern Ocean, sea ice, and deep-sea hydrothermal vents: a review. *Mar. Biotechnol.* **7:**253–271.

Mancuso Nichols, C., S. Garon, J. P. Bowman, G. Raguénès, and J. Guezennec. 2004. Production of exopolysaccharides by Antarctic marine bacterial isolates. *J. Appl. Microbiol.* **96:**1057–1066.

Maranger, R., D. F. Bird, and S. K. Juniper. 1994. Viral and bacterial dynamics in Arctic sea ice during the spring algal bloom near Resolute, N.W.T., Canada. *Mar. Ecol. Prog. Ser.* **111:**121–127.

Meiners, K., R. Gradinger, J. Fehling, G. Civitarese, and M. Spindler. 2003. Vertical distribution of exopolymer particles in sea ice of Fram Strait (Arctic) during autumn. *Mar. Ecol. Prog. Ser.* **248:**1–13.

Meiners, K., R. Brinkmeyer, M. A. Granskog, and A. Lindfors. 2004. Abundance, size distribution and bacterial colonization of exopolymer particles in Antarctic sea ice (Bellingshausen Sea). *Aquat. Microb. Ecol.* **35:**283–296.

Methé, B. A., K. E. Nelson, J. W. Deming, B. Momen, E. Melamud, X. Zhang, J. Moult, R. Madupa, W. C. Nelson, R. J. Dodson, L. M. Brinkac, S. C. Daugherty, A. S. Durkin, R. T. DeBoy, J. F. Kolonay, S. A. Sullivan, L. Zhou, T. M. Davidsen, M. Wu, A. L. Huston, M. Lewis, B. Weaver, J. F. Weidman, H. Khouri, T. R. Utterback, T. V. Feldblyum, and C. M. Fraser. 2005. The psychrophilic lifestyle as revealed by the genome sequence of

Colwellia psychrerythraea 34H through genomic and proteomic analyses. *Proc. Natl. Acad. Sci. USA* **102:**10913–10918.

Miteva, V. I., P. P. Sheridan, and J. E. Brenchley. 2004. Phylogenetic and physiological diversity of microorganisms isolated form a deep Greenland glacier ice core. *Appl. Environ. Microbiol.* **70:**202–213.

Mock, T., and D. N. Thomas. 2005. Recent advances in sea-ice microbiology. *Environ. Microbiol.* **7:**605–619.

Murray, J. L. S., and P. A. Jumars. 2002. Clonal fitness of attached bacteria predicted by analog modeling. *BioScience* **52:**343–355.

Murray, J. B., J.-P. Muller, G. Neukum, S. C. Werner, S. van Gasselt, E. Hauber, W. J. Markiewicz, J. W. Head III, B. H. Foing, D. Page, K. L. Mitchell, G. Portyankina, and the HRSC Co-Investigator Team. 2005. Evidence from the Mars Express High Resolution Stereo Camera for a frozen sea close to Mars' equator. *Nature* **434:**352–356.

Nichols, C. A. M., J. Guezennec, and J. P. Bowman. 2005a. Bacterial exopolysaccharides from extreme marine environments with special considerations of the Southern Ocean, sea ice, and deep-sea hydrothermal vents: a review. *Mar. Biotechnol.* **7:**253–271.

Nichols, C. M., S. G. Lardiére, J. P. Bowman, P. D. Nichols, J. A. E. Gibson, and J. Guézennec. 2005b. Chemical characterization of exopolysaccharides from Antarctic marine bacteria. *Microb. Ecol.* **49:**578–589.

Oelke, C., and T. Zhang. 2004. A model study of circum-Arctic soil temperatures. *Permafrost and Periglac. Process.* **15:**103–121.

Osterkamp, T. E. 2003. Establishing long-term permafrost observatories for active-layer and permafrost investigations in Alaska: 1977–2002. *Permafrost and Periglac. Process.* **14:**331–342.

Price, P. B. 2000. A habitat for psychrophiles in deep Antarctic ice. *Proc. Natl. Acad. Sci. USA* **97:**1247–1251.

Price, P. B., and T. Sowers. 2004. Temperature dependence of metabolic rates for microbial growth, maintenance, and survival. *Proc. Natl. Acad. Sci. USA* **101:**4631–4636.

Price, P. B., O. V. Nagornov, R. Bay, D. Chirkin, Y. He, P. Miocinovic, A. Richards, K. Woschnagg, B. Koci, and V. Zagorodnov. 2002. Temperature profile for glacial ice at the South Pole: implications for life in a nearby subglacial lake. *Proc. Natl. Acad. Sci. USA* **99:**7844–7847.

Priscu, J. C., and B. C. Christner. 2004. Earth's ice biosphere, p. 130–145. *In* A. Bull (ed.), *Microbial Diversity and Bioprospecting*, ASM Press, Washington, DC.

Priscu, J. C., B. C. Christner, C. M. Foreman, and G. Royston-Bishop. Biological material in ice cores. *Encyclopedia Quaternary Sci.*, in press.

Richardson, M. I., and M. A. Mischna. 2005. Long-term evolution of transient liquid water on Mars. *J. Geophys. Res.* **110:**1–21.

Rivkina, E., E. I. Friedmann, C. P. McCay, and D. A. Gilichinsky. 2000. Metabolic activity of permafrost bacteria below the freezing point. *Appl. Environ. Microbiol.* **66:**3230–3233.

Rivkina, E., K. Laurinavichius, J. McGrath, J. Tiedje, V. Shcherbakova, and D. Gilichinsky. 2004. Microbial life in permafrost. *Adv. Space Res.* **33:**1215–1221.

Rogers, S. O., V. Theraisnathan, L. J. Ma, Y. Zhao, G. Zhang, S.-G. Shin, J. D. Castello, and W. T. Starmer. 2004. Comparison of protocols for decontamination of environmental ice samples for biological and molecular examinations. *Appl. Environ. Microbiol.* **70:**2540–2544.

Sheridan, P. P., V. I. Miteva, and J. E. Brenchley. 2003. Phylogenetic analysis of anaerobic psychrophilic enrichment cultures obtained from a Greenland glacier ice core. *Appl. Environ. Microbiol.* **69:**2153–2160.

Shur, Y., K. M. Hinkel, and F. E. Nelson. 2005. The transient layer: implications for geocryology and climate-change science. *Permafrost and Periglac. Process.* **16:**5–17.

Smith, S. L., M. M. Burgess, D. Riseborough, and F. M. Nixon. 2005. Recent trends from Canadian permafrost thermal monitoring network sites. *Permafrost and Periglac. Process.* **16:**19–30.

Sowers, T. 2001. The N_2O record spanning the penultimate deglaciation from the Vostok ice core. *J. Geophys. Res. Atmos.* **106:**31903–31914.

Sowers, T., V. Miteva, and J. Brenchley. 2006. Assessing N_2O anomalies in the Vostok ice core in terms of in-situ N_2O production by nitrifying microorganisms, p. 43. *In* R. Margesin and F. Schinner (ed.), *Book of Abstracts, International Conference on Alpine and Polar Microbiology*, Innsbruck, March 27–31, 2006, Institute of Microbiology, Leopold-Franzens-University, Innsbruck, Austria.

Stierle, A. P., and H. Eicken. 2002. Sediment inclusions in Alaskan coastal sea ice: spatial distribution, interannual variability and entrainment requirements. *Arctic Antarctic Alpine Res.* **34:**103–114.

Stroeve, J. C., M. C. Serreze, F. Fetterer, T. Arbetter, W. Meier, J. Maslanik, and K. Knowles. 2005. Tracking the Arctic's shrinking ice cover: another extreme minimum in 2004. *Geophys. Res. Lett.* **32:**1–4.

Sullivan, W. T., and J. A. Baross. Prologue. *In* W. T. Sullivan and J. A. Baross (ed.), *Planets and Life: the Emerging Science of Astrobiology*, Cambridge University Press, in press.

Thomas, D. N., and G. S. Dieckmann. 2002. Antarctic sea ice—a habitat for extremophiles. *Science* **295:**641–644.

Thomas, D. N., and S. Papadimitriou. 2003. Biogeochemistry of sea ice, p. 267–302. *In* D. N. Thomas and G. S. Dieckmann (ed.), *Sea Ice: An Introduction to Its Physics, Chemistry, Biology and Geology*, Blackwell Science, Oxford, United Kingdom.

Thomas, D. N., R. J. Lara, H. Eicken, G. Kattner, and A. Skoog. 1995. Dissolved organic matter in Arctic multi-year sea ice during winter: major components and relationship to ice characteristics. *Polar Biol.* **15:**477–483.

Vetter, Y.-A., J. W. Deming, P. A. Jumars, and B. B. Krieger-Brockett. 1998. A predictive model of bacterial foraging by means of freely released extracellular enzymes. *Microb. Ecol.* **36:**75–92.

Vorobyova, E., V. Soina, M. Gorlenko, N. Minkovskaya, N. Zalinova, A. Mamukelashvili, D. Gilichinsky, E. Rivkina, and T. Vishnivetskaya. 1997. The deep cold biosphere: facts and hypothesis. *FEMS Microbiol. Rev.* **20:**277–290.

Washburn, A. L. 1980. *Geocryology: a Survey of Periglacial Processes and Environments*. Halsted Press, New York, 406pp.

Wells, L. E., and J. W. Deming. 2006. Modeled and measured dynamics of viruses in Arctic winter sea-ice brines. *Environ. Microbiol.* **8:**1115–1121.

ZoBell, C. E. 1934. Microbiological activities at low temperatures with particular reference to marine bacteria. *Quart. Rev. Biol.* **9:**460–466.

Physiology and Biochemistry of Extremophiles
Edited by C. Gerday and N. Glansdorff
© 2007 ASM Press, Washington, D.C.

Chapter 11

Lake Vostok and Subglacial Lakes of Antarctica: Do They Host Life?

GUIDO DI PRISCO

INTRODUCTION—LAKE VOSTOK

Seismic traces collected from 1955 to 1964 on the East Antarctic plateau near the Russian station Vostok (78°S 106°E, elevation 3,488 m, mean temperature −55°C) revealed a large body of liquid water under the ice sheet, possibly occupying a structural depression within the East Antarctica Precambrian craton. The existence of a lake, also indicated by the unusually smooth ice surface observed from the air, was confirmed in 1973–1975 by radio-echo sounding. Owing to its location, it was named Lake Vostok.

Airborne 60-MHz radio-echo sounding records showed that Lake Vostok (Kapitsa et al., 1996; Siegert et al., 1996) is the largest of 77 (at that time) subglacial lakes that have existed for millions of years beneath the East Antarctic ice sheet and appear to be part of a cycle of ice melt/freeze. The European Research Satellite-1 provided radar altimeter data, which allowed to define its dimensions and predict the density of water, which appears to be fresh. Lake Vostok is one of the largest in the world ($\sim$14,000 km^2, comparable to Lake Ontario); its depth reaches values exceeding 1,100 m. It is clearly visible from the air, because the area corresponding to the lake appears 10-fold flatter than the surrounding regions.

Lakes are now known to exist in both West and East Antarctica; a recent inventory (very likely not exhaustive) places their number at 145 (Siegert et al., 2005). No lake has been sampled yet; the current knowledge comes from ice cores and aerial observations. Biologists, chemists, geologists, glaciologists, climatologists, and limnologists share great expectations to reach liquid water.

Information on the basic aspects of the knowledge gathered on Lake Vostok and Antarctic subglacial lakes and on the first proposals for further action is summarized in proceedings of two workshops. The first one (Lake Vostok: A Curiosity or A Focus for Interdisciplinary Study?) took place in Washington, D.C. (Bell and Karl, 1998), the second (SCAR International Workshop on Subglacial Lake Exploration) in Cambridge, United Kingdom (Ellis-Evans, 1999).

In 2000, the Scientific Committee on Antarctic Research (SCAR) established a Group of Specialists on Subglacial-Lake Exploration. The group launched the program Subglacial Antarctic Lake Environments (SALE), one of five wide-ranging programs endorsed by SCAR in 2004.

The lake has a surface slope and lies beneath an ice sheet ranging from 4,150 m at the northern end to 3,750 m at the southern end, under an average pressure of around 400 bar; it is reached by water and particles $\sim$1 million years after deposition on the ice. Studinger et al. (2004) have provided a bathymetry map covering the entire lake, which is made of two subbasins, separated by a ridge with very shallow water. The larger, deeper southern basin is twice the spatial area of the northern one. The calculated water volume is 5,400 ±1,600 km^3. The sediment is thick, a few hundred meters; it almost certainly contains meteorites and other cosmic material, being an extraordinary source of information on extraterrestrial flux over many million years. Despite the ice-sheet temperature (about −55°C), the water is kept liquid by the pressure of the ice and by the uniform heat rising from the interior of Earth. The latter keeps the water at a mild temperature of −2.65°C. There is no light. Knowledge on salinity, availability and concentration of nutrients and gases, energy sources, possible geothermal vents, and so on is indirect.

Owing to temperature and pressure, Lake Vostok should be supersaturated with dissolved gases in

---

G. di Prisco • Institute of Protein Biochemistry, National Research Council, I-80131 Naples, Italy.

equilibrium with a solid solution of gas hydrate known as clathrate (Miller, 1969), formed by combination of ice and gas. McKay et al. (2003) analyzed clathrate formation and discussed the fate and concentration of gases. As the ice melts into the lake, trapped air is brought in; the total air in the water increases linearly with the age of the lake. Owing to depth, the air is pressurized, so that clathrate is formed. Under the ice transition zone (see below), the nitrogen/oxygen composition of the clathrate is similar to that in the air (Ikeda et al., 1999). In situ measurements of dissolved gases, in particular nitrogen and oxygen, will be important to elucidate physical and biological processes in the lake. The oxygen concentration, which sets the lake redox state, will be high (50-fold that in air-equilibrated water) and may generate a biological stress. Putative living organisms should cope with the need to deal with abundance of oxygen radicals and thus may have high levels of superoxide dismutase, catalase, and peroxidase.

The ice sheet has existed for 15 million years or more (Barrett, 2003), isolating the lake from the atmosphere. The origins of Lake Vostok may thus date back to the Miocene. The residence time of the lake water has been calculated to be in the range 55,000 to 110,000 years (Petit et al., 2003; Studinger et al., 2004). Estimates for the age of the water body range from 1 million years (Kapitsa et al., 1996) to 15 million years (Siegert et al., 2003); it is a "fossil ambient" which evolved during a very long time span in almost total isolation from the biosphere.

The importance of Lake Vostok in terms of glaciology, tectonics, and biology was rapidly acknowledged.

In 1989, an international team began drilling the ice above the lake at the thinner southern end (78°28′ S, 106°48′ E). In 1998, Russian, British, French, and Belgian scientists and engineers pushed the drilling to 3,623 m, ~120 m above the liquid surface and then stopped because of concerns regarding contamination. The ice core is the deepest one ever recovered.

The isotopic analysis of the ice cores allows to define three zones:

1. The ice of the upper 3,310 m (*glacial ice*) is an extensive environmental clock of Earth's paleoclimate over the past 420,000 years (Petit et al., 1999), covering four complete ice-age climate cycles of the late Quaternary. Concentrations of carbon dioxide and methane appear strongly correlated with Antarctic paleotemperatures, suggesting a significant contribution of greenhouse gases to the glacial–interglacial change, in keeping with the current debate on the future of Earth's climate. It contains nutrients of Aeolian origin (formic, sulfuric, nitric, methylsulfuric acids; marine salts, minerals).

2. The ice between 3,310 and 3,539 m (*transition ice*) is part of the continuous ice column but is disrupted by deformations caused by sliding movements of the sheet, which impair deciphering the climatic record (Souchez et al., 2002), and is of relatively poor quality, because of fractures.

3. Below 3,539 m, there are ~200 m of younger *accretion ice*, produced by slow melting and freezing equilibrium; namely, there is water exchange from the base of ice to the lake and vice versa (Siegert et al., 2000). Accretion ice is probably generated by a mechanism similar to frazil ice, formed by small crystals floating on water, followed by consolidation through interstitial water freezing. The ice features change radically: it is made of very large (10 to 100 cm) crystals, it has low electric conductivity and gas content, and the content of stable isotopes shifts (Jouzel et al., 1999). The ionic content is 5- to 50-fold lower than that of glacial ice, suggesting that the salinity of the lake water is <0.01% (Priscu et al., 2005). Such shift enriches the water in clathrates, causes sedimentation phenomena, and (together with ice motion and, possibly, geothermal heating) stimulates water circulation and geochemical gradients; constituents in the accretion ice are thought to reflect those in the liquid water in a proportion equal to the partitioning occurring when the water freezes (Siegert et al., 2001).

From X-ray diffraction experiments, performed on ice monocrystals taken along the 3,623-m ice core, including its deepest part, lattice distortion was observed, related to the bending of the basal plane and torsion of the lattice around the *c*-axis of the crystal. The observed curvature supports basal slip as the predominant deformation mode, accommodated by recrystallization processes (Montagnat et al., 2003).

The two subbasins may have different chemical and biological composition (Studinger et al., 2004). Sediments released by basal melting are likely to accumulate in the northern basin, whereas preglacial sediments are more likely located at the bottom of the southern, deeper basin. Thus, the sediment sampling strategy depends on the type of sediment targeted.

SUBGLACIAL LAKES: SCIENTIFIC OBJECTIVES

The study of subglacial lakes requires a multidisciplinary approach, because basic questions are addressed to life, Earth, atmosphere, and climate sciences. In view of the large geological time span, paleosciences are at the base of the study.

The scientific objectives comprise (i) understanding formation and evolution of subglacial processes and environments; (ii) determining the existence, origins, evolution, and maintenance of life; and (iii) understanding limnology and paleoclimate history recorded in lake sediments.

Climatology, besides studying the climate cycles covered by the ice core referring to the paleoclimate over the past 420,000 years (Petit et al., 1999), seeks for further indications on Antarctica's past climate changes in the lake sediments.

Geophysics seeks for identification and measurement of lake surfaces by radio-echo sounding, bathymetry of water accumulation and sediment thickness by seismic measurements, and geological data analysis to understand tectonics and ice-sheet setting of lake environments.

Glaciology needs insight into ice flow over lake environments by direct surface measurements and satellite data and relationship between ice-sheet processes and water circulation; it will establish numerical models of ice-sheet history to identify the formation and evolution of lake environments.

Geology and *Cenozoic paleoclimate* seek to study origin, transport, and deposition of sediment and correlate surface sediment in each subenvironment to extant process; it will use paleoenvironmental data to determine water and ice-sheet histories and to evaluate temporal changes in Cenozoic paleoclimate relative to histories determined from Antarctic marginal sequences and global Cenozoic records; it will analyze sediment minerals and specimens of geological bedrock to establish the tectonic setting and its temporal evolution.

Functional genomics and *phylogeny* will investigate genomes, gene function, and phylogenetic relationships of organisms, extinct and extant life signatures in the ice sheet (glacial and accretion ice) to determine the possible origins of biotic constituents, genetic diversity in the water accumulations and benthic sediments. Tools will comprise molecular approaches to biodiversity and to evolutionary adaptive strategies and development of laboratory techniques and conditions mimicking the native ones as much as possible (e.g., growth and biochemical assays at high pressures). Characterization of macromolecules and membranes, identification of preferential metabolic pathways, gene expression and transfer, and DNA persistence will be pursued.

Limnology will study vertical density gradients and their use to model water motion, biogeochemical processes, and metabolic activities in the sediments, relationship to genomic data, role of geochemical and isotopic composition of selected water constituents in biological processes, water column stability, age of lake water, and hydrological links among environments.

Targeting access to subglacial lakes is a huge venture. Searching for concerted efforts, SCAR is establishing links among the program SALE and at least two of the other four wide-ranging SCAR programs, namely Antarctic Climate Evolution (ACE) and Evolution and Biodiversity in the Antarctic: The Response of Life to Change (EBA). One of the objectives of ACE is to develop paleoclimate models for the Antarctic; the paleoclimatic record contained in lake sediments will provide important new information from the interior of the continent. Moreover, because ACE will offer information on the formation and development of subglacial-lake environments, it will provide SALE with model results to establish the history of lakes in the context of ice-sheet and climate evolution. Because "[n]othing in biology makes sense except in the light of evolution" (Dobzhansky, 1973), EBA aims to further understand the evolutionary history, biology, and biodiversity of Antarctic biota, in a climatic and tectonic context. Subglacial lakes offer a unique opportunity to examine biodiversity and evolutionary responses in isolated systems that provide analogs for life on early Earth and other planetary bodies. Novel responses to the environment are expected to be found in these lake systems, important end members for biodiversity and polar-community dynamics.

ARE SUBGLACIAL ENVIRONMENTS TOO EXTREME TO HOST LIFE?

In any environment on Earth, no matter how extreme, life is present. Lately, the dogma that life on Earth evolved in a hot environment has become an open question for research. Although the cryosphere had been traditionally viewed as being devoid of life, it has become clear that, on the contrary, it supports some of the most unusual and extreme microbial ecosystems on Earth (and maybe also life in extraterrestrial bodies). Microorganisms have been found in a great diversity of icy environments (where they stay viable for very long periods of time), e.g., permafrost, polar oceans, and snow, sea ice, glacial ice, and cryoconite holes. Examples include ice-covered hypersaline and other lakes (Priscu et al., 1998) and cryptoendolithic communities colonizing the pore spaces of exposed rocks in the Dry Valleys (de la Torre et al., 2003) and other locations of Antarctica; methanogenic *Archaea* (Tung et al., 2005); and ultrasmall microorganisms found in the deepest part of a 3,053-m ice core in Greenland (Miteva and Brenchley, 2005).

Consequently, the question arises, can Lake Vostok and other Antarctic subglacial lakes support life? We have no direct evidence, as liquid water has not been reached yet. The difficulties associated with reaching the surface without contaminating the lake will be discussed below.

If forms of life do exist in subglacial lakes, microorganisms will most likely be dominant. They have existed on Earth since at least 3.7 billion years, dominating the three domains of life: *Bacteria*, *Archaea*, and the less developed forms of *Eukarya*. Of the estimated 3×10^6 species of prokaryotes, <5,000 have been described (Colwell, 1997). During evolution, microorganisms have developed biochemical, physiological, and morphological diversities and have been able to colonize every environment on Earth, including the most extreme and hostile, thanks to (i) mechanisms of adaptation to temperature, salt, and pH extremes; (ii) new redox couples for energy production; (iii) new mechanisms of energy acquisition; and (iv) survival strategies against lack of nutrients. They are extremely resistant. Halophiles have been found in an inclusion of a 250-million-year-old salt crystal (Vreeland et al., 2000); a bacterial spore has been extracted from a 40-million-year-old amber, placed into a culture medium and identified (Cano and Borucki, 1995). There is bacterial life in Earth's crust at depths of several kilometers, in the ocean bottom near "black smokers," at very high temperatures and pressures, and (as mentioned above) at extremely low temperatures.

On these grounds, it is reasonable to expect life also in subglacial lakes. Originating from soil before ice-sheet formation, and/or from ice trapping (after being carried by air circulation) and slow transport to the water, microbes may thrive in the lake water and sediment. The latter, provided there are suitable energy sources, may also shelter protozoa and micrometazoa (e.g., rotifers, tardigrades, nematodes). In both cases, in Lake Vostok and other lakes, these organisms have probably remained in isolation for several million years. This time span is short in evolutionary terms [taking the mutation frequency in prokaryotes into account, species divergence may require up to 100 million years (Lawrence and Ochman, 1998)], yet sufficient to at least produce adaptive changes in the genotype, bound to be unique on Earth. To date, we know of no organism capable of withstanding the exceptional suite of extreme conditions of Lake Vostok at the same time (low temperature, high pressure, darkness, pH, low salinity/availability of nutrients/energy, and high concentrations of gases). If lake-floor biota exists, the time of isolation may have been long enough to produce evolutionary divergence, in view of the evolutionary pressure in such an extreme environment. The mechanisms responsible for phenotype alterations might also be linked to recombination phenomena (Tiedje, 1998).

According to several hypotheses on the origin of life in cold environments (Woese, 1987; Russell et al., 1988; Pace, 1991; Levy and Miller, 1998), the discovery of bacterial strains stemming from these ancestral cells would be of great importance for studies on evolution. Unique strategies of molecular, biochemical, and physiological adaptations must have gradually developed to allow metabolic compatibility.

Thus, there is no doubt that it is worthwhile to invest funds in a concerted way for adequate (albeit expensive) logistics and to provide science with tools necessary to explore the existence of, and gain access to, life in subglacial lakes. To date, the ice core of Lake Vostok yields an extensive overview of life in—and, although indirectly, under—the ice. The water being as yet inaccessible, accretion ice, in view of the water exchange because of the melting–freezing equilibrium at the lake water interface, is the most suitable template for the Vostok ecosystem. Indeed, living conditions for microorganisms trapped into the ice and carried to the Lake do exist.

LIFE IN THE VOSTOK ICE CORE

The only information on living organisms in the Lake Vostok area is coming from the ice core. The findings are reported in several publications in issues of top journals of the past decade.

The possibility of contamination by foreign bacteria (mostly from the drilling/recovery processes and subsequent laboratory procedures) is indeed a major problem. It calls for extreme cautiousness in drawing conclusions. Controversial interpretations have been offered; however, the ensemble of findings constitutes an important background for the debate that will undoubtedly continue until the liquid water of the lake is reached and sampled (and even after).

At least one energy source is essential for the survival of microorganisms. If such source came from the ice sheet, Lake Vostok would be one of the most oligotrophic habitats on Earth. Although unlikely, connections of the lake (which is below sea level) with the ocean cannot be excluded. Recent work (Bulat et al., 2004) shows that geothermal energy sources are indeed present.

In the lake, clathrates are one of the most interesting factors for physiological characterization. Methane clathrates are known to exist at great marine depths, but their microbiological characterization is scanty; clathrates of other gases are even less known. There is evidence of sulfate-reducing archaebacteria that utilize methane clathrate as electron source (Hinrichs et al., 1999; Pancost et al., 2000). The existence and role of clathrates would be very interesting, especially for biotechnological applications in the area of extremophiles.

Based on conductivity values, a model of glacial habitat was proposed (Price, 2000). The model is based on a network of channels in glacial ice and accretion ice, containing a liquid phase, in which bacteria are able to move and get energy and carbon from ions in solution. To maintain neutral pH, bacteria need much energy to either keep strong proton pumps active or maintain the integrity of membranes having low permeability to protons. The liquid medium contains highly concentrated acidic solutions, e.g., up to 2.5 M sulfuric acid, which lower the freezing point of the inclusions and keep the concentrations of nutrients and dissolved organic carbon (DOC) high. Cells, in hypothesized amounts of at least one per milliliter, would be metabolically active but unable to reproduce. The electron acceptors are sulfuric and nitric acids; the carbon and energy sources (electron donors) for biosynthetic activity are methylsulfuric, formic, and acetic acids. The postulated concentration of methylsulfuric acid (~0.084 M) would be sufficient to keep alive up to seven biologically functional (but unable to reproduce) cells per milliliter of thawed ice for 4×10^5 years in glacial ice and 10^2 cells/ml for 10^4 years in accretion ice.

Lake Vostok is seeded regularly with microorganisms from the ice sheet. By ice-core analyses of glacial ice, between 1,500 and 2,750 m (covering 110,000 to 240,000 years), Abyzov (1993) and Abyzov et al. (1998) found (i) prokaryotic and eukaryotic microorganisms (800 to 11,000 cells/ml thawed ice), (ii) proportionality between dust particles and cell number, and (iii) mesophilic microorganisms capable of metabolic activity (as shown by consumption of ^{14}C-labeled organic substrates). Actinomycetes, fungi, yeasts, diatoms and other microalgae, and pollen of higher plants (Table 1) were also detected. Consumption of ^{14}C-labeled organic compounds showed the existence of metabolically active cells, which tend to decrease at increasing depths. Further information on bacterial life in the basal zone of the transition ice and in the accretion ice is reviewed in Abyzov et al. (2001).

Evidence from at least four independent laboratories indicates that accretion ice contains bacteria. Samples from the depth range 3,551 to 3,607 m were analyzed (see Table 1). At 3,590 m, 2.8×10^3 to 3.6×10^4 cells/ml thawed ice were found (Priscu et al., 1999). The techniques used were epifluorescence microscopy (following staining of DNA), scanning electron microscopy (SEM), and DNA amplification. Under conditions remote from real, no growth was observed, but under extreme conditions substrates could merely be used for maintenance activities. Thus, metabolic results remain equivocal with respect to viability. Biotite (73%), quartz (13%), feldspar (K; 9%), muscovite (2%), plagioclasium (2%), and ferric oxide (1%) were detected. DOC was 0.51 mg/l. From molecular taxonomic analysis of genomic DNA and small-subunit ribosomal RNA-encoding DNA molecules (16S rRNA genes), and other analyses, the largest number of the sequences belongs to α- and β-*Proteobacteria*. *Actinomyces* was also found but no *Archaea*. From the partition coefficients for the water- to ice-phase change, according to tentative estimates (Priscu and Christner, 2004), applying findings at Lake Bonney (Dry Valleys),

Table 1. Microorganisms in the ice core above Lake Vostok

Microorganism	Method	Depth (m)	Reference
Bacteria: cocci, diplococci, rods, oval cells *Actinomyces*, *Saccharomyces* Presumably: *Fusarium*, *Trichotecium*, *Mucor*, *Penicillium*, *Aspergillus* Pollen, microalgae	Epifluorescence, SEM	1,500–2,750	Abyzov et al., 1998, 2001
α-, β-*Proteobacteria* (*Acidovorax*, *Afipia*, and *Comamonas*) *Actinomyces*	Epifluorescence, SEM, genomic DNA amplification	3,590	Priscu et al., 1999
Predominance of gram-positive bacteria	Epifluorescence, SEM, dual-laser flow cytometry, biomarkers	3,603	Karl et al., 1999
Brachybacteria, *Sphingomonas*, *Paenibacillus*, *Methylobacterium*, *Cytophaga/Flavobacterium/Bacteroides* α-, β-*Proteobacteria* Gram-positive bacteria	Colony isolation, amplification of 16S rRNA genes	3,593	Christner et al., 2001
H. thermoluteolus	Amplification of 16S rRNA genes	3,551–3,607	Bulat et al., 2004

Lake Vostok may contain 10^5 to 10^6 cells/ml, and 1.2 mg/l of DOC, compatible with heterotrophic microbial growth. On the basis of volume of the ice sheet and subglacial lakes (Siegert et al., 2000) and bacterial volume to carbon conversion factors (Riemann and Søndergaard, 1986), it was estimated that the concentration of cells and DOC in all subglacial lakes are similar to those postulated for Lake Vostok. Thus, the cell number of Antarctic subglacial lakes plus ice sheet, 1.0×10^{26} and containing 12% of the total cell number and carbon associated with Antarctica, is similar to that (1.3×10^{26}) of Earth's surface freshwater lakes and rivers (Whitman et al., 1998), suggesting that such an organic-carbon reservoir in Antarctica is significant in global carbon storage and dynamics. Based on several assumptions, of course these values are very tentative and need to be refined on the basis of future information.

At 3,603 m, in the accretion ice containing no dust inclusions [which makes the results not directly comparable to those of Priscu et al. (1999)], Karl et al. (1999) found bacterial cells at concentrations of 200 to 300 cells/ml thawed ice. These values are 1 order of magnitude lower than in deep marine environments with scarce amount of nutrients. The techniques used were epifluorescence microscopy, SEM, and dual-laser flow cytometry. Biomass was estimated using two biosensors: (i) ATP (an essential link between energy production and biosynthesis, regulating and driving cellular metabolism with other nucleotides, and a precursor for DNA and RNA biosynthesis; this test was negative, perhaps because of the low number of cells) and (ii) lipopolysaccharide (a marker of walls of gram-negative cells; this test was positive). Metabolic activity was measured by the formation of $^{14}CO_2$ and ^{14}C incorporation in macromolecules during incubation with exogenous ^{14}C-organic substrates. Incubations with ^{14}C-acetate and ^{14}C-glucose produced $^{14}CO_2$, indicating metabolically active cells, supporting the hypothesis that Lake Vostok may contain viable microorganisms. Because organic and inorganic nutrients are the key to survival in any Earth's habitat, the presence of potential C and N substrates for growth was also estimated. DOC is fivefold lower than in deep marine environments; hence, Lake Vostok is an oligotrophic habitat (low levels of nutrients, biomass, and energy fluxes); nitrate and nitrite ions are present; most of the free oxygen is probably sequestered as clathrate.

In search for viable bacteria in the accretion ice, Christner et al. (2001) analyzed a sample taken from 3,593 m below the surface, adopting a suite of precautions to minimize contamination and the usual additional risk of artifactually generated DNA molecules during PCR amplifications. Isolates belonging to the *Brachybacteria*, *Methylobacterium*, *Paenibacillus*, and *Sphingomonas* lineages were obtained (see Table 1). Populations of 16S rRNA genes were amplified; they originate from a member of the *Cytophaga/Flavobacterium/Bacteroides* lineage, α- and β-*Proteobacteria*, and gram-positive bacteria. Some rRNA genes sequences of these microorganisms appear closely related (although not identical) to isolates found in other cold environments (Christner et al., 2000).

Important results were recently obtained from analyses of accretion-ice samples at 3,551 and 3,607 m (Bulat et al., 2004), located at ~200 and 150 m from the water–ice interface, both containing sediment inclusions and 20,000 and 15,000 years old. A sample from glacial ice, at 3,001 m, ~300,000 years old (Petit et al., 1999), was used for comparison. Rigorous precautions were taken and suitable databases were used to minimize contamination; the advantages and drawbacks inherent in each procedure and criterion were highlighted. The clones were short in length because it was impossible to amplify full-sized 16S rRNA genes, suggesting the existence of very small amounts of DNA, possibly because of damage by high oxygen content. Of 16 bacterial phylotypes recovered from the accretion ice, only one 16S rRNA genes phylotype in one of the samples at 3,607 m successfully passed the contaminant database and screening criteria, yielding relevance to the lake environment. The phylotype represents a β-proteobacterium, the extant thermophile *Hydrogenophilus thermoluteolus*, recently encountered only in hot springs in Japan (a closely related species, *H. hirschii*, thrives in hot springs at Yellowstone). In addition to limited distribution, its noncontaminant status was supported by the absence of *Hydrogenophilus* sp. present in known contaminant databases. Its optimal growth temperature is ~50°C, favoring the hypothesis that its occurrence in ice is due to the freezing process, rather than to thriving in the ice channels, as suggested by Price (2000). The accretion ice has very low levels of organic carbon and nutrients; these conditions preferably sustain chemolithoautotrophic rather than heterotrophic biota.

These findings are consistent with the existence of a geothermal environment beneath Lake Vostok. Although there is no indication of *Archaea* so far, they may well be present in such environment. Bulat et al. (2004) also discuss the relevant geophysical implications. There is no ^3He enrichment in accretion ice (Jean Baptiste et al., 2001), thus excluding a contribution of mantle-derived hydrothermal fluids such as "black smokers." Several indicators (geological setting, long-term tectonic activity, and ^{18}O) support the existence of deep faults, where water at great depths may heat up and rise back into the lake, causing hydrothermal circulation. The hydrothermal plumes may flush out bacteria and sediment toward the freezing zone, where accretion may trap them in ice.

The upper layer of accreted ice contains visible sediment inclusions (trapped where the ice sheet enters the lake in a shallow area), whereas the lower part is clean and is probably formed over the deep part of the lake where inclusion of particles is no longer possible (Jouzel et al., 1999). Flow cytometry indicated that the number of cells in the upper layer, considering the ratio to contaminants, is so low that accreted ice approaches sterile conditions. Consequently, Bulat et al. (2004) suggest that the water body should host minimal levels, if any, of living microorganisms. However, this hypothesis is essentially based on a single tool (PCR and DNA signature), known to be subject to shortcomings. In addition, it seems to merely refer to the upper layer of the lake water: but what is the situation in the sediment and in the deeper bulk of liquid water? Support is definitely needed from other independent lines of evidence.

PERSPECTIVES, PROBLEMS, AND PRECAUTIONS

What will happen in the future is tightly linked to finding solutions to the serious problem of reaching liquid water.

The participation of biology in subglacial-lake research implies highlighting the dramatic problem of contamination, especially in conditions of low ambient biomass. Some of the lakes might have interconnections; in such a case, nonsterile drilling procedures would introduce contaminants into more than one lake (Price et al., 2002). The acknowledgement of this difficulty has led to a recent comparison of the most commonly used decontamination protocols (Rogers et al., 2004) and to a very detailed description of appropriate conditions to ensure contaminant-free conditions, including monitoring the disappearance of intentional contamination by *Serratia marcescens*, in sections of the Vostok ice core taken at two depths (Christner et al., 2005).

The debate on the question whether Lake Vostok does host life and in which forms highlights the importance of sampling the liquid water. Future research depends on the quality of the samples that will be extracted; thus, entry and sampling should not be attempted until having established contamination control. This is one of the most important tasks of SCAR–SALE, to be pursued in the long-standing tradition of multidisciplinary and international cooperation. Underestimating the dangers may lead to disaster, and special precautions are imperative since the very first stages in order not to spoil the lake.

At present, reasons for serious concern do exist. For example, the original borehole near the Russian station contains 60 tons of a mixture of kerosene (aircraft fuel) and Freon, which had been introduced in order to keep the hole open, and is now stacked, with all its pollutants (including foreign bacteria) on the bottom of the hole. The mixture is already well within the accretion ice, and the possibility that it will slowly leak through the portion which has not been drilled yet, eventually reaching the water, is not a remote danger. If such a process takes place, it will not take long to contaminate the whole lake. A rational approach suggests not to use the old hole, until procedures to thoroughly clean it up have been devised. As far as the author is aware, this is not the case. From the available information, the Russian Antarctic organization is planning to reach the water by completing the drill of the remaining 120 m of accretion ice in three yearly stages, expressing confidence that the drilling-fluid extraction technique will be adequate. This project is cause for great concern (Inman, 2005).

The claim that the environment is almost sterile (Bulat et al., 2004) does not appear conclusive. Other teams working on glacial and accretion ice, very conscious of the contamination issue and using extremely careful precautions and several (not a single one) methods of analysis, do not share the view that the lake hosts very little or no life, a view supported by findings in other parts of the cryosphere. A large part of the continuous input from the ice sheet received by such a huge lake may well consist of contaminating material, but it would indeed be surprising and exceptional if contaminants accounted for the whole input. The sterility hypothesis also weakens one of the most important reasons to do things slowly but surely. We all know that political arguments tend to overcome scientific reasoning when huge financial investments are needed. The argument that expensive efforts to cleanly reach the water are not worthwhile after all would weaken the rationale of biologically safe sampling and speed up actions without adequate precautions. But impatience is a very bad advisor, and hurriedly pushing the process implies taking heavy responsibilities toward the whole scientific community.

Irreversibly contaminating one of the last pristine habitats on Earth and spoiling a unique and invaluable ecosystem is a price that the scientific community cannot afford to pay. Biologists and experts in the other disciplines are certainly aware of the implications of either a promising success or a frustrating failure. After reaching the water, it would be sad to see studies on Lake Vostok deprived of the essential contribution of biology, or—even worse—to see the beginning of lengthy disputes on whether a microorganism found in the lake is a contaminant or not. However, we are confident that the science community will succeed in preserving this scientific treasure.

After clean and sterile field and laboratory conditions become routine, it will be necessary to solve additional problems, e.g., to devise experimental conditions approaching those that lake organisms have to face in their environment. However, compared with the contamination issue, this appears a relatively minor concern.

Several alternative strategies have been proposed to bypass the difficulties. Some radical suggestions of some years ago (e.g., to leave Lake Vostok forever as it is; to postpone drilling for generations) have been set aside. National Antarctic institutions are endeavoring to experiment with new drilling technologies (e.g., hot-water drilling) in smaller subglacial lakes. An interesting choice is Lake Ellsworth, a much smaller lake (10-km long), located across the subglacial foothills of the Ellsworth Mountains, West Antarctica (Siegert et al., 2004). The ice sheet is 4-km thick, but the temperature is higher (−30°C). This lake is within easier logistic coverage, and the base of the West Antarctica ice sheet has been attained and sampled several times in the past by the British Antarctic Survey. Hot-water drilling essentially consists of shower heads spraying water at high temperature and pressure, melting the way down the ice sheet. The hole would be kept open for 24 to 36 h; during this time, probes would be deployed for water and sediment sampling. This UK project, which has raised the interest of several countries, aims at making Lake Ellsworth the best characterized subglacial lake within a reasonable time, namely 2009. Another alternative, proposed by the Italian Antarctic organization, is Lake Concordia, in the subglacial system at Dome C, a few hundred kilometers from Lake Vostok (Tabacco et al., 2003).

These strategies imply that reaching the liquid water of Lake Vostok will perhaps see a few years' delay. It will occur only after achieving success in ecosystem preservation by means of previous work elsewhere. To many of us, this approach seems rational and safe. True, the ecosystems of a couple of subglacial lakes will be at risk, but this seems acceptable if Lake Vostok (by far the most valuable complex, also in view of the wealth of information that the scientific community has been able to gather in several science fields in a couple of decades) becomes accessible with the highest possible degree of safety.

Perspectives in Astrobiology

Lake Vostok and other subglacial lakes are a unique test area for studying life existence, evolution, and persistence in icy moons and planets (Jouzel et al., 1999; Price et al., 2002; Priscu and Christner, 2004; Bulat et al., 2004; Christner et al., 2005). This is a further compelling reason for investing resources and energy in studying life in subglacial-lake environments.

On Europa, one of the moons of Jupiter, 3- to 4-km-thick ice appears to cover a 50- to 100-km-deep liquid ocean (Kivelson et al., 2000; Turtle and Pierazzo, 2001; Chyba and Phillips, 2001). Europa's surface appears similar to Earth's polar ice floes, suggesting periodic exchange between the ice shell and the liquid ocean. National Aeronautics and Space Administration (NASA) and European Space Agency (ESA) are planning missions (perhaps in 2014) to icy Mars (Wharton et al., 1995) in order to search for liquid water, ice shells, and evidence of extinct or extant life. Water ice exists at the poles and below the surface (Malin and Carr, 1999), and Martian meteorites (bearing microfossils and chemical signatures of potential biological origin) have suggested that prokaryotes were once present (Thomas-Keprta et al., 2002).

The danger of irreversibly contaminating extraterrestrial bodies with biological material coming from Earth is an even major concern for space missions, and again, this calls for extreme care. Consequently, safe drilling and appropriate experimental procedures not only are a must for accessing the ecosystems of subglacial lakes, but will have the added value of providing essential codes of action in astrobiology.

Besides contamination control, extraterrestrial missions will require vehicles and equipment (for drilling and many other purposes) of weight and dimensions as small as possible. Robotics will be needed. All of these great challenges are in common with subglacial-lake explorations and will entail enormous logistic efforts and are hard for a single country to afford, again calling for multinational and multidisciplinary projects. The collaborative philosophy will indeed yield added value to these exciting ventures.

Acknowledgments. This study is in the framework of the Italian National Programme for Antarctic Research (PNRA). I thank S. Bulat, B. Christner, J.-R. Petit, J. Priscu, and M. Siegert for fruitful discussion.

REFERENCES

Abyzov, S. S. 1993. Micro-organisms in the Antarctic ice, p. 265–295. *In* E. I. Friedman (ed.), *Antarctic Microbiology*. Wiley-Liss, New York, NY.

Abyzov, S. S., I. N. Mitskevich, and M. N. Poglazova. 1998. Microflora at the deep glacier horizons of central Antarctica. *Mikrobiologiâ* 67:451–458.

Abyzov, S. S., I. N. Mitskevich, M. N. Poglazova, N. I. Barkov, V. Y. Lypenkov, N. E. Bobin, B. B. Koudryashov, V. M. Pashkevich, and M. V. Ivanov. 2001. Microflora in the basal strata at Antarctic ice core above the Vostok lake. *Adv. Space Res.* 28:701–706.

Barrett, P. 2003. Palaeoclimatology: cooling a continent. *Nature* 421:221–223.

Bell, R. E., and D. M. Karl (ed.). 1998. Lake Vostok: a curiosity or a focus for interdisciplinary study? p. 1–83. *Lake Vostok Workshop Final Report*. NSF Press, Washington, D.C.

Bulat, S. A., I. A. Alekhina, M. Blot, J.-R. Petit, M. de Angelis, D. Wagenbach, V. Ya. Lipenkov, L. P. Vasilyeva, D. M. Wloch, D. Raynaud, and V. V. Lukin. 2004. DNA signature of thermophilic bacteria from the aged accretion ice of Lake Vostok, Antarctica: implications for searching for life in extreme icy environments. *Int. J. Astrobiol.* 3:1–12.

Cano, R. J., and M. K. Borucki. 1995. Revival and identification of bacterial spores in 25–40 million-year old Dominican amber. *Science* 268:1060–1064.

Christner, B. C., J. A. Mikucki, C. M. Foreman, J. Denson, and J. C. Priscu. 2005. Glacial ice cores: a model system for developing extraterrestrial decontamination protocols. *Icarus* 174:572–584.

Christner, B. C., E. Mosley-Thompson, L. G. Thompson, and J. N. Reeve. 2001. Isolation of bacteria and 16S rDNAs from Lake Vostok accretion ice. *Environ. Microbiol.* 3:570–577.

Christner, B. C., E. Mosley-Thompson, L. G. Thompson, V. Zagorodnov, K. Sandman, and J. N. Reeve. 2000. Recovery and identification of viable bacteria immured in glacial ice. *Icarus* 144:479–485.

Chyba, C. F., and C. B. Phillips. 2001. Possible ecosystems and the search for life on Europa. *Proc. Natl. Acad. Sci. USA* 98:801–804.

Colwell, R. R. 1997. Microbial biodiversity and biotechnology, p. 279–288. *In* M. L. Reaka-Kudla, D. E. Wilson, and E. O. Wilson (ed.), *Biodiversity II: Understanding and Protecting Our Biological Resources*. Joseph Henry Press, National Academy of Sciences, Washington, D.C.

de la Torre, J. R., B. M. Goebel, E. I. Friedmann, and N. R. Pace. 2003. Microbial diversity of cryptoendolithic communities from the McMurdo Dry Valleys, Antarctica. *Appl. Environ. Microbiol.* 69:3858–3867.

Dobzhansky, T. 1973. Nothing in biology makes sense except in the light of evolution. *Am. Biol. Teacher* 35:125–129.

Ellis-Evans, J. C. (ed.). 1999. International Workshop on Subglacial Lake Exploration. *Workshop Report and Recommendations—Supporting Material*. SCAR, Cambridge, United Kingdom.

Hinrichs, K.-U., J. M. Hayes, S. Sylva, P. G. Brewer, and E. DeLong. 1999. Methane-consuming archaebacteria in marine sediments. *Nature* 398:802–805.

Ikeda, T., H. Fukazawa, S. Mae, L. Pepin, P. Duval, B. Champagnon, V. Ya. Lipenkov, and T. Hondoh. 1999. Extreme fractionation of gases caused by formation of clathrate hydrates in Vostok, Antarctica. *Geophys. Res. Lett.* 26:91–94.

Inman, M. 2005. The plan to unlock Lake Vostok. *Science* 310:611–612.

Jean Baptiste, P., J.-R. Petit, V. Ya. Lipenkov, D. Raynaud, and N. I. Barkov. 2001. Constraints on hydrothermal processes and water exchange in Lake Vostok from helium isotopes. *Nature* 411:460–462.

Jouzel, J., J.-R. Petit, R. Souchez, N. I. Barkov, V. Ya. Lipenkov, D. Raynaud, M. Stievenard, N. I. Vassiliev, V. Verbeke, and F. Vimeux. 1999. More than 200 meters of lake ice above subglacial Lake Vostok, Antarctica. *Science* 286:2138–2141.

Kapitsa, A. P., J. K. Ridley, G. de Q. Robin, M. J. Siegert, and I. A. Zotikov. 1996. A large deep freshwater lake beneath the ice of central East Antarctica. *Nature* 381:684–686.

Karl, D. M., D. F. Bird, K. Björkman, T. Houlihan, R. Shackelford, and L. Tupas. 1999. Microorganisms in the accreted ice of lake Vostok, Antarctica. *Science* 286:2144–2147.

Kivelson, M. G., K. K. Khurana, C. T. Russell, M. Volwerk, R. J. Walker, and C. Zimmer. 2000. Galileo magnetometer measurements: a stronger case for a subsurface ocean at Europa. *Science* 289:1340–1343.

Lawrence, J. G., and H. Ochman. 1998. Molecular archaeology of the *Escherichia coli* genome. *Proc. Natl. Acad. Sci. USA* 95:9413–9417.

Levy, M., and S. L. Miller. 1998. The stability of the RNA bases: implications for the origin of life. *Proc. Natl. Acad. Sci. USA* 95:7933–7938.

Malin, M. C., and M. H. Carr. 1999. Groundwater formation of martian valleys. *Nature* 397:589–591.

McKay, C. P., K. P. Hand, P. T. Doran, D. T. Andersen, and J. C. Priscu. 2003. Clathrate formation and the fate of noble and biologically useful gases in Lake Vostok, Antarctica. *Geophys. Res. Lett.* 30:1702. doi:10.1029/2003GL017490.

Miller, S. L. 1969. Clathrate hydrates of air in Antarctic ice. *Science* 165:489–490.

Miteva, V. I., and J. E. Brenchley. 2005. Detection and isolation of ultrasmall microorganisms from a 120,000-year-old Greenland glacier iced core. *Appl. Environ. Microbiol.* 71:7806–7818.

Montagnat, M., P. Duval, P. Bastie, B. Hamelin, and V. Y. Lipenkov. 2003. Lattice distortion in ice crystals from the Vostok core (Antarctica) revealed by hard X-ray diffraction: implication in the deformation of ice at low stresses. *Earth Planet. Sci. Lett.* 214:369–378.

Pace, N. R. 1991. Origin of life—facing up to the physical setting. *Cell* 65:531–533.

Pancost, R. D., J. S. S. Damsté, S. De Lint, M. J. E. C. Van Der Maarel, J. C. Gottschal, and Medinaut Shipboard Scientific Party. 2000. Biomarker evidence for widespread anaerobic methane oxidation in Mediterranean sediments by a consortium of methanogenic Archaea and Bacteria. *Appl. Environ. Microbiol.* 66:1126–1132.

Petit, J.-R., M. Blot, and S. Bulat. 2003. Lac Vostok: A la decouverte d'un environnement sous glaciaire et de son contenu biologique, p. 273–316. *In* M. Gargaud and J. P. Parisot (eds.), *Environnement de la Terre Primitive*. Presses Universitaires, Bordeaux, France.

Petit, J.-R., J. Jouzel, D. Raynaud, N. I. Barkov, J. M. Barnola, I. Basile, M. Bender, J. Chappellaz, M. Davis, G. Delaygue, M. Delmotte, V. M. Kotlyakov, M. Legrand, V. Y. Lipenkov, C. Lorius, L. Pepin, C. Ritz, E. Saltzman, and M. Stievenard. 1999. Climate and atmospheric history of the past 420,000 years from the Vostok ice record, Antarctica. *Nature* 399:429–436.

Price, P. B. 2000. A habitat for psychrophiles in deep Antarctic ice. *Proc. Natl. Acad. Sci. USA* 97:1247–1251.

Price, P. B., O. V. Nagornov, R. Bay, D. Chirkin, Y. He, P. Miocinovic, A. Richards, K. Woschnagg, B. Koci, and V. Zagorodnov. 2002. Temperature profile for glacial ice at the South Pole: implications for life in a nearby subglacial lake. *Proc. Natl. Acad. Sci. USA* 99:7844–7847.

Priscu, J. C., and B. C. Christner. 2004. Earth's icy biosphere, p. 130–145. *In* A. Bull (ed.), *Microbial Diversity and Bioprospecting*. ASM Press, Washington, D.C.

Priscu, J. C., E. E. Adams, W. B. Lyons, M. A. Voytek, D. W. Mogk, R. L. Brown, C. P. McKay, C. D. Takacs, K. A. Welch, C. F. Wolf, J. D. Kirshtein, and R. Avci. 1999. Geomicrobiology of subglacial ice above Lake Vostok, Antarctica. *Science* 286:2141–2144.

Priscu, J. C., C. H. Fritsen, E. E. Adams, S. J. Giovannoni, H. W. Paerl, C. P. McKay, P. T. Doran, D. A. Gordon, B. D. Lanoil, and J. L. Pinckney. 1998. Perennial Antarctic lake ice: an oasis for life in a polar desert. *Science* 280:2095–2098.

Priscu, J. C., M. C. Kennicutt II, R. E. Bell, S. A. Bulat, J. C. Ellis-Evans, V. V. Lukin, J.-R. Petit, R. W. Powell, M. J. Siegert, and I. Tabacco. 2005. Exploring subglacial Antarctic lake environments. *EOS* 86:193–200.

Riemann, B., and M. Søndergaard. 1986. *Carbon Dynamics in Eutrophic, Temperate Lakes*. Elsevier Science B.V., Amsterdam, Netherlands.

Rogers, S. O., V. Theraisnathan, L. J. Ma, Y. Zhao, G. Zhang, S.-G. Shin, J. D. Castello, and W. T. Starmer. 2004. Comparison of protocols for decontamination of environmental ice samples for biological and molecular examinations. *Appl. Environ. Microbiol.* 70:2540–2544.

Russell, M. J., A. J. Hall, A. G. Cairns-Smith, and P. S. Braterman. 1988. Submarine hot springs and the origin of life. *Nature* 336:117.

Siegert, M. J., S. Carter, I. Tabacco, S. Popov, and D. D. Blankenship. 2005. A revised inventory of Antarctic subglacial lakes. *Antarc. Sci.* 17:453–460.

Siegert, M. J., J. A. Dowdeswell, M. R. Gorman, and N. F. McIntyre. 1996. An inventory of Antarctic subglacial lakes. *Antarc. Sci.* 8:281–286.

Siegert, M. J., J. C. Ellis-Evans, M. Tranter, C. Mayer, J.-R. Petit, A. Salamatin, and J. C. Priscu. 2001. Physical, chemical and biological processes in Lake Vostok and other Antarctic subglacial lakes. *Nature* 414:603–609.

Siegert, M. J., R. Hindmarsh, H. Corr, A. Smith, J. Woodward, E. C. King, A. J. Payne, and I. Joughin. 2004. Subglacial Lake Ellsworth: a candidate for in situ exploration in West Antarctica. *Geophys. Res. Lett.* 31:L23403. doi:10.1029/2004GL021477.

Siegert, M. J., R. Kwok, C. Mayer, and B. Hubbard. 2000. Water exchange between the subglacial lake Vostok and the overlying ice sheet. *Nature* 403:643–646.

Siegert, M. J., M. Tranter, J. C. Ellis-Evans, J. C. Priscu, and W. B. Lyons. 2003. The hydrochemistry of Lake Vostok and the potential for life in Antarctic subglacial lakes. *Hydrol. Process.* 17:795–814.

Souchez, R., P. Jean-Baptiste, J. R. Petit, V. Y. Lipenkov, and J. Jouzel. 2002. What is the deepest part of the Vostok ice core telling us? *Earth-Sci. Rev.* 60:131–146.

Studinger, M., R. E. Bell, and A. A. Tikku. 2004. Estimating the depth and shape of subglacial Lake Vostok's water cavity from aerogravity data. *Geophys. Res. Lett.* 31:L12401. doi:10.1029/2004GL019801.

Tabacco, I., E. A. Forieri, A. Della Vedova, A. Zirizzotti, C. Bianchi, P. De Michelis, and A. Passerini. 2003. Evidence of 14 new subglacial lakes in Dome C-Vostok area. *Terra Antarc. Rep.* 8:175–179.

Thomas-Keprta, K. L., S. J. Clemett, D. A. Bazylinski, J. L. Kirschvink, D. S. McKay, S. J. Wentworth, H. Valli, E. K. Gibson, Jr., and C. S. Romanek. 2002. Magnetofossils from ancient Mars: a robust biosignature in the martian meteorite ALH84001. *Appl. Environ. Microbiol.* 68:3663–3672.

Tiedje, J. M. 1998. Exploring microbial life in Lake Vostok, p. 19–21. *In* R. Bell and D. M. Karl (ed.), *Lake Vostok Workshop Final Report*. NSF Press, Washington, D.C.

Tung, H. C., N. E. Bramall, and P. B. Price. 2005. Microbial origin of excess methane in glacial ice and implications for life on Mars. *Proc. Natl. Acad. Sci. USA* 102:18292–18296.

Turtle, E. P., and E. Pierazzo. 2001. Thickness of a Europan ice shell from impact crater simulations. *Science* 294:1326–1328.

Vreeland, R. H., W. D. Rosenzweig, and D. W. Powers. 2000. Isolation of a 250 million-year old halotolerant bacterium from a primary salt crystal. *Nature* 407:897–900.

Wharton, R. A., Jr., R. A. Jamison, M. Crosby, C. P. McKay, and J. W. Rice, Jr. 1995. Paleolakes on Mars. *J. Paleolimnol.* 13:267–283.

Whitman, W. B., D. C. Coleman, and W. J. Wiebe. 1998. Prokaryotes: the unseen majority. *Proc. Natl. Acad. Sci. USA* 95:6578–6583.

Woese, C. R. 1987. Microbial Evolution. *Microbiol. Rev.* 51:221–271.

Chapter 12

Psychrophiles: Membrane Adaptations

NICHOLAS J. RUSSELL

INTRODUCTION

All of the major groups of microorganisms, including both prokaryotes (bacteria) comprising the *Bacteria* plus *Archaea*, as well as eukaryotes (protists) including the yeasts and fungi, microalgae and lichens, and protozoa, have representatives that are adapted to life at low temperatures. Because of their small size and consequent inability to insulate themselves against the cold, all microorganisms must adapt the composition of their cellular membranes to enable them to function in psychrophilic habitats. This chapter considers those membrane adaptations, and the emphasis will be on the adaptive changes occurring in prokaryotes; where relevant, the distinctive changes in eukaryotes will be compared.

It is important that a distinction is made between phenotypic and genotypic changes in relation to differences in growth temperature. Some researchers, particularly those working with higher organisms such as fish or animals, refer to these changes as acclimation and adaptation, respectively. Phenotypic adaptation is the change in cellular composition of microbial populations brought about by alterations in enzyme activity due directly to temperature changes per se or by more complex changes in regulation. This may well involve shifts in gene expression with some genes being switched on and others off, so that the cellular proteome is altered by thermal shifts with increases or decreases in the levels of specific enzymes of lipid metabolism. Genotypic adaptation refers to adaptive changes on an evolutionary timescale (usually longer than that for phenotypic adaptation), which involve an alteration in genetic structure, i.e., mutations occur and are positively selected if favorable to become established as part of the genome. Of particular relevance to the membrane adaptations of psychrophiles are the phenotypic and genotypic adaptations in lipid composition (the cellular "lipiome"), for which there is much information. Much less is known about the corresponding changes in membrane proteins in response to low temperature.

Despite their small size and generally unicellular nature, microorganisms are metabolically not only extraordinarily diverse but also extremely capable: this is particularly true of bacteria. Taking temperature as the environmental parameter, microbes like any other organisms must be able to sense and respond quickly (over a timescale relative to their cellular doubling times) to thermal changes. Indeed, their small size means that not only are they unable to insulate themselves, but they are also unable to move sufficient distance to use avoidance as an adaptive strategy. Therefore, they have only one option, namely to alter their biochemical make-up in the face of temperature changes.

Cold, particularly extreme persistent cold at subzero temperatures, is usually accompanied by a reduction in water activity (a_w), caused by freezing of water. This has the effect of increasing the solute concentration in both residual liquid water (e.g., brine channels in sea ice) and ice (e.g., glaciers). The combined stresses of cold, lowered a_w, and raised solute concentration, all have a direct effect on membrane lipid conformation and hence membrane properties. In addition, there may be other accompanying stresses, such as low nutrient concentrations in oligotrophic habitats, which will impinge on adaptive changes in solute uptake systems of the cytoplasmic membrane for maintaining cellular activity. Thermal changes may well be cyclical and involve freeze–thaw cycles that can damage membranes through osmosis or even by the formation of physically disrupting ice crystals if the changes are rapid and extreme.

MEMBRANE STABILITY AND TEMPERATURE

Like the membranes of higher organisms, those of microorganisms are comprised mainly of proteins and lipids, together with a smaller amount of carbohydrate in the form of glycoproteins, glycolipids, or other molecules, organized as described originally in the Fluid-Mosaic Model of membrane structure. There are some specialized microbial exceptions, such as gas vesicle membranes that are comprised only of proteins. Microbial membrane lipids consist mainly of phospholipids together with glyco(phospho)lipids (the latter are found particularly in gram-positive bacteria); the outer membrane of gram-negative bacteria contains lipopolysaccharide (LPS) in the outer leaflet, which has an amphiphilic structure comparable to that of phospholipids with fatty acyl chains, including hydroxy and hydroxyacyl fatty acids that contribute to the hydrophobic core of the membrane and alter in response to temperature changes. Some bacteria contain hopanoids that are saturated triterpenoid equivalents of sterols found in eukaryotic microorganisms. Sterols play an important role in cold adaptation and the regulation of membrane fluidity in eukaryotes, generally acting to stiffen membranes, but there is less direct evidence for the same role for hopanoids in prokaryotes, which seem to be distributed randomly amongst species and genera (Ourisson et al., 1987). Molecules such as tetraterpenoid carotenoids, common in pigmented microbes, may also reinforce membranes. For correct cellular function, the lipid bilayer should be in the liquid-crystalline phase, in which the lipid molecules are highly mobile and undergo rapid rotational and vibrational motions that allow integral membrane proteins to undergo necessary conformational changes and to diffuse laterally within the plane of the membrane. As temperature falls, lipids will lose fluidity and eventually pass through a transition to form a gel phase, in which the molecules are packed much more tightly with greatly reduced motions; proteins are squeezed out of gel phase domains and may become crowded in the liquid-crystalline domains. The temperature of this transition (T_m) depends largely on the fatty acyl composition of the lipid and can be most readily measured in model systems containing a single type of lipid.

However, in natural membranes, there is usually a mixture of lipid types with different combinations of fatty acyl chains so that no abrupt transition to a gel phase is seen. Instead as the temperature is lowered, domains of gel phase lipid will form that will increase in size as the temperature continues to fall. Experiments with mutants of *Escherichia coli* and with the wall-less bacterium *Acholeplasma laidlawii*, in both of which the fatty acyl composition of lipids can be controlled and varied widely, have shown that cells remain viable, and their membranes function normally with up to one-quarter of the lipid in the gel phase. If more gel phase forms, then cells become leaky, the proteins cannot function so well, and cellular viability decreases. This is relevant to life in the cold because if the temperature falls suddenly, then gel phase domains may form and membrane functions and cellular viability will be impaired. In eukaryotic yeast, fungi, and algae, the liquid-crystalline to gel phase transition temperature of the membrane lipids depends on the phospholipid/sterol ratio, as well as on the fatty acyl composition. The rigid planar ring system of sterols interacts with the approximately 10 carbons of an acyl chain at its carbonyl end, lowering the fluidity above T_m but increasing it below T_m; the molecular packing density of the remainder of the acyl chain at the methyl end is lowered. Therefore, sterols have a complex effect on bilayer stability but generally act to strengthen natural membranes within the growth temperature range of an organism. Whether hopanoids in bacteria that contain them perform a similar function in thermal regulation of membrane fluidity has not been demonstrated unequivocally; they do reinforce membranes, but several growth parameters besides temperature influence their proportions in membranes (Kannenberg and Poralla, 1999).

Together, the thermally dependent changes in lipid composition, including those necessary for growth at low temperatures, are often referred to as "homeoviscous adaptation" (Sinensky, 1974). This implies that lipid viscosity and membrane fluidity are maintained at constant levels across the growth temperature range. However, this is not the case, as demonstrated by the formation of domains of lipid with different phase behavior. Therefore, a better terminology might be to regard thermal adaptation of membrane lipid composition as a "homeophasic adaptation"; the use of such a term also accommodates another structural feature of membrane lipids, as discussed below.

When tested as single types, some lipids such as phosphatidylglycerol or phosphatidylcholine will generally form a bilayer, whereas others will form non-bilayer phases—e.g., hexagonal phases. What phase a particular lipid will form depends on its molecular shape: those with a cylindrical shape form bilayers, whereas those with a (truncated) conical shape form non-bilayer phases (Israelachvili et al., 1980; Goldfine, 1984; de Kruiff, 1997). Of the common microbial membrane lipids, notably phosphatidylethanolamines containing unsaturated fatty acyl chains and diphosphatidylglycerol (cardiolipin) in the presence of a divalent cation such as Ca^{2+} form hexagonal phases, whereas phosphatidylglycerol, phosphatidylcholine, and phophatidylserine all form a bilayer (lamellar) phase. However, microbial membranes always contain a mixture of lipids so that the overall structure is a bilayer. The presence of lipids that have a tendency to form non-bilayer phases gives a certain tension to the membrane and may be important in helping to drive processes such as sporulation and cell division that involve segregation of membranes. Non-bilayer-forming lipids may also have special architectural roles in helping to pack irregularly shaped proteins within the bilayer. The balance of bilayer and non-bilayer lipids is important in relation to transport for two reasons: first it is the spatial organization of membrane lipids that is crucial to the passive permeability properties and second the transport proteins are integral membrane proteins so they interact with and their activity is influenced by membrane lipids. Significantly in the context of psychrophilic microorganisms, low temperatures favor the formation of non-bilayer phases, so the cellular response to a decrease in growth temperature has to not only counteract a lowering of membrane fluidity but also must prevent the formation of non-bilayer phases. Indeed, sometimes a simple analysis of the temperature dependence of fatty acid composition in terms of fluidity might indicate that the change is in the "wrong" direction until lipid phase behavior is taken into consideration.

THERMAL CHANGES IN LIPID FATTY ACYL COMPOSITION

Microorganisms modify their membrane lipid fatty acyl composition in response to thermal changes by altering unsaturation, (methyl) branching, or chain length. Many bacteria also form cyclopropane fatty acids by cyclization of a *cis*-double bond. However, although cyclopropane chains reduce membrane fluidity and their proportion in membrane lipids is altered by changes in growth temperature, they are not always in the correct direction and do not seem to be part of a coordinated thermal response; rather the changes are probably more dependent on growth phase and may be an adaptation to strengthen the membrane and protect it from, for example, oxidative stress. This raises the general point that in many studies of the thermal dependence of fatty acyl composition, insufficient attention is paid to the stage at which batch cultures are harvested, which has a large influence on lipid composition.

The thermally dependent changes in fatty acyl unsaturation, branching, and chain length are now considered in turn.

Unsaturation

As in higher organisms, the most common response to a decrease in temperature is an increase in the proportion of unsaturated fatty acyl chains. The double bond in unsaturated lipids can be introduced by two different mechanisms. The most widespread of these mechanisms amongst the different microbial groups is the use of a desaturase enzyme, which is part of a complex (usually membrane bound) enzyme system that transfers two hydrogen atoms from fatty acyl chains via a "mini respiratory chain" to a terminal acceptor, which is usually oxygen. Therefore, the process is an aerobic one, but in some *Bacteria*, alternative terminal acceptors such as Fe(III) or nitrite are used so the process is anaerobic (Bowman et al., 1997). The substrate is generally the fatty acyl chains of an intact membrane acyl lipid (e.g., phospholipids), but there are also examples of acyl coenzyme A or acyl-ACP being used as the substrate. As in higher organisms, the desaturase component of the enzyme complex inserts the double bond at a specific position relative to the carboxyl end of the molecule, which is most commonly at the $\Delta 9$ position; however, a wide range of other positional specificities are found amongst bacteria, including the $\Delta 10$ or $\Delta 5$ positions. In eukaryotic microorganisms such as yeast, and in two groups of prokaryotic cyanobacteria, polyunsaturated fatty acids (PUFA) are made by repetitive desaturations of $\Delta 9$-unsaturated substrates, commonly at the $\Delta 12$ and $\Delta 15$ positions and at the $\Delta 6$ position in some. A recent study shows that the *Archaea*, which are abundant in cold environments, can also use unsaturation as a way of regulating membrane fluidity and that the mechanism may involve incomplete saturation of precursors rather than desaturation (Nichols et al., 2004). It is unclear as to how widespread such regulation is among *Archaea*, since relatively few species have been investigated in this respect.

The unsaturated acyl lipids in *Bacteria* are most commonly *sn*-1 saturated, *sn*-2 monounsaturated (Harwood and Russell, 1984), so the major response to a decrease in growth temperature is to increase their proportion or to convert them to diunsaturated phospholipids in which each fatty acyl chain is monounsaturated. Whichever alternative predominates, the overall effect is to lower the liquid-crystalline to gel phase transition temperature and thereby to maintain the membrane lipid bilayer in a fluid state. In some bacteria (e.g., certain *Bacillus* spp. after a sudden decrease in temperature) and in two groups of cyanobacteria, as well as in eukaryotic microorganisms, polyunsaturated fatty acyl chains may be synthesized by multiple desaturations: in yeast, there may be up to three double bonds, cyanobacteria contain three or four, whilst fungi and algae may have up to four, five, or six. The introduction of multiple double bonds has the same effect of lowering the phase transition temperature, although the influence of second, third, and subsequent double bonds is successively less with the introduction of the first double bond, giving by far the greatest increase in fluidity (Russell, 1989).

A number of cold-adapted bacteria, such as *Shewanella* and *Colwellia* spp., commonly found in the Southern Ocean and other cold waters, contain PUFA in their membrane lipids (Russell and Nichols, 1999). They are also present in some other bacteria, for instance the gram-positive *Kineococcus radiotolerans*, which are neither psychrophilic nor marine (Phillips et al., 2002), confirming the view of Russell and Nichols (1999) that PUFA distribution is more related to phylogeny than psychrophilicity. These PUFA are (somewhat surprisingly) identical to those found in higher organisms such as fish and humans and include arachidonic acid (20:4 all *cis* $\Delta 5,8,11,14$), eicosapentaenoic acid (20:5 all *cis* $\Delta 5,8,11,14,17$), and docosahexaenoic acid (22:6 all *cis* $\Delta 4,7,10,13,16,19$), which are precursors of regulatory compounds such as prostaglandins, thromboxanes, and leukotrienes. As a result, their biosynthesis in eukaryotic microorganisms is well understood, proceeding via a series of interacting elongation and desaturation steps that, depending on the nature of the starting fatty acid and the sequence of the reactions, produce two major groups of PUFAs known as the $\omega 3$ and $\omega 6$ series (Certik and Shimizu, 1999). Therefore, it was assumed that they would be made via the same mechanism in bacteria, particularly since lateral gene transfer of the DNA responsible for encoding the elongation/desaturation enzymes has been demonstrated in the laboratory between *Shewanella* and a cyanobacterium (Takeyama et al., 1997). However, it has been possible neither to prove the natural existence of this pathway in bacteria containing such PUFA nor to demonstrate the predicted intermediates even using sensitive methods involving radiolabelled precursors (N.J. Russell and D. S. Nichols, unpublished results).

Instead, an alternative mechanism for PUFA synthesis has been proposed that involves a polyketide synthase (PKS) enzyme complex (Metz et al., 2001). Such PKS enzymes have been well characterized in a wide range of organisms where they are responsible for the biosynthesis of a variety of molecules including antibiotics and other secondary metabolites. The PKS systems carry out an abbreviated set of reactions compared to those of fatty acid synthase (FAS) with some *trans*-unsaturated products that can be isomerized to *cis*-unsaturated analogs (Staunton and Weissman, 2001). Overall, the set of reactions can be achieved anaerobically, thus avoiding the necessity of the involvement of an oxygen-dependent desaturase. Metz et al. (2001) have constructed transgenic strains of *E. coli* that could synthesize eicosapentaenoic acid and showed that it was independent of fatty acid synthesis via the anaerobic pathway. However, whilst such experiments show that it is possible to produce PUFA using the PKS pathway, it does not prove it is the mechanism in bacteria that produce PUFA naturally. Smith (2002) has produced genetic evidence that there has been horizontal gene transfer of PUFA synthesis genes between marine bacteria but that the mechanism of biosynthesis in *Colwellia* may well be different to that in *Shewanella*.

Not all the psychrophilic bacteria of cold marine waters have the metabolic capacity to make PUFAs, and some species that are not cold adapted also contain them. Thus the question arises "Do PUFA-containing phospholipids have any role in cold adaptation?" A problem of answering this question is that it is still unclear as to what exactly is the molecular conformation of such lipids in the membrane. The presence of multiple *cis*-unsaturated double bonds shortens the fatty acyl chain (which may well "double back on itself" to form a hairpin-like conformation), which influences the packing of the molecule that in turn alters the balance between bilayer and non-bilayer phase-forming tendencies. These ideas have been incorporated in a number of models of membrane lipid organization, but no direct experiments have been performed on the lipids from *Shewanella* or *Colwellia* spp. Some doubt is cast on the necessity of PUFAs for cold adaptation by the results of Allen et al. (1999), who showed that in the psychrotolerant (and piezotolerant) *Photobacterium profundum* SS9, monounsaturated fatty acids rather than PUFA were essential for growth at high pressure and low temperature.

There are also some bacteria that make monounsaturated fatty acids via the so-called anaerobic pathway. Since *E. coli* uses this pathway, it is well characterized biochemically and genetically: *cis*-unsaturated double bonds are introduced as part of a modified Type II FAS enzyme system, involving specific dehydration and isomerization reactions (Harwood and Russell, 1984). Recently, a variation of the "*E. coli*" pathway has been identified in *Streptococcus pneumoniae*, and on the basis of genomic sequence analysis, other anaerobes such as clostridia will also have a different mechanism of introducing the double bond into fatty acids anaerobically (Marrakchi et al., 2002). Such FAS systems synthesize both saturated and monounsaturated fatty acids, in contrast to the usual form of FAS that makes saturated fatty acids exclusively. It used to be thought that the presence of either system (i.e., the anaerobic pathway or the desaturase complex) was mutually exclusive, but there are examples of bacteria in which more than one system is present, such as in those cold-adapted bacteria that contain "animal-like" PUFA discussed above (Russell and Nichols, 1999). These bacteria contain monounsaturated fatty acids synthesized via the anaerobic pathway in addition to their PUFA made (it is assumed) by a polyketide pathway. However, in the majority of microorganisms, a single system is used to make unsaturated fatty acids, and a single system always predominates.

For many years, it was assumed that *trans*-unsaturated fatty acyl chains were not normal constituents of the membrane lipids in any organism. They were known to be formed as byproducts of the microbial hydrogenation of *cis*-unsaturated fatty acids by anaerobic bacteria in the rumen of ruminants, and small amounts could be found incorporated into the lipids of both the bacteria and their hosts. More recently, they have been proved as being bona fide membrane acyl lipid constituents in a number of cold-adapted gram-negative aerobic bacteria, particularly *Vibrio* and *Pseudomonas* spp. Although these bacteria contain predominantly *cis*-unsaturated lipids, the proportion of *trans*-unsaturated fatty acyl chains in their lipids increases when the cells are stressed by starvation, the presence of toxicants such as phenols or solvents, or by increases in temperature (Heipieper et al., 2003). The *trans*-unsaturated fatty acids are synthesized by direct and non-reversible isomerization of *cis*-unsaturated fatty acyl chains without a saturated intermediate. The gene for the *cis/trans* isomerase enzyme has been cloned and the enzyme purified. Although the purified enzyme is only active in vitro on free (i.e., non-esterified) fatty acids, it does perform the isomerization in membrane vesicle preparations, particularly when organic solvents are present (Pedrotta and Witholt, 1999). Therefore, it is assumed that in vivo the substrates are membrane acyl lipids, which is consistent with the fact that the enzyme does not use ATP or other cofactors: it is located in the periplasm and it contains a haem-binding site of the cytochrome c type from the predicted protein sequence (Heipieper et al., 2003). This mirrors the fact that many bacteria containing *trans*-unsaturated membrane lipids are resistant to organic solvents, and the presence of these lipids is an adaptive mechanism to counteract the disruptive (including fluidizing) effects of such compounds. *Trans*-unsaturated lipids have a fluidity that is intermediate between that of saturated and *cis*-unsaturated lipids, so the action of the *cis/trans* isomerase enzyme in forming *trans*-unsaturated lipids by direct isomerization of membrane lipids in situ would effectively stabilize the membrane against disruption by either organic solvent or elevated temperatures. The latter is supported by data on the psychrophile *Pseudomonas* E-3, which when grown isothermally at 4 or 30°C contains 2 to 3% of 16:1Δ9 *trans* in its membrane lipids, but when cultures are shifted suddenly from 4 to 30°C, this value rises to 14% within 2 h after the shift and then decreases to 5% after 24 h (Okuyama et al., 1997). This suggests that the conversion of *cis*- to *trans*-unsaturated lipid is a rapid response mechanism of some psychrophiles to deal with the immediate effects of membrane disruption by solvent or raised temperature. It cannot play a part in adaptation to low temperature, because the reaction is not reversible and *trans*-unsaturated fatty acids are not converted to their *cis*-counterparts. However, the presence of *trans*-unsaturated fatty acids may well be a psychrophilic adaptation, particularly in psychrotolerant species to enable them to grow at mesophilic temperatures as well as at or close to zero; given their wide growth temperature range, psychrotolerants require a mechanism to avoid hyperfluidity of their membranes at the upper end of their thermal range. Indeed, recent data for the psychrotolerant *Pseudomonas putida* indicate that sudden increases in temperature, including the thawing of frozen stored cells for analysis, stimulate the production of *trans*-unsaturated fatty acids (Härtig et al., 2005).

Methyl Branching

Many bacteria, particularly gram-positives, contain methyl-branched fatty acids, which may be iso-branched (methyl group on first methylene carbon atom from the methyl end) or anteiso-branched (methyl group on second methylene carbon). Compared with *cis*-unsaturated lipids, those with methyl-branched acyl chains have a relatively more ordered liquid-crystalline phase but a more disordered gel phase, and the effect of anteiso-branches is greater than that of iso-branches—i.e., they have an "intermediate" fluidity (Macdonald et al., 1983, 1985). The response to a decrease in growth temperature is complex in that increases in the proportions of methyl-branched fatty acids are usually accompanied by changes in the relative proportions of iso- and anteiso-branched fatty acids; in addition, there is often a change in the average fatty acyl chain length (see below).

Anteiso-branched fatty acids seem to be particularly associated with growth at low temperatures. For example, in the psychrotolerant *Listeria monocytogenes*, which contains predominantly branched chain fatty acids in its lipids, cold-sensitive mutants have been isolated that are deficient in the synthesis of anteiso-15:0, and the wild-type phenotype can be restored by adding its precursor 2-methylbutyric acid to cultures (Annous et al., 1997), which also restores the correct level of membrane fluidity (Jones et al., 2002). We have found that gram-positive bacteria isolated from Antarctic soils generally have a predominance of anteiso-branched chain fatty acids, whether they are psychrotolerant or psychrophilic, but the psychrotolerant isolates generally have smaller proportions, with the balance being made up with iso-branched fatty acids (White, 1999). However, there are a number of exceptions, and the situation is complicated by differences in fatty acyl chain length (which also alter membrane fluidity, see below) and by the presence in some isolates of small amounts of unsaturated fatty acids, which have a relatively greater influence on fluidity compared with branched acids. There may well be hopanoids in some strains, but their presence was not tested; these membrane-stiffening lipids are distributed apparently randomly amongst bacterial species. Our experience with Antarctic bacterial isolates serves to emphasize that it is usually difficult, if not impossible, to predict the cold hardiness of a microorganism from its fatty acid composition.

Chain Length

The fluidity of membranes can be increased by shortening the average fatty acyl chain length of lipids, and changes in chain length often accompany other alterations in unsaturation and/or methyl branching. The psychrotolerant gram-negative bacterium *Psychrobacter oleovorans* (formerly *Micrococcus cryophilus*) is somewhat distinctive in that it uses acyl chain length exclusively to adjust membrane fluidity over its growth temperature range (Russell, 1984b). This bacterium is also unusual in that it has a very simple fatty acid composition in which 16:1ω7 plus 18:1ω9 comprise 97% of the total fatty acids, and it regulates membrane fluidity through decreasing the C18/C16 ratio by up to fourfold in response to a decrease in growth temperature. In the majority of microorganisms where there are temperature-dependent changes in fatty acyl chain length, they occur in concert with alterations of unsaturation and/or methyl branching.

REGULATION OF ADAPTIVE CHANGES IN LIPID COMPOSITION

On the basis of experiments largely with *E. coli*, it has been proposed that thermal control of membrane lipids operates to maintain the composition within a "window" between gel and non-lamellar phases (Morein et al., 1996). This statement embodies the twin themes of conserving both fluidity and phase properties of membrane lipids in the face of a temperature change, as well as accommodating the fact that a proportion of gel phase lipid can coexist with the predominant liquid-crystalline phase. For psychrophiles to adapt to growth at low temperatures, they must counteract both the lowering of fluidity and the increased propensity to form non-bilayer phases. The changes in fatty acyl composition that are mainly responsible for restoring and maintaining fluidity (headgroup changes play a minor role) in the cold will have to be consistent with avoiding the formation of non-bilayer phases.

It might be argued that psychrophiles would have recognizably different lipid (headgroup and fatty acid) compositions compared with those of mesophiles and thermophiles. Whilst it is true that psychrophiles often have large amounts of unsaturated or branched chain fatty acids, so also do many mesophiles. Even when phylogenetic differences are minimized by comparing species of the same genus, there may be large differences in fatty acid composition and the response to temperature shifts (e.g., see Suutari and Laakso, 1992). The reason is that lipid composition varies widely between different phylogenetic groups, and there are many different ways of achieving the same level of appropriate membrane fluidity and phase behavior at low temperatures. We do not understand the reason for the complexities of lipid diversity amongst microorganisms (Dowhan, 1997), and Cronan (2003) has pointed out that we know less about the regulation of phospholipid synthesis compared with fatty acid synthesis in bacteria, and our understanding of how the mix of different phospholipids is regulated is particularly poor.

Therefore, to study the regulation of cold adaptation of lipid composition, it is necessary to perform temperature shift experiments on specific organisms and to determine phenotypic changes. Such experiments have focused largely on changes of fatty acyl composition, which is very easy to measure using gas chromatographic techniques. There do not appear to be unique mechanisms used by psychrophiles, i.e., they utilize the same mechanisms of altering fatty acyl unsaturation, methyl branching, and chain length as used by mesophiles and thermophiles. Much less is known about lipid headgroup composition, and there are few reports of the kinetics of change in either headgroup or fatty acyl composition following a change to low temperature in psychrophilic microorganisms. The time-course of changes in fatty acid synthesis in response to temperature shifts is very rapid, occurring within minutes or even seconds. However, the changes in membrane lipid synthesis are much slower, not least because cultures usually stop growing for up to several hours when they are subjected to sudden decreases of temperature. The length of the lag period is proportional to both the extent of the shock and the final temperature (Gounot and Russell, 1999). During this period, the expression of housekeeping genes is reduced (but not in some psychrophiles, a fact that may distinguish them from psychrotolerants) and that for cold shock proteins (CSP) is increased. CSP have three broad functions (Phadtare et al., 2000). They ensure that translation continues by acting as chaperones to mRNA and by stabilizing the interaction of initiation and elongation factors with the ribosome; cellular metabolism is modulated to conserve resources vital for cell viability (e.g., the expression of glycolytic enzymes is increased); and membrane lipids are modified to regulate fluidity (e.g., the Δ5 desaturase in *Bacillus* is a CSP) (Aguilar et al., 1998). However, our understanding of the integration of changes in membrane lipid and fatty acid composition with cell growth and division is poorly understood. For bacteria that use the anaerobic pathway to synthesize unsaturated fatty acids and those that regulate fluidity by changes in methyl branching or acyl chain length, there has to be de novo synthesis of new membrane lipid ("replacement synthesis"), and the timescale of changes is slower than those microorganisms which increase membrane lipid unsaturation in situ using desaturase ("modification synthesis") (Russell, 1984a). Cronan (2003) has noted that the anaerobic pathway is much less common than desaturase as a mechanism for increasing lipid unsaturation. One might speculate further that one of the reasons for this distribution is the inability of bacteria with such an anaerobic mechanism to rapidly adapt to decreases in temperature.

Another aspect of cold adaptation that requires more research is the coordination of membrane events (thermal sensing, phospholipid synthesis, and fluidity regulation) with those occurring intracellularly (gene expression, translation, and biosynthesis of fatty acids). More than 25 years ago, Kates and coworkers speculated that temperature could modulate membrane desaturases by altering their membrane penetration and thereby change substrate accessibility (Kates and Pugh, 1980); and Fulco's group identified a temperature-sensitive modulator of Δ5 desaturase hyperinduction and synthesis in *Bacillus megaterium* (Fulco and Fujii, 1980). Proof of how difficult it is to elucidate the thermal regulation of desaturases is demonstrated by the fact that it is only very recently that de Mendoza and coworkers, studying the membrane topology of the Δ5 desaturase of *B. subtilis*, have identified a new transmembrane domain in this phospholipid desaturase, together with a two-component regulatory system: they have proposed a model for regulation of desaturase activity and thermal sensing (Mansilla and de Mendoza, 2005). The desaturase uses membrane phospholipids as its substrate, and the integral membrane protein kinase component (DesK) responds to lipid fluidity (e.g., caused by a temperature down-shift) by phosphorylating the cytoplasmic DesR response regulator that is a transcriptional activator of the Δ5 desaturase gene; subsequent desaturation of membrane phospholipids restores correct membrane fluidity and DesK resumes its dominant phosphatase state.

It should be noted that desaturases can only regulate membrane fluidity in response to a decrease in temperature, because their activity is not reversible. The process of unsaturated fatty acid hydrogenation to form saturated derivatives, which would be required to change unsaturation after a thermal shift-up, is rare in bacteria, occurring in a few anaerobes in the rumen of ruminants (Harwood and Russell, 1984). Unsaturated membrane lipids must be either removed and replaced (turnover) or more usually in bacteria simply replaced by dilution as newly synthesized saturated lipids are inserted into the membrane during growth. Even closely related species that use similar mechanisms may have quite different time-courses for shift-up and shift-down: for instance, *B. megaterium* responds rapidly to shift-up in environmental temperature, whereas *B. licheniformis* responds more slowly and makes the "wrong" fatty acid profile for at least one generation after the thermal shift, although both species modulate their membrane fluidity by altering the de novo biosynthetic pattern of branched chain fatty acids (Thomas, 1989). Neither species has the capacity to isomerize *cis*-unsaturated lipids to *trans*-unsaturated lipids, which can occur rapidly because existing membrane lipids are modified, so the distinction must be due to differences in regulatory properties of their fatty acid synthesis systems. In comparison, bacilli have a rapid response mechanism involving lipid desaturation after shift-down of temperature, followed by a longer term slower mechanism of replacement synthesis of lipids with altered branching (Russell, 1984a).

WHAT CAN WE LEARN FROM GENOMICS?

Recently, the genomes of several psychrophilic bacteria, including *Desulfotalea psychrophila* (Rabus et al., 2004), *Pseudoalteromonas haloplanktis* (Medique et al., 2005), and *Colwellia psychrerythraea* (Methe et al., 2005) have been sequenced fully, whilst those from two methanogens *Methanogenium frigidum* and *Methanococcoides burtonii* (Saunders et al., 2003) have been sequenced partially. Genomic and proteomic analyses have revealed a number of enzymes involved in lipid and fatty acid biosynthesis, including those such as fatty acyl desaturases and *cis/trans* isomerases that will play a role in thermal adaptation of membranes. However, such enzymes are also present in mesophiles, and hence their presence cannot be regarded as a psychrophilic adaptation per se. What is needed are studies of their gene dosage and regulation of expression at low temperature. Medique et al. (2005) point out that desaturases not only increase membrane lipid fluidity at low temperatures but also consume oxygen, which is more soluble at low temperatures and so has a greater capacity for generating reactive oxygen species that damage cells; *P. haloplanktis* is also unusual in that it lacks the otherwise ubiquitous molybdopterin-dependent metabolism, thereby eliminating a group of reactions capable of generating reactive oxygen species.

Proteomic studies addressing this aspect have been reported recently. For example, more than 30 overexpressed proteins have been identified in the psychrotolerant *Bacillus psychrosaccharolyticus* grown at low temperatures (Seo et al., 2004). However, in that study, soluble cytoplasmic but not membrane proteins were investigated. Goodchild et al. (2004) identified a wide range of proteins involved in cold adaptation in *Methanococcoides burtonii*, not only from the proteome but also on the basis of mRNA levels and enzyme assays. The major groups of cold-related proteins were those involved in energy production, transcription, and translation. Interestingly, cultures grown at the apparent optimum growth temperature expressed heat-shock proteins, indicating that growing at their fastest rate under laboratory conditions was stressful for cells. The same may not be true in their native habitat, in which they will have had much longer to adapt to the particular physical and chemical conditions, compared with the artificial media and conditions (e.g., shaken glass flasks) that are used in the laboratory. Proteomic/genomic studies are in their infancy. They are useful for answering "big questions" and have begun to provide us with a global view of cold adaptation. However, thus far, they have not answered specific questions about membrane adaptation, such as the identity of thermal sensors or the feedback control of membrane lipid composition.

SOLUTE TRANSPORT ACROSS MEMBRANES

Membrane lipid composition and specifically the lipid fluidity and phase behavior influences the uptake of ions and solutes in two ways. First, it is the lipid composition of membranes that determines their passive permeability, which is particularly important for the uptake of small neutral molecules including gases. It is also responsible for the low proton permeability of membranes in not only psychrophiles but also mesophiles and (hyper)thermophiles (as measured at their respective growth temperature). If membranes were not relatively impermeable to protons, the proton motive force would dissipate and cells would de-energize. One of the functions of temperature-dependent changes in membrane lipid composition is to regulate passive permeability. Such adaptation has been termed "homeo-proton permeability adaptation" by Konings and coworkers (van de Vossenberg et al., 1999).

The movement of most molecules into and out of cells is mediated by transport proteins, which are integral membrane proteins that interact with lipids in the bilayer. This interaction influences their activity, as demonstrated by experiments with auxotrophic mutants of *E. coli* that require fatty acids for growth and *Acholeplasma laidlawii* that naturally requires lipid supplementation for growth. Changing lipid composition alters the activation energy of transport, as revealed by inflections on Arrhenius plots of solute uptake rates across the growth temperature range (Russell, 1989). Lipids immediately in contact with integral membrane proteins exchange rapidly with bulk lipids, but there remains the possibility that in native microbial membranes, transporters for certain substrates (e.g., the phosphotransferase system and the lactose permease of bacteria) require specific lipids for activity. Such specific lipid–protein associations are found in all thermal groups of bacteria, but in psychrophiles, the requirement for a liquid-crystalline bilayer is particularly acute due to the low temperatures at which they operate and the accompanying loss of kinetic energy. The transport systems of psychrophiles appear to be adapted to increase their substrate affinity relative to those of mesophiles, and the temperature-dependent changes in lipid composition will have evolved in tandem with protein structure to optimize transport activity (Russell, 2003b).

FUTURE PROGRESS

For many years, the focus of cold adaptation of membranes (and thermal adaptation in general) was on compositional changes, with a heavy emphasis on the fatty acid composition of membranes. Less attention has been given to biosynthetic aspects of such changes, and virtually all of our knowledge of the enzymology of lipid synthesis in bacteria is based on experiments with *E. coli*. No extremophile has been investigated rigorously in any depth, which would be worthwhile doing since it is now becoming apparent that there is probably a greater diversity of mechanisms of, for example, unsaturated fatty acid biosynthesis than hitherto realized (e.g., Marrakchi et al., 2002).

Bacteria possess an interactive set of stress responses, with common elements including two-component sensor/effector systems and sigma factors for coordinate transcriptional control of adaptive biochemical and physiological changes to shifts in extreme conditions. The common elements of these responses go some way toward explaining the coordinated regulation of adaptive changes, and the application of the new genomics/proteomics technology has begun to show just how extensive is the coordination of such cellular stress responses (e.g., Goodchild et al., 2004). Nonetheless, we still know very little of how intracellular events are linked to the regulation of membrane lipid composition, but progress is now being made in understanding the regulation of membrane lipid desaturation (Mansilla and de Mendoza, 2005). It is to be hoped that the same progress can be made for the control of fluidity by changes in fatty acid methyl branching as used by the majority of gram-positive bacteria, for which the cold-sensitive enzyme has not been identified. Similarly, regulation of fluidity by cis/trans isomerization and the synthesis of PUFA remain to be elucidated (Russell, 2003a). More physiological/biochemical studies of microbial lipid metabolism and its thermal regulation, in combination with a genomics/proteomics approach, are required if we are to understand cold-adaptive changes in membrane lipid composition.

REFERENCES

Aguilar, P. S., J. E. Cronan, and D. de Mendoza. 1998. A *Bacillus subtilis* gene induced by cold shock encodes a membrane phospholipid desaturase. *J. Bacteriol.* 180:2194–2200.

Allen, E. E., D. Facciotti, and D. H. Bartlett. 1999. Monounsaturated but not polyunsaturated fatty acids are required for growth of the deep-sea bacterium *Photobacterium profundum* SS9 at high pressure and low temperature. *Appl. Environ. Microbiol.* 65:1710–1720.

Annous, B. A., L. A. Becker, D. O. Bayles, D. P. Labeda, and B. J. Wilkinson. 1997. Critical role of anteiso-C15:0 fatty acid in the growth of *Listeria monocytogenes* at low temperatures. *Appl. Environ. Microbiol.* 63:3887–3894.

Bowman, J. P., S. A. McCammon, D. S. Nichols, et al. 1997. *Shewanella gelidimarina* sp. nov. and *Shewanella frigidimarina* sp. nov., novel Antarctic species with the ability to produce eicosapentaenoic acid (C20:5ω3) and grow anaerobically by dissimilatory Fe(III) reduction. *Int. J. Syst. Bacteriol.* 47:1040–1047.

Certik, M., and S. Shimizu S. 1999. Biosynthesis and regulation of microbial polyunsaturated fatty acid production. *J. Biosci. Bioeng.* 87:1–14.

Cronan, J. E. 2003. Bacterial membrane lipids: where do we stand? *Annu. Rev. Microbiol.* 57:203–224.

de Kruiff, B. 1997. Lipid polymorphism and biomembrane function. *Curr. Opin. Chem. Biol.* 1:564–569.

Dowhan, W. 1997. Molecular basis for membrane phospholipids diversity: Why are there so many lipids? *Annu. Rev. Biochem.* 66:157–165.

Fulco, A. J., and D. K. Fujii. 1980. Adaptive regulation of membrane lipid biosynthesis in bacilli by environmental temperature, p. 77–98. *In* M. Kates and A. Kuksis (ed.), *Membrane Fluidity. Biophysical Techniques and Cellular Regulation*. The Humana Press, Clifton, NJ.

Goldfine, H. 1984. Bacterial membranes and lipid packing theory. *J. Lipid Res.* 25:1501–1507.

Goodchild, A., N. F. Saunders, H. Ertan, et al. 2004. A proteomic determination of cold adaptation in the Antarctic archaeon, *Methanococcoides burtonii*. *Mol. Microbiol.* 53:309–321.

Gounot, A.-M., and N. J. Russell. 1999. Physiology of cold-adapted microorganisms, p. 33–55. *In* R. Margesin and F. Schinner (ed.), *Cold-Adapted Microorganisms*. Springer-Verlag, Berlin, Germany.

Härtig, C., N. Loffhagen, and H. Harms. 2005. Formation of *trans* fatty acids is not involved in growth-linked membrane adaptation of *Pseudomonas putida*. *Appl. Environ. Microbiol.* 71:1915–1922.

Harwood, J. L., and N. J. Russell. 1984. *Lipids in Plants and Microbes*. George Allen and Unwin, London, United Kingdom.

Heipieper, H. J., F. Meinhardt, and A. Segura. 2003. The cis–trans isomerase of unsaturated fatty acids in Pseudomonas and Vibrio: biochemistry, molecular biology and physiological function of a unique stress adaptive mechanism. *FEMS Microbiol. Lett.* 229:1–7.

Israelachvili, J. N., S. Marčela, and R. G. Horn. 1980. Physical principles of membrane organization. *Quart. Rev. Biophys.* 13:121–200.

Jones, S. L., P. Drouin, B. J. Wilkinson, and P. D. Morse II. 2002. Correlation of long-range membrane order with temperature-dependent growth characteristics of parent and a cold-sensitive, branched-chain-fatty-acid-deficient mutant of *Listeria monocytogenes*. *Arch. Microbiol.* 177:217–222.

Kannenberg, E. L., and K. Poralla. 1999. Hopanoid biosynthesis and function. *Naturwissenschaften* 86:168–176.

Kates, M., and E. Pugh. 1980. Role of phospholipid desaturases in control of membrane fluidity, p. 153–170. *In* M. Kates and A. Kuksis (ed.), *Membrane Fluidity. Biophysical Techniques and Cellular Regulation*. The Humana Press, Clifton, NJ.

Macdonald, P. M., B. McDonough, B. D. Sykes, and R. N. McElhaney. 1983. Fluorine-19 nuclear magnetic resonance studies of lipid fatty acyl chain order and dynamics in *Acholeplasma laidlawii* B membranes. Effects of methyl-branch substitution and of trans unsaturation upon membrane acyl-chain orientational order. *Biochemistry* 22:5103–5111.

Macdonald, P. M., B. D. Sykes, and R. N. McElhaney. 1985. Fluorine-19 nuclear magnetic resonance studies of lipid fatty acyl chain order and dynamics in *Acholeplasma laidlawii* B membranes. Gel-state disorder in the presence of methyl iso- and anteiso-branched-chain substituents. *Biochemistry* 24:2412–2419.

Mansilla, M. C., and D. de Mendoza. 2005. The *Bacillus subtilis* desaturase: a model to understand phospholipid modification and temperature sensing. *Arch. Microbiol.* 183:229–235.

Marrakchi, H., K.-H. Choi, and C. O. Rock. 2002. A new mechanism for anaerobic unsaturated fatty acid formation in *Streptococcus pneumoniae*. *J. Biol. Chem.* 277:44809–44816.

Medique, C., E. Krin, G. Pascal, et al. 2005. Coping with cold: The genome of the versatile marine Antarctica bacterium *Pseudoalteromonas haloplanktis* TAC125. *Genome Res.* 15:1325–1335.

Methe, B. A., K. E. Nelson, J. W. Deming, et al. 2005. The psychrophilic lifestyle as revealed by the genome sequence of *Colwellia psychrerythraea* 34H through genomic and proteomic analyses. *Proc. Natl. Acad. Sci. USA*. 102:10913–10918.

Metz, J. G., P. Roessler, D. Facciotti, et al. 2001. Production of polyunsaturated fatty acids by polyketide synthases in both prokaryotes and eukaryotes. *Science* **293**:290–293.

Morein, S., A. Andersson, L. Rilfors, and G. Lindblom. 1996. Wild-type *Escherichia coli* cells regulate the membrane lipid composition in a "window" between gel and non-lamellar structures. *J. Biol. Chem.* **271**:6801–6809.

Nichols, D. S., M. R. Miller, N. W. Davies, A. Goodchild, M. Rafferty, and R. Cavicchioli. 2004. Cold adaptation in the archaeon *Methanococcoides burtonii* involves membrane lipid unsaturation. *J. Bacteriol.* **186**:8508–8515.

Okuyama, H., D. Enari, and N. Morita. 1997. Identification and characterization of 9-*cis*-hexadecenoic acid *cis–trans* isomerase from a psychrotolerant bacterium *Pseudomonas* sp. strain E-3. *Proc. NIPR Symp. Polar Biol. Microbiol.* **10**:153–162.

Ourisson, G., M. Rohmer, and K. Poralla. 1987. Prokaryotic hopanoids and other polyterpenoid sterol surrogates. *Annu. Rev. Microbiol.* **41**:301–333.

Pedrotta, V., and B. Witholt. 1999. Isolation and characterization of the *cis–trans*-unsaturated fatty acid isomerase of *Pseudomonas oleovorans* GPo12. *J. Bacteriol.* **181**:3256–3261.

Phadtare, S., K. Yamanaka, and M. Inouye. 2000. The cold shock response, p. 33–45. *In* G. Storz and R. Hengge-Aronis (ed.), *Bacterial Stress Responses*. American Society for Microbiology Press, Washington, DC.

Phillips, R. W., J. Wiegel, C. J. Berry, C. Fliermans, A. D. Peacock, D. C. White, and L. J. Shimkets. 2002. *Kineococcus radiotolerans* sp. nov., a radiation-resistant, gram-positive bacterium. *Int. J. Syst. Evol. Microbiol.* **52**:933–938.

Rabus, R., A. Ruepp, T. Frickey, et al. 2004. The genome of *Desulfotalea psychrophila*, a sulfate-reducing bacterium from permanently cold Arctic sediments. *Environ. Microbiol.* **6**: 887–902.

Russell, N. J. 1984a. Mechanisms of thermal adaptation in bacteria: blueprints for survival. *Trends Biochem. Sci.* **9**:108–112.

Russell, N. J. 1984b. The regulation of membrane fluidity in bacteria by acyl chain length, p. 329–347. *In* L. A. Manson and M. Kates (ed.), *Biomembranes*. Humana Press, Clifton, NJ.

Russell, N. J. 1989. Functions of lipids: structural roles and membrane functions, p. 279–365, vol. 2. *In* C. Ratledge and S. G. Wilkinson (ed.), *Microbial Lipids*. Academic Press, London, United Kingdom.

Russell, N. J. 2003a. Psychrophily and resistance to low temperatures. *In Encyclopaedia of Life Support Systems*. EOLSS Publishers Co Ltd. Published electronically, contribution number 6-73-03-00@http://www.eolss.com.

Russell, N. J. 2003b. Membrane adaptation and solute uptake systems. *In Encyclopaedia of Life Support Systems*. EOLSS Publishers Co Ltd. Published electronically, contribution number 6-73-03-02@http://www.eolss.com.

Russell, N. J., and D. S. Nichols. 1999. Polyunsaturated fatty acids in marine bacteria—a dogma rewritten. *Microbiology* **145**: 767–779.

Saunders, N. F., T. Thomas, P. M. Curmi, et al. 2003 Mechanisms of thermal adaptation revealed from the genomes of the Antarctic Archaea *Methanogenium frigidum* and *Methanococcoides burtonii*. *Genome Res.* **13**:1580–1588.

Seo, J. B., H. S. Kim, G. Y. Jung, et al. 2004. Psychrophilicity of *Bacillus psychrosaccharolyticus*: A proteomic study. *Proteomics* **4**:3654–3659.

Sinensky, M. 1974. Homeoviscous adaptation: a homeostatic process that regulates the viscosity of membrane lipids in *Escherichia coli*. *Proc. Natl. Acad. Sci. USA* **71**:522–525.

Smith, M. C. 2002. Molecular genetics of polyunsaturated fatty acid biosynthesis in Antarctic bacteria. Ph.D. Thesis. University of Tasmania, Australia.

Staunton, J., and K. J. Weissman. 2001. Polyketide biosynthesis: a millennium review. *Nat. Prod. Rep.* **18**:380–416.

Suutari, M., and S. Laakso. 1992. Unsaturated and branched chain-fatty acids in temperature adaptation of *Bacillus subtilis* and *Bacillus megaterium*. *Biochim. Biophys. Acta* **1126**:119–124.

Takeyama, H., D. Takeda, K. Yazawa, et al. 1997. Expression of the eicosapentanoic acid synthesis gene cluster from *Shewanella* sp. in a transgenic marine cyanobacterium, *Synechococcus* sp. *Microbiology* **143**:2725–2731.

Thomas, A. 1989. Thermal adaptation of bacterial membrane lipids. Ph.D. Thesis. University of Wales, United Kingdom.

van de Vossenberg, J. C. L. M., A. J. M. Driessen, M. S. da Costa, and W. N. Konings. 1999. Homeostasis of the membrane proton permeability in *Bacillus subtilis* grown at different temperatures. *Biochim. Biophys. Acta* **1419**:97–104.

White, P. L. 1999. The effects of environmental warming on Antarctic soil microbial communities. Ph.D. Thesis. University of London, United Kingdom.

Chapter 13

Cold-Adapted Enzymes

TONY COLLINS, SALVINO D'AMICO, JEAN-CLAUDE MARX, GEORGES FELLER,
AND CHARLES GERDAY

INTRODUCTION

Cold-adapted enzymes are produced by microorganisms living at permanently low temperature, which constitutes the major environment on planet Earth and includes deep sea, polar, and mountain regions. To enable growth and development of these so-called psychrophilic microorganisms, adaptation of the enzymatic repertoire to allow for appropriate reaction rates in the low temperature environment is necessary. Indeed, chemical and enzyme–catalyzed reactions generally obey the Arrhenius law: $k = A \exp(-E_a/RT)$ (Arrhenius, 1889). Hence any decrease in temperature (T) should induce an exponential decrease in the reaction rate (k). Here, E_a is the empirical activation energy which determines the relative dependency of the reaction to temperature; the higher the activation energy the lower the reaction rate. In some cases, the thermodependence of a reaction can be high and can for example display a Q_{10} value (the ratio of the rates of the reaction measured at a temperature interval of 10°C) close to or higher than 3. In other cases, when the activation energy is close to zero, the exponential term of the Arrhenius relation tends to 1 and the reaction rate becomes nearly temperature independent and no adaptation is therefore needed.

One can reasonably expect that most enzymes produced by psychrophilic microorganisms require molecular adaptation so as to secure an appropriate activity at low temperatures. In contrast, especially in the case of intracellular enzymes catalyzing reactions close to a thermodynamic equilibrium and characterized by a weak control coefficient in a cascade of transformations, one would expect that a decrease in activity induced by lowering the temperature would not significantly modify the overall flux. These would be quite resistant to the selective pressure exerted on their k_{cat} and/or K_m and could therefore display catalytic parameters similar to those of their mesophilic counterparts. These factors could explain why some enzymes produced by permanently cold-adapted microorganisms do not show the adaptations typical of extracellular or intracellular control enzymes.

In this chapter, we will only consider those enzymes significantly adapted to low temperatures, that is, displaying a high specific activity at low temperatures. A certain number of reviews have already been published on cold-adapted enzymes and some of the most comprehensive ones published by the main teams working in this field include the following: Russell, 1990, 2000; Jaenike, 1990; Feller et al., 1996; Feller and Gerday, 1997; Gerday et al., 1997, 2000; Smalas et al., 2000; Zecchinon et al., 2001; D'Amico et al., 2002; Gianese et al., 2002; Deming, 2002; Cavicchioli et al., 2002; Glansdorff and Xu, 2002; Feller and Gerday, 2003; Hoyoux et al., 2004; Georlette et al., 2004 and Marx et al., 2004.

GENERAL PROPERTIES

Many enzymes produced by cold-adapted microorganisms have now been fully characterized in terms of their physical, chemical, and kinetic properties but still only 11 structures have been solved by X-ray crystallography: α-amylase (Aghajari et al., 1998), citrate synthase (Russell et al., 1998), malate dehydrogenase (Sun-Yong et al., 1999), triosephosphate isomerase (Alvarez et al., 1998), Ca^{2+}-Zn^{2+} protease (Aghajari et al., 2003), xylanase (Van Petegem et al., 2003), adenylate kinase (Bae and Phillips, 2004), cellulase (Violot et al., 2005), subtilisin-like protease (Arnorsdottir et al., 2005), tyrosine phosphatase (Hiroki et al., 2005), and β-galactosidase (Skalova

---

Tony Collins, Salvino D'Amico, Jean-Claude Marx, Georges Feller' and Charles Gerday • Institute of Chemistry, B6, University of Liege, Sart-Tilman, B-4000 Liege, Belgium.

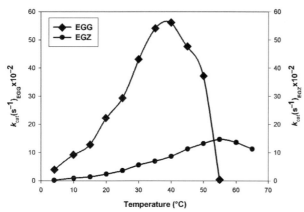

Figure 1. Thermodependence of the activity of the cold-adapted cellulase from *Pseudoalteromonas haloplanktis* (♦) compared to that of the mesophilic counterpart from *Erwinia chrysanthemi* (●).

et al., 2005). In addition some three-dimensional structures have also been described for enzymes from cold-adapted eukaryotic organisms: salmon trypsin (Helland et al., 1998), Atlantic cod uracil-DNA glycosylase (Leiros et al., 2003), and shrimp alkaline phosphatase (de Backer et al., 2002). From these various observations it was deduced that, in general, the cold-adapted enzymes display three general properties: first, they have a high specific activity at low and moderate temperatures, meaning often much higher than that of their mesophilic counterparts over a temperature range from 0 to 40°C (Fig. 1), second, their apparent optimum temperature is significantly shifted towards low temperatures (see Fig. 1), and third, their specific activity around their "environmental temperature" is in general lower than that of their closely related counterparts around their own environmental temperature (see Fig. 1). This could mean that the adaptation is not fully complete and hence another challenge is to understand why.

Enzymes from psychrophiles which do not have a higher specific activity at low to moderate temperatures when compared to their mesophilic counterparts have been reported in a few instances, most notably a triosephosphate isomerase (Alvarez et al., 1998) and an aspartate aminotransferase (Birolo et al., 2000). These are certainly not isolated cases: first for the reasons already mentioned above—all enzymes of a psychrophile do not need to be cold adapted; second, because sometimes no closely related mesophilic counterpart has been characterized and hence an accurate comparative analysis is impossible; and third, because in some cases measurement of the specific activity is not a trivial experiment. In this context, the importance of growth temperature on the molecular integrity of cold-adapted enzymes is clearly demonstrated in the case of the triosephosphate isomerase where the recombinant form of the enzyme was obtained from *E. coli* cultivated at 37°C and at the same time it was shown that the cold-adapted enzyme displayed a half-life of 10 min. at 25°C (Alvarez et al., 1998). Clearly, at 37°C the enzyme is a mixture of native and unfolded forms. This has also been underlined by Gerike et al. (1997) with a cold-active citrate synthase; indeed, when the host cells were grown at 27°C the specific activity of the enzyme was 98 mU/mg and this dropped to 21 mU/mg when the cells were grown at 37°C.

Psychrophilic microorganisms are usually capable of growing over a broad range of temperatures but some are confined within very narrow limits, e.g., *Moritella profunda*, which displays a maximal growth rate at 2°C and a maximum temperature of growth of only 12°C (Xu et al., 2003c). This narrow range of growth temperatures is probably related to a particular sensitivity of certain proteins or macromolecular structures near and above the apparent optimum temperature of growth. Indeed, it has been recently shown that the physical chemistry of a microorganism closely resembles that of globular proteins and that a growth rate model can be directly derived from the temperature behavior of proteins (Ratkowsky et al., 2005). Much broader ranges of growth temperatures, even up to and above 40°C (Jay, 1986), have been reported for other microorganisms, e.g., temperature ranges of 2 to 44°C (Astwood and Wais, 1998). Clearly, there is a continuum in the adaptation of microorganisms found in cold environments and any distinction in categories is futile. Therefore, to designate microorganisms adapted to cold—psychrotrophic, psychrotolerant or psychrophilic—we will use the appellation psychrophiles, sticking to the definition proposed a long time ago by Ingraham and Stokes (1959): "psychrophiles are microorganisms that grow well at 0°C."

The in vitro growth temperature of these psychrophilic microorganisms is very important for enzyme production, especially for extracellular enzymes, since the production is highly dependent on temperature. For example, in the case of lipase secretion by Antarctic *Moraxella* strains, it has been shown that when grown at 25°C, the so-called optimum temperature of the strains, lipase production is reduced to 10% of the amount produced at 3°C, and at 17°C the production level is intermediate to these (Feller et al., 1990). In a similar experiment, the production of α-amylase by an Antarctic *Pseudoalteromonas* strain was very high when the strain was grown at 4°C, significantly lower at 18°C, and almost completely abolished at 25°C (Fig. 2) (Feller et al., 1997). In this case also the density of cells at the end of the exponential phase was several orders of magnitude larger at 4°C than at 18°C, which itself was much larger than that obtained at 25°C, a temperature also

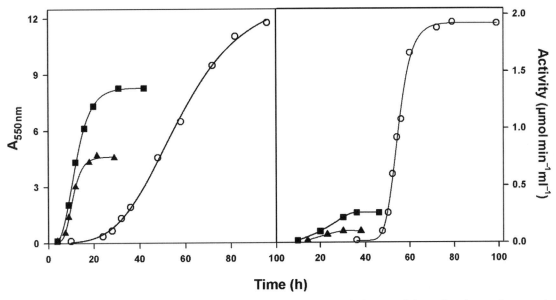

Figure 2. Growth, in rich medium, of the Antarctic strain *Pseudoalteromonas haloplanktis* A23 (left panel) and α-amylase secretion (right panel) at 4°C (○), 18°C (■), and 25°C (▲). There is a net inhibition of exoenzyme production and of cell density when the temperature is raised. Adapted from Feller et al., 1994.

close to the apparent optimum. This does mean that the optimum temperature as defined by the shortest doubling time is not a true optimum temperature but rather a critical temperature at which the bacterium is in a situation of physiological stress. Similar data were observed for protease production by psychrotrophic bacteria, where it was shown that the optimum temperature for this was in general considerably lower than that for cell growth (Margesin and Schinner, 1992). In another systematic investigation, the production of various extracellular enzymes such as cellulases, pectate lyases, chitinases, and chitobiases by several strains permanently or seasonally exposed to cold temperatures was followed as a function of growth temperature. Several patterns were observed; sometimes the enzyme production decreased as a function of growth temperature but in other cases this increased up to the optimum, so that no general conclusion could be drawn (Buchon et al., 2000). One has to add, however, that enzyme production can be significantly obscured by the fact that different growth temperatures can give rise to the expression of different isoforms (homologous or not) of the same enzyme family. Indeed, it was recently shown that a cold-adapted *Pseudomonas fluorescens* strain grown at 7°C expressed high levels of a lipase 1 and a protease 2 whereas at 37°C, a lipase 2 and a protease 1 were produced in greater amounts. The two proteases were found to be homologous enzymes whereas the two lipases were not homologous and displayed quite different apparent optimum temperatures: 28°C for lipase 1 and 40°C for lipase 2 (Mostafa et al., 2003). Thus, this underlines the necessity when screening for cold-adapted enzymes to cultivate the psychrophilic microorganisms at a temperature as close as possible to the temperature of their environment since the above-mentioned phenomenon could be at the origin of artifactual abnormal properties of cold-adapted enzymes.

CATALYSIS AT LOW TEMPERATURES

Like the microorganisms from which they originate, cold-adapted enzymes display a large diversity of temperature dependences. In some cases the range of temperatures is very limited and their apparent optimum temperature is not much higher than that of the environmental temperatures, such as in Arctic sediments in which the apparent optimum temperatures of four glycoside hydrolases did not exceed 15 to 18°C (Arnosti and Jorgensen, 2003). Similarly, in sea ice samples from the Arctic, temperature optima of 15°C were recorded for chitobiase and leucine amino peptidase (Huston et al., 2000). It is also interesting to note that here the greatest release of proteases by one of the psychrophilic isolates was obtained after growing the strain at −1°C, a temperature very close to that of the environment.

In some cases, the catalytic activity of cold-adapted enzymes encompasses a broad range of temperatures, e.g., an aspartate aminotransferase from an Antarctic strain is active from low temperatures up to an apparent optimum of 64°C (Birolo et al., 2000) while the optimum was found to be 57°C for a glucoamylase from *Candida antarctica* (Demot and Verachter, 1987) and 50°C for a pullulanase from *Micrococcus* sp. (Kimura and Horikoshi, 1990). Similar examples can be found in the literature and therefore the apparent optimum temperature of an enzyme does not clearly reflect the psychrophily of their host. More important is the fact that a cold-adapted enzyme retains high activity at low temperatures. This is notably striking in the case of a xylanase which retains, at 5°C, 60% of its maximal activity (Collins et al., 2002), in the case of a chitinase from Antarctic sea water which also displays, at 5°C, 40% of its maximum activity found at 35°C (Bendt et al., 2001), and in the case of OTCase showing at 5°C more than 30% of the maximal activity found around 25°C (Xu et al., 2003a). Normally, if one assumes a Q_{10} of 2, the activity at 5°C should only be around 10% of the maximum activity. The figures are of course variable but indicate that, in general, cold-adapted enzymes are less sensitive to temperature changes than their mesophilic counterparts and this will be tentatively explained in the following sections.

As already mentioned, the thermodependence of a chemical reaction was first described by Arrhenius on the basis of the thermodependence of the rate of sucrose inversion with the well known equation: $k = A \exp(-E_a/RT)$, in which A is a pre-exponential factor also known as the frequency factor, E_a the activation energy, R the gas constant, and T the temperature in Kelvin. Indeed, Arrhenius was among the first to postulate the existence of an activated state of the reactant in equilibrium with the inactive form and able to react. But what is the physical and thermodynamic meaning of A and E_a? In the Arrhenius equation, the exponential factor, $\exp(-E_a/RT)$, represents the fraction of the total number of molecules that possess the appropriate energy to take part in the reaction and the pre-exponential factor A has the dimension of frequency and the product $A \exp(-E_a/RT)$ gives the specific reaction rate. To better understand the significance of these terms, a first approach is to invoke the collision theory as applied to a bimolecular reaction of the form: B + C → D. In this reaction, the specific rate is dependent on the frequency of collisions between B and C and this, in favorable cases, provides an energy high enough for the reaction to proceed. In this case, the pre-exponential factor A corresponds to the frequency of collisions between the particles. However, colliding is not enough to force the reactants to react, and to better understand the process an excellent analogy was proposed by Haynie (2001) in his valuable book entitled *Biological Thermodynamics*: "it is not just a collision between an enzyme and substrate that brings about an enzymatic reaction, it is a collision that brings the substrate into the active site with the correct orientation. For instance, a hand can collide with a glove in a huge variety of ways but only a relatively small number of collisions will result in a glove going on the correct hand." An additional factor is therefore necessary to take into account this restriction, and the rate constant of the reaction between two gases should be written as: $k = Zp \exp(-E_a/RT)$, in which p is the steric or probability factor referring to the requirement that the reactants should be in the appropriate orientation and Z is the collision frequency. One can immediately see the analogy between this equation and the Arrhenius equation since $A = Zp$ and A is indeed a frequency factor.

In 1935, Eyring introduced the theory of an absolute reaction rate. In this theory, the rate of a chemical reaction at a given temperature depends only on the concentration of the energy-rich activated complex in equilibrium with inactive forms which breaks down at a rate given by the relation: $k = k_B T/hK^*$. The activated state is seen as an intermediate form critical for the process, since on attainment of this configuration and provided that the motion of this activated state occurs along the reaction coordinate (one degree of vibrational freedom is replaced by a translational motion along the reaction coordinate) there is a high probability that the reaction will go to completion. Evans and Polanyi (1935) suggested the term "transition state" for this intermediate state. The rate of a reaction expressed in the above-mentioned equation is therefore equal to the concentration of the activated complex at the top of the barrier multiplied again by the frequency of crossing the barrier. k_B is the Boltzmann constant, the gas constant per molecule, h is the Planck constant, and K^* is the equilibrium constant between the activated complex and the un-activated molecules. Ordinary thermodynamic relations can be applied to this equilibrium and since: $\Delta G^* = -RT \ln K^*$

$$k = k_B \left(\frac{T}{h}\right) \exp\left(\frac{-\Delta H^*}{RT}\right) \exp\left(\frac{+\Delta S^*}{R}\right) \quad (1)$$

Assuming that ΔS^* does not vary with temperature and after transformation one gets: $d(\ln k)/dT = 1/T + \Delta H^*/RT^2$, and comparing this equation with that of Arrhenius we obtain $E_a = \Delta H^* + RT$. E_a is in fact a measure of the activation enthalpy and can be easily determined from an Arrhenius plot; $\ln k$ as a function of $1/T$, which usually gives, within a certain temperature range, a straight line of slope $-E_a/R$ (see also Lonhienne et al., 2000) As $E_a = \Delta H^* + RT$, equation (1) can be rewritten as $k = k_B \cdot T/h \cdot \exp(-E_a/RT) \cdot \exp(+\Delta S^*/R) \cdot \exp(1)$ and comparison of this with the other form of the Arrhenius equation gives $A = k_B(T/h) \exp(+\Delta S^*/R) \exp(1)$. Thus, the factor A, or frequency factor, is in fact a measure of the activation entropy. In these developments it was assumed that all activated molecules passing over the energy barrier will lead to products, but for various reasons which will be tentatively developed later, there is the possibility that the activated state returns to its initial state (re-crossing). Therefore, it is necessary to introduce another coefficient to take into account this possibility and equation (1) should be rather written as $k = \kappa k_B(T/h) \exp(-\Delta H^*/RT) \exp(+\Delta S^*/R)$ in which κ is known as the transmission coefficient and, in certain cases, is lower than unity.

The discussion up to now has focused on chemical reactions essentially in the gas phase so as to avoid the possible influence of the solvent or of other solutes that may be present. To try to understand how these theories apply to enzyme-catalyzed reactions occurring in the condensed phase, we can start with the Henri-Michaelis equation related to a single enzyme interacting with a single substrate and with only one activated state: $E+S \leftrightarrow ES \leftrightarrow ES^* \rightarrow E+P$. The free energy of activation corresponds to the difference between the energy of the activated complex and that of the reactants $E+S$. One has to note, however, that if the free energy of the ES complex is lower than that of the reactants $E+S$, as in the case of a low K_M, the activation energy will be the difference between the free energy of ES^* and that of ES. One of the weaknesses of the transition state theory, as developed initially by Pelzer and Wigner (1932) and Eyring (1935) for chemical reactions, is that the coupling between the solvent and the reactants was considered to be strong enough to maintain an equilibrium between chemical species. Moreover, for enzyme-catalyzed reactions, the possible influence of the solvent is generally neglected and the transmission coefficient κ is always considered to be equal to one. This coefficient initially represented the possibility of re-crossing the energy barrier but a more generalized expression takes into account other possible effects specific to any enzyme-catalyzed reaction (Garcia-Viloca et al., 2004). Indeed, a generalized expression of the reaction rate has been proposed:

$$k_{cat} = \gamma(T) k_B \left(\frac{T}{h}\right) \exp\left(\frac{-\Delta G^*}{RT}\right) \quad (2)$$

with $\gamma_{(T)} = \Gamma_{(T)} \kappa_{(T)} g_{(T)}$. In this equation, $\Gamma_{(T)}$ arises from possible re-crossing and should be equal to or less than 1; $\kappa_{(T)}$ takes into account possible tunnelling effects, signifying that some of the complexes having a free energy lower than the activation energy can give rise to the product, so $\kappa_{(T)}$ should be greater than or equal to 1 and is different from the κ found in equation (1) and finally, $g_{(T)}$ expresses the possible deviation of the system from the equilibrium distribution, it can be less than or greater than one.

The assumption that the transmission coefficient, as expressed by κ in equation (1) or by $\gamma_{(T)}$ in equation (2), is equal to 1 is certainly questionable in the case of cold-adapted enzymes since the low temperature of the environment induces an important increase in the solvent viscosity, in particular in the intracellular space. Indeed, the viscosity of the cytoplasm at 20°C is on average 2.5 cP (Mastro and Keith, 1984), a value which is comparable to a 25% sucrose solution whereas at 0°C the internal viscosity is close to 5 cP. Demchenko et al. (1989) examined the influence of the viscosity of the solvent on the rates of the chemical reaction catalyzed by a lactate dehydrogenase at 25°C. They showed that the maximal rate, V_{max}, for lactate oxidation using NAD^+ as a cofactor decreased from 8.5 relative units in low viscosity buffer to 1.5 in a 44% sucrose solution (viscosity = ~6 cP). Similar effects were obtained with other solvent additives such as glycerol and ethylene glycol as well as for pyruvate reduction reaction. Indeed, although a certain strengthening of the enzyme-substrate complexes was observed in these solutions under the form of a K_M reduction, the most pronounced effect of viscosity was at the level of the catalytic steps. As an inverse relationship between the reaction rate and the solvent viscosity was observed, the use of Kramers theory (Kramers, 1940) for interpretation of the data was suggested: $k_{cat} = (A/\eta_r) \exp(-E_a/k_B T)$, in which η_r differs from the viscosity η by a factor δ which describes the coupling between the dynamics of the protein and that of the solvent molecules. Therefore, η_r is a parameter that takes into account the dissipation of energy (internal friction) in the process of activation. In a recent theoretical analysis of the possible influence of the solvent viscosity on the rate of enzyme-catalyzed reactions (Siddiqui et al., 2004), the following relation was proposed: $k_{cat} = [1+(\eta/\eta_0)^2-(\eta/\eta_0)] \exp(-E_a/RT)]^{1/2}$ where η is the relative viscosity of the medium and η_0 a factor correlated to the vibrational frequency at the top of the energy barrier and to the mass and diameter of the particle. This relationship has still to be verified experimentally.

The behavior of cold-adapted enzymes can therefore significantly differ from that of their mesophilic counterparts in the sense that the solvent can also play a negative effect due to its elevated viscosity at low temperatures. Cold-adapted enzymes, in particular those found intracellularly, have to face simultaneously two challenges: one is related to the exponential reverse relationship between the rate and temperature and the other to the similarly reverse relationship between rate and viscosity.

THERMODYNAMICS OF ACTIVITY

The rate of an enzyme-catalyzed reaction is exponentially related to the activation energy, $\Delta G^* = \Delta H^* - T\Delta S^*$. Hence, one way to improve the performance of cold-adapted enzymes would be to reduce the activation energy, either by decreasing ΔH^* or by increasing ΔS^*. The sign of ΔS^* can be positive or negative but in many cases the activation process corresponds to an increase of order and ΔS^* is therefore often negative and tends to increase the value of ΔG^*. In other cases, however, ΔS^* is positive as a result of changes in the distribution of water molecules during the activation process. The activation parameters have been calculated for some cold-adapted enzymes and are compared to those of mesophilic enzymes in Table 1. It can be seen that, except for the chitinase, the activation energy, ΔG^*, of the cold-adapted enzymes is always lower than that of their mesophilic counterparts and these figures are related to the much higher specific activity of cold-adapted enzymes. The lower activation energy is essentially the result of a much lower activation enthalpy ΔH^* and is probably the reason for the lower sensitivity of cold-adapted enzymes to temperature changes as compared to their mesophilic counterparts. The low value of ΔH^* is, however, always compensated by a less favorable activation entropy ΔS^*. Indeed, as in the case of the enthalpy terms, the difference between the activation entropy of cold-adapted enzymes and that of their mesophilic counterpart is always negative whatever the sign, positive or negative, of the entropic parameter. This enthalpy-entropy compensation can be associated to the higher plasticity of the catalytic site of the cold-adapted enzyme or of the whole protein. Indeed, for a defined catalytic reaction, the lower activation enthalpy can be related to the amount of heat required to secure an appropriate complementarity between the enzyme and substrate, e.g., a lesser number of weak bonds have to be broken in the case of cold-adapted enzymes.

As far as the entropic term is concerned and for a positive value of the activation entropy, it is clear that, as the entropy of the cold-adapted enzyme is already elevated, the additional entropy needed for the activation will be lower than in the case of the more rigid mesophilic counterpart. If, on the other hand, the entropic change is negative, the more flexible cold-adapted enzyme will need to undergo a more negative entropic change than the mesophilic enzyme so as to secure the appropriate order corresponding to the activated state, i.e., stabilization of the transition state of the substrate. Looking again at Table 1, one can see that if the entropic changes during the activation process of psychrophilic enzymes were kept similar to those of their mesophilic homologues, the activation energy of the cold-adapted enzymes would be quite low. In the case of the α-amylase for example, the decrease of the activation enthalpy observed (about 12 kJ/mole) would increase the k_{cat} about 140-fold at 10°C. However, due to the entropy compensation only a threefold increase is recorded and this explains why the cold-adapted enzyme cannot achieve a specific activity at the low temperature of its environment comparable to that of the mesophilic counterpart at its own environmental temperature (Lonhienne et al., 2000).

Table 1. Catalytic properties and activation parameters of various psychrophile enzymes compared to those of their mesophilic counterparts

Enzyme	Source	T (°C)	k_{cat} (s^{-1})	T_{opt} (°C)	ΔG^* (kJ/mole)	ΔH^* (kJ/mole)	$T\Delta S^*$ (kJ/mole)	$\Delta\Delta H^*$ (p-m) (kJ/mole)	$T\Delta\Delta S^*$ (p-m) (kJ/mole)	References
Amylase	Psychro	10	294	28	57.68	34.7	−23.0	−11.7	−10.9	D'Amico et al., 2003
	Meso		97	54	58.52	46.4	−12.1			
Chitinase	Psychro	15	1.7	−	69.2	60.2	−9.0	−14.1	−16.1	Lonhienne et al., 2001a
	Meso		3.9	−	67.2	74.3	+7.1			
Chitobiase	Psychro	15	3.8	−	59.5	44.7	−14.8	−26.8	−22.8	Lonhienne et al., 2001b
	Meso		0.9	−	63.5	71.5	+8.0			
Cellulase	Psychro	4	0.18	37	71.6	46.2	−25.4	−19.6	−13.0	Garsoux et al., 2004
	Meso		00.1	56	78.2	65.8	−12.4			
Subtilisin	Psychro	15	25.4	40	62.0	36.0	−26.5	−10.0	−6.3	Davail et al., 1994
	Meso		5.4	60	66.0	46.0	−20.2			
Xylanase (bacterium)	Psychro	10	515.5	35	54.0	21.0	−3.3	−37.0	−3.1	Collins et al., 2003
	Meso		59.5	62	60.0	58.0	−2.0			
Xylanase (yeast)	Psychro	5	14.8		52.3	45.4	−7.0	−4.5	−2.3	Petrescu et al., 2000
	Meso		4.9		54.6	49.9	−4.7			

From Table 1 it can also be seen that the activity of the chitinase from the gram-positive Antarctic strain *Arthrobacter* TAD20 is lower than that of the homologous mesophilic enzyme at 15°C. Here, a soluble preparation of chitin from crab shells was used as substrate, and differences in substrate specificity may play a role in this unexpected data as chitins from different origins are structurally quite different. (Lonhienne et al., 2001a)

INACTIVATION AND CONFORMATIONAL STABILITY

From the few examples cited in Table 1, one can see that the apparent optima of cold-adapted enzymes are significantly lower than those of mesophilic enzymes; this being due to the rapid inactivation of the psychrophilic enzymes as the temperature is raised. It is, however, worth mentioning that in many cases this inactivation does not necessarily correspond to the unfolding of the enzyme. Indeed, as shown in Fig. 3, the inactivation of a cold-adapted α-amylase (Feller et al., 1999, D'Amico et al., 2003a), DNA ligase (Georlette et al., 2003), and xylanase (Collins et al., 2003) precedes any significant structural change as recorded by fluorescence spectroscopy or differential scanning calorimetry. In contrast, for mesophilic and thermophilic enzymes the loss of activity is due to the unfolding of the molecular edifice. Interestingly, this also introduces the concept of real temperature optima since the inactivation does not correspond to structural changes as it is usually observed and the optima are therefore not dependent on the assay duration. An interesting discussion of the significance of temperature optima can be found in Daniel et al. (2001). This inactivation of cold-adapted enzymes has been attributed to the higher flexibility of the catalytic site since an increase in temperature above a certain threshold gives rise to an excessive number of conformations, in rapid equilibrium with each other, which are unable to properly accommodate the substrate.

Numerous examples of the structural instability of cold-adapted enzymes can be found in the literature and apart from those already mentioned in the text and in Table 1 other examples include: a lipase

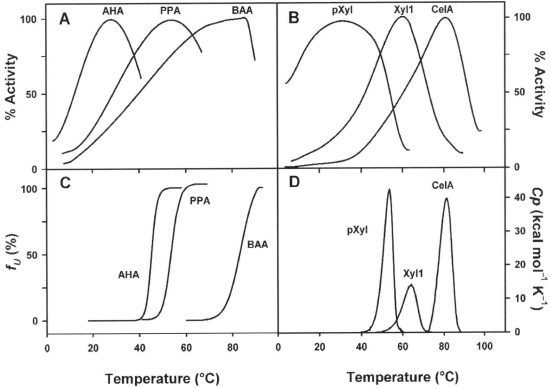

Figure 3. Thermodependence of activity (A, B), upper panel, and unfolding process, as recorded by fluorescence spectroscopy (C), and differential micro-calorimetry (D), of cold-adapted α-amylase (AHA), mesophilic α-amylase from pig pancreas (PPA), thermostable α-amylase from *Bacillus amyloliquefaciens* (BAA), family 8 cold-adapted xylanase from Antarctic *Pseudoalteromonas haloplanktis* (pXyl), family 11 mesophilic xylanase from *Streptomyces* sp. S38(Xyl1), and family 8 thermostable endoglucanase from *Clostridium thermocellum* (CelA). The apparent maximal activities of cold-adapted enzymes, AHA, and pXyl are reached well before any significant structural changes. Adapted from Georlette et al., 2004.

(Arpigny et al., 1997), β-lactamase (Feller et al., 1997), alanine-dehydrogenase (Galkin et al., 1999), alanine racemase (Okubo et al., 1999), phosphoglycerate kinase (Bentahir et al., 2000), catalase (Yumoto et al., 2000), elongation factor (Thomas and Cavicchioli, 2000), alkaline phosphatase (Hauksson et al., 2000), isocitrate lyase and malate synthase (Watanabe et al., 2001), pectate lyase (Van Truong et al., 2001), β-galactosidase (Hoyoux et al., 2001; Fernandes et al., 2002; Cieslinski et al., 2005), isopropyl malate dehydrogenase (Svingor et al., 2001), esterase (Suzuki et al., 2002a), subtilase (Pazgier et al., 2003), ornithine carbamoyltransferase (Xu et al., 2003a), dihydrofolate reductase (Xu et al., 2003b) isocitrate lyase (Watanabe and Takada, 2004), and an isocitrate dehydrogenase (Mizuho et al., 2004, Watanabe et al., 2005).

Thermodynamic stability curves have notably been obtained for the α-amylase from *Pseudoalteromonas haloplanktis* (Feller et al., 1999; D'Amico et al., 2003a) and a deep-sea dihydrofolate reductase from *Moritella profunda* (Hata et al., 2004). These enzymes, unexpectedly for proteins of such a large size, show a reversible thermal unfolding. The maximum conformational stability, ΔG_{stab} which is about 34 kJ/mol., for the α-amylase is observed around 20°C, well above its environmental temperature, whereas that for dihydrofolate reductase, about 16 kJ/mol., is found around 15°C. These figures can be compared to the values observed for mesophilic or thermophilic counterparts: about 71 kJ/mol. for the mesophilic α-amylase from pig pancreas at 30°C and about 145 kJ/mol for the dihydrofolate reductase from *Thermotoga maritima* at 35°C. Thus, these two cold-adapted enzymes display a marginal stability over the whole temperature range.

At the environmental temperature, around 0°C, the stabilization energy drops to about 10 kJ/mol for the α-amylase and is close to 15 kJ/mole for the dihydrofolate reductase which shows a rather flat stability curve from 10 to 25°C. This low stability at low temperatures has been attributed to the preferential hydration of charged and hydrophobic residues inducing a negative value of the stabilization enthalpy which, in the case of α-amylase, is compensated for by the negative value of the stabilization entropy. This low stability, driven by an unfavorable enthalpy, presumably provides an appropriate flexibility of the molecular edifice so as to facilitate substrate accommodation at low temperatures. Hence, the low value of the activation energy of cold-adapted enzymes appears to be associated with a low stability of the molecular edifice and that is presumably necessary to limit the energy cost of the accommodation of the substrates by the enzymes. It is also interesting to note that the maximum stability of these cold-adapted enzymes as well as of their mesophilic and thermophilic counterparts is found around room temperature despite the large differences of the apparent optimum temperatures. This has already been observed for the majority of proteins and is attributed to the hydrophobic effect maximum around room temperature (Kumar et al., 2002).

STRUCTURAL MODIFICATIONS

Analysis of the available three-dimensional structures and site-directed mutagenesis experiments has shown that the low stability of cold-adapted enzymes has been achieved through a reduction of the number and/or strength of weak interactions (hydrogen bonds, and ionic and hydrophobic interactions), by increased interactions with the solvent, a decrease in the number and/or strength of hydrophobic internal clusters and by entropic effects tending to increase the entropy of the unfolded form and to lower its free energy. Indeed, some cold-adapted enzymes display a higher number of glycine residues and a reduced number of proline residues. Other entropic effects, such as a higher proportion of hydrophobic residues at the surface of the enzyme, increase the free energy of the native state through the organization of water molecules around the exposed hydrophobic groups. Some psychrophilic enzymes also show an increase in surface charges, mainly of negative charges, e.g., in subtilisin (Narinx et al., 1997; Arnorsdottir et al., 2005), β-lactamase (Feller et al., 1997), citrate synthase (Russell et al., 1998), malate dehydrogenase (Sun-Yong et al., 1999), and cellulase (Garsoux et al., 2004). This excess of negative charges induces repulsion between groups and prevents an inappropriate compacting of the surface of the protein. The implication of charged residues in the stability of psychrophilic, mesophilic, and thermophilic citrate synthases has been analyzed by Kumar and Nussinov (2004), and different factors contributing to the stability and instability of proteins have been discussed extensively by Vieille and Zeikus (2001) and more specifically for the cold-adapted enzymes by Feller and Gerday (1997). This will not be re-discussed here. Each cold-adapted enzyme is modulated using a specific strategy, probably as a function of structural requirements and makes a selection among the above-mentioned factors to improve the flexibility at the level of the catalytic site. Shoichet et al. (1995) have shown that the configuration of the active sites of cold-adapted enzymes is not optimized for stability, since numerous mutations of the two catalytic residues and of the substrate binding sites in T4 lysozyme give systematically rise to a reduced enzymatic activity and an increased thermal stability. In cold-adapted enzymes, the catalytic and substrate binding sites are almost always identical to their mesophilic counterparts, but indirect changes such as an improved access to the catalytic cavities and improved flexibility of the sites through long distance effects participate in the adaptation process.

MUTAGENESIS STUDIES

The structural modifications believed to be involved in cold-adaptation have been examined in some limited cases using site-directed mutagenesis and directed evolution approaches. The first microbial cold-adapted enzyme submitted to genetic engineering was a subtilisin from an Antarctic bacterial strain (Narinx et al., 1997), in which it was notably shown that the reinforcement of one Ca^{2+} binding site by a single mutation improved both the specific activity and the thermostability, while a double mutation further improved the specific activity but reduced the thermostability. A triosephosphate isomerase (Alvarez et al., 1998) was also engineered and it was shown that the replacement of an alanine by a serine considerably increased the stability of the enzyme while the specific activity was drastically reduced. More-recent experiments concerned an alkaline phosphatase (Tsigos et al., 2001; Mavromatis et al., 2002) and showed that very discrete amino acid substitutions have a considerable impact on the active site flexibility and catalytic performance. For example, the replacement of two glycine residues by alanine gave rise to a complete loss or severe decrease of the enzyme activity. In isocitrate lyase from *Colwellia maris* two glutamine residues were replaced by a histidine and lysine commonly found in the amino acid sequence of homologous enzymes from many other organisms (Watanabe and Takada, 2004). These site-directed mutations led to a decrease of the specific activity of the enzyme, especially between 10 and 25°C, suggesting that the two glutamine residues contribute significantly to cold-adaptation. Several mutations were also introduced in a cold-adapted chitinase (Mavromatis et al., 2003) with the aim of increasing its thermal stability. The introduction of a disulfide bridge found in the *Serratia marcescens* chitinase resulted in a complete loss of activity whereas the introduction of a Gly93Pro mutation produced not only an enzyme with a higher melting temperature (T_m) but also a higher activation energy, shifting the general properties of this enzyme toward those of the mesophilic counterpart. In a cold-active citrate synthase (Gerike et al., 2001), point mutations were introduced with the aim of reducing the accessibility of the active site region and a 50% reduction of activity was recorded while mutations tending to increase the stability of the enzyme did not induce a loss of specific activity. Another systematic investigation using essentially a rational approach was carried out on a cold-adapted α-amylase (D'Amico et al., 2003b) in an endeavor to render the psychrophilic enzyme similar to its mesophilic counterpart in terms of specific activity and thermal stability. The introduction of five single mutations in association with a disulfide bridge resulted in a mutant with a stabilization energy raised from 50 to 80% of that of the mesophilic enzyme while the melting temperature was increased by 8°C and the specific activity was comparable to that of the mesophilic enzyme and was, at 25°C, half of that of the wild type form.

From the above discussion, it is clear that a high specific activity at low temperatures is always associated with a decrease in thermal stability, suggesting that in a naturally cold environment, the flexibility of the molecular edifice is the main determinant of the catalytic efficiency and, in general, any attempt to stabilize a cold-adapted enzyme leads to a decrease in the specific activity at low temperature. However, it is possible in vitro, at least to a certain extent and in a limited number of cases, to increase the thermostability while retaining or further increasing the specific activity of a cold-adapted enzyme. Nevertheless, this is probably more appropriate to enzymes interacting with small size substrates or when small-size synthetic substrates are used rather that natural large-sized macromolecular substrates since the flexibility is probably less crucial in these cases.

Another approach used to investigate psychrophilic enzymes is directed evolution. Miyazaki et al. (2000) applied this to a protease (Davail et al., 1994) so as to improve the thermostability while retaining a high catalytic efficiency at low temperatures. After several cycles, mainly using error prone PCR and recombination, a seven-mutant variant was produced displaying a much higher thermoresistance at 60°C and also a threefold increase of the specific activity at 10°C. This prompted the authors to suggest that the low stability of cold-adapted enzymes was a consequence of random genetic drift due to the absence of selective pressure and not a property related to the high activity of these enzymes at low temperatures. A similar explanation was proposed to explain why thermophilic enzymes never display a high specific activity at room temperature. However, the small chromogenic substrate used for the characterization of the mutants, s-AAPF-pNa, is quite different from the macromolecular substrates encountered by the enzyme in natural environments and it is possible that an increase in the rigidity of the enzyme shifted the specificity of the multi-substrate protease toward this unusual substrate. Interestingly, similar data, tending to demonstrate that activity and stability could be independent properties, were also shown for an esterase from a mesophilic *B. subtilis* (Giver et al., 1998) using in vitro evolution. However, according to the authors, the mutations that improved both the stability and activity were very rare, and the more-stable variants showed a decrease in k_{cat} at 30°C. Only the k_{cat}/K_M ratio was increased by a factor of two for the more stable mutant and again, in these experiments, a small chromogenic substrate, *p*-nitrophenyl acetate, was used. Indeed, the importance of the substrate in evaluating the relative performance of wild type and mutated enzymes has been demonstrated during attempts to transform mesophilic subtilisin BPN′ into a cold-adapted enzyme (Taguchi et al., 2000). Here, mutants with a reduced stability and enhanced catalytic properties on small synthetic substrates at low temperatures were isolated while no improvement of the caseinolytic activity was observed. Random mutagenesis consisting of replacement of a valine residue by various other amino acid residues was applied to the single substrate enzyme 3-isopropyl malate dehydrogenase from *Thermus thermophilus* (Suzuki et al., 2002b) and several mutants showing a significant increase of k_{cat} at low temperatures were found to be less stable than the original thermophilic enzyme. Here, the enhancement of the catalytic properties was associated with an increase of the steric hindrance causing a destabilization of the ternary complex. It is also worth mentioning that mutants of *P. furiosus* OTCase selected in vivo to complement, at 30 and 15°C, yeast mutants lacking OTCase displayed a much higher K_m value for ornithine than the wild type, a lower optimal temperature, a higher k_{cat} value (up to fivefold), and a markedly increased thermolability. This example is one of the rare cases where it could be directly observed that thermolability was not a property acquired during genetic drift at a low temperature independent of the changes responsible for kinetic adaptation (Roovers et al., 2001).

Therefore, we believe that the high activity of psychrophilic enzymes at low temperatures does indeed require an improved flexibility of the molecular edifice that, in vivo, is achieved by a decreased stability of the enzymes. This does not mean that it is not possible to improve in vitro, at least to a certain extent, the stability of a cold-adapted enzymes while maintaining a high catalytic efficiency. It is clear that in Nature the constraints that contribute to modulate the properties of an enzyme are much more diversified than in a test tube, in which one can select only one or two properties.

COLD DENATURATION

From Fig. 3 one can see that psychrophilic enzymes are much more sensitive to low temperatures than their mesophilic and thermophilic counterparts and hence are more susceptible to cold denaturation. The enthalpy and entropy contributions to conformational stability have been measured for an α-amylase which displays a reversible thermal unfolding (D'Amico et al., 2003a). It was shown that at 0°C the weak residual stabilization energy is due to an unfavorable negative stabilization enthalpy which is compensated for by a negative entropy term. In contrast, the residual stability at 37°C is driven by a favorable enthalpy term partially compensated for by an unfavorable positive entropy contribution as expected due to the dissipative forces exerted on the structure by heat. This negative stabilization enthalpy observed at 0°C is apparently due to the hydration of polar and hydrophobic groups and especially aromatic groups. Indeed, the hydration of charged and polar residues occurs with a highly negative free energy change which is mainly driven by a negative enthalpy and which decreases with decreasing temperature (Makhtadze and Privalov, 1994). For the aliphatic hydrophobic groups, the free energy change associated with hydration is positive and contributes to weakly stabilize the native and compact structure of the protein (ΔG_{stab} is positive but close to zero). In contrast, the hydration of aromatic groups has a negative free energy and therefore contributes to destabilize the native structure, especially at low temperature, although this effect is much more limited (ΔG_{stab} is negative but close to zero) than with polar and charged groups. This has been attributed to the fact that, contrarily to aliphatic groups, aromatic groups can form hydrogen bonds with water molecules, because of the partial positive charge of their surface.

THE IMPORTANCE OF FLEXIBILITY

In cold-adapted enzymes, the high catalytic activity is always associated with a low stability of the protein edifice and infers that the former is the consequence of a high flexibility of the three-dimensional structure. In support of this, flexibility indexes showed that the overall flexibility of proteins increases when thermostability is reduced and that the flexibility of proteins displaying similar catalytic activities around their apparent temperature optima is also similar. However, the situation is not always simple since the variation of the indexes expresses the average flexibility that can be quite small. In many cases only specific crucial parts of the molecular structure display a change in flexibility as a function of the usual working temperature of the enzymes and the variation of the indexes does not reflect the real change in local flexibility (Vihinen, 1987). In addition, the flexibility may be related to the structural modifications occurring as a function of time and hence corresponds to the dynamics of the structure, or may be related to the amplitude of the conformational changes which are crucial for the accommodation of the macromolecular substrates at low temperatures.

Different techniques have been used to evaluate the flexibility of protein structures. If the crystallographic structure is known the atomic temperature factors or B-factors, representing the spread of electron densities, can be used for comparisons of the relative flexibility of proteins provided that the resolution of the crystal structures, the crystallization temperature used, and the packing and solvent content are comparable. A significant correlation between the B-factors and the respective stabilities has been found in adenylate kinases from psychrophilic, mesophilic, and thermophilic *Bacillus* sp. with values of 49.5, 27.7, and 10.9 Å^2 respectively (Bae and Phillips, 2004). The situation is more complicated in malate dehydrogenases, however, since the enzyme from *Aquaspirillium arcticum* shows an average B value of 15.10 Å^2 surprisingly well below the value of 23.56 Å^2 for the enzyme from *Thermus flavus* (Sun-Yong et al., 1999). However, when the relative B-factors were used, corresponding to the ratio between the B-factors of individual atoms and the average B-factor, it was found that the temperature factors of all the atoms of the residues interacting with NADH and oxaloacetate in the psychrophilic malate dehydrogenase show a twofold increase when compared to their mesophilic counterparts, thereby reflecting an increase in the relative flexibility of this part of the protein edifice. Similar data have been observed for citrate synthases (Russell et al., 1998).

Molecular dynamics simulations (Bransdal et al., 1999; Tindbaek et al., 2004) can also be very useful in the search for a possible correlation between the catalytic activity and the flexibility of enzymes. In a recent study (Tindbaek et al., 2004) it was shown that when a region of a cold-adapted subtilisin, predicted to be flexible, was transferred to the mesophilic counterpart savinase, the mutant displayed a broader substrate specificity, a significant increase in specific activity at low temperatures, an increase in global and binding region flexibilities, and a global destabilization as measured by differential scanning calorimetry (DSC) experiments. So, clearly the activity of the enzyme is related to the flexibility of the structure.

Experimental techniques can also be used to estimate the relative flexibility of protein structures. For example, amide hydrogen-deuterium exchange as measured by Fourier-transformed infra-red spectroscopy (FTIR), nuclear magnetic resonance (NMR), or mass spectrometry (Yamamoto and Kunihiko, 2003) has been used but has provided only mitigated success. When applied to a psychrophilic alcohol dehydrogenase, it was shown that only those parts of the cold-adapted enzyme involved in substrate and cofactor binding showed a greater flexibility as compared to a thermophilic homologue (Liang et al., 2004). Using this technique a good correlation between the stability and flexibility of thermophilic and mesophilic 3-isopropylmalate dehydrogenases was also found (Zavodszky et al., 1998) but when extended to the psychrophilic enzyme it was shown that this appears to be more rigid than its mesophilic homologue (Svingor et al., 2001). In other cases the data obtained were also puzzling; a thermophilic ribonuclease displayed a flexibility similar to that of its mesophilic counterpart (Hollien and Marqusee, 2002) while a thermophilic α-amylase from *B. licheniformis* showed a higher flexibility than that of its mesophilic counterpart from *B. amyloliquefaciens* (Fitter and Heberle, 2000). Furthermore, no inverse correlation between the stability of the extremely heat stable rubredoxin from *Pyrococcus furiosus* and the flexibility as measured by amide hydrogen exchange was found (Hernandez et al., 2000). These discrepancies may be due to the fact that H/D exchange rates depend on the temperature at which the measurement is carried out and on the time scale used, since not all time scales are relevant to stability. Moreover, one has to keep in mind that the correlation between thermal stability and structural fluctuations is rather complex. If indeed an increase in thermostability is generally associated with an increase in rigidity, the thermodynamic stability of a protein at a given temperature can be achieved not only by an increase of the stabilization enthalpy (increase in the number and/or strength of weak bonds) but also through an increase in entropy of the native form resulting in a smaller difference between the entropy of the unfolded and native forms to the benefit of the stability (Aguilar et al., 1997).

A successful correlation between thermostability and flexibility has been obtained for a certain number of psychrophilic enzymes using acrylamide quenching of tryptophan fluorescence, namely for a Ca^{2+}-Zn^{2+} protease (Chessa et al., 2000), α-amylase (D'Amico et al., 2003a), DNA ligase (Georlette et al., 2003), and xylanase (Collins et al., 2003). Here, the quenching effect was also found to be much more temperature dependent in psychrophilic enzymes.

Another means of evaluating the relative flexibility of proteins is by measurement of neutron scattering. This has been recently applied to measure the macromolecular dynamics in bacteria from different thermal environments where the measured dynamics is dominated by protein conformational fluctuations since these represent about 70% of the dry weight of the cell

(Tehei et al., 2004). In this very interesting paper it was clearly shown that the resilience, taken here as an index of rigidity, was similar at the respective environmental temperatures and that the effective force constants determining the mean macromolecular resilience increased with the physiological temperature from 0.2 N/m in psychrophiles to 0.6 N/m in hyperthermophiles. Thus this supports the hypothesis that thermal adaptation is achieved through evolution towards an appropriate resilience of the macromolecular structures and that flexibility is related to the catalytic efficiency and plays a central role in temperature adaptation. These data further indicate that the difference in stabilization free energy is dominated by enthalpy terms rather than entropic terms. It is therefore clear that the flexibility plays a central role in temperature adaptation and that this flexibility is also related to catalytic efficiency as further recently demonstrated in cold-adapted uracil DNA glycosylase compared to a mesophilic homologue (Olufsen et al., 2005).

CONCLUSION

Cold-adapted enzymes from psychrophilic microorganisms have evolved structural peculiarities which, in general, lead to a decreased stability of the protein edifice mainly through a decrease in the stabilization enthalpy corresponding to a reduction in the number and/or strength of weak bonds. This destabilization seems to be necessary so as to achieve an appropriate flexibility of the whole or crucial parts of the molecular structure and enables reduction of the energy cost of induced fit enzyme mechanisms and an easy accommodation of the substrate at low temperatures. Each enzyme adopts its own strategy compatible with the preservation of the structural characteristics required to secure adequate substrate specificity. In many cases, the adaptation does not seem to be complete since the specific activities do not attain those of the corresponding mesophilic enzymes at their own environmental temperatures. This is probably because the "freezing effect" of low temperatures on the structure cannot be completely compensated for by the low-stability-induced flexibility. Indeed, the stability of the structures has reached some sort of limit as illustrated by the unusually low thermodynamic stability, ΔG_{stab}, of the cold-adapted enzymes investigated so far. The two main properties of these enzymes—a high specific activity at low and moderate temperatures and a low thermostability enabling their rapid inactivation in a complex mixture—render these enzymes particularly suitable for various low to moderate temperature biotechnological processes.

REFERENCES

Aghajari, N., G. Feller, C. Gerday, and R. Haser. 1998. Structures of the psychrophilic *Alteromonas haloplanktis* alpha-amylase give insight into cold adaptation at a molecular level. *Structure* 6:1503–1516.

Aghajari, N., F. Van Petegem, V. Villeret, J.-P. Chessa, C. Gerday, R. Haser, and J. Van Beeumen. 2003. Crystal structure of a psychrophilic metalloprotease reveals new insights into catalysis by cold-adapted proteases. *Proteins: Struct. Funct. Genet.* 50:636–647.

Aguilar, C. F., I. Sanderson, M. Moracci, M. Ciaramella, R. Nucci, M. Rossi, and L. H. Pearl. 1997. Crystal structure of the β-glycosidase from the hyperthermophilic archaeon *Sulfolobus solfataricus*: resilience as a key factor in thermostability. *J. Mol. Biol.* 271:789–802.

Alvarez, M., J.-P. Zeelen, V. Mainfroid, F. Rentier-Delrue, J. Martial, and L. Wyns. 1998. Triosephosphate isomerase(TIM) of the psychrophilic bacterium Vibrio *marinus*. Kinetic and structural properties. *J. Biol. Chem.* 273:2199–2206.

Arnorsdottir, J., M. M., Kristjansson, and R. Ficner. 2005. Crystal structure of a subtilisin-like serine proteinase from a psychrotrophic *Vibrio* species reveals structural aspects of cold adaptation. *FEBS J.* 272:832–845.

Arnosti, C., and B. Jorgensen. 2003. High activity and low temperature optima of extracellular enzymes in Arctic sediments: implication for carbon cycling by heterotrophic microbial communities. *Mar. Ecol. Prog. Ser.* 249:15–24.

Arpigny, J.-L., J. Lamotte, and C. Gerday. 1997. Molecular adaptation to cold of an Antarctic bacterial lipase. *J. Mol. Catal. B: Enzymatic.* 3:29–35.

Arrhenius, S. 1889. Uber die reaktionsgeschwindigkeit bei der inversion von rohrzucker durch sauren. *Z. Physik. Chem.* 4:226–248.

Astwood, A. C., and A. C. Wais. 1998. Psychrotrophic bacteria isolated from a constantly warm tropical environment. *Curr. Microbiol.* 36:148–151.

Bae, E., and G. N. Phillips. 2004. Structures and analysis of highly homologous psychrophilic, mesophilic, and thermophilic adenylate kinases. *J. Biol. Chem.* 279:28202–28208.

Bendt, A., H. Huller, U. Kammel, E. Helmke, and T. Schweder. 2001. Cloning, expression and characterization of a chitinase gene from the Antarctic psychrotolerant bacterium *Vibrio* sp.strain Fi:7. *Extremophiles.* 5:119–126.

Bentahir, M., G. Feller, M. Aittaleb, J. Lamotte-Brasseur, T. Himri, J.-P. Chessa, and C. Gerday. 2000. Structural, kinetic, and calorimetric characterization of a cold-active phosphoglycerate kinase from the Antarctic *Pseudomonas* sp. TACII 18. *J. Biol. Chem.* 275:11147–11153.

Birolo, L., M. L. Tutino, B. Fontanella, C. Gerday, K. Mainolfi, S. Pascarella, G. Sannia, F. Vinci, and G. Marino. 2000. Aspartate aminotransferase from the Antarctic bacterium Pseudoalteromonas haloplanktis TAC 125. *Eur. J. Biochem.* 267:2790–2802.

Bransdal, B. O., E. S. Heimstad, I. Sylte, and A. O. Smalas. 1999. Comparative molecular dynamics of mesophilic and psychrophilic protein homologues studied by 1.2 ns simulations. *J. Biomol. Struct.* 3:493–506.

Buchon, L., P. Laurent, A. M. Gounot, and J. F. Guespin-Michel. 2000. Temperature dependence of extracellular enzymes production by psychrotrophic and psychrophilic bacteria. *Biotechnol. Lett.* 22:1577–1581.

Cavicchioli, R., K. S. Siddiqui, D. Andrews, and K. R. Somers. 2002. Low-temperature extremophiles and their applications. *Curr. Opin. Biotechnol.* 13:253–261.

Cieslinski, H., J. Kur, A. Bialkowska, I. Baran, K. Makowski, and M. Turkiewicz. 2005. Cloning, expression and purification of a recombinant cold-adapted β-galactosidase from Antarctic bacterium *Pseudoalteromonas* sp.22B. *Prot. Expr. Purif.* **39**: 27–34.

Collins, C., M-A Meuwis, I. Stals, M. Claeyssens, G. Feller, and C. Gerday. 2002. A novel family & xylanase, functional and physicochemical characterization. *J. Biol. Chem.* **277**: 35133–35139.

Collins, T., M.-A. Meuwis, C. Gerday, and G. Feller. 2003. Activity, stability and flexibility in glycosidases adapted to extreme thermal environments. *J. Mol. Biol.* **328**:419–428.

Chessa, J.-P., I. Petrescu, M. Bentahir, J. Van Beeumen, and C. Gerday. 2000. Purification, physico-chemical characterization and sequence of the heat labile alkaline metalloprotease isolated from a psychrophilic *Pseudomonas* species. *Biochim. Biophys. Acta.* **1479**:265–274.

D'Amico, S., P. Claverie, T. Collins, D. Georlette, E. Gratia, A. Hoyoux, M.-A. Meuwis, G. Feller, and C. Gerday. 2002. Molecular basis of cold adaptation. *Philos. Trans. R. Soc. Lond. Series B.: Biol. Sci.* **357**:917–925.

D'Amico, S., C. Gerday, and G. Feller. 2003a. Activity-stability relationships in extremophilic enzymes. *J. Biol. Chem.* **278**: 7891–7896.

D'Amico, S., C. Gerday, and G. Feller. 2003b. Temperature adaptation of proteins: engineering mesophilic-like activity and stability in a cold-adapted α-amylase. *J. Mol. Biol.* **332**:981–988.

Daniel, R. M., M. J. Danson, and R. Eisenthal. 2001. The temperature optima of enzymes: a new perspective on an old phenomenon. *Trends Biochem. Sci.* **26**:223–225.

Davail, S., G. Feller, E. Narinx, and C. Gerday. 1994. Cold adaptation of proteins. Purification, characterization, and sequence of the heat-labile subtilisin from the Antarctic psychrophile *Bacillus* TA41.1994. *J. Biol. Chem.* **269**:17448–17453.

De Backer, M., S. McSweeney, H. B. Rasmussen, B. W. Riise, P. Lindley, and E. Hough. 2002. The 1.9A crystal structure of heat-labile shrimp alkaline phosphatase. *J. Mol. Biol.* **318**:1265–1274.

Demchenko, A. P., O. I. Rusyn, and E. A. Saburova. 1989. Kinetics of the lactate dehydrogenase reaction in high-viscosity media. *Biochim. Biophys. Acta.* **998**:196–203.

Deming, J. W. 2002. Psychrophiles and polar regions. *Curr. Opin. Microbiol.* **5**:301–309.

Demot, R., and H. Verachtert. 1987. Purification and characterization of extracellular α-amylase and glucosamylase from the yeast *Candida antarctica* CBS 6678. *Eur. J. Biochem.* **4**:643–654.

Evans, M. G., and M. Polanyi. 1935. Some applications of the transition state method to the calculation of reaction velocities especially in solution. *Trans. Faraday Soc.* **31**:875–894.

Eyring, H. 1935. The activated complex in chemical reactions. *J. Chem. Phys.* **3**:107–115.

Feller, G., and C. Gerday. 1997. Psychrophilic enzymes: molecular basis of cold adaptation. *CMLS, Cell. Mol. Life Sci.* **53**:830–841.

Feller, G., and C. Gerday. 2003. Psychrophilic enzymes: hot topics in cold adaptation. *Nature Rev. Microbiol.* **1**:200–207.

Feller, G., E. Narinx, J.-L. Arpigny, Z. Zekhnini, and C. Gerday. 1994. Temperature dependence of growth, enzyme secretion and activity of psychrophilic Antarctic bacteria. *Appl. Microbiol. Biotechnol.* **41**:477–479.

Feller, G., E. Narinx, J.-L. Arpigny, M. Aittaleb, E. Baise, S. Genicot, and C. Gerday. 1996. Enzymes from psychrophilic organisms. *FEMS Microbiol. Rev.* **18**:189–202.

Feller, G., D. d'Amico, and C. Gerday. 1999. Thermodynamic stability of a cold-active α-amylase from the Antarctic bacterium *Alteromonas haloplanktis*. *Biochemistry.* **38**:4613–4619.

Feller, G., M. Thiry, J.-L. Arpigny, M. Mergeay, and C. Gerday. 1990. Lipases from psychrotrophic Antarctic bacteria. *FEMS Microbiol. Lett.* **66**:239–244.

Feller, G., Z. Zekhnini, J. Lamotte-Brasseur, and C. Gerday. 1997. Enzyme from cold-adapted microorganisms. The class C β-lactamase from the Antarctic psychrophile *Psychrobacter immobilis* A5. *Eur. J. Biochem.* **244**:186–191.

Fernandes, S., B. Geueke, O. Delgado, J. Coleman, and R. Hatti-Kaul. 2002. β-galactosidase from a cold-adapted bacterium: purification, characterization and application for lactose hydrolysis. *Appl. Microbiol. Biotechnol.* **58**:313–321.

Fitter, J., and J. Heberle. 2000. Structural equilibrium fluctuations in mesophilic and thermophilic α-amylase. *Biophys. J.* **79**: 1629–1636.

Galkin, A., L. Kulakova, H. Ashida, Y. Sawa, and N. Esaki. 1999. Cold-adapted alanine dehydrogenase from two Antarctic bacterial strains: gene cloning, protein characterization, and comparison with mesophilic and thermophilic counterparts. *Appl. Environ. Microbiol.* **65**:4014–4020.

Garcia-Viloca, M., J. Gao, M. Karplus, and D. G. Truhlar. 2004. How enzymes work: analysis by modern rate theory and computer simulations. *Science* **303**:186–195.

Garsoux, G., J. Lamotte-Brasseur, C. Gerday, and G. Feller. 2004. Kinetic and structural optimisation to catalysis at low temperatures in a psychrophilic cellulase from the Antarctic bacterium *Pseudoalteromonas haloplanktis*. *Biochem. J.* **384**:247–253.

Georlette, D., V. Blaise, T. Collins, S. D'Amico, E. Gratia, A. Hoyoux, J.-C. Marx, G. Sonan, G. Feller, and C. Gerday. 2004. Some like it cold: biocatalysis at low temperatures. *FEMS Microbiol. Rev.* **28**:25–42.

Georlette, D., B. Damien, V. Blaise, E. Depiereux, V. N. Uversky, C. Gerday, and G. Feller. 2003. Structural and functional adaptations to extreme temperatures in psychrophilic, mesophilic and thermophilic DNA ligases. *J. Biol. Chem.* **278**:37015–37023.

Gerday, C., M. Aittaleb, J.-L. Arpigny, E. Baise, J.-P. Chessa, G. Garsoux, I. Petrescu, and G. Feller. 1997. Psychrophilic enzymes: a thermodynamic challenge. *Biochim. Biophys. Acta.* **1342**:119–131.

Gerday, C., M. Aittaleb, M. Bentahir, J.-P. Chessa, P. Claverie, T. Collins, S. D'Amico, J. Dumont, G. Garsoux, D. Georlette, A. Hoyoux, T. Lonhienne, M.-A. Meuwis, and G. Feller. 2000. Cold-adapted enzymes: from fundamentals to biotechnology. *Trends Biotechnol.* **18**:103–107.

Gerike, U., M. Danson, N. J. Russell, and D. Hough. 1997. Sequencing and expression of the gene encoding a cold-active citrate synthase from an Antarctic bacterium, strain DS2-3R. *Eur. J. Biochem.* **248**:49–57.

Gerike, U., M. J. Danson, and D. W. Hough. 2001. Cold-active citrate synthase: mutagenesis of active-site residues. *Protein. Engng.* **14**:655–661.

Gianese, G., F. Bossa, and S. Pascarella. 2002. Comparative structural analysis of psychrophilic and mesophilic and thermophilic enzymes. *Proteins.* **47**:236–249.

Giver, L., A. Gershenson, P. Freskgard, and F. Arnold. 1998. Directed evolution of a thermostable esterase. *Proc. Natl. Acad. Sci. USA* **95**:12809–12813.

Glansdorff, N., and Y. Xu. 2002. Microbial life at low temperatures/mechanisms of adaptation and extreme biotopes. Implications for exobiology and the origin of life. *Rec. Res. Develop. Microbiol.* **6**:1–21.

Hata, K., R. Hono, M. Fujisawa, R. Kitahara, Y. Kamatari, K. Akasaka, and X. Ying. 2004. High pressure NMR study of dihydrofolate reductase from a deep-sea bacterium *Moritella profunda*. *Cell. Mol. Biol.* **50**:311–316.

Hauksson, J. B., O. S. Andresson, and B. Asgeirsson. 2000. Heat-labile bacterial alkaline phosphatase from a marine *Vibrio* sp. *Enz. Microb. Technol.* **27**:66–73.

Haynie, D. T. 2001. Collision theory, p. 261–262. *In* D. T. Haynie (ed.), *Biological Thermodynamics*, 1st ed. Cambridge University Press, Cambridge United Kingdom.

Helland, R., I. Leiros, G. Berglund, N. P. Willassen, and A. O. Smalas. 1998. The crystal structure of anionic salmon trypsin in complex with bovine pancreatic trypsin inhibitor. *Eur. J. Biochem.* **256**:317–324.

Hernandez, G., F. E. Jenney, M. W. Adams, and D. M. Lemaster. 2000. Millisecond time scale conformational flexibility in a hyperthermophile at ambient temperature. *Proc. Natl. Acad. Sci. USA* **97**:2962–2964.

Hiroki, T., B. Mikami, and A. Yasuo. 2005. Crystal structure of cold-active protein–tyrosine phosphatase from a psychrophile, *Shewanella* sp. *J. Biochem.* (Tokyo) **137**:69–77.

Hollien, J., and S. Marqusee. 2002. Comparison of the folding processes of *T. Thermophilus* and *E. coli* ribonucleases H. *J. Mol. Biol.* **316**:327–340.

Hoyoux, A., V. Blaise, T. Collins, S. D'Amico, E. Gratia, A. L. Huston, J.-C. Marx, G. Sonan, Y. Zeng, G. Feller, and C. Gerday. 2004. Extreme catalysts from low-temperature environments. *J. Biosci. Bioeng.* **98**:317–330.

Hoyoux, A., I. Jennes, P. Dubois, S. Genicot, F. Dubail, J.-M. Francois, E. Baise, G. Feller, and C. Gerday. 2001. Cold-adapted β-galactosidase from the Antarctic psychrophile *Pseudoalteromonas haloplanktis*. *Appl. Environ. Microbiol.* **67**:1529–1535.

Huston, A. L., B. B. Krieger-Brockett, and J. Deming. 2000. Remarkably low temperature optima for extracellular enzyme activity from Artic bacteria and sea ice. *Environ. Microbiol.* **2**:383–388.

Ingraham, J. L., and J. L. Stokes. 1959. Psychrophilic bacteria. *Bacteriol. Rev.* **23**:97–108.

Jaenike, R. 1990. Protein structure and function at low temperatures. *Philos. Trans. R. Soc. Lond. Series B.: Biol. Sci.* **326**:535–551.

Jay, J. M. 1986. Characteristics and growth of psychrotrophic microorganisms. p. 579–592. *In* D. Van Nostrand (ed.), *Modern Food Microbiology*, 3rd ed. Reinhold Company, New-York.

Kimura, T., and K. Horikoshi. 1990. Characterization of pullulan-hydrolysing enzyme from an alkali-psychrotrophic *Micrococcus* sp. *Appl. Microbiol. Biotechnol.* **34**:52–56.

Kramers, H. A. 1940. Brownian motion in a field of force and the diffusion model of chemical reactions. *Physica.* **7**:284–304.

Kumar, S., and R. Nussinov. 2004. Different roles of electrostatics in heat and in cold: adaptation by citrate synthase. *Chem. Bio. Chem.* **5**:280–290.

Kumar, S., T. Chung-Jung, and R. Nussinov. 2002. Maximal stabilities of reversible two-state proteins. *Biochemistry* **41**:5359–5374.

Leiros, I., E. Moe, O. Lanes, A. O. Smalas, and N. P. Willassen. 2003. The structure of uracil-DNA glycosylase from Atlantic cod (*Gadus morhua*) reveals cold-adaptation features. *Acta Crystallog. Sect.D.* **59**:1357–1365.

Liang, Z., I. Tsigos, T. Lee, V. Bouriotis, K. Resing, N. Ahn, and J. Klinman. 2004. Evidence for increased local flexibility in psychrophilic alcohol dehydrogenase relative to its thermophilic homologue. *Biochemistry* **43**:14676–14683.

Lonhienne, T., C. Gerday, and G. Feller. 2000. Psychrophilic enzymes: revisiting the thermodynamic parameters of activation may explain local flexibility. *Biochim. Biophys. Acta.* **1543**:1–10.

Lonhienne, T., E. Baise, G. Feller, V. Bouriotis, and C. Gerday. 2001a. Enzyme activity determination on macromolecular substrates by isothermal titration calorimetry: application to mesophilic and psychrophilic chitinases. *Biochim. Biophys. Acta.* **1545**:349–356.

Lonhienne, T., J. Zoidakis, C. E. Vorgias, G. Feller, C. Gerday, and V. Bouriotis. 2001b. Modular structure, local flexibility and cold-activity of a novel chitobiase from a psychrophilic Antarctic bacterium. *J. Mol. Biol.* **310**:291–297.

Makhtadze, G. I., and P. L. Privalov. 1994. Hydration effects in protein unfolding. *Biophys. Chem.* **51**:291–309.

Margesin, R., and F. Schinner. 1992. Extracellular protease production by psychrotrophic bacteria from glaciers. *Intern. Biodet. Biodegr.* **29**:177–189.

Marx, J.-C., V. Blaise, T. Collins, S. D'Amico, D. Delille, E. Gratia, A. Hoyoux, A. L. Huston, G. Sonan, G. Feller, and C. Gerday. 2004. A perspective on cold enzymes: current knowledge and frequently asked questions. *Cell. Mol. Biol.* **50**:643–655.

Mastro, A. M., and A. D. Keith. 1984. Diffusion in the aqueous compartment. *J. Cell Biol.* **99**:180–187.

Mavromatis, K., I. Tsigos, M. Tzanodaskalaki, M. Kokkinidis, and V. Bouriotis. 2002. Exploring the role of a glycine cluster in cold adaptation of an alkaline phosphatase. *Eur. J. Biochem.* **269**:2330–2335.

Mavromatis, K., G. Feller, M. Kokkinidis, and V. Bouriotis. 2003. Cold adaptation of a psychrophilic chitinase: a mutagenesis study. *Protein Engng.* **16**:497–503.

Miyazaki, K., P. L. Wintrode, R. A. Grayling, D. N. Rubingh, and F. Arnold. 2000. Directed evolution study of temperature adaptation in a psychrophilic enzyme. *J. Mol. Biol.* **297**:1015–1026.

Mizuho, Y., S. Takehiko, N. Katsutoshi, and Y. Takada. 2004. Characterization of chimeric isocitrate dehydrogenase of a mesophilic nitrogen-fixing bacterium, *Azobacter vinelandii*, and a psychrophilic bacterium, *Colwellia maris*. *Curr. Microbiol.* **48**:383–388.

Mostafa, W. H., H. H. Radman, and A. M. Hashem. 2003. Expression of cold adaptative enzymes of *Pseudomonas fluorescens*. *New Egypt. J. Microbiol.* **6**:162–187.

Narinx, E., E. Baise, and C. Gerday. 1997. Subtilisin from psychrophilic Antarctic bacteria: characterization and site-directed mutagenesis of residues possibly involved in the adaptation to cold. *Protein Engng.* **10**:1271–1279.

Okubo, Y., K. Yokoigawa, N. Esaki, K. Soda, and H. Kawai. 1999. Characterization of psychrophilic alanine racemase from *Bacillus psychrosaccharolyticus*. *Biochem. Biophys. Res. Comm.* **256**:333–340.

Olufsen, M., A. O. Smalas, E. Moe, and B. O. Brandsdal. 2005. Increased flexibility as a strategy for cold adaptation. A comparative molecular dynamics study of cold- and warm-active uracil DNA glycosylase. *J. Biol. Chem.* **280**:18042–18048.

Pazgier, M., M. Turkiewicz, H. Kalinowska, and S. Bielicki. 2003. The unique cold-adapted extracellular subtilase from psychrophilic yeast *Leucosporidium antarcticum*. *J. Molecul. Cat. B.* **21**:39–42.

Pelzer, H., and E. Wigner. 1932. Velocity coefficient of interchanges reactions. *Z. Phys. Chem.* **B15**:445–471.

Petrescu, I., J. Lamotte-Brasseur, J.-P. Chessa, P. Ntarima, M. Claeyssens, B. Devreese, G. Marino, and C. Gerday. 2000. Xylanase from psychrophilic yeast *Cryptococcus adeliae*. *Extremophiles* **4**:137–144.

Ratkowsky, D. A., J. Olley, and T. Ross. 2005. Unifying temperature effects on the growth rate of bacteria and the stability of globular proteins. *J. Theor. Biol.* **233**:351–362.

Roovers, M., R. Sanchez, C. Legrain, and N. Glansdorff. 2001. Experimental evolution of enzyme temperature activity profile selection in vivo and characterization of low-temperature adapted mutants of *Pyrococcus furiosus* ornithine carbamoyl transferase. *J. Bacteriol.* **183**:1101–1105.

Russell, N. J. 1990. Cold-adaptation of microorganisms. *Philos. Trans. R. Soc. Lond. B. Biol. Sci.* **326**:595–611.

Russell, N. J. 2000. Toward a molecular understanding of cold activity of enzymes from psychrophiles. *Extremophiles* 4:83–90.

Russell, R. J., U. Gerike, M. J. Danson, D. W. Hough, and G. L. Taylor. 1998. Structural adaptations of the cold-active citrate synthase from an Antarctic bacterium. *Structure* 6:351–361.

Shoichet, B. K., W. A. Baase, R. Kuroki, and B. W. Matthews. 1995. A relationship between protein stability and protein function. *Proc. Natl. Acad. Sci. USA* 92:452–456.

Smalas, A. O., H. K. Leiros, V. Os, and N. P. Willassen. 2000. Cold-adapted enzymes. *Biotechnol. Ann. Rev.* 6:1–57.

Siddiqui, K. S., S. A. Bokhari, A. J. Afzal, and S. Singh. 2004. A novel thermodynamic relationship based on Kramers theory for studying enzyme kinetics under high viscosity. *IUBMB Life* 56:403–407.

Skalova, T., J. Dohnalek, V. Spiwok, P. Lipovova, E. Vondrackova, H. Petrokova, J. Duskova, H. Strnad, B. Kralova, and J. Hasek. 2005. Cold-active β-galactosidase from *Arthrobacter* sp. C2-2 forms compact 660kDa hexamers: crystal structure at 1.9 Å resolution. *J. Mol. Biol.* 353:282–294.

Sun-Yong, K., H. Kwang-Yeon, K. Sung-Hou, S. Ha-Chin, H. Ye-Sun, and C. Yunge. 1999. Structural basis for cold adaptation. Sequence, biochemical properties, and crystal structure of malate dehydrogenase from a psychrophile *Aquaspirillium arcticum*. *J. Biol. Chem.* 274:11761–11767.

Suzuki, T., T. Nakayama, T. Kurihara, T. Nishino, and N. Esaki. 2002a. Primary structure and catalytic properties of a cold-active esterase from a psychrotroph, *Acinetobacter* sp. Strain no.6. isolated from Siberian soil. *Biosci. Biotechnol. Biochem.* 66:1682–1690.

Suzuki, T., M. Yasugi, F. Arisaka, T. Oshima, and A. Yamagishi. 2002b. Cold-adaptation mechanism of mutant enzymes of 3-isopropylmalate dehydrogenase from *Thermus thermophilus*. *Protein Engng.* 15:471–476.

Svingor, A., J. Kardos, I. Hadju, A. Nemeth, and P. Zavodszky. 2001. A better enzyme to cope with cold. Comparative flexibility studies on psychrotrophic, mesophilic, and thermophilic IPMDHS. *J. Biol. Chem.* 276:28121–28125.

Taguchi, S., S. Komada, and H. Momose. 2000. The complete amino acid substitutions at position 131 that are positively involved in cold adaptation of subtilisin BPN'. *Appl. Environ. Microbiol.* 66:1410–1415.

Tehei, M., B. Franzetti, D. Madern, M. Ginzburg, B. Z. Ginzburg, M.-T., Giudici-Orticoni, M. Bruschi, and G. Zaccai. 2004. Adaptation to extreme environments: macromolecular dynamics in bacteria compared in vivo by neutron scattering. *EMBO Rep.* 5:66–70.

Thomas, T., and R. Cavicchioli. 2000. Effect of temperature on stability and activity of elongation factor 2 proteins from Antarctic and thermophilic methanogens. *J. Bacteriol.* 182:1328–1332.

Tindbaek, N., A. Svendsen, P. Oestergaard, and H. Draborg. 2004 Engineering a substrate-specific cold-adapted subtilisin. *Protein Engng.* 17:149–156.

Tsigos, I., K. Mavromatis, M. Tzanodaslaki, C. Pozidis, M. Kokkinidis, and V. Bouriotis. 2001. Engineering the properties of a cold-active enzyme through rational redesign of the active site. *Eur. J. Biochem.* 268:5074–5080.

Van Petegem, F., T. Collins, M.-A. Meuwis, C. Gerday, G. Feller, and J. Van Beeumen. 2003. The structure of a cold-adapted family 8 xylanase at 1.3 A resolution. Structural adaptations to cold and investigation of the active site. *J. Biol. Chem.* 278:7531–7539.

Van Truong, L., H. Tuyen, E. Helmke, L. Tran Binh, and T. Schweder. 2001. Cloning of two pectate lyase genes from the marine Antarctic bacterium *Pseudoalteromonas haloplanktis* strain ANT/505 and characterization of the enzymes. *Extremophiles* 5:35–44.

Vieille, C., and G. J. Zeikus. 2001. Hyperthermophilic enzymes: sources, uses and molecular mechanisms for thermostability. *Microbiol. Mol. Rev.* 65:1–43.

Vihinen, M. 1987. Relationship of protein flexibility to thermostability. *Protein Engng.* 1:477–480.

Violot, S., N. Aghajari, M. Czjzek, G. Feller, G. K. Sonan, P. Gouet, C. Gerday, R. Haser, and V. Receveur-Bréchot. 2005. Structure of a full length psychrophilic cellulase from *Pseudoalteromonas haloplanktis* revealed by X-ray diffraction and small angle X-ray scattering. *J. Mol. Biol.* 348:1211–1224.

Watanabe, S., Y. Takada, and N. Fukunaga. 2001. Purification and characterization of a cold-adapted isocitratelyase and a malate synthase from *Colwellia maris*, a psychrophilic bacterium. *Biosci. Biotechnol. Biochem.* 65:1095–1103.

Watanabe, S., and Y. Takada. 2004. Amino acid residues involved in cold-adaptation of isocitrate lyase from a psychrophilic bacterium, *Colwellia maris*. *Microbiology* (Reading, UK). 150: 3393–3403.

Watanabe, S., Y. Yasutake, Y. I. Tanaka, and Y. Takada. 2005. Elucidation of stability determinant of cold-adapted monomeric isocitratedehydrogenase from a psychrophilic bacterium *Colwellis maris*, by construction of chimeric enzymes. *Microbiology* 151:1083–1094.

Xu, Y., G. Feller, C. Gerday, and N. Glansdorff. 2003a. Metabolic enzymes from psychrophilic bacteria: challenge of adaptation to low temperatures in ornithine carbamoyltransferase from *Moritella abyssi*. *J. Bacteriol.* 185:2161–2168.

Xu, Y., G. Feller, C. Gerday, and N. Glansdorff. 2003b. *Moritella* cold-active dihydrofolate reductase: are there natural limits to optimisation of catalytic efficiency at low temperature? *J. Bacteriol.* 185:5519–5526.

Xu, Y., Y. Nogi, C. Kato, Z. Liang, H. J. Ruger, D. De Kegel, and N. Glansdorff. 2003c. *Moritella profunda* sp. Nov.and *Moritella abysi* sp. Nov., two psychropiezophilic organisms isolated from deep Atlantic sediments. *Int. J. Syst. Evol. Microbiol.* 53:533–538.

Yamamoto, T., and G. Kunihiko. 2003. Study of flexibility of protein by mass spectrometry. *J. Mass Spectro. Soc. Jap.* 51:412–414.

Yumoto, I., D. Ichihashi, H. Iwata, A. Istokovics, N. Ichise, H., Matsuyama, H. Okuyama, and K. Kawasaki. 2000. Purification and characterization of a catalase from the facultatively psychrophilic bacterium *Vibrio rumiensis* S-1 exhibiting high catalase activity. *J. Bacteriol.* 182:1903–1909.

Zavodszky, P., J. Kardos, A. Svingor, and G. A. Petsko. 1998. Adjustment of conformational flexibility is a key event in the thermal adaptation of proteins. *Proc. Natl. Acad. Sci. USA* 95:7406–7411.

Zecchinon, L., P. Claverie, T. Collins, S. D'Amico, D. Delille, G. Feller, D. Georlette, E. Gratia, A. Hoyoux, M.-A. Meuwis, G. Sonan, and C. Gerday. 2001. Did psychrophilic enzymes really win the challenge? *Extremophiles* 5:313–321.

Chapter 14

The Cold-Shock Response

MASAYORI INOUYE AND SANGITA PHADTARE

INTRODUCTION

When a bacterial culture growing exponentially at a temperature optimum for its growth is shifted to low temperature, it exhibits cold-shock response. This is irrespective of the preferred optimum growth temperature; thus all types of bacteria such as psychrotrophic, psychrophilic, mesophilic, and thermophilic bacteria possess cellular machinery to elicit this response.

COLD-SHOCK RESPONSE OF MESOPHILIC BACTERIA

The mesophilic cold-shock response has been studied extensively (for review see Ermolenko and Makhatadze, 2002; Phadtare, 2004; Phadtare et al., 1999, 2000; Weber and Marahiel, 2003; Yamanaka et al., 1998) in recent years using two bacteria, Escherichia coli and Bacillus subtilis as model organisms. Many of the key features of cold-shock response discovered during these studies apply well to all the classes of bacteria mentioned above. E. coli, an enterobacterium with an optimum temperature of 37°C, often encounters sudden drastic temperature changes as a result of excretion from animals. Therefore, having an efficient cold-shock response system is critical for this organism to quickly adapt to the new environment. The main features of cold-shock response with emphasis on E. coli are discussed below.

The change in temperature is sensed at the levels of membrane, nucleic acids, and proteins. The membranes lose their flexible, crystalline state at low temperature, which affects the membrane-associated functions such as transport. Cold-shock adaptation involves restoring the flexibility of membranes by various mechanisms such as increasing the proportion of unsaturated fatty acids (UFAs), shortening the fatty acid chain length, or altering fatty acid branching from iso to anteiso. Phospholipids with UFAs have lower melting points and greater degree of flexibility than phospholipids containing saturated fatty acids, and the proportion of UFAs increases upon temperature downshift. This is known as homeoviscous adaptation (Sinensky, 1974). In E. coli, the UFA produced after temperature downshift is cis-vaccenic acid (cis-11-octadecenic acid). The enzyme β-ketoacyl-acyl carrier protein (ACP) synthase II converts palmitoleic acid to cis-vaccenic acid. The synthesis of this enzyme is not induced upon cold shock, but the enzyme is activated at low temperature (Garwin and Cronan, 1980; Garwin et al., 1980). In contrast, the desaturation system in B. subtilis is cold inducible. Bacillus is the only nonphotosynthetic bacterium in which the presence and cold induction of desaturase is reported (Aguilar et al., 1998, 2001), and des is the strongest cold-inducible gene (Beckering et al., 2002; Kaan et al., 2002). The desaturase system of Bacillus has been studied in detail (Aguilar et al., 2001), and it has been shown that deletion of des gene does not cause cold sensitivity in Bacillus growing in a rich medium (Aguilar et al., 1998). However, its deletion causes a severe cold-sensitive phenotype in the absence of isoleucine. In addition, B. subtilis (Kaneda, 1967, 1991) and Lactobacillus monocytogenes (Annous et al., 1997) show prominent presence of anteiso-branched fatty acids that have lower melting points than iso-fatty acids at low temperature.

DNA is another temperature-sensing cellular component, and negative supercoiling of DNA increases in response to temperature downshift and affects DNA-related functions, such as replication, transcription, and recombination (Krispin and Allmansberger, 1995; Mizushima et al., 1997; Wang and Syvanen, 1992). The changes in supercoiling influence relative orientation of the −35 and −10 regions, which in turn influences the recognition of some σ^{70} promoters by RNA polymerase (Wang and Syvanen, 1992). A shift in

---

M. Inouye and S. Phadtare • Department of Biochemistry, Robert Wood Johnson Medical School, 675 Hoes Lane, Piscataway, NJ 08854.

temperature is also sensed at level of proteins, and certain proteins such as the aspartate chemoreceptor (Tar) of *E. coli* sense changes in temperature by reversible methylation of its cytoplasmic signaling/adaptation domain (Nishiyama et al., 1999).

Cold-shock response of many bacteria including *E. coli* is characterized by a transient arrest of cell growth immediately following the temperature downshift, during which a number of genes are induced, in contrast to a severe inhibition of general protein synthesis. CspA (cold-shock protein A) is the major cold-shock protein in *E. coli*. CspA belongs to a family of nine homologous proteins. In fact, many bacteria have multiple CspA homologs, suggesting that these play essential role(s) in cellular adaptation to temperature changes. CspA homologs are widespread in *Bacteria* including hyperthermophilic bacteria but are not present in *Archaea*. CspA homologs are also found in higher organisms, for example, the putative wheat nucleic acid-binding protein WCSP1. Induction of this protein is cold specific (Karlson et al., 2002). However, CspA homologs have not been identified in yeasts, but genes encoding cold-shock-inducible proteins such as *TIP1* (temperature-inducible protein), *TIR1*, and *TIR2* are induced in budding yeast *Saccharomyces cerevisiae* (Kondo and Inouye, 1991). *NSR1*, another cold-induced gene, is involved in ribosomal RNA processing and ribosomal synthesis at low temperature in yeast (Kondo and Inouye, 1992). Of the nine *E. coli* CspA homologs, four [CspA (Goldstein et al., 1990), CspB (Lee et al., 1994), CspG (Nakashima et al., 1996), and CspI (Wang et al., 1999)] are cold-shock inducible. CspC and CspE are constitutively produced at 37°C (Phadtare and Inouye, 2001), while CspD is induced in the stationary phase and under nutrient-limiting conditions and is involved in replication inhibition (Yamanaka and Inouye, 1997; Yamanaka et al., 2001). Interestingly, none of the cold-shock-inducible homologs are singularly responsible for cold adaptation, and single, double, or triple deletion mutants are not cold sensitive. On the other hand, a quadruple deletion mutant of *cspA*, *cspB*, *cspG*, and *cspE* exhibits cold sensitivity (Xia et al., 2001). Based on this observation, it was suggested that function(s) of CspA homologs is redundant and they compensate for each other. The other Csps that are significantly induced upon temperature downshift include CsdA (cold-shock domain A) (ribosome-associated RNA helicase) (Toone et al., 1991), RbfA (ribosome-binding factor) (Dammel and Noller, 1995), NusA (protein involved in transcription) (Friedman et al., 1984), and PNP (ribonuclease) (Donovan and Kushner, 1986). On the other hand, cold-shock induction of proteins such as RecA (recombination factor) (Walker, 1984), IF-2 (initiation factor) (Gualerzi and Pon, 1990), H-NS (nucleoid-associated, DNA-binding protein) (Dersch et al., 1994), GyrA (subunit of topoisomerase DNA gyrase) (Sugino et al., 1977), Hsc66 and HscB (presumed cold-shock molecular chaperones) (Lelivelt and Kawula, 1995), dihydrolipoamide transferase, and pyruvate dehydrogenase (Jones and Inouye, 1994) is moderate. Other cold-shock-inducible proteins include: (i) trigger factor (TF) that acts as a molecular chaperone and presumably helps protein synthesis and folding to continue at low temperature and may also help in refolding of cold-damaged proteins (Kandror and Goldberg, 1997), (ii) a ribosome-associated protein, pY, that inhibits translation at the elongation stage by blocking binding of aminoacyl-tRNA to the ribosomal A site (Agafonov et al., 2001), and (iii) trehalose-6-phosphate synthase (OtsA) and trehalose-6-phosphate phosphatase (OtsB) involved in the synthesis of trehalose that probably has protective effect during cold shock (Kandror et al., 2002).

The CspA homologs are reported from a number of other mesophilic bacteria such as *Salmonella enterica* serovar Typhimurium (Craig et al., 1998), *Lactococcus lactis* (Chapot-Chartier et al., 1997), *Lactobacillus plantarum* (Derzelle et al., 2003), and *Enterococcus faecalis* (Panoff et al., 1997). Unlike *E. coli*, in case of some bacteria such as *L. lactis*, there is no lag period in cell growth after the temperature downshift. In case of *B. subtilis*, the growth continues at low temperature at a reduced doubling time without apparent growth lag. The three CspA homologs, CspB, CspC, and CspD, of *B. subtilis* were shown to be essential for efficient growth at optimal temperature, for efficient adaptation to low temperatures, and for survival during stationary phase (Graumann et al., 1997). In addition to these, other proteins such as CheY (chemotaxis), Hpr (sugar uptake), ribosomal proteins S6 and L7/L12 (translation), peptidyl propyl *cis/trans* isomerase (protein folding), cysteine synthase, ketol acid reductoisomerase, glyceraldehyde dehydrogenase, and triosephosphate isomerase (general metabolism) were induced by temperature downshift in this bacterium (Graumann et al., 1996; Graumann and Marahiel, 1999).

CspA FAMILY OF *E. COLI*

The cold-shock induction of CspA and its homologs is presumed to be regulated at the levels of transcription, mRNA stability, and translation and does not need any additional transcription factors.

Transcription

There is a divided opinion regarding the contribution of cold-shock transcription of *cspA* in its dramatic induction upon cold shock (Gualerzi et al., 2003). The various observations reported regarding this aspect are listed here. (i) Cold-shock induction of *cspA* does not need additional transcription factors. (ii) Moderate (approximately threefold) increase in activity was seen at 15°C by studying the expression profiles of reporter genes such as *lacZ* and *cat* fused to the *cspA* promoter (Goldenberg et al., 1997; Mitta et al., 1997). (iii) Increase in *cspA* transcript (four- to fivefold) upon cold shock was seen by primer extension, Northern blot analysis, and DNA microarray analysis of the global transcript profile of cold-shocked cells (Phadtare and Inouye, 2004). (iv) *cspA* can be induced at 15°C independently of its promoter; however, cold-shock induction of *cspA* from a heterologous promoter is significantly less than that from *cspA* promoter. (v) Cold-shock-inducible CspA homologs contain an unusually long 5'-untranslated region (5'-UTR), with a highly conserved, 11-base sequence termed the "cold box." It is presumed that this is a transcriptional pausing site that is bypassed by RNA polymerase immediately upon temperature downshift, but at higher concentration of CspA, it binds to its own mRNA to destabilize the elongation complex and thus regulates its own expression. Thus, overproduction of 5'-UTR leads to prolonged synthesis of CspA, an effect suppressed by the coproduction of CspA (Fang et al., 1998; Jiang et al., 1996). (vi) *cspA* transcription is enhanced by an AT-rich sequence (UP element) immediately upstream of the −35 region (Goldenberg et al., 1997; Mitta et al., 1997); deletion of this region leads to diminished activity of the *cspA* promoter. (vii) *cspA* promoter region includes an extended −10 box (a TGn motif preceding the −10 box), another important factor that may contribute toward high promoter activity. It is reported that in the presence of this motif, the −35 region is dispensable. (viii) CspE negatively regulates the expression of CspA at the level of transcription at 37°C by binding to the cold box region and increasing the efficiency of pausing by RNA polymerase (Bae et al., 1999). (ix) Production of CspA is also observed at 37°C during early exponential growth phase, although only at one-sixth level as compared with its cold-shock induction level. This induction is presumed to be due to its position near *oriC* resulting in higher gene dosage effect, high concentration of its transcription activator Fis, and higher stability of its mRNA at this phase due to lower RNase activity (Brandi et al., 1999). Later, this production was shown to be due to nutritional upshift (Yamanaka and Inouye, 2001).

All these analyses could not distinguish if the increase in *cspA* transcript is due to cold-shock transcription alone or is a combined effect of transcription and stabilization of its mRNA upon cold shock. In order to address this question, recently studies were carried out to evaluate promoter activity of *cspA* during cold shock in a cell-free system. It was observed that: (i) the extended −10 motif is critical for high-level expression of *cspA*; however it does not contribute to its low-temperature expression, (ii) transcription from the *cspA* promoter is cold-sensitive and appears to play little or no role in low-temperature induction of *cspA* expression, (iii) presence of any cellular factor that may act as transcription enhancer for *cspA* in vivo was not detected (Phadtare and Severinov, 2005).

Stabilization of mRNA

The *cspA* mRNA is transiently and dramatically stabilized immediately following cold shock, which probably plays a vital role in its cold-shock production. With the help of deletion analysis of its 5'-UTR, it was shown that this region is responsible for the extreme instability of *cspA* mRNA at 37°C (half-life 12 s) and has a positive effect on its stabilization upon cold shock (half-life more than 20 min) (Mitta et al., 1997). The *cspA* promoter is active at 37°C, but CspA production is low at this temperature due to extreme instability of its mRNA. This was supported by an observation that artificial stabilization of *cspA* mRNA at 37°C achieved by introducing a three-base substitution mutation within the 159-base 5'-UTR that disrupts the target of degradation by RNase E results in constitutive expression of *cspA* at 37°C (Fang et al., 1997). As mentioned above, recent data show that contribution of transcription in cold-shock induction of *cspA* is marginal, and stabilization of its mRNA probably is a major factor that leads to its dramatic induction at low temperature (Phadtare and Severinov, 2005).

Translation

The cold-shock induction of *cspA* is also regulated at the level of translation. The mRNAs for *cspA*, *cspB*, *cspG*, *cspI*, *csdA*, and *rbfA* contain a sequence located 14-bases downstream of their initiation codons. This element is presumed to enhance translation initiation in cold-shock mRNAs and is called translation-enhancing element (TEE, originally termed as downstream box). With the help of toeprinting experiments and polysome profile analysis of the fusion constructs of *lacZ*, it was shown that A/T richness of this region plays a significant role in *E. coli* gene expression (Qing et al., 2004). As this sequence is complementary to a region in the penultimate stem of 16S rRNA, it was initially thought to enhance translation initiation by facilitating the formation of translation preinitiation complex through binding to

16S rRNA; however this view is now disputed, and the exact mechanism of the enhancing effect on translation initiation by TEE is unknown at present (Mitta et al., 1997; Moll et al., 2001).

Structure of CspA

Rapid and two-state folding seem to be intrinsic properties of CspA and its homologs (Jacob et al., 1997; Perl et al., 1998; Reid et al., 1998). The three-dimensional structure of CspA from *E. coli* and CspB from *B. subtilis* has been resolved by X-ray crystallography and nuclear magnetic resonance analysis (Feng et al., 1998; Newkirk et al., 1994; Schindelin et al., 1994, 1993; Schnuchel et al., 1993). The protein consists of five antiparallel β-strands (β1 to β5) that form a β-barrel structure with two β-sheets. The two putative RNA-binding motifs, RNP1 and RNP2, are located on the β2 and β3 strands, respectively. The surface of CspA has a surface patch of aromatic residues (RNP1 W^{11}, F^{18} and F^{20}, and RNP2 F^{31}, H^{33}, and F^{34}) that are involved in the hydrophilic interactions between protein and nucleic acid. Presumably, the aromatic patch residues contribute to nucleic acid binding by intercalating between DNA or RNA bases. Mutations of the three phenylalanine residues from the aromatic cluster adversely affected the RNA binding in the case of CspA from *E. coli* (Hillier et al., 1998). Similarly, in the case of CspB from *B. subtilis*, the nucleic acid binding as well as the protein stability was abolished by the mutations in the two RNP sites (Schindler et al., 1998; Schroder et al., 1995).

Functions of CspA Homologs

A number of functions have been attributed to CspA and its homologs. In the mesophilic lactic acid bacterium *L. plantarum*, overproduction of its three Csps CspL, CspP, and CspC leads to improved adaptation to cold shock, stationary phase, and freezing stresses (Derzelle et al., 2003). A cold-shock like gene in *Streptomyces* has possible effects on its antibiotic biosynthesis (Martinez-Costa et al., 2003). In *Staphylococcus aureus*, CspA is involved in susceptibility of the organism to an antimicrobial peptide derived from human neutrophil cathepsin G. A disruption mutation in *cspA* decreased the capacity of *S. aureus* to respond to cold-shock stress and increased resistance to this antimicrobial peptide. It was suggested that genes regulated by CspA encode proteins that influence the susceptibility of this organism to the antimicrobial peptide (Katzif et al., 2003). In *Shewanella putrefaciens*, an alkyl hydroperoxide reductase (AhpC) is induced by both osmotic stress and cold shock (Leblanc et al., 2003). Cold-shock pretreatment of sporulating *B. subtilis* cells increases the heat resistance of the spores formed from these cells (Movahedi and Waites, 2002). The activity of sigma B in *L. monocytogenes* is stimulated by high osmolarity and is necessary for efficient uptake of osmoprotectants. It was shown that during cold shock, sigma B contributes to adaptation in a growth phase-dependent manner and is necessary for efficient accumulation of betaine and carnitine as cryoprotectants (Becker et al., 2000). CspE influences UV sensitivity in *E. coli* K-12 cells (Mangoli et al., 2001). CspC and CspE from *E. coli* regulate the expression of number of proteins such as OsmY, Dps, ProP, and KatG. OsmY is induced in response to osmotic stress and upon stationary phase, Dps is induced by osmotic, oxidative stress and upon stationary phase, ProP is induced by osmotic stress, and KatG is induced by oxidative stress. These proteins are regulated by RpoS, and there is a possibility that CspC and CspE may regulate these through RpoS. CspE and CspC also regulate the expression of universal protein A, UspA, a protein responding to numerous stresses (Phadtare and Inouye, 2001). CspE has also shown to be involved in various functions such as camphor resistance and chromosome condensation (Hu et al., 1996; Sand et al., 2003), downregulation of λ Q-mediated transcription antitermination (Hanna and Liu, 1998), and downregulation of poly(A)-mediated 3′ to 5′ exonucleolytic decay by PNPase (Feng et al., 2001). CspD in *E. coli* is induced by starvation and upon stationary phase (Yamanaka and Inouye, 1997). It inhibits the initiation and elongation steps of minichromosome replication in vitro and is presumed to function as a novel inhibitor of DNA replication and to play a regulatory role in chromosomal replication in nutrient-depleted cells (Yamanaka et al., 2001). Recently, in the case of *Pseudomonas putida*, it was shown that CspA plays a role in tolerance to solvents such as toluene (Segura et al., 2005). Even though primary function of CspA homologs may be to counteract the detrimental effect of cold shock, they seem to be involved in a number of other stresses as well. This suggests that regulation and functions of these proteins are complex.

Function of CspA homologs as RNA chaperones has been studied in detail (Bae et al., 2000; Jiang et al., 1997; Phadtare et al., 2002). One of the effects of low temperatures is the stabilization of the secondary structures of RNA. This would hinder both transcription elongation and ribosomal movement on RNA and thus translation. Based on the observations that CspA homologs: (i) can bind and melt secondary structures in nucleic acids, (ii) are induced significantly immediately upon cold shock, and (iii) are important for adaptation of cells to low temperature, it was speculated that CspA homologs function as

"RNA chaperones" by destabilizing the secondary structures in nucleic acids, thus facilitating transcription and translation at low temperature. The structure of CspA homologs is suitable for their role as RNA chaperones. These proteins have overall negative surface charge with a positively charged aromatic patch on the surface. After binding to RNA by virtue of stacking of the aromatic side chains with RNA bases, the approach of other RNA for intramolecular or intermolecular base pairing will be prevented by charge repulsion (Graumann and Marahiel, 1998). CspA from *E. coli* binds RNA and single-stranded (ss) DNA without sequence specificity and with low binding affinity (Jiang et al., 1997). CspB, CspC, and CspE from *E. coli* selectively bind RNA/ss DNA (the preferred sequences being UUUUU, AGGGAGGGA, and AU-rich regions, respectively), while CspB from *B. subtilis* binds to T-rich regions preferentially; however the magnitude of this selectivity is small (Lopez et al., 2001; Phadtare and Inouye, 1999). It has been shown that by virtue of the nucleic acid melting activity, certain CspA homologs act as transcription antiterminators and thus aid in cold acclimation of cells (Bae et al., 2000; Phadtare et al., 2002). CspE, a prototypical member of this family, was used to analyze these activities. CspE caused transcription antitermination at *rho*-independent terminators for several genes. Mutants of CspE that fully retain their nucleic acid-binding activity, but have decreased nucleic acid melting activity, lose their ability to cause transcription antitermination and cold acclimation of cells (Phadtare et al., 2002).

PROTECTIVE EFFECT OF SUGARS DURING COLD SHOCK

Trehalose is known to exert protection to cells undergoing heat, osmotic stress, and cold adaptation (Kandror et al., 2002). Mutant cells deficient in trehalose production show reduced viability at 4°C, which can be complemented by genes involved in trehalose synthesis (*otsA* and *otsB*). Cellular levels of trehalose increase significantly (eightfold) upon temperature downshift from 37 to 16°C. Although the mechanism of protection is not known at present, it is suggested that trehalose prevents denaturation and aggregation of proteins, protects against oxidative damage, functions by acting as a free radical scavenger, and stabilizes cellular membrane (Kandror et al., 2002). It has also been shown that trehalose stabilizes a cold-adapted protease by preventing its autolysis (Pan et al., 2005). The protective effect is also seen in higher organisms such as *Drosophila*, *Caenorhabditis elegans*, and yeasts.

Another sugar, maltose, was also shown to accumulate and confer protection during cold stress in plants (Kaplan and Guy, 2004). Recent global transcript profiling of *E. coli* cells undergoing cold shock showed that several genes encoding proteins involved in sugar transport and metabolism were induced by cold shock (Phadtare and Inouye, 2004). These sugars include maltose along with mannose, ribose, xylose, and trehalose. It is interesting that cold-shock induction of mannose and maltose transport systems was prominently repressed in a cold-sensitive *csp* quadruple deletion mutant that has significantly prolonged (4 h) lag period as opposed to 1-h lag period of the wild-type strain (Xia et al., 2001) at 15°C.

PROTEIN FOLDING AT LOW TEMPERATURE

Recent reports suggest that proper folding of proteins as well as refolding of cold-damaged proteins is important after cold shock (Kandror and Goldberg, 1997). Global transcript profiling of *E. coli* cells undergoing cold shock showed that certain molecular chaperones were induced upon cold shock. These include caseinolytic proteases (Clps), TF, and GroEL and GroES (Phadtare and Inouye, 2004). Cold-shock chaperone role has been reported for the Clps in cyanobacteria and for TF in *E. coli*. It has been suggested that ClpB may renature and solubilize aggregated proteins at low temperatures at which translation is repressed (Porankiewicz and Clarke, 1997). On the other hand, TF (peptidyl prolyl isomerase) catalyzes the *cis/trans* isomerization of peptide bonds N-terminal to the proline residue. It is moderately induced upon cold shock and is important for cellular viability during storage at 4°C. TF is presumed to help protein synthesis and folding to continue at low temperature (Kandror and Goldberg, 1997), accelerate proline-limited steps in protein folding and associate with ribosomes, and help folding of newly formed protein chains (Maier et al., 2003). Transient high levels of GroELS was also seen at 37°C in *E. coli* cells overproducing CspC and CspE (Phadtare and Inouye, 2001); thus it needs to be seen if transient induction of these chaperones is significant for cold shock or is a mere consequence of high CspA homolog levels immediately following cold shock.

COLD-SHOCK RESPONSE OF COLD-ADAPTED BACTERIA

Cold-adapted bacteria are either psychrophilic or psychrotrophic. Both are capable of growing at or near zero temperatures, but the optimum and upper temperature limits for growth are lower for psychrophiles

Table 1. Comparison between cold-shock response of *E. coli* and psychrotrophs

E. coli	Psychrotrophs
Synthesis of housekeeping genes is repressed immediately upon temperature downshift	Synthesis of housekeeping genes is not repressed at low temperature
Few cold-shock proteins are present	Large number of cold-shock proteins are present
CspA homologs are induced at significant levels	Cold-shock proteins are induced at moderate levels
Cold-shock induction of CspA homologs is regulated at various levels	mRNAs for cold-shock proteins preexist—mostly post-translational regulation is responsible for their cold-shock induction
Cold-shock induction of CspA homologs is transient	Cold-shock proteins persist during growth at low temperature
Does not produce cold acclimation proteins (Caps) during continued growth at low temperature	Caps are produced during continued growth at low temperature

compared with psychrotrophs. Psychrophiles often inhabit permanently cold habitats, whereas psychrotrophs are predominantly in the environments that experience thermal fluctuations. Cold-shock response of cold-adapted bacteria is similar to that of mesophiles in aspects such as in many cases a lag phase of growth precedes acclimation to low temperature, specific proteins are induced by temperature downshift, membranes undergo adaptive changes, and enzymes are adapted to function at low temperature. On the other hand, unlike mesophiles, cold-adapted bacteria are capable of growing over a wide range of temperature including near or even below freezing; thus it is expected that specific mechanisms exist in the cold acclimation of cold-adapted bacteria to cope with these extreme low temperatures, which are absent in mesophiles. One of the main differences in the cold-shock response of these two types of bacteria is the presence of cold acclimation proteins (Caps) in cold-adapted bacteria. Mesophiles with the exception of *E. faecalis* do not produce Caps in response to continuous growth at low temperature (Panoff et al., 1997). Table 1 lists the major differences in the cold-shock response of *E. coli* and psychrotrophs (Berger et al., 1997; Hebraud and Potier, 1999).

As mentioned above, for survival and growth at cold temperature, cold-adapted bacteria have evolved a number of adaptations to their cellular components, including membranes, energy-generating and protein synthesis machinery, biodegradative enzymes, and the components responsible for nutrient uptake (Russell, 1998). The details of membrane adaptations and cold-adapted enzymes are described separately in this book and will be only briefly mentioned here.

MEMBRANE-ASSOCIATED CHANGES

The changes occurring in the membrane composition in these bacteria are basically the same as those observed in mesophilic bacteria and involve increasing the ratio of unsaturated fatty acyl residues, *cis* double bonds, chain shortening, and methyl branching. These changes are mediated by modification of preexisting lipids by cold-shock-activated enzymes and by de novo synthesis of specific enzymes upon cold shock (Russell, 1990).

COLD-ADAPTED ENZYMES

Unlike mesophiles, structural and metabolic proteins of psychrotrophs are functional at low temperature and are active over a wide range of temperatures, and the low-temperature growth also involves the synthesis of new proteins upon cold shock. Compared to their mesophilic counterparts, enzymes from psychrotrophic bacteria are more thermolabile but are more active at low temperature (Hebraud and Potier, 1999). Flexibility of proteins with respect to temperature is accomplished by reduction of electrostatic noncovalent weak interactions and reduced hydrophobicity (Feller and Gerday, 1997). Recently, the crystal structure of a subtilisin-like serine proteinase from the psychrotrophic marine bacterium, *Vibrio* sp., was resolved. Comparison of this protease with its mesophilic and thermophilic homologs proteinase K and thermitase showed that the cold-adapted enzyme has more of its apolar surface exposed. In addition, it has a strong anionic character arising from the high occurrence of uncompensated negatively charged residues at its surface. The data also suggested a dual role of electrostatic interactions in the adaptation of enzymes to both high and low temperatures (Arnorsdottir et al., 2005).

TRANSLATION

Psychrotrophic bacteria, similar to *E. coli*, exhibit a lag period of growth immediately following the cold shock; however, in the latter, this phase is characterized by the inhibition of general protein synthesis, while in psychrotrophs, the synthesis of housekeeping proteins continues after the temperature downshift. It has been suggested that the initiation of protein synthesis is more cold tolerant in psychrotrophic bacteria such as *Pseudomonas fluorescens* than that in *E. coli* (Broeze et al., 1978).

ADENYLATE METABOLISM

Recently, it has been shown that in case of mesophiles and thermophiles, rapid loss of ATP occurs upon cold shock. On the other hand, in case of cold-loving bacteria such as *Pseudomonas*, ATP levels increase (approximately 20 to 50%) upon cold shock, even at sub-zero temperatures (Napolitano and Shain, 2005). These results suggest that unlike mesophiles and thermophiles, in case of psychrotrophs, elevated adenylate nucleotides may compensate for maintaining biochemical processes at low temperature, and key enzymes such as F_1 ATP synthase and AMP phosphatase/deaminase involved in adenylate metabolism may be implicated. It was also observed that cold shock-induced increase in ATP was maintained for a prolonged period of time and was followed by corresponding increase in ADP and ATP after approximately 24 h, leading to continued increase in the adenylate pool size.

COLD-SHOCK PROTEINS OF PSYCHROTROPHIC BACTERIA

One of the important features of cold-shock response of mesophiles, thermophiles, psychrophiles, and psychrotrophs is the production of Csps. There are distinct differences (Hebraud and Potier, 1999) in the level, number, and amount of Csps produced in mesophiles and cold-loving bacteria (Table 1). CspA homologs are found in a number of psychrotrophic bacteria; however these did not reveal the presence of a specific domain that can differentiate them from their mesophilic counterparts. Csps of few well-studied psychrotrophic bacteria are described below.

Arthrobacter globiformis

A. globiformis is a psychrotrophic bacterium capable of growth between −5 and +32°C. Upon transfer from 25 to 4°C, this bacterium exhibits a 14-h lag for cell growth, followed by growth with a generation time of 25 h (Berger et al., 1997). Proteomic analysis of its cold-shock response revealed that four types of proteins are produced upon temperature downshift: (i) Csps that are overexpressed only after a large temperature downshift, (ii) Csps with optimal expression after mild shocks, (iii) proteins overexpressed irrespective of magnitude of temperature downshift—these were considered to be early Caps, and (iv) proteins which were present at high concentrations only in 4°C steady-state cells and appeared to be late Caps (Berger et al., 1996). A gene termed *capA* that is homologous to *E. coli cspA* was identified in this bacterium (Berger et al., 1997). Proteins encoded by both genes showed a high degree of sequence identity, and a particular residue or domain that may be significant for cold-adapted bacteria was not detected. CapA was induced immediately upon temperature downshift and was still expressed during prolonged growth at low temperature. The mRNA for *capA* preexists in the cells prior to cold shock, and its cold-shock induction is regulated at the level of translation. It plays a crucial role in cold adaptation of this bacterium.

Pseudomonas fragi

P. fragi is a psychrotrophic bacterium capable of growing over a wide range of temperature (from 2 to 35°C). Study of its cold-shock response is important, as it is the main food-spoiling bacterium in refrigerated foods. Its cold-shock response (after temperature downshift from 30 or 20°C to 5°C) includes induction of Csps that are transiently induced and Caps that show prolonged induction (Michel et al., 1997). Among the cold-induced proteins, four low-molecular-weight proteins are prominent. These include Caps—CapA and CapB and TapA and TapB. These are homologs of *E. coli* CspA. Unlike CapA and CapB, TapA and TapB are transiently expressed. CapA and CapB are optimally expressed at low temperatures (4 to 10°C), while TapA and TapB are optimally produced at 25 to 30°C and are also heat-shock inducible. Similar to *Arthrobacter*, persistent expression of Caps at low temperature presumably plays a critical role in cold adaptation of *Pseudomonas* (Fig. 1).

Aeromonas hydrophila

Aeromonas is a psychrotrophic food-spoilage bacterium. Upon temperature downshift from 30 to 5°C, levels of very few proteins were altered in this bacterium (Imbert and Gancel, 2004). Most housekeeping proteins were similarly expressed at these two temperatures. Its cold-shock proteins are of two types, Csps and Caps. Interestingly, no CspA homolog was found in this bacterium. It was speculated that in case of *Vibrio vulnificus*, lack of a CspA homolog is related to the ability of this bacterium to enter into a viable but not culturable state (VBNC) when the temperature drops below 15°C (Carroll et al., 2001). A similar

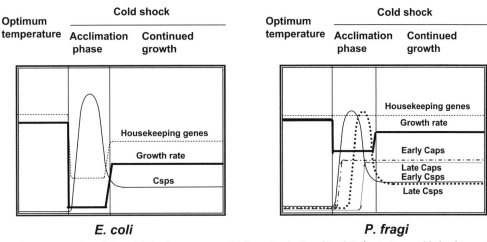

Figure 1. Schematic representation of the cold-shock response and adaptation in *E. coli* and *P. fragi*. Csps, cold-shock proteins; Caps, cold acclimation proteins.

assumption was made for *Aeromonas*, although it can grow at 5°C.

COLD-SHOCK RESPONSE OF OTHER BACTERIA

Cold-shock response has been studied from other groups of bacteria as well.

Thermophilic Bacteria

Thermotoga maritima is a hyperthermophilic bacterium, named so as it has an optimum growth temperature of 80°C. No cold-shock response has been demonstrated for *T. maritima*; however, the presence of CspA homologs [*Tm*CspB and *Tm*CspL (Nelson et al., 1999)] that are capable of performing functions of their *E. coli* counterparts (Phadtare et al., 2003) and the presence of 5'-UTR in their mRNA, which is essential for the regulation of expression upon cold shock, suggest that this organism may exhibit cold-shock response if subjected to temperature downshift of adequate magnitude. CspA homologs present in this bacterium may play an important role in cold acclimation of cells. *Tm*CspB and *Tm*CspL are the most thermostable CspA homologs known to date with melting temperatures above 80°C (Wassenberg et al., 1999). *Tm*CspB has been extensively studied (Frankenberg et al., 1999; Kremer et al., 2001; Martin et al., 2001; Perl and Schmid, 2001; Perl et al., 1998; Wassenberg et al., 1999; Welker et al., 1999) and is shown to have a β-barrel structure similar to that of *E. coli* CspA. Its thermostability is presumed to be due to the presence of an Arg residue in the penultimate N-terminal position that forms an ion cluster with several centrally and C-terminally located residues (Kremer et al., 2001).

Recently, a proteomic study of cold-shock response of a thermophilic bacterium *B. stearothermophilus* after downshift in the temperature from 65°C to 37 and 25°C has been reported (Topanurak et al., 2005). Eight cold-inducible proteins such as glucosyltransferase, anti-sigma B factor, Mrp protein homolog, dihydroorotase, hypothetical transcriptional regulator in FeuA-SigW intergenic region, RibT protein, phosphoadenosine phosphosulfate reductase, and prespore-specific transcriptional activator RsfA showed significant change in their levels. Interestingly, six of these proteins are correlated with the signal transduction pathway of bacterial sporulation.

Cyanobacteria

The cold-shock response machinery of cyanobacteria is different from that of *E. coli*. The two main differences are: (i) the absence of CspA homologs and (ii) the presence of desaturases. In fact, desaturases are present only in photosynthetic bacteria with the exception of *Bacillus*. These photosynthetic bacteria lack CspA homologs, but instead have cold-inducible, RNA-binding proteins (Rbps) (Sato, 1995). These proteins belong to the RNA-binding domain family and are structurally distinct from the Csd family of proteins. In addition to these, cold shock also elicits induction of CrhC, an RNA helicase, Clp proteins, and S21 protein in cyanobacteria (Chamot et al., 1999; Chamot and Owttrim, 2000; Los and Murata, 1999; Schelin et al., 2002). *Synechocystis* has 41 histidine kinases, and one of these, Hik33, is suggested to act as a cold sensor by sensing the rigidification of membrane lipids (Suzuki et al., 2000). Mutations in the gene encoding Hik33 diminished the extent of induction of several cold-inducible genes, such as *desD* encoding a desaturase and *crh* encoding an

RNA helicase. This suggested a role of Hik33 in perceiving and transducing cold signal to regulate the expression of cold-inducible genes.

Desaturases play an important role in cold-shock response of cyanobacteria. They regulate the change in membrane composition in response to temperature downshift to maintain the membrane-related functions. These enzymes introduce double bonds into fatty acids that have been esterified to glycerolipids and are bound to the thylakoid membrane in these bacteria (Murata and Wada, 1995). Cold-shock induction of two of the *Synechococcus* desaturases encoded by *desA* and *desB* is closely regulated by synthesis and stabilization of their mRNAs (Sakamoto and Bryant, 1997). In cyanobacteria, desaturation of lipids is connected to acclimatization of photosynthetic activity at low temperature; therefore desaturases are essential for the cold-shock adaptation of these bacteria. These enzymes are also important from the commercial point of view as introduction of the *desC* gene for the acyl-lipid desaturase from the thermophilic cyanobacterium *Synechococcus vulcanus* into *Nicotiana tabacum* significantly increased lipid content, the extent of fatty acid unsaturation, and importantly, chilling tolerance of these plants (Ishizaki-Nishizawa et al., 1996; Orlova et al., 2003).

GLOBAL TRANSCRIPT PROFILING OF COLD-SHOCK RESPONSE OF VARIOUS BACTERIA

With the advent of DNA microarray technology, several groups have carried out global transcript profiling of cold-shock response of different bacteria. Array technology is perhaps the most advanced and comprehensive technique available that allows the analysis of the entire complement of expressed genes in a cell. Its advantages include quantitative gene expression as measured by mRNA abundance in parallel expression profiles and rapid determination of complete gene-expression patterns. Our analysis of cold-shock response of *E. coli* revealed that in addition to the known cold-shock-inducible genes, new genes such as those encoding proteins involved in sugar transport and metabolism and remarkably genes encoding certain heat-shock proteins are induced by cold shock. In light of strong reduction in metabolic activity of the cell after temperature downshift, the induction of sugar metabolism machinery was unexpected. The deletion of four *csp*s (*cspA*, *cspB*, *cspG*, and *cspE*) affected cold-shock induction of mostly those genes that are transiently induced in the acclimation phase, emphasizing that CspA homologs are essential in the acclimation phase, which is relevant to the observation that CspA homologs function as RNA chaperones (Phadtare and Inouye, 2004). Analysis of cold-shock response of *E. coli* from Deho's group (Polissi et al., 2003) showed that extracytoplasmic stress response regulators *rpoE* and *rseA* were induced by cold shock. They also included a PNPase-deficient strain in their analyses and showed that PNPase both negatively and positively modulates the transcript abundance of certain genes and thus plays a complex role in controlling cold adaptation. Cold-shock response of *Yersinia pestis* (Han et al., 2005) also revealed upregulation of a specific set of genes whose protein products are designed to prevent or eliminate cold-induced DNA or RNA structuring, to remodel cell membrane components for maintenance of normal functions, to elevate the energy generation for ensuring ATP-dependent responses during cold adaptation, and to synthesize or transport compatible solutes such as cryoprotectants, and at the same time to cause repression of certain genes that encode major heat-shock proteins.

A similar analysis was carried out for *B. subtilis* by two groups (Beckering et al., 2002; Kaan et al., 2002). The analysis showed induction of known cold-shock-inducible genes such as CspB, CspC, and CspD. The strongest cold-inducible gene was *des* encoding fatty acid desaturase. However, the analysis also showed that the two-component system DesKR exclusively controls the desaturase gene *des* and is not the cold-triggered regulatory system of global relevance. Few genes encoding ribosomal proteins and genes belonging to the operon *ptb-bcd-buk-lpd-bkdA1-bkdA2-bkdB*, which encode enzymes involved in degradation of branched chain amino acid, were upregulated. Two new cold-inducible genes, the elongation factor homolog *ylaG* and the sigma (L)-dependent transcriptional activator homolog *yplP*, were also identified. Many of the repressed genes included those that encode enzymes involved in the biosynthesis of amino acids, nucleotides, and coenzymes, indicating metabolic adaptation of the cells to the decreased growth rate at the lower temperature.

Clarks' group studied the effect of cold shock on the hyperthermophilic methanarchaeon *Methanococcus jannaschii* from its optimal growth temperature of 85 to 65°C (Boonyaratanakornkit et al., 2005). The gene encoding an RNA helicase and genes involved in transcription and translation, and proteases and transport proteins were upregulated. A gene that codes for an 18-kDa FKBP (FK 506-binding protein)-type PPIase (peptidylprolyl cis-trans isomerase), which may facilitate protein folding at low temperatures, was also upregulated. Cold-shock response was also studied in another hyperthermophilic archaeon *Pyrococcus furiosus* (Weinberg et al., 2005). The cold shock of temperature downshift from 95 to 72°C resulted in a 5-h lag phase. The DNA microarray analysis showed that cells undergo three very different responses at 72°C such as an early shock (1 to 2 h), a late shock (5 h), and an adapted response (occurring after many generations at 72°C). Genes unique to each of these responses were identified. These included proteins involved in translation, solute transport, amino acid biosynthesis, and tungsten and intermediary carbon metabolism, as well as numerous membrane-associated proteins, such as two major membrane proteins termed CipA (PF0190) and CipB (PF1408). The *Archaea* do not contain members of CspA homologs, instead they all contain homologs of CipA and CipB.

Recently, DNA microarray analysis was carried out to study cold shock response of a cyanobacterium *Synechocystis* sp. (Inaba et al., 2003). Cold inducibility of gene expression in this bacterium was enhanced by the rigidification of membrane lipids which was engineered by disruption of genes for fatty acid desaturases. Double mutation of the *desA* and *desD* genes encoding desaturases rigidified the plasma membrane of *Synechocystis* at physiological temperatures. DNA microarray analysis was thus performed to examine the effects of the membrane rigidification on the regulation of gene expression upon exposure of cells to cold and heat stress. The analysis revealed that cold-inducible genes are of three types based on the effects of the rigidification of membrane lipids: (i) genes whose expression was not induced by cold in wild-type cells but became strongly cold inducible upon rigidification of membrane lipids, for example, certain heat-shock genes, genes for subunits of the sulfate transport system, and the *hik34* gene for a histidine kinase, (ii) genes whose cold inducibility was moderately enhanced by the rigidification of membrane lipids; most of genes in this group encoded proteins of as yet unknown function, and (iii) genes whose cold inducibility was unaffected by the rigidification of membrane lipids, such as genes for an RNA helicase and an Rbp.

CONCLUSION AND PERSPECTIVE

Study of cold-shock response from mesophilic, thermophilic, and cold-loving bacteria has shown that although there are certain common features that are present across all these groups, each group distinctively maintains unique features that allow its growth at a particular temperature. Study of these features is also enlightening from evolution point of view. The cold-shock response is executed at various levels such as cell membrane, transcription, translation, and metabolism. Csps play a key role in dealing with the initial detrimental effect of cold shock and maintaining the continued growth of the organism at low temperature. Study of these aspects of cold-shock response has provided insights into the general principles underlying cellular functions including those of RNA, membranes, and metabolism. Recently, the study of cold-shock response has acquired a new perspective. Many of the features that are found to be critical for cold-shock response and adaptation of bacteria are now being exploited for biotechnological applications. Cellular events occurring during cold-shock response are used in applications such as in food and agricultural industry and in research. Integration of structural elements (such as TEE) of the *E. coli* major cold-shock protein CspA into vectors allows expression of high level of proteins at low temperature with improved solubility and stability. The structural elements from cold-loving bacteria are fast gaining importance in biotechnology such as use of cold-active enzymes in the detergent and food industries, in dairy industry to decrease the lactose content of milk, and in the manufacture of ice-cream, dietary supplements in the form of polyunsaturated fatty acids from certain Antarctic marine psychrophiles. It can be thus safely speculated that with the new exciting discoveries, the study of cold-shock response and adaptation is rapidly gaining importance in both basic research and biotechnology.

REFERENCES

Agafonov, D. E., V. A. Kolb, and A. S. Spirin. 2001. Ribosome-associated protein that inhibits translation at the aminoacyl-tRNA binding stage. *EMBO Rep.* **2:**399–402.

Aguilar, P. S., J. E. Cronan, Jr., and D. de Mendoza. 1998. A *Bacillus subtilis* gene induced by cold shock encodes a membrane phospholipid desaturase. *J. Bacteriol.* **180:**2194–2200.

Aguilar, P. S., A. M. Hernandez-Arriaga, L. E. Cybulski, A. C. Erazo, and D. de Mendoza. 2001. Molecular basis of thermosensing: a two-component signal transduction thermometer in *Bacillus subtilis*. *EMBO J.* **20:**1681–1691.

Annous, B. A., L. A. Becker, D. O. Bayles, D. P. Labeda, and B. J. Wilkinson. 1997. Critical role of anteiso-C15:0 fatty acid in the growth of *Listeria monocytogenes* at low temperatures. *Appl. Environ. Microbiol.* **63:**3887–3894.

Arnorsdottir, J., M. M. Kristjansson, and R. Ficner. 2005. Crystal structure of a subtilisin-like serine proteinase from a psychrotrophic *Vibrio* species reveals structural aspects of cold adaptation. *FEBS. J.* **272:**832–845.

Bae, W., S. Phadtare, K. Severinov, and M. Inouye. 1999. Characterization of *Escherichia coli cspE*, whose product negatively regulates transcription of *cspA*, the gene for the major cold shock protein. *Mol. Microbiol.* **31:**1429–1441.

Bae, W., B. Xia, M. Inouye, and K. Severinov. 2000. *Escherichia coli* CspA-family RNA chaperones are transcription antiterminators. *Proc. Natl. Acad. Sci. USA* **97:**7784–7789.

Becker, L. A., S. N. Evans, R. W. Hutkins, and A. K. Benson. 2000. Role of sigma(B) in adaptation of *Listeria monocytogenes* to growth at low temperature. *J. Bacteriol.* **182:**7083–7087.

Beckering, C. L., L. Steil, M. H. Weber, U. Volker, and M. A. Marahiel. 2002. Genomewide transcriptional analysis of the cold shock response in *Bacillus subtilis*. *J. Bacteriol.* **184:**6395–6402.

Berger, F., N. Morellet, F. Menu, and P. Potier. 1996. Cold shock and cold acclimation proteins in the psychrotrophic bacterium *Arthrobacter globiformis* SI55. *J. Bacteriol.* **178:**2999–3007.

Berger, F., P. Normand, and P. Potier. 1997. capA, a cspA-like gene that encodes a cold acclimation protein in the psychrotrophic bacterium *Arthrobacter globiformis* SI55. *J. Bacteriol.* **179:**5670–5676.

Boonyaratanakornkit, B. B., A. J. Simpson, T. A. Whitehead, C. M. Fraser, N. M. El-Sayed, and D. S. Clark. 2005. Transcriptional profiling of the hyperthermophilic methanarchaeon *Methanococcus jannaschii* in response to lethal heat and nonlethal cold shock. *Environ. Microbiol.* **7:**789–797.

Brandi, A., R. Spurio, C. O. Gualerzi, and C. L. Pon. 1999. Massive presence of the *Escherichia coli* 'major cold-shock protein' CspA under non-stress conditions. *EMBO J.* **18:**1653–1659.

Broeze, R. J., C. J. Solomon, and D. H. Pope. 1978. Effects of low temperature on in vivo and in vitro protein synthesis in *Escherichia coli* and *Pseudomonas fluorescens*. *J. Bacteriol.* **134:**861–874.

Carroll, J. W., M. C. Mateescu, K. Chava, R. R. Colwell, and A. K. Bej. 2001. Response and tolerance of toxigenic *Vibrio cholerae* O1 to cold temperatures. *Antonie Van Leeuwenhoek* **79:**377–384.

Chamot, D., W. C. Magee, E. Yu, and G. W. Owttrim. 1999. A cold shock-induced cyanobacterial RNA helicase. *J. Bacteriol.* **181:**1728–1732.

Chamot, D., and G. W. Owttrim. 2000. Regulation of cold shock-induced RNA helicase gene expression in the Cyanobacterium *anabaena* sp. strain PCC 7120. *J. Bacteriol.* **182:**1251–1256.

Chapot-Chartier, M. P., C. Schouler, A. S. Lepeuple, J. C. Gripon, and M. C. Chopin. 1997. Characterization of cspB, a coldshock-inducible gene from *Lactococcus lactis*, and evidence for a family of genes homologous to the *Escherichia coli* cspA major cold shock gene. *J. Bacteriol.* **179:**5589–5593.

Craig, J. E., D. Boyle, K. P. Francis, and M. P. Gallagher. 1998. Expression of the cold-shock gene cspB in *Salmonella typhimurium* occurs below a threshold temperature. *Microbiology* **144:**697–704.

Dammel, C. S., and H. F. Noller. 1995. Suppression of a coldsensitive mutation in 16S rRNA by overexpression of a novel ribosome-binding factor, RbfA. *Genes Dev.* **9:**626–637.

Dersch, P., S. Kneip, and E. Bremer. 1994. The nucleoid-associated DNA-binding protein H-NS is required for the efficient adaptation of *Escherichia coli* K-12 to a cold environment. *Mol. Gen. Genet.* **245:**255–259.

Derzelle, S., B. Hallet, T. Ferain, J. Delcour, and P. Hols. 2003. Improved adaptation to cold-shock, stationary-phase, and freezing stresses in *Lactobacillus plantarum* overproducing cold-shock proteins. *Appl. Environ. Microbiol.* **69:**4285–4290.

Donovan, W. P., and S. R. Kushner. 1986. Polynucleotide phosphorylase and ribonuclease II are required for cell viability and mRNA turnover in *Escherichia coli* K-12. *Proc. Natl. Acad. Sci. USA* **83:**120–124.

Ermolenko, D. N., and G. I. Makhatadze. 2002. Bacterial coldshock proteins. *Cell. Mol. Life Sci.* **59:**1902–1913.

Fang, L., Y. Hou, and M. Inouye. 1998. Role of the cold-box region in the 5' untranslated region of the cspA mRNA in its transient expression at low temperature in *Escherichia coli*. *J. Bacteriol.* **180:**90–95.

Fang, L., W. Jiang, W. Bae, and M. Inouye. 1997. Promoterindependent cold-shock induction of cspA and its derepression at 37 degrees C by mRNA stabilization. *Mol. Microbiol.* **23:**355–364.

Feller, G., and C. Gerday. 1997. Psychrophilic enzymes: molecular basis of cold adaptation. *Cell. Mol. Life Sci.* **53:**830–841.

Feng, W., R. Tejero, D. E. Zimmerman, M. Inouye, and G. T. Montelione. 1998. Solution NMR structure and backbone dynamics of the major cold-shock protein (CspA) from *Escherichia coli*: evidence for conformational dynamics in the single-stranded RNA-binding site. *Biochemistry* **37:**10881–10896.

Feng, Y., H. Huang, J. Liao, and S. N. Cohen. 2001. *Escherichia coli* poly(A)-binding proteins that interact with components of degradosomes or impede RNA decay mediated by polynucleotide phosphorylase and RNase E. *J. Biol. Chem.* **276:**31651–31656.

Frankenberg, N., C. Welker, and R. Jaenicke. 1999. Does the elimination of ion pairs affect the thermal stability of cold shock protein from the hyperthermophilic bacterium *Thermotoga maritima*? *FEBS Lett.* **454:**299–302.

Friedman, D. I., E. R. Olson, C. Georgopoulos, K. Tilly, I. Herskowitz, and F. Banuett. 1984. Interactions of bacteriophage and host macromolecules in the growth of bacteriophage lambda. *Microbiol. Rev.* **48:**299–325.

Garwin, J. L., and J. E. Cronan, Jr. 1980. Thermal modulation of fatty acid synthesis in *Escherichia coli* does not involve de novo enzyme synthesis. *J. Bacteriol.* **141:**1457–1459.

Garwin, J. L., A. L. Klages, and J. E. Cronan, Jr. 1980. Betaketoacyl-acyl carrier protein synthase II of *Escherichia coli*. Evidence for function in the thermal regulation of fatty acid synthesis. *J. Biol. Chem.* **255:**3263–3265.

Goldenberg, D., I. Azar, A. B. Oppenheim, A. Brandi, C. L. Pon, and C. O. Gualerzi. 1997. Role of *Escherichia coli* cspA promoter sequences and adaptation of translational apparatus in the cold shock response. *Mol. Gen. Genet.* **256:**282–290.

Goldstein, J., N. S. Pollitt, and M. Inouye. 1990. Major cold shock protein of *Escherichia coli*. *Proc. Natl. Acad. Sci. USA* **87:**283–287.

Graumann, P., K. Schroder, R. Schmid, and M. A. Marahiel. 1996. Cold shock stress-induced proteins in *Bacillus subtilis*. *J. Bacteriol.* **178:**4611–4619.

Graumann, P., T. M. Wendrich, M. H. Weber, K. Schroder, and M. A. Marahiel. 1997. A family of cold shock proteins in *Bacillus subtilis* is essential for cellular growth and for efficient protein synthesis at optimal and low temperatures. *Mol. Microbiol.* **25:**741–756.

Graumann, P. L., and M. A. Marahiel. 1998. A superfamily of proteins that contain the cold-shock domain. *Trends Biochem. Sci.* **23:**286–290.

Graumann, P. L., and M. A. Marahiel. 1999. Cold shock response in *Bacillus subtilis*. *J. Mol. Microbiol. Biotechnol.* **1:**203–209.

Gualerzi, C. O., A. M. Giuliodori, and C. L. Pon. 2003. Transcriptional and post-transcriptional control of cold-shock genes. *J. Mol. Biol.* **331:**527–539.

Gualerzi, C. O., and C. L. Pon. 1990. Initiation of mRNA translation in prokaryotes. *Biochemistry* **29:**5881–5889.

Han, Y., D. Zhou, X. Pang, L. Zhang, Y. Song, Z. Tong, J. Bao, E. Dai, J. Wang, Z. Guo, J. Zhai, Z. Du, X. Wang, J. Wang, P. Huang, and R. Yang. 2005. DNA microarray analysis of the heat- and cold-shock stimulons in *Yersinia pestis*. *Microbes Infect.* **7:**335–348.

Hanna, M. M., and K. Liu. 1998. Nascent RNA in transcription complexes interacts with CspE, a small protein in *E. coli* implicated in chromatin condensation. *J. Mol. Biol.* **282:**227–239.

Hebraud, M., and P. Potier. 1999. Cold shock response and low temperature adaptation in psychrotrophic bacteria. *J. Mol. Microbiol. Biotechnol.* **1:**211–219.

Hillier, B. J., H. M. Rodriguez, and L. M. Gregoret. 1998. Coupling protein stability and protein function in *Escherichia coli* CspA. *Fold Des.* **3:**87–93.

Hu, K. H., E. Liu, K. Dean, M. Gingras, W. DeGraff, and N. J. Trun. 1996. Overproduction of three genes leads to camphor resistance and chromosome condensation in *Escherichia coli*. *Genetics* **143:**1521–1532.

Imbert, M., and F. Gancel. 2004. Effect of different temperature downshifts on protein synthesis by *Aeromonas hydrophila*. *Curr. Microbiol.* **49:**79–83.

Inaba, M., I. Suzuki, B. Szalontai, Y. Kanesaki, D. A. Los, H. Hayashi, and N. Murata. 2003. Gene-engineered rigidification of membrane lipids enhances the cold inducibility of gene expression in synechocystis. *J. Biol. Chem.* **278:**12191–12198.

Ishizaki-Nishizawa, O., T. Fujii, M. Azuma, K. Sekiguchi, N. Murata, T. Ohtani, and T. Toguri. 1996. Low-temperature resistance of higher plants is significantly enhanced by a nonspecific cyanobacterial desaturase. *Nat. Biotechnol.* **14:**1003–1006.

Jacob, M., T. Schindler, J. Balbach, and F. X. Schmid. 1997. Diffusion control in an elementary protein folding reaction. *Proc. Natl. Acad. Sci. USA* **94:**5622–5627.

Jiang, W., L. Fang, and M. Inouye. 1996. The role of the 5′-end untranslated region of the mRNA for CspA, the major cold-shock protein of *Escherichia coli*, in cold-shock adaptation. *J. Bacteriol.* **178:**4919–4925.

Jiang, W., Y. Hou, and M. Inouye. 1997. CspA, the major cold-shock protein of *Escherichia coli*, is an RNA chaperone. *J. Biol. Chem.* **272:**196–202.

Jones, P. G., and M. Inouye. 1994. The cold-shock response—a hot topic. *Mol. Microbiol.* **11:**811–818.

Kaan, T., G. Homuth, U. Mader, J. Bandow, and T. Schweder. 2002. Genome-wide transcriptional profiling of the *Bacillus subtilis* cold-shock response. *Microbiology* **148:**3441–3455.

Kandror, O., A. DeLeon, and A. L. Goldberg. 2002. Trehalose synthesis is induced upon exposure of *Escherichia coli* to cold and is essential for viability at low temperatures. *Proc. Natl. Acad. Sci. USA* **99:**9727–9732.

Kandror, O., and A. L. Goldberg. 1997. Trigger factor is induced upon cold shock and enhances viability of *Escherichia coli* at low temperatures. *Proc. Natl. Acad. Sci. USA* **94:**4978–4981.

Kaneda, T. 1967. Fatty acids in the genus *Bacillus*. I. Iso- and anteiso-fatty acids as characteristic constituents of lipids in 10 species. *J. Bacteriol.* **93:**894–903.

Kaneda, T. 1991. Iso- and anteiso-fatty acids in bacteria: biosynthesis, function, and taxonomic significance. *Microbiol. Rev.* **55:**288–302.

Kaplan, F., and C. L. Guy. 2004. β-Amylase induction and the protective role of maltose during temperature shock. *Plant Physiol.* **135:**1674–1684.

Karlson, D., K. Nakaminami, T. Toyomasu, and R. Imai. 2002. A cold-regulated nucleic acid-binding protein of winter wheat shares a domain with bacterial cold shock proteins. *J. Biol. Chem.* **277:**35248–35256.

Katzif, S., D. Danavall, S. Bowers, J. T. Balthazar, and W. M. Shafer. 2003. The major cold shock gene, *cspA*, is involved in the susceptibility of *Staphylococcus aureus* to an antimicrobial peptide of human cathepsin G. *Infect. Immun.* **71:**4304–4312.

Kondo, K., and M. Inouye. 1991. TIP 1, a cold shock-inducible gene of *Saccharomyces cerevisiae*. *J. Biol. Chem.* **266:**17537–17544.

Kondo, K., and M. Inouye. 1992. Yeast NSR1 protein that has structural similarity to mammalian nucleolin is involved in pre-rRNA processing. *J. Biol. Chem.* **267:**16252–16258.

Kremer, W., B. Schuler, S. Harrieder, M. Geyer, W. Gronwald, C. Welker, R. Jaenicke, and H. R. Kalbitzer. 2001. Solution NMR structure of the cold-shock protein from the hyperthermophilic bacterium *Thermotoga maritima*. *Eur. J. Biochem.* **268:**2527–2539.

Krispin, O., and R. Allmansberger. 1995. Changes in DNA supertwist as a response of *Bacillus subtilis* towards different kinds of stress. *FEMS Microbiol. Lett.* **134:**129–135.

Leblanc, L., C. Leboeuf, F. Leroi, A. Hartke, and Y. Auffray. 2003. Comparison between NaCl tolerance response and acclimation to cold temperature in *Shewanella putrefaciens*. *Curr. Microbiol.* **46:**157–162.

Lee, S. J., A. Xie, W. Jiang, J. P. Etchegaray, P. G. Jones, and M. Inouye. 1994. Family of the major cold-shock protein, CspA (CS7.4), of *Escherichia coli*, whose members show a high sequence similarity with the eukaryotic Y-box binding proteins. *Mol. Microbiol.* **11:**833–839.

Lelivelt, M. J., and T. H. Kawula. 1995. Hsc66, an Hsp70 homolog in *Escherichia coli*, is induced by cold shock but not by heat shock. *J. Bacteriol.* **177:**4900–4907.

Lopez, M. M., K. Yutani, and G. I. Makhatadze. 2001. Interactions of the cold shock protein CspB from *Bacillus subtilis* with single-stranded DNA. Importance of the T base content and position within the template. *J. Biol. Chem.* **276:**15511–15518.

Los, D. A., and N. Murata. 1999. Responses to cold shock in cyanobacteria. *J. Mol. Microbiol. Biotechnol.* **1:**221–230.

Maier, R., B. Eckert, C. Scholz, H. Lilie, and F. X. Schmid. 2003. Interaction of trigger factor with the ribosome. *J. Mol. Biol.* **326:**585–592.

Mangoli, S., V. R. Sanzgiri, and S. K. Mahajan. 2001. A common regulator of cold and radiation response in *Escherichia coli*. *J. Environ. Pathol. Toxicol. Oncol.* **20:**23–26.

Martin, A., V. Sieber, and F. X. Schmid. 2001. In-vitro selection of highly stabilized protein variants with optimized surface. *J. Mol. Biol.* **309:**717–726.

Martinez-Costa, O. H., M. Zalacain, D. J. Holmes, and F. Malpartida. 2003. The promoter of a cold-shock-like gene has pleiotropic effects on *Streptomyces* antibiotic biosynthesis. *FEMS Microbiol. Lett.* **220:**215–221.

Michel, V., I. Lehoux, G. Depret, P. Anglade, J. Labadie, and M. Hebraud. 1997. The cold shock response of the psychrotrophic bacterium *Pseudomonas fragi* involves four low-molecular-mass nucleic acid-binding proteins. *J. Bacteriol.* **179:**7331–7342.

Mitta, M., L. Fang, and M. Inouye. 1997. Deletion analysis of *cspA* of *Escherichia coli*: requirement of the AT-rich UP element for *cspA* transcription and the downstream box in the coding region for its cold shock induction. *Mol. Microbiol.* **26:**321–335.

Mizushima, T., K. Kataoka, Y. Ogata, R. Inoue, and K. Sekimizu. 1997. Increase in negative supercoiling of plasmid DNA in *Escherichia coli* exposed to cold shock. *Mol. Microbiol.* **23:**381–386.

Moll, I., M. Huber, S. Grill, P. Sairafi, F. Mueller, R. Brimacombe, P. Londei, and U. Blasi. 2001. Evidence against an Interaction between the mRNA downstream box and 16S rRNA in translation initiation. *J. Bacteriol.* **183:**3499–3505.

Movahedi, S., and W. Waites. 2002. Cold shock response in sporulating *Bacillus subtilis* and its effect on spore heat resistance. *J. Bacteriol.* **184:**5275–5281.

Murata, N., and H. Wada. 1995. Acyl-lipid desaturases and their importance in the tolerance and acclimatization to cold of cyanobacteria. *Biochem. J.* **308:**1–8.

Nakashima, K., K. Kanamaru, T. Mizuno, and K. Horikoshi. 1996. A novel member of the *cspA* family of genes that is induced by cold shock in *Escherichia coli*. *J. Bacteriol.* **178:**2994–2997.

Napolitano, M. J., and D. H. Shain. 2005. Distinctions in adenylate metabolism among organisms inhabiting temperature extremes. *Extremophiles* **9:**93–98.

Nelson, K. E., R. A. Clayton, S. R. Gill, M. L. Gwinn, R. J. Dodson, D. H. Haft, E. K. Hickey, J. D. Peterson, W. C. Nelson, K. A. Ketchum, L. McDonald, T. R. Utterback, J. A. Malek, K. D. Linher, M. M. Garrett, A. M. Stewart, M. D. Cotton, M. S., Pratt, C. A. Phillips, D. Richardson, J. Heidelberg, G. G. Sutton, R. D. Fleischmann, J. A. Eisen, O. White, S. L. Salzberg, H. O. Smith, J. C. Venter, and C. M. Fraser. 1999. Evidence for lateral gene transfer between archaea and bacteria from genome sequence of *Thermotoga maritima*. *Nature* **399:**323–329.

Newkirk, K., W. Feng, W. Jiang, R. Tejero, S. D. Emerson, M. Inouye, and G. T. Montelione. 1994. Solution NMR structure of the major cold shock protein (CspA) from *Escherichia coli*: identification of a binding epitope for DNA. *Proc. Natl. Acad. Sci. USA* **91:**5114–5118.

Nishiyama, S. I., T. Umemura, T. Nara, M. Homma, and I. Kawagishi. 1999. Conversion of a bacterial warm sensor to a cold sensor by methylation of a single residue in the presence of an attractant. *Mol. Microbiol.* **32:**357–365.

Orlova, I. V., T. S. Serebriiskaya, V. Popov, N. Merkulova, A. M. Nosov, T. I. Trunova, V. D. Tsydendambaev, and D. A. Los. 2003. Transformation of tobacco with a gene for the thermophilic acyl-lipid desaturase enhances the chilling tolerance of plants. *Plant Cell. Physiol.* **44:**447–450.

Pan, J., X. L. Chen, C. Y. Shun, H. L. He, and Y. Z. Zhang. 2005. Stabilization of cold-adapted protease MCP-01 promoted by trehalose: prevention of the autolysis. *Protein Pept. Lett.* **12:**375–378.

Panoff, J. M., D. Corroler, B. Thammavongs, and P. Boutibonnes. 1997. Differentiation between cold shock proteins and cold acclimation proteins in a mesophilic gram-positive bacterium, *Enterococcus faecalis* JH2-2. *J. Bacteriol.* **179:**4451–4454.

Perl, D., and F. X. Schmid. 2001. Electrostatic stabilization of a thermophilic cold shock protein. *J. Mol. Biol.* **313:**343–357.

Perl, D., C. Welker, T. Schindler, K. Schroder, M. A. Marahiel, R. Jaenicke, and F. X. Schmid. 1998. Conservation of rapid two-state folding in mesophilic, thermophilic and hyperthermophilic cold shock proteins. *Nat. Struct. Biol.* **5:**229–235.

Phadtare, S. 2004. Recent developments in bacterial cold-shock response. *Curr. Issues Mol. Biol.* **6:**125–136.

Phadtare, S., J. Alsina, and M. Inouye. 1999. Cold-shock response and cold-shock proteins. *Curr. Opin. Microbiol.* **2:**175–180.

Phadtare, S., J. Hwang, K. Severinov, and M. Inouye. 2003. CspB and CspL, thermostable cold-shock proteins from *Thermotoga maritima*. *Genes Cells* **8:**801–810.

Phadtare, S., and M. Inouye. 1999. Sequence-selective interactions with RNA by CspB, CspC and CspE, members of the CspA family of *Escherichia coli*. *Mol. Microbiol.* **33:**1004–1014.

Phadtare, S., and M. Inouye. 2001. Role of CspC and CspE in regulation of expression of RpoS and UspA, the stress response proteins in *Escherichia coli*. *J. Bacteriol.* **183:**1205–1214.

Phadtare, S., and M. Inouye. 2004. Genome-wide transcriptional analysis of the cold shock response in wild-type and cold-sensitive, quadruple-*csp*-deletion strains of *Escherichia coli*. *J. Bacteriol.* **186:**7007–7014.

Phadtare, S., M. Inouye, and K. Severinov. 2002. The nucleic acid melting activity of *Escherichia coli* CspE is critical for transcription antitermination and cold acclimation of cells. *J. Biol. Chem.* **277:**7239–7245.

Phadtare, S., and K. Severinov. 2005. Extended −10 motif is critical for activity of the *cspA* promoter but does not contribute to low-temperature transcription. *J. Bacteriol.* **187:**6584–6589.

Phadtare, S., Yamanaka, K, and Inouye, M. 2000. The cold shock response, p. 33–45. *In* G. Storz, and R. Hengge-Aronis (ed.), *The Bacterial Stress Responses*. ASM Press, Washington, D.C.

Polissi, A., W. De Laurentis, S. Zangrossi, F. Briani, V. Longhi, G. Pesole, and G. Deho. 2003. Changes in *Escherichia coli* transcriptome during acclimatization at low temperature. *Res. Microbiol.* **154:**573–580.

Porankiewicz, J., and A. K. Clarke. 1997. Induction of the heat shock protein ClpB affects cold acclimation in the cyanobacterium *Synechococcus* sp. strain PCC 7942. *J. Bacteriol.* **179:**5111–5117.

Qing, G., B. Xia, and M. Inouye. 2004. Enhancement of translation initiation by A/T-rich sequences downstream of the initiation codon in *Escherichia coli*. *J. Mol. Microbiol. Biotechnol.* **6:**133–144.

Reid, K. L., H. M. Rodriguez, B. J. Hillier, and L. M. Gregoret. 1998. Stability and folding properties of a model beta-sheet protein, *Escherichia coli* CspA. *Protein Sci.* **7:**470–479.

Russell, N. J. 1990. Cold adaptation of microorganisms. *Philos. Trans. R. Soc. Lond., B., Biol. Sci.* **326:**595–608.

Russell, N. J. 1998. Molecular adaptations in psychrophilic bacteria: potential for biotechnological applications. *Adv. Biochem. Eng. Biotechnol.* **61:**1–21.

Sakamoto, T., and D. A. Bryant. 1997. Temperature-regulated mRNA accumulation and stabilization for fatty acid desaturase genes in the cyanobacterium *Synechococcus* sp. strain PCC 7002. *Mol. Microbiol.* **23:**1281–1292.

Sand, O., Gingras, M., Beck, N., Hall, C., and N. Trun. 2003. Phenotypic characterization of overexpression or deletion of the *Escherichia coli crcA*, *cspE* and *crcB* genes. *Microbiology* **149:**2107–2117.

Sato, N. 1995. A family of cold-regulated RNA-binding protein genes in the cyanobacterium *Anabaena variabilis* M3. *Nucleic Acids Res.* **23:**2161–2167.

Schelin, J., F. Lindmark, and A. K. Clarke. 2002. The *clpP* multigene family for the ATP-dependent Clp protease in the cyanobacterium *Synechococcus*. *Microbiology* **148:**2255–2265.

Schindelin, H., W. Jiang, M. Inouye, and U. Heinemann. 1994. Crystal structure of CspA, the major cold shock protein of *Escherichia coli*. *Proc. Natl. Acad. Sci. USA* **91:**5119–5123.

Schindelin, H., M. A. Marahiel, and U. Heinemann. 1993. Universal nucleic acid-binding domain revealed by crystal structure of the *B. subtilis* major cold-shock protein. *Nature* **364:**164–168.

Schindler, T., D. Perl, P. Graumann, V. Sieber, M. A. Marahiel, and F. X. Schmid. 1998. Surface-exposed phenylalanines in the RNP1/RNP2 motif stabilize the cold-shock protein CspB from *Bacillus subtilis*. *Proteins* **30:**401–406.

Schnuchel, A., R. Wiltscheck, M. Czisch, M. Herrler, G. Willimsky, P. Graumann, M. A. Marahiel, and T. A. Holak. 1993. Structure in solution of the major cold-shock protein from *Bacillus subtilis*. *Nature* **364:**169–171.

Schroder, K., P. Graumann, A. Schnuchel, T. A. Holak, and M. A. Marahiel. 1995. Mutational analysis of the putative nucleic acid-binding surface of the cold-shock domain, CspB, revealed an essential role of aromatic and basic residues in binding of single-stranded DNA containing the Y-box motif. *Mol. Microbiol.* **16:**699–708.

Segura, A., P. Godoy, P. van Dillewijn, A. Hurtado, N. Arroyo, S. Santacruz, and J. L. Ramos. 2005. Proteomic analysis reveals the participation of energy- and stress-related proteins in the response of *Pseudomonas putida* DOT-T1E to toluene. *J. Bacteriol.* **187:**5937–5945.

Sinensky, M. 1974. Homeoviscous adaptation—a homeostatic process that regulates the viscosity of membrane lipids in *Escherichia coli*. *Proc. Natl. Acad. Sci. USA* **71:**522–525.

Sugino, A., C. L. Peebles, K. N. Kreuzer, and N. R. Cozzarelli. 1977. Mechanism of action of nalidixic acid: purification of *Escherichia coli nalA* gene product and its relationship to DNA gyrase and a novel nicking-closing enzyme. *Proc. Natl. Acad. Sci. USA* **74:**4767–4771.

Suzuki, I., D. A. Los, Y. Kanesaki, K. Mikami, and N. Murata. 2000. The pathway for perception and transduction of low-temperature signals in *Synechocystis*. *EMBO J.* **19:**1327–1334.

Toone, W. M., K. E. Rudd, and J. D. Friesen. 1991. *deaD*, a new *Escherichia coli* gene encoding a presumed ATP-dependent RNA helicase, can suppress a mutation in *rpsB*, the gene encoding ribosomal protein S2. *J. Bacteriol.* **173:**3291–3302.

Topanurak, S., S. Sinchaikul, B. Sookkheo, S. Phutrakul, and S. T. Chen. 2005. Functional proteomics and correlated signaling pathway of the thermophilic bacterium *Bacillus stearothermophilus* TLS33 under cold-shock stress. *Proteomics* **5:**4456–4471.

Walker, G. C. 1984. Mutagenesis and inducible responses to deoxyribonucleic acid damage in *Escherichia coli*. *Microbiol. Rev.* **48:**60–93.

Wang, J. Y., and M. Syvanen. 1992. DNA twist as a transcriptional sensor for environmental changes. *Mol. Microbiol.* **6:**1861–1866.

Wang, N., K. Yamanaka, and M. Inouye. 1999. CspI, the ninth member of the CspA family of *Escherichia coli*, is induced upon cold shock. *J. Bacteriol.* **181:**1603–1609.

Wassenberg, D., C. Welker, and R. Jaenicke. 1999. Thermodynamics of the unfolding of the cold-shock protein from *Thermotoga maritima*. *J. Mol. Biol.* **289:**187–193.

Weber, M. H., and M A. Marahiel. 2003. Bacterial cold shock responses. *Sci. Prog.* **86:**9–75.

Weinberg, M. V., G. J. Schut, S. Brehm, S. Datta, and M. W. Adams. 2005. Cold shock of a hyperthermophilic archaeon: *Pyrococcus furiosus* exhibits multiple responses to a suboptimal growth temperature with a key role for membrane-bound glycoproteins. *J. Bacteriol.* **187:**336–348.

Welker, C., G. Bohm, H. Schurig, and R. Jaenicke. 1999. Cloning, overexpression, purification, and physicochemical characterization of a cold shock protein homolog from the hyperthermophilic bacterium *Thermotoga maritima*. *Protein Sci.* **8:**394–403.

Xia, B., H. Ke, and M. Inouye. 2001. Acquirement of cold sensitivity by quadruple deletion of the cspA family and its suppression by PNPase S1 domain in *Escherichia coli*. *Mol. Microbiol.* **40:**179–188.

Yamanaka, K., L. Fang, and M. Inouye. 1998. The CspA family in *Escherichia coli*: multiple gene duplication for stress adaptation. *Mol. Microbiol.* **27:**247–255.

Yamanaka, K., and M. Inouye. 1997. Growth-phase-dependent expression of cspD, encoding a member of the CspA family in *Escherichia coli*. *J. Bacteriol.* **179:**5126–5130.

Yamanaka, K., and M. Inouye. 2001. Induction of CspA, an *E. coli* major cold-shock protein, upon nutritional upshift at 37 degrees C. *Genes Cells* **6:**279–290.

Yamanaka, K., W. Zheng, E. Crooke, Y. H. Wang, and M. Inouye. 2001. CspD, a novel DNA replication inhibitor induced during the stationary phase in *Escherichia coli*. *Mol. Microbiol.* **39:**1572–1584.

Physiology and Biochemistry of Extremophiles
Edited by C. Gerday and N. Glansdorff
© 2007 ASM Press, Washington, D.C.

Chapter 15

Perception and Transduction of Low Temperature in Bacteria

S. Shivaji, M. D. Kiran, and S. Chintalapati

INTRODUCTION

Bacteria are known to survive under a wide range of temperatures, varying from subzero (<0°C), as in the polar caps (Price, 2000), to temperatures above the boiling point of water (>100°C), as in hydrothermal vents (Stetter, 1996). Among these, the psychrophilic bacteria—which are capable of growing in the temperature range of 0°C to 30°C, with optimum growth ~15°C to 25°C—constitute a sizeable proportion of the bacterial diversity, because of the fact that a good proportion of Earth's biosphere (~85%) experiences temperatures of <5°C for at least a few months in a year (Baross and Morita, 1978) and >75% are permanently cold (<5°C throughout the year). These cold-loving extremophiles could serve as a model system to unravel the molecular basis of survival and multiplication at low temperature (Shivaji and Ray 1995; Ray et al., 1998; Cavicchioli et al., 2002; Deming, 2002; Feller and Gerday, 2003; Xu et al., 2003: Georlette et al., 2004; Goodchild et al., 2004; Marx et al., 2004; Shivaji et al., 2004; Chintalapati et al., 2005; Shivaji, 2005).

The ability of microorganisms to adapt to low-temperature conditions would primarily depend on the ability to sense changes in temperature, and such a response appears to be all the more important when cells are shifted abruptly to low temperature. This ability has been attributed to cell membrane and biomolecules, namely DNA, RNA, and proteins (Eriksson et al., 2002). In fact, the membrane that acts as an interface between the external and internal environments of the cell and that which undergoes changes in membrane fluidity at low temperatures could thus form one of the primary sensors of cold (Rowbury, 2003). These changes in the membrane's structural and dynamic characteristics could in turn affect the functions of membrane proteins, such as proteins involved in respiration and transport (Beney and Gervais, 2000; Bond and Sansom, 2003; Kasamo, 2003), and as a consequence the cell adapts to the change in its environment. Therefore, there is a need to understand the strategies by which a bacterium modulates the physical state of the membrane (membrane fluidity). With this in view, this chapter focuses on the various mechanisms by which membrane fluidity is modulated in bacteria vis-à-vis its importance in cold adaptation. A detailed update on the perception and transduction of low-temperature signals in bacteria is also included.

LOW-TEMPERATURE-INDUCED CHANGES IN MEMBRANE FLUIDITY

Bacteria modulate membrane fluidity by various methods. For instance, changes in the size and charge of the polar head groups are known to alter the packing of glycerophospholipids in the membrane bilayer and thus could change the fluidity of the membrane. However, such changes in the polar head groups are less frequent and less effective in modifying the membrane fluidity (Hasegawa et al., 1980; Suutari and Laakso, 1994). Changes in the proportion of short- and long-chain fatty acids depending on the growth temperature are also known to modulate the fluidity of the membrane, as in the marine bacterium *Shewanella putrefaciens* SCRC-2738 (Akimoto et al., 1990). Longer chains span the width of the bilayer and decrease membrane fluidity, whereas shorter chains, with <12 carbons, are unable to span the bilayer and thus maintain the fluid state of the membrane (Quinn, 1981). However, this mode of adaptation through chain-length modification is possible only in growing cells (Denich et al., 2003) and therefore may not be the universal method of modulating membrane fluidity. Proteins, by their ability to interact with lipids, also contribute to the overall stability of the membrane bilayer (Epand, 1998; Heipieper et al., 1994; Takeuchi

---

S. Shivaji, M. D. Kiran, and S. Chintalapati • Centre for Cellular and Molecular Biology, Uppal Road, Hyderabad 500 007, Andhra Pradesh, India.

et al., 1978, 1981), but the interaction itself is dependent on head group acylation, membrane fluidity, and membrane thickness, implying that it does not cause the primary effect on fluidity. Compared to the above strategies, changes in fatty acid desaturation, changes in fatty acid isomerization, and changes in the composition of carotenoids appear to be the common modes of modulation of membrane fluidity in cells growing at or exposed to low temperatures.

Changes in Fatty Acid Desaturation

The membrane normally has both saturated and unsaturated fatty acids, and therefore altering the proportion of these fatty acids could effectively alter the fluidity of the membrane. The rule of thumb is that saturated fatty acids decrease, whereas unsaturated fatty acids increase, the fluidity of the membrane because the acyl chains of the saturated fatty acids pack tightly, but the unsaturated fatty acid acyl chains exhibit poor packing because of the kink caused by the *cis* double bound (Quinn, 1981; McElhaney, 1982; Mendoza and Cronan, 1983; Mansilla et al., 2004). It is a well-established fact that psychrophilic bacteria, as compared with mesophilic bacteria, have a greater proportion of unsaturated fatty acids (Prabagaran et al., 2005; Reddy et al., 2002a, 2003a, 2003b, 2003c, 2004) (Table 1) and also that mesophilic or psychrophilic bacteria when exposed to low temperatures respond by increasing the level of unsaturated fatty acids in the membrane phospholipids (Russell, 1984, 1990; Murata and Wada, 1995). This phenomenon of thermal regulation of membrane fluidity has been studied in detail in *Escherichia coli*, *Bacillus subtilis*, *Synechocystis* sp. PCC 6803, and other cyanobacteria.

Temperature-Dependent Changes in the Unsaturated Fatty Acids of *E. coli*

Marr and Ingraham (1962) were the first to demonstrate that *E. coli*, which has palmitic acid (C16:0), palmitoleic acid (C16:1(9)), and *cis*-vaccenic acid (C18:1(11)), responds to a decrease in temperature by increasing the amount of C18:1(11) and reducing palmitic acid C16:0 (Cronan and Gelmann, 1973; Baldassare et al., 1976). In *E. coli*, the synthesis of these unsaturated fatty acids is brought about by three enzymes: FabA, FabB, and FabF (Bloch, 1963; Heath and Rock, 2002; Albanesi et al., 2004). It was also shown that FabF, which is required to convert C16:1(9) to C18:1(11) (Garwin et al., 1980), at low temperature in *E. coli* is independent of de novo synthesis of mRNA and protein (Garwin and Cronan, 1980), thus implying that low-temperature regulation of membrane fluidity in *E. coli* is dependent on the activity of the enzyme and not the synthesis of FabF.

Temperature-Dependent Changes in the Unsaturated Fatty Acids of Various *Bacillus* Species

Using desaturase enzymes, when transferred to low temperature, *B. subtilis* and *Bacillus megaterium* convert already synthesized saturated fatty acids to unsaturated fatty acids (Fulco, 1969; Fujii and Fulco, 1977; Diaz et al., 2002). Subsequent studies also indicated that in both *B. subtilis* and *B. megaterium* the desaturating system required de novo synthesis of RNA and protein (Grau and de Mendoza, 1993; Grau et al., 1994). Aguilar et al. (1998) identified the *des* gene encoding the sole desaturase of *B. subtilis* and demonstrated that the gene is not detectable at 37°C but is induced transiently upon downshift in temperature (Aguilar et al., 1999). Increase in anteiso-branched fatty acids (which exhibit methyl branching at the last but two carbons of the fatty acid) on lowering the growth temperature, with concomitant decrease in iso-branched fatty acids (methyl branching at the last but one carbon of the fatty acid), was also observed in *B. subtilis* and *Brevibacterium fermentans* (Suutari and Laakso, 1992; Klein et al., 1999). These anteiso-branched-chain fatty acids increase the fluidity of membranes compared with the corresponding iso-branched-chain fatty acids (Kaneda, 1991; Okuyama et al., 1991). As of yet, the biochemical mechanism for modulating the ratio of iso- to anteiso-branched-chain fatty acids is still unknown (Mansilla et al., 2004).

Temperature-Dependent Changes in the Unsaturated Fatty Acids of Cyanobacteria

Cyanobacteria are divided into four groups based on their fatty acid composition, the position of double bonds in the fatty acids, and the distribution of fatty acids at the *sn*-1 and the *sn*-2 positions of the glycerolipids (Kenyon, 1972; Kenyon et al., 1972; Murata et al., 1992). Therefore, changes that occur in the fatty acid composition, when cyanobacteria are exposed to low temperature, would be different, depending on the group to which the cyanobacterium belongs. For instance, with alterations of growth temperature, species belonging to group I change both the extent of unsaturation and the chain length of fatty acids. All other cyanobacteria respond to low temperature by modulating only the extent of unsaturation of fatty acids (Sato et al., 1979), such as a decrease in the levels of $C18:1_{(9)}$ and $C18:2_{(9,12)}$ and increase in the levels of $C18:3_{(9,12,15)}$ (Sato and Murata, 1980; Sato et al., 1979; Murata et al., 1992; Wada and Murata, 1990).

Table 1. Comparison of the fatty acid composition of psychrophilic bacteria (boldface) with their nearest mesophilic phylogenetic neighbor[a]

Fatty acid	Halomonas variabilis	H. glaciei (DD39[T])	Pseudomonas orientalis	Pseudomonas antarctica (CMS35[T])	Leifsonia poae	Leifsonia rubra (CMS 76or[T])	Leifsonia aurea (CMS 81y[T])	Kocuria rosea (ATCC 186[T])	Kocuria polaris (CMS 76or[T])	Planococcus okeanokoites	Planococcus antarcticus (CMS 26or[T])	Planococcus psychrophilus (CMS 53or[T])
C12:0	—	0.5	2.3	0.8	—	—	—	—	—	—	—	—
C14:0	—	0.3	0.4	0.3	—	1.3	0.2	0.5	1.1	—	—	—
C14:1	—	2.9	—	—	—	—	—	—	—	—	Trace	—
Iso-C14:0	—	—	—	—	—	—	—	1.3	0.7	Trace	1.0	3.2
C15:0	—	0.2	0.2	0.3	<1.0	0.5	0.2	0.8	0.6	33.9	14.2	—
C15:1	—	—	—	0.8	<1.0	3.6	1.6	—	—	Trace	9.8	—
Iso-C15:0	—	—	—	—	1.0	7.8	3.3	5.1	3.2	2.9	1.3	5.6
Anteiso-C15:0	—	—	—	—	36.3	37.5	43.7	65.0	70.6	14.0	43.2	41.3
C16:0	17.5	11.6	30.4	24.8	<1.0	1.8	1.4	3.2	2.4	4.7	4.2	4.4
Iso-C16:0	—	—	—	—	15.6	11.5	5.0	3.0	0.9	28.1	4.0	8.1
C16:1	—	—	—	—	—	—	—	3.5	2.2	2.8	3.0	3.2
C16:1 Δ7c	2.4	13.5	32	29.6	—	—	—	—	—	—	—	—
C16:1 Δ9t	—	—	—	5.8	—	—	—	—	—	—	—	—
C16:1 Δ9c	—	—	—	2.3	—	—	—	—	—	—	—	—
Iso-C16:1	—	—	—	—	—	—	—	—	—	11.7	1.2	7.2
C17:0	—	—	0.3	0.7	—	—	—	—	—	—	1.0	6.1
Cyclo-C17:0	11.4	5.5	4.9	1.4	—	—	—	—	—	—	—	—
Iso-C17:0	—	—	—	—	<1.0	1.8	6.3	1.5	2.9	Trace	0.3	—
Anteiso-C17:0	—	—	—	—	45.4	20.3	28.2	7.1	5.6	Trace	9.5	7.3
Anteiso-C17:1	—	—	—	—	—	—	—	5.0	4.8	—	—	—
Iso + anteiso-C17:1	—	—	—	—	—	—	—	—	—	—	4.2	11.3
C18:0	—	5	1.1	1.4	<1.0	3.7	4.3	2.9	1.2	1.8	0.3	1.6
Iso-C18:0	—	—	—	—	—	—	—	0.9	3.0	Trace	—	—
C18:1	15.1	31.2	25.3	—	<1.0	5.0	4.1	—	0.9	—	1.0	1.0
Cyclo-C19:0	50.9	—	0.7	—	—	—	—	—	—	—	—	—
A	—	2.4	—	—	—	—	—	—	—	—	—	—
B	—	10.0	—	—	—	—	—	—	—	—	—	—
C	—	10.3	—	—	—	—	—	—	—	—	—	—

[a] —, fatty acid not detected; A, hexadecanoic acid 9,10-dimethoxy methyl ester; B, cyclopropane octadecanoic acid 2-octyl methyl ester; C, octadecanoic acid 11-methoxy methyl ester; trace, <0.1%. Data from Reddy et al. (2002a, 2003a, 2003b, 2003c, 2004).

In cyanobacteria, the unsaturation of the fatty acids is catalyzed by four acyl-lipid desaturases: DesA, DesB, DesC, and DesD, which introduce a double bond in the Δ12, Δ15, Δ9, and Δ6 position of the fatty acid, respectively. It has been demonstrated that polyunsaturated fatty acids and fatty acyl-lipid desaturases are essential for the acclimation of cyanobacteria to low temperatures (Murata and Wada, 1995; Nishida and Murata, 1996). At 34°C, *Fad12*, a mutant of *Synechocystis* sp. PCC 6803 (Wada and Murata, 1989), grew almost at the same rate as the wild type, but at 22°C the *Fad12* mutant grew three times slower (Wada and Murata, 1989). These results indicated that the *Fad12* mutant was cold sensitive, and this was attributed to the inability of the mutant to convert $C18:1_{(9)}$ to $C18:2_{(9,12)}$, an activity associated with Δ12 desaturase. These results were subsequently confirmed by targeted mutagenesis of the gene for the Δ12 desaturase (*desA*) in *Synechocystis* sp. PCC 6803 (Murata and Wada, 1995; Tasaka et al., 1996) and *Synechococcus* sp. PCC 7002 (Sakamoto et al., 1998). Both the mutants showed a considerable increase in the level of $C18:1_{(9)}$ at the expense of the polyunsaturated fatty acid C18:3 in the membrane lipids (Table 2), and the strains were cold sensitive at 25°C compared to the wild-type strain (Tasaka et al., 1996). Another mutant of *Synechocystis* sp. PCC 6803 (Tasaka et al., 1996), in which both the Δ12 and Δ6 desaturases (*desA* and *desD*) were mutagenized, also showed a cold-sensitive phenotype, and in this mutant, an increase in the level of $C18:1_{(9)}$ at the expense of the polyunsaturated fatty acid C18:3 was observed in the membrane lipids (Table 2) at 25°C (Tasaka et al., 1996). These observations confirmed that polyunsaturated fatty acids are essential for growth at low temperatures and demonstrated that replacement of polyunsaturated fatty acids by monounsaturated fatty acids has a significant effect on the growth of cells. In accordance with this observation, it was confirmed that *Anacystis nidulans*, which contains only saturated (C16:0 and C18:0) and monounsaturated fatty acids ($C16:1_{(9)}$ and $C18:1_{(9)}$), is sensitive to growth at low temperature (Murata, 1989), but when it was transformed with the Δ12 desaturase from *Synechocystis* sp. PCC 6803, the transformed cells synthesized diunsaturated fatty acids ($C16:2_{(9,12)}$ and $C18:2_{(9,12)}$) at the expense of monounsaturated fatty acids and were able to tolerate lower temperatures, unlike the wild-type cells (Wada et al., 1990). Further, when the *desC* gene for the acyl-lipid Δ9 desaturase of the thermophilic cyanobacterium *Synechococcus vulcanus* was introduced into *Nicotiana tabacum* (Orlova et al., 2003), the chilling tolerance of these plants increased, indicating that introduction of unsaturated fatty acids increases cold tolerance.

Changes in Proportion of *cis* and *trans* Fatty Acids

Geometrical isomers such as the *cis–trans* fatty acids also effect membrane fluidity (Morita et al., 1993; Okuyama et al., 1990). Increased levels of *cis*-unsaturated fatty acids lower the phase-transition temperature because it provokes an unmovable kink of 30° in the acyl chain (Heipieper et al., 2003), whereas *trans*-unsaturated fatty acids that have a lesser kink do not lower the phase transition to the same extent. The *trans*-monounsaturated fatty acids were first discovered in *Methylosinus trichosporium*, a methane-utilizing bacterium (Makula, 1978). Subsequently, it was found that *trans*-monounsaturated are predominant in gram-negative bacteria (Keweloh and Heipieper, 1996) and are synthesized by direct isomerization of *cis*-unsaturated fatty acids to *trans*-unsaturated fatty acids without shifting of a double bond (Heipieper et al., 1992; Diefenbach and Keweloh, 1994; von Wallbrunn et al., 2003). Further, the conversion of *cis*- to *trans*-monounsaturated fatty acids is not dependent on growth (Diefenbach et al., 1992; Heipieper and de Bont, 1994) and is unaffected by chloramphenicol, implying that it does not require de novo protein biosynthesis (Heipieper et al., 1992).

In psychrophilic *Vibrio* sp. (strain ABE-1) and *Pseudomonas* sp. (strain E-3), an increase in *trans*-fatty acid content with increase in growth temperature was observed (Okuyama et al., 1990, 1991). Subsequent studies also confirmed an increase in *trans*-monounsaturated fatty acids when bacteria were subjected to solvent and thermal stress (Diefenbach et al., 1992; Heipieper and de Bont, 1994; Junker and Ramos, 1999; Okuyama et al., 1991). This increase in *trans*-unsaturated fatty acid could effectively reduce the fluidity of the membrane (Okuyama et al., 1991). In a recent study, a psychrophilic bacterium, *Pseudomonas syringae* (Lz4W) from Antarctica, was used as a model system to establish a correlation, if any, between thermal adaptation, *trans*-fatty acid content, and membrane fluidity (Kiran et al., 2004, 2005). When grown at 28°C, the bacterium showed increased proportion of saturated and *trans*-monounsaturated fatty acids compared with cells grown at 5°C and the membrane fluidity decreased with growth temperature. Further, when the *cti* gene was mutated in *P. syringae* (Lz4W), it was observed that the growth of the mutant at 5°C was not altered but was arrested at 28°C, thus implying that the synthesis of *trans*-fatty acid and modulation of membrane fluidity to levels comparable to the wild-type cells are essential for growth at higher temperatures. In fact, the *cti*-null mutant-complemented strain of *P. syringae* (Lz4W–C30b), which was capable of synthesizing the *trans*-fatty acid, could grow at 28°C, thus confirming

Table 2. Fatty acid composition of total glycerolipids from wild type and mutant cells of Synechocystis sp. PCC 6803 and Synechococcus sp. PCC 7002[a]

Strain	Growth temperature (°C)	Fatty acid (mole %)								
		$C16:0$	$C16:1_{(9)}$	$C18:0$	$C18:1_{(9)}$	$C18:1_{(11)}$	$C18:2_{(6,9)}$	$C18:2_{(9,12)}$	$C18:3_{(9,12,15)}$	$C18:3_{(6,9,12)}$	$C18:4_{(6,9,12,15)}$

Strain	Growth temp	C16:0	C16:1(9)	C18:0	C18:1(9)	C18:1(11)	C18:2(6,9)	C18:2(9,12)	C18:3(9,12,15)	C18:3(6,9,12)	C18:4(6,9,12,15)
Synechocystis sp. PCC 6803											
Wild type	25	53	4	1	9	Trace	Trace	10	5	14	4
desA	25	55	4	1	28	Trace	12	0	0	0	0
desB	25	51	5	2	6	Trace	Trace	13	0	22	0
desD	25	52	6	1	8	Trace	0	23	11	0	0
desBdesD	25	53	6	1	12	Trace	0	29	0	0	0
desAdesD	25	51	7	Trace	41	Trace	0	0	0	0	0
Wild type	22	51	3	1	2	Trace	Trace	6	8	21	8
Fad6	22	55	3	1	3	2	Trace	17	10	5	4
Fad12	22	53	3	1	25	Trace	16	Trace	Trace	0	0
Synechococcus sp. PCC 7002											
Wild type	22	51	7	1	7	0	0	15	19	0	0
desA	22	49	6	1	43	0	0	0	0	0	0
desB	22	49	8	1	8	0	0	34	0	0	0

[a] Trace, <0.1%. Data from Tasaka et al. (1996) and Sakamoto et al. (1998).

that psychrophilic *P. syringae* requires *trans*-monounsaturated fatty acids for growth at higher temperatures (Kiran et al., 2004, 2005).

Role of Carotenoids in Modulating Membrane Fluidity

In vitro studies based on the interaction of carotenoids with model membranes have clearly demonstrated that more polar carotenoids stabilize the membrane to a greater extent than less polar carotenoids (Gabrielska and Gruszecki 1996; Subczynski et al., 1992). This is an interesting hypothesis because several Antarctic bacteria have been found to contain carotenoid type of pigments, which are associated with the cell membrane (Chauhan and Shivaji, 1994; Jagannadham et al., 1996b, 2000; Chattopadhyay et al., 1997; Gupta et al., 2004; Reddy et al., 2000, 2002a, 2002b, 2003a, 2003b; Shivaji et al., 1988, 1992, 2005). Further, the synthesis of the type of carotenoid (polar versus less polar) appears to be dependent on the growth temperature (Chattopadhyay et al., 1997; Jagannadham et al., 2000). Therefore, it could be speculated that in these bacteria temperature-dependent synthesis of carotenoids may be a strategy to modulate membrane fluidity by altering the levels of the polar and nonpolar carotenoids. With this in view, the carotenoid pigments of a psychrophilic bacterium *Sphingobacterium antarcticum* were characterized with respect to their in vivo localization, temperature-dependent biosynthesis, and in vitro interaction with synthetic membranes (Jagannadham et al., 2000). The results indicated an increase in the levels of two polar carotenoids (zeaxanthin and an unidentified carotenoid PS1) from 53 to 85% and a concomitant decrease in the levels of three less polar carotenoids (β-cryptoxanthin, β-carotene, and an unidentified carotenoid PS5) from 47 to 15% (Table 3). This confirmed the findings of an earlier study on psychrophilic *Micrococcus roseus*, in which the relative amount of polar carotenoids was higher in cells grown at 5°C than in cells grown at 25°C (Chattopadhyay et al., 1997). Simultaneously, in *S. antarcticum*, an increase in the biosynthesis of unsaturated and branched-chain fatty acids in cells grown at 5°C compared to cells grown at 25°C was observed. Taken together, these results suggested that in cells grown at 5°C, unsaturated and branched-chain fatty acids, which increase quantitatively at low temperature, would increase the fluidity of the membrane, whereas polar carotenoids that stack well in the membrane would facilitate membrane stabilization and thus overcome the fluidizing effect of the unsaturated fatty acids (Jagannadham et al., 1991, 1996a, 2000). This indeed could be a mechanism by which cells maintain

Table 3. Quantitative changes in the carotenoid content of *S. antarcticum* grown at 5°C and 25°C[a]

Pigment	Relative quantity (%)	
	5°C	25°C
Unidentified	6.12 ± 2.29	3.69 ± 1.75
Zeaxanthin	79.23 ± 3.63	49.28 ± 2.45
β-Cryptoxanthin	12.84 ± 1.88	30.19 ± 0.82
Unidentified	0.61 ± 0.22	1.78 ± 0.53
β-Carotene	1.71 ± 0.20	15.04 ± 2.80

[a] Data from Jagannadham et al. (2000).

homeoviscous adaptation, namely an optimum membrane fluidity. In fact, it was observed that the membrane fluidity of intact cells of *S. antarcticum* was similar in cells grown at 5°C and 25°C (Jagannadham et al., 2000). As of yet it is not known whether polar carotenoids are required for the survival of pigmented chemotrophic bacteria at low temperatures. However, because psychrophilic bacteria synthesize higher amounts of polar carotenoids when grown at 5°C compared with 25°C, it is possible that these pigments modulate membrane fluidity depending on the growth temperature. Thus, by switching over the synthesis of carotenoids from one type (polar) to another type (nonpolar), a bacterium could alter the fluidity of its membrane and thus influence homeoviscous adaptation.

Role of Cold-Inducible Genes in Low-Temperature Adaptation

Membrane fluidity modulation in response to low temperature is thus crucial to the bacterium, but two questions still need to be addressed: What is the nature of the sensor? and What are the downstream events that follow the sensing of the change in temperature? Proteomics and DNA microarray approaches have provided some insight into this aspect; a number of genes and proteins have been identified, and many of these have been implicated in cold adaptation. A discussion on these genes and proteins needs to be dealt with separately in another review. However, it may be worthwhile to mention the key genes and proteins that are induced following downshift in temperature in *Synechocystis* (Suzuki et al., 2001), *E. coli* (Goldstein et al., 1990; Lee et al., 1994; Jones and Inouye, 1994; Newkirk et al., 1994; Sommerville and Ladomery, 1996; Jiang et al., 1997; Graumann and Marahiel, 1999; Phadtare et al., 1999), *B. subtilis* (Graumann et al., 1997; Graumann and Marahiel, 1999; Schindler et al., 1999; Kaan et al., 2002), *Vibrio cholerae* (Datta and Bhadra, 2003), and *Methanococcoides burtonii* (Goodchild et al., 2004). They could be categorized as follows: (i) genes for fatty acid desaturases; (ii) genes that serve as RNA chaperones similarly to the Csp proteins of *E. coli* and *B. subtilis*; (iii) genes involved in replication, such as genes for DnaA, RecA, NusA, and HNS (Jones et al., 1987; Graumann and Marahiel, 1999; Jones and Inouye, 1994; Brandi et al., 1994); (iv) genes involved in transcription such as *rpoA* and *sigD*; (v) genes involved in translation such as *fus* gene for elongation factor EF-G, genes for RNA helicases that destabilize the secondary structures of mRNAs so as to facilitate initiation of translation, and genes for ribosomal proteins, such as *S6*, *L7/L12* (Graumann and Marahiel, 1999), *TF*, *CsdA*, *RbfA* (Xia et al., 2003), *IF2*, *IF2α*, and *IF2β* (Jones et al., 1987); (vi) the *hliA*, *hliB*, and *hliC* genes that encode high-light-inducible proteins (Suzuki et al., 2001); (vii) the *cytM* gene for an alternative form of cytochrome *c*; (viii) the *ndhD2* gene for subunit of 4-NADH dehydrogenases and genes for a number of enzymes such as caseinolytic proteases, polynucleotide phosphorylase (Jones and Inouye, 1994; Graumann and Marahiel, 1999), peptidyl-prolyl *cis/trans* isomerase, (Goodchild et al., 2004), enzymes of oxidative decarboxylation of pyruvate (Graumann and Marahiel, 1999; Jones and Inouye, 1994), enzymes involved in methanogenesis, and amino acid synthesis, and (ix) various other genes that do not fall in any of the above categories. Thus, it is evident that cold stress enhances the expression of many genes whose products control membrane fluidity, transcription, translation, and the energy status of the cell.

Perception of Temperature and Signal Transduction

One of the predominant signal transduction mechanisms employed by bacteria is the phosphotransfer pathway commonly referred to as the two-component signal transduction system, which consists of a sensor kinase (histidine kinase) and a response regulator, found in bacteria, *Archaea*, and *Eukarya* (Hoch, 2000; Mizuno, 1997; Mizuno et al., 1996; Aguilar et al., 2001; Weber and Marahiel, 2003). The sensor is normally an integral membrane protein consisting of membrane-spanning domains, a signal-recognition domain, and an autokinase domain. Recognition of signal causes autophosphorylation of a histidine by the autokinase, and the phosphate is then transferred to an aspartate on the response regulator (Fig. 1). However, in contrast to the above simple two-component signal transduction systems, multistep phosphorelay systems

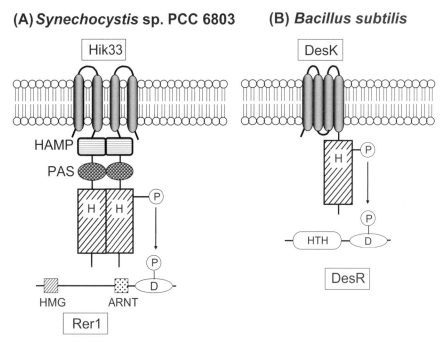

Figure 1. Diagrammatic representation of the two-component signal transduction pathway for the perception and transduction of low-temperature signals in *Synechocystis* (A) and *B. subtilis* (B). In panel (A), Hik33 and Rer1 are the histidine kinase (sensor) and the response regulators, respectively, of *Synechocystis*. The four transmembrane domains (shaded cylinders), the histidine kinase domain and the histidine residue (vertical rectangles with an H), the HAMP domain (histidine–kinase–adenylyl–cyclase–methyl-binding protein phosphatase), and the PAS (PER–ARNF–SIM) domain are indicated. In Rer1, the receiver domain with an aspartate residue (D), the DNA-binding domain (HMG), and the transcriptional activation domain (ARNT) are indicated. In panel (B), DesK and DesR are the histidine kinase and response regulators, respectively, of *B. subtilis*. The four transmembrane domains (shaded cylinders) and the histidine kinase domain (vertical rectangle with an H) of DesK are indicated. In DesR, HTH refers to the DNA-binding domain and the receiver domain with an aspartate residue (D) are also shown. (The data for this figure has been taken from Sakamoto and Murata (2002) and Kuntz et al. (1997)).

have been discovered in both prokaryotes and eukaryotes (Appleby et al., 1996; Zhang and Shi, 2005). These multiple-step-type signal transduction pathways have a hybrid-type histidine kinase in which both the histidine kinase and the response regulator receiver domains are present in the same protein. Further, the histidine-containing phosphotransfer domain could exist either as a domain in the hybrid-type histidine kinase or as a separate protein. In bacteria, quorum sensing is also achieved by a two-component regulatory system, but in this system the signaling molecule is taken up into the cells, unlike in the two-component system for low-temperature sensing (Hellingwerf et al., 1998).

Sensors and Transducers of Low Temperature in B. subtilis

The first direct evidence for the two-component signal transduction mechanism involved in sensing cold has come from studies on B. subtilis (see Fig. 1). At low temperature, the des gene in B. subtilis was induced, and it appeared logical to assume that the des gene was regulated by the two-component system (Hoch, 2000). The assumption was based on the earlier observation that indicated that B. subtilis genome has a two-gene operon composed of desK (yocF) and desR (yocG) located downstream to des (Kunst et al., 1997); further, desK and desR exhibited structural similarity to histidine kinases and response regulators of the two-component system, respectively. Inactivation of desK or desR inhibited induction of des (Aguilar et al., 1999, 2001). The des pathway, upon decrease in constant growth temperature or temperature downshift, is activated by DesK, which senses the decrease in membrane fluidity (Mansilla et al., 2003). Evidence that membrane fluidity, rather than growth temperature, controls the transcription of the des gene was obtained by experiments in which the proportion of anteiso-branched-chain fatty acids of B. subtilis membranes was varied (Cybulski et al., 2002).

The DesK protein of B. subtilis, which acts as the sensor, has four transmembrane segments that define the sensor domain and a long cytoplasmic C-terminal tail harboring the histidine kinase (see Fig. 1). This C-terminal kinase of DesK undergoes autophosphorylation in the presence of ATP in the conserved His 188 (Albanesi et al., 2004) and serves as a phosphodonor of the effector protein DesR, which becomes phosphorylated at Asp54 (Albanesi et al., 2004). Recently, it was shown that only the phosphorylated form of DesR binds to a DNA sequence extending from the −28 to −77 positions relative to the start site of the temperature-regulated des genes (Aguilar et al., 2001; Cybulski et al., 2004). It was also demonstrated that DesR, which is a transcriptional regulator, is directly involved in the activation of the des genes at low temperatures and that unsaturated fatty acids act as negative regulators of des expression (Aguilar et al., 2001). Thus, a regulatory loop consisting of DesK–DesR and unsaturated fatty acids constitutes a novel mechanism for the sensing and transduction of low-temperature signal in B. subtilis (Aguilar et al., 2001). In E. coli, it was reported that the high-temperature signal was transduced via the CpxA–CpxR relay system, where CpxA is a histidine kinase containing transmembrane regions and CpxR a response regulator. Similar CpxA–CpxR phosphorelay systems have been observed also in Salmonella enterica serovar Typhimurium and Yersinia pestis (de Wulf et al., 2000).

Sensors and Transducers of Low Temperature in Cyanobacteria

On the basis of the observation that catalytic hydrogenation of fatty acids of plasma membrane lipids in Synechocystis sp. PCC 6803 causes a decrease in membrane fluidity and enhanced transcription of the desA gene, Murata and Wada (1995) proposed a scheme for the acclimatization of cyanobacterial cells to low temperature (Vigh et al., 1993). According to this model, upon a downward shift in temperature, the fluidity of the membrane decreases (see Fig. 1). This signal is detected by a membrane-bound sensor of low temperature and is then transmitted to regulatory molecules that directly or indirectly interact with the regulatory regions of genes for desaturases, leading to their activation. As a result, the levels of desaturases increase and fatty acids are desaturated (see Fig. 1). Finally, the accelerated accumulation of unsaturated fatty acids leads to the recovery of membrane fluidity and to the restoration of the activity of membrane-associated enzymes (Murata and Wada, 1995; Nishida and Murata, 1996).

To identify the temperature sensor in cyanobacterial cells, Suzuki et al. (2000a) systematically inactivated each of the putative histidine kinases (corresponding to 43 open reading frames) in Synechocystis and identified two histidine kinases, namely a membrane-bound Hik33 and a soluble Hik19, which affected the inducibility of the desB gene and a putative response regulator Rer1 (Suzuki et al., 2000a, 2000b). Hik33 has a highly conserved histidine kinase domain at its carboxyl terminus, two membrane-spanning domains at its amino terminus (Sakamoto and Murata, 2002), and a type-P linker and a leucine zipper in the middle region, characteristic of several membrane-bound histidine kinases from various organisms (Park et al., 1998). It was hypothesized that rigidification of membrane lipids around Hik33 might alter the conformation of the membrane-spanning domains of Hik33. These conformational changes might generate an active form of Hik33 (Taylor and Zhulin, 1999; Suzuki et al., 2000a), which would play the role of a global cold sensor involved in cold induction (Inaba et al., 2003). It was suggested that cold stress promotes a conformational change in the HAMP region of Hik33 (see Fig. 1) and that, as a consequence, Hik33 would dimerize and become activated (Mikami et al., 2003). As of yet, the cross talk between Hik33 and Rer1 and the subsequent regulation of transcription of the *desB* gene by Rer1 remain to be characterized. Mutations in *Hik33* and *Hik19* genes greatly reduced (but did not eliminate) the cold induction of *desB*, *desD*, and *crh* (cold-inducible RNA helicase) genes but had no effect on the cold induction of the *desA* gene (Suzuki et al., 2001). Thus, a separate sensor apparently operates to regulate the cold induction of the *desA* gene and to partially control *desB*, *desD*, and *crh* expression (Browse and Xin, 2001; Suzuki et al., 2001). Further, Rer1, which functions downstream of Hik33 and Hik19, only regulates the cold induction of *desB* and not of *desD* or *crh*, which presumably are regulated by additional response elements (Suzuki et al., 2001).

Results of Fourier transform infrared spectroscopy (FTIR) analysis (Szalontai et al., 2000) strongly suggest that Hik33 might recognize a change in membrane fluidity at low temperature (Murata and Los, 1997; Los and Murata, 2000). This possibility was examined by mutation of the *hik33* gene in desA-/desD- cells. In the resultant desA-/desD-/hik33- cells, the expression of Hik33-regulated genes, such as the *hliA*, *hliB*, and *sigD*, was no longer inducible by cold (Inaba et al., 2003). Thus, it appears that Hik33 perceives a decrease in membrane fluidity, which depends on an increase in the extent of unsaturation of fatty acids, as a primary signal of cold stress.

Sensing of Temperature through Alteration in DNA Conformation

Eriksson et al. (2002) have suggested that temperature-induced conformational or physicochemical changes in DNA, RNA, and proteins could form the basis for sensing temperature. In bacteria, the expression of many genes is dependent on DNA conformation, which in turn depends on temperature-dependent changes in DNA supercoiling (Grau et al., 1994; Hurme and Rhen, 1998). Topoisomerases II and I (Drlica, 1992; Tse-Dinh et al., 1997) and the nucleoid-associated proteins such as H–NS (Williams and Rimsky 1997; Dorman et al., 1999) regulate supercoiling. For instance, in *Shigella* the expression of the virulence regulator, virF, at low temperature is suppressed by H–NS (Tobe et al., 1993) and the ability of H–NS to mediate transcriptional repression is dependent on the superhelical state of the promoter region (Falconi et al., 1998). However, when temperature is increased to 37°C, the ability of H–NS to bind cooperatively to its target sequence at the *virF* promoter sequences decreases, owing to a conformational shift in the local DNA topology, allowing transcription of *virF* (Falconi et al., 1998). Furthermore, *E. coli* and many other bacteria express the protein StpA (Dorman et al., 1999; Sonnenfield et al., 2001), which also has a function similar to that of H–NS.

Sensing of Temperature through Alteration in RNA Conformation

RNA molecules, because of their ability to form pronounced secondary and tertiary structures (Andersen and Delihas, 1990) and their ability to form intermolecular RNA:RNA hybrids (Lease and Belfort, 2000), have a strong potential to act as temperature sensors (Lai, 2003; Narberhaus et al., 2005). Excellent examples are the protein LcrF, which in *Y. pestis* is a virulence regulator (Hoe and Gougen 1993; Straley and Perry 1995), and the virulence-activating transcription factor prfA from *Listeria monocytogenes*, which are thermoregulated (Romby and Ehresmann, 2003). In these cases, the *LcrF* mRNA and *prfA* mRNA at 25°C are folded into a secondary structure, thus preventing translation (Hoe and Gougen, 1993). However, at elevated temperature, the stem–loop structure melts, thus allowing translation. Thus, *lcrF* mRNA and *prfA* mRNA would serve as both the messenger and the thermosensor. In *E. coli* and serovar Typhimurium, cold expression of RpoS is dependent on the expression of the *dsrA* gene encoding a small regulatory RNA (Sledjeski et al., 1996; Majdalani et al., 1998). The temperature-dependent expression of these small regulatory RNAs can modify the activity of proteins and the stability and translation of mRNAs and thus are critical to regulation of genes (Majdalani et al., 2005).

Sensing of Temperature through Alteration in Protein Conformation

Temperature-induced changes in the conformation of proteins observed in bacteria could also form the basis for sensing changes in temperature. In serovar Typhimurium TlpA, a 371-amino-acid protein (Gulig et al., 1993; Hurme et al., 1996, 1997) has the capacity to sense temperature variations and to regulate gene expression. TlpA at 28°C has the capacity to oligomerize and bind DNA and suppress its own expression. But when temperature increases, the oligomerization decreases, leading to a derepression of the target gene. The sensory capacity of TlpA is dependent on the coiled–coil structure of TlpA, which illustrates sensing of temperature through changes in protein conformation.

Earlier studies had suggested that membrane proteins that undergo temperature-dependent phosphorylation–dephosphorylation in bacteria might act as sensors (Ray et al., 1994a, 1994b). A correlation was observed between phosphorylation and dephosphorylation of a set of membrane proteins in response to upshift and downshift of temperature in the psychrophile P. syringae (Ray et al., 1994a). In P. syringae, it was also observed that the phosphorylated membrane protein could induce phosphorylation of a cytosolic 66-kDa protein (Ray et al., 1994b). In the same psychrophilic bacterium, differential phosphorylation of lipopolysaccharides associated with the outermost layer of the outer membrane of the bacterium was also observed (Ray et al., 1994c).

CONCLUSION

Modulation in membrane fluidity appears to be crucial for low-temperature sensing in bacteria, and this is normally achieved by the conversion of saturated fatty acids to unsaturated fatty acids. This conversion brings about a physical change in the packing of the fatty acids in membranes, and this change is sufficient to activate a sensor. The sensor perceives and transduces the signal to a response regulator, which then induces the upregulation of genes involved in membrane fluidity modulation. Aspects of this two-component signal transduction system have been understood, to a limited extent, with respect to a few bacteria. Yet we are far from understanding many key aspects of bacterial signal transduction in response to low temperature. For instance, there is a need to identify sensor molecule(s) and response regulators in various bacteria; a need to understand the cross talk between the above two components; and ultimately the need to understand the function of the response regulator as an inducer of gene expression. Further, how does the cell choose the specific sensor that needs to be activated when one or more sensors perceive the same signal and under such conditions how does it choose the specific response regulator? The domains in the sensor that perceive the change in membrane fluidity and the lipid molecules that specifically interact with the sensor are yet not known. Therefore, studies directed toward finding answers to the above questions would ultimately unravel the molecular basis of low-temperature perception, signal transduction, and survival at low temperature.

Acknowledgments. This work was supported by a grant from the India–Japan Cooperative Science Programme of the Department of Science and Technology, Government of India and the Japanese Society for the Promotion of Science, Government of Japan to S.S. M.D.K. and S.C. thank C.S.I.R. and U.G.C. Government of India, respectively, for research fellowships.

REFERENCES

Aguilar, P. S., J. E. Cronan, Jr., and D. de Mendoza. 1998. A *Bacillus subtilis* gene induced by cold shock encodes a membrane phospholipids desaturase. *J. Bacteriol.* 180:2194–2200.

Aguilar, P. S., P. Lopez, and D. de Mendoza. 1999. Transcriptional control of the low temperature inducible des gene, encoding the delta 5 desaturase of *Bacillus subtilis*. *J. Bacteriol.* 181: 7028–7033.

Aguilar, P. S., A. M. Hernandez-Arriaga, L. E. Cybulski, A. C. Erazo, and D. de Mendoza. 2001. Molecular basis of thermosensing: a two-competent signal transduction thermometer in *Bacillus subtilis*. *EMBO J.* 20:1681–1691.

Akimoto, M., T. Ishii, K. Yamagaki, K. Ohtaguchi, K. Koide, and Y. Kazunaga. 1990. Production of eicosapentaenoic acid by a bacterium isolated from mackerel intestines. *J. Am. Oil Chem. Soc.* 67:911–915.

Albanesi, D., M. C. Mansilla, and D. de Mendoza. 2004. The membrane fluidity sensor DesK of *Bacillus subtilis* controls the signal decay of its cognate response regulator. *J. Bacteriol.* 186:2655–2663.

Andersen, J., and N. Delihas. 1990. micF RNA binds to the 5′ end of ompF mRNA and to a protein from *Escherichia coli*. *Biochemistry* 29:9249–9256.

Appleby, J. L., J. S. Parkinson, and R. B. Bourret. 1996. Signal transduction via the multi-step phosphorelay: not necessarily a road less traveled. *Cell* 86:845–848.

Baldassare, J. J., K. B. Rhinehart, and D. F. Silbert. 1976. Modifications of membrane lipid: physical properties in relation to fatty acid structure. *Biochemistry* 15:2986–2994.

Baross, J. A., and R. Y. Morita. 1978. Microbial life at low temperatures: ecological aspects, p. 9–71. *In* D. J. Kushner (ed.), *Microbial Life in Extreme Environments*. Academic Press, New York, NY.

Beney, L., and P. Gervais. 2000. Influence of the fluidity of the membrane on the response of microorganisms to environmental stresses. *Appl. Microbiol. Biotechnol.* 57:34–42.

Bloch, K. 1963. The biological synthesis of unsaturated fatty acids. *Biochem. Soc. Symp.* 24:1–16.

Bond, P. J., and M. S. Sansom. 2003. Membrane protein dynamics versus environment: simulations of OmpA in a micelle and in a bilayer. *J. Mol. Biol.* 329:1035–1053.

Brandi, A., C. L. Pon, and C. O. Gualerzi. 1994. Interaction of the main cold shock protein CS 7.4 (CspA) of *Escherichia coli* with the promoter region of HNS. *Biochimie* 76:1090–1098.

Browse, J., and Z. Xin. 2001. Temperature sensing and cold acclimation. *Curr. Opin. Plant Biol.* 4:241–246.

Cavicchioli, R., Siddiqui, K. S., Andrews, D, and K. R. Sowers. 2002. Low temperature extremophiles and their applications. *Curr. Opin. Biotechnol.* 13:253–261.

Chattopadhyay, M. K., M. V. Jagannadham, M. Vairamani, and S. Shivaji. 1997. Carotenoid pigments of an Antarctic psychrotrophic bacterium *Micrococcus roseus*: temperature dependent biosynthesis, structure and interaction with synthetic membranes. *Biochem. Biophys. Res. Commun.* 239:85–90.

Chauhan, S., and S. Shivaji. 1994. Growth and pigmentation in *Sphingobacterium antarcticus*, a psychrotrophic bacterium from Antarctica. *Polar Biol.* **14:**31–36.

Chintalapati, S., M. D. Kiran, and S. Shivaji. 2005. Role of membrane lipid fatty acids in cold adaptation. *Cell. Mol. Biol.* **50:**631–642.

Cronan, Jr., J. E., and E. P. Gelmann. 1973. An estimate of the minimum amount of unsaturated fatty acid required for growth of *Escherichia coli*. *J. Biol. Chem.* **248:**1188–1195.

Cybulski, L. E., D. Albanesi, M. C. Mansilla, S. Altabe, P. S. Aguilar, and D. de Mendoza. 2002. Mechanism of membrane fluidity optimization: isothermal control of the *Bacillus subtilis* acyl lipid desaturase. *Mol. Microbiol.* **45:**1379–1388.

Cybulski, L. E., G. del Solar, P. O. Craig, M. Espinosa, and D. de Mendoza. 2004. *Bacillus subtilis* DesR functions as a phosphorylation-activated switch to control membrane lipid fluidity. *J. Biol. Chem.* **17:**39340–39347.

Datta, P. P., and R. K. Bhadra. 2003. Cold shock response and major cold shock proteins of *Vibrio cholerae*. *Appl. Environ. Microbiol.* **69:**6361–6369.

De Mendoza, D., and J. E. Cronan, Jr. 1983. Thermal regulation of membrane lipid fluidity in bacteria. *Trends Biochem. Sci.* **8:**49–52.

de Wulf, P., B. J. Akerley, and E. C. Lin. 2000. Presence of the Cpx system in bacteria. *Microbiology* **146:**247–248.

Deming, J. W. 2002. Psychrophiles and polar regions. *Curr. Opin. Microbiol.* **5:**301–309.

Denich, T. J., L. A. Beaudette, H. Lee, and J. T. Trevors. 2003. Effect of selected environmental and physico-chemical factors on bacterial cytoplasmic membranes. *J. Microbiol. Methods* **52:**149–182.

Diaz, A. R., M. C. Mansilla, A. J. Vila, and D. de Mendoza. 2002. Membrane topology of the acyl-lipid desaturase from *Bacillus subtilis*. *J. Biol. Chem.* **277:**48099–48106.

Diefenbach, R., and H. Keweloh. 1994. Synthesis of *trans* unsaturated fatty acids in *Pseudomonas putida* P8 by direct isomerization of the double bond of lipids. *Arch. Microbiol.* **162:**120–125.

Diefenbach, R., H. J. Heipieper, and H. Keweloh. 1992. The conversion of *cis*- into *trans*-unsaturated fatty acids in *Pseudomonas putida* P8: evidence for a role in the regulation of membrane fluidity. *Appl. Microbiol. Biotechnol.* **38:**382–387.

Dorman, C. J., J. C. D. Hinton, and A. Free. 1999. Domain organization and oligomerization among H–NS-like nucleoid-associated proteins in bacteria. *Trends Microbiol.* **7:**124–128.

Drlica, K. 1992. Control of bacterial DNA supercoiling. *Mol. Microbiol.* **6:**425–433.

Epand, R. M. 1998. Lipid polymorphism and protein–lipid interactions. *Biochim. Biophys. Acta* **1376:**353–368.

Eriksson, S., R. Hurme, and M. Rhen. 2002. Low temperature sensors in bacteria. *Philos. Trans. R. Soc. Lond. B.* **357:**887–893.

Falconi, M., B. Colonna, G. Prosseda, G. Micheli, and C. O. Gualerzi. 1998. Thermoregulation of *Shigella* and *Escherichia coli* EIEC pathogenicity. A temperature dependent structural transition of DNA modulates accessibility of virF promoter to transcriptional repressor H–NS. *EMBO J.* **17:**7033–7043.

Feller, G., and C. Gerday. 2003. Psychrophilic enzymes: hot topics in cold adaptation. *Nat. Rev. Microbiol.* **1:**200–208.

Fujii, D. K., and A. J. Fulco. 1977. Biosynthesis of unsaturated fatty acids by Bacilli. Hyperinduction and modulation of desaturase synthesis. *J. Biol. Chem.* **252:**3660–3670.

Fulco, A. J. 1969. The biosynthesis of unsaturated fatty acids by Bacilli. I. Temperature induction of the desaturation reaction. *J. Biol. Chem.* **244:**889–895.

Gabrielska, J., and W. I. Gruszecki. 1996. Zeaxanthin (dihydroxy-beta-carotene) but not beta-carotene rigidifies lipid membranes: a 1H-NMR study of carotenoid-egg phosphatidylcholine liposomes. *Biochim. Biophys. Acta* **1285:**167–174.

Garwin, J. L., and J. E. Cronan, Jr. 1980. Thermal modulation of fatty acid synthesis in *Escherichia coli* does not involve de novo enzyme synthesis. *J. Bacteriol.* **141:**1457–1459.

Garwin, J. L., A. L. Klages, and J. E. Cronan, Jr. 1980. Structural, enzymatic and genetic studies of beta-ketoacyl-acyl carrier protein synthases I and II of *Escherichia coli*. *J. Biol. Chem.* **255:**11949–11956.

Georlette, D., V. Blaise, T. Collins, S. D'Amico, E. Gratia, A. Hoyoux, J. C. Marx, G. Sonan, G. Feller, and C. Gerday. 2004. Some like it cold: biocatalysis at low temperatures. *FEMS Microbiol. Rev.* **28:**25–42.

Goldstein, J., N. S. Pollitt, and M. Inouye. 1990. Major cold-shock protein of *Escherichia coli*. *Proc. Natl. Acad. Sci. USA* **87:**283–287.

Goodchild, A., N. F. W. Saunders, H. Ertan, M. Raftery, M. Guilhaus, P. M. G. Curmi, and R. Cavicchioli. 2004. A proteomic determination of cold adaptation in the Antarctic archaeon, *Methanococcoides burtonii*. *Mol. Microbiol.* **53:**309–321.

Grau, R., and D. de Mendoza. 1993. Regulation of the synthesis of unsaturated fatty acids by growth temperature in *Bacillus subtilis*. *Mol. Microbiol.* **8:**535–542.

Grau, R., D. Gardiol, G. C. Glikin, and D. de Mendoza. 1994. DNA supercoiling and thermal regulation of unsaturated fatty acid synthesis in *Bacillus subtilis*. *Mol. Microbiol.* **11:**933–941.

Graumann, P. L., and M. A. Marahiel. 1999. Cold shock response in *Bacillus subtilis*. *J. Mol. Microbiol. Biotechnol.* **1:**203–209.

Graumann, P., T. M. Wendrich, M. H. Weber, K. Schröder, and M. A. Marahiel. 1997. A family of cold shock proteins in *Bacillus subtilis* is essential for cellular growth and for efficient protein synthesis at optimal and low temperatures. *Mol. Microbiol.* **25:**741–756.

Gulig, P. A., H. Danbara, D. G. Guiney, A. J. Lax, F. Norel, and M. Rhen. 1993. Molecular analysis of *spv* virulence genes of the *Salmonella* virulence plasmid. *Mol. Microbiol.* **7:**825–830.

Gupta, P., G. S. N. Reddy, D. Delille, and S. Shivaji. 2004. *Arthrobacter gangotriensis* sp. nov. and *Arthrobacter kerguelensis* sp. nov. from Antarctica. *Int. J. Syst. Evol. Microbiol.* **54:**2375–2378.

Hasegawa, Y., N. Kawada, and Y. Nosho. 1980. Change in chemical composition of membrane of *Bacillus caldotenax* after shifting the growth temperature. *Arch. Microbiol.* **126:**103–108.

Heath, R. J., and C. O. Rock. 2002. The Claisen condensation in biology. *Nat. Prod. Rep.* **19:**581–596.

Heipieper, H. J., and J. A. M. De Bont. 1994. Adaptation of *Pseudomonas putida* S12 to ethanol and toluene at the level of fatty acid composition of membranes. *Appl. Environ. Microbiol.* **60:**4440–4444.

Heipieper, H. J., R. Diefenbach, and H. Keweloh. 1992. Conversion of *cis* unsaturated fatty acids to *trans*, a possible mechanism for the protection of phenol degrading *Pseudomonas putida* P8 from substrate toxicity. *Appl. Environ. Microbiol.* **58:**1847–1852.

Heipieper, H. J., F. J. Weber, J. Sikkema, H. Keweloh, J. A. M. de Bont. 1994. Mechanisms of resistance of whole cells to toxic organic solvents. *Trends Biotechnol.* **12:**409–414.

Heipieper, H. J., F. Meinhardt, and A. Segura. 2003. The *cis–trans* isomerase of unsaturated fatty acids in *Pseudomonas* and *Vibrio*: biochemistry, molecular biology and physiological function of a unique stress adaptive mechanism. *FEMS Microbiol. Lett.* **229:**1–7.

Hellingwerf, K. J., W. C. Crielaard, M. J. Teixeira de Mattos, W. D. Hoff, R. Kort, D. T. Verhamme, and C. Avignone-Rossa. 1998. Current topics in signal transduction in bacteria. *Antonie Leeuwenhoek* **74:**211–227.

Hoch, J. A. 2000. Two component and phosphorelay signal transduction. *Curr. Opin. Microbiol.* **3:**165–170.

Hoe, N. P., and J. D. Gougen. 1993. Temperature sensing in *Yersinia pestis*: translation of the LerF activator protein is thermally regulated. *J. Bacteriol.* 175:7901–7909.

Hurme, R., and M. Rhen. 1998. Temperature sensing in bacterial gene regulation—what it all boils down to. *Mol. Microbiol.* 30:1–6.

Hurme, R., K. Berndt, D. Namork, and M. Rhen. 1996. DNA binding exerted by a bacterial gene regulator with extensive coiled coil domains. *J. Biol. Chem.* 272:12626–12631.

Hurme, R., K. Berndt, S. J. Normark, and M. Rhen. 1997. A proteinaceous gene regulatory thermometer in *Salmonella*. *Cell* 90:55–64.

Inaba, M., I. Suzuki, B. Szalontai, Y. Kanesaki, D. A. Los, H. Hayashi, and N. Murata. 2003. Gene-engineered rigidification of membrane lipids enhances the cold inducibility of gene expression in *Synechocystis*. *J. Biol. Chem.* 278:12191–12198.

Jagannadham, M. V., V. Jayathirtha Rao, and S. Shivaji. 1991. The major carotenoid pigment of a psychrotrophic *Micrococcus roseus*: purification, structure and interaction of the pigment with synthetic membranes. *J. Bacteriol.* 173:7911–7917.

Jagannadham, M. V., M. K. Chattopadhyay, and S. Shivaji. 1996a. The major carotenoid pigment of a psychrotrophic *Micrococcus roseus* strain: fluorescence properties of the pigment and its binding to membranes. *Biochem. Biophys. Res. Commun.* 220:724–728.

Jagannadham, M. V., K. Narayanan, Ch. Mohan Rao, and S. Shivaji. 1996b. In vivo characteristics and localisation of carotenoid pigments in psychrotrophic and mesophilic *Micrococcus roseus* using photoacoustic spectroscopy. *Biochem. Biophys. Res. Commun.* 227:221–226.

Jagannadham, M. V., M. K. Chattopadhyay, C. Subbalakshmi, M. Vairamani, K. Narayanan, Ch. Mohan Rao, and S. Shivaji. 2000. Carotenoids of an Antarctic psychrotolerant bacterium *Sphingobacterium antarcticus* and a mesophilic bacterium *Sphingobacterium multivorum*. *Arch. Microbiol.* 173:418–424.

Jiang, W., Y. Hou, and M. Inouye. 1997. CspA, the major cold shock protein of *Escherichia coli* is an RNA chaperone. *J. Biol. Chem.* 272:196–202.

Jones, P. G., and M. Inouye. 1994. The cold shock response: a hot topic. *Mol. Microbiol.* 11:811–818.

Jones, P. G., R. A. VanBogelan, and F. C. Neidhardt. 1987. Induction of proteins in response to low temperature in *Escherichia coli*. *J. Bacteriol.* 169:2092–2095.

Junker, F., and J. L. Ramos. 1999. Involvement of the *cis/trans* isomerase Cti in solvent resistance of *Pseudomonas putida* DOT-T1E. *J. Bacteriol.* 181:5693–5700.

Kaan, T., G. Homuth, U. Mäder, J. Bandow, and T. Schweder. 2002. Genome-wide transcriptional profiling of the *Bacillus subtilis* cold-shock response. *Microbiology* 148:3441–3455.

Kaneda, T. 1991. Iso- and anteiso-fatty acids in bacteria: biosynthesis, function and taxonomic significance. *Microbiol. Rev.* 55:288–302.

Kasamo, K. 2003. Regulation of plasma membrane H+-ATPase activity by the membrane environment. *J. Plant Res.* 116:517–523.

Kenyon, C. N. 1972. Fatty acid composition of unicellular strains of blue-green algae. *J. Bacteriol.* 109:827–834.

Kenyon, C. N., R. Rippka, and R. Y. Stanier. 1972. Fatty acid composition and physiological properties of some filamentous blue-green algae. *Arch. Mikrobiol.* 83:216–236.

Keweloh, H., and H. J. Heipieper. 1996. *Trans*-unsaturated fatty acids in bacteria. *Lipids* 31:129–137.

Kiran, M. D., J. S. S. Prakash, S. Annapoorni, S. Dube, T. Kusano, H. Okuyama, N. Murata, and S. Shivaji. 2004. Psychrophilic *Pseudomonas syringae* required *trans* monounsaturated fatty acid for growth at higher temperature. *Extremophiles* 8:401–410.

Kiran, M. D., S. Annapoorni, I. Suzuki, N. Murata, and S. Shivaji. 2005. *Cis-trans* isomerase gene in psychrophilic *Pseudomonas syringae* is constitutively expressed during growth and under conditions of temperature and solvent stress. *Extremophiles* 9:117–125.

Klein, W., M. H. Weber, and M. A. Marahiel. 1999. Cold shock response of *Bacillus subtilis*: isoleucine dependent switch in the fatty acid branching pattern for membrane adaptation to low temperatures. *J. Bacteriol.* 181:5341–5349.

Kunst, F., N. Ogasawara, I. Moszer, A. M. Albertini, G. Alloni, V. Azevedo, M. G. Bertero, P. Bessieres, A. Bolotin, S. Borchert, R. Borriss, L. Boursier, A. Brans, M. Brauwn, S. C. Brignell, S. Born, S. Brouillet, C. V. Bruschi, B. Caldwell, V. Capuano, N. M. Carter, S. K. Choi, J. J. Codani, I. F. Connerton, A. Danchin et al. 1997. The complete genome sequence of the gram-positive bacterium *Bacillus subtilis*. *Nature* 390:249–256.

Lai, E. C. 2003. RNA sensors and riboswitches: self-regulating messages. *Curr. Biol.* 13:R285–R291.

Lease, R. A., and M. Belfort. 2000. A *trans*-acting RNA as a control switch in *Escherichia coli*: DsrA modulates function by forming alternative structures. *Proc. Natl. Acad. Sci. USA* 97:9919–9924.

Lee, S. J., A. Xie, W. Jiang, J. P. Etchegaray, P. G. Jones, and M. Inouye. 1994. Family of the major cold-shock protein, CspA (CS7.4) of *Escherichia coli*, whose members show a high sequence similarity with the eukaryotic Y-box binding proteins. *Mol. Microbiol.* 11:833–839.

Los, D. A., and N. Murata. 2000. Regulation of enzymatic activity and gene expression by membrane fluidity. *Science's STKE* 2000:pe1. doi: 10.1126/stke.2000.62.pe1.

Majdalani, N., C. Cunning, D. Sledjeski, T. Elliot, and S. Gottesman. 1998. DsrA RNA regulates translation of RpoS message by an anti-antisense mechanism independent of its action as an antisilencer of transcription. *Proc. Natl. Acad. Sci. USA* 95:12462–12467.

Majdalani, N., C. K. Vanderpool, and S. Gottesman. 2005. Bacterial small RNA regulators. *Crit. Rev. Biochem. Mol. Biol.* 40:93–113.

Makula, R. A. 1978. Phospholipid composition of methane-utilizing bacteria. *J. Bacteriol.* 134:771–777.

Mansilla, M. C., P. S. Aguilar, D. Albanesi, L. E. Cybulski, S. Altabe, and D. de Mendoza. 2003. Regulation of fatty acid desaturation in *Bacillus subtilis*. *Prostaglandins Leukot. Essent. Fatty Acids* 68:187–190.

Mansilla, M. C., L. E. Cybulski, D. Albanesi, and D. de Mendoza. 2004. Control of membrane lipid fluidity by molecular thermosensors. *J. Bacteriol.* 186:6681–6688.

Marr, A. G., and J. J. Ingraham. 1962. Effect of temperature on the composition of fatty acids in *Escherichia coli*. *J. Bacteriol.* 84:1260–1267.

Marx, J. C., V. Blaise, T. Collins, S. D'Amico, D. Delille, E. Gratia, A. Hoyoux, A. L. Huston, G. Sonan, G. Feller, and C. Gerday. 2004. A perspective on cold enzymes: current knowledge and frequently asked questions. *Cell. Mol. Biol.* 50:643–655.

McElhaney, R. N. 1982. Effects of membrane lipids on transport and enzymic activities. *Curr. Top. Membr. Transp.* 17:317–380.

Mikami, K., I. Suzuki, and N. Murata. 2003. 4 sensors of abiotic stress in *Synechocystis*. *Topics Curr. Gen.* 4:103–119.

Mizuno, T. 1997. Compilation of all genes encoding two-component phosphotransfer signal transducers in the genome of *Escherichia coli*. *DNA Res.* 4:161–168.

Mizuno, T., T. Kaneko, and S. Tabata. 1996. Compilation of all genes encoding bacterial two-component signal transducers in the genome of the cyanobacterium, *Synechocystis* sp. strain PCC 6803. *DNA Res.* 3:407–414.

Morita, N., A. Shibahara, K. Yamamoto, K. Shinkai, G. Kajimoto, and H. Okuyama. 1993. Evidence for *cis-trans* isomerization of a double bond in the fatty acids of the psychrophilic bacterium *Vibrio* sp. strain ABE-1. *J. Bacteriol.* 175:916–918.

Murata, N. 1989. Low temperature effects of cyanobacterial membranes. *J. Bioenerg. Biomembr.* 21:61–75.

Murata, N., and H. Wada. 1995. Acyl lipid desaturases and their importance in the tolerance and acclimatization to cold of cyanobacteria. *Biochem. J.* 308:1–8.

Murata, N., and L. A. Los. 1997. Membrane fluidity and temperature perception. *Plant Physiol.* 115:875–879.

Murata, N., H. Wada, and Z. Gombos. 1992. Modes of fatty-acid desaturation in cyanobacteria. *Plant Cell Physiol.* 33:933–941.

Narberhaus, F., T. Waldminghaus, and S. Chowdhury. 2005. RNA thermometers. *FEMS Microbiol. Rev.* 20:1–14.

Newkirk, K., W. Feng, W. Jiang, R. Tejero, S. D. Emerson, M. Inouye, and G. Montelione. 1994. Solution NMR structure of the major cold-shock proteins (CspA) from *Escherichia coli*: identification of a binding epitope for DNA. *Proc. Natl. Acad. Sci. USA* 91:5114–5118.

Nishida, I., and N. Murata. 1996. Chilling sensitivity in plants and cyanobacteria: the crucial contribution of membrane lipids. *Annu. Rev. Plant Physiol. Plant Mol. Biol.* 47:541–568.

Okuyama, H., S. Sasaki, S. Higashi, and N. Murata. 1990. A *trans*-unsaturated fatty acid in a psychrophilic bacterium, *Vibrio* sp. strain ABE-1. *J. Bacteriol.* 172:3515–3518.

Okuyama, H., N. Okajima, S. Sasaki, S. Higashi, and N. Murata. 1991. The *cis*/*trans* isomerization of the double bond of a fatty acid as a strategy for adaptation to changes in ambient temperature in the psychrophilic bacterium, *Vibrio* sp. strain ABE-1. *Biochim. Biophys. Acta* 1084:13–20.

Orlova, I. V., T. S. Serebriiskaya, V. Popov, N. Merkulova, A. M. Nosov, T. I. Trunova, V. D. Tsyendambaev, and D. A. Los. 2003. Transformation of tobacco with a gene for the thermophilic acyl-lipid desaturase enhances the chilling tolerance of plants. *Plant Cell Physiol.* 44:447–450.

Park, H., S. K. Saha, and M. Inouye. 1998. Two-domain reconstitution of a functional protein histidine kinase. *Proc. Natl. Acad. Sci. USA* 95:6728–6732.

Phadtare, S., J. Alsina, and M. Inouye. 1999. Cold-shock response and cold-shock proteins. *Curr. Opin. Microbiol.* 2:175–180.

Prabagaran, S. R., K. Suresh, R. Manorama, D. Delille, and S. Shivaji. 2005. *Marinomonas ushuaiensis* sp. nov., isolated from coastal seawater in Ushuaia, Argentina, Sub-Antarctica. *Int. J. Syst. Evol. Microbiol.* 55:309–313.

Price, P. B. 2000. A habitat for psychrophiles in deep Antarctic ice. *Proc. Natl. Acad. Sci. USA* 97:1247–1251.

Quinn, P. J. 1981. The fluidity of cell membranes and its regulation. *Prog. Biophys. Mol. Biol.* 38:1–104.

Ray, M. K., G. Seshu Kumar, and S. Shivaji. 1994a. Phosphorylation of membrane proteins in response to temperature in an Antarctic *Pseudomonas syringae*. *Microbiology* 140:3217–3223.

Ray, M. K., G. Seshu Kumar, and S. Shivaji. 1994b. Tyrosine phosphorylation of a cytosolic protein from the antarctic psychrotrophic bacterium *Pseudomonas syringae*. *FEMS Microbiol. Letts.* 122:49–54.

Ray, M. K., G. Seshu Kumar, and S. Shivaji. 1994c. Phosphorylation of lipopolysaccharides in the Antarctic psychrotroph *Pseudomonas syringae*: a possible role in temperature adaptation. *J. Bacteriol.* 176:4243–4249.

Ray, M. K., G. Seshukumar, K. Janiyani, K. Kannan, P. Jagtap, M. K. Basu, and S. Shivaji. 1998. Adaptation to low temperature and regulation of gene expression in Antarctic psychrotrophic bacteria. *J. Biosci.* 23:423–435.

Reddy, G. S. N., R. K. Aggarwal, G. I. Matsumoto, and S. Shivaji. 2000. *Arthrobacter flavus* sp. nov., a psychrophilic bacterium isolated from a pond in McMurdo Dry Valley, Antarctica. *Int. J. Syst. Evol. Microbiol.* 50:1553–1561.

Reddy, G. S. N., J. S. S. Prakash, M. Vairamani, S. Prabhakar, G. I. Matsumoto, and S. Shivaji. 2002a. *Planococcus antarcticus* and *Planococcus psychrophilus* spp. nov. isolated from cyanobacterial mat samples collected from ponds in Antarctica. *Extremophiles* 6:253–261.

Reddy, G. S. N., J. S. S. Prakash, G. I. Matsumoto, E. Stackebrandt, and S. Shivaji. 2002b. *Arthrobacter roseus* sp. nov., a psychrophilic bacterium isolated from an Antarctic cyanobacterial mat sample. *Int. J. Syst. Evol. Microbiol.* 52:1017–1021.

Reddy, G. S. N., P. U. M. Raghavan, N. B. Sarita, J. S. S. Prakash, N. Nagesh, D. Delille, and S. Shivaji. 2003a. *Halomonas glaciei* sp. nov. isolated from fast ice of Adelie Land, Antarctica. *Extremophiles* 7:55–61.

Reddy, G. S. N., J. S. S. Prakash, R. Srinivas, G. I. Matsumoto, and S. Shivaji. 2003b. *Leifsonia rubra* sp. nov. and *Leifsonia aurea* sp. nov. psychrophilic species isolated from a pond in Antarctica. *Int. J. Syst. Evol. Microbiol.* 53:977–984.

Reddy, G. S. N., J. S. S. Prakash, V. Prabahar, G. I. Matsumoto, E. Stackebrandt, and S. Shivaji. 2003c. *Kocuria polaris* sp. nov., an orange pigmented psychrophilic bacterium isolated from an Antarctic cyanobacterial mat sample. *Int. J. Syst. Evol. Microbiol.* 53:183–187.

Reddy, G. S. N., G. I. Matsumoto, P. Schumann, E. Stackebrandt, and S. Shivaji. 2004. Psychrophilic pseudomonads from Antarctica: *Pseudomonas antarctica* sp. nov., *Pseudomonas meridiana* sp. nov. and *Pseudomonas proteolytica* sp. nov. *Int. J. Syst. Evol. Microbiol.* 54:713–719.

Romby, P., and C. Ehresmann. 2003. At the flick of a switch: a *Listeria* mRNA turns on and off its own expression in response to temperature. *The ELSO Gazette*, http://www.the-elso-gazette.org/magazines/issue14/mreviews/mreviews1.asp.

Rowbury, R. J. 2003. Temperature effects on biological systems: introduction. *Sci. Prog.* 86:1–8.

Russell, N. J. 1984. Mechanisms of thermal adaptation in bacteria: blueprints for survival. *Trends Biochem. Sci.* 9:108–112.

Russel, N. J. 1990. Cold adaptation of micro-organisms. *Philos. Trans. R. Soc. Lond. B.* 326:595–611.

Sakamoto, T., and N. Murata. 2002. Regulation of the desaturation of fatty acids and its role in tolerance to cold and salt stress. *Curr. Opin. Microbiol.* 5:206–210.

Sakamoto, T., G. Shen, S. Higashi, N. Murata, and D. A. Bryant. 1998. Alteration of low-temperature susceptibility of the cyanobacterium *Synechococcus* sp. PCC 7002 by genetic manipulation of membrane lipid unsaturation. *Arch. Microbiol.* 169:20–28.

Sato, N., and N. Murata. 1980. Temperature shift-induced responses in lipids in the blue-green alga, *Anabaena variabilis*. The central role of diacylmonogalactosylglycerol in thermo-adaptation. *Biochim. Biophys. Acta* 619:353–366.

Sato, N., N. Murata, Y. Miura, and N. Ueta. 1979. Effect of growth temperature on lipid and fatty acid compositions in the blue-green algae, *Anabaena variabilis* and *Anacystis nidulans*. *Biochim. Biophys. Acta* 572:19–28.

Schindler, T., P. L. Graumann, D. Perl, S. Ma, F. X. Schmid, and M. A. Marahiel. 1999. The family of cold shock proteins of *Bacillus subtilis*. Stability and dynamics in vitro and in vivo. *J. Biol. Chem.* 274:3407–3413.

Shivaji, S. 2005. Microbial diversity and molecular basis of cold adaptation in Antarctic bacteria, p. 3–24. *In* T. Satyanarayana and B. N. Johri (ed.), *Microbial Diversity: Current Perspectives and Potential Applications*. I.K. International Pvt. Ltd., New Delhi, India.

Shivaji, S., and M. K. Ray. 1995. Survival of strategies of psychrotrophic bacteria and yeasts of Antarctica. *Indian J. Microbiol.* 35:263–281.

Shivaji, S., N. S. Rao, L. Saisree, V. Sheth, G. S. N. Reddy, and P. M. Bhargava. 1988. Isolation and identification of *Micrococcus roseus* and *Planococcus* sp. from Schirmacher Oasis, Antarctica. *J. Biosci.* **13**:409–414.

Shivaji, S., M. K. Ray, N. Shyamala Rao, L. Saisree, M. V. Jagannadham, G. Seshu Kumar, G. S. N. Reddy, and P. M. Bhargava. 1992. *Sphingobacterium antarcticus* sp. nov. a psychrotrophic bacterium from the soils of Schirmacher Oasis, Antarctica. *Int. J. Syst. Bacteriol.* **42**:102–116.

Shivaji, S., G. S. N. Reddy, P. U. M. Raghavan, N. B. Sarita, and D. Delille. 2004. *Psychrobacter salsus* sp. nov. and *Psychrobacter adeliensis* sp. nov. isolated from fast ice from Adelie Land, Antarctica. *Syst. Appl. Microbiol.* **27**:628–635.

Shivaji, S., G. S. N. Reddy, R. P. Aduri, R. Kutty, and K. Ravenschlag. 2005. Bacterial diversity of a soil sample from Schirmacher Oasis, Antarctica. *Cell. Mol. Biol.* **50**:525–536.

Sledjeski, D. D., A. Gupta, and S. Gottesman. 1996. The small RNA, DsrA, is essential for the low temperature expression of RpoS during exponential growth in *Escherichia coli*. *EMBO J.* **15**:3993–4000.

Sommerville, J., and M. Ladomery. 1996. Masking of mRNA by Y-box proteins. *FASEB J.* **10**:435–443.

Sonnenfield, J. M., C. M. Burns, C. F. Higgins, and J. Hinton. 2001. The nucleoid-associated protein StpA binds curved DNA, has a greater DNA-binding affinity than H-NS and is present in significant levels in hns mutants. *Biochimie* **83**:243–249.

Stetter, K. O. 1996. Hyperthermophilic prokaryotes. *FEMS Microbiol. Rev.* **18**:49–158.

Straley, S., and R. D. Perry. 1995. Environmental modulation of gene expression and pathogenesis in *Yersinia*. *Trends Microbiol.* **3**:310–317.

Subczynski, W. K., E. Markowska, W. I. Gruszecki, and J. Sielewiesiuk. 1992. Effect of polar carotenoids on dimyristoylphosphatidylcholine membranes: a spin-label study. *Biochim. Biophys. Acta* **1105**:97–108.

Suutari, M., and S. Laakso. 1992. Changes in fatty acid branching and unsaturation of *Streptomyces griseus* and *Brevibacterium fermentans* as a response to growth temperature. *Appl. Environ. Microbiol.* **58**:2338–2343.

Suutari, M., and S. Laakso. 1994. Microbial fatty acids and thermal adaptation. *Crit. Rev. Microbiol.* **20**:285–328.

Suzuki, I., D. A. Los, Y. Kanesaki, K. Mikami, and N. Murata. 2000a. The pathway for perception and transduction of low-temperature signals in *Synechocystis*. *EMBO J.* **19**:1327–1334.

Suzuki, I., D. A. Los, and N. Murata. 2000b. Perception and transduction of low temperature signals to induce desaturation of fatty acids. *Biochem. Soc. Trans.* **28**:626–630.

Suzuki, I., Y. Kanasaki, K. Mikami, M. Kanehisa, and N. Murata. 2001. Cold regulated genes under the control of cold sensor hik33 in *Synechocystis*. *Mol. Microbiol.* **40**:235–245.

Szalontai, B., Y. Nishiyaina, Z. Gombos, and N. Murata. 2000. Membrane dynamics as seen by Fourier transform infrared spectroscopy in a cyanobacterium, *Synechocystis* PCC 6803. The effects of lipid unsaturation and the protein-to-lipid ratio. *Biochim. Biophys. Acta* **1509**:409–419.

Takeuchi, Y., S. I. Ohnishi, M. Ishinaga, and M. Kito. 1978. Spin-labeling of *Escherichia coli* membrane by enzymatic synthesis of phosphatidylglycerol and divalent cation-induced interaction of phosphatidylglycerol with membrane proteins. *Biochim. Biophys. Acta* **506**:54–63.

Takeuchi, Y., S. I. Ohnishi, M. Ishinaga, and M. Kito. 1981. Dynamic states of phospholipids in *Escherichia coli* B membrane. Electron spin resonance studies with biosynthetically generated phospholipid spin labels. *Biochim. Biophys. Acta* **646**:119–125.

Tasaka, Y., Z. Gombos, Y. Nishiyama, P. Mohanty, T. Ohba, K. Ohki, and N. Murata. 1996. Targeted mutagenesis of acyl-lipid desaturases in *Synechocystis*: evidence for the important roles of polyunsaturated membrane lipids in growth, respiration and photosynthesis. *EMBO J.* **15**:6416–6425.

Taylor, B. L., and I. B. Zhulin. 1999. PAS domains: internal sensors of oxygen, redox potential and light. *Microbiol. Mol. Biol. Rev.* **63**:479–506.

Tobe, T., M. Yoshokawa, T. Mizuno, and C. Sasakawa. 1993. Transcriptional control of the invasion regulatory gene *virB* of *Shigella exneri*: activation by VirF and repression by H–NS. *J. Bacteriol.* **175**:6142–6149.

Tse-Dinh, Y.-C., H. Qi, and R. Menzel. 1997. DNA supercoiling and bacterial adaptation: thermotolerance and thermoresistance. *Trends Microbiol.* **5**:323–326.

Vigh, L., D. A. Los, I. Horvath, and N. Murata. 1993. The primary signal in the biological perception of temperature: Pd-catalyzed hydrogenation of membrane lipids stimulated the expression of the *desA* gene in *Synechocystis* PCC6803. *Proc. Natl. Acad. Sci. USA* **90**:9090–9094.

von Wallbrunn, A., H. H. Richnow, G. Neumann, F. Meinhardt, and H. J. Heipieper. 2003. Mechanism of *cis–trans* isomerization of unsaturated fatty acids in *Pseudomonas putida*. *J. Bacteriol.* **185**:1730–1733.

Wada, H., and N. Murata. 1989. *Synechocystis* sp. PCC6803 mutants defective in desaturation of fatty acids. *Plant Cell Physiol.* **30**:971–978.

Wada, H., and N. Murata. 1990. Temperature-induced changes in the fatty acid composition of the cyanobacterium, *Synechocystis* PCC6803. *Plant Physiol.* **92**:1062–1069.

Wada, H., Z. Gombos, and N. Murata. 1990. Enhancement of a chilling tolerance of a cyanobacterium by genetic manipulation of fatty acid desaturation. *Nature* **347**:200–203.

Weber, M. H. W., and A. M. Marahiel. 2003. Bacterial cold shock responses. *Sci. Prog.* **86**:9–75.

Williams, R. M., and S. Rimsky. 1997. Molecular aspects of the *E. coli* nucleoid protein, H-NS: a central controller of gene regulatory networks. *FEMS Microbiol. Lett.* **156**:175–185.

Xia, B., H. Ke, U. Shinde, and M. Inouye. 2003. The role of RbfA in 16S rRNA processing and cell growth at low temperature in *Escherichia coli*. *J. Mol. Biol.* **332**:575–584.

Xu, Y., G. Feller, C. Gerday, and N. Glansdorff. 2003. Metabolic enzymes from psychrophilic bacteria: challenge of adaptation to low temperatures in ornithine carbamoyltransferase from *Moritella abyssi*. *J. Bacteriol.* **185**:2161–2168.

Zhang, W., and L. Shi. 2005. Distribution and evolution of multiple-step phosphorelay in prokaryotes: lateral domain recruitment involved in the formation of hybrid-type histidine kinases. *Microbiology* **151**:2159–2173.

Chapter 16

An Interplay between Metabolic and Physicochemical Constraints: Lessons from the Psychrophilic Prokaryote Genomes

ANTOINE DANCHIN

INTRODUCTION

In a ratchet-driven evolutionary process, the selection pressure resulting from the environment together with that from the biochemical objects that make a cell operates on the phenotype. Several levels, nested in a top-down way, cooperate to build up the phenotype: the population, the organism, the cell. In contrast, the genetic program transmitted from generation to generation operates at the DNA level and influences the phenotype in a bottom-up way. To survive and reproduce, the organisms need to display strong coupling processes associating the phenotype and the chromosome structure, taking into account both bottom-up physicochemical constraints and top-down constraints deriving from the process of evolution itself. The usual way to consider evolution is to contemplate the action of the global adaptation of the organism to its environment, rather than consider, in the cell, the organism or the population, the influence of the lowest grain of life, the basic cell components. The DNA structure, for example, is constrained not only by its physicochemical buildup as a charged polymer but also by the very nature of the building blocks making the cell and by their interconversions, metabolism in short. Presence or absence of a single metabolic gene will considerably alter the nucleotide content of the genome, independently of the environment (see for example the G+C increase of *Escherichia coli* DNA when *mutT* [Cox and Yanofsky, 1967] or *ndk* is absent [Miller et al., 2002]). Comparing organisms growing in very diverse environmental conditions gives us a handle to tackle the problem. Temperature, in particular, is an excellent probe of the effect of all kinds of selective constraints as, except perhaps in homeothermic animals, no process, even deeply imbedded in the cell, can escape its influence. Microbes, which can develop in liquid water from −10 to 110°C, provide us with the largest range of concrete responses to temperature.

While the astronomical position of the Earth should make it a frozen planet, the greenhouse effect mediated by carbon dioxide keeps water mostly in liquid form at its surface. Most of it, however, remains almost permanently below 15°C, and the water temperature of the deep oceans stays between 2 and 4°C. The discovery by Carl Woese that many of the prokaryotes living under hot conditions constituted a new domain of life (Woese and Fox, 1997), in parallel with challenges to the primeval soup model of the origin of life in rich broth at medium temperature (see e.g. Danchin, 1989), prompted much interest for thermophilic and even hyperthermophilic organisms, resulting in a lack of interest for organisms living under cold conditions. The interest for high temperature was further supported by the name "*Archaea*," which suggested that those conditions had something very ancient and were indeed linked to the origin of life. The feeling was so prevalent that nobody challenged immediately the claim by Baross and Deming that living organisms multiplied at temperatures as high as 300°C, under very high pressure (Baross and Deming, 1983). It was soon recognized that many building blocks of biomolecules would only stand for a few minutes at such high temperature (in particular aspartate) (Trent et al., 1984; White, 1984). The question of the existence of nucleotides remained particularly intriguing since their carbohydrate moiety, ribose in particular, is quite unstable even at ordinary temperature (Larralde et al., 1995). Despite these drawbacks, the constraints posed by relatively high temperature (when lower than that of boiling water) do not considerably alter living processes. The situation in the cold would be much worse. Indeed, while the structure of water does not change very much until it boils, (pure) water undergoes

---

A. Danchin • Genetics of Bacterial Genomes URA 2171 CNRS, Institut Pasteur, 28, rue du Docteur Roux, 75724 Paris Cedex 15, France.

dramatic changes below 10 to 12°C, and this can be related to cold denaturation of macromolecules (Tsai et al., 2002; Wernet et al., 2004). Organisms living in the cold are therefore of special interest.

In this chapter, I summarize the knowledge we have about the genomes of psychrophilic bacteria, a subclass of the cold-adapted bacteria, with emphasis on the specific selective features relevant to cold adaptation. Known psychrophilic *Bacteria* mostly belong to five phylogenetic groups, the alpha, gamma and delta subdivisions of *Proteobacteria*, the *Cytophaga-Flavobacterium-Bacteroides* clade and the gram-positive branch. Isolates from cold environments include *Arthrobacter* sp., *Psychrobacter* sp., and members of the genera *Halomonas, Pseudomonas, Hyphomonas,* and *Sphingomonas* (Bowman et al., 1997). In the case of *Archaea*, several studies have shown that this kingdom is well represented in cold environments (Schleper et al., 1998; Nozhevnikova et al., 2001; Saunders et al., 2003; Sheridan et al., 2003; Simankova et al., 2003; Cavicchioli, 2006). However, only a few cold growing *Bacteria* and *Archaea* have yet been submitted to genomic studies. Despite this, we shall see that interesting rules begin to emerge about the way living organisms cope with cold. Furthermore, genes and gene regions from psychrophiles have been studied for some time, and when this informs us about some of their adaptive properties, this will be discussed. Finally, there is some overlap between cold-adapted bacteria and aquatic conditions, as most of the cold regions on the globe are aquatic (mostly marine). While there are only a few completely sequenced genomes at the present time, several are in progress and their number is expected to rise steadily in the few next years.

PSYCHROPHILY: A MATTER OF OPINION?

Cold-tolerant bacteria are classified into psychrotrophs (from the Greek ψυχρος, cold, and τροφειν, to eat) and psychrophiles (φιλειν, to like). It is generally accepted that psychrotrophic organisms can grow at low temperature, even at or below 0°C, but that their "optimum" growth temperature is higher than 20°C. In contrast, psychrophilic organisms would be those with an optimum growth temperature lower than 15°C, and they would not be supposed to grow at temperatures higher than 20 to 25°C. Unfortunately, as always in biology, as soon as a rule is established, a counter-example is found. The definition of psychrophily given above assumes that one knows what an optimum growth temperature is. This, of course, will depend on the growth medium and not only on the temperature (Ron, 1975), while the conditions needed to establish the fastest growth of bacteria are far from being generally established. The domain of bacteria isolation and identification of growth factors is still highly empirical. Furthermore, the existence of an optimum assumes that organisms can grow as isolated entities, which is not the case of the vast majority of living organisms. The very concept of "temperature optimum" is fuzzy, as this refers to bacteria that are cultivatable in the laboratory, and depends considerably on the culture medium used.

In summary, we will retain as a rule of thumb the following classification and retain in this article only organisms that are generally considered as psychrophiles. First, psychrophiles have an optimum growth temperatures of 8 to 10 to 20°C and are capable of growth at temperatures less than 7°C. Second, mesophiles have an optimum growth at temperatures of 25 to 40°C and do no grow below 7 to 10°C. Third, thermophiles have an optimum growth at temperatures higher than 55° C. The paradigm of a fourth category, the poorly defined psychrotrophs, is *Listeria monocytogenes*, which, although developing well at 37°C can grow in refrigerators, where it becomes a dangerous food pathogen. Because these organisms are in fact a sub-category of mesophiles, they will not be discussed here. It is also worth mentioning that at the borders of this temperature range, there exists hyper-extremophiles: in our case, Glansdorff and Xu have reviewed the existence of prokaryotes living in polar permafrost or in other permanently cold conditions with remarkably low optimum growth temperatures (Glansdorff and Xu, 2002). In spite of their considerable interest, they cannot be analyzed yet in the absence of enough genome data.

While this discussion may appear to be simply semantic, delineating the true conditions defining psychrophily is important when one tries to find rules governing growth at low temperature: the constraints suffered by organisms that can grow in a large band of temperatures are obviously different from those which can only sustain growth in a narrow window (Phadtare et al., 1999). We shall discuss in this review chapter constraints that may be revealed by genome sequences, referring to other chapters which will provide the reader with a detailed analysis of each specific topic. Namely we can recognize six major constraints:

1. *Catalysis*. The constraint is discussed by Gerday, and as it probably does not have an important contribution to genome organization, we shall not discuss it further (Chapter 13).

2. *Shaping of macromolecules*. The second constraint is linked to the first one. However, because it

interferes globally with the overall properties of the cell, we shall discuss what has already been observed in genome studies in this domain.

3. *Stability of macromolecules (aging processes).* Stability of macromolecules is linked to the first two constraints. It has also specific features, in particular related to aging processes, which will be discussed.

4. *Changes in solubility at low temperature (gases and radicals).* Gases and radicals are more soluble and stable at low temperature; this has consequences which are already visible in the features extracted from genome annotations.

5. *Membrane fluidity.* Cold conditions are known to rigidify membrane structures: we can see in genomes how this has been dealt with.

6. *Viscosity.* Viscosity is discussed by Gerday (Lobry and Necsulea, 2006) for its role in catalysis. It may also have implications in the genome organization. However, in the absence of enough data on gene organization in genomes, we shall not discuss in details.

Before analyzing the constraints 2 to 5 we shall briefly summarize what was known at the onset of genome programs.

EARLY STUDIES

As early as in 1991, the sequences of large bacterial genome continuous sections (100 kb or more) were known, and when the first complete bacterial genome sequences were published in 1995, continuous sequences of microbial genomes' sections longer than those of the complete genomes then published had already permitted considerable in silico analysis of global genome properties. Ten genome sequences were known by the end of 1997, one of a member of the *Eukarya* (*Saccharomyces cerevisiae*), two members of the *Archaea* (*Methanocaldococcus jannaschii*, *Methanothermobacter thermautotrophicus*) and seven members of the *Bacteria* (*Bacillus subtilis, Escherichia coli, Haemophilus influenzae, Helicobacter pylori, Mycoplasma genitalium, Mycoplasma pneumoniae, Synechocystis* PCC6803) covering the three kingdoms of life. According to the GOLD site (http://www.genomesonline.org/), we are witnessing today 460 complete genomes from prokaryotes, while 1139 projects were under way as on 4 February 2007. In this fast evolving context, psychrophiles genome studies began early on. They were, however, much less fashionable than studies of organisms growing at high temperature. Nevertheless, attempts to correlate DNA composition and specific gene sequences to growth under cold conditions were undertaken. A typical example of these early studies is found in the work of Bernardet and Kerouault in 1989, who isolated and characterized bacteria of the *Cytophaga psychrophila* complex (Bernardet and Kerouault, 1989), which causes disease in salmon and trout at 10°C. These bacteria did not use carbohydrates as carbon supply, secreted proteases, and hydrolyzed tyrosine. While psychrophilic, they could not grow below 3°C. Their DNA G+C content was approximately 34%, in line with what was usually expected for organisms growing in the cold. The follow-up of this type of work is likely to provide us the sequence of the genomes of psychrophilic *Aeromonas* genus as well as that of *Vibrio* genus, of which we have already genomes from mesophilic species.

Interestingly, as studies became more numerous, no correlation could be found between the genomic G+C content and optimal growth temperature in prokaryotes. Galtier and Lobry showed that the genomic G+C content varied between 25 and 77% in mesophilic genera (mean 54.2% G+C), and between 31 and 67% in thermophilic genera (optimum growth temperature around 45°C, mean 48.8% G+C), with no significant correlation with optimum growth temperature (Galtier and Lobry, 1997). In contrast, they observed that the expected relationship between G+C content in RNA that has some secondary structure and optimal growth temperature was present in prokaryotes. This pattern was found in all the molecules and all the genera examined, suggesting that any secondary structure that must endure a high temperature requires a high G+C content (Galtier and Lobry, 1997). The same pattern (a correlation between RNA G+C content and optimum growth temperature, but no correlation between genomic G+C content and optimum growth temperature) was found by intrageneric analyses of 9 *Bacillus* species, 64 *Clostridium* species, 22 *Cytophaga* species, 8 *Methanobacterium* species, 7 *Methanococcus* species, and 31 *Pseudomonas* species. The trend seems to hold at low temperature, as witnessed by the G+C content of the ribosomal RNA genes. More work, however, is needed to substantiate this point as A+T-rich gram positives tend to have RNAs richer in A+Ts, as do bacteria which have undergone genome-reducing evolution (*Buchnera* sp, *Mollicutes* etc.).

Archaea, which have generally been studied mostly for their ability to survive and develop at hot temperatures, also thrive in cold conditions and make up a significant fraction of the picoplankton biomass in the vast habitats encompassed by cold and deep marine waters (DeLong et al., 1994; Stein et al., 1996; Murray et al., 1998; Vetriani et al., 1999). *Archaea* from the kingdom *Crenarchaeota* were also recovered from anaerobic freshwater-lake sediments, showing that they represented a lineage diverging prior to all recently identified crenarchaeotes isolated from temperate environments (Schleper et al., 1997a). At that time (during 1997), all uncultured crenarchaeotal sequences recovered from moderate or cold environments formed a distinct, monophyletic group separate from the thermophilic crenarchaeotes. This demonstrated that crenarchaeotes, thought until the mid-1990s to be solely extreme thermophiles, had radiated into a large variety of temperate environments (Schleper et al., 1997a). This type of observation will certainly be considerably extended as more metagenomics studies are performed (Cottrell et al., 2005). In 1997, moreover, first studies of the uncultivated psychrophilic archaeon *Cenarchaeum symbiosum* substantiated the ubiquitous nature of *Archaea* in cold-environments (Schleper et al., 1997b; 1998).

All these studies paved the way for true genome programs involving psychrophiles, which began to be undertaken just after the turn of the millennium.

GENOME PROGRAMS

The sequences of several genomes of cold-tolerant organisms have been published (typically that of *Listeria* species [Glaser et al., 2001]), but the first genome sequences of psychrophilic organisms only appeared in 2004 to 2005. This is the beginning of a fast growing list.

Bacteria

The first five genome sequences deciphered from psychrophilic bacteria belonged to the *Proteobacteria* family (Table 1). This precludes generalization to the whole *Bacteria* kingdom. Nevertheless, rules observed in this large clade can be used as references for further exploration of psychrophile-specific properties. Cold-tolerant microorganisms have been discovered in large quantity in glaciated and permanently frozen environments, as well as on the deep-sea floor of oceans, but their growth patterns are far from being generally understood. Some controversy is associated with the discovery of many minutest bacteria (less than 0.1 μm in size) in glacier ice core (Miteva and Brenchley, 2005) that genome studies might help to solve. As a consequence, most of the information we have at present on these organisms is structural in essence: it corresponds to genome sequence identification and in silico annotation of gene sequences, usually with little or no experimental validation.

Desulfotalea psychrophila strain LSv54 is a rod-shaped δ-proteobacterium, as are most sulfate-reducing bacteria. It was isolated from permanently cold Arctic marine sediments. It has a 3,523,383-bp circular chromosome with 3,118 predicted genes and two circular plasmids of 121,586 and 14,663 bp (Rabus et al., 2004). The genome comprises seven rRNA gene clusters three being clustered together, but only 64 tRNA genes. This organism makes selenocysteine, as it has all the required components to incorporate this twenty-first amino acid into proteins.

Photobacterium profundum SS9 was isolated at a depth of 2,500 m. This barophilic (piezophilic) organism belongs to the γ-proteobacteria, family *Vibrionaceae*. As seems to be the rule in vibrios (Okada et al., 2005), it has two circular chromosomes, 4,085,301- and 2,237,950-bp long, and a circular 80-kbp plasmid; while emphasis has been placed on its

Table 1. List of cold-growing bacteria for which complete genomes are available

Species	Size (Mb)	Nb chromosomes	Division	Ecosystem	Temperature
Escherichia coli K-12 for reference	4.45	1	γ-Proteobacteria	Mammalian gut/contaminated water effluents	Mesophilic (range, 16 to 46°C; optimum growth, 37°C)
Desulfotalea psychrophila LSv54	3.52	1	δ-Proteobacteria	Cold marine sediments	Psychrophilic (growth at <0°C)
Photobacterium profundum SS9	4.09 2.23	2	γ-Proteobacteria	Deep sea/high pressure	Psychrophilic (range, 2 to 20°C; optimum growth, 15°C)
Colwellia psychrerythraea 34H	5.37		γ-Proteobacteria	Cold marine sediments	Psychrophilic (range, <0 to 19°C; optimum growth, 8°C)
Pseudoalteromonas haloplanktis TAC125	3.21 0.64	2	γ-Proteobacteria	Antarctic costal seawater	Psychrophilic (range, <0 to 25°C; optimum growth, 15°C)
Psychrobacter arcticus sp. 253–4	2.65	1	γ-Proteobacteria	Soil sea-ice/Siberian permafrost core	Psychrophilic (optimum growth, unknown)
Pelagibacter ubique HTCC1062	1.31	1	α-Proteobacteria	Seawater columns	Mesophilic (optimum growth, unknown)

ability to withstand extremely high pressure (up to 90 MPa), not much has been described about its life in the cold. It has a very large number of rRNA gene clusters (14 in chromosome I and 1 in chromosome II), which stand out by their remarkable polymorphism. Indeed there seems to be 5 and 2% variation between the various copies of the 23S and 16S RNAs, respectively (Vezzi et al., 2005). Another unexpected feature of the genome is the presence of two complete operons for F_1F_0 ATP synthase, one on each chromosome (Vezzi et al., 2005). The full complement for selenocysteine metabolism is present in the genome.

Colwellia psychrerythraea 34H was also isolated from Arctic marine sediments. The type strain is strictly barophilic (Deming et al., 1988) and belongs to the γ-proteobacteria, family *Colwelliaceae*. All characterized members are psychrophilic and have been obtained from stably cold marine environments, including deep sea and Arctic and Antarctic sea ice (Deming, 2002). Its genome is 5,373,180-bp long with 4,937 predicted coding DNA sequences (CDSs) (Methe et al., 2005). It has nine rRNA gene clusters and 88 tRNA genes. The genes needed for selenocysteine incorporation into proteins have not been described, suggesting that the organism does not use this amino acid.

Pseudoalteromonas haloplanktis TAC125 was isolated from coastal seawater from Antarctica. It has two circular chromosomes, of 3,214,944 and 635,328 bp, and a plasmid 4,081 bp long. Chromosome I has 9 rRNA gene clusters, and 118 tRNA genes, in line with the very fast growth of the organism at low temperature (Medigue et al., 2005). It also belongs to the γ-proteobacteria. It does not make or use selenocysteine. The *Pseudoalteromonas* genus is ubiquitous in aqueous and marine environments, with specific genomic signatures in the 16S RNA, differentiating it in particular from *Colwellia* species (Ivanova et al., 2004).

Psychrobacter arcticus 273-4 was isolated from a 20- to 40-thousand-year-old Siberian permafrost core (Ponder et al., 2005). *Psychrobacter* are halotolerant, psychrophilic to mesophilic, aerobic, nonmotile, gram-negative coccobacilli often found in pairs also belonging to the γ-proteobacteria. It has one chromosome 2,650,701 bp long, with 2,147 CDSs, 4 rRNA gene clusters and only 49 tRNA genes. It does not seem to incorporate selenocysteine into proteins [https://maple.lsd.ornl.gov/microbial/psyc/].

A sixth genome, that of *Pelagibacter ubique* HTCC1062, belonging to the α-proteobacteria, may perhaps be added to the list, as these bacteria constitute a huge proportion of planktonic bacteria and may develop well at low temperature. These bacteria have the smallest genome (1,308,759 bp) for almost autonomous cells (it cannot make thiamine or biotin, however, making it an auxotrophic organism). The genome harbors 1 rRNA gene cluster, 32 tRNA genes including, curiously, 4 tRNA[met]. This cosmopolitan oceanic organism has not been explicitly characterized as a psychrophile, but its features might be used as references for the core genes of bacteria, including that particular category (Giovannoni et al., 2005).

A detailed analysis of the general features of genomes and proteomes from psychrophilic bacteria is presented after the following summary of what has been collected in the case of psychrophilic *Archaea*, before we possessed complete genome sequences of psychrophilic organisms from this kingdom.

Draft Genome Sequences of *Archaea*

First genomic analyses of *Cenarchaeum symbiosum*, a non-cultivated organism living in specific association with a marine sponge, revealed typical features of *Archaea* (*Crenarchaeote*) (Schleper et al., 1998). The sequence analysis of 28 kbp of contiguous DNA from two *C. symbiosum* variants confirmed the phylogenetic affiliation inferred from the analysis of the 16S rRNA sequence. Interestingly, the rRNA gene order, spacer region, and structure were highly related to those of hyperthermophilic crenarchaeota. The glutamate-1-semialdehyde aminotransferase gene, which is located directly upstream of the ribosomal operon, was further found in the same relative location on a fosmid derived from a planktonic marine crenarchaeote, confirming the phylogenetic affiliation (Schleper et al., 1998). Several other ubiquitous genes were also analyzed in psychrophilic *Archaea*: for example, DNA polymerases of *Methylosphaera hansonii*, or uncultivated *C. symbiosum* had more cysteines and lower isoelectric point in the mesophilic/psychrophilic versus thermophilic DNA polymerases, indicating that composition of proteins in some amino acid might be a signature of temperature adaptation (Schleper et al., 1997b). However, in the case of cysteine, metabolic constraints due to sulfur assimilation might have a dominating influence (see below).

The methanogens stand out as a group of organisms that have species capable of growth at 0°C (*Methylobacter psychrophilus*, *Methylocella* spp., *Methylocapsa acidiphila* and *Methylomonas scandinavica* [Trotsenko and Khmelenina, 2005], and *Methanosarcina* spp., [Simankova et al., 2003]) and 110°C (*Methanopyrus kandleri* (Slesarev et al., 2002)). The genome sequence of several mesophilic methanogens is known: *Methanosarcina acetivorans* (37°C; Galagan et al., 2002), *Methanosarcina mazei* (37°C; Deppenmeier et al., 2002), *Methanococcus maripaludis* (37°C; Hendrickson et al., 2004), and a draft sequence of *Methanosarcina barkeri* (35°C). The most remarkable observation coming from these organisms is perhaps the presence of the 22nd amino acid pyrrolysine, encoded by the UAG codon (Srinivasan et al., 2002). Equally important is the fact that cysteine is not directly loaded onto tRNA^{cys}, but that serine phosphate is, with subsequent homeotopic transformation (Danchin, 1989) into cysteinyl-tRNA^{cys}, in a process which constitutes the cysteine biosynthetic pathway (Sauerwald et al., 2005). This has certainly a considerable impact on the cysteine content of proteins in the cognate organisms.

We also have good draft sequences of *Methanogenium frigidum* and *Methanococcoides burtonii*, allowing some understanding of their genomic and proteomic features (Saunders et al., 2003). Five genes were identified as unique to *M. frigidum* and *Methanococcoides burtonii*, including one likely to code for a 3-helical bundle DNA/RNA-binding protein. Growth in the cold requires exquisite control of RNA folding. Interestingly S1-box RNA-binding domain of polynucleotide phosphorylase was identified as the best threading template for two sequences from *M. burtonii*. There was also a gene coding for an ArsR-like transcriptional regulator similar to SmtB from *Synechococcus*. Cold-shock-protein (Csp) genes were expected. Counterparts were found in *M. frigidum* but the absence of a Csp homolog in *M. burtonii* was remarkable (Saunders et al., 2003). Peptidyl prolyl *cis-trans* isomerases (PPIases) are present in *Archaea*, but with a distribution that differs from those of *Bacteria* (Maruyama et al., 2004). No rule about the situation in psychrophiles can be derived yet.

The study showed that the bulk amino acid composition in proteins from cold-adapted *Archaea* was sufficiently different to distinguish them from both mesophilic and (hyper)thermophilic organisms. Strong evidence for decreased Gln and increased Glu content, and moderate evidence for a decrease in Thr, His, and Ser content for thermophiles had been previously reported for those organisms (Kreil and Ouzounis, 2001; Tekaia et al., 2002). The study of Saunders and co-workers showed that in the *Archaea*, there was an almost linear trend in the content of Gln, Thr, and Leu over the complete range of optimal growth temperatures from psychrophiles to hyperthermophiles (Saunders et al., 2003). This study, however, is somewhat ambiguous as it studies the overall amino acid content of the proteomes and not the way individual proteins are distributed within the proteome (see below). Also it is highly sensitive to the G+C content of the organisms and therefore might not be related to the adaptation of the proteins to particular temperature conditions, cold conditions in particular (Saunders et al., 2003), but, rather, reflect an overlap of several independent constraints.

BACTERIA UNDER COLD CONDITIONS

All investigators involved in sequencing the genomes of psychrophilic *Bacteria* looked for common features which would account for cold-adaptation. Among those, the resistance to oxygen, membrane-increased fluidity at low temperature, and folding of macromolecules were investigated. The folding of proteins was a feature of major interest, as it was assumed that growth in the cold might help fold proteins in heterologous systems that would be hard to fold at medium or high temperature.

GENERAL RULES OF BACTERIAL GENOME ORGANIZATION

As stated above, there is no clear relationship between the overall G+C content of the organism and psychrophily. However, the small sample we have does not allow us to draw too firm conclusions, and there is some tendency in the genomes of psychrophiles to be more A+T rich. It is important, however, to note that metabolic constraints play a role which is at least as important as temperature in the genome composition as most limitations tend to drive the composition towards A+T (in general (Rocha and Danchin, 2002) and in the particular example of *Pelagibacter ubique* (Giovannoni et al., 2005)). There seems to exist a compositional adaptation in G+C content of the rRNA genes, which are submitted to considerable selection pressure for the proper folding of ribosomal RNA (Nakashima et al., 2003). There is significant polymorphism in the rRNA genes as noted for *Pseudoalteromonas haloplanktis* TAC125 and in particular for *Photobacterium profundum* SS9 where the polymorphism is unusually large. The genomes have one or two chromosomes, and plasmids may also be present. An interesting case of explicit recruitment of a unidirectional replication plasmid as an authentic chromosome has been discovered in *Pseudoalteromonas haloplanktis* TAC125 (Medigue et al., 2005). The driving force for that particular recruitment is not understood (in particular the whole metabolism of histidine, including histidine tRNA synthetase, is coded from the plasmid-like chromosome). The split of the genome into two chromosomes might increase the speed of replication and allow faster growth at low temperature and/or when abundant nutrient sources become suddenly available. There is indeed some correlation with a high number of tRNA genes and the presence of two chromosomes in marine bacteria (*Vibrios* in particular). Finally, these organisms possess a standard counterpart of core "persistent" genes (Fang et al., 2005), suggesting that adaptation to cold would only be superimposed on a minimal genome structure of the type found in *Pelagibacter ubique*, for example (Giovannoni et al., 2005).

The taming of energy by living organisms derives mostly from the vectorial transport of protons through membranes. As protons leak in more easily through lipid bilayers as the temperature increases, this should apparently not become a big challenge for growth in the cold. However, protons scarcely travel free: they are either associated with one or several water molecules or they titrate basic chemical groups (such as the amido or guanidinium group) while creating large networks of hydrogen bonds. Ultimately, when the temperature approaches 0°C, these networks culminate in the formation of solid ice, with both a set of preferred orientations and a change in the overall volume. The direction of the flow of protons inside the cell is organized along the surface of membranes, in "columns" of water molecules or traveling along nucleic acid or protein fibers, and this flow is therefore very important to distribute the available protons from their "sources" (the ATP synthase molecules) to the places where they participate in reactions or are expelled from the cell (in respiration centers). This must be highly organized as the number of free protons inside an average bacterium of one cubic micrometer in volume, with a pH of 7.6 (if this meant something under those conditions), would approximately be 15, as anyone can easily calculate. The management of protons inside the cell in the cold is therefore most probably an important challenge since the cell has not only to fluidify its water content (as it can do with solutes such as glycerol or trehalose) but also to preserve the internal proton network (which it can do with a network of protein or nucleic acid fibers, a domain that has been only scarcely explored until now), as it plays a major role in the shaping of macromolecules.

THE RNA WORLD

In response to temperature downshift, a number of changes occur in the cellular physiology of model bacteria such as *E. coli* and *B. subtilis* (Phadtare, 2004). RNAs are affected at two levels: stabilization of secondary structures of nucleic acids leading to reduced efficiency of mRNA translation and transcription, and altered ribosome function (often arrest at the level of translation initiation). A number of cold-shock proteins are meant to cope with these alterations.

A large fraction of the complexes present in cells is made of RNA-protein complexes. This has long been known for the paradigm of the cell nanomachines, the ribosome. This is also true of the secreting machinery. But it is only recently that one has uncovered a large number of RNA-coding genes, with a variety of RNA molecules acting as such, or in interaction with proteins. RNAs are now recognized as extremely important both for their catalytic activity (ribozymes and RNA machines such as ribosomal RNAs), as structural components (tRNAs, tmRNA, etc.), and for *cis*- (riboswitches; Tucker and Breaker, 2005) and *trans*- (small non-coding RNAs; Gottesman, 2005) regulation of gene expression. These RNAs are folded, often in very complex and subtle structures, which can be so precise that they can recognize and interact with very small molecules in a highly specific way: one of the recently discovered riboswitches interacts specifically with the smallest amino acid, glycine (Mandal et al., 2004). RNA genes have not been labelled in most genomes of psychrophiles, except those coding for the major species (rRNA and tRNA in particular), but several riboswitches and the *csrB* gene (E. Krin, AD, unpublished) have been found in *P. haloplanktis* for example, suggesting that many regulatory RNAs are encoded in the genomes of the psychrophiles as in the genome of their mesophilic counterparts.

Because RNA folding is heavily dependent on hydrogen bonds, it seems likely that they suffer considerable constraints due to variations of the structure of water as temperature decreases. Folding and unfolding of RNA must be a major challenge below 10 to 12°C. Moreover, indeed, the genome of *Pseudoalteromonas haloplanktis* TAC125 revealed a considerable number of helicases (most of them RNA helicases) (Medigue et al., 2005) of the DEA[DH] family (Cordin et al., 2005). Many cold-shock proteins are also likely to be involved in folding/unfolding of RNA (Phadtare et al., 1999), but there is no considerable difference between the number of these proteins in mesophiles and psychrophiles. The situation of the H-NS protein is interesting as it is a major cold-shock protein in *E. coli* but absent from *P. haloplanktis* TAC125 and perhaps *C. psychrerythraea* 34H. This protein is, however, present in *Photobacterium profundum* SS9, *Desulfotalea psychrophila* LSv54 and probably in *Psychrobacter arcticus* 273-4.

Ribosomal RNA is significantly polymorphic in a large proportion of genomes (note that this conclusion requires sequencing of rRNA regions individually, and not direct assembly from a shotgun sequencing approach as may have been the case in some cases) (Cilia et al., 1996). This is an intriguing observation as the ribosome is the first molecular chaperone shaping proteins: one may wonder whether there is not some kind of choice depending on the growth conditions among rRNA molecules in the genome of *Photobacterium profundum* SS9, where polymorphism is very high (Vezzi et al., 2005). This organism is an extreme barophilic organism, and it may be more important to adapt protein folding as a function of pressure than of temperature.

The Chaperonin Enigma

Escherichia coli is the most frequent bacterial host for heterologous protein expression. Unfortunately many interesting proteins make inclusion bodies (Huang et al., 2001; Klosch et al., 2005), and several studies have tried to remedy that fact by co-expressing molecular chaperones or lowering the growth temperature (Bera and Bernhardt, 1999). An obvious choice would be to express proteins in psychrophiles (Tutino et al., 2001). Their genomes possess the three major PPIase families: cyclophilins, FK506 binding proteins (FKBPs), and parvulins (Galat, 2003; Golbik et al., 2005). However, how this possession plays a role in adaptation to cold is not known: the number of members of each family differs from genome to genome. The genomes of psychrophilic bacteria also have the counterpart of major chaperonins such as the essential GroES GroEL complex. A remarkable observation poses interesting questions about the role of this complex. Indeed, studies using the GroES GroEL complex from the psychrophile *Oleispira antarctica*—the genome of which is still unknown—expressed in *E. coli* revealed that these bacteria, which normally do not grow at low temperature could grow at temperature as low as 4°C (Strocchi et al., 2006; Ferrer et al., 2004). An attempt using the GroES GroEL counterpart from *P. haloplanktis* TAC125 did not restore growth of *E. coli* (Medigue et al., 2005). The observation is most remarkable as the GroEL protein of *O. antarctica* is highly similar to that of *E. coli*. The researchers identified two residues (K468 and S471) as potentially important for allowing the protein, after ATP binding, to dissociate easily at low temperature, whereas at moderate ones, at which the affinity for nucleotides is lower, the interaction involving both residues is not influenced (Ferrer et al., 2004). Further work will be needed to substantiate this interpretation as these amino acid changes exist in GroEL counterparts of other non-psychrophilic organisms, while they are absent from those of the sequenced psychrophiles. In any event, this observation has to be reconciled with other observations that suggest that the bulk proteome of psychrophiles displays particular features that are not entirely overlapping with those of mesophiles, as we shall see.

The Proteome

We shall now explore the global features of the proteome of psychrophiles compared to that of mesophiles and thermophiles. As already alluded to above, the analysis of the bulk proteome of prokaryotes has long indicated that some important biases existed in the distribution of amino acids in proteins (Tekaia et al., 2002; Pe'er et al., 2004). These studies, however, did not consider the individual proteins of the proteome but only a concatenate or a subset of all protein sequences, precluding fine analyses of the causes of the biases observed (as well as creating possible artifacts due to overrepresentation of some proteins in the proteome). Earlier studies have suggested a bias in proline and arginine in psychrophilic enzymes (Oikawa et al., 2001; 2005): the genome studies do not substantiate that this would be the rule. In fact, a recent study using correspondence analysis of the individual proteins making the proteome of model prokaryotes has found that three major biases drove the overall compositions of the proteins, and that these biases spanned throughout the *Bacteria* and the *Archaea* (Pascal et al., 2005). The first bias is driven by the content in charged amino acids (arginine, aspartate, glutamate, and lysine) opposed to that of hydrophobic amino acids (preferentially the large ones: leucine and phenylalanine). The second one is driven by the G+C content of the first codon position of amino acids, separating between Asn, Cys, Ile, Lys, Met, Phe, Ser, Thr, Trp, Tyr and Ala, Asp, Gln, Glu, Gly, His, Pro, Val. The driving force behind the structure of the genetic code is a matter of much controversy (Di Giulio, 2005), but it may be interesting to note that sulfur-containing amino acids as well as aromatics cluster together in this list, whereas the acidic amino acids are in the second cluster. The third bias is driven by the aromatics, and the most biased proteins it contains are mostly orphan proteins. This led the authors of the study to suggest that they might participate in a process of gain of function during evolution. They proposed that many of these proteins could be considered as "gluons" stabilizing multicomponent complexes, thus labelling the "self" of a species (Pascal et al., 2005). Remarkably, when this study was extended to the comparison between psychrophiles, mesophiles, and thermophiles, it appeared that a supplementary bias, driven by an opposition between asparagine and cysteine separated the psychrophiles from the mesophiles (and the thermophiles). In that particular study, the role of cysteine, although consistent with high reactivity towards reactive oxygen species (ROS, see below), could not be considered statistically significant, because of the low cysteine content of the corresponding proteins (Medigue et al., 2005). In contrast, the contribution of asparagine was certainly significant. The authors proposed that asparagine is less counterselected at low temperature because its spontaneous cyclization and deamidation tendency is lower at low temperature (Pascal et al., 2005). If this is substantiated, this would provide an excellent reason for expressing foreign proteins into fast-growing psychrophiles.

Reactive Oxygen Species

Gases are more soluble at low than at high temperature, while radicals are more stable in the former conditions. In the presence of molecular oxygen (dioxygen), this has the consequence that ROS are more frequent and stable for a longer time. Hence it is expected that oxygen is particularly toxic at low temperature. Chemically, in the presence of oxygen, amino acids display very different stability, either as isolated molecules or within the proteins. Dioxygen would mostly affect both sulfur-containing amino acids (cysteine, irreversibly and methionine reversibly), as well as histidine and tryptophane. Does this have consequences in the gene counterparts of the genome of psychrophiles? Cysteine is oxidized to several forms: sulfenic, sulfinic, and sulfonic acid. In *Eukarya*, the sulfenic and the sulfinic acid derived from cysteine can be reduced while sulfonic derivatives cannot. The situation is not well studied in prokaryotes, and the equivalents of sulfinate reductase (sulfiredoxin, Srx) have not yet been demonstrated biochemically. However, there may exist counterparts of the eukaryotic enzyme in cyanobacteria and in *Mycobacterium tuberculosis*, and it has been proposed that the enzyme evolved from the division protein ParB (Basu and Koonin, 2005). We did not find clear counterparts of the gene in the available genomes of psychrophiles. Srx's are supposed to reduce the hyperoxidized sulfinic acid form of the catalytic cysteine of 2-Cys peroxiredoxins.

Peroxiredoxins (Prxs) are a superfamily of thiol-specific antioxidant proteins, also known as thioredoxin peroxidase or alkyl hydroperoxidase reductase. They have an essential protective properties against hydrogen peroxide, organic hydroperoxide, and peroxynitrite. Prxs are abundant in most *Bacteria*, *Archaea*, and *Eukarya*, and exist in several isoforms in all these organisms. Prxs contain a conserved N-terminal catalytic cysteine called the peroxidatic cysteine. According to the occurrence, or not, of a second conserved cysteine, the Prxs are divided into two groups, the 1-Cys and 2-Cys Prxs, respectively. The mechanism of reduction is only partially known. As expected, all psychrophiles have Prxs, sometimes as multiple paralogs: there are, for example, two *ahpC* Prx genes in *P. haloplanktis* as compared to one in *E. coli*. This is consistent with a need to repair oxidized residues and/or protect against some forms of ROS. Consistent with cysteine sensitivity at least under particular conditions, there is a relative decrease in cysteine content in some of the organisms. Furthermore, the analog selenocysteine is present as the 21st amino acid in only two of the five psychrophiles, in *Desulfotalea psychrophila*, which is sulfate reducing and thus poised to cope with oxidized forms of sulfur and its higher mass counterpart and in *Photobacterium profundum*, which thrives at the bottom of deep ocean. However, while stability of the ROS is increased at low temperature, their reactivity in increased at higher temperature and the balance between these contradictory effects is not known. There is indication that psychrophilic prokaryotes living under high pressure conditions may have proteins enriched in cysteine (Schleper et al., 1997b; Xu et al., 1998). It will be most interesting to see the behavior of cysteine-rich proteomes in future studies, in relation with the metabolic capacities of the corresponding organisms.

The sulfonates are scavenged by sulf(on)atases, with sulfate recycled into sulfur assimilation or dissimilated in sulfate-reducing bacteria. Methionine is prone to oxidation and is easily sulfoxidated into two different enantiomers (the highly oxidized form methionine sulfone is relatively rare): the S form reduced by MsrA and the R form reduced by MsrB (Ezraty et al., 2005). Both types of peptide methionine sulfoxide reductases exist, often as multiple copies in psychrophiles, consistent with the increased toxicity of oxygen. As this modification is reversible, it has been proposed that methionine acts as a protectant against oxygen deleterious effects, and crystallographic data support well this contention (Sundby et al., 2005).

In contrast, the histidine and tryptophane ROS adducts cannot be repaired and are presumably recycled after proteolysis, but the pathways are not well known.

Finally, nucleotides can also be affected by ROS and, indeed, enzymes meant to scavenge oxidized purines are coded in the genome, such as *mutM*, *mutT*, *mutY*, and a complex DNA repair system is involved in that correction. It is likely that many conserved genes of unknown function are related to these processes.

Metabolic Constraints

A remarkable way to lower the incidence of ROS in *P. haloplanktis* has been to evolve by eliminating in a concerted way the metabolism of molybdopterin, as several molybdoenzymes are producing ROS (Medigue et al., 2005). This is not the case in the other psychrophiles, and even the compact genome of *Pelagibacter ubique* codes for this metabolism. In contrast, showing that this pathway is not essential even though extremely frequent, *Saccharomyces cerevisiae* lacks this metabolism, increasing its basal respiration rate to lower ROS production.

Basic thermophysics states that reaction speed will increase as temperature increases. This is a general rule and does not show any particular difference for a particular temperature. Chemical reactions will be slow at low temperature and fast at high temperature. However, an important feature of the reactions involved in living organisms is the fine balance that must be maintained between synthesis and degradation, anabolism and catabolism. The optimum discovered by homeothermic organisms (around 37°C) is probably the temperature best integrating this balance (we shall see below that the structure of membranes has also its word in the selection pressure, which leads to that particular temperature range). It is therefore expected that metabolic fluxes are altered in psychrophilic organisms as compared to their mesophilic counterparts. However, the very small sample of genomes we possess at the moment preclude to find rules of metabolic organization in psychrophiles, although they do possess the complete complement of persistent genes, showing no difference with Bacteria growing under different conditions (Fang et al., 2005). Furthermore, we are witnessing the situation of marine organisms, and this introduces a specific constraint in terms of formation of osmoprotectants (such as the ubiquitous potassium glutamate) and limitation in nitrogen, phosphorus, and iron availability. Most do not have a phospho*enol*pyruvate-dependent phosphotransferase system (PTS) to transport carbon sources (but *Photobacterium profundum* has several) but have a PTS-dependent nitrogen assimilation regulatory pathway. This goes well with an efficient Entner-Doudoroff pathway that can be used to equilibrate the NADH/NADPH supply under conditions of easy availability of dioxygen.

The main prominent metabolic feature of these bacteria is indeed that they have a large number of putative dioxygenases, allowing them to easily dispose of molecular oxygen (Medigue et al., 2005).

Membrane Fluidity, Protein Export, and Secretion

Bacterial membranes are constructed around core lipid bilayer. Such structures are known to be prone to spontaneous organization, becoming locally rigid structures. Much speculation has been devoted to the role (if any) of that organization [Russell, this volume], and it has even been speculated that the temperature of the homeotherms was driven by the need for fluid membranes with standard phosphatidyldiglycerides, with saturated fatty acids. While this does not seem to be the real cause of the ubiquity of the 37°C optimum, it is clear that as temperature falls, the lipid bilayer becomes more and more rigid, asking for some sort of adaptation. Membrane fluidity can be increased in two ways: either by incorporating unsaturated fatty acids or by including branched-chain fatty acids in the diglycerides (Cybulski et al., 2002). There is considerable evidence correlating the production of increased proportions of membrane unsaturated fatty acids with bacterial growth at low temperature (Sakamoto and Murata, 2002). Interestingly, desaturating saturated fatty acids can be easily obtained by desaturases which use dioxygen as a substrate: this has the remarkable advantage to couple elimination of toxic oxygen at low temperature while increasing membrane fluidity (Sakamoto and Bryant, 2002). This is typical of a concerted evolution feature that would have been presented as a mystery by opponents to the general scenario of selective stabilization as the driving force of evolution. The fatty acids produced by *P. profundum* SS9 grown at various temperatures and pressures were characterized, and differences in fatty acid composition as a function of phase growth and between inner and outer membranes noted. *Photobacterium profundum* SS9 was found to exhibit enhanced proportions of both monounsaturated and polyunsaturated fatty acids when grown at a decreased temperature or elevated pressure (Bartlett, 1999). The drawback to the presence of unsaturated fatty acids in the membrane lipids is that they are sensitive to ROS: this places another constraint to ROS scavenging for psychrophiles.

The psychrophiles described in this review chapter have standard export and secretion systems of the universal types, adapted to the particular structure of the envelope of gram-negative organisms, which possess an outer membrane. However, there may exist special feature for import of solutes (Russell, 1990), and in the case of a *P. haloplanktis* isolate, experiments have suggested the presence of an unusual secretion pathway in these organisms (Tutino et al., 2002). Further work is needed to substantiate this potentially interesting observation.

TENTATIVE CONCLUSIONS

Genome programs have only fairly recently been dealing with life in the cold. It is therefore impossible to draw firm conclusions at this point about specific constraints for sustained life at temperatures prevailing at the bottom of oceans or near the major ice cores of the Earth. However, it seems clear that prokaryotes thrive and develop well under those conditions provided they are supplied with sufficient nutrients. Some seem even to have developed ways to sustain cycles of feast and famine, with the onset of rapid exponential growth when nutrients are suddenly available. Three main features can be observed in the genomes and proteomes of these organisms: a variety of means to cope with ROS, a multiplicity of nucleic acid folding and unfolding devices, and, finally, a bias in the amino acid composition of their proteome. To these features many individual solutions have been implemented to cope with cold conditions. Whether conserved features will withstand sequencing of many new partners of the psychrophiles will be known in a few years time.

REFERENCES

Baross, J., and J. Deming. 1983. Growth of "black smoker" bacteria at temperatures of at least 250°C. *Nature* 303:423–426.

Bartlett, D. H. 1999. Microbial adaptations to the psychrosphere/piezosphere. *J. Mol. Microbiol. Biotechnol.* 1:93–100.

Basu, M. K., and E. V. Koonin. 2005. Evolution of eukaryotic cysteine sulfinic acid reductase, sulfiredoxin (Srx), from bacterial chromosome partitioning protein ParB. *Cell Cycle* 4:947–952.

Bera, A. K., and R. Bernhardt. 1999. GroEL-assisted and -unassisted refolding of mature and precursor adrenodoxin: the role of the precursor sequence. *Arch. Biochem. Biophys.* 367:89–94.

Bernardet, J. F., and B. Kerouault. 1989. Phenotypic and genomic studies of "Cytophaga psychrophila" isolated from diseased rainbow trout (*Oncorhynchus mykiss*) in France. *Appl. Environ. Microbiol.* 55:1796–1800.

Bowman, J. P., S. A. McCammon, M. V. Brown, D. S. Nichols, and T. A. McMeekin. 1997. Diversity and association of psychrophilic bacteria in Antarctic sea ice. *Appl. Environ. Microbiol.* 63:3068–3078.

Cavicchioli, R. 2006. Cold-adapted archaea. *Nat. Rev. Microbiol.* 4:331–343.

Cilia, V., B. Lafay, and R. Christen. 1996. Sequence heterogeneities among 16S ribosomal RNA sequences, and their effect on phylogenetic analyses at the species level. *Mol. Biol. Evol.* 13:451–461.

Cordin, O., J. Banroques, N. K. Tanner, and P. Linder. 2005. The DEAD-box protein family of RNA helicases. *Gene.* (in press)

Cottrell, M. T., L. Yu, and D. L. Kirchman. 2005. Sequence and expression analyses of cytophaga-like hydrolases in a Western arctic metagenomic library and the Sargasso Sea. *Appl. Environ. Microbiol.* 71:8506–8513.

Cox, E. C., and C. Yanofsky. 1967. Altered base ratios in the DNA of an *Escherichia coli* mutator strain. *Proc. Natl. Acad. Sci. USA* 58:1895–1902.

Cybulski, L. E., D. Albanesi, M. C. Mansilla, S. Altabe, P. S. Aguilar, and D. de Mendoza. 2002. Mechanism of membrane fluidity optimization: isothermal control of the *Bacillus subtilis* acyl-lipid desaturase. *Mol. Microbiol.* 45:1379–1388.

Danchin, A. 1989. Homeotopic transformation and the origin of translation. *Prog Biophys Mol Biol* 54:81–86.

DeLong, E. F., K. Y. Wu, B. B. Prezelin, and R. V. Jovine. 1994. High abundance of *Archaea* in Antarctic marine picoplankton. *Nature* 371:695–697.

Deming, J., L. Somers, W. Straube, D. Swartz, and M. MacDonell. 1988. Isolation of an obligately barophilic bacterium and description of a new genus, *Colwellia* gen. nov. *Syst. Appl. Microbiol.* 10:152–160.

Deming, J. W. 2002. Psychrophiles and polar regions. *Curr Opin Microbiol* 5:301–309.

Deppenmeier, U., A. Johann, T. Hartsch, R. Merkl, R. A. Schmitz, R. Martinez-Arias, A. Henne, A. Wiezer, S. Baumer, C. Jacobi, H. Bruggemann, T. Lienard, A. Christmann, M. Bomeke, S. Steckel, A. Bhattacharyya, A. Lykidis, R. Overbeek, H. P. Klenk, R. P. Gunsalus, H. J. Fritz, and G. Gottschalk. 2002. The genome of *Methanosarcina mazei*: evidence for lateral gene transfer between *Bacteria* and *Archaea*. *J. Mol. Microbiol. Biotechnol.* 4:453–461.

Di Giulio, M. 2005. The origin of the genetic code: theories and their relationships, a review. *Biosystems* 80:175–184.

Ezraty, B., L. Aussel, and F. Barras. 2005. Methionine sulfoxide reductases in prokaryotes. *Biochim Biophys Acta* 1703:221–229.

Fang, G., E. Rocha, and A. Danchin. 2005. How essential are nonessential genes? *Mol. Biol. Evol.* 22:2147–2156.

Ferrer, M., H. Lunsdorf, T. N. Chernikova, M. Yakimov, K. N. Timmis, and P. N. Golyshin. 2004. Functional consequences of single:double ring transitions in chaperonins: life in the cold. *Mol. Microbiol.* 53:167–182.

Galagan, J. E., C. Nusbaum, A. Roy, M. G. Endrizzi, P. Macdonald, W. FitzHugh, S. Calvo, R. Engels, S. Smirnov, D. Atnoor, A. Brown, N. Allen, J. Naylor, N. Stange-Thomann, K. DeArellano, R. Johnson, L. Linton, P. McEwan, K. McKernan, J. Talamas, A. Tirrell, W. Ye, A. Zimmer, R. D. Barber, I. Cann, D. E. Graham, D. A. Grahame, A. M. Guss, R. Hedderich, C. Ingram-Smith, H. C. Kuettner, J. A. Krzycki, J. A. Leigh, W. Li, J. Liu, B. Mukhopadhyay, J. N. Reeve, K. Smith, T. A. Springer, L. A. Umayam, O. White, R. H. White, E. Conway de Macario, J. G. Ferry, K. F. Jarrell, H. Jing, A. J. Macario, I. Paulsen, M. Pritchett, K. R. Sowers, R. V. Swanson, S. H. Zinder, E. Lander, W. W. Metcalf, and B. Birren. 2002. The genome of *M. acetivorans* reveals extensive metabolic and physiological diversity. *Genome Res.* 12:532–542.

Galat, A. 2003. Peptidylprolyl *cis/trans* isomerases (immunophilins): biological diversity—targets—functions. *Curr. Top. Med. Chem.* 3:1315–1347.

Galtier, N., and J. R. Lobry. 1997. Relationships between genomic G+C content, RNA secondary structures, and optimal growth temperature in prokaryotes. *J. Mol. Evol.* 44:632–636.

Giovannoni, S. J., H. J. Tripp, S. Givan, M. Podar, K. L. Vergin, D. Baptista, L. Bibbs, J. Eads, T. H. Richardson, M. Noordewier, M. S. Rappe, J. M. Short, J. C. Carrington, and E. J. Mathur. 2005. Genome streamlining in a cosmopolitan oceanic bacterium. *Science* 309:1242–1245.

Glansdorff, N., and Y. Xu. 2002. Microbial life at low temperatures: mechanisms of adaptation and extreme biotopes. Implications for exobiology and the origin of life. *Recent Res. Devel. Microbiol.* 6:1–21.

Glaser, P., L. Frangeul, C. Buchrieser, C. Rusniok, A. Amend, F. Baquero, P. Berche, H. Bloecker, P. Brandt, T. Chakraborty, A. Charbit, F. Chetouani, E. Couve, A. de Daruvar, P. Dehoux, E. Domann, G. Dominguez-Bernal, E. Duchaud, L. Durant, O. Dussurget, K. D. Entian, H. Fsihi, F. Garcia-del Portillo, P. Garrido, L. Gautier, W. Goebel, N. Gomez-Lopez, T. Hain, J. Hauf, D. Jackson, L. M. Jones, U. Kaerst, J. Kreft, M. Kuhn, F. Kunst, G. Kurapkat, E. Madueno, A. Maitournam, J. M. Vicente, E. Ng, H. Nedjari, G. Nordsiek, S. Novella, B. de Pablos, J. C. Perez-Diaz, R. Purcell, B. Remmel, M. Rose, T. Schlueter, N. Simoes, A. Tierrez, J. A. Vazquez-Boland, H. Voss, J. Wehland, and P. Cossart. 2001. Comparative genomics of *Listeria* species. *Science* 294:849–852.

Golbik, R., C. Yu, E. Weyher-Stingl, R. Huber, L. Moroder, N. Budisa, and C. Schiene-Fischer. 2005. Peptidyl Prolyl cis/trans-Isomerases: comparative reactivities of Cyclophilins, FK506-Binding Proteins, and Parvulins with Fluorinated Oligopeptide and Protein Substrates. *Biochemistry* 44:16026–16034.

Gottesman, S. 2005. Micros for microbes: non-coding regulatory RNAs in bacteria. *Trends Genet.* 21:399–404.

Hendrickson, E. L., R. Kaul, Y. Zhou, D. Bovee, P. Chapman, J. Chung, E. Conway de Macario, J. A. Dodsworth, W. Gillett, D. E. Graham, M. Hackett, A. K. Haydock, A. Kang, M. L. Land, R. Levy, T. J. Lie, T. A. Major, B. C. Moore, I. Porat, A. Palmeiri, G. Rouse, C. Saenphimmachak, D. Soll, S. Van Dien, T. Wang, W. B. Whitman, Q. Xia, Y. Zhang, F. W. Larimer, M. V. Olson, and J. A. Leigh. 2004. Complete genome sequence of the genetically tractable hydrogenotrophic methanogen *Methanococcus maripaludis*. *J. Bacteriol.* 186:6956–6969.

Huang, C. C., L. T. Li, M. C. Shen, J. Y. Chen, and S. W. Lin. 2001. Domain specific monoclonal anti-factor VIII antibodies generated by inclusion body-renatured factor VIII peptides. *Thromb. Res.* 101:405–415.

Ivanova, E. P., S. Flavier, and R. Christen. 2004. Phylogenetic relationships among marine *Alteromonas*-like proteobacteria: emended description of the family *Alteromonadaceae* and proposal of *Pseudoalteromonadaceae* fam. nov., *Colwelliaceae* fam. nov., *Shewanellaceae* fam. nov., *Moritellaceae* fam. nov., *Ferrimonadaceae* fam. nov., *Idiomarinaceae* fam. nov. and *Psychromonadaceae* fam. nov. *Int. J. Syst. Evol. Microbiol.* 54:1773–1788.

Klosch, B., W. Furst, R. Kneidinger, M. Schuller, B. Rupp, A. Banerjee, and H. Redl. 2005. Expression and purification of biologically active rat bone morphogenetic protein-4 produced as inclusion bodies in recombinant *Escherichia coli*. *Biotechnol. Lett.* 27:1559–1564.

Kreil, D. P., and C. A. Ouzounis. 2001. Identification of thermophilic species by the amino acid compositions deduced from their genomes. *Nucleic Acids Res* 29:1608–1615.

Larralde, R., M. P. Robertson, and S. L. Miller. 1995. Rates of decomposition of ribose and other sugars: implications for chemical evolution. *Proc. Natl. Acad. Sci. USA* 92:8158–8160.

Lobry, J. R., A. Necsulae 2006. Synonymous codon usage and its potential link with optimal growth temperature in prokaryotes. *Gene* 385:128–136.

Mandal, M., M. Lee, J. E. Barrick, Z. Weinberg, G. M. Emilsson, W. L. Ruzzo, and R. R. Breaker. 2004. A glycine-dependent riboswitch that uses cooperative binding to control gene expression. *Science* 306:275–279.

Maruyama, T., R. Suzuki, and M. Furutani. 2004. Archaeal peptidyl prolyl cis-trans isomerases (PPIases) update 2004. *Front. Biosci.* 9:1680–1720.

Medigue, C., E. Krin, G. Pascal, V. Barbe, A. Bernsel, P. N. Bertin, F. Cheung, S. Cruveiller, S. D'Amico, A. Duilio, G. Fang, G. Feller, C. Ho, S. Mangenot, G. Marino, J. Nilsson, E. Parrilli, E. P. Rocha, Z. Rouy, A. Sekowska, M. L. Tutino, D. Vallenet, G. von Heijne, and A. Danchin. 2005. Coping with cold: the genome of the versatile marine Antarctica bacterium *Pseudoalteromonas haloplanktis* TAC125. *Genome Res.* 15:1325–1335.

Methe, B. A., K. E. Nelson, J. W. Deming, B. Momen, E. Melamud, X. Zhang, J. Moult, R. Madupu, W. C. Nelson, R. J. Dodson, L. M. Brinkac, S. C. Daugherty, A. S. Durkin, R. T. DeBoy, J. F. Kolonay, S. A. Sullivan, L. Zhou, T. M. Davidsen, M. Wu, A. L. Huston, M. Lewis, B. Weaver, J. F. Weidman, H. Khouri, T. R. Utterback, T. V. Feldblyum, and C. M. Fraser. 2005. The psychrophilic lifestyle as revealed by the genome sequence of *Colwellia psychrerythraea* 34H through genomic and proteomic analyses. *Proc. Natl. Acad. Sci. USA* 102:10913–10918.

Miller, J. H., P. Funchain, W. Clendenin, T. Huang, A. Nguyen, E. Wolff, A. Yeung, J. H. Chiang, L. Garibyan, M. M. Slupska, and H. Yang. 2002. *Escherichia coli* strains (*ndk*) lacking nucleoside diphosphate kinase are powerful mutators for base substitutions and frameshifts in mismatch-repair-deficient strains. *Genetics* 162:5–13.

Miteva, V. I., and J. E. Brenchley. 2005. Detection and isolation of ultrasmall microorganisms from a 120,000-year-old greenland glacier ice core. *Appl. Environ. Microbiol.* 71:7806–7818.

Murray, A. E., C. M. Preston, R. Massana, L. T. Taylor, A. Blakis, K. Wu, and E. F. DeLong. 1998. Seasonal and spatial variability of bacterial and archaeal assemblages in the coastal waters near Anvers Island, Antarctica. *Appl. Environ. Microbiol.* 64:2585–2595.

Nakashima, H., S. Fukuchi, and K. Nishikawa. 2003. Compositional changes in RNA, DNA and proteins for bacterial adaptation to higher and lower temperatures. *J. Biochem. (Tokyo)* 133:507–513.

Nozhevnikova, A. N., M. V. Simankova, S. N. Parshina, and O. R. Kotsyurbenko. 2001. Temperature characteristics of methanogenic archaea and acetogenic bacteria isolated from cold environments. *Water Sci. Technol.* 44:41–48.

Oikawa, T., N. Yamamoto, K. Shimoke, S. Uesato, T. Ikeuchi, and T. Fujioka. 2005. Purification, characterization, and overexpression of psychrophilic and thermolabile malate dehydrogenase of a novel Antarctic psychrotolerant, *Flavobacterium frigidimaris* KUC-1. *Biosci. Biotechnol. Biochem.* 69:2146–2154.

Oikawa, T., K. Yamanaka, T. Kazuoka, N. Kanzawa, and K. Soda. 2001. Psychrophilic valine dehydrogenase of the Antarctic psychrophile, *Cytophaga* sp. KUC-1: purification, molecular characterization and expression. *Eur. J. Biochem.* 268:4375–4383.

Okada, K., T. Iida, K. Kita-Tsukamoto, and T. Honda. 2005. *Vibrios* commonly possess two chromosomes. *J. Bacteriol.* 187:752–757.

Pascal, G., C. Medigue, and A. Danchin. 2005. Universal biases in protein composition of model prokaryotes. *Proteins* 60:27–35.

Pe'er, I., C. E. Felder, O. Man, I. Silman, J. L. Sussman, and J. S. Beckmann. 2004. Proteomic signatures: amino acid and oligopeptide compositions differentiate among phyla. *Proteins* 54:20–40.

Phadtare, S. 2004. Recent developments in bacterial cold-shock response. *Curr. Issues Mol. Biol.* 6:125–136.

Phadtare, S., J. Alsina, and M. Inouye. 1999. Cold-shock response and cold-shock proteins. *Curr. Opin. Microbiol.* 2:175–180.

Ponder, M. A., S. J. Gilmour, P. W. Bergholz, C. A. Mindock, R. Hollingsworth, M. F. Thomashow, and J. M. Tiedje. 2005. Characterization of potential stress responses in ancient Siberian permafrost psychroactive bacteria. *FEMS Microbiol. Ecol.* 53:103–115.

Rabus, R., A. Ruepp, T. Frickey, T. Rattei, B. Fartmann, M. Stark, M. Bauer, A. Zibat, T. Lombardot, I. Becker, J. Amann, K. Gellner, H. Teeling, W. D. Leuschner, F. O. Glockner, A. N. Lupas, R. Amann, and H. P. Klenk. 2004. The genome of *Desulfotalea psychrophila*, a sulfate-reducing bacterium from permanently cold Arctic sediments. *Environ. Microbiol.* 6:887–902.

Rocha, E. P., and A. Danchin. 2002. Base composition bias might result from competition for metabolic resources. *Trends Genet.* 18:291–294.

Ron, E. Z. 1975. Growth rate of *Enterobacteriaceae* at elevated temperatures: limitation by methionine. *J. Bacteriol.* 124:243–246.

Russell, N. J. 1990. Cold adaptation of microorganisms. *Philos. Trans. R. Soc. Lond. B Biol. Sci.* 326:595–608, discussion 608–511.

Sakamoto, T., and D. A. Bryant. 2002. Synergistic effect of highlight and low temperature on cell growth of the Delta12 fatty acid desaturase mutant in *Synechococcus* sp. PCC 7002. *Photosynth. Res.* 72:231–242.

Sakamoto, T., and N. Murata. 2002. Regulation of the desaturation of fatty acids and its role in tolerance to cold and salt stress. *Curr. Opin. Microbiol.* **5:**208–210.

Sauerwald, A., W. Zhu, T. A. Major, H. Roy, S. Palioura, D. Jahn, W. B. Whitman, J. R. Yates, III, M. Ibba, and D. Soll. 2005. RNA-dependent cysteine biosynthesis in archaea. *Science* **307:**1969–1972.

Saunders, N. F., T. Thomas, P. M. Curmi, J. S. Mattick, E. Kuczek, R. Slade, J. Davis, P. D. Franzmann, D. Boone, K. Rusterholtz, R. Feldman, C. Gates, S. Bench, K. Sowers, K. Kadner, A. Aerts, P. Dehal, C. Detter, T. Glavina, S. Lucas, P. Richardson, F. Larimer, L. Hauser, M. Land, and R. Cavicchioli. 2003. Mechanisms of thermal adaptation revealed from the genomes of the Antarctic Archaea *Methanogenium frigidum* and *Methanococcoides burtonii*. *Genome Res.* **13:**1580–1588.

Schleper, C., E. F. DeLong, C. M. Preston, R. A. Feldman, K. Y. Wu, and R. V. Swanson. 1998. Genomic analysis reveals chromosomal variation in natural populations of the uncultured psychrophilic archaeon *Cenarchaeum symbiosum*. *J. Bacteriol.* **180:**5003–5009.

Schleper, C., W. Holben, and H. P. Klenk. 1997a. Recovery of crenarchaeotal ribosomal DNA sequences from freshwater-lake sediments. *Appl. Environ. Microbiol.* **63:**321–323.

Schleper, C., R. V. Swanson, E. J. Mathur, and E. F. DeLong. 1997b. Characterization of a DNA polymerase from the uncultivated psychrophilic archaeon *Cenarchaeum symbiosum*. *J. Bacteriol.* **179:**7803–7811.

Sheridan, P. P., V. I. Miteva, and J. E. Brenchley. 2003. Phylogenetic analysis of anaerobic psychrophilic enrichment cultures obtained from a Greenland glacier ice core. *Appl. Environ. Microbiol.* **69:**2153–2160.

Simankova, M. V., O. R. Kotsyurbenko, T. Lueders, A. N. Nozhevnikova, B. Wagner, R. Conrad, and M. W. Friedrich. 2003. Isolation and characterization of new strains of methanogens from cold terrestrial habitats. *Syst. Appl. Microbiol.* **26:**312–318.

Slesarev, A. I., K. V. Mezhevaya, K. S. Makarova, N. N. Polushin, O. V. Shcherbinina, V. V. Shakhova, G. I. Belova, L. Aravind, D. A. Natale, I. B. Rogozin, R. L. Tatusov, Y. I. Wolf, K. O. Stetter, A. G. Malykh, E. V. Koonin, and S. A. Kozyavkin. 2002. The complete genome of hyperthermophile *Methanopyrus kandleri* AV19 and monophyly of archaeal methanogens. *Proc. Natl. Acad. Sci. USA* **99:**4644–4649.

Srinivasan, G., C. M. James, and J. A. Krzycki. 2002. Pyrrolysine encoded by UAG in Archaea: charging of a UAG-decoding specialized tRNA. *Science* **296:**1459–1462.

Stein, J. L., T. L. Marsh, K. Y. Wu, H. Shizuya, and E. F. DeLong. 1996. Characterization of uncultivated prokaryotes: isolation and analysis of a 40-kilobase-pair genome fragment from a planktonic marine archaeon. *J. Bacteriol.* **178:**591–599.

Strocchi, M., M. Ferrer, K. N. Timmis, and P. N. Golyshin. 2006. Low temperature-induced systems failure in *Escherichia coli*: insights from rescue by cold-adapted chaperones. *Proteomics* **6:**193–206.

Sundby, C., U. Harndahl, N. Gustavsson, E. Ahrman, and D. J. Murphy. 2005. Conserved methionines in chloroplasts. *Biochim. Biophys. Acta* **1703:**191–202.

Tekaia, F., E. Yeramian, and B. Dujon. 2002. Amino acid composition of genomes, lifestyles of organisms, and evolutionary trends: a global picture with correspondence analysis. *Gene* **297:**51–60.

Trent, J. D., R. A. Chastain, and A. A. Yayanos. 1984. Possible artefactual basis for apparent bacterial growth at 250 degrees C. *Nature* **307:**737–740.

Trotsenko, Y. A., and V. N. Khmelenina. 2005. Aerobic methanotrophic bacteria of cold ecosystems. *FEMS Microbiol. Ecol.* **53:**15–26.

Tsai, C. J., J. V. Maizel, Jr., and R. Nussinov. 2002. The hydrophobic effect: a new insight from cold denaturation and a two-state water structure. *Crit. Rev. Biochem. Mol. Biol.* **37:**55–69.

Tucker, B. J., and R. R. Breaker. 2005. Riboswitches as versatile gene control elements. *Curr Opin Struct Biol* **15:**342–348.

Tutino, M. L., A. Duilio, R. Parrilli, E. Remaut, G. Sannia, and G. Marino. 2001. A novel replication element from an Antarctic plasmid as a tool for the expression of proteins at low temperature. *Extremophiles* **5:**257–264.

Tutino, M. L., E. Parrilli, L. Giaquinto, A. Duilio, G. Sannia, G. Feller, and G. Marino. 2002. Secretion of alpha-amylase from *Pseudoalteromonas haloplanktis* TAB23: two different pathways in different hosts. *J. Bacteriol.* **184:**5814–5817.

Vetriani, C., H. W. Jannasch, B. J. MacGregor, D. A. Stahl, and A. L. Reysenbach. 1999. Population structure and phylogenetic characterization of marine benthic Archaea in deep-sea sediments. *Appl. Environ. Microbiol.* **65:**4375–4384.

Vezzi, A., S. Campanaro, M. D'Angelo, F. Simonato, N. Vitulo, F. M. Lauro, A. Cestaro, G. Malacrida, B. Simionati, C. Cannata, C. Romualdi, D. H. Bartlett, and G. Valle. 2005. Life at depth: *Photobacterium profundum* genome sequence and expression analysis. *Science* **307:**1459–1461.

Wernet, P., D. Nordlund, U. Bergmann, M. Cavalleri, M. Odelius, H. Ogasawara, L. A. Naslund, T. K. Hirsch, L. Ojamae, P. Glatzel, L. G. Pettersson, and A. Nilsson. 2004. The structure of the first coordination shell in liquid water. *Science* **304:**995–999.

White, R. H. 1984. Hydrolytic stability of biomolecules at high temperatures and its implication for life at 250°C. *Nature* **310:**430–432.

Woese, C. R., and G. E. Fox. 1977. Phylogenetic structure of the prokaryotic domain: the primary kingdoms. *Proc. Natl. Acad. Sci. USA* **74:**5088–5090.

Xu, Y., Y. Zhang, Z. Liang, M. Van de Casteele, C. Legrain, and N. Glansdorff. 1998. Aspartate carbamoyltransferase from a psychrophilic deep-sea bacterium, Vibrio strain 2693: properties of the enzyme, genetic organization and synthesis in *Escherichia coli*. *Microbiology* **144:**1435–1441.

IV. HALOPHILES

Chapter 17

Biodiversity in Highly Saline Environments

AHARON OREN

INTRODUCTION

About 70% of the surface of planet Earth is covered by seawater: a salty environment that contains approximately 35 g of total dissolved salts per liter, 78% of which is NaCl. Although many microorganisms are unable to cope with life at seawater salinity, the marine environment cannot be considered "extreme": the seas are populated by a tremendous diversity of micro- and macroorganisms, at least as diverse as the world of freshwater organisms.

However, there are also environments with salt concentrations much higher than those found in the sea. When salt concentrations increase, the biological diversity decreases, and at concentrations about 150 to 200 g/liter, macroorganisms no longer survive. On the other hand, highly salt-tolerant and often even highly salt-requiring microorganisms can be found up to the highest salt concentrations: NaCl-saturated brines that contain salt concentrations of over 300 g/liter. Halophilic *Archaea*, *Bacteria*, and eukaryotic unicellular algae live in the Dead Sea, in the Great Salt Lake, in saltern crystallizer ponds, and in other salt-saturated environments, and they often reach high densities in such environments.

This chapter explores the world of high salt environments worldwide and the diversity of microorganisms that inhabit these environments.

DIVERSITY OF HYPERSALINE ENVIRONMENTS

Highly saline environments can be encountered on all continents. They include natural salt lakes with highly diverse chemical compositions, artificial salt lakes such as solar salterns for the production of NaCl from seawater, underground deposits of rock salt, as well as salted food products, highly saline soils, and others (Javor, 1989; Oren, 2002a).

The two largest truly hypersaline inland salt lakes are the Great Salt Lake, Utah, and the Dead Sea. The Great Salt Lake, a remnant of the ice-age saline Lake Bonneville that has largely dried out, has a salt composition that resembles that of seawater ("thalassohaline" brines). Owing to climatic changes and to human interference (division of the lake into a northern and a southern basin by a rockfill railroad causeway in the 1950s), the salinity of the lake has been subject to strong fluctuations in the past century. The northern basin is nowadays saturated with respect to NaCl. It is unfortunate that we know so little about the microbiology of the Great Salt Lake: after the pioneering studies by Fred Post in the 1970s (Post, 1977), the study of the microbial communities in the lake has been sadly neglected. However, a recent renewed interest in the biology of the lake is expected to change the picture, so that we soon may expect to get a much better picture of the diversity of microorganisms in the largest of all hypersaline lakes, their properties, and their dynamics (Baxter et al., 2005).

The Dead Sea, with its present-day salt concentration of over 340 g/liter, is an example of an "athalassohaline" brine, which has an ionic composition greatly different from that of seawater. Magnesium, not sodium, is the dominant cation, calcium is present as well in very high concentrations, and the pH is relatively low: around 6, as compared with 7.5 to 8 in thalassohaline brines. Indeed, the present-day Dead Sea is a remnant of the Pleistocene Lake Lisan, whose salts were of marine origin, but massive precipitation of halite and other geological phenomena have greatly changed the chemical properties of the brine. Yet, a few types of microorganisms can survive even in the

---

A. Oren • The Institute of Life Sciences and the Moshe Shilo Minerva Center for Marine Biogeochemistry, The Hebrew University of Jerusalem, Jerusalem 91904, Israel.

waters of the Dead Sea. However, the increase in salt concentration and relative increase in divalent cation concentrations in the past decades have made the Dead Sea environment too extreme for massive development of even the most salt-adapted microorganisms. Only when the upper water layers become diluted as a result of winter rain floods do dense microbial communities develop in the lake. A 10 to 15% dilution is sufficient to trigger massive blooms of the green alga *Dunaliella* and different types of red halophilic *Archaea* (Oren, 1988, 1999a).

Other natural hypersaline lakes are highly alkaline. Mono Lake, California (total salt concentration of around 90 g/liter; pH about 9.7 to 10), is an example of such a soda lake. Even more extreme are some of the soda lakes of the East African Rift Valley such as Lake Magadi, Kenya, as well as the lakes of Wadi Natrun, Egypt, and some soda lakes in China: here dense communities of halophilic *Archaea* and other prokaryotes are found in salt-saturated brines at pH values above 10. This illustrates that some halophilic microorganisms are true "polyextremophiles" (Rothschild and Mancinelli, 2001), organisms that can simultaneously cope with more than one type of environmental stress. The discovery of a truly thermophilic halophile, *Halothermothrix orenii*, isolated from a salt lake in Tunisia, shows that also life at high temperatures is compatible with life at high salt. This anaerobic fermentative bacterium grows up to salt concentrations of 200 g/liter (optimum: 100 g/liter) at temperatures up to 68°C (optimum 60°C) (Cayol et al., 1994).

Coastal solar salterns, found worldwide in dry tropical and subtropical climates, are man-made, thalassohaline hypersaline environments in which seawater is evaporated for the production of salt. Such saltern systems are operated as a series of ponds of increasing salinity, enabling controlled sequential precipitation of different minerals (calcite, gypsum, and halite). As a result, these saltern ecosystems present us with a more or less stable gradient of salt concentrations, from seawater salinity to NaCl precipitation and beyond, with each pond enabling the growth of those microbial communities adapted to the specific salinity of its brines. Dense and varied microbial communities generally develop both in the water and in the surface sediments of the saltern ponds (Oren, 2005). It is therefore not surprising that these saltern ecosystems have become popular objects for the study of microbial biodiversity and community dynamics at high salt concentrations, and much of our understanding of the biology of halophilic microorganisms is based on studies of the saltern environment and in-depth studies of microorganisms isolated from such salterns.

Another hypersaline aquatic habitat that appears to harbor interesting communities of halophilic microorganisms is the highly saline anoxic brines found in several sites near the bottom of the sea. Owing to the fact that these deep-sea anoxic hypersaline basins are not easily accessible for sampling, little is known thus far on their microbiology. However, a preliminary exploration of such brines from the bottom of the Red Sea, using culture-independent techniques, yielded evidence for the presence of a wealth of novel types of halophiles (Eder et al., 1999). A comprehensive multidisciplinary research program was recently launched, aimed at the elucidation of the biology of the deep-sea hypersaline anoxic basins in the Eastern Mediterranean Sea. The first published data that emerged from this program (van der Wielen et al., 2005) prove that we may expect many surprises from this previously unexplored type of hypersaline environment.

An overview of the biology of natural and man-made hypersaline lakes, as related to their chemical and physical properties, can be found in a recent monograph (Oren, 2002a).

Halophilic and halotolerant microorganisms are not only found in aquatic habitats. They can be recovered from many other environments in which high salt concentrations and/or low water activities occur. Halophilic and highly halotolerant bacteria can easily be recovered from saline soils. Some plants that grow on saline soils in arid areas actively excrete salt from their leaves, and the phylloplane of these plants thus appeared to be an interesting novel environment for halophiles (Simon et al., 1994), an environment that deserves to be investigated in further depth. Salted food products—especially when crude solar salt is used for salting—can be an excellent growth substrate for halophilic or halotolerant microorganisms. In fact, the production of some traditionally fermented food products in the Far East is based on the activity of halophilic bacteria.

Maybe the most surprising environment in which halophilic microorganisms have been found is the rock salt deposits found in many places worldwide. Live bacteria (endospore-forming organisms of the genus *Bacillus*) have even been recovered from rock salt crystals that had been buried for 250 million years (Vreeland et al., 2000), while viable *Archaea* of the family *Halobacteriaceae* or their 16S ribosomal RNA genes were recovered from ancient salt deposits as well (Fish et al., 2002; Leuko et al., 2005). These microorganisms appear to survive within small liquid inclusions within the solid rock salt. Although the claim that these organisms indeed had survived within the crystals for millions of years is not uncontested, it is now well established that indeed halophilic *Bacteria* and *Archaea* can retain their viability for long times in such brine inclusions within salt crystals.

PHYLOGENETIC DIVERSITY OF HALOPHILIC MICROORGANISMS

The ability to grow at salt concentrations exceeding those of seawater is widespread in the tree of life (Oren, 2000, 2002a, 2002b). Figure 1 presents the three-domain *Archaea–Bacteria–Eukarya* phylogenetic tree, based on small subunit rRNA gene comparisons, indicating those branches that contain representatives able to grow at salt concentrations above 100 g/liter.

Halophiles are thus found in all three domains of life. Among the *Eukarya*, we find relatively few representatives. Halophilic macroorganisms are rare; one of the few existing ones is the brine shrimp (genus *Artemia*) found in many salt lakes worldwide at salt concentrations up to 330 g/liter (Javor, 1989). The most widespread representative of the *Eukarya* in hypersaline ecosystems is the algal genus *Dunaliella*. *Dunaliella* is a unicellular green alga that is present as the major or sole primary producer in the Great Salt Lake, the Dead Sea, and salterns. Some species can accumulate massive amounts of β-carotene, and their cells are therefore orange-red rather than green. Some *Dunaliella* species prefer the low-salt marine habitat, but others, notably *D. salina*, can still grow in the NaCl-saturated brines of saltern crystallizer ponds. Also among the protozoa, we find halophilic and halotolerant types. Different ciliate, flagellate, and amoeboid protozoa can be observed in the biota of saltern evaporation ponds of intermediate salinity (Oren, 2005). Although predation of the halophilic microbial communities is possible up to the highest salt concentrations (Hauer and Rogerson, 2005; Park et al., 2003), protozoa do not appear to be very abundant in most hypersaline ecosystems. Another, often neglected, group of eukaryal halophiles is that of the fungi. Fungi are generally not abundantly found in environments of high salt concentrations. However, it was recently ascertained that certain fungi, notably the halophilic black yeasts, find their natural ecological niches in the hypersaline waters of solar salterns (Gunde-Cimerman et al., 2000). More recent surveys have shown that the role that fungi may play in high salt environments has been grossly underestimated thus far (Gunde-Cimerman et al., 2004; Butinar et al., 2005).

Within the domain *Archaea*, we find halophiles in two major branches of *Euryarchaeota*: the *Halobacteriales* and the methanogens. The branch of extremely halophilic, generally red pigmented, aerobic *Archaea* of the order *Halobacteriales* consists entirely of halophiles (Oren, 2001a). These are the organisms that dominate the heterotrophic communities in the Dead Sea, in the northern basin of the Great Salt Lake, in the crystallizer ponds of solar salterns, and also in many soda lakes. Their massive presence is generally obvious by the red coloration of the brines, caused mainly by 50-carbon carotenoids (α-bacteroruberin and derivatives), but retinal based protein pigments (the light-driven proton pump bacteriorhodopsin and the light-driven primary chloride pump halorhodopsin) may also contribute to the coloration of the cells.

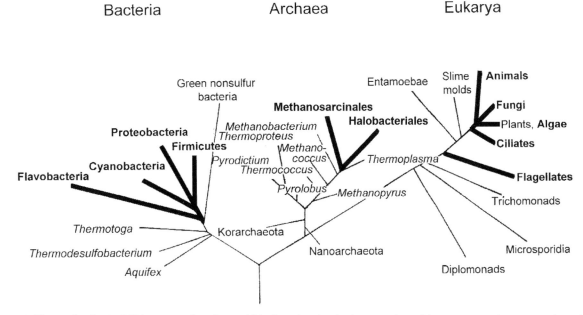

Figure 1. The small subunit rRNA sequence-based tree of life. Branches that harbor organisms able to grow at salt concentrations above 100 g/liter are highlighted. Based in part on Fig. 11.13 in Madigan et al. (2003).

There are also obligatory anaerobic halophilic methanogenic *Archaea*. Here, the halophiles do not form a separate phylogenetic branch, but they appear interspersed between non-halophilic relatives.

Most known halophilic and halotolerant prokaryote species belong to the domain *Bacteria*. Microscopic examination of water and sediment samples of saltern evaporation ponds of intermediate salinity shows an abundance of forms of bacteria. Diverse communities of cyanobacteria, unicellular as well as filamentous, are conspicuously found in the microbial mats that cover the bottom sediments of salterns at salt concentrations up to 200 to 250 g/liter. Below the cyanobacterial layer, massive development of photosynthetic purple sulfur bacteria (*Halochromomatium*, *Halorhodospira*, and related organisms, belonging to the Proteobacteria branch of the domain *Bacteria*) is often seen as well (Oren, 2005). The domain *Bacteria* contains many aerobic heterotrophic organisms of widely varying phylogenetic affiliation (Ventosa et al., 1998). The recent discovery of the genus *Salinibacter* (*Bacteroidetes* phylum), a genus abundant in saltern crystallizer ponds (see below), shows that the domain *Bacteria* contains some microorganisms that are no less salt tolerant and salt dependent than the most halophilic among the archaeal order *Halobacteriales*, which was thus far considered to contain the best salt adapted of all microorganisms. There is one lineage within the *Bacteria* that appears to consist entirely of halophiles: the group of obligatory anaerobic bacteria of the order *Halanaerobiales* (families *Halanaerobiaceae* and *Halobacteroidaceae*) (Oren, 2001b). These fermentative organisms, which typically grow optimally at salt concentrations between 50 and 200 g/liter, may well be responsible for much of the anaerobic degradation of carbohydrates and other compounds in the anaerobic sediments of hypersaline lakes.

Last but not least, this survey of microbial diversity at high salt concentrations should also mention the occurrence of viruses. Many halophilic *Bacteria* and *Archaea* have bacteriophages that attack them and may cause their lysis. Free virus-like particles as well as lysing cells releasing large number of mature bacteriophages have been observed during electron microscopic examination of the biomass of saltern crystallizer ponds (Guixa-Boixareu et al., 1996) and the Dead Sea (Oren et al., 1997), and the viral assemblage in Spanish saltern pond has been partially characterized by pulsed-field gel electrophoresis (Diez et al., 2000). It was calculated that lysis by viruses is quantitatively far more important than bacterivory by protozoa in regulating the prokaryotic community densities of saltern ponds at the highest salinities (Guixa-Boixareu et al., 1996).

METABOLIC DIVERSITY OF HALOPHILIC MICROORGANISMS

As salt concentrations increase, the number of physiological types of microorganisms encountered in hypersaline lakes and other high salt ecosystems decreases. To give a few examples: we do not know any methanogenic *Archaea* growing at salt concentrations above 100 g/liter and using hydrogen plus carbon dioxide or acetate as their substrates. Methanogenesis at higher salt concentrations does occur, but it is mainly based on degradation of methylated amines. No truly halophilic dissimilatory sulfate-reducing bacteria are known to oxidize acetate, while sulfate reduction with lactate as electron donor can proceed up to salt concentrations of 200 to 250 g/liter at least. Other metabolic activities that are notably absent at the highest salt concentrations are the two stages of autotrophic nitrification: oxidation of ammonium ions to nitrite and oxidization of nitrite to nitrate. Microbial activities that are possible up to the highest salt concentrations are aerobic respiration and oxygenic photosynthesis. Denitrification, anoxygenic photosynthesis with sulfide as electron donor and fermentations are processes that have been documented to proceed in environments at or close to salt saturation, as well as in cultures of isolated microorganisms grown at salt concentrations of 200 g/liter and higher (Oren, 1999b, 2000, 2002a).

A possible explanation has been brought forward for the apparent absence of certain metabolic types of microorganisms at the highest salt concentrations. This explanation was based on the balance between the energetic cost of osmotic adaptation and the amount of energy made available to the organisms in the course of their dissimilatory metabolism (Oren, 1999b). Life at high salt concentrations is energetically costly as the cells have to accumulate high concentrations of solutes to provide osmotic balance between their cytoplasm and the brines in which they live. No microorganism uses NaCl to balance the NaCl outside, and therefore osmotic balance is always accompanied by the establishment of concentration gradients across the cell membrane, and this can only be done at the expense of energy.

Two fundamentally different modes of osmotic adaptation are known in the microbial world: accumulation of KCl, i.e., inorganic ions, to provide the osmotic equilibrium, or synthesis of accumulation of organic osmotic solutes. The "high salt-in" strategy, based on the accumulation of potassium and chloride ions up to molar concentrations in the cytoplasm, is used by a few groups of microorganisms only. The aerobic halophilic *Archaea* of the order *Halobacteriales* use this mode of osmotic adaptation.

Not all halophilic *Archaea* use this strategy: the halophilic members of the methanogens accumulate organic osmotic solutes. Within the domain *Bacteria*, we thus far know only two groups of halophiles that use the "high salt-in" strategy. One is the fermentative anaerobes of the order *Halanaerobiales* (low G+C branch of the *Firmicutes*) (Oren, 2002a). The second is the only recently discovered red aerobic *Salinibacter* (*Bacteroidetes* branch) (Oren et al., 2002). It is interesting to note that both the halophilic *Archaea* and *Salinibacter* possess halorhodopsin, a light-driven primary chloride pump, to facilitate the uptake of chloride into the cells. Calculations have shown that the "high salt-in" strategy of osmotic adaptation is energetically favorable (Oren, 1999b). However, this mode of life depends on the complete adaptation of the intracellular enzymatic machinery to function in the presence of high ionic concentration. Special adaptations of the protein structure are necessary to achieve this, and as a result, those microorganisms that use KCl as their osmotic solute have become strictly dependent on the presence of high salt concentrations. Such organisms are generally restricted to life at a narrow range of extremely high salt concentrations. They lack the flexibility to adapt to a wide range of salt concentrations and to changes in the salt concentration of their medium, a flexibility that is so characteristic of many microorganisms that use the second strategy of osmotic adaptation.

That second strategy is based on the exclusion of inorganic ions from the cytoplasm to a large extent while balancing the osmotic pressure exerted by the salts in the environment with simple uncharged or zwitterionic organic solutes. A tremendous variety of such organic solutes have been detected in different halophilic and halotolerant microorganisms. Thus, algae of the genus *Dunaliella* produce and accumulate molar concentrations of glycerol while regulating the intracellular glycerol in accordance with the outside salinity. Glycerol is never found as an osmotic solute in the prokaryote world. Osmotic, "compatible" solutes produced by different groups of prokaryotes include simple sugars (sucrose and trehalose), amino acid derivatives [glycine betaine, ectoine (1,4,5,6-tetrahydro-2-methyl-4-pyrimidine carboxylic acid), and others], and other classes of compounds (Oren, 2002a). In many cases, more than one solute may be produced by a single organism. For example, photosynthetic sulfur bacteria of the genus *Halorhodospira* (γ-Proteobacteria) typically contain cocktails of glycine betaine, ectoine, and trehalose. De novo biosynthesis of such organic osmotic solutes is energetically expensive. However, most "low salt-in" organisms are also able to accumulate suitable organic solutes when such compounds are present in the medium, thus enabling the cells to save considerable amounts of energy. The great advantage of the "low salt-in" strategy of life at high salt concentration is that no or little adaptation of the intracellular enzymatic machinery is necessary. Cells that use organic osmotic solutes to provide osmotic balance generally display a large extent of adaptability to a wide range of salinities and can rapidly adjust to changes in medium salinity.

Integration of the available information on the energetic cost of osmotic adaptation and information on the amount of energy generated by the different types of dissimilatory metabolism has enabled the establishment of a coherent model that may explain which types of metabolism can occur at the highest salt concentrations and which cannot (Oren, 1999b). Processes that provide plenty of energy (e.g., aerobic respiration and denitrification) can function at high salt concentrations, independent of the mode of osmotic adaptation of the organisms that perform them. On the other hand, dissimilatory processes that yield little energy only (e.g., autotrophic nitrification and production of methane from acetate) are problematic at the highest salt concentrations unless the cells can economize on the amount of energy required to produce or accumulate osmotic solutes. There, the "high salt-in" strategy appears to be advantageous, and this is therefore the strategy adopted by the Halanaerobiales, the specialized group of halophilic fermentative *Bacteria*. The model explains why for example autotrophic nitrification is not likely to occur at high salt concentrations: only very little energy is gained in the process and (most) nitrifying bacteria belong to the Proteobacteria, a group that uses organic osmotic solutes rather than KCl to provide osmotic balance. Also, the apparent lack of certain types of methanogens and sulfate-reducing bacteria becomes understandable: those reactions that yield little energy do not occur at the highest salinities and those reactions that are energetically more favorable do. Both groups depend on organic osmotic solutes for growth at high salt concentrations (Oren, 1999b, 2002a).

SALINIBACTER RUBER, AN EXTREMELY HALOPHILIC MEMBER OF THE *BACTERIA*

The recently discovered *Salinibacter ruber*, a species of red, extremely halophilic *Bacteria* isolated from saltern crystallizer ponds, presents us with an interesting model for the study of the adaptation of microorganisms to life at the highest salt concentrations (Oren, 2004; Oren et al., 2004).

In the past, *Archaea* of the order *Halobacteriales*, family *Halobacteriaceae*, were always considered to be the extreme halophiles par excellence, being the sole heterotrophs active at the highest salinities such as those that occur in saltern crystallizer ponds and other NaCl-saturated environments. All known heterotrophs representatives of the domain *Bacteria* could be classified as moderate halophiles. Those few that were still able to grow at salt concentrations above 300 g/liter did so at very slow rates only and had their optimum growth at far lower salt concentrations (Ventosa et al., 1998). However, evidence for the presence of significant number of extremely halophilic representatives of the domain *Bacteria* in saltern crystallizer ponds, sometimes representing up to 15 to 20% and more of the prokaryotic community, was first obtained in the late 1990s on the basis of molecular ecological, culture-independent studies (Antón et al., 2000). When soon afterward the organism, a rod-shaped red aerobic bacterium, was brought into culture (Antón et al., 2002), the organism appeared to be extremely interesting, and its study has deepened our understanding of phylogenetic as well as physiological and metabolic diversity in the world of halophiles.

Salinibacter ruber, as the organism was named, belongs phylogenetically to the *Salinibacter Bacteroidetes* branch of the *Bacteria*. Its closest relative as based on 16S rRNA sequence comparison is the genus *Rhodothermus*, red, aerobic thermophiles isolated from marine hot springs. *Salinibacter* is no less halophilic than the most salt-requiring and salt-tolerant organisms within the *Halobacteriaceae*: it is unable to grow at salt concentrations below 150 g/liter, it thrives optimally at 200 to 250 g/liter, and it grows in media saturated with NaCl as well. Examination of the mode of osmotic adaptation and the properties of the intracellular enzymes showed a great similarity between *Salinibacter* and the *Halobacteriaceae*: in contrast to all earlier examined aerobic halophilic or halotolerant members of the *Bacteria*, *Salinibacter* did not contain organic osmotic solutes but was found to use KCl to provide osmotic balance (Oren et al., 2002). Accordingly, the intracellular enzymatic systems were found to be salt tolerant, and in many cases salt dependent. The finding of a gene coding for halorhodopsin, the light-driven inward chloride pump known thus far from halophilic *Archaea* only, made the similarity between the two even greater. We may here have an example of convergent evolution, in which two, phylogenetically disparate organisms have obtained highly similar adaptations that have enabled them to grow at the highest salt concentrations but have also restricted their possibility to survive at lower salinities (Mongodin et al., 2005; Oren, 2004).

We still know little about the interrelationships between *Salinibacter* and halophilic *Archaea* in the habitat they share: the brines of saltern crystallizer ponds and probably other salt lakes as well. Being very similar in their physiological properties, *Salinibacter* should be expected to compete with the *Halobacteriaceae* for the same substrates and other resources. What selective advantages either group has to ensure its coexistence with the other remains to be determined.

THE MICROBIAL COMMUNITY STRUCTURE IN HYPERSALINE ENVIRONMENTS— CULTURE-DEPENDENT AND CULTURE-INDEPENDENT APPROACHES

As described in the previous section, it was the application of culture-independent studies of the microbial diversity, using small subunit rRNA gene sequence-based techniques that presented the first evidence of the existence of *Salinibacter* (Antón et al., 2000), an organism that was until that time completely overlooked, even when it probably had been present as colonies on agar plates inoculated with saltern brines in the past. Microbiologists working with halophiles silently assumed that red colonies that developed on plates with salt concentrations of 200 to 250 g/liter can only belong to members of the *Halobacteriaceae*. After the molecular approach had indicated what to look for, the isolation of the organism harboring the novel 16S rRNA gene sequence followed rapidly (Antón et al., 2002).

The application of molecular biological techniques to the study of the microbial diversity in hypersaline ecosystems started in the mid-1990s with the studies by Benlloch et al. (1995) in the salterns of Santa Pola, Alicante, Spain. Sequencing of 16S rRNA genes amplified from DNA extracted from the biomass showed that the dominant phylotype in this environment indeed belonged to a member of the *Halobacteriaceae*, but differed from all thus far isolated members of the family at the genus level. Fluorescence in situ hybridization experiments then showed that this phylotype belongs to a highly unusually shaped prokaryote: extremely thin, flat, perfectly square, or rectangular cells that contain gas vesicles (Antón et al., 1999). This type of cell was first detected during microscopic examination of water from a coastal brine pool on the Sinai Peninsula (Walsby, 1980). The abundance of such cells in the salterns had become well known in subsequent years (Guixa-Boixareu et al., 1996; Oren et al., 1996). However, until recently, this intriguing microorganism defied all attempts toward its isolation.

The elusive flat square halophilic *Archaea* were brought into culture in 2004, independently by two groups of investigators, working in salterns in Spain (Bolhuis et al., 2004) and in Australia (Burns et al., 2004a). Using appropriate growth media (preferentially low in nutrients) and in addition a large amount of patience (incubation times of 8 to 12 weeks), Burns et al. (2004b) showed that in fact the majority of prokaryotes that can be detected in the saltern crystallizer ponds using 16S rRNA gene sequence-based, culture-independent techniques can also be cultured. In most non-extreme ecosystems, there still is a tremendous difference, generally of many orders of magnitude, between the numbers of prokaryotes observed microscopically and the numbers that can be grown as colonies on plates. Thanks to the recently developed new approaches, the saltern crystallizer environment is probably the first ecosystem for which the "great plate count anomaly," as the phenomenon is often designated, has ceased to exist.

More extensive molecular ecological studies have been made in the Alicante salterns along the salt gradient, to obtain a more complete picture of the development of the microbial diversity as the salinity increases during the gradual evaporation of seawater (Benlloch et al., 2001, 2002; Casamajor et al., 2002; see also Oren, 2002c). Benthic cyanobacterial mats that develop on the bottom of saltern ponds of intermediate salinity have been the subject of molecular ecological studies as well (Mouné et al., 2002). Similar techniques have been used to characterize the microbial diversity in the athalassohaline alkaline Mono Lake, California (Humayoun et al., 2003). These studies make it clear that many of the microorganisms that dominate the communities before NaCl saturation is reached still await isolation and characterization.

EPILOGUE

Although only few groups of macroorganisms have learned to live at salt concentrations much higher than those of seawater, many types of microorganisms have developed the adaptations necessary for life in hypersaline environments. Many can even live at the salinity of saturated solutions of NaCl, the salt concentration encountered in some natural salt lakes as well as in saltern crystallizer ponds. It has been suggested that the ability to live at high salt concentrations may have appeared very early in prokaryote evolution and that life may even have emerged in a hypersaline environment—a concentrated solution of organic compounds in tidal pools of partially evaporated seawater (Dundas, 1998). The theory of a hypersaline origin of life is, however, not supported by phylogenetic evidence: most halophiles are located on distant, relatively "recent" branches of the small subunit rRNA gene sequence-based phylogenetic tree. Moreover, the great variety in strategies used by the present-day halophiles to cope with the high salinity in their environment shows that adaptation to life at high salt concentrations has probably arisen many times during the evolution of the three domains of life (Oren, 2002a).

The world of the halophilic microorganisms is highly diverse. We find halophiles dispersed all over the phylogenetic tree of life. Metabolically, they are almost as diverse as the "non-extremophilic" microbial world: we know halophilic autotrophs as well as heterotrophs, aerobes as well as anaerobes, phototrophs as well as chemoautotrophs. Thus, hypersaline ecosystems can function to a large extent in the same way as "conventional" freshwater and marine ecosystems. Owing to the absence of macroorganisms and the generally low levels of predation by protozoa, the microbial community densities of halophiles in hypersaline environments may be extremely high: counts of 10^7 to 10^8 red halophilic Archaea per ml of brine are not exceptionally high in Great Salt Lake, the Dead Sea, and in saltern crystallizer ponds, and they often impart a bright red color to the brines. The presence of such dense communities makes such environments ideal model systems for the study of the functioning of microorganisms in nature.

While osmotic equilibrium of the cell's cytoplasm with the salinity of the environment is essential for any halophilic or halotolerant microorganism to function, there are multiple ways in which this osmotic equilibrium can be achieved. There is therefore a considerable diversity within the world of the halophilic microorganisms with respect to the way the cells cope with the salt outside. Notably, there are two basically different approaches toward the solution of the problem: keeping the salt out or allowing massive amounts of salt (KCl rather than NaCl) to enter the cytoplasm. There is no clear correlation between the phylogenetic position of a halophilic microorganism and the strategy it uses to obtain osmotic balance. As the case of *Salinibacter* clearly shows, similar solutions have turned up in completely unrelated microorganisms.

Culture-independent techniques have taught us how diverse the microbial communities in salt lakes really are. A few recent breakthroughs have enabled the cultivation of a number of halophiles (the flat square gas-vacuolated Archaea and *Salinibacter*) that are among the dominant forms of life in many hypersaline environments. An in-depth study of such ecologically relevant organisms will undoubtedly deepen our understanding of the functioning of the highly saline ecosystems, as well as shed more light on the nature of the adaptation of life to function at the highest salt concentrations.

REFERENCES

Antón, J., E. Llobet-Brossa, F. Rodríguez-Valera, and R. Amann. 1999. Fluorescence in situ hybridization analysis of the prokaryotic community inhabiting crystallizer ponds. *Environ. Microbiol.* **1:**517–523.

Antón, J., R. Rosselló-Mora, F. Rodríguez-Valera, and R. Amann. 2000. Extremely halophilic bacteria in crystallizer ponds from solar salterns. *Appl. Environ. Microbiol.* **66:**3052–3057.

Antón, J., A. Oren, S. Benlloch, F. Rodríguez-Valera, R. Amann, and R. Rosselló-Mora. 2002. *Salinibacter ruber* gen. nov., sp. nov., a novel extreme halophilic member of the bacteria from saltern crystallizer ponds. *Int. J. Syst. Evol. Microbiol.* **52:**485–491.

Baxter, B. K., C. D. Litchfield, K. Sowers, J. D. Griffith, P. Arora DasSarma, and S. DasSarma. 2005. Microbial diversity of Great Salt Lake, p. 11–25. *In* N. Gunde-Cimerman, A. Oren, and A. Plemenitaš (ed.), *Adaptation to Life at High Salt Concentrations in Archaea, Bacteria, and Eukarya.* Springer, Dordrecht, The Netherlands.

Benlloch, S., A. J. Martínez-Murcia, and F. Rodríguez-Valera. 1995. Sequencing of bacterial and archaeal 16S rRNA genes directly amplified from a hypersaline environment. *Syst. Appl. Microbiol.* **18:**574–581.

Benlloch, S., S. G. Acinas, J. Antón, A. López-López, S. P. Luz, and F. Rodríguez-Valera. 2001. Archaeal biodiversity in crystallizer ponds from a solar saltern: culture versus PCR. *Microb. Ecol.* **41:**12–19.

Benlloch, S., A. López-López, E. O. Casamajor, L. Øvreas, V. Goddard, F. L. Dane, G. Smerdon, R. Massana, I. Joint, F. Thingstd, C. Pedrós-Alió, and F. Rodríguez-Valera. 2002. Prokaryotic genetic diversity throughout the salinity gradient of a coastal solar saltern. *Environ. Microbiol.* **4:**349–360.

Bolhuis, H., E. M. te Poele, and F. Rodríguez-Valera. 2004. Isolation and cultivation of Walsby's square archaeon. *Environ. Microbiol.* **6:**1287–1291.

Burns, D. G., H. M. Camakaris, P. H. Janssen, and M. L. Dyall-Smith. 2004a. Cultivation of Walsby's square haloarchaeon. *FEMS Microbiol. Lett.* **238:**469–473.

Burns, D. G., H. M. Camakaris, P. H. Janssen, and M. L. Dyall-Smith. 2004b. Combined use of cultivation-dependent and cultivation-independent methods indicates that members of most haloarchaeal groups in an Australian crystallizer pond are cultivable. *Appl. Environ. Microbiol.* **70:**5258–5265.

Butinar, L., I. Spencer-Martins, S. Santos, A. Oren, and N. Gunde-Cimerman. 2005. Yeast diversity in hypersaline habitats. *FEMS Microbiol. Lett.* **244:**229–234.

Casamajor, E. O., R. Massana, S. Benlloch, L. Øvreas, B. Diez, V. J. Goddard, J. M. Gasol, I. Joint, F. Rodríguez-Valera, and C. Pedrós-Alió. 2002. Changes in archaeal, bacterial and eukaryal assemblages along a salinity gradient by comparison of genetic fingerprinting methods in a multipond solar saltern. *Environ. Microbiol.* **4:**338–348.

Cayol, J.-L., B. Ollivier, B. K. C. Patel, G. Prensier, J. Guezennec, and J.-L. Garcia. 1994. Isolation and characterization of *Halothermothrix orenii* gen. nov., sp. nov., a halophilic, thermophilic, fermentative, strictly anaerobic bacterium. *Int. J. Syst. Bacteriol.* **44:**534–540.

Diez, B., J. Antón, N. Guixa-Boixereu, C. Pedrós-Alió, and F. Rodríguez-Valera. 2000. Pulsed-field gel electrophoresis analysis of virus assemblages present in a hypersaline environment. *Int. Microbiol.* **3:**159–164.

Dundas, I. 1998. Was the environment for primordial life hypersaline? *Extremophiles* **2:**375–377.

Eder, W., W. Ludwig, and R. Huber. 1999. Novel 16S rRNA gene sequences retrieved from highly saline brine sediments of Kebrit Deep, Red Sea. *Arch. Microbiol.* **172:**213–218.

Fish, S. A., T. J. Shepherd, T. J. McGenity, and W. D. Grant. 2002. Recovery of 16S ribosomal RNA gene fragments from ancient halite. *Nature* **417:**432–436.

Guixa-Boixareu, N., J. I. Caldérón-Paz, M. Heldal, G. Bratbak, and C. Pedrós-Alió, C. 1996. Viral lysis and bacterivory as prokaryotic loss factors along a salinity gradient. *Aquat. Microb. Ecol.* **11:**213–227.

Gunde-Cimerman, N., P. Zalar, G. S. de Hoog, and A. Plemenitaš. 2000. Hypersaline water in salterns—natural ecological niches for halophilic black yeasts. *FEMS Microbiol. Ecol.* **32:**235–240.

Gunde-Cimerman, N., P. Zalar, U. Petrovič, M. Turk, T. Kogej, S. de Hoog, and A. Plemenitaš. 2004. Fungi in the salterns, p. 103–111. *In* A. Ventosa (ed.), *Halophilic Microorganisms.* Springer-Verlag, Berlin, Germany.

Hauer, G., and A. Rogerson. 2005. Heterotrophic protozoa from hypersaline environments. *In* N. Gunde-Cimerman, A. Oren, and A. Plemenitaš (ed.), *Adaptation to Life at High Salt Concentrations in Archaea, Bacteria, and Eukarya,* p. 521–539. Springer, Dordrecht, The Netherlands.

Humayoun, S. B., N. Bano, and J. T. Hollibaugh. 2003. Depth distribution of microbial diversity in Mono Lake, a meromictic soda lake in California. *Appl. Environ. Microbiol.* **69:**1030–1042.

Javor, B. 1989. *Hypersaline Environments. Microbiology and Biogeochemistry.* Springer-Verlag, Berlin, Germany.

Leuko, S., A. Legat, S. Fendrihan, H. Wieland, C. Radax, C. Gruber, M. Pfaffenhuemer, G. Weidler, and H. Stan-Lotter. 2005. Isolation of viable haloarchaea from ancient salt deposits and application of fluorescent stains for *in situ* detection of halophiles in hypersaline environmental samples and model fluid inclusions, p. 93–104. *In* N. Gunde-Cimerman, A. Oren, and A. Plemenitaš (ed.), *Adaptation to Life at High Salt Concentrations in Archaea, Bacteria, and Eukarya.* Springer, Dordrecht, The Netherlands.

Madigan, M. T., J. M. Martinko, and J. Parker. 2003. *Brock Biology of Microorganisms.* Pearson Education., Inc., Upper Saddle River, NJ.

Mongodin, E. F., K. E. Nelson, S. Daugherty, R. T. deBoy, J. Wister, H. Khouri, J. Weidman, D. A. Walsh, R. T. Papke, G. Sanchez Perez, A. K. Sharma, C. L. Nesbó, D> MacLeod, E. Bapteste, W. F. Doolittle, R. L. Charlebois, B. Legault, and F. Rodríguez-Valera. 2005. The genome of *Salinibacter ruber*: convergence and gene exchange among hyperhalophilic bacteria and archaea. *Proc. Natl. Acad. Sci. USA* **102:**18147–18152.

Mouné, S., P. Caumette, R. Matheron, and J. C. Willison. 2002. Molecular sequence analysis of prokaryotic diversity in the anoxic sediments underlying cyanobacterial mats of two hypersaline ponds in Mediterranean salterns. *FEMS Microbiol. Ecol.* **44:**117–130.

Oren, A. 1988. The microbial ecology of the Dead Sea, p. 193–229. *In* K. C. Marshall (ed.), *Advances in Microbial Ecology,* vol. 10. Plenum Publishing Company, New York, NY.

Oren, A. 1999a. Microbiological studies in the Dead Sea: future challenges toward the understanding of life at the limit of salt concentrations. *Hydrobiologia* **405:**1–9.

Oren, A. 1999b. Bioenergetic aspects of halophilism. *Microbiol. Mol. Biol. Rev.* **63:**334–348.

Oren, A. 2000. Life at high salt concentrations. *In* M. Dworkin, S. Falkow, E. Rosenberg, K.-H. Schleifer, and E. Stackebrandt (ed.), *The Prokaryotes: An Evolving Electronic Resource for the Microbiological Community,* 3rd ed., release 3.1, 20 January 2000. Springer-Verlag, New York, NY, http://link.springer-ny.com/link/service/books/10125.

Oren, A. 2001a. The order Halobacteriales. *In* M. Dworkin, S. Falkow, E. Rosenberg, K.-H. Schleifer, and E. Stackebrandt (ed.), *The Prokaryotes: An Evolving Electronic Resource for the Microbiological Community,* 3rd ed., release 3.2, 25 July

2001. Springer-Verlag, New York, NY, http://link.springer-ny.com/link/ service/books/10125.

Oren, A. 2001b. The order Haloanaerobiales. *In* M. Dworkin, S. Falkow, E. Rosenberg, K.-H. Schleifer, and E. Stackebrandt (ed.), *The Prokaryotes: An Evolving Electronic Resource for the Microbiological Community*, 3rd ed., release 3.2, 25 July 2001. Springer-Verlag, New York, NY, http://link.springer-ny.com/link/service/books/10125.

Oren, A. 2002a. *Halophilic Microorganisms and their Environments*. Kluwer Scientific Publishers, Dordrecht, The Netherlands.

Oren, A. 2002b. Diversity of halophilic microorganisms: environments, phylogeny, physiology, and applications. *J. Indust. Microbiol. Biotechnol.* **28**:56–63.

Oren, A. 2002c. Molecular ecology of extremely halophilic Archaea and Bacteria. *FEMS Microbiol. Ecol.* **39**:1–7.

Oren, A. 2004. The genera *Rhodothermus*, *Thermonema*, *Hymenobacter* and *Salinibacter*. *In* M. Dworkin, S. Falkow, E. Rosenberg, K.-H. Schleifer, and E. Stackebrandt (ed.), *The Prokaryotes: An Evolving Electronic Resource for the Microbiological Community*, 3rd ed., release 3.17, 31 August 2004. Springer-Verlag, New York, NY, http://link.springer-ny.com/link/service/books/10125/.

Oren, A. 2005. Microscopic examination of microbial communities along a salinity gradient in saltern evaporation ponds: a 'halophilic safari', p. 43–57. *In* N. Gunde-Cimerman, A. Oren, and A. Plemenitaš (ed.), *Adaptation to Life at High Salt Concentrations in Archaea, Bacteria, and Eukarya*. Springer, Dordrecht, The Netherlands.

Oren, A., S. Duker, and S. Ritter. 1996. The polar lipid composition of Walsby's square bacterium. *FEMS Microbiol. Lett.* **138**:135–140.

Oren, A., G. Bratbak, and M. Heldal. 1997. Occurrence of virus-like particles in the Dead Sea. *Extremophiles* **1**:143–149.

Oren, A., M. Heldal, S. Norland, and E. A. Galinski. 2002. Intracellular ion and organic solute concentrations of the extremely halophilic Bacterium *Salinibacter ruber*. *Extremophiles* **6**:491–498.

Oren, A., F. Rodríguez-Valera, J. Antón, S. Benlloch, R. Rosselló-Mora, R. Amann, J. Coleman, and N. J. Russell. 2004. Red, extremely halophilic, but not archaeal: the physiology and ecology of *Salinibacter ruber*, a bacterium isolated from saltern crystallizer ponds, p. 63–76. *In* A. Ventosa (ed.), *Halophilic Microorganisms*. Springer-Verlag, Berlin, Germany.

Park, J. S., H. Kim, D. H. Choi, and B. C. Cho. 2003. Active flagellates grazing on prokaryotes in high salinity waters of a solar saltern. *Aquat. Microb. Ecol.* **33**:173–179.

Post, F. J. 1977. The microbial ecology of the Great Salt Lake. *Microb. Ecol.* **3**:143–165.

Rothschild, L. J., and R. L. Mancinelli. 2001. Life in extreme environments. *Nature* **409**:1092–1101.

Simon, R. D., A. Abeliovich, and S. Belkin. 1994. A novel terrestrial halophilic environment: the phylloplane of *Atriplex halimus*, a salt-excreting plant. *FEMS Microbiol. Ecol.* **14**:99–110.

van der Wielen, P. W. J. J., H. Bolhuis, S. Borin, D. Daffonchio, C. Corselli, L. Giuliano, G. D'Auria, G. J. de Lange, A. Huebner, S. P. Varnavas, J. Thomson, C. Tamburini, D. Marty, T. J. McGenity, K. N. Timmis, and the BioDeep Scientific Party. 2005. The enigma of prokaryotic life in deep hypersaline anoxic basins. *Science* **307**:121–123.

Ventosa, A., J. J. Nieto, and A. Oren. 1998. Biology of aerobic moderately halophilic bacteria. *Microbiol. Mol. Biol. Rev.* **62**:504–544.

Vreeland, R. H., W. D. Rosenzweig, and D. W. Powers. 2000. Isolation of a 250 million-year-old halotolerant bacterium from a primary salt crystal. *Nature* **407**:897–900.

Walsby, A. E. 1980. A square bacterium. *Nature* **283**:69–71.

Chapter 18

Response to Osmotic Stress in a Haloarchaeal Genome: a Role for General Stress Proteins and Global Regulatory Mechanisms

GUADALUPE JUEZ, DAVID FENOSA, AITOR GONZAGA, ELENA SORIA, AND
FRANCISCO J. M. MOJICA

INTRODUCTION

Halophilic *Archaea* (haloarchaea) inhabit hypersaline environments, such as solar salterns and salty lakes, with very high salt concentrations, where salt precipitation is commonplace and where relatively high temperatures (up to 55°C) are frequently reached (Rodríguez-Valera, 1988; Oren, 1999). Haloarchaea are highly specialized for life under these extreme conditions. They are able to grow in saturated sodium chloride concentrations, and most of them require a minimum of 1.5 to 3 M NaCl and 0.005 to 0.04 M magnesium salts for growth (Tindall and Trüper, 1986; Juez, 1988). To compensate for the osmotic pressure, haloarchaea accumulate high concentrations of potassium as their main compatible solute. This intracellular ionic content varies according to the salinity of the medium and can reach up to 5 M potassium (Christian and Waltho, 1962; Ginzburg et al., 1970). These organisms are therefore subject to extreme environmental salinity as well as to extreme intracellular ionic concentrations. The intracellular ionic concentrations compensate for the excess of acidic amino acids typical of haloarchaeal proteins, which are destabilized in the absence of proper cation concentration (Lanyi, 1974; Danson and Hough, 1997). The halophilic nature of the haloarchaeal proteins is accompanied by a cation-dependent character. Indeed, a low-salt challenge may have a drastic effect on protein stability and function. Hypoosmotic stress by low salinities or after dilution with water is, in fact, a frequent event in their habitat which, in addition to implying protein aggregation, could commonly promote cell lysis. Whilst in other organisms the cell wall counteracts turgor pressure under hypoosmotic conditions, in the case of most haloarchaea, in particular rod and pleomorphic shapes, the cell wall is composed of a halophilic glycoprotein whose stability depends on the ionic concentration in the external medium, particularly of NaCl and also magnesium (Mescher and Strominger, 1976). For haloarchaea such as *Haloferax* and *Halobacterium* requiring a minimum of approximately 1.5 and 3 M NaCl for growth (10 and 20% of total salts corresponding in proportions to those found in seawater) and a minimum of 0.02 to 0.04 M and 0.005 to 0.01 M, respectively, of magnesium, the lowest limit at which cell lysis may be prevented is at 0.5 to 1 M NaCl (3 to 5% of total salts, which in the case of the different *Haloferax* members seems to coincide with the minimal magnesium requirements) (Juez, 1982, 1988; Torreblanca et al., 1986). Haloarchaeal cocci, such as *Halococcus*, may require high salinities for growth (2.5 M NaCl or about 15% of total salts as minimal salinities) but are more resistant to lysis upon salt dilution than rods and can even survive after exposure to distilled water due to their heteropolysaccharidic cell wall (Gibbons, 1974; Steber and Schleifer, 1975). Nevertheless, a common fact for haloarchaea is the effect that hyposaline conditions have on proteins, which to a certain degree may resemble the effect of high temperatures. In summary, while haloarchaea are particularly specialized for life under hypersaline conditions, withstanding harsh dehydration or low water activity, in these organisms hypoosmotic stress is a really harsh and usually lethal condition (Juez, 2004).

In order to counteract osmotic challenge, haloarchaea have had to evolve particularly effective mechanisms. On the one hand, osmotic balance seems to be the main limiting factor in adaptation to changing salinities (Mojica et al., 1997; Juez, 2004). Adaptation after a shift from low to high salinity (10 to 30% of

---

G. Juez, D. Fenosa, A. Gonzaga, E. Soria, and F. J. M. Mojica • División de Microbiología, Campus de San Juan, Universidad Miguel Hernandez, Sant Joan d'Alacant 03550 Alicante, Spain.

salts) involves a long lag period during which potassium is gradually accumulated in cells and the high-salt-related proteins are synthesized. Meanwhile, after a shift from high to low salinity (30 to 10% of salts), there is a drastic decrease in intracellular potassium content and an immediate induction of the newly required proteins together with a fast recovery of cells after the osmotic downshift has been overcome (Mojica et al., 1997). Adaptation to hypoosmotic conditions must therefore require a fast response and effective protection. On the other hand, the stability of haloarchaeal proteins might be a critical point to take into account under osmotic stress, and, in this respect, molecular chaperones could contribute to the proper folding of other proteins as protective machineries (Juez, 2004). However, the possible role of the different haloarchaeal molecular chaperone systems in stress response networks is currently a matter for debate and has yet to be clarified. In this context, it must be mentioned that very few osmoregulated genes have been reported to date. Amongst the previously described genes with differential expression depending on the salinity of the medium are those corresponding to the gas vesicles (Englert et al., 1990), a protein with chaperone activity (Franzetti et al., 2001), and certain membrane and DNA-binding proteins (Mojica et al., 1993). Finally, global regulatory mechanisms in response to environmental stimuli are also currently a topical issue which should be studied in detail.

OSMOREGULATION IN HALOARCHAEA: ARE THERE FUNCTIONAL DOMAINS RELATED TO THE RESPONSE TO OSMOTIC CONDITIONS IN THE HALOARCHAEAL GENOME?

The description of different haloarchaeal genomes has created new means of understanding the biology of this group of organisms (Ng et al., 2000; Baliga et al., 2004b; Falb et al., 2005). Comparative genomic transcription analyses are providing useful information regarding the behavior of haloarchaeal systems in response to environmental perturbations (Baliga et al., 2004a; Muller and DasSarma, 2005). However, the environmental adaptation processes, in particular as regards osmotic stress, are poorly understood. In order to contribute to the knowledge of osmoadaptation mechanisms in haloarchaea, we have analyzed the global expression in the *Haloferax volcanii* genome and attempted to identify osmoregulated genes and osmoregulatory mechanisms (Mojica et al., 1993; Ferrer et al., 1996; Juez, 2004; E. Soria and G. Juez, unpublished data). The transcriptional response to different osmotic conditions appears to be quite widespread over the *H. volcanii* genome (Fig. 1). We have been able to distinguish specific high-salt and low-salt responses, as well as more general stress behaviors such as responses to both low and high salt and to both osmotic stress and heat shock (Fig. 1), which may help to understand the osmoadaptation processes and the connection between different networks of adaptation to environmental conditions. A general overview of differential transcription in the *H. volcanii* genome clearly reflects the fact that adaptation to hyposaline conditions involves much more widespread transcriptional activity than adaptation to hypersaline conditions (see Fig. 1). This extensive and strong expression in adaptation to low salt is in accordance with the severe effect of hypoosmotic challenge for haloarchaea, which requires as fast and as effective a response as possible (Juez, 2004).

It can be noticed that, as a global overview, differential transcription in the *H. volcanii* genome reveals clear gene clusters and large genomic regions with coordinated expression (Ferrer et al., 1996; Juez, 2004; Soria and Juez, unpublished). Some genomic regions may show transcription profiles ranging from a high diversity of responses to different environmental stimuli to a clearly homogeneous response pattern to the environment (Fig. 1). Clustering of osmoregulated genes in the haloarchaeal genome may reflect coordinated transcription regulation mechanisms. As previously suggested, certain homogeneous and alternating responses to salinity in adjacent regions could be related to osmoregulatory mechanisms (Ferrer et al., 1996). At this time, we may conclude that global regulation of the osmotic response could be achieved through DNA topology (Soria and Juez, unpublished). Organization of genes in gene clusters, not necessarily cotranscribed nor organized in operons, may allow global regulatory mechanisms such as DNA topology to play an effective role in adaptation to the environment. The role of Z-DNA structures in transcription regulation as a response to environmental stimuli in haloarchaea has already been suggested (Yang and DasSarma, 1990; Mojica et al., 1993; Yang et al., 1996; Juez, 2004). The presence of gene clusters and large genomic regions with a simultaneous response to the environment, the effect of gyrase inhibitors on the transcription levels of these genomic regions, as well as the presence of sequences susceptible of non-B DNA configuration within the regulatory regions of the osmoregulated genes suggest that DNA structure might be an important global regulatory mechanism in the haloarchaeal genome, being able to coordinate the response to the environment of even large genomic domains (Soria and Juez, unpublished).

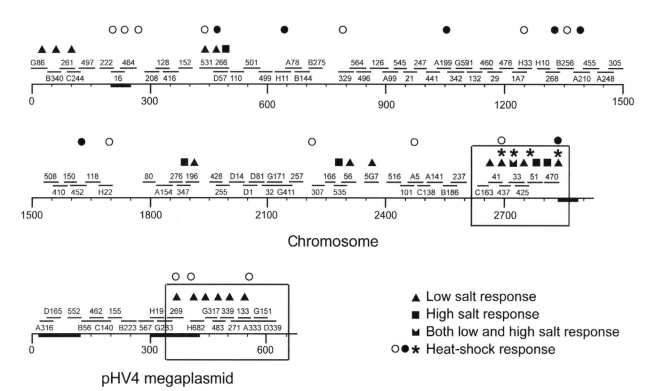

Figure 1. Transcriptional map of the *Haloferax volcanii* genome. The figure shows an overview of differentially transcribed regions in the chromosome and the pHV4 megaplasmid. Symbols are not drawn to scale and represent a summary of the most representative responses. Genome transcription analysis was mainly based on the use of cDNA probes to hybridize against restriction fragments of the cosmid clones of a genomic library of the organism (Charlebois et al., 1991). Transcriptionally induced regions, over the whole genome, in cells growing in low (12% salts) and high (30% salts) salinity conditions were described previously (Ferrer et al., 1996; Juez, 2004). Two genomic stretches, indicated by boxes, have been the subject of a more extensive analysis through the detection of transcripts arising from genomic regions (by Northern blot hybridization) and including the long-term response in cultures growing at different salinities (8, 10, 12, 15, 20, 25, 30, and 35% salt medium), as well as the immediate response after a downshift (30 to 10% salt medium), an upshift (10 to 30% salt medium) and a heat shock (37 to 55°C in 20% salt medium, indicated by asterisks) (Juez, 2004; Soria and Juez, unpublished). A mixture of salts in the proportions found in seawater (30% salts containing in w/v: 23.4% NaCl, 1.95% $MgCl_2$, 2.9% $MgSO_4$, 0.12% $CaCl_2$, 0.6% KCl, 0.03% $NaHCO_3$, and 0.075% NaBr) was used, as described previously (Rodríguez-Valera et al., 1980; Mojica et al., 1997). The map also includes minor and major signals (indicated as empty and solid circles, respectively) of heat-shock responses, as well as FII AT-rich regions containing IS elements (indicated by solid black bars below the distance scale), previously reported by Trieselmann and Charlebois (1992). A kilobase-pair distance scale and cosmid clones representing the genome are shown.

In this respect, there is a significant presence of a domain of about 200 kb within the largest of the extrachromosomal replicons of *H. volcanii*, the replicon pHV4, which could be related to adaptation to hypoosmotic conditions (see Fig. 1). This region shows extensive and coordinated transcription enhancement under low salinities (Ferrer et al., 1996; Juez, 2004; Soria and Juez, unpublished). Similar low-salt induction was also observed within the probably homologous replicon pHM500 from *Haloferax mediterranei* (Ferrer et al., 1996). We have previously pointed out the possibility of this genomic region being responsible for the ability of members of the genus *Haloferax* to grow at lower salinities (NaCl concentrations) than other haloarchaeal groups (Ferrer et al., 1996; Juez, 2004). The recent detection within this pHV4 stretch of sequences codifying for several membrane proteins, particularly different cation transport systems, or several transcription regulators, strengthens the hypothesis of its involvement in the adaptation to osmotic challenge. On the other hand, the presence of a peculiar structure of short tandem repeats for which a possible role in replicon stability was previously described (Mojica et al., 1995, 2000) would provide a stable character to this replicon, or at least to this genomic region, perhaps an essential genomic element for the organism. This large pHV4 region related to adaptation to hyposaline conditions appears to be under transcription regulation by DNA topology and may constitute a clearly defined functional domain within the *H. volcanii* genome (E. Soria and G. Juez, unpublished). Our interest is currently focused on the nature, origin, and evolution of sequences within this peculiar genomic domain (A. Gonzaga and G. Juez, unpublished data).

Completely different behavior is shown by a chromosome region that appears to participate in adaptation to different stressing conditions (Fig. 1, position 2650 to 2850). This stretch of about 200 kb seemed to concentrate responses to either low- or to high-salt conditions (Ferrer et al., 1996; Juez, 2004), as well as to heat shock (Trieselmann and Charlebois, 1992). A more recent and extensive analysis (Soria and Juez, unpublished) has revealed a complex transcriptional profile (Fig. 1). Specific responses to particular environmental conditions as well as general stress responses can be distinguished. Particular regions or transcripts are specifically induced by low salt, by high salt, or by heat shock (Fig. 1) and could help to clarify the mechanisms specifically involved in adaptation to hypoosmotic versus hyperosmotic conditions or to temperature shock. Other sequences show expression enhancement at both low- and high-salt conditions (a U-type response) and could be considered to be related to general stress. Frequently, transcripts with this U-type response to salt are also induced after heat shock, showing a clear general stress nature. A general stress behavior, with response to heat shock, has also been observed for certain sequences responding to low-salt conditions, while it has not been observed for specific high-salt responses. Furthermore, the overlap of responses to heat shock and osmotic stress, particularly hypoosmotic stress, seems to be a frequent feature within the haloarchaeal genome (Juez, 2004; see also Fig. 1). This fact may reflect a connection between different response networks but overall suggests the relevance of general stress proteins in adaptation to hyposaline challenge for the haloarchaeal cell. Within this chromosome region, we have detected sequences codifying for transcriptional regulators and certain general stress proteins such as an oxydoreductase and several proteases, which could correspond to some of the general stress responses observed. This genomic region certainly seems to be highly involved in adaptation to the environment and offers the possibility of distinguishing specific adaptation processes as well as general stress mechanisms.

A lengthy chromosomal region (around position 900 to 1400), which is the most transcriptionally active region under optimal conditions, also seems to include some of the strongest responses to heat shock, among which those of the chaperonin subunit genes *cct1* and *cct2*, located at positions 1037 to 1058 and 1318 to 1330, respectively, can be noticed (Trieselmann and Charlebois, 1992; Kuo et al., 1997). This region harbors essential genes, such as different RNA polymerase subunit genes or chaperonin subunit genes, related to transcription or protein synthesis and stabilization (Charlebois et al., 1991; Kuo et al., 1997). This large chromosome stretch does not seem to play a significant role in osmoadaptation, at least in the long-term response of cells growing under low- or high-salt conditions (Ferrer et al., 1996). However, chaperonin genes may also be induced after salt dilution, although not as dramatically as after heat shock (Kuo et al., 1997). In fact, in haloarchaea, both hypoosmotic stress and heat shock would promote haloarchaeal protein destabilization and aggregation. Both types of stressing conditions might require certain common protection mechanisms, among which molecular chaperones might be key elements (Juez, 2004).

MOLECULAR CHAPERONES AND OTHER STRESS PROTEINS MUST PLAY AN IMPORTANT ROLE IN ADAPTATION TO OSMOTIC STRESS

Haloarchaea must have evolved effective protection mechanisms in order to withstand the harsh environmental conditions in their natural habitat, such as extremely high salinity, moderately high temperature, or the lethal stress, which may be implied by salt dilution. Apart from specific mechanisms of adaptation to different conditions, other general stress responses must be essential for survival under different types of stress. Transcriptional behavior in the *H. volcanii* genome supports this idea. According to the transcriptional patterns, previous protein synthesis analysis revealed proteins specifically related to adaptation to high or to low osmotic conditions, as well as general stress proteins overexpressed under both hypo- and hyperosmotic conditions (Mojica et al., 1997). In addition, molecular chaperones may be involved in the response to osmotic stress, besides the expected heat shock, in these extreme halophiles. Some heat-shock proteins, among them the Cct family chaperonins, are also slightly induced upon salt dilution (Daniels et al., 1984; Kuo et al., 1997). A novel haloarchaeal protein with chaperone activity was found to participate in the response to hyposaline conditions (Franzetti et al., 2001). On the basis of the expression pattern and molecular mass of certain *H. volcanii* proteins, we suggested previously that general stress proteins and molecular chaperones, as the DnaK chaperone system, might play an important role in the adaptation to osmotic stress, particularly hypoosmotic stress, in haloarchaea (Mojica et al., 1997). Nevertheless, the DnaK system was not yet described in this organism, neither was it detected among heat-shock proteins nor transcriptional responses. In fact, its origin and universal presence in haloarchaea and other archaeal groups has been a controversial matter to date (Gupta and Singh, 1992; Gupta, 1998; Gribaldo et al., 1999; Philippe et al., 1999). We have corroborated the universal presence of this chaperone system among haloarchaea and have evidence suggesting that it could be involved in the response to general stress, in particular to hyposaline stress (D. Fenosa and G. Juez, unpublished data. The role that the different molecular chaperone machineries must play in haloarchaea is a subject of current interest yet to be clarified.

In *Archaea*, chaperonins (Hsp60 family) are similar to CCT eukaryal type chaperonins, differing clearly from bacterial chaperonins (for a review see Trent, 1996). Archaeal chaperonins have aroused great interest, and their activity in stabilizing other proteins under denaturing conditions has been proven. While bacterial chaperonins are assisted by the DnaK system, in the case of *Archaea*, at least in thermophilic Archaea, an eukaryal like prefoldin system seems to fulfill the DnaK function, cooperating with the chaperonin machinery in the proper folding of other proteins under thermal destabilization (Leroux et al., 1999; Okochi et al., 2002). It is in this scenario that the role of the DnaK system in *Archaea* is currently confusing. Its discontinuous presence among *Archaea* and the current uncertainty about its origin (Gupta and Singh, 1992; Gupta, 1998; Gribaldo et al., 1999; Philippe et al., 1999) has obscured the knowledge of its possible biological role in these organisms. The DnaK cluster genes have not been detected in the genomes of several *Archaea*, in particular in hyperthermophilic *Archaea*. Nevertheless, the cluster is present in members of the Thermoplasmatales (*Thermoplasma*, *Ferroplasma*, and *Picrophilus*), in *Methanothermobacterium*, *Methanosarcina*, and *Methanococcoides*, and, as recently observed (Fenosa and Juez, unpublished), in all haloarchaeal groups. Horizontal transfer from bacteria might have happened in the case of *Methanosarcina*, where two different DnaK gene clusters are present and one of them is clearly related to gram-positive bacteria of the *Clostridium* group (Gribaldo et al., 1999; Macario et al., 1999; Deppenmeier et al., 2002). As already pointed by Philippe et al. (1999), the DnaK protein is present in a coherent set of archaeal branches with a common origin, but horizontal transfer may confuse the Hsp70 family phylogeny. Contrary to previous hypotheses that suggested an origin of the DnaK protein in haloarchaea from high G+C gram-positive bacteria (Gupta and Singh, 1992; Gupta, 1998), we have evidence supporting its vertical origin. Moreover, the haloarchaeal chaperone system appears to be related to that of other archaeal groups and could probably be an essential element in adaptation to stressing conditions in these organisms (Fenosa and Juez, unpublished).

Among other attempts to corroborate this hypothesis, we have analyzed the DnaK system gene cluster (including the genes codifying for the DnaK chaperone and its cochaperones DnaJ and GrpE) from haloarchaea and other *Archaea* by means of protein phylogenetic relationships, as well as analyzing the surrounding and intergenic regions (Fenosa and Juez, unpublished). Several different approaches suggest a typical archaeal nature of the DnaK system gene cluster and a common origin for the system in haloarchaea and other archaeal groups. In this respect, a detailed analysis of protein alignments will contribute toward clarifying its origin and possible role in haloarchaea and other *Archaea*. The DnaK, DnaJ, and GrpE proteins show characteristic or distinctive amino acid substitutions for haloarchaea, even within highly conserved regions or protein domains (Fig. 2 and 3). These amino acid substitutions are frequently related to the halophilic character of the protein, and, as described for other haloarchaeal proteins (Dennis and Shimmin, 1997), certain highly conserved residues are substituted for glutamate (E) or aspartate (D) in haloarchaea (see Fig. 2). The haloarchaeal chaperone system seems to have evolved as other haloarchaeal proteins and to have a much more ancient origin than previously thought (Fenosa and Juez, unpublished). Needless to say, the former is not the overall substitution pattern, particular signatures are more likely to be related to phylogenetic divergence, and the degree of conservation within phylogenetic groups supports this idea. It should be mentioned that haloarchaea present characteristic amino acid substitutions which frequently coincide with substitutions in other archaeal groups (see Fig. 2 and 3). The different archaeal lineages are connected by coincident specific residues, suggesting a common origin for the archaeal DnaK system. On the other hand, the presence of consistent archaeal substitutions within functional protein domains might have a phylogenetic or functional significance (see Fig. 2). Haloarchaea and other *Archaea* share particular amino acid positions with thermophilic bacteria, such as the *Thermus-Deinococcus* group (Fig. 2), a fact that could be understood as a reflection of a common ancient origin or of protein stability and function under extreme conditions. However, the most relevant fact is that the haloarchaeal and other archaeal DnaK system proteins contain all the functional domains described in other organisms.

CHAPTER 18 • RESPONSE TO OSMOTIC STRESS IN A HALOARCHAEAL GENOME

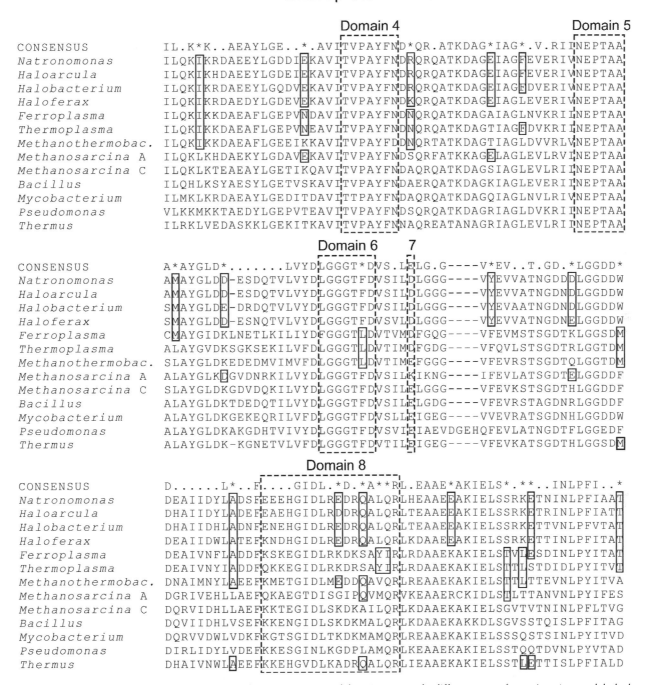

Figure 2. Protein sequence alignment of the DnaK chaperone. Conserved domains among the different types of organisms (external dashed boxed) and distinctive amino acid substitutions for haloarchaea and other *Archaea* (internal boxes) are indicated. A consensus sequence is also shown. For simplicity, a limited central region of the protein and sequences from representatives of different archaeal and bacterial genera are shown. Conserved domains (domains 4 to 8) correspond to Hsp70 signature (TVPAYFND), connect 1 (NEPTAA), phosphate 2 (LGGGTFD), Hinge residue (E), and nuclear localization signal (NLS), respectively.

The DnaK protein, a highly conserved protein among the different types of organisms, is a clear example of the conservation of these functional domains in the archaeal lineages where it has been identified (Fig. 2). In the case of the DnaJ cochaperone, a much more variable protein with significant diversity of sequence even within phylogenetic groups, the presence in haloarchaea and other *Archaea* of the N-terminal J domain, the glycine-rich region, or the four zinc-fingers might be significant (see Fig. 3). The only

DnaJ protein

```
CONSENSUS          ......A*.G............C..C.G.G....**...C..C.G.G............
Haloarcula         EIDLEEAYNGATKQLNVTRPEACDDCDGAGHPPGADSETCPECNGQGQTTQVQQTPM
Haloferax          TIDLEDAYHGVERDVTIRRREVCPECDGEGHPADADVNTCSECNGSGQQTTVQQTPF
Ferroplasma        NVTMEDAYYGASKPIKFKRNAMCEVCNGTGAE-GGVLVTCPTCHGTGQERISRGQGF
Thermoplasma       DISLEEAYYGTEKRIKFRRNAICPDCKGTGAK-NGKLITCPTCHGTGQQRVVRGQGF
Methanosarcina C   YITFEEAAFGVRKDIDIPRTERCSTCSGTGAKPGTSPKRCPTCGGTGQVRTTRSTLG
Bacillus           TLSFEDAAFGKETTIEIPREETCETCKGSGAKPGTNPETCSHCGGSGQLNVEQNTPF
Mycobacterium      ELDFVEAAKGVAMPLRLTSPAPCTNCHGSGARPGTSPKVCPTCNGSGVINRNQG---
Pseudomonas        DLDLEEAVRGTTVTIRVPTLVGCKTCNGSGAKPGTTPVTCTTCGGIGQVRMQQ----
Thermus            PLTLEEAFHGGERVVEVAG---|--------|----------|---------------

CONSENSUS          .........C..C.G.G......C..C.G.G............P.G...*...R....
Haloarcula         GRVQQRTTCRRCDGEGTLYDETCSTCRGNGVVQNDASLEVEIPSGIADGQTLRMERE
Haloferax          GRVQQTTTCRACGGEGKTYSEDCSECRGSGRVRRTRDVTITIPAGFRDGQRLRYRGE
Ferroplasma        FNFVQVIVCRTCMGKGKIPKTPCKACAGKGYIPKMENISITIPKGVDTNTRLRTQKL
Thermoplasma       FRMVTVTTCNTCGGKGRIPEEKCPRCNGTGTIVVDEDITVKIPRGATDNLRLRVSGK
Methanosarcina C   MQFISTTTCSTCHGRGQIIESPCPVCGGAGRVRNKRTITVNPAGADSGMSLRLSGE
Bacillus           GKVVNRRVCHHCEGTGKIIKNKCADCGGKGKIKKRKINVTIPAGVDDGQQLRLSGQ
Mycobacterium      -AFGFSEPCTDCRGSGSIIEHPCEECKGTGVTTRTRTINVRIPPGVEDGQRIRLAGQ
Pseudomonas        GFFSVQQTCPRCHGTGKMISDPCGSCHGQGRVEEQKTLSVKVPAGVDTGDRIRLTGE
Thermus            -------|--------|----------|----RRVSVRIPPGVREGSVIRVPGM
```

Figure 3. Protein sequence alignment of the DnaJ cochaperone. A central region of the protein, including the zinc-finger sequences (CxxCxGxG) (indicated by external dashed boxes), is shown. Distinctive amino acid substitutions for haloarchaea and other *Archaea* (internal boxes) and a consensus sequence are also indicated.

exception would be the case of the highly degenerated sequence of the DnaJ protein from the second gene cluster identified in *Methanosarcina* (Deppenmeier et al., 2002) (named here as cluster A or *Methanosarcina* A in Fig. 2), where the zinc-fingers and glycine-rich regions have been lost, but the *dnaJ* gene copy imported from *Clostridium* group (named here as gene cluster C or *Methanosarcina* C in Fig. 2 and 3) could replace its function. The consistent conservation of functional domains, such as the repeated zinc-finger signature (CxxCxGxG) lying within such a variable stretch, can be explained as results of selective pressure for protein functionality. In summary, the evolution of the DnaK system in haloarchaea, as well as in other archaeal lineages where it has been detected, strongly suggests an essential role for this chaperone machinery in these organisms. New frontiers are currently opening up as regards our understanding of the function and interaction of the different molecular chaperone machineries in *Archaea*.

Acknowledgments. The authors thank F. Rodríguez-Valera and thank W.F. Doolittle for providing the *Haloferax volcanii* genomic library.

This research was supported by grants GV97-VS-25-82 and Grupos03/060 from the "Generalitat Valenciana" and PB96-0330 and BMC2000-0948-C02 from the Spanish Ministry of Science and Technology (MCYT).

REFERENCES

Baliga, N. S., S. J. Bjork, R. Bonneau, M. Pan, C. Iloanusi, M. C. H. Kottemann, L. Hood, and J. DiRuggiero. 2004a. Systems level insights into the stress response to UV radiation in the halophilic archaeon *Halobacterium* NRC-1. *Genome Res.* 14:1025–1035.

Baliga, N. S., R. Bonneau, M. T. Facciotti, M. Pan, G. Glusman, E. W. Deutsch, P. Shannon, Y. Chiu, R. S. Weng, R. R. Gan, P. Hung, S. V. Date, E. Marcotte, L. Hood, and W. V. Ng. 2004b. Genome sequence of *Haloarcula marismortui*: A halophilic archaeon from the Dead Sea. *Genome Res.* 14:2221–2234.

Charlebois, R. L., L. C. Schalkwyk, J. D. Hofman, and W. F. Doolittle. 1991. Detailed physical map and set of overlapping clones covering the genome of the archaebacterium *Haloferax volcanii* DS2. *J. Mol. Biol.* 222:509–524.

Christian, J. H. B., and J. A. Waltho. 1962. Solute concentrations within cells of halophilic and non-halophilic bacteria. *Biochim. Biophys. Acta* 65:506–508.

Daniels, C. J., A. H. Z. McKee, and W. F. Doolittle. 1984. Archaebacterial heat-shock proteins. *EMBO J.* 3:745–749.

Danson, M. J., and D. W. Hough. 1997. The structural basis of halophilicity. *Comp. Biochem. Physiol.* 117A:307–312.

Dennis, P. P., and L. C. Shimmin. 1997. Evolutionary divergence and salinity-mediated selection in halophilic archaea. *Microbiol. Mol. Biol. Rev.* 61:90–104.

Deppenmeier, U., A. Johann, T. Hartsch, R. Merkl, R. A. Schmitz, R. Martinez-Arias, A. Henne, A. Wiezer, S. Baumer, C. Jacobi, H. Bruggemann, T. Lienard, A. Christmann, M. Bomeke, S. Steckel, A. Bhattacharyya, A. Lykidis, R. Overbeek, H. P. Klenk, R. P. Gunsalus, H. J. Fritz, and G. Gottschalk. 2002. The genome of *Methanosarcina mazei*: evidence for lateral gene transfer between bacteria and archaea. *J. Mol. Microbiol. Biotechnol.* **4**:453–461.

Englert, C., M. Horne, and F. Pfeifer. 1990. Expression of the major gas vesicle protein gene in the halophilic archaebacterium *Haloferax mediterranei* is modulated by salt. *Mol. Gen. Genet.* **222**:225–232.

Falb, M., F. Pfeiffer, P. Palm, K. Rodewald, V. Hickmann, J. Tittor, and D. Oesterhelt. 2005. Living with two extremes: conclusions from the genome sequence of *Natronomonas pharaonis*. *Genome Res.* **15**:1336–1343.

Ferrer, C., F. J. M. Mojica, G. Juez, and F. Rodriguez-Valera. 1996. Differentially transcribed regions of *Haloferax volcanii* genome depending on the medium salinity. *J. Bacteriol.* **178**:309–313.

Franzetti, B., G. Schoehn, C. Ebel, J. Gagnon, R. W. Ruigrok, and G. Zaccai. 2001. Characterization of a novel complex from halophilic archaebacteria, which displays chaperone-like activities in vitro. *J. Biol. Chem.* **276**:29906–29914.

Gibbons, N. E. 1974. *Halobacteriaceae* fam.nov., p. 269–272. *In* R. E. Buchanan and N. E. Gibbons (ed.), *Bergey's Manual of Determinative Bacteriology*, 8th ed. Williams and Wilkins, Baltimore, MD.

Ginzburg, M., L. Sachs, and B. Z. Ginzburg. 1970. Ion metabolism in a halobacterium. I. Influence of age of culture on intracellular concentrations. *J. Gen. Physiol.* **55**:187–207.

Gribaldo, S., V. Lumia, R. Creti, E. C. de Macario, A. Sanangelantoni, and P. Cammarano. 1999. Discontinuous occurrence of the *hsp70* (*dnaK*) gene among Archaea and sequence features of HSP70 suggest a novel outlook on phylogenies inferred from this protein. *J. Bacteriol.* **181**:434–443.

Gupta, R. S. 1998. Protein phylogenies and signature sequences: A reappraisal of evolutionary relationships among archaebacteria, eubacteria, and eukaryotes. *Microbiol. Mol. Biol. Rev.* **62**:1435–1491.

Gupta, R. S., and B. Singh. 1992. Cloning of the HSP70 gene from *Halobacterium marismortui*: relatedness of archaebacterial HSP70 to its eubacterial homologs and a model for the evolution of the HSP70 gene. *J. Bacteriol.* **174**:4594–4605.

Juez, G. 1982. Aislamiento, estudio taxonómico, ultraestructural y molecular de nuevos grupos de halófilos extremos. Ph.D. thesis. University of Alicante, Alicante, Spain.

Juez, G. 1988. Taxonomy of extremely halophilic archaebacteria, vol. II, p. 3–24. *In* F. Rodríguez-Valera (ed.), *Halophilic Bacteria*. CRC Press, Boca Raton, FL.

Juez, G. 2004. Extremely halophilic Archaea: Insights into their response to environmental conditions, p. 243–253. *In* A. Ventosa (ed.), *Halophilic Microorganisms*. Springer-Verlag, Heidelberg, Germany.

Kuo, Y. P., D. K. Thompson, A. St Jean, R. L. Charlebois, and C. J. Daniels. 1997. Characterization of two heat shock genes from *Haloferax volcanii*: a model system for transcription regulation in the Archaea. *J. Bacteriol.* **179**:6318–6324.

Lanyi, J. K. 1974. Salt dependent properties of proteins from extremely halophilic bacteria. *Bacteriol. Rev.* **38**:272–290.

Leroux, M. R., M. Fändrich, D. Klunker, K. Siegers, A. N. Lupas, J. R. Brown, E. Schiebel, C. M. Dobson, and F. U. Hartl. 1999. MtGimC, a novel archaeal chaperone related to the eukaryotic chaperonin cofactor GimC/prefoldin. *EMBO J.* **18**:6730–6743.

Macario, A. J. L., M. Lange, B. K. Ahring, and E. Conway de Macario. 1999. Stress genes and proteins in the Archaea. *Microbiol. Mol. Biol. Rev.* **63**:923–967.

Mescher, M. F., and J. L. Strominger. 1976. Structural (shape-maintaining) role of the cell surface glycoprotein of *Halobacterium salinarium*. *Proc. Natl. Acad. Sci. USA* **73**:2687–2691.

Mojica, F. J. M., E. Cisneros, C. Ferrer, F. Rodriguez-Valera, and G. Juez. 1997. Osmotically induced response in representatives of halophilic prokaryotes: the Bacterium *Halomonas elongata* and the Archaeon *Haloferax volcanii*. *J. Bacteriol.* **179**:5471–5481.

Mojica, F. J. M., C. Diez-Villaseñor, E. Soria, and G. Juez. 2000. Biological significance of a family of regularly spaced repeats in the genomes of Archaea, Bacteria and mitochondria. *Mol. Microbiol.* **36**:244–246.

Mojica, F. J. M., C. Ferrer, G. Juez, and F. Rodriguez-Valera. 1995. Long stretches of short tandem repeats are present in the largest replicons of the Archaea *Haloferax mediterranei* and *Haloferax volcanii* and could be involved in replicon partitioning. *Mol. Microbiol.* **17**:85–93.

Mojica, F. J. M., G. Juez, and F. Rodriguez-Valera. 1993. Transcription at different salinities of *Haloferax mediterranei* sequences adjacent to partially modified *Pst*I sites. *Mol. Microbiol.* **9**:613–621.

Muller, J. A., and S. DasSarma. 2005. Genomic analysis of anaerobic respiration in the archaeon *Halobacterium sp.* strain NRC-1: dimethyl sulfoxide and trimethylamine N-oxide as terminal electron acceptors. *J. Bacteriol.* **187**:1659–1667.

Ng, W. V., S. P. Kennedy, G. G. Mahairas, B. Berquist, M. Pan, H. D. Shukla, S. R. Lasky, N. S. Baliga, V. Thorsson, J. Sbrogna, S. Swartzell, D. Weir, J. Hall, T. A. Dahl, R. Welti, Y. A. Goo, B. Leithauser, K. Keller, R. Cruz, M. J. Danson, D. W. Hough, D. G. Maddocks, P. E. Jablonski, M. P. Krebs, C. M. Angevine, H. Dale, T. A. Isenbarger, R. F. Peck, M. Pohlschroder, J. L. Spudich, K.-H. Jung, M. Alam, T. Freitas, S. Hou, C. J. Daniels, P. P. Dennis, A. D. Omer, H. Ebhardt, T. M. Lowe, P. Liang, M. Riley, L. Hood, and S. DasSarma. 2000. Genome sequence of *Halobacterium* species NRC-1. *Proc. Natl. Acad. Sci. USA* **97**:12176–12181.

Okochi, M., T. Yoshida, T. Maruyama, Y. Kawarabayasi, H. Kikuchi, and M. Yohda. 2002. *Pyrococcus* prefoldin stabilizes protein-folding intermediates and transfer them to chaperonins for correct folding. *Biochem. Biophys. Res. Commun.* **291**:769–774.

Oren, A. (ed). 1999. *Microbiology and Biogeochemistry of Hypersaline Environments*. CRC Press, Boca Raton, FL.

Philippe, H., K. Budin, and D. Moreira. 1999. Horizontal transfers confuse the prokaryotic phylogeny based on the HSP70 protein family. *Mol. Microbiol.* **31**:1007–1009.

Rodríguez-Valera, F. (ed). 1988. *Halophilic Bacteria*. CRC Press, Boca Raton, FL.

Rodríguez-Valera, F., F. Ruiz-Berraquero, and A. Ramos-Cormenzana. 1980. Behaviour of mixed populations of halophilic bacteria in continuous cultures. *Can. J. Microbiol.* **26**:1259–1263.

Steber, J., and K. H. Schleifer. 1975. *Halococcus morrhuae*: a sulfated heteropolysaccharide as the structural component of the bacterial cell wall. *Arch. Microbiol.* **105**:173–177.

Tindall, B. J., and H. G. Trüper. 1986. Ecophysiology of the aerobic halophilic archaeabacteria. *Syst. Appl. Microbiol.* **7**:202–212.

Torreblanca, M., F. Rodríguez-Valera, G. Juez, A. Ventosa, M. Kamekura, and M. Kates. 1986. Classification of non-alkaliphilic halobacteria based on numerical taxonomy and polar lipid composition, and description of *Haloarcula* gen. nov. and *Haloferax* gen. nov. *Syst. Appl. Microbiol.* **8**:89–99.

Trent, J. D. 1996. A review of acquired thermotolerance, heat-shock proteins and molecular chaperones in archaea. *FEMS Microbiol. Rev.* **18**:249–258.

Trieselmann, B. A., and R. L. Charlebois. 1992. Transcriptionally active regions in the genome of the archaebacterium *Haloferax volcanii*. *J. Bacteriol.* **174**:30–34.

Yang, C.-F., and S. DasSarma. 1990. Transcriptional induction of purple membrane and gas vesicle synthesis in the archaebacterium *Halobacterium halobium* is blocked by a DNA gyrase inhibitor. *J. Bacteriol.* **172**:4118–4121.

Yang, C.-F., J.-M. Kim, E. Molinari, and S. DasSarma. 1996. Genetic and topological analysis of the bop promoter of *Halobacterium halobium*: Stimulation by DNA supercoiling and non-B-DNA structure. *J. Bacteriol.* **178**:840–845.

Chapter 19

Molecular Adaptation to High Salt

FREDERIC VELLIEUX, DOMINIQUE MADERN, GIUSEPPE ZACCAI, AND CHRISTINE EBEL

INTRODUCTION

Halophilic organisms inhabit extremely saline environments up to NaCl saturation. Halophiles are found in all three domains of life: *Bacteria*, *Archaea*, and *Eukarya*. One of the motivations for their study is the hope to reach an understanding of the molecular and cellular mechanisms underlying their ability to cope with these hostile conditions. Another motivation comes from the fact that a large number of halophilic microorganisms are *Archaea*, and some macromolecular machineries from *Archaea* share similarities with those from *Eukarya*. Examples are complexes involved in translation, proteolysis, or protein folding (Langer et al., 1995; Maupin-Furlow et al., 2004). *Archaea*, therefore, offer simple macromolecular models to describe systems that are more complex in *Eukarya*.

Extreme halophiles require multimolar salt for growth. Their study has shown that they have developed a wide variety of strategies to thrive in media that are hostile to other life forms (for a complete review, see Oren, 2002). In order to counterbalance the external osmotic pressure, extreme halophiles accumulate salt—mainly KCl—close to saturation in their cytosol. All biochemical reactions occur in this extreme medium.

In this chapter, we first present briefly the insights into adaptation provided by the study of the four genomes of extreme halophiles sequenced to date. The focus will then shift to molecular adaptation of halophilic proteins, defined as proteins isolated from extreme halophiles. We shall not address membrane proteins or ribosomes. The starting point of the analysis will be the high-resolution structures of halophilic proteins available at this time. DNA–protein interactions will be considered with the only example described so far, which concerns DNA binding by a protein from a nonextreme halophile. Structural information has been combined with complementary phylogenetic analysis and solution studies. Different aspects concerning solvation, stabilization of the folded and associated assemblies of proteins, and salt effect will be presented. Molecular evolution also has to select appropriate solubility and dynamics in order to permit and favor halophilic protein activity at high salt.

WHAT DID WE LEARN FROM HALOPHILIC GENOME SEQUENCES?

For a long time, it was thought that halophilic *Archaea* (*Halobacteriaceae*) were the only prokaryotic cells adapted to extreme salt environments. Phylogenetic studies have revealed that more than 11 genera belong to the clade of halophilic *Archaea* (Oren, 2002). They have all developed a unique adaptive strategy to counterbalance the strong osmotic pressure induced by the high sodium chloride content of the external medium: they accumulate and maintain a high potassium chloride concentration inside their cytosol. Recently, the discovery of the eubacterium *Salinibacter ruber* in saltern crystallizer ponds indicated that eubacteria are also able to accumulate KCl in their cytosol (Anton et al., 2002; Oren et al., 2002). Extremely halophilic bacterial species are very difficult to identify because of their strong phenotypic similarities with haloarchaea. The genome sequences of three *Halobacteriaceae*—*Halobacterium* sp. NRC-1, *Haloarcula marismortui*, and *Natronomonas pharaonis*—and of the halophilic eubacterium *S. ruber* are now available (Ng et al., 2000; Baliga et al., 2004; Falb et al., 2005; Mongodin et al., 2005). Genomic analyses helped us to understand the metabolic strategies and physiological responses they have developed to live in their specific environments.

F. Vellieux, D. Madern, and C. Ebel • Laboratoire de Biophysique Moléculaire, IBS, Institut de Biologie Structurale Jean-Pierre Ebel, 41 rue Jules Horowitz, F-38027 Grenoble France; CEA; CNRS; Universite Joseph Fourier. G. Zaccai • Institut Laue Langevin, 6 rue Jules Horowitz, BP156, 38042 Grenoble Cedex 9, France.

Archaea

The analysis of the genomes of *Halobacterium* sp. NRC-1 (Ng et al., 2000), *H. marismortui* (Baliga et al., 2004), and *N. pharaonis* (Falb et al., 2005) reveals a common organization in multiple replicons. In *H. marismortui*, some of these replicons might be considered as (small) chromosomes, because they encode essential functions. Some replicons have a low G+C content and can be seen as reservoirs for insertion sequences.

Owing to strong similarities in their respective physiologies, *H. marismortui* and *Halobacterium* sp. NRC-1 possess a common set of metabolic enzymes. In both strains, glucose catabolism is achieved by a modified Entner–Doudoroff pathway. However, striking differences exist in other metabolic pathways of the two organisms, which, for example, use distinct pathways for arginine breakdown: the arginine deaminase pathway is used by *Halobacterium* sp. NRC-1, whereas *H. marismortui* degrades arginine via the arginase pathway. In both strains, the genes coding for arginine synthesis and degradation functions are segregated on different replicons (Ng et al., 2000; Baliga et al., 2004). *N. pharaonis* is a chemoorganotrophic microorganism that normally uses amino acids of the environment as sole carbon source. It misses several genes encoding key glycolytic enzymes, suggesting that the microorganism is not able to degrade glucose. *N. pharaonis* grows in highly salty and alkaline conditions (pH ~11). Such pH conditions cause reduced levels of ammonia. According to the genome analysis, *N. pharaonis* has various mechanisms to supply ammonia, which is then converted to glutamate (Falb et al., 2005). Ammonia can enter the cell by direct uptake or can be provided from the uptake and reduction of nitrate. A third mechanism involves uptake and hydrolysis of urea.

In order to maintain the osmotic equilibrium between cytosol and external medium, *Halobacterium* sp. NRC-1 and *H. marismortui* possess multiple sets of genes coding for active K^+ transporters and Na^+ antiporters. In addition to the high salt concentration, the natural habitats of *Halobacterium* sp. NRC-1, *H. marismortui*, and *N. pharaonis* have similar characteristics of low oxygen solubility and high light intensity. All strains have a set of genes encoding opsins, which use light energy to maintain physiological ion concentrations and to generate chemical energy in the form of a proton gradient. *Halobacterium* sp. NRC-1 contains genes coding for the ion pumps halorhodopsin and bacteriorhodopsin. For *N. pharaonis*, genes coding for the chloride pump halorhodopsin and sensory rhodopsin II have been identified (Falb et al., 2005). Genes encoding protein-binding sensory pigments that absorb blue light and halocyanin precursor-like proteins have been identified in both *Halobacterium* sp. NRC-1 and *H. marismortui*. The detection of genes encoding circadian clock regulator-like proteins suggests that *Halobacterium* sp. NRC-1 and *H. marismortui* are able to regulate their metabolism in response to the circadian cycle (Ng et al., 2000; Baliga et al., 2004). In the genomes of *Halobacterium* sp. NRC-1, *H. marismortui*, and *N. pharaonis*, a large number of genes coding for transducers and motility proteins have been identified. A large family of multidomain proteins, which can act both as sensors and as transcriptional regulators, are encoded in each genome. It should be pointed out that *H. marismortui* contains five times more copies of this type of gene than *Halobacterium* sp. NRC-1, suggesting that *H. marismortui* has an enhanced capability to adapt to a fluctuating environment. In haloalkaliphilic *N. pharaonis*, analysis of the genes coding for electron transport chain proteins indicates that protons, rather than sodium, are the coupling ions between respiratory chain and ATP synthase in this organism.

As was reported in the study of individual halophilic proteins (Madern et al., 2000a), the proteomes of both *Halobacterium* sp. NRC-1 and *H. marismortui* are highly acidic, with an average pI ~5. The average pI of the proteome has not yet been computed for *N. pharaonis*.

Eubacteria

The complete genome sequence of *S. ruber* reveals that lateral gene transfer (LGT) from haloarchaea has played an important role in the evolutionary fate of this bacterium (Mongodin et al., 2005). A phylogenetic analysis of 16S rRNA sequences indicates that *S. ruber* is rooted within the *Bacteroides/Chlorobium* group of bacteria, while the analysis of most *S. ruber* open reading frames confirms that *S. ruber* and *Chlorobium tepidum* are closely related. However, similarity sequence analysis of individual open reading frames from *S. ruber* indicates that part of its genes are found to be phylogenetically related to specific genes from halophilic *Archaea*. As an example, the LGT phenomenon is well established in *S. ruber* genes coding for the K^+ uptake/efflux systems and the cationic amino acid transporters, for which >50% of the genes are recruited from haloarchaea. The most striking observation is the presence of four genes coding for rhodopsins in the *S. ruber* genome. These rhodopsins are linked with halorhodopsin, suggesting that they function as inward-directed chloride pumps. In addition, two genes encoding putative sensory rhodopsin molecules have been detected by sequence similarity. It should be pointed out that, at the genome scale, the total number of these manifest gene-transfer events is small.

S. ruber possesses a complete set of genes for the transport and degradation of organic compounds. In addition, all the components required for fermentation have also been identified. The analysis of the genes involved in glucose catabolism indicates that *S. ruber* uses the Embden–Meyerhoff pathway.

The calculated normalized distribution of pI values for the open reading frames of *S. ruber* has a mean value of 5.2, which is very close to that computed for the haloarchaeal proteome.

TWO EVOLUTIONARY MECHANISMS

Genome analysis of extremely halophilic microorganisms revealed that there are at least two evolutionary mechanisms that have driven adaptation to high salinity.

The first mechanism, which is common to *Halobacterium* sp. NRC-1, *H. marismortui*, and *S. ruber*, consists in a series of amino acid substitutions that replace neutral amino acids by acidic ones. The crystallographic structures described below reveal the very acidic surfaces of halophilic proteins. Molecular adaptation to high salt has been studied in great detail at the protein level, using as a model system the malate dehydrogenase from *H. marismortui*. High-resolution information was used in conjunction with phylogenetic, functional, stability, solubility, and dynamics studies (for reviews, see Madern et al., 2000a; Mevarech et al., 2000; Ebel and Zaccai, 2004; Tehei and Zaccai, 2005). The role of acidic amino acids and of other structural features in the solubility, stability, and dynamics of halophilic proteins will be discussed further in the chapter.

The second mechanism that is operative for the adaptation of *S. ruber* to high salt is LGT from haloarchaeal species that thrive in the same saline environment. LGT was demonstrated by the analysis of archaeal and hyperthermophilic bacterial genomes to take place between microbial communities (Nelson et al., 1999; Ruepp et al., 2000).

HIGH-RESOLUTION STRUCTURAL INFORMATION

Crystallographic Studies of Halophilic Proteins

Tables 1 and 2 summarize the characteristics of halophilic proteins whose crystallographic structures have been solved to date. These are presented in Color Plate 2. Historically, the first halophilic protein to be crystallized and have its three-dimensional structure solved is the malate dehydrogenase from the halophilic archaeon *H. marismortui* (*Hm* MalDH), a homotetrameric enzyme (Dym et al., 1995). MalDH had been first described as a dimer (Pundak et al., 1981), but further solution studies have established that it was a tetramer (Bonneté et al., 1993). Its analysis contributed to the establishment of the lactate dehydrogenase-like MalDH family of enzymes (Madern, 2002).

The purified enzyme was crystallized using an original modification of the classical vapor diffusion method (Richard et al., 1995; Costenaro et al., 2001): addition of the organic solvent 2-methyl-pentane-diol to *Hm* MalDH in NaCl causes phase separation in the crystallization drop. The MalDH enzyme segregates in the salt-containing phase. Water evaporates from the reservoir to reach the protein-containing droplet, leading to an increase in the volume of the crystallization drop, a process during which nucleation and crystal growth occur.

Table 1. Halophilic proteins with available high-resolution structures

Protein Data Bank code	Protein	Name	Resolution[a] (Å)	Reference
1HLP	Malate dehydrogenase, holo	*H. marismortui* MalDH	3.2	Dym et al., 1995
1D3A	Malate dehydrogenase, apo		2.95	Richard et al., 2000
2HLP	Malate dehydrogenase, E267R, apo		2.6	Richard et al., 2000
1O6Z	Malate dehydrogenase, R207S, R292S, holo		1.95	Irimia et al., 2003
1DOI	2Fe–2S Ferredoxin	*H. marismortui* Fd	1.9	Frolow et al., 1996
1EOZ	2Fe–2S Ferredoxin	*H. salinarum* Fd	NMR	Marg et al., 2005
1E10	2Fe–2S Ferredoxin		NMR	Marg et al., 2005
1VDR	Dihydrofolate reductase	*H. volcanii* DHFR	2.6	Pieper et al., 1998
1ITK	Catalase-peroxidase	*H. marismortui* CP	2	Yamada et al., 2002
1MOG	Dodecin	*H. salinarum* dodecin	1.7	Bieger et al., 2003
1TJO	DNA-protecting protein during starvation A	*H. salinarum* DpsA	1.6	Zeth et al., 2004
2AZ1	Nucleoside diphosphate kinase, apo	*H. salinarum* NDK	2.35	Besir et al., 2005
2AZ3	Nucleoside diphosphate kinase, CDP complex		2.2	Besir et al., 2005

[a] Numbers are given for structures obtained by crystallography. For a same publication, only the crystal form with best resolution is quoted. NMR, structure obtained from nuclear magnetic resonance spectroscopy.

Table 2. Ions detected and net charge of halophilic proteins with an X-ray structure

Organism	Detected ions	Amino acids	Subunits	Negative charges[a]
H. marismortui MalDH	2 Na$^+$; 8 Cl$^-$	1,212	4	152
H. marismortui Fd[b]	6 K$^+$	128	1	28
H. volcanii DHFR	3 PO$_3^-$	162	1	14
H. marismortui CP	6 SO$_4^-$, 16 Cl$^-$, 6 K$^+$	1,462	2	150
H. salinarum dodecin	12 Mg^{2+}, 2 Cl$^-$, 1 Na$^+$	804	12	147
H. salinarum DpsA[b]	4 SO$_4^-$, 6 Mg^{2+}, 4 Na$^+$	2,184	12	368
H. salinarum NDK	5 Ca^{2+}, 4 Mg^{2+}	960	6	132

[a] At pH 7. All numbers are given per biologically active protein.
[b] Neglecting iron.

Initially, the structure was solved and published at 3.2 Å resolution (Dym et al., 1995). Later, additional information was gathered as the resolution gradually increased (2.9 Å for the native enzyme, 2.65 Å for the *E242R* mutant, eventually reaching 1.9 Å for the *R207S* and *R292S* mutant) (Richard et al., 2000; Irimia et al., 2003). Different features were observed, which are associated with the halophilic character of *Hm* MalDH: the surface of the enzyme displays a large negative isoelectric potential, resulting from a marked excess of negatively charged residues over positively charged side chains. Color Plate 2 presents, for comparison, the representation of the structure of a nonhalophilic homolog of *Hm* MalDH. This negatively charged surface is assumed to effectively recruit a large number of solvent components in a salt-rich intracellular medium, where the salt ions are also hydrated (see below). The second structural feature is the presence, detected at subunit interfaces, of specific ion-binding sites (Color Plate 3). The incorporated ions are integral components of the protein's three-dimensional structure. In addition, the presence of a large number of salt bridges and salt-bridge networks was noticed between subunits, a feature that is usually associated with thermostable proteins.

The three-dimensional structures of several halophilic proteins (Color Plate 2) later confirmed these initial findings, while allowing additional insight into alternative means to adapt to high-salt environments at the molecular level. Thus, the halophilic ferredoxin from *H. marismortui* (*Hm* Fd) also showed a negatively charged surface and numerous specific cation-binding sites (Frolow et al., 1996). Most interestingly, the structure revealed the presence of a hyperacidic insertion, in the form of two amphipathic surface helices. Such "halophilic addition," i.e., the incorporation of a single negative domain, can be thought of as a very straightforward means for a protein to adapt to high salt. The same features were observed in the structure of Fd from *Halobacterium salinarum* (Marg et al., 2005). The other process of "halophilic substitution," leading to the acquisition of a surface with evenly distributed negative side chains, is assumed to take considerably longer during molecular evolution than the acquisition of an additional domain with the requested characteristics.

The structure of *Haloferax volcanii* dihydrofolate reductase (*Hv* DHFR) (Pieper et al., 1998) features a negatively charged surface. A highly acidic C-terminal segment is reminiscent of the added acidic domain seen in *Hm* FD. However, the negative character is only slightly more pronounced than that of nonhalophilic dihydrofolate reductases (DHFRs), which are exceptionally acidic. Although it has an optimal activity at 3 to 4 M salt, *Hv* DHFR is stable at rather low monovalent salt (0.5 M). The three-dimensional structure suggests that two adjacent aspartate residues (D54 and D55) allow the essential conformational transitions necessary for enzyme activity to take place in a salted environment. Contrary to halophilic *Hm* MalDH, the three-dimensional structure did not reveal any trends in salt-bridge contents (or clusters thereof). A second *Hv* DHFR has been discovered in *H. volcanii* (Ortenberg et al., 2000). The first *Hv* DHFR described here is very likely the result of an LGT event from a nonhalophilic organism. The second corresponds to the true functional DHFR. Such an observation helps to explain why the properties of the first *Hv* DHFR differ from those generally observed with halophilic proteins.

H. marismortui catalase peroxidase (*Hm* CP) exhibits dual activities; the two activities are modulated by the solvent composition (Cendrin et al., 1994). The 2.0-Å resolution structure of *Hm* CP (Yamada et al., 2002) reveals the halophilic character of the enzyme. As is the case with *Hm* MalDH, the surface of this bidomain homodimeric protein is acidic, with a large excess (54%) of acidic Asp and Glu side chains over basic side chains (8% of Arg, Lys, and His side chains). The crystalline enzyme binds numerous ions, with 6 sulfate ions, 16 chlorides, and 6 ions of unknown type. Thus, *Hm* CP possesses specific ion-binding sites, where the ions are connected to basic side chains or, by hydrogen bonds, to amide groups and to water molecules. A large fraction of these ions is found at the dimer interface, and it is assumed that the presence of the bound ions in their specific binding sites in the protein is essential to maintaining the integrity of the three-dimensional structure and, thus, enzymatic activity.

H. salinarum dodecin is a small polypeptide (68 residues) with the property of coassembling with flavin cofactors to form homododecamers. The three-dimensional structure of the dodecameric assembly has been solved by X-ray crystallography (Bieger et al., 2003). The molecule is a hollow sphere with outer diameter ~60 Å. Both the outer and the inner surfaces of the 12-mer are negatively charged, with a large excess of acidic side chains over basic ones (24% versus 6%, such excesses thus appear to be a hallmark of halophilic proteins or enzymes). In the structure, numerous ions are observed bound to the protein: 12 magnesium ions are located in the inner compartment of the hollow sphere (one per polypeptidic chain). These are bound to aspartate side chains and to water molecules. The two types of channels linking the inner compartment to the outside are plugged by chloride ions, one of which is in direct interaction with a sodium cation. An additional magnesium ion is present on the external surface, where it is linked to a glutamate side chain. This ion is involved in the crystal lattice-forming contacts between dodecameric molecules. In addition, the structure revealed the presence of important salt-bridge interactions, in particular each dodecin monomer being involved in four intersubunit salt bridges that are reminiscent of the salt-bridge networks located in the *Hm* MalDH intersubunit interfaces.

The three-dimensional structure of the iron uptake and storage ferritin DpsA from *H. salinarum* (*Hs* DpsA) has been obtained in three forms with increasing iron contents (Zeth et al., 2004). *Hs* DpsA is a homododecameric protein shell (outer diameter ~90 Å) that surrounds a central iron storage cavity. The iron-binding sites and iron-binding properties of this protein will not be discussed here because they are the raison d'être of the protein, thus not to be related to the halophilic character of the protein. In addition to the iron ions, the crystallographic structures showed the presence of sulfate, magnesium, and sodium ions (some of which are associated with the iron-binding sites). When compared with nonhalophilic ferritins, halophilic DpsA comprises an elongated N-terminal tail enriched in acidic residues, reminiscent of the additional acidic domain of *Hm* Fd. Another difference concerns the iron translocation pathway, which involves histidine residues in *Hs* DpsA when carboxylate side chains are the participating elements in nonhalophilic ferritins: in the KCl-enriched cytosol of *H. salinarum*, the iron ions would compete with potassium ions for binding to carboxylate side chains, thus reducing the efficiency of iron translocation. Molecular adaptation of the iron translocation function to high salt would be obtained by the replacement of acidic side chains by the more basic His residues. Nonetheless, the outer surface of *Hs* DpsA shares with other halophilic proteins a marked acidic character.

The three-dimensional structure of *H. salinarum* nucleoside diphosphate kinase (*Hs* NDK) has recently appeared in the literature (Besir et al., 2005): crystals were obtained both for the native homohexameric enzyme and for a (His_6)-tagged construct. The latter was studied to investigate the effect of basic tag addition on the halophilic properties of the enzyme. Like all halophilic proteins investigated thus far, the surface of the *Hs* NDK protein has a marked acidic character. Although no electron density is found for the hexa-His tag in the crystals of the modified NDK construct, the addition of this short stretch of basic residues is sufficient to confer low salt-folding ability to the protein.

With the availability of an increasing number of structures of halophilic proteins, the molecular features related to haloadaptation and protein stability in high salt are gradually emerging: the surfaces of high-salt-adapted proteins all show a marked excess of negative over positive charges, except in regions where the presence of basic amino acid side chains is required for proper biological function. The acquisition of this negative amino acid sequence character appears to have taken place in two different ways (which are not mutually exclusive): either by halophilic addition of an acidic domain or stretch of residues, seen as a means to readily confer a halophilic character to a nonhalophilic protein, or by the less expeditious halophilic replacement of side chains throughout the sequence to confer the requested negative surface. An appealing evolutionary scenario can be put forward, in which the first step would be the rapid acquisition of negatively charged stretches of residues. This would rapidly confer at least a partial halophilic character to the adapting protein. Afterward, haloadaptation could proceed over a longer period of time by the gradual replacement of protein side chains, conferring the full haloadapted character to the protein.

Another feature detected in the crystallographic structures is the presence in the proteins of specific anion- or cation-binding sites. When considering these, the limitations of X-ray crystallography for the visualization of bound ions should be kept in mind: solvent density peaks are first assigned as water molecules. Only very well-ordered ions can be distinguished from bound water, and sodium (which contains the same number of electrons as an H_2O molecule) can only be distinguished from water on the basis of its coordination pattern (unless high enough resolution, under 1 Å, allows the assignment of water hydrogen atoms, a situation not encountered so far for halophilic proteins). In addition protein–solvent interactions could be modified upon crystallization. It is thus likely that halophilic proteins interact with more ions than viewed in crystal structures. High-resolution structures of halophilic proteins are nonetheless seen to comprise specific ion-binding sites (Color Plate 3, Table 2), which are integral components of the macromolecule and thought to be essential for stability: these ion-binding sites are often observed at subunit interfaces, where they mediate intersubunit contacts. Removal of ions from these sites, for example, by lowering the salt content of the buffer leads to the disruption of the subunit interface and thus to protein instability (see the complementary studies on *Hm* MalDH described below).

The third feature observed in the three-dimensional structures of halophilic proteins is the presence of an increased number of salt bridges and networks at the interface between subunits. This observation can be associated with the increased number of ion pairs and salt-bridge networks observed in the three-dimensional structures of thermostable and hyperthermostable proteins.

PROTEIN–DNA INTERACTIONS IN A HALOPHILIC CONTEXT

A number of hyperthermophilic *Archaea* accumulate moderate salt concentration (0.5 to 1 M) in their cytosol. The three-dimensional structures of their proteins share some of the features emphasized above for halophilic proteins (from organisms that require salt for growth). Table 3 summarizes some references to structures of salt-adapted proteins from nonextreme halophiles. Although a detailed description of these three-dimensional structures is out of the scope of this chapter, it seems interesting to report the features concerning the crystallographic structures and subsequent solution studies on the TATA-box-binding protein (TBP)—wild type, mutants, and complexes—of the hyperthermophilic *Pyrococcus woesei*. *P. woesei* grows optimally at 95°C to 100°C and 0.6 M NaCl.

The crystallographic structure of *P. woesei* TBP (DeDecker et al., 1996) (Color Plate 2) was compared with the models of eukaryotic TATA-binding proteins. All models have very similar folds, as each TBP monomer is composed of two similar substructures (N and C terminal) related by diad symmetry. However, the archaeal TBP contains a C-terminal acidic additional tail with six glutamate residues, which is absent in the eukaryotic TBPs. Another difference in the structure concerns the electrostatic potential surrounding the proteins, which has a more pronounced negative character in *P. woesei* TBP (*Pw* TBP) because of the presence of a higher number of acidic residues on the surface, in particular with negatively charged stirrups. Several of the acidic side chains participate to ion pairs. Thus, the archaeal TATA-box-binding protein possesses two of the characteristics usually associated with halophilic proteins: a negative surface and the presence of a higher number of surface ion pairs than nonhalophilic proteins. However, a positively charged surface is expected for areas that bind the cognate DNA, in order to neutralize the negative charges of the DNA's sugar-phosphate backbone. Later, the same group solved the structures of two

Table 3. Salt-adapted proteins from nonextreme halophile with available high-resolution structures

PDB code	Protein	Name	Resolution (Å)	Reference
1FTR	Formylmethanofuran: tetrahydromethanopterin formyltransferase	*Methanopyrus kandleri* FTR	1.73	Ermler et al., 1997
1QLM	Methenyltetrahydromethanopterin cyclohydrolase	*M. kandleri* MCH	2	Grabarse et al., 1999
1EZW	Coenzyme F_{420}-dependent methylenetetrahydromethanopterin reductase	*M. kandleri* MER	1.65	Shima et al., 2000
1E6V	Methyl-coenzyme M reductase	*M. kandleri* MCR	2.7	Grabarse et al., 2000
1QV9	Coenzyme F_{420}-dependent methylenetetrahydromethanopterin dehydrogenase	*M. kandleri* MTD	1.54	Hagemeier et al., 2003
1JR9	Manganese superoxide dismutase	*Bacillus halodenitrificans* SOD	2.8	Liao et al., 2002
1Y7W	Carbonic anhydrase	*Dunaliella salina* CA	1.86	Premkumar et al., 2005
1PCZ	TATA-box-binding protein	*P. woesei* PDB	2.2	DeDecker et al., 1996
1AIS	TATA-box-binding protein/transcription factor (II)B/TATA-box		2.1	Kosa et al., 1997
1D3U	TATA-box binding protein/transcription factor B/extended TATA-box promoter		2.4	Littlefield et al., 1999

ternary complexes. These comprise the C-terminally truncated (Δ182 to Δ191) archaeal TBP, the C-terminal core (TFBc) of the transcription factor B (TFB is a homolog of the eukaryotic transcription factor TFIIB), and a DNA molecule containing either the TATA element (Kosa et al., 1997) or the B recognition element in addition to the TATA box (Littlefield et al., 1999). The second structure indicated that the orientation of the archaeal TBP bound to the TATA box element was artifactual in the first structure.

In parallel with the crystallographic work and because the *Pw* TBP protein originates from a archaeon accumulating salt in the molar range, studies using this model system and aimed at revealing the influence of salt on protein–DNA interactions have been initiated (O'Brien et al., 1998). This work reported the first experimental demonstration of an increase in the protein–DNA association in parallel with increasing salt concentration, thus revealing the halophilic nature of the specific *Pw* TBP–TATA-box element interaction. This interaction can only be measured at high salt concentrations. In addition, the experimental data suggested that this high-salt protein–DNA interaction is accompanied by the removal of a large number of water molecules from the hydrophobic buried surface. The situation is opposite to that observed with nonhalophilic protein–DNA interactions, where the binding constant decreases with increasing salt. From the *Pw* TBP crystal structure available at the time, it was expected that the direct interaction of the negative stirrups on the TBP molecule with the DNA element (which also has an acidic character) would be unfavorable. However, the presence of neutralizing cations around the protein–DNA interface region is yet to be verified. The crystal structures of *Pw* TBP and of its complexes were obtained at too-low resolutions to allow the unambiguous assignment of density peaks as corresponding to salt cations (in particular Na^+, indistinguishable from electron density corresponding to water molecules).

Later, Bergqvist et al. (2001, 2002, 2003) showed that the halophilic character of the *Pw* TBP–DNA interaction was due principally to three or four interface glutamate residues (E12, E41, E42, and E128): site-directed mutagenesis on these residues and studies of the DNA-binding properties of the resulting mutants indicated the additive effects of the mutations, which gradually reduce and eventually lose the original salt dependence of the halophilic protein–DNA interaction. With the wild-type protein, the DNA-binding event is accompanied by the net uptake of two ions. With the E12AE128A mutant, the net ion uptake is zero. The triple mutant E12AE41KE128A releases one ion and the quadruple mutant E12AE41KE42KE128A two ions. The latter two mutants exhibit the characteristic decrease in DNA-binding affinity with increasing salt concentrations found for nonhalophilic proteins. Thus, the halophilic character of the *Pw* TBP–DNA interaction is totally reversed by the accumulated effects of three or four mutations. This finding indicates that, for *Pw* protein–DNA interaction the halophilic phenotype could be rapidly acquired in evolutionary time by a limited number of point mutations.

BEYOND THE STRUCTURE

Solvation

If crystallization allows the identification of a limited number of ions in the structure of halophilic proteins (Table 2), only solution studies can permit an estimate of the composition of the solvation shell. Density measurements, small-angle X-ray and neutron scattering, as well as analytical ultracentrifugation can be used to evaluate a "density" (in different scales for the different techniques) for the solvated particle (Eisenberg, 1976, 1981). From measurements at different solvent salt concentrations, a solvation shell composition can be calculated. This characterization was made for different halophilic proteins. Elongation factor EF1α—previously named EF-Tu—from *H. marismortui* (Ebel et al., 1992), glyceraldehyde-3-phosphate dehydrogenase from *H. volcanii* (Ebel et al., 1995), *Hm* MalDH (Bonneté et al., 1993) were characterized in KCl and/or NaCl solutions as solvated by 0.2 to 0.4 g of water and 0.1 to 0.2 g of NaCl or KCl per gram of protein. This corresponds to large but not unusual amounts of water—acidic residues are expected to be highly solvated—and exceptionally large amounts of salt. In particular, the amount of salt is much larger than the amount detected in the crystallographic structures and reported in Table 2.

Further investigations were performed on *Hm* MalDH in the presence of various salts and at different salt and protein concentrations (Ebel et al., 2002). Well-defined experimental protocols were established in these studies. The results indicated that the solvation shell composition was little affected by the type of anion in the solvent salt. This feature was rather surprising: it is generally considered that the stabilizing effect of cosolvents is related to their propensity to increase the preferential hydration of the macromolecule (Timasheff, 1993) and that the effect of anions is generally predominant when compared with that of the cations (Collins, 1997). For *Hm* MalDH, the modulation of preferential hydration by the type of anion is masked: the global content of the solvation shell is not affected significantly. However, anions of high charge density stabilize the active folded protein at much lower salt concentration when compared with chloride-containing salts (Ebel et al., 1999). The large stabilizing effects of fluoride and sulfate anions are thus related to a limited number of "strong" binding sites. These can be those that are detected by crystallography at the interface between the subunits (Color Plate 3). As will be detailed below, they stabilize the folded dimer in addition to the tetramer assembly. *H. volcanii* NADP (*Hv* NADP)-citrate dehydrogenase displays stability that is little (or not) sensitive to anions (Madern et al., 2004). This corroborates the fact that anions stabilize folded active *Hm* MalDH through specific binding sites and not through nonspecific effects.

The quantification of the solvation shell composition of *Hm* MalDH was derived from combined density and small-angle neutron-scattering measurements (Ebel et al., 2002). These data justified an analysis in terms of an invariant particle made up of protein and solvation shell. Thus, one can consider the solvation shell around the protein to have the same composition when the salt concentration is varied within the range of particle stability. An alternative model of analysis, which considers binding sites that can exchange water and ion, gives essentially the same description of the solvation shell: there is saturation of the solvent-binding sites by the solvent salt ions at the lowest solvent salt concentration that stabilizes the folded tetramer. The results are given in Table 4. Clearly, changing the solvent salt cation determines different protein solvations. This is logical in view of the very acidic character of the protein. The hydration is highly variable, with values in $MgCl_2$ that are twice those in NaCl, $NaCH_3CO_2$, or $(NH_4)_2SO_4$. The numbers of associated salt molecules are also variable, from 85 mol/mol in $MgCl_2$, to ~50 mol/mol in Na salts, and to 0 mol/mol in $(NH_4)_2SO_4$. Positive values correspond to accumulation of salt ions that exceeds the presence of the counter ions. For nonhalophilic proteins such as bovine serum albumin, β-lactoglobulin, and lysozyme in NaCl and $MgCl_2$, salt binding is very low but measurable. Ammonium sulfate in general produces preferential hydration (Ebel et al., 2002 and references therein). Thus, the diversity of behavior observed for solvent interactions of *Hm* MalDH in different solvents is a general protein feature. The solvation of halophilic proteins appears to be exceptional in quantitative rather than qualitative terms.

Salt and Halophilic Protein Stability

Electrostatic contributions to the stability of halophilic proteins

The energetics of halophilic proteins is expected to be strongly dependent on the large number of negative charges on their surfaces. The electrostatic contribution to stability was addressed in a theoretical approach in 1998 by Elcock and McCammon, using the available structures of *Hm* MalDH at 3.2 Å, that of *Hm*, and that of a nonhalophilic homolog 2Fe–2S ferredoxin (Elcock and McCammon, 1998). The Poisson–Boltzmann equation of classical electrostatics was used. Electrostatic interactions between acidic residues, which remain repulsive even at high salt concentrations, were found to be a major factor in the low-salt destabilization of the proteins. An analysis as a function of pH also showed increased stability at low pH and upward shifts in the pK_a upon protein folding, in agreement with experimental data on pH/salt concentration effects obtained for *Hv* DHFR (see Bohm and Jaenicke, 1994). However, *Hm* MalDH is markedly stabilized at pH 8 and 9 when compared with pH 6 and 7 (Madern and Zaccai, 1997), a feature that was not modeled. Specific effects of salts—salt binding and hydration effects—were out of the scope of such analysis, because only the valence was considered.

Salt bridges, ion binding, and stabilization of the active dimer and tetramer of *Hm* MalDH by salt

Crystallographic analysis of *Hm* MalDH shows a tetramer made up of two dimers interacting mainly via complex salt-bridge clusters. In the R207S/R292S *Hm* MalDH mutant, these salt bridges are partially

Table 4. Composition of the solvation shell of *Hm* MalDH[a]

Salt	Solvent salt (M)	Associated water (mole/mole tetramer)	Associated salt (mole/mole tetramer)	$c_{salt,\ salvation}$ (M)
$MgCl_2$	0.2–1.3	4100	85	1.1–1.4
NaCl	2–5	2100	55	1.5–3.4
$NaCH_3CO_2$	1–3.5	1800	40	1.1–3.1
$(NH_4)_2SO_4$	1–3	2100	0	0–1.2

[a] The value given for the number of associated salt molecules corresponds to salt in addition to the counter ion. The two values given for $c_{salt,\ salvation}$, the salt concentration in the solvation shell, correspond to the two limiting cases concerning the dissociation of the counter ions from the polypeptide chains.

disrupted. The protein was found to be an active tetramer at high salt concentrations. At lower salt concentrations, stable oligomeric intermediates, including an active dimer, could be trapped at given pH, temperature, or solvent conditions (Madern et al., 2000b). The crystallographic structure of this mutant revealed the location of the chloride anion between two monomeric subunits (Irimia et al., 2003). This result suggested that ion binding could be important for the stabilization of the active dimer.

Indeed, in the presence of ammonium sulfate and sodium or potassium fluoride, the active dimer could be identified as a stable species for the wild-type apoprotein when lowering the salt concentration, although in the presence of NaCl or KCl the tetramer low-salt dissociation, unfolding, and inactivation are unresolved events (Irimia et al., 2003). These three salts were selected because it was noticed that salts with anions of high charge density stabilize a folded and active form of Hm MalDH at rather low salt concentration (Ebel et al., 1999), although the global solvation of the protein is not strongly dependent on the type of anions of the solvent salt (Ebel et al., 2002). It is thus likely that fluoride and sulfate ions stabilize the active dimer by their association in place of chloride at the interface between monomers (see Color Plate 3).

The autoassociation of the dimers into tetramers for the (R207S and R292S) mutant was shown to increase with increasing solvent salt concentration. On the basis of changes in the association constant with salt concentration, the formation of the tetramer is accompanied by binding of 3 to 10 moles of salt (6 to 20 ions) and up to ~100 moles of water per mole of tetramer. Because probably not all the solvent molecules at the interface are related to the association event, these values are in fair agreement with the 220 water and 8 Cl$^-$ detected in the structure at the interface between the dimers (Color Plate 3).

Salts and stabilization of halophilic proteins

The requirement of multimolar salt concentration for the stabilization of the folded active state of halophilic proteins—for example, 2.5 M KCl at pH 7, 20°C for long-term stability of Hm MalDH (Pundak et al., 1981)—cannot be considered as an intrinsic character of halophilic proteins. A number of halophilic proteins are stable and active at moderate salt concentrations. It appears that large oligomeric complexes such as the 20S proteasome or the P45 chaperone from Hm display little salt dependence for their stability and activity (Franzetti et al., 2001, 2002). Changing the salt type, pH, and temperature and adding cofactors can drastically modify the stability of halophilic proteins. When the pH is raised from 7 to 8, Hm MalDH low-salt inactivation is shifted to lower salt by 0.4 M salt in NaCl (Madern and Zaccaï, 1997). The cofactor NADH has a huge stabilizing effect; in the presence of 1.5 mM NADH, the dimer is stable above 0.1 M NaCl and the tetramer above 0.5 M at pH 8 and 4°C (Madern and Zaccaï, 1997; Irimia et al., 2003). The relative role of anions and cations in the stabilization of Hm MalDH was estimated from the comparison of the effects of various salts in a large range of concentrations (Ebel et al., 1999). Increasing the salt concentration stabilizes, first, the folded form and then in several cases destabilizes it. The latter effect was described also for nonhalophilic proteins. The effects of anions and cations were found to superimpose. For very stabilizing cations or anions, the effect of the salt ion of opposite charge is, however, limited. For the low-salt transition, anions and cations with the highest charge density are the most efficient to stabilize the folded form: $Ca^{2+} \approx Mg^{2+} > Li^+ \approx NH_4^+ \approx Na^+ > K^+ > Rb^+ > Cs^+$, and $SO_4^{2-} \approx OAc^- \approx F^- > Cl^-$. Temperature studies in ammonium sulfate showed cold unfolding, which was not detected in NaCl (Bonneté et al., 1994).

Numerous weak interactions involving protein–protein, protein–ion, protein–water, and protein–ligand contacts are involved in the stabilization of the folded and unfolded states of proteins. Macromolecules interacting in the more efficient way with the cosolvent and/or less efficiently with water are stabilized by increasing the cosolvent content of the solvent. Stabilization by solvent can be described by the superimposition of specific and nonspecific effects (Collins, 1997; Moelbert et al., 2004; Dill et al., 2005). Specific effects of salts depend on the characteristics of a binding site or on the chemical nature of the surface of the macromolecule, whereas nonspecific ones depend mainly on the properties of the solvent itself and poorly on the characteristics of the macromolecule. Ions of high charge density order water in comparison with the state of pure water. This leads to the hydration of exposed surfaces. This nonspecific effect favors the folded compact state at higher salt concentrations (Timasheff, 1993). Stability is also affected by salt through electrostatic unspecific interactions of the ions with the protein, but also through specific weak or strong interactions with ion-binding sites. The stabilizing effect of anions is in general very well defined, but not that of cations. This is because cations of high charge not only have water-structuring effects (favoring the folded form at high salt) but also interact with the peptide bonds (stabilizing the unfolded form at high salt concentrations) and with potential binding sites defined by acidic residues at the surface of most of the proteins (Ebel et al., 1999 and references therein).

The fact that Mg^{2+} and Na^+ salts accumulate in the solvation shell of the folded *Hm* MalDH suggests that weak and strong specific interactions of these cations with the folded protein are a stabilizing feature, which superimposes on their water-ordering effect. Ammonium salts in general promote the nonspecific preferential hydration of proteins, as found also for *Hm* MalDH. Cold unfolding observed for *Hm* MalDH in this salt indicates a stabilization process dominated by entropy, a feature often related to nonspecific water accumulation at the interfaces. Concerning anions, their stabilizing efficiency for *Hm* MalDH is related to the presence of specific binding sites detected in the crystallographic structures (see above). The large variety of solvation patterns found for *Hm* MalDH in different salts and the related diversity of temperature dependency for protein stability demonstrate clearly that the protein can adapt in a versatile way to its environment.

The stabilizing effects of salts in the presence of urea were compared for the DHFR from *Escherichia coli* (*Ec* DHFR) and for two isoforms of DHFR from *H. volcanii*, in the presence of CsCl, KCl, and NaCl (Wright et al., 2002). Extrapolation to infinite dilution of urea provides a value for the free energy of unfolding $\Delta G°$ in the absence of urea, at various salt concentrations. The absolute value of $\Delta G°$ and variation with KCl, CsCl, and sucrose concentrations are similar for the three enzymes. Researchers concluded that these enzymes are stabilized by the same mechanisms, which are not specific to halophilism. However, *Ec* DHFR is by itself extremely acidic. In addition, the behavior of the three enzymes in the presence of NaCl differs, which is interpreted in terms of ion binding. Here again, versatility appears in the details of the mechanisms leading to protein stability in various environments.

A study on the malate dehydrogenase from the extremely halophilic eubacterium *S. ruber* (*Sr* MalDH) has shown that at least one alternate strategy for microorganisms is to cope with proteins that are not specifically adapted to high salt concentrations but tolerate it (Madern and Zaccai, 2004). The biochemical characterization of *Sr* MalDH indicates that, unlike other halophilic proteins, the enzyme is not enriched in acidic residues and that it behaves as a nonhalophilic protein. From these observations, it can be hypothesized that the enzyme, which is functional in the cytoplasm of an extremely halophilic organism, can bypass the evolutionary mechanism—acidification of the protein surfaces—described for halophilic *Archaea*. This would make sense during the time course of evolution, if the ancestor protein (in an ancestral cellular lineage) were able to sustain sufficient enzymatic activity in nonoptimal conditions (in this case, a high concentration of KCl) in order to support the metabolic processes required for survival and growth.

Solubility

The solubility of proteins is affected by high concentrations of salts. The highly acidic surface of halophilic proteins was proposed to play an important role in maintaining protein solubility at high salt concentrations. In solution studies such as sedimentation or scattering, the colligative properties of the solution are measured, i.e., the numbers of macromolecules in solution are counted. When increasing the weight concentration, c, the apparent number of macromolecules does not increase strictly linearly with c. The deviation from the expected linear behavior is characterized by the second virial coefficient, A_2. It indicates nonideality of solution arising from the macromolecular concentration and tells about weak interparticle interactions and solubility. Positive values of A_2 correspond to particle distributions, indicating overall repulsion between the macromolecules, favoring high solubility. Negative values of A_2 correspond to particle distributions, indicating overall attraction between the macromolecules, favoring low solubility. Moderately negative values are related to mild attraction between macromolecules and were found to statistically correlate with favorable conditions for protein crystallization. This was also found for halophilic proteins and, particularly, for *Hm* MalDH in its original crystallization condition (dilution in the presence of organic solvents) (Costenaro et al., 2001). Weak interactions between macromolecules in solution were characterized for *Hm* MalDH in its native tetrameric state in a number of solvent salt conditions (Costenaro et al., 2002). Positive values were found, for example, in 3 M KCl, as in the whole range of NaCl concentrations, allowing the stabilization of the tetramer. In these solvents, the protein solubility is very high (>100 mg ml^{-1}). The values of A_2 were found to decrease at low salt (0.2 M) for *Hm* MalDH solubilized in magnesium chloride and at high salt (3M) in ammonium sulfate. This complex behaviour can be understood by considering the effect of solvation. It has been mentioned above that, for a given type of salt in the solvent, the composition of the solvation shell can be considered as invariant in the range of salt concentration where the protein is a stable tetramer. At low magnesium chloride concentrations, the salt concentration is larger in the solvation shell than in the bulk solvent. Increasing $MgCl_2$ in the solvent leads to an equalization of the salt concentration in the solvation shell and in the bulk solvent and even to a slight preferential hydration of the protein at higher salt. In the presence of ammonium sulfate, the solvation shell is always enriched in water, and the difference in salt concentrations is more marked at high salt. In the presence of NaCl or KCl, in the multimolar range where the protein is stable, the salt concentrations in the solvation shell and in the bulk solvent are always close to each other. A_2 is lowered in conditions where the composition of the solvent in the local domain differs from that in the bulk solvent, a feature that contributes as a negative entropic terms to A_2 (Costenaro and Ebel, 2002). The two situations of cosolvent depletion and water depletion (cosolvent accumulation) at the macromolecular interface have the same consequence: an effective macromolecular attraction. It is thought that the macromolecular surfaces of halophilic proteins have evolved in order to develop interactions with solvent ions and thus to avoid water enrichment at their surfaces. This preserves their solubility in the crowded and salted cytosol of halophilic cells (Costenaro and Ebel, 2002; Costenaro et al., 2002; Ebel and Zaccai, 2004).

Neutron-scattering studies of molecular dynamics in extreme halophiles

The dynamics of soluble and membrane proteins from extreme halophiles has been studied extensively with the aim of understanding the molecular mechanisms leading to stability, solubility, and activity in highly concentrated salt environments. It is now well accepted that appropriate conformational flexibility is essential for enzyme catalysis and for biological molecular activity in general. The conformational stability and flexibility of a protein structure results from a balance of known intramolecular and protein–solvent forces. In other words, these forces maintain biological structure and govern atomic motions. They include hydrogen bonding, electrostatic, and van der Waals interactions as well as effective forces arising from the hydrophobic effect. They are weak forces because their associated energies are similar to thermal energy at usual temperatures. This is why biological matter is "soft."

We can picture a neutron spectroscopy experiment as one in which neutrons of known energy and momentum are bounced off protein atoms (Gabel et al., 2002). By measuring the energy and momentum of neutrons after collisions, it is possible to compute the energy and momentum of the protein atomic motions, under the conditions of the experiment with respect to temperature, solvent composition, and protein state. In particular, neutron-scattering experiments provide quantitative measurements of the amplitudes and frequencies of thermal atomic fluctuations in a protein. These are fast motions on the picosecond to nanosecond time scale. They are, nevertheless, essential for biological activity because they act as the lubricant that enables conformational changes on physiological time scales. The dynamical behavior of proteins over a broad temperature range is well described by the conformational substate model of Frauenfelder et al. (1988). At physiological temperatures, protein flexibility arises from fluctuations between different protein states with small differences in structure. At very low temperatures, proteins are inflexible and biologically inactive. They behave like hard, solid materials, their atoms held tightly in the structure; the protein is trapped within one conformational substate, and atomic thermal motions are represented by harmonic vibrations about equilibrium positions. At higher temperatures, activation energy becomes available for the atoms to sample different conformational substates; the protein becomes "soft" and active. Relations between dynamics and stability were established in neutron-scattering experiments on Hm MalDH under different solvent conditions.

Studies reported in other sections of this chapter have shown that soluble proteins from extremely halophilic *Archaea* are active and soluble in a wide range of solvent salt conditions with varying stability. Hm MalDH in the apo form requires molar solvent salt concentrations for stability and solubility. In NaCl or KCl solutions, it binds exceptional amounts of water and salt ions. The crystal structure of Hm MalDH shows intersubunit salt-bridge clusters, stabilized by chloride ion binding. Kinetic inactivation of the protein in low-salt solvents occurs as a first-order reaction and is due to concomitant dissociation of the tetramer and unfolding of monomers. In molar NaCl or KCl in H_2O, enthalpic terms dominate the kinetics of the process. The protein is more stable in NaCl than in KCl, probably because of the higher hydration and binding energies of Na^+ compared with K^+. Hm MalDH is also more stable in D_2O (^2H_2O, heavy water) than in H_2O solutions. Solvent effects on Hm MalDH were well characterized, and protein dynamics was measured in corresponding conditions, by neutron scattering, to explore the correlation between dynamics and stability.

The mean square fluctuations, $\langle u^2 \rangle$, of Hm MalDH atoms on the picosecond-to-nanosecond time scale, in 2 M NaCl in D_2O, 2 M KCl in D_2O, and 2 M NaCl in H_2O solutions were measured as a function of temperature, T (Tehei et al., 2001) (Fig. 1). Starting at similar values (~1.5 Å^2) for the three solvent conditions at 280 K (~7°C), the mean square fluctuations rise at different rates as the temperature is increased. The *resilience* of the structure, expressed as an effective force constant $\langle k' \rangle$, is calculated from

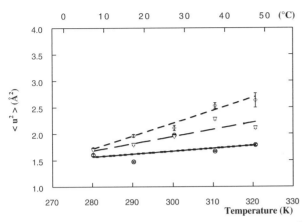

Figure 1. Mean-square fluctuations $\langle u^2 \rangle$ in Hm MalDH measured by neutron scattering. $\langle u^2 \rangle$ values are plotted as a function of temperature for three different solvent conditions: 2 M NaCl D$_2$O (circles), 2 M KCl D$_2$O (triangles), and 2 M NaCl H$_2$O (diamonds). (Modified from Tehei et al., 2001, with permission from the publisher.)

the slope of the $\langle u^2 \rangle$ versus T line (Zaccai, 2000). A more resilient structure is more rigid; the $\langle u^2 \rangle$ value rises less steeply with temperature and vice versa. The two independent parameters ($\langle u^2 \rangle$, $\langle k' \rangle$) calculated from the neutron data provide information on the flexibility of the protein structure and on its rigidity, respectively. The resilience values calculated from the data are 0.1, 0.2, and 0.5 N/m for 2 M NaCl in H$_2$O, 2 M KCl in D$_2$O, and 2 M NaCl in D$_2$O, respectively. For this halophilic protein in these solvent conditions, stability is directly correlated with resilience, showing the dominance of the enthalpy term in the activation free energy of stabilization. It is interesting to note that in the NaCl, KCl, H$_2$O, and D$_2$O series, Hm MalDH is expected to be the least resilient and shows the largest fluctuations in KCl/H$_2$O, the condition closest to physiological conditions.

KCl is selected universally as the dominant cytoplasmic salt, and considerable energy is consumed pumping Na$^+$ ions out of cells. The selection of K$^+$ solvation is a molecular dynamic adaptation mechanism operative in the extreme halophiles. Is it because structures stabilized by Na$^+$-solvation would be too resilient and are not sufficiently flexible for efficient biological activity? Further work is needed to address this question. A molecular dynamic adaptation mechanism was also suggested by results from neutron-scattering experiments on bacteria adapted to various temperatures (psychrophiles, mesophiles, thermophiles, and hyperthermophiles). The resilience and mean-square fluctuation values are such that they permit not only stability at the physiological temperature but also appropriate flexibility to favor biological activity (Tehei et al., 2004).

CONCLUSION

Extreme halophilic organisms require high salt concentrations for growth. They accumulate multimolar salt concentrations in their cytosol to counterbalance the high osmotic pressure of their environment. They have developed strategies to adapt to their environment, and specific cellular responses have been inferred from the information given by the halophilic genomes. Their proteomes have adapted to allow proper stability and function in high salt conditions. Acidic surfaces, specific ion-binding sites, and multiple salt bridges have been observed in the seven high-resolution structures of halophilic proteins available to date. These features allow the recruitment of solvent components for the stabilization of folded active enzymes and assemblies. Large amounts of perturbed solvent have been measured in solution studies of halophilic proteins. The composition of the solvation shell—salt ions and water—depends on solvent composition. The negatively charged molecular surface allows the protein solubility required in the crowded salt-rich cytosol of the organism. Experimental studies have shown that in the presence of physiological K+ salts, halophilic protein dynamics permits not only stability at physiological temperatures but also appropriate flexibility to favor biological activity. From the versatile behavior of halophilic enzymes in different solvent environments and their robustness toward hostile environments, biotechnological perspectives can be considered.

REFERENCES

Anton, J., A. Oren, S. Benlloch, F. Rodriguez-Valera, R. Amann, and R. Rossello-Mora. 2002. *Salinibacter ruber* gen. nov., sp. nov., a novel, extremely halophilic member of the Bacteria from saltern crystallizer ponds. *Int. J. Syst. Evol. Microbiol.* 52:485–491.

Baliga, N. S., R. Bonneau, M. T. Facciotti, M. Pan, G. Glusman, E. W. Deutsch, P. Shannon, Y. Chiu, R. S. Weng, R. R. Gan, P. Hung, S. V. Date, E. Marcotte, L. Hood, and W. V. Ng. 2004. Genome sequence of *Haloarcula marismortui*: a halophilic archaeon from the Dead Sea. *Genome Res.* 14:2221–2234.

Bergqvist, S., R. O'Brien, and J. E. Ladbury. 2001. Site-specific cation binding mediates TATA binding protein–DNA interaction from a hyperthermophilic archaeon. *Biochemistry* 40:2419–2425.

Bergqvist, S., M. A. Williams, R. O'Brien, and J. E. Ladbury. 2002. Reversal of halophilicity in a protein–DNA interaction by limited mutation strategy. *Structure* 10:629–637.

Bergqvist, S., M. A. Williams, R. O'Brien, and J. E. Ladbury. 2003. Halophilic adaptation of protein–DNA interactions. *Biochem. Soc. Trans.* 31:677–680.

Besir, H., K. Zeth, A. Bracher, U. Heider, M. Ishibashi, M. Tokunaga, and D. Oesterhelt. 2005. Structure of a halophilic nucleoside diphosphate kinase from *Halobacterium salinarum*. *FEBS Lett.* 579:6595–6600.

Bieger, B., L. O. Essen, and D. Oesterhelt. 2003. Crystal structure of halophilic dodecin: a novel, dodecameric flavin binding protein from *Halobacterium salinarum*. *Structure* 11:375–385.

Bohm, G., and R. Jaenicke. 1994. A structure-based model for the halophilic adaptation of dihydrofolate reductase from *Halobacterium volcanii*. *Protein Eng.* 7:213–220.

Bonneté, F., C. Ebel, H. Eisenberg, and G. Zaccai. 1993. Biophysical study of halophilic malate dehydrogenase in solution: revised subunit structure and solvent interactions in native and recombinant enzyme. *J. Chem. Soc. Faraday Trans.* 89:2659–2666.

Bonneté, F., D. Madern, and G. Zaccai. 1994. Stability against denaturation mechanisms in halophilic malate dehydrogenase "adapt" to solvent conditions. *J. Mol. Biol.* 244:436–447.

Cendrin, F., H. M. Jouve, J. Gaillard, P. Thibault, and G. Zaccai. 1994. Purification and properties of a halophilic catalase-peroxidase from *Haloarcula marismortui*. *Biochim. Biophys. Acta* 1209:1–9.

Collins, K. D. 1997. Charge density-dependent strength of hydration and biological structure. *Biophys. J.* 72:65–76.

Costenaro, L., and C. Ebel. 2002. Thermodynamic relationships between protein–solvent and protein–protein interactions. *Acta Crystallogr. D* 58:1554–1559.

Costenaro, L., G. Zaccai, and C. Ebel. 2001. Understanding protein crystallisation by dilution: the ternary NaCl–MPD–H_2O system. *J. Cryst. Growth* 232:102–113.

Costenaro, L., G. Zaccai, and C. Ebel. 2002. Link between protein–solvent and weak protein–protein interactions gives insight into halophilic adaptation. *Biochemistry* 41:13245–13252.

DeDecker, B. S., R. O'Brien, P. J. Fleming, J. H. Geiger, S. P. Jackson, and P. B. Sigler. 1996. The crystal structure of a hyperthermophilic archaeal TATA-box binding protein. *J. Mol. Biol.* 264:1072–1084.

Dill, K. A., T. M. Truskett, V. Vlachy, and B. Hribar-Lee. 2005. Modeling water, the hydrophobic effect, and ion solvation. *Annu. Rev. Biophys. Biomol. Struct.* 34:173–199.

Dym, O., M. Mevarech, and J. L. Sussman. 1995. Structural features that stabilize halophilic malate dehydrogenase from an archaebacterium. *Science* 267:1344–1346.

Ebel, C., W. Altekar, J. Langowski, C. Urbanke, E. Forest, and G. Zaccai. 1995. Solution structure of glyceraldehyde-3-phosphate dehydrogenase from *Haloarcula vallismortis*. *Biophys. Chem.* 54:219–227.

Ebel, C., L. Costenaro, M. Pascu, P. Faou, B. Kernel, F. Proust-De Martin, and G. Zaccai. 2002. Solvent interactions of halophilic malate dehydrogenase. *Biochemistry* 41:13234–13244.

Ebel, C., P. Faou, B. Kernel, and G. Zaccai. 1999. Relative role of anions and cations in the stabilisation of halophilic malate dehydrogenase. *Biochemistry* 38:9039–9047.

Ebel, C., F. Guinet, J. Langowski, C. Urbanke, J. Gagnon, and G. Zaccai. 1992. Solution studies of elongation factor Tu from the extreme halophile Halobacterium marismortui. *J. Mol. Biol.* 223:361–371.

Ebel, C., and G. Zaccai. 2004. Crowding in extremophiles: linkage between solvation and weak protein–protein interactions, stability and dynamics, provides insight into molecular adaptation. *J. Mol. Recognit.* 17:382–389.

Eisenberg, H. 1976. *Biological Macromolecules and Polyelectrolytes in Solution*. Clarendon Press, Oxford, United Kingdom.

Eisenberg, H. 1981. Forward scattering of light, X-rays and neutrons. *Q. Rev. Biophys.* 14:141–172.

Elcock, A. H., and J. A. McCammon. 1998. Electrostatic contributions to the stability of halophilic proteins. *J. Mol. Biol.* 280:731–748.

Ermler, U., W. Grabarse, S. Shima, M. Goubeaud, and R. K. Thauer. 1997. Crystal structure of methyl-coenzyme M reductase: the key enzyme of biological methane formation. *Science* 278:1457–1462.

Falb, M., F. Pfeiffer, P. Palm, K. Rodewald, V. Hickmann, J. Tittor, and D. Oesterhelt. 2005. Living with two extremes: conclusions from the genome sequence of *Natronomonas pharaonis*. *Genome Res.* 15:1336–1343.

Franzetti, B., G. Schoehn, C. Ebel, J. Gagnon, R. W. Ruigrok, and G. Zaccai. 2001. Characterization of a novel complex from halophilic archaebacteria, which displays chaperone-like activities in vitro. *J. Biol. Chem.* 276:29906–29914.

Franzetti, B., G. Schoehn, D. Garcia, R. W. Ruigrok, and G. Zaccai. 2002. Characterization of the proteasome from the extremely halophilic archaeon *Haloarcula marismortui*. *Archaea* 1:53–61.

Frauenfelder, H., F. Parak, and R. D. Young. 1988. Conformational substates in proteins. *Annu. Rev. Biophys. Biophys. Chem.* 17:451–479.

Frolow, F., M. Harel, J. L. Sussman, M. Mevarech, and M. Shoham. 1996. Insights into protein adaptation to a saturated salt environment from the crystal structure of a halophilic 2Fe–2S ferredoxin. *Nat. Struct. Biol.* 3:452–458.

Gabel, F., D. Bicout, U. Lehnert, M. Tehei, M. Weik, and G. Zaccai. 2002. Protein dynamics studied by neutron scattering. *Q. Rev. Biophys.* 35:327–367.

Grabarse, W., F. Mahlert, S. Shima, R. K. Thauer, and U. Ermler. 2000. Comparison of three methyl-coenzyme M reductases from phylogenetically distant organisms: unusual amino acid modification, conservation and adaptation. *J. Mol. Biol.* 303:329–344.

Grabarse, W., M. Vaupel, J. A. Vorholt, S. Shima, R. K. Thauer, A. Wittershagen, G. Bourenkov, H. D. Bartunik, and U. Ermler. 1999. The crystal structure of methenyltetrahydromethanopterin cyclohydrolase from the hyperthermophilic archaeon *Methanopyrus kandleri*. *Structure* 7:1257–1268.

Hagemeier, C. H., S. Shima, R. K. Thauer, G. Bourenkov, H. D. Bartunik, and U. Ermler. 2003. Coenzyme F420-dependent methylenetetrahydromethanopterin dehydrogenase (Mtd) from *Methanopyrus kandleri*: a methanogenic enzyme with an unusual quarternary structure. *J. Mol. Biol.* 332:1047–1057.

Irimia, A., C. Ebel, D. Madern, S. B. Richard, L. W. Cosenza, G. Zaccai, and F. M. Vellieux. 2003. The oligomeric states of *Haloarcula marismortui* malate dehydrogenase are modulated by solvent components as shown by crystallographic and biochemical studies. *J. Mol. Biol.* 326:859–873.

Kosa, P. F., G. Ghosh, B. S. DeDecker, and P. B. Sigler. 1997. The 2.1-A crystal structure of an archaeal preinitiation complex: TATA-box-binding protein/transcription factor (II)B core/TATA-box. *Proc. Natl. Acad. Sci. USA* 94:6042–6047.

Langer, D., J. Hain, P. Thuriaux, and W. Zillig. 1995. Transcription in archaea: similarity to that in eucarya. *Proc. Natl. Acad. Sci. USA* 92:5768–5772.

Liao, J., M. Y. Liu, T. Chang, M. Li, J. Le Gall, L. L. Gui, J. P. Zhang, T. Jiang, D. C. Liang, and W. R. Chang. 2002. Three-dimensional structure of manganese superoxide dismutase from *Bacillus halodenitrificans*, a component of the so-called "green protein". *J. Struct. Biol.* 139:171–180.

Littlefield, O., Y. Korkhin, and P. B. Sigler. 1999. The structural basis for the oriented assembly of a TBP/TFB/promoter complex. *Proc. Natl. Acad. Sci. USA* 96:13668–13673.

Madern, D. 2002. Molecular evolution within the L-malate and L-lactate dehydrogenase super-family. *J. Mol. Evol.* 54:825–840.

Madern, D., M. Camacho, A. Rodriguez-Arnedo, M. J. Bonete, and G. Zaccai. 2004. Salt-dependent studies of NADP-dependent isocitrate dehydrogenase from the halophilic archaeon *Haloferax volcanii*. *Extremophiles* 8:377–384.

Madern, D., C. Ebel, and G. Zaccai. 2000a. Halophilic adaptation of enzymes. *Extremophiles* 4:91–98.

Madern, D., C. Ebel, M. Mevarech, S. B. Richard, C. Pfister, and G. Zaccai. 2000b. Insights into the molecular relationships between malate and lactate dehydrogenases: structural and biochemical properties of monomeric and dimeric intermediates of a mutant of tetrameric L-[LDH-like] malate dehydrogenase from the halophilic archaeon *Haloarcula marismortui*. *Biochemistry* 39:1001–1010.

Madern, D., and G. Zaccai. 1997. Stabilisation of halophilic malate dehydrogenase from *Haloarcula marismortui* by divalent cations—effects of temperature, water isotope, cofactor and pH. *Eur. J. Biochem.* 249:607–611.

Madern, D., and G. Zaccai. 2004. Molecular adaptation: the malate dehydrogenase from the extreme halophilic bacterium *Salinibacter ruber* behaves like a non-halophilic protein. *Biochimie* 86:295–303.

Marg, B. L., K. Schweimer, H. Sticht, and D. Oesterhelt. 2005. A two-alpha-helix extra domain mediates the halophilic character of a plant-type ferredoxin from halophilic archaea. *Biochemistry* 44:29–39.

Maupin-Furlow, J. A., M. A. Gil, I. M. Karadzic, P. A. Kirkland, and C. J. Reuter. 2004. Proteasomes: perspectives from the Archaea. *Front. Biosci.* 9:1743–1758.

Moelbert, S., B. Normand, and P. De Los Rios. 2004. Kosmotropes and chaotropes: modelling preferential exclusion, binding and aggregate stability. *Biophys. Chem.* 112:45–57.

Mongodin, E. F., K. E. Nelson, S. Daugherty, R. T. Deboy, J. Wister, H. Khouri, J. Weidman, D. A. Walsh, R. T. Papke, G. Sanchez Perez, A. K. Sharma, C. L. Nesbo, D. MacLeod, E. Bapteste, W. F. Doolittle, R. L. Charlebois, B. Legault, and F. Rodriguez-Valera. 2005. The genome of *Salinibacter ruber*: convergence and gene exchange among hyperhalophilic bacteria and archaea. *Proc. Natl. Acad. Sci. USA* 102:18147–18152.

Nelson, K. E., R. A. Clayton, S. R. Gill, M. L. Gwinn, R. J. Dodson, D. H. Haft, E. K. Hickey, J. D. Peterson, W. C. Nelson, K. A. Ketchum, L. McDonald, T. R. Utterback, J. A. Malek, K. D. Linher, M. M. Garrett, A. M. Stewart, M. D. Cotton, M. S. Pratt, C. A. Phillips, D. Richardson, J. Heidelberg, G. G. Sutton, R. D. Fleischmann, J. A. Eisen, O. White, S. L. Salzberg, H. O. Smith, J. C. Venter, and C. M. Fraser. 1999. Evidence for lateral gene transfer between Archaea and bacteria from genome sequence of *Thermotoga maritima*. *Nature* 399:323–329.

Ng, W. V., S. P. Kennedy, G. G. Mahairas, B. Berquist, M. Pan, H. D. Shukla, S. R. Lasky, N. S. Baliga, V. Thorsson, J. Sbrogna, S. Swartzell, D. Weir, J. Hall, T. A. Dahl, R. Welti, Y. A. Goo, B. Leithauser, K. Keller, R. Cruz, M. J. Danson, D. W. Hough, D. G. Maddocks, P. E. Jablonski, M. P. Krebs, C. M. Angevine, H. Dale, T. A. Isenbarger, R. F. Peck, M. Pohlschroder, J. L. Spudich, K. W. Jung, M. Alam, T. Freitas, S. Hou, C. J. Daniels, P. P. Dennis, A. D. Omer, H. Ebhardt, T. M. Lowe, P. Liang, M. Riley, L. Hood, and S. DasSarma. 2000. Genome sequence of *Halobacterium* species NRC-1. *Proc. Natl. Acad. Sci. USA* 97:12176–12181.

O'Brien, R., B. DeDecker, K. G. Fleming, P. B. Sigler, and J. E. Ladbury. 1998. The effects of salt on the TATA binding protein–DNA interaction from a hyperthermophilic archaeon. *J. Mol. Biol.* 279:117–125.

Oren, A. 2002. *Halophilic Microorganisms and their Environments*. Kluwer Academic, Dordrecht, Netherlands/Boston, MA.

Oren, A., M. Heldal, S. Norland, and E. A. Galinski. 2002. Intracellular ion and organic solute concentrations of the extremely halophilic bacterium *Salinibacter ruber*. *Extremophiles* 6:491–498.

Ortenberg, R., O. Rozenblatt-Rosen, and M. Mevarech. 2000. The extremely halophilic archaeon *Haloferax volcanii* has two very different dihydrofolate reductases. *Mol. Microbiol.* 35:1493–1505.

Pieper, U., G. Kapadia, M. Mevarech, and O. Herzberg. 1998. Structural features of halophilicity derived from the crystal structure of dihydrofolate reductase from the Dead Sea halophilic archaeon, *Haloferax volcanii*. *Structure* 6:75–88.

Premkumar, L., H. M. Greenblatt, U. K. Bageshwar, T. Savchenko, I. Gokhman, J. L. Sussman, and A. Zamir. 2005. Three-dimensional structure of a halotolerant algal carbonic anhydrase predicts halotolerance of a mammalian homolog. *Proc. Natl. Acad. Sci. USA* 102:7493–7498.

Pundak, S., H. Aloni, and H. Eisenberg. 1981. Structure and activity of malate dehydrogenase from the extreme halophilic bacteria of the Dead Sea. 2. Inactivation, dissociation and unfolding at NaCl concentrations below 2 M. Salt, salt concentration and temperature dependence of enzyme stability. *Eur. J. Biochem.* 118:471–477.

Richard, S., F. Bonneté, O. Dym, and G. Zaccai. 1995. Protocol 21: the MPD–NaCl–H_2O system for the crystallization of halophilic proteins, p. 149–153. *In* S. Dassarma (ed.), *Archaea: A Laboratory Manual*. Cold Spring Harbor Laboratory Press, New York, NY.

Richard, S. B., D. Madern, E. Garcin, and G. Zaccai. 2000. Halophilic adaptation: novel solvent protein interactions observed in the 2.9 and 2.6 Å resolution structures of the wild type and a mutant of malate dehydrogenase from *Haloarcula marismortui*. *Biochemistry* 39:992–1000.

Ruepp, A., W. Graml, M. L. Santos-Martinez, K. K. Koretke, C. Volker, H. W. Mewes, D. Frishman, S. Stocker, A. N. Lupas, and W. Baumeister. 2000. The genome sequence of the thermoacidophilic scavenger *Thermoplasma acidophilum*. *Nature* 407:508–513.

Shima, S., E. Warkentin, W. Grabarse, M. Sordel, M. Wicke, R. K. Thauer, and U. Ermler. 2000. Structure of coenzyme F(420) dependent methylenetetrahydromethanopterin reductase from two methanogenic archaea. *J. Mol. Biol.* 300:935–950.

Tehei, M., B. Franzetti, D. Madern, M. Ginzburg, B. Z. Ginzburg, M. T. Giudici-Orticoni, M. Bruschi, and G. Zaccai. 2004. Adaptation to extreme environments: macromolecular dynamics in bacteria compared in vivo by neutron scattering. *EMBO Rep.* 5:66–70.

Tehei, M., and G. Zaccai. 2005. Adaptation to extreme environments: macromolecular dynamics in complex system. *Biochim. Biophys. Acta* 1724:404–410.

Tehei, M., D. Madern, C. Pfister, and G. Zaccai. 2001. Fast dynamics of halophilic malate dehydrogenase and BSA measured by neutron scattering under various solvent conditions influencing protein stability. *Proc. Natl. Acad. Sci. USA* 98:14356–14361.

Timasheff, S. N. 1993. The control of protein stability and association by weak interactions with water: how do solvents affect these processes? *Annu. Rev. Biophys. Biomol. Struct.* 22:67–97.

Wright, D. B., D. D. Banks, J. R. Lohman, J. L. Hilsenbeck, and L. M. Gloss. 2002. The effect of salts on the activity and stability of *Escherichia coli* and *Haloferax volcanii* dihydrofolate reductases. *J. Mol. Biol.* 323:327–344.

Yamada, Y., T. Fujiwara, T. Sato, N. Igarashi, and N. Tanaka. 2002. The 2.0 Å crystal structure of catalase-peroxidase from *Haloarcula marismortui*. *Nat. Struct. Biol.* 9:691–695.

Zaccai, G. 2000. How soft is a protein? A protein dynamics force constant measured by neutron scattering. *Science* 288:1604–1607.

Zeth, K., S. Offermann, L. O. Essen, and D. Oesterhelt. 2004. Iron-oxo clusters biomineralizing on protein surfaces: structural analysis of *Halobacterium salinarum* DpsA in its low- and high-iron states. *Proc. Natl. Acad. Sci. USA* 101:13780–13785.

V. ACIDOPHILES

Chapter 20

Physiology and Ecology of Acidophilic Microorganisms

D. BARRIE JOHNSON

INTRODUCTION

The general definition of an acidophile is that it is an organism that displays a pH optimum for growth at a value significantly <7. There is no common agreement on the pH boundary which delineates acidophily, but a useful guide is that extreme acidophiles have optimum pH for growth of <3.0 and that moderate acidophiles grow optimally at pH 3 to 5. Although many moderately acidophilic microorganisms can grow at pH < 3, including mesophiles (e.g., *Thiomonas* spp. and the *Acidobacteriacae*), moderate thermophiles (e.g., some *Alicyclobacillus* spp.), and extreme thermophiles (e.g., some *Sulfolobus* spp.), these will not be considered in this chapter, where the focus will be the physiological and phylogenetic diversities of extreme acidophiles and how these microorganisms interact with and adapt to their environment.

Extremely acidophilic organisms are exclusively microbial and comprise both prokaryotes and eukaryotes, and the axiom that as an environmental parameter becomes more extreme, biodiversity declines holds true for both groups. Although some angiosperms (e.g., *Juncus bulbosus*) have been observed to grow in highly acidic lakes, their root systems grow in sediments in which the pH is usually much higher than that of the water body itself. Many eukaryotic microorganisms that have been observed in extremely low-pH environments are acid tolerant rather than truly acidophilic and may grow equally well, or better, in circumneutral pH environments. Acidophilic/acid-tolerant eukaryotes include some yeasts and fungi, e.g., *Acontium velatum*, a copper-tolerant (14 mM) mitosporic fungus that grows between pH 0.2 and 0.7, and *Scytalidium acidophilum*, which tolerates 140 mM copper and grows at pH 0 (though not at pH > 7) and optimally between pH 1 and 2. Protozoa (flagellates, ciliates, and amoeba) have also commonly been observed in acid mine drainage (AMD) (acidic and metal-rich effluents arising from abandoned mines and mine spoils) and have been demonstrated to graze chemolithotrophic and heterotrophic acidophiles in vitro, in cultures of pH < 2 (Johnson and Rang, 1993). Microalgae, such as *Chlamydomonas acidophila*, *Euglena mutabilis*, and the moderate thermoacidophiles *Cyanidium caldarium* and *Galdieria sulphuraria* may be major contributors to net primary production in extremely acidic environments. Multicellular life forms tend to be rare in these situations, a notable exception being some species of rotifers (e.g., *Cephalodella hoodi*). Reviews of eukaryotic diversity in low-pH environments include Gross and Robbins (2000; fungi and yeasts), Packroff and Woelfl (2000; heterotrophic protists), Gross (2000; microalgae), and Deneke (2000; rotifers and crustaceans). In situ and laboratory studies of eukaryotic microorganisms have mostly utilized traditional (nonbiomolecular) approaches. More recently, the biodiversity of eukaryotic microbial communities in extremely acidic (pH 0.8 to 1.38) metal-rich and moderately thermal (30°C to 50°C) waters in the Richmond mine at Iron Mountain, California, was studied using biomolecular techniques (Baker et al., 2004). Fungal, algal, and protozoan representatives were identified in clone libraries and novel acidophilic fungi isolated.

EXTREMELY ACIDIC ENVIRONMENTS: ORIGINS AND GENERAL CHARACTERISTICS

There are a number of biological processes that generate acidity, such as fermentation and nitrification. Oxidation of elemental sulfur and sulfide minerals can result in the production of sufficient sulfuric acid to overwhelm the pH-buffering capacity of terrestrial and aquatic environments, and it is this process, which may be abiotic but which is mostly biologically accelerated, that is responsible for generating most of the

---

D. B. Johnson • School of Biological Sciences, University of Wales, Bangor LL57 2UW, United Kingdom.

extremely acidic niches found on our planet. Such sites may form naturally, e.g., in volcanic areas where elemental sulfur (formed by the condensation of hydrogen sulfide and sulfur dioxide in volcanic gases) is oxidized by sulfur-oxidizing archaea and bacteria, often at elevated temperatures. The resulting high levels of acidity cause the partial or complete destruction of minerals in the vicinity and the formation of acidic mud pots (e.g., in solfatara fields found in Yellowstone National Park, Wyoming; Whakarewarewa, New Zealand; and Krisuvik, Iceland). The measured pH values of these acidic mud pots and associated acid springs and streams is frequently <3 (e.g., Johnson et al., 2003), and they may show a wide range of temperatures. Submarine hydrothermal systems also discharge large quantities of sulfidic minerals, though here the buffering capacity of seawater limits the net amount of acidity produced when these oxidize, though the pH of water in the immediate vicinity of hydrothermal vents tends to be significantly lower than that of bulk marine waters.

Natural exposure of sulfide mineral-rich rock strata often results in the oxidation of these minerals and the deposition of highly colored, ferric iron-rich secondary deposits (gossans) in locations ranging from the high arctic to the tropics. In the past, gossans have served as important guides for prospectors to the presence and nature of metal-rich ores to be found below these iron caps. The process of excavation, comminution, and disposal of mineral ores (and coals) has inadvertently led to the generation of acidic and extremely acidic environments in areas throughout the world that have, or have had, active mining industries. Degradation of surface and ground waters caused by AMD is acknowledged as one of the most severe forms of environmental pollution worldwide. The ways in which extreme acidity may arise in these situations can be demonstrated by reference to the most ubiquitous and abundant sulfide mineral, pyrite. Like other sulfide minerals, pyrite (FeS_2) is stable in situations where either oxygen or water is absent (e.g., in buried, undisturbed ore bodies). In moist environments, however, pyrite is susceptible to attack by molecular oxygen or by a suitable chemical oxidant, the most significant of which is ferric iron (Fe^{3+}). The relative importance of oxygen and ferric iron in this respect depends on pH, with ferric iron assuming a more important role in acidic liquors. Ferric iron attack on pyrite results in the oxidative dissolution of the mineral (Rohwerder et al., 2003), though the products of this initial reaction are not fully oxidized:

$$FeS_2 + 6Fe^{3+} + 3H_2O \rightarrow 7Fe^{2+} + S_2O_3^{2-} + 6H^+ \quad (1)$$

Though this reaction may proceed in both aerated and anoxic environments, its proliferation depends on either the continued input of extraneous ferric iron or the regeneration of ferric iron, which is an oxygen-dependent reaction:

$$4Fe^{2+} + O_2 + 4H^+ \rightarrow 4Fe^{3+} + 2H_2O \quad (2)$$

The rate at which ferrous iron oxidizes abiotically is also highly dependent on pH and is very slow at pH < 4 at ambient temperatures, even in oxygen-saturated solutions (Stumm and Morgan, 1981). Some chemolithotrophic bacteria and archaea exploit this situation by using energy from ferrous iron oxidation to generate ATP. Thiosulfate, formed as the initial sulfur product of pyrite dissolution by reaction (1), is unstable in acidic, ferric iron-containing media and oxidizes to form a variety of polythionates, which are in turn exploited as energy sources by sulfur-oxidizing acidophiles. Some sulfide minerals, such as sphalerite (ZnS), are oxidized by ferric iron but are also inherently unstable in acidic liquors. Proton attack of acid-soluble sulfides results in the production of large amounts of elemental sulfur, but only in the absence of sulfur oxidizers (Rohwerder et al., 2003). Even though the "primary" oxidizers of sulfide minerals are often iron-oxidizing acidophiles, the greater energy yields that are available from the oxidation of elemental sulfur and reduced sulfur compounds compared to ferrous iron (Kelly, 1978) results in numbers of sulfur-oxidizing chemolithotrophs often exceeding those of iron oxidizers in mineral leaching environments (e.g., Okibe et al., 2003). This association is, however, of benefit to both groups of chemolithotrophs—the sulfur oxidizers are supplied with an energy source from (in the case of acid-insoluble sulfides) otherwise inaccessible substrates, while acid production due to sulfur oxidation counterbalances proton consumption associated with ferrous iron oxidation, maintaining low pH conditions at which ferric iron is soluble and most of the iron oxidizers thrive.

An important consequence of the production of extreme acidity coupled with accelerated dissolution of metallic sulfides and other minerals is that the concentrations of most (cationic) metals in these environments tend to be far greater than those in neutral pH ecosystems. This imposes a potential additional stress on indigenous microorganisms and is probably the major reason why AMD and similar acidic wastewaters are highly toxic to most life forms.

Extremely acidic environments vary greatly in major physicochemical parameters, such as temperature, redox potentials, dissolved solutes, and oxygen concentrations, as illustrated in Table 1, which

Table 1. Water chemistries of representative acidic geothermal and mine waters

Ecosystem	pH	T (°C)	E_h (mV)	E_c (µS cm^{-1})	SO$_4$ (mg liter^{-1})	Fe (mg liter^{-1})
Yellowstone (Gibbon River area)[a]	3.3	78	+553			
Yellowstone (Frying Pan hot spring)[a]	2.7	70	+647			
Galways Soufriere (Montserrat)[b]	1.7–7.4	30–98	−200 to +700	1,450–1,902	400 to >1,600	<1–250
Gages Soufriere (Montserrat)[b]	1.0–2.5	65–97	+405 to +630	1,791–1,856	800 to >1,600	100–500
Parys mine (Wales)[c]	2.7	9.0	+420	2,995	1,550	490
Richmond mine (California)[d]	−3.6 to +1.5	29–47			14,000–760,000	2,470–79,700

[a] Data from Johnson et al. (2003).
[b] Data from Atkinson et al. (2000).
[c] Data from Coupland and Johnson (2004).
[d] Data from Nordstrom et al. (2000).

includes data from natural (geothermal) ecosystems and anthropogenic (mine-impacted) environments.

PHYLOGENETIC DIVERSITY AND PHYSIOLOGICAL CHARACTERISTICS OF EXTREME ACIDOPHILES

Acidophilic microorganisms are widely distributed throughout the three domains of living organisms. Within the archaeal domain, both extremely acidophilic *Euryarchaeota* and *Crenarchaeota* are known, and a number of different bacterial phyla (*Firmicutes*, *Actinobacteria*, and *Proteobacteria* [α, β, and γ subphyla], *Nitrospira*, and *Aquifex*) also contain extreme acidophiles. Acidophilic prokaryotes frequently have been subdivided on the basis of their temperature optima as mesophiles (temperature optima 20°C to 40°C), moderate thermophiles (40°C to 60°C), or (extreme) thermophiles (>60°C). Conveniently, mesophilic acidophiles are mostly gram-negative bacteria, moderate thermophiles gram-positive bacteria, and (extreme) thermophiles exclusively archaea. However, as increasing numbers of novel acidophiles are characterized, more exceptions to this pattern are emerging, such as mesophilic *Ferroplasma* (an archaeon) and the thermophilic gram-negative bacterium *Acidicaldus*.

Adaptations and Consequences of Living at Low pH

Acidophilic microorganisms live in low-pH liquors but maintain intracellular pH values close to neutral (Norris and Ingledew, 1992; Matin, 1999). The majority of their cytoplasmic enzymes that have been studied have pH optima close to 7, though several intracellular enzymes from the euryarchaeote *Ferroplasma* have been shown to have extremely low pH optima (Golyshina and Timmis, 2005). Extracellular enzymes and redox-active proteins (e.g., rusticyanin) located in the periplasms of gram-negative acidophiles are also, by necessity, active at low pH. Large pH gradients (ΔpH) are maintained in deenergized as well as metabolically active cells, and the respiratory chains of acidophiles have to excrete protons against these gradients. The proton motive force (PMF), which is used to drive the conversion of ATP from ADP via membrane-bound ATPases as well as transport of substrates across membranes and the rotation of flagella, derives from both the membrane potential (Δω, generated by the transport of electric charge) and the ΔpH:

$$\text{PMF(mV)} = \Delta\omega - 2.3\left(\frac{RT}{F}\right)\Delta\text{pH} \quad (3)$$

The ways in which acidophiles have solved the problem of maintaining pH integrity and the bioenergetic conundrum of pumping protons against a concentration gradient are by (i) possessing cell membranes that are highly impermeable to hydronium ions (H_3O^+) and (ii), in contrast to neutrophilic microorganisms, generating positive inside membrane potentials. Interestingly, cell membranes of acidophilic bacteria have been reported to have no obvious features that distinguish them from those of neutrophiles (Norris and Ingledew, 1992), though *Ferroplasma* cell membranes contain novel caldarchaetidylglycerol tetraether lipids that have extremely low proton permeabilities (Golyshina and Timmis, 2005). Possession of these ether lipids has been postulated to be a major reason for the survival of these archaea, which do not possess cell walls, in extremely low pH (<1) environments. Positive membrane potentials derive from the active influx of cations such as K^+ (Alpers et al., 2001). In energized cells, the PMF is positive, whereas it collapses to zero when bacteria are deenergized. Because ΔpH values are similar in energized and deenergized acidophiles, it is clear that the Δω component varies with energy status of the cells, becoming increasingly large as the PMF declines. Indeed, the positive membrane potential has a protective role in acidophiles when their metabolism declines, as the increased magnitude counteracts the impelling force of proton influx, thereby preventing acidification of the cytoplasm and consequent death of the microorganism (Matin, 1999).

One important consequence of maintaining large transmembrane proton gradients is that acidophiles are highly sensitive to low molecular-weight organic acids, such as acetic and pyruvic acids (Norris and Ingledew, 1992). The pK_a values of many of these acids are greater than the pH of liquors in which extreme acidophiles typically live, and therefore they occur as protonated (and lipophilic) acids. These permeate microbial membranes and dissociate in the cell cytoplasm, and the consequent disequilibrium results in continued influx of undissociated acid. The resulting acidification of the cytoplasm can reach lethal proportions even in the presence of micromolar concentrations of some aliphatic acids. Possession of positive membrane potentials has positive and negative consequences for acidophiles. On the one hand, their tolerance of cationic metals such as copper is, in general, far greater than that of neutrophilic bacteria, though this may also have a specific genetic basis in some acidophiles, such as *Acidocella* (Ghosh et al., 1997). Conversely, most acidophiles are unusually sensitive to anions, with the exception of sulfate, particularly when they are deenergized and $\Delta\omega$ values are consequently large (Norris and Ingledew, 1992).

One of the seeming advantages of living in low-pH environments is that acidophilic prokaryotes have a ready-made pH gradient that might be used to generate ATP using membrane-bound ATPases. Proton influx has to be controlled, however, to avoid acidification of the cytoplasm, and respiratory systems that result in protons being pumped into the periplasm (gram-negative bacteria) or external milieu (gram-positive bacteria and archaea) operate in acidophiles as in neutrophilic prokaryotes. In ferrous iron oxidation, however, proton influx can be counterbalanced by uptake of electrons (Fig. 1). Oxidation of ferrous iron has to necessarily take place outside the cytoplasm in order to avoid the precipitation of ferric iron, which is highly insoluble above pH 2.5. In gram-negative iron oxidizers, ferrous iron oxidation is mediated by redox-active proteins (such as rusticyanin in *Acidithiobacillus (At.) ferrooxidans*) retained in the acidic periplasm. Electrons are transferred via soluble and/or membrane-bound cytochromes ultimately to cytochrome oxidase, where they are used, together with protons, to reduce oxygen to water.

Metabolic Diversity in Acidophiles: Energy Transformations and Carbon Assimilation

Acidophilic microorganisms exhibit a similar range of energy-transforming reactions and means of

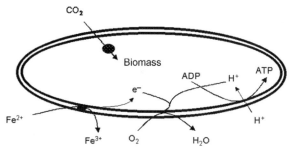

Figure 1. Schematic representation of iron oxidation and CO_2 fixation by a gram-negative acidophilic chemolithotrophic bacterium. Iron oxidation is mediated by a redox-active protein retained within the acidic periplasm. Electrons are transferred via soluble and/or membrane-bound cytochromes ultimately to cytochrome oxidase, where they are used, together with protons, to reduce oxygen to water. Carbon dioxide fixation is mediated by cytoplasmic ribisco.

assimilating carbon as neutrophiles. For example, both chemical and solar energy sources are exploited by acidophiles, and both inorganic and organic forms of carbon may be assimilated. Phototrophic acidophiles appear to be exclusively eukaryotic; no acidophilic cyanobacteria or anaerobic photosynthetic bacteria have been described. Inorganic electron donors are particularly significant energy sources in many extremely acidic environments, as the abundance of ferrous iron and reduced forms of sulfur often greatly exceeds that of organic carbon, for reasons described previously. Other inorganic electron donors used by neutrophilic chemolithotrophs [ammonium, nitrite, and manganese(II)] have so far not been shown to be used by acidophiles (manganese oxidation at low pH would not be expected on thermodynamic grounds). However, hydrogen can be used as an electron donor by some acidophilic bacteria (e.g., *At. ferrooxidans*) and archaea (e.g., *Acidianus* spp.). There have been occasional reports of other metals acting as energy sources; for example, copper(I), uranium(IV), and molybdenum(V) have been reported to be oxidized, and coupled to CO_2 fixation, by *At. ferrooxidans*. However, it is often difficult to demonstrate unequivocally the biological oxidation of reduced metals as opposed to chemical oxidation by ferric iron (which is also produced by this iron oxidizer). The potential also exists that arsenite [As(III)] can serve as an alternative electron donor for extremely acidophilic organisms as it is for the moderate acidophile, *Thiomonas arsenivorans* (Battaglia-Brunet et al., 2006).

The range of organic electron donors used by most acidophilic heterotrophic and mixotrophic prokaryotes is relatively narrow, though there are some notable exceptions. Most mesophilic acidophilic heterotrophs use low molecular-weight monomeric compounds, such as sugars, alcohols, and some amino acids, but not polymeric substances. Some organic acids are also catabolized (e.g., by *Acidocella* spp.), though, as noted, these tend to be highly toxic to acidophiles. Catabolism of aromatic compounds appears also to be restricted in heterotrophic acidophiles, though one mesophilic isolate (*Acidocella aromatica*) has been found to grow on a wide range of aromatics (including phenol, substituted phenols, benzoic acid, and naphthalene) (Hallberg et al., 1999), and the moderate thermophile *Acidicaldus organivorans* can also grow on phenol (Johnson et al., 2006).

The electron acceptor most commonly used by acidophiles is molecular oxygen. Indeed, for iron-oxidizing acidophiles, the high redox potential of the ferrous/ferric couple (+770 mV at pH 2) means that, on thermodynamic grounds, oxygen is the only feasible electron acceptor. On the other hand, this high redox potential means that ferric iron is an attractive alternative electron acceptor to oxygen (the redox potential of the oxygen–water couple is only slightly greater, at +840 mV, than the ferrous–ferric couple), particularly because ferric iron tends to be very abundant in extremely acidic environments and is soluble (and therefore more bioavailable) at pH < 2.3. Growth coupled with ferric iron reduction has been demonstrated in autotrophic (e.g., *At. ferrooxidans*, using sulfur or hydrogen as electron donors), mixotrophic (e.g., *Sulfobacillus* spp.), and heterotrophic (e.g., *Acidiphilium* spp.) acidophiles. Anaerobic growth by sulfur reduction has been demonstrated in some acidophilic archaea (e.g., *Acidianus* spp.). Proof that extreme acidophiles can catalyze the dissimilatory reduction of sulfate has, however, proved more elusive, though this has been demonstrated for a consortium of moderate acidophiles (Kimura et al., 2006).

In terms of carbon acquisition, acidophiles comprise obligate autotrophs, obligate heterotrophs, and others that can fix CO_2, but which do this only when there is little or no available organic carbon. Many mesophilic and moderately thermophilic autotrophic acidophiles fix CO_2 by the Calvin reductive pentose phosphate cycle, using the enzyme ribulose 1,5-biphosphate carboxylase (RuBPCase or Rubisco). In some cases (e.g., the moderate thermophile *Sulfobacillus thermosulfidooxidans*), CO_2-enriched air is required for rapid autotrophic growth. Some acidophilic archaea also grow autotrophically, though details of the pathway(s) are less well known. There is evidence for the CO_2 fixation via the acetyl coenzyme A (acetyl-CoA) pathway in *Sulfolobus* spp. and for a modified 3-hydroxypropionate pathway in *Acidianus brierleyi*.

The ability to use either inorganic C or organic C as a carbon source is widespread amongst gram-positive acidophiles (e.g., *Sulfobacillus* and *Acidimicrobium* spp.). Owing to the energy requirement for fixing CO_2, these acidophiles use organic carbon when this is available, though their abilities to switch to autotrophic mode in organic carbon-depleted environments that contain large potential energy reserves (in terms of reduced sulfur and iron) give them a competitive advantage in fluctuating environments. The term mixotrophs has sometimes been used to describe bacteria that use either autotrophic or heterotrophic metabolisms. However, this term is ambiguous as it has also been used to describe prokaryotes, like *Ferrimicrobium acidiphilum*, that use an inorganic energy source (ferrous iron, in this case) and an organic carbon source (Bacelar-Nicolau and Johnson, 1999).

CHARACTERISTICS OF CLASSIFIED ACIDOPHILIC PROKARYOTES

The list of characterized extreme acidophiles continues to expand. Entire genomes have been sequenced of the iron/sulfur-oxidizing bacterium *Acidithiobacillus ferrooxidans* and of the archaea *Thermoplasma acidophilum*, *Picrophilus torridus*, *Sulfolobus tokodaii*, and *Ferroplasma acidarmanus*, and more genome sequences of acidophiles are due to be completed in the near future.

Iron- and Sulfur-Oxidizing Aerobes

Leptospirillum spp.

Leptospirillum spp. form a distinct lineage within the deep-rooted *Nitrospira* phylum, which also includes nitrifying, sulfate-reducing, and magnetotactic bacteria. However, *Leptospirillum* spp. are unique in this phylum in being both obligately acidophilic and capable of the dissimilatory oxidation of ferrous iron. Three species of *Leptospirillum* are currently recognized, and a fourth species has been proposed (Table 2). These have very similar physiological characteristics, the most notable of which is that they have been shown to utilize only one electron donor (ferrous iron) and a single electron acceptor (molecular oxygen), though they vary in their tolerance to temperature and pH and in their abilities to fix nitrogen. Ferric iron produced by *Leptospirillum* spp. chemically oxidizes sulfide minerals and is in turn reduced to ferrous iron, which is then reoxidized by the bacterium. Thus, although *Leptospirillum* (*L.*) *ferrooxidans* has only been found to use a single electron donor (Fe^{2+}), it is able, indirectly, to exploit the energy available from the oxidation of sulfide

Table 2. Phenotypic characteristics of *Leptospirillum* spp.

Characteristic	*L. ferrooxidans*	*L. thermoferrooxidans*	*L. ferriphilum*	*L. ferrodiazotrophum*
Oxidation of Fe^{2+}	+	+	+	+
N_2 fixation	+	ND	–	+
G+C (mol%)	51–52	56	55–58	~57.5
Cell length (µm)	0.9–1.1	1.5–2.0	0.9–3.5	0.5–3.5
Endospores	–	–	–	–
Motility	+	+	+	+
rrn gene operon copy number	3	ND	2	1
$T_{optimum}$ (°C)	30–37	45–50	35–45	ND
$pH_{minimum}$	1.1	1.3	0.8	<1.0

minerals (sulfane sulfur, in pyrite) using ferric iron as an abiotic electron shuttle.

The first *Leptospirillum* to be described, *L. ferrooxidans*, was originally isolated from a copper mine in Armenia in 1972 and was, at the time, only the second iron-oxidizing acidophile to be characterized. Later, a novel thermotolerant (temperature optimum 45°C to 50°C) species, *Leptospirillum thermoferrooxidans*, was described by Golovacheva et al. (1992); however, there is some uncertainty regarding the viability of the single strain of this bacterium. For some time before *L. thermoferrooxidans* was described, it had been recognized that bacteria categorized as *L. ferrooxidans* could be separated into two groups on the basis of the G+C contents of their chromosomal DNA. Coram and Rawlings (2002) showed that analysis of 16S rRNA genes and of the 16S–23S rRNA gene spacer regions of over 16 strains of *Leptospirillum* clearly separated these two groups (I and II) and that these represented different species. The novel species, *Leptospirillum ferriphilum*, was shown to include thermotolerant as well as mesophilic species. Interestingly, all of the *Leptospirillum*-like bacteria isolated from commercial mineral processing tank operations in South Africa by Coram and Rawlings (2002) and by Okibe et al. (2003) were *L. ferriphilum*, rather than *L. ferrooxidans*. A representative of a third group of *Leptospirillum* was identified in extremely acidic and metal-rich waters within the abandoned Richmond mine at Iron Mountain, California (Tyson et al., 2004). This bacterium was found in a biofilm, where it was greatly outnumbered by group II (*L. ferriphilum*) representatives. Random shotgun cloning of the biofilm showed that a single *nif* operon was present and that this was found on a genome fragment of group III *Leptospirillum*. A cultivation technique, using nitrogen-free liquid media, was used to enrich the novel bacteria, which resulted in the successful isolation of a group III representative, with the proposed name *Leptospirillum ferrodiazotrophum* (Tyson et al., 2005).

Acidithiobacillus spp.

Currently, there are four recognized species of this genus that grow autotrophically on sulfur, sulfide, and reduced inorganic sulfur compounds (RISCs). One of these, the iron-oxidizing and -reducing acidophile, *Acidithiobacillus ferrooxidans*, is a facultative anaerobe and is described in a later section. The other three species, *Acidithiobacillus thiooxidans*, *Acidithiobacillus albertensis*, and *Acidithiobacillus caldus*, are obligate aerobes and share many physiological traits, with the most significant differentiating characteristic being that strains of *At. thiooxidans* (and *At. albertensis*) are all mesophilic (most strains appear to be fairly temperature sensitive and do not grow above 35°C), while *At. caldus* is a moderate thermophile, with a temperature optimum of ~45°C and maximum of ~55°C. Phylogenetically, these acidothiobacilli fall into a monophyletic group within the γ-Proteobacteria, close to the cusp between the β and γ subgroups.

At. thiooxidans was, in 1921, the first microorganism with a pH optimum of <3 to be isolated and described. It couples the oxidation of elemental sulfur and a variety of RISCs to sulfate, with concomitant production of protons, and is one of the most acidophilic of all known bacteria, being capable of growing at pH 0.5. It is likely that many strains of *At. caldus* have, in the past, been mistakenly identified as *At. thiooxidans*, as the two share many physiological characteristics. Phylogenetically, the two acidophilic bacteria are also closely related. Although, as noted earlier, *At. caldus* is a moderate thermophile, it also grows within the upper temperature range of *At. thiooxidans* and, at 30°C to 35°C, will outcompete the mesophile. This appears to be a major reason why *At. caldus* has been identified as the dominant sulfur-oxidizing microorganism in the leachates of many commercial mineral solubilization processes. *At. caldus* is also more metabolically versatile than its mesophilic counterpart, in that it can grow mixotrophically in yeast extract- or glucose-containing media, while *At. thiooxidans* is a strict autotroph.

The situation concerning *At. albertensis* is less clear (Kelly and Wood, 2000), as the original isolate may have been lost from culture without being sequenced. It was originally reported to produce a glycocalyx, to possess a tuft of flagella, but to be nearly identical to *At. thiooxidans* in most other respects, apart from a significantly higher G+C content of its chromosomal DNA (61.5 mol%, as opposed to 51 mol% for *At. thiooxidans*).

Hydrogenobaculum acidophilum

H. acidophilum is an obligate autotroph and grows aerobically using either hydrogen or elemental sulfur as an electron donor. Although it is capable of growing at pH 2, its optimum pH for growth is 3 to 4. It is, however, the most thermotolerant of all characterized acidophilic bacteria, with a temperature optimum at 65°C and maximum at ~70°C. Phylogenetically, it belongs to an early branching division of the domain *Bacteria* and is most closely related to the extreme thermophile, *Aquifex pyrophilus*. Recently, a novel *Hydrogenobaculum* (H55) isolated from an acidic sulfate- and chloride-rich geothermal spring in Yellowstone National Park, Wyoming, has been described (Donahoe-Christiansen et al., 2004). Isolate H55 shares 98% 16S rRNA gene identity with *H. acidophilum* and has a similar pH optimum but is less thermotolerant. This bacterium was shown to oxidize As(III) to As(V), though appeared unable to use As(III) as the sole energy source for chemolithotrophic growth. In contrast to *H. acidophilum*, isolate H55 used hydrogen, but not elemental sulfur, as the electron donor.

Thiobacillus spp.

The genus *Thiobacillus* was thoroughly revised by Kelly and Wood (2000) on the basis of the phylogenetic diversity of this group that had been revealed by biomolecular (16S rRNA gene sequence) analysis. This resulted in several acidophilic species being reclassified as *Acidithiobacillus* spp (γ-*Proteobacteria*), whilst the original genus title was retained for the type species (*Thiobacillus thioparus*), which is a β-proteobacterium. However, there were insufficient data available at the time to assign the halotolerant acidophilic iron/sulfur oxidizer *Thiobacillus prosperus* to the existing or newly described genera. Previously, it had been shown that another iron oxidizer, isolate m-1 (which did not oxidize sulfur), was erroneously classified as a strain of *Thiobacillus ferrooxidans* (Lane et al., 1992), and again this acidophile is awaiting formal redesignation.

Sulfur-oxidizing archaea

Several genera of sulfur-oxidizing archaea (all of which are thermoacidophilic crenarchaeotes) have been described; one of these (*Acidianus*) can also respire anaerobically on sulfur, as is described in a later section. *Sulfolobus (S.) acidocaldarius* was the first thermoacidophilic prokaryote to be described and is the most thoroughly studied acidophilic archaeon. Some *Sulfolobus* spp. (all of which are obligate aerobes) can grow autotrophically on sulfur, though only one (*Sulfolobus metallicus*) is an obligate chemoautotroph and is apparently also unique amongst this genus in catalyzing the dissimilatory oxidation of ferrous iron; however, there is some evidence for iron oxidation by *S. tokodaii*. Other facultatively chemolithotrophic sulfur-oxidizing archaea are *Metallosphaera* spp. and *Sulfurococcus yellowstonensis*. Like *S. metallicus* and *A. brierleyi* (see below), *Metallosphaera* spp. are recognized as being important microorganisms involved in the commercial processing of mineral ores at high temperatures in stirred tank bioreactors (Norris et al., 2000).

Facultative Anaerobic Acidophiles That Catalyze the Dissimilatory Oxidoreduction of Iron

Acidithiobacillus ferrooxidans

Acidithiobacillus ferrooxidans is the most well studied of all acidophilic microorganisms and has often, though erroneously, been regarded as an obligate aerobe. While its most well-known physiological characteristic (the ability to catalyze the dissimilatory oxidation of ferrous iron in acidic milieu) has necessarily to be coupled with the reduction of molecular oxygen, it is also able to use other electron donors (elemental sulfur and sulfide, RISCs, hydrogen, and formic acid), all of which can be coupled to the reduction of ferric iron, facilitating its growth in the absence of oxygen (e.g., Pronk et al., 1991). The fact that *At. ferrooxidans* is a facultative anaerobe at least partially accounts for the fact that it is frequently present as the numerically dominant acidophile in anoxic, extremely acidic mine waters at their point of discharge from underground mine sites (e.g., Coupland and Johnson, 2004) but is increasingly displaced by other iron oxidizers (notably *L. ferrooxidans*) as the drainage streams become aerated and redox potentials increase. *At. ferrooxidans* is also less acidophilic (pH minimum of ~1.3 to 1.5, dependent on strain) than iron-oxidizing *Leptospirillum*, *Ferroplasma*, and *Sulfobacillus* spp.

It has become increasingly apparent that strains and isolates classified as *At. ferrooxidans* exhibit considerable genetic variation (Kelly and Wood, 2000). Even discounting isolates such as *T. ferrooxidans* m-1, which clearly represents a different genus (see above), Karavaiko et al. (2003) showed that strains that comprised a monophyletic cluster (together with *At. thiooxidans*) fell into four phylogenetic groups and that some groups were more distantly related to each other than they were to *At. thiooxidans*. There is a strong case for reconsidering the status of some bacteria currently named as *At. ferrooxidans* and, as in *Leptospirillum* spp., reclassifying these as different species. Whilst all *At. ferrooxidans* strains contained within this monophyletic cluster have the core characteristics of being acidophilic iron and sulfur oxidizers, some physiological differences have been noted. For example, while many strains (including the type strain) produce large amounts of copper-containing protein rusticyanin when respiring on ferrous iron, other *At. ferrooxidans* appear not to do so (Blake and Johnson, 2000).

Sulfobacillus spp.

Sulfobacillus (Sb.) spp. fall into the phylum *Firmicutes* (low G+C mol% gram-positive, sporulating eubacteria). Two of the currently recognized species (*Sb. thermosulfidooxidans* and *Sulfobacillus acidophilus*) and the proposed species *Sulfobacillus montserratensis* share several important physiological traits, including a remarkable metabolic flexibility in terms of their mechanisms of acquiring energy and carbon (Norris et al., 1996; Yahya et al., 1999). They can grow autotrophically via the oxidation of ferrous iron, sulfur, and sulfidic minerals; anaerobically using ferric iron as electron acceptor; and heterotrophically on a variety of organic carbon sources. *Sb. thermosulfidooxidans* and *Sb. acidophilus* are both moderate thermophiles, with temperature optima of ~47°C and maxima of 55°C to 60°C, while *Sb. montserratensis* is mesophilic, with an optimum growth temperature at 37°C. *Sb. montserratensis* is also significantly more tolerant of extreme acidity than the moderately thermophilic species and can grow at pH 0.8 (Yahya and Johnson, 2002). *Sulfobacillus* spp. appear to have poorly developed systems for the assimilatory reduction of sulfate, in that autotrophic growth on ferrous iron requires the provision of a reduced form of sulfur, such as tetrathionate. Optimum growth of axenic cultures is observed when organic carbon (typically yeast extract) is included in growth media or when *Sulfobacillus* spp. are grown in coculture with autotrophic acidophiles, such as *Leptospirillum* (Okibe and Johnson, 2004). A recently described *Sulfobacillus* sp. (*Sulfobacillus sibiricus*) was shown to grow between 17°C and 60°C (optimum 55°C) and to grow reproducibly only in the presence of both an inorganic energy source (ferrous iron, sulfur, or sulfide minerals) and organic carbon (e.g., yeast extract or glucose) (Melamud et al., 2003). Whether or not *Sb. sibiricus* could grow anaerobically by ferric iron respiration was not ascertained.

Alicyclobacillus spp. and other acidophilic Firmicutes

The first *Alicyclobacillus* spp. to be described (*Alicyclobacillus acidocaldarius*, *Alicyclobacillus acidoterrestris*, *Alicyclobacillus cycloheptanicus*, and *Alicyclobacillus hesperidum*) were obligately heterotrophic *Firmicutes*, not all of which grew optimally at pH 3 or below, and none of which were able to catalyze the dissimilatory oxidation of iron or sulfur. A characteristic and diagnostic feature of these bacteria is that they contain ω-alicyclic fatty acids as major components of their cell membranes. More recently, the genus has been revised to include iron- and sulfur-oxidizing acidophiles (*Alicyclobacillus tolerans* and *Alicyclobacillus disulfidooxidans*) (Karavaiko et al., 2005). One isolate, closely related to *Alb. tolerans* (strain GSM), was shown to oxidize ferrous iron (and sulfur), though this was not coupled with growth in the presence of organic carbon, and also shown to grow anaerobically using ferric iron as the electron acceptor (Johnson et al., 2001). Interestingly, the isolate GSM did not produce ω-alicyclic fatty acids.

Other currently unclassified acidophiles of the phylum *Firmicutes* have been described that are phylogenetically distinct from both *Sulfobacillus* and *Alicyclobacillus* (Johnson et al., 2001). Although not fully characterized, these (e.g., isolate SLC66) appear to be obligately heterotrophic iron oxidizers that do not oxidize sulfur. These distinct *Firmicutes* have been detected in a number of mineral-rich environments, including weathering sulfidic regoliths and tailings.

Acidimicrobium/Ferrimicrobium spp.

Two acidophilic genera of *Actinobacteria*, each currently represented by a single species, have been described, one of which (*Ferrimicrobium*) awaits validation. These are both capable of oxidizing ferrous iron (though not sulfur) and can reduce ferric iron in the absence of oxygen (Clark and Norris, 1996; Johnson et al., 2001). A major difference between the two bacteria is that *Acidimicrobium ferrooxidans* is a moderate thermophile (temperature optima and maxima, 48°C and 55°C, respectively), while *Ferrimicrobium (Fm.) acidiphilum* is mesophilic. Production of endospores has not been described for either species.

The original strain of *Acidimicrobium ferrooxidans* (isolated from New Mexico) was considered to be an obligate heterotroph, though a second (and designated type) strain (from Iceland) was shown also to fix CO_2. In contrast, *Fm. acidiphilum*, which is widely distributed in metal-rich, acidic environments, often in close association with chemolithotrophic acidophiles, is apparently unable to fix CO_2.

Ferroplasma spp.

Ferroplasma spp. are extremely acidophilic iron-oxidizing, mesophilic, and thermotolerant *Euryarchaeota* (Golyshina and Timmis, 2005; Dopson et al., 2004). These prokaryotes lack cell walls and grow as pleomorphic or irregular cocci. The type strain of *Ferroplasma (Fp.) acidiphilum* was isolated from a pilot plant bioreactor in Kazakhstan, in which a pyrite/arsenopyrite was being bioleached (Golyshina et al., 2000); since then, other strains have been isolated from similar environments (Dopson et al., 2004). Around the same time that *Fp. acidiphilum* was characterized, a second species, *F. acidarmanus*, was isolated from the abandoned Richmond mine at Iron Mountain, where it was shown to be the dominant acidophile in the most extremely acidic (pH < 1) niches (Edwards et al., 2000). *Ferroplasma acidarmanus* is the most acidophilic of all known iron oxidizers and is capable of growing at pH 0. Recently, a third *Ferroplasma* sp. has been described (Hawkes et al., 2005). *Ferroplasma cupricumulans* was isolated from a chalcocite heap bioleaching operation and grows aerobically in ferrous iron/yeast extract medium or anaerobically using either yeast extract or tetrathionate as electron donors and ferric iron as an electron acceptor.

There has been some dispute concerning carbon acquisition by *Ferroplasma* spp. *Fp. acidiphilum* was originally described as an autotroph, though it required yeast extract for growth (Golyshina et al., 2000). However, Dopson et al. (2004) were unable to demonstrate CO_2 fixation in the three strains of *Fp. acidiphilum* (including the original type strain) or in cultures of *Fp. acidarmanus* and concluded that *Ferroplasma* spp. were not autotrophic. All four strains tested by Dopson et al. (2004) were able to grow heterotrophically on yeast extract and various sugars in the absence of ferrous iron, and none was able to oxidize reduced sulfur tetrathionate. Ferric iron reduction by two strains of *Fp. acidiphilum*, isolated from mineral bioleachate, was reported by Okibe et al. (2003), and Dopson et al. (2004) later confirmed that both *Fp. acidiphilum* and *Fp. acidarmanus* could grow anaerobically on yeast extract using ferric iron as electron donor.

Other Iron-Reducing Heterotrophs

Acidiphilium spp.

Acidiphilium spp. are probably the most widely distributed heterotrophic, acidophilic bacteria. Many of these α-*Proteobacteria* are adept scavengers, presumably reflecting the conditions of their natural aquatic environments, which tend to contain little dissolved organic carbon. Although six different species are currently recognized, 16S rRNA gene sequence data suggest that some of these are strains of the same species. One *Acidiphilium* sp. (*Acidiphilium (A.) acidophilum*) is a facultative autotroph and can use elemental sulfur as the sole electron donor. All classified *Acidiphilium* spp. catalyze the dissimilatory reduction of ferric iron to ferrous; in some species (including the type species, *Acidiphilium cryptum*) this is constitutive, while in others (including *A. acidophilum*) iron reduction is inducible (Johnson and Bridge, 2002). These bacteria can induce the reductive dissolution of a wide range of ferric iron minerals, including jarosites and goethite, as well as soluble ferric iron (Bridge and Johnson, 2000). In the majority of strains, iron reduction occurs optimally under microaerobic rather than in strictly anoxic conditions and is coupled with growth. Several *Acidiphilium* spp. have been shown to produce bacteriochlorophyll-*a*, though they cannot grow as phototrophs. One such is *Acidiphilium rubrum*, which grows as strongly pigmented pink-purple colonies on solid media; its bacteriochlorophyll-*a* has been shown to have a zinc-containing, rather than a magnesium-containing, porphyrin (Kishimoto et al., 1995).

Acidocella spp.

Like *Acidiphilium* spp., *Acidocella* are α-*Proteobacteria* commonly found in environments impacted by acid mine waters. In general, *Acidocella* spp. are less tolerant of extreme acidity and heavy metals than *Acidiphilium* spp., though there are exceptions (e.g. strain GS19h, which grows in the presence of molar concentrations of cadmium or zinc) (Ghosh et al., 1997). *Acidocella* spp. grow readily in liquid and on solid media, where they tend to form bright white colonies, as these heterotrophic acidophiles do not produce bacteriochlorophyll. This helps differentiate them from colonies of *Acidiphilium*, which tend to be cream or pink/purple colored. The range of organic compounds utilized by both *Acidocella* and

Acidiphilium spp. is similar (mostly sugars, low-molecular-weight alcohols, and some amino acids), though a notable exception is *Acidocella aromatica*, which has been reported to grow both on aliphatic acids (at concentrations that are lethal to most acidophiles) (Gemmell and Knowles, 2000) and on a variety of aromatic compounds (Hallberg et al., 1999). Iron reduction has been demonstrated with several strains of *Acidocella* using the ferric iron mineral schwertmannite, a major ferric iron mineral present in mine drainage-impacted streams (Coupland, 2005).

Acidicaldus organivorans

Acidicaldus (Acd.) organivorans is one of the most thermotolerant of all known extremely acidophilic bacteria (growing at up to 65°C) and was isolated from geothermal areas in Yellowstone National Park (Johnson et al., 2006). *Acd. organivorans* is an obligate heterotroph that grows on a variety of carbon sources, including phenol, and also oxidizes elemental sulfur to sulfuric acid. It is also a facultative anaerobe, capable of growing by ferric iron respiration.

Sulfur-Reducing Acidophiles

Four acidophilic genera (all archaea) have been described that can grow anaerobically using elemental sulfur as an electron acceptor. *Acidianus* spp. are so called after the Roman god Janus, who is usually portrayed as a figure with two faces looking in opposite directions, as these crenarchaeotes are able to use elemental sulfur either as an electron donor/energy source in the presence of oxygen or as an electron acceptor in anoxic environments (using hydrogen as electron donor). Three species of *Acidianus (Ad.)* have been described: *Acidianus infernus* and *Acidianus ambivalens* are both obligate autotrophs, while *Acidianus brierleyi* is a facultative autotroph and can use ferrous iron as well as sulfur as an electron donor. A fourth species, *Acidianus tengchonensis*, has been proposed, which does not use ferrous iron as an energy source, though it is closely related (99.8% 16S rRNA gene sequence identity) to *Ad. brierleyi* (He et al., 2004). *Stygiolobus (Sg.) azoricus* is, in contrast, an obligately anaerobic thermoacidophile; this crenarchaeote also grows by coupling the oxidation of hydrogen to the reduction of sulfur. An archaeon closely related to *Sg. azoricus*, *Sulfurisphaera ohwakuensis*, has phenotypic traits similar to those of *Acidianus* spp. (e.g., in being a facultative anaerobe) and is one of the most thermotolerant of characterized acidophiles, growing at up to 92°C. The other known sulfur-reducing acidophilic archaeon is the euryarchaeote *Thermoplasma*. Two species of this moderate thermophile (*Thermoplasma acidophilum* and *Thermoplasma volcanium*) have been described; both are facultative anaerobes and obligately heterotrophic. *Thermoplasma* spp. lack cell walls and may grow as cocci or disk- and clubbed-shaped cells, sometimes as filaments (Segerer et al., 1988).

Picrophilus spp. are also moderately thermophilic euryarchaeotes, though unlike *Thermoplasma* spp. they are obligate aerobes (and do not reduce or oxidize sulfur), and unlike both *Ferroplasma* and *Thermoplasma*, they possess a wall-like outer structure. Two species have been characterized: *Picrophilus oshimae* and *Picrophilus torridus*; both are obligate heterotrophs. *Picrophilus* spp. are the most acid tolerant of all life forms that have been described, with pH optima of about 0.7 and minima of pH < 0 (Schleper et al., 1995), and the internal pH of these archaea (~4.6) is significantly lower than other acidophiles that have been studied.

ACIDOPHILIC COMMUNITIES

Relationships and Interactions

In both natural and anthropogenic environments, acidophilic microorganisms live in communities that range from relatively simple (two to three dominant members) to highly complex, and within these, acidophiles interact positively or negatively with each other (Johnson, 1998, 2001). Reference has already been made to the interplay between iron- and sulfur-oxidizing chemolithotrophs during the oxidation of sulfidic minerals, such as pyrite, which is an example of a mutualistic relationship (an association in which both partners derive benefit). A more complex synergistic relationship has been described for the heterotrophic iron oxidizer *Fm. acidophilum* and the autotrophic sulfur oxidizer *At. thiooxidans*, involving the transfer of carbon as well as interacting iron/sulfur transformations (Bacelar-Nicolau and Johnson, 1999). Neither of these bacteria can grow on pyrite in pure culture, yet mixed cultures are highly effective at degrading the mineral. Unlike *Leptospirillum* spp. and *At. ferrooxidans*, *Fm. acidophilum* is unable to fix CO_2, but the low amounts of soluble organic carbon leaked by the autotroph are sufficient to support its growth (like many other acidophilic heterotrophs, *Fm. acidophilum* is an adept scavenger). This is the key to the success of the partnership—*Fm. acidophilum* accelerates ferrous iron oxidation, ferric iron attacks pyrite generating sulfur and RISCs, which are oxidized by *At. thiooxidans* and provide the energy required for CO_2 fixation by this autotroph.

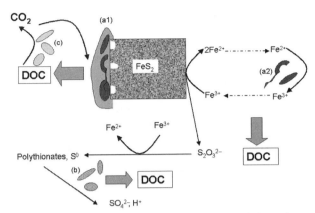

Figure 2. Oxidation of pyrite by consortia of acidophilic prokaryotes. The initial attack on the mineral is by ferric iron generated by iron-oxidizing prokaryotes that are either attached to the mineral within a biofilm [population (a1)] or free swimming [population (a2)]. Elemental sulfur and RISCs formed as intermediates of pyrite oxidation are oxidized to sulfuric acid by sulfur-oxidizing prokaryotes [population (b)]. A significant proportion of the carbon dioxide fixed by the chemoautotrophs is lost from cells as exudates and lysates. This dissolved organic carbon is metabolized by heterotrophic acidophiles [population (c)] and the carbon dioxide released may be reassimilated by the primary producers.

In a more general context, the relationship between chemolithotrophic and heterotrophic acidophiles in acidic mineral leaching is often mutualistic. Organic compounds released from iron and sulfur oxidizers as exudates, cell lysis products, and so on can accumulate and inhibit the growth of these bacteria, many of which (e.g., *Leptospirillum* spp.) are highly sensitive to organic materials in general and carboxylic acids in particular. Catabolism of these materials by heterotrophic and mixotrophic prokaryotes such as *Acidiphilium*, *Acidimicrobium*, and *Ferroplasma* spp. therefore has positive feedback on the autotrophs by relieving potential inhibition as well as generating CO_2 that can be fixed by the autotrophs (Fig. 2). In both environmental and industrial situations, heterotrophic (or mixotrophic) acidophiles coexist with chemolithotrophs (e.g., Goebel and Stackebrandt, 1994; Bond et al., 2000a). The significance of interactions involving transfer of organic and inorganic carbon and transformations of iron and sulfur, in terms of optimizing the rates of pyrite dissolution by moderately thermophilic microbial cultures, has been highlighted by Okibe and Johnson (2004).

Because of their relative abundance in many extremely acidic environments, iron and sulfur are extremely important primary sources of energy, which is reflected in the large diversity of bacteria and archaea that exploit either or both of these elements in their reduced forms, as described previously. Chemolithotrophic iron and/or sulfur oxidizers are the sole or major primary producers in many extremely acidic environments (a situation analogous to submarine hydrothermal vents), particularly in subterranean sites such as deep mines, though wooden props may contribute to the net carbon budget in these locations. In AMD streams and other surface environments, photosynthesis by acidophilic microalgae (e.g., *E. mutabilis*) can also be an important component of net primary production, though the presence and activity of eukaryotic life forms may be restricted in ecosystems where physicochemical parameters (e.g., temperatures, concentrations of soluble metals, as well as pH) are particularly extreme.

Macroscopic Growths and Streamer Communities

Acidophilic microbial communities can achieve massive proportions, most notably as streamer-like growths in underground abandoned mine chambers. These may take the form of thick biofilms on moist surfaces, slime-like growths in ponds, stalactite-like forms hanging from pit props and mine roofs (smaller growths of this type sometimes being referred to as "snotites"), or long gelatinous filaments in flowing mine waters ("acid streamers"; Color Plate 4). Although these growths were first described in the scientific literature over 70 years ago, it is only within the last decade, with the advent of biomolecular techniques, that the microbial communities that make up acid streamers and slimes have been elucidated. Using a combination of fluorescent in situ hybridization (FISH) and analysis of clone libraries of extracted 16S rRNA genes, Bond et al. (2000b) and Bond and Banfield (2001) examined slime biofilms and snotites that had developed on the exposed surface of a pyrite ore within the abandoned Richmond mine at Iron Mountain. The major microorganisms identified were *Leptospirillum* spp. (*L. ferriphilum* and smaller numbers of *L. ferrodiazotrophum*) and *Fp. acidarmanus*, *Sulfobacillus*, and *Acidimicrobium/Ferrimicrobium*-related species. Some δ-proteobacterial gene sequences (possibly sulfate-reducing bacteria) were also detected in clone libraries. In contrast, Lopez-Archilla et al. (2004) found that large dendritic filaments growing in the extremely acidic (pH 1.7 to 2.2) Rio Tinto, Spain, were composed primarily of γ- and α-*Proteobacteria* and smaller numbers of *Firmicutes* and β-*Proteobacteria*. Although most of the phylotypes identified in this study were related to previously characterized bacteria, no 16S rRNA gene sequence obtained was related to any known acidophile. In a third study, Hallberg et al. (2006) found that acid streamers growing in AMD draining abandoned copper mine (pH ~2.5) and in a chalybeate spring (pH ~2.8) in North Wales were very different from both the Iron Mountain and Rio Tinto growths and, in both cases, displayed very limited biodiversity. Using a combination of cultivation-dependent and cultivation-independent approaches (terminal restriction fragment length polymorphism, clone library analysis, and FISH), Hallberg et al. (2006) showed that the streamer communities at both sites were entirely bacterial and that >90% of the bacteria in the copper mine streamers were dominated by a single uncharacterized β-proteobacterium (previously identified in clone libraries obtained from some extremely acidic environments, including Iron Mountain), while the same prokaryote and another related to the iron-oxidizing neutrophile *Gallionella ferruginea* accounted for most of the bacteria in the chalybeate acid streamers. Using a modified plating technique, the uncharacterized β-proteobacterium was isolated in pure culture and shown to be a novel iron-oxidizing acidophile. These three studies have highlighted the fact that macroscopic streamer growths, although often displaying similar gross morphologies, are highly heterogeneous in their microbial compositions. The precise reasons for these differences have not been determined, but differences in physicochemical characteristics (such as temperature and dissolved solutes) are likely to be important factors.

REFERENCES

Alpers, S.-V., J. L. C. M. van de Vossenberg, A. J. M. Driessen, and W. N. Konings. 2001. Bioenergetics and solute uptake under extreme conditions. *Extremophiles* **5:**285–294.

Atkinson, T., S. Cairns, D. A. Cowan, M. J. Danson, D. W. Hough, D. B. Johnson, P. R. Norris, N. Raven, R. Robson, C. Robinson, and R. J. Sharp. 2000. A microbiological survey of Montserrat island hydrothermal biotopes. *Extremophiles* **4:**305–313.

Bacelar-Nicolau, P., and D. B. Johnson. 1999. Leaching of pyrite by acidophilic heterotrophic iron-oxidizing bacteria in pure and mixed cultures. *Appl. Environ. Microbiol.* **65:**585–590.

Baker, B. J., M. A. Lutz, S. C. Dawson, P. L. Bond, and J. F. Banfield. 2004. Metabolically active eukaryotic communities on extremely acidic mine drainage. *Appl. Environ. Microbiol.* **70:**6264–6271.

Battaglia-Brunet, F., C. Joulian, F. Garrido, M.-C. Dictor, D. Morin, K. Coupland, D. B. Johnson, K. B. Hallberg, and P. Baranger. 2006. Oxidation of arsenite by *Thiomonas* strains and characterization of *Thiomonas arsenivorans* sp. nov. *Antonie Leeuwenhoek* **89:**99–108.

Blake, R., and D. B. Johnson. 2000. Phylogenetic and biochemical diversity among acidophilic bacteria that respire on iron, p. 53–78. *In* D. R. Lovley (ed.), *Environmental Microbe–Metal Interactions*. ASM Press, Washington, DC.

Bond, P. L., G. K. Druschel, and J. F. Banfield. 2000a. Comparison of acid mine drainage communities in physically and geochemically distinct ecosystems. *Appl. Environ. Microbiol.* **66:**4962–4971.

Bond, P. L., S. P Smriga, and J. F. Banfield. 2000b. Phylogeny of microorganisms populating a thick, subaerial, predominantly lithotrophic biofilm at an extreme acid mine drainage site. *Appl. Environ. Microbiol.* **66:**3842–3849.

Bond, P. L., and J. F. Banfield. 2001. Design and performance of rRNA targeted oligonucleotide probes for in situ detection and phylogenetic identification of microorganisms inhabiting acid mine drainage environments. *Microb. Ecol.* **41:**149–161.

Bridge, T. A. M., and D. B. Johnson. 2000. Reductive dissolution of ferric iron minerals by *Acidiphilium* SJH. *Geomicrobiol. J.* **17:**193–206.

Clark, D. A., and P. R. Norris. 1996. *Acidimicrobium ferrooxidans* gen. nov., sp. nov.: mixed culture ferrous iron oxidation with *Sulfobacillus* species. *Microbiology* **141:**785–790.

Coram, N. J., and D. E. Rawlings. 2002. Molecular relationship between two groups of the genus *Leptospirillum* and the finding that *Leptospirillum ferriphilum* sp. nov. dominates South African commercial bioleaching tanks that operate at 40°C. *Appl. Environ. Microbiol.* **68:**838–845.

Coupland, K. 2005. A study of the geomicrobiology of acid mine drainage-impacted environments. Ph.D. thesis. University of Wales, United Kingdom.

Coupland, K., and D. B. Johnson. 2004 Geochemistry and microbiology of an impounded subterranean acidic water body at Mynydd Parys, Anglesey, Wales. *Geobiology* **2:**77–86.

Deneke, R. 2000. Review of rotifers and crustaceans in highly acidic environments of pH values ≤3. *Hydrobiologia* **433:**167–172.

Donahoe-Christiansen, J., S. D'Imperio, C. R. Jackson, W. P. Innskeep, and T. R. McDermott. 2004. Arsenite-oxidizing *Hydrogenobaculum* strain isolated from an acid-sulfate-chloride geothermal spring in Yellowstone National Park. *Appl. Environ. Microbiol.* **70:**1865–1868.

Dopson, M., C. Baker-Austin, A. Hind, J. P. Bowman, and P. L. Bond. 2004. Characterization of *Ferroplasma* isolates and *Ferroplasma acidarmanus* sp. nov., extreme acidophiles from acid mine drainage and industrial bioleaching environments. *Appl. Environ. Microbiol.* **70:**2079–2088.

Edwards, K. J., P. L. Bond, T. M. Gihring, and J. F. Banfield. 2000. An archaeal iron-oxidising extreme acidophile important in acid mine drainage. *Science* **287:**1796–1799.

Gemmell, R. T., and C. J. Knowles. 2000. Utilisation of aliphatic compounds by acidophilic heterotrophic bacteria: the potential for bioremediation of acidic wastewaters contaminated with toxic organic compounds and heavy metals. *FEMS Microbiol. Lett.* **192:**185–190.

Ghosh, S., N. R. Mahapatra, and P. C. Banerjee. 1997. Metal tolerance in *Acidocella* strains and plasmid-mediated transfer of this characteristic to *Acidiphilium multivorum* and *Escherichia coli*. *Appl. Environ. Microbiol.* **63:**4523–4527.

Goebel, B. M., and E. Stackebrandt. 1994. Cultural and phylogenetic analysis of mixed microbial populations found in natural and commercial bioleaching environments. *Appl. Environ. Microbiol.* **60:**1614–1621.

Golovacheva, R. S., O. V. Golyshina, G. I. Karavaiko, A. G. Dorofeev, T. A. Pivovarova, and N. A. Chernykh. 1992. A new iron-oxidizing bacterium, *Leptospirillum thermoferrooxidans*, sp. nov. *Microbiology* **61:**1056–1065 (English translation of *Mikrobiologiya*).

Golyshina, O. V., and K. N. Timmis. 2005. *Ferroplasma* and relatives, recently discovered cell wall-lacking archaea making a living in extremely acid, heavy metal-rich environments. *Environ. Microbiol.* **7:**1277–1288.

Golyshina, O. V., T. A. Pivovarova, G. I. Karavaiko, T. F., Kondrat'eva, E. R. B., Moore, W. R. Abraham, H. Lunsdorf, K. N. Timmis, M. M. Yakimov, and P. N. Golyshin. 2000. *Ferroplasma acidiphilum* gen. nov., sp. nov., an acidophilic, autotrophic, ferrous-iron-oxidizing, cell-wall-lacking, mesophilic member of the *Ferroplasmaceae* fam. nov., comprising a distinct lineage of the Archaea. *Int. J. Syst. Evol. Microbiol.* **50:**997–1006.

Gross, S., and E. I. Robbins. 2000. Acidophilic and acid-tolerant fungi and yeasts. *Hydrobiologia* **433:**91–109.

Gross, W. 2000. Ecophysiology of algae living in highly acidic environments. *Hydrobiologia* **433:**31–37.

Hallberg, K. B., D. B. Johnson, and P. A. Williams. 1999. A novel metabolic phenotype among acidophilic bacteria: aromatic degradation and the potential use of these microorganisms for the treatment of wastewater containing organic and inorganic pollutants, p. 719–728. *In* R. Amils, and A. Ballester (ed.), *Biohydrometallurgy and the Environment Toward the Mining of the 21st Century, Process Metallurgy 9A*. Elsevier, Amsterdam, The Netherlands.

Hallberg, K. B., K. Coupland, S. Kimura, and D. B. Johnson. 2006. Macroscopic streamer growths in acidic, metal-rich mine waters in North Wales consist of novel and remarkably simple bacterial communities. *Appl. Environ. Microbiol.* **72:**2022–2030.

Hawkes, R. B., P. D. Franzmann, and J. J. Plumb. 2005. Moderate thermophiles including "*Ferroplasma cyprexacervatum*" sp. nov., dominate an industrial scale chalcocite heap bioleaching operation, p. 657–666. *In* S. T. L. Harrison, D. E. Rawlings, and J. Petersen (ed.), *Proceedings of the 16th International Biohydrometallurgy Symposium*. University of Cape Town, South Africa.

He, Z.-G., H. Zhong, and Y. Li. 2004. *Acidianus tengchongensis* sp. nov., a new species of acidothermophilic archaeon isolated from an acidothermal spring. *Curr. Microbiol.* **48:**159–163.

Johnson, D. B. 1998. Biodiversity and ecology of acidophilic microorganisms. *FEMS Microbiol. Ecol.* **27:**307–317.

Johnson, D. B. 2001. Importance of microbial ecology in the development of new mineral technologies. *Hydrometallurgy* **59:**147–158.

Johnson, D. B., and L. Rang. 1993. Effects of acidophilic protozoa on populations of metal-mobilising bacteria during the leaching of pyritic coal. *J. Gen. Microbiol.* **139:**1417–1423.

Johnson, D. B., and T. A. M. Bridge. 2002. Reduction of ferric iron by acidophilic heterotrophic bacteria: evidence for constitutive and inducible enzyme systems in *Acidiphilium* spp. *J. Appl. Microbiol.* **92:**315–321.

Johnson D. B., P. Bacelar-Nicolau, N. Okibe, A. Yahya, and K. B. Hallberg. 2001. Role of pure and mixed cultures of Gram-positive eubacteria in mineral leaching, p. 461–470. *In* V. S. T. Ciminelli, and O. Garcia, Jr. (ed.), *Biohydrometallurgy: Fundamentals, Technology and Sustainable Development, Process Metallurgy 11A*. Elsevier, Amsterdam, The Netherlands.

Johnson, D. B., N. Okibe, and F. F. Roberto. 2003. Novel thermo-acidophiles isolated from geothermal sites in Yellowstone National Park: physiological and phylogenetic characteristics. *Arch. Microbiol.* **180:**60–68.

Johnson, D. B., B. Stallwood, S. Kimura, and K. B. Hallberg. 2006. Isolation and characterization of *Acidicaldus organovorus*, gen. nov., sp. nov.; a novel sulfur-oxidizing, ferric iron-reducing thermo-acidophilic heterotrophic *proteobacterium*. *Arch. Microbiol.* **185:**212–221.

Karavaiko, G. I., T. P Tourova, T. F. Kondrat'eva, A. M. Lysenko, T. V. Kolganova, S. N. Ageeva, L. N. Muntyan, and T. A. Pivovarova. 2003. Phylogenetic heterogeneity of the species *Acidithiobacillus ferrooxidans*. *Int. J. Syst. Evol. Microbiol.* **53:**113–119.

Karavaiko, G. I., T. I. Bogdanova, T. P. Tourova, T. F. Kondrat'eva, I. A. Tsaplina, M. A. Egorova, E. N. Krasnol'nikova, and L. M. Zakharchuk. 2005. Reclassification of "*Sulfobacillus thermosulfidooxidans* subsp. *thermotolerans*" strain K1 as *Alicyclobacillus tolerans* sp. nov. and *Sulfobacillus disulfidooxidans* Dufresne et al 1996 as *Alicyclobacillus disulfidooxidans* comb. nov., and amended description of the genus *Alicyclobacillus*. *Int. J. Syst. Evol. Microbiol.* **55:**941–947.

Kelly, D. P. 1978. Bioenergetics of chemolithotrophic bacteria, p. 363–386. *In* A. T. Bull, and P. M. Meadow (ed.), *Companion to Microbiology*. Longman, London, United Kingdom.

Kelly, D. P., and A. P. Wood. 2000. Reclassification of some species of *Thiobacillus* to the newly designated genera *Acidithiobacillus* gen. nov., *Halothiobacillus* gen. nov., and *Thermothiobacillus* gen. nov. *Int. J. Syst. Evol. Microbiol.* **50:**511–516.

Kimura, S., K. B. Hallberg, and D. B. Johnson. 2006. Sulfidogenesis in low pH (3.8–4.2) media by a mixed population of acidophilic bacteria. *Biodegradation* **17:**57–65.

Kishimoto, N., F. Fukaya, K. Inagaki, T. Sugio, H, Tanaka, and T. Tano. 1995. Distribution of bacteriochlorophyll *a* among aerobic and acidophilic bacteria and light-enhanced CO_2-incorporation in *Acidiphilium rubrum*. *FEMS Microbiol. Ecol.* **16:**291–296.

Lane, D. J., A. P. Harrison, Jr., D. Stahl, B. Pace, S. J. Giovannoni, G. J. Olsen, and N. R. Pace. 1992. Evolutionary relationships among sulfur- and iron-oxidizing eubacteria. *J. Bacteriol.* **174:**269–278.

Lopez-Archilla, A. I., E. Gerard, D. Moreira, and P. Lopez-Garcia. 2004. Macrofilamentous microbial communities in the metal-rich and acidic River Tinto, Spain. *FEMS Microbiol. Lett.* **235:**221–228.

Matin, A. 1999. pH homeostasis in acidophiles, p. 152–166. *In Bacterial Responses to pH*. Novartis Foundation Symposium 221. Wiley, Chichester, United Kingdom.

Melamud, V. S., T. A. Pivovarova, T. P. Tourova, T. V. Kolganova, G. A. Osipov, A. M. Lysenko, T. F. Kondrat'eva, and G. Karavaiko. 2003. *Sulfobacillus sibiricus* sp. nov., a new moderately thermophilic bacterium. *Microbiology* **72:**605–612 (English translation of *Mikrobiologiya*).

Nordstrom, D. K., C. N. Alpers, C. J. Ptacek, and D. W. Blowes. 2000. Negative pH and extremely acidic minewaters from Iron Mountain, California. *Environ. Sci. Technol.* **34:**254–258.

Norris, P. R., and W. J. Ingledew. 1992. Acidophilic bacteria: adaptations and applications, p. 115–142. *In* R. A. Herbert, and R. J. Sharp (ed.), *Molecular Biology and Biotechnology of Extremophiles*. Blackie, Glasgow, United Kingdom.

Norris, P. R., D. A. Clark, J. P. Owen, and S. Waterhouse. 1996. Characteristics of *Sulfobacillus acidophilus* sp. nov. and other moderately thermophilic mineral sulphide-oxidizing bacteria. *Microbiology* **142:**775–783.

Norris, P. R., N. P. Burton, and N. A. M. Foulis. 2000. Acidophiles in bioreactor mineral processing. *Extremophiles* **4:**71–76.

Okibe, N., and D. B. Johnson. 2004. Biooxidation of pyrite by defined mixed cultures of moderately thermophilic acidophiles in pH-controlled bioreactors: the significance of microbial interactions. *Biotechnol. Bioeng.* **87:**574–583.

Okibe, N., M. Gericke, K. B. Hallberg, and D. B. Johnson. 2003. Enumeration and characterization of acidophilic microorganisms isolated from a pilot plant stirred tank bioleaching operation. *Appl. Environ. Microbiol.* **69:**1936–1943.

Packroff. G., and S. Woelfl. 2000. A review on the occurrence and taxonomy of heterotrophic protists in extreme acidic environments of pH values ≤3. *Hydrobiologia* **433:**153–156.

Pronk, J. T., K. Liem, P. Bos, and J. G. Kuenen. 1991. Energy transduction by anaerobic ferric iron reduction in *Thiobacillus ferrooxidans*. *Appl. Environ. Microbiol.* **57:**2063–2068.

Rohwerder, T., T. Gehrke, K. Kinzler, and W. Sand. 2003. Bioleaching review part A: progress in bioleaching: fundamentals and mechanisms of bacterial metal sufide oxidation. *Appl. Microbiol. Biotechnol.* **63:**239–248.

Schleper, C., G. Puehler, B. Kuhlmorgen, and W. Zillig. 1995. Life at extremely low pH. *Nature* **375:**741–742.

Segerer, A. H., T. A. Langworthy, and K. O. Stetter. 1988. *Thermoplasma acidophilum* and *Thermoplasma volcanium* sp. nov. from solfatara fields. *Syst. Appl. Microbiol.* **10:**161–171.

Stumm, W., and J. J. Morgan. 1981. Aquatic chemistry: an introduction emphasizing chemical equilibria in natural waters. Wiley, New York, NY.

Tyson, G. W., J. Chapman, P. Hugenholtz, E. E. Allen, R. J. Ram, P. M. Richardson, V. V. Solovyev, E. M. Rubin, D. S. Rokhsar, and J. F. Banfield. 2004. Community structure and metabolism through reconstruction of microbial genomes from the environment. *Nature* **428:**37–43.

Tyson, G. W., I. Lo, B. J. Baker, E. E. Allen, E. E., P. Hugenholtz, and J. F. Banfield. 2005. Genome-directed isolation of the key nitrogen-fixer *Leptospirillum ferrodiazotrophum* sp. nov. from an acidophilic microbial community. *Appl. Environ. Microbiol.* **71:**6319–6324.

Yahya, A., and D. B. Johnson. 2002. Bioleaching of pyrite at low pH and low redox potentials by novel mesophilic Gram-positive bacteria. *Hydrometallurgy* **63:**181–188.

Yahya, A., F. F. Roberto, and D. B. Johnson. 1999. Novel mineral-oxidising bacteria from Montserrat (W.I.): physiological and phylogenetic characteristics, p. 729–740. *In* R. Amils, and A. Ballester (ed.), *Biohydrometallurgy and the Environment Toward the Mining of the 21st Century, Process Metallurgy 9A*. Elsevier, Amsterdam, The Netherlands.

Chapter 21

Acidophiles: Mechanisms To Tolerate Metal and Acid Toxicity

SYLVIA FRANKE AND CHRISTOPHER RENSING

INTRODUCTION

There are many theories on how life evolved (Nisbet and Sleep, 2001). Most view an origin of microbial life in hot and chemotrophic conditions probable (Woese, 1987), but this has also been challenged (Galtier et al., 1999), and others proposed a cold (Bada and Lazcano, 2002) or alkaliphilic beginning (Russell and Hall, 1997). Nevertheless, it is possible that life arose in conditions that were hot, acidic, and relatively rich in metals (Nisbet and Sleep, 2001; Di Giulio, 2005). Therefore, early organisms had to overcome these conditions, perhaps utilize them. Acidophiles are microorganisms belonging to eubacteria, archaea, and eukaryotes that need to grow in environments of low pH (<3). Other microorganisms are highly tolerant to extremely acidic conditions but can also grow at neutral pH. As early as 1943, two extremely acid- and copper-tolerant fungi were isolated by Starkley and Waksman (1943). More recently, fungi and protists, able to tolerate pH values between 0.8 and 2 and high metal concentrations, could be isolated from Iron Mountain in California and the Tinto River in Spain (Amaral-Zettler et al., 2002, 2003; Baker et al., 2004). Some acidophiles, such as *Ferroplasma acidarmanus*, are extremely adapted to acidic conditions and can even grow around pH 0 (Futterer et al., 2004; Macalady et al., 2004). In addition to their acid tolerance, many acidophiles are extremely resistant to metals and metalloids (Dopson et al., 2003) (Table 1). The reason for this ability to tolerate high concentrations of metals is currently not known, although acidic mine waters often contain elevated concentrations of metals such as copper and zinc and metalloids such as arsenic (Johnson and Hallberg, 2003). There are a few reports mostly in *Acidithiobacillus ferrooxidans* (formerly *Thiobacillus ferrooxidans*) describing copper, mercury, and arsenic resistance in acidophiles (Das et al., 1997; Leduc et al., 1997; Butcher et al., 2000; Butcher and Rawlings, 2002; Barreto et al., 2003; Sugio et al., 2003; Takeuchi et al., 2005). However, these metal/metalloid resistance determinants are not significantly different from similar determinants in neutrophiles (Mukhopadhyay et al., 2002; Barreto et al., 2003; Mergeay et al., 2003; Rensing and Grass, 2003). It has become clear that there can be no sole cause for this high metal resistances, but it is likely that they have arisen from an accumulation of evolutionary changes leading to the current phenotypes. Therefore, we intend to explore different possibilities in this chapter. Consider an experimental population, which could be either a single species or a mixture of strains, grown under increasingly acidic conditions over several generations. Over time, this will lead to adaptation and increased fitness for growth under acidic conditions. The basis for this adaptation is variation in the genotype that can (i) arise by mutation or expression of preexisting genetic potential or (ii) be due to horizontal gene transfer. This in turn will lead to competition among strains with different genotypes for available resources and will lead to a change in the genetic makeup of a population (Travisano and Rainey, 2000; Kassen and Rainey, 2004).

Ecological success in any given environment is termed fitness, a measure of the ability of one genotype to reproduce in comparison with another. The underlying change in fitness resulting from the current genetic makeup of any organism in an acidophilic environment is therefore dependent not only on the physical and chemical environment but also on the interaction and competition among strains with different genotypes. This includes both inter- and intraspecies competition. For example, if a single culture is repeatedly grown at more and more acidic conditions, mutations arise spontaneously and affect fitness randomly, leading to diversification. The beneficial mutations and some of these strains with more fit genotypes will dominate the population. The increase in fitness is

---

S. Franke • Department of Biology, Skidmore College, 815 North Broadway, Saratoga Springs, NY 12866. C. Rensing • Department of Soil, Water, and Environmental Science, University of Arizona, Tucson, AZ 85718.

Table 1. Upper-level concentrations of some metals in a variety of acidophiles where metabolic activity has been recorded (from Dopson et al., 2003)

Organism	Metal concentration whereby metabolic activity occurs (mM)				
	As(III)	Cu(II)	Zn(II)	Cd(II)	Ni(II)
Neutrophilic bacterium					
E. coli	4	1	1	0.5	1
Acidophilic bacteria					
Acidiphilium cryptum	ND[a]	15	125	700	20
A. multivorum	30	10	40	20	350
Acidiphilium symbioticum KM2	ND	20	150	700	20
A. symbioticum H8	ND	15	150	700	30
Acidiphilium angustum	ND	5	8	<0.2	10
Acidiphilium strain GS18h	ND	15	60	10	30
Acidocella aminolytica	ND	30	500	200	150
Acidocella facilis	ND	<1	100	<1	1
Acidocella strain GS19h	ND	15	900	700	150
A. ferrooxidans	84	800	1071	500	1000
Sulfobacillus thermosulfidooxidans	ND	6	43	ND	5
Acidophilic archaea					
F. acidarmanus	13	16	ND	9	ND
Metallosphaera sedula	1	16	150	1	ND
S. acidocaldarius	ND	1	10	10	1
S. solfataricus	ND	1	10	10	0.1

[a] ND, not determined.

significant at the beginning of the experiment and then slows down until an evolutionary equilibrium is reached. This is because an increase in fitness is often accompanied by a loss of capacity in respect of other environments. For example, acid-adapted organisms might no longer be able to grow under the neutral conditions that were satisfactory for growth of the original culture.

Most important, this excursion into a simple experiment is meant to acknowledge the continuous nature of the evolutionary process. Therefore, the resulting genotypes offer the chance that the selected strains, which have arisen in a laboratory environment, may be very different from the genotypes in a possible acidic, metal-rich early Earth. Similar observations have been made for the forced evolution of antibiotic-resistant organisms in the laboratory compared with the natural environment; different solutions pertain and in some instance the selection of different amino acid changes in the same protein are seen in the two contrasting environments. Secondly, it proposes that acidophily is dependent on a variety of factors that can be reinvented numerous times, leading to convergent evolution.

FEATURES OF ACIDOPHILES AIDING PROTECTION FROM METAL TOXICITY

Hot, acidic, and metal-rich environments are often dominated by archaeal members of the *Thermoplasmales* family. Members of this family include *F. acidarmanus*, *Picrophilus torridus*, and *Thermoplasma acidophilum* (Golyshina and Timmis, 2005). All members possess tetraether-linked monolayer membrane lipids as their main lipid component (Macalady et al., 2004). In addition, *Crenarchaeota*, such as *Sulfolobus solfataricus*, also contain tetraether monolayers (Hopmans et al., 2000). Because a consistent relationship between membrane core lipid composition and pH optimum could be established, a selective advantage of this membrane structure was proposed (Macalady et al., 2004). The ether linkage is less sensitive to acid hydrolysis in comparison with the ester linkage of eubacteria. Archaeal membranes would also be less permeable to protons and metals because of their isoprenoid core. In eubacteria, these kinds of membranes are not present. However, novel lipid membranes are not an essential requirement for acid tolerance because many gram-negative acidophiles (e.g., *Acidiphilium acidophilum*, formerly *Thiobacillus acidophilum*) have conventional lipid membranes. It may be the combination of high temperature and acidity that has combined to select for novel membrane structures.

Another factor that could aid acidophiles in combating high external concentrations of metals is their inside-positive membrane potential. The capacity to invert $\Delta\psi$ is vital for acidophiles in the generation and maintenance of a large pH gradient. In essence, increased net influx of K^+ will generate inside-positive membrane potential and stabilize the proton motif force (PMF) at values typical for neutrophiles. The PMF consists of two components: the membrane potential ψ and the proton gradient ΔpH, expressed in mV at 37°C as PMF = $\Delta\psi$ – 61.5 ΔpH. It is obvious that a positive membrane potential will decrease the PMF. Another way to look at the membrane potential is:

$$\Delta\mu_H^+ = 2.3RT(pH_o - pH_i) + nF\Delta V$$

Electrogenic H$^+$ pumps such as quinines extrude protons into the medium or periplasm, thus creating a high membrane potential that is negative inside. The pH difference would be minor so the main component of the PMF would be the membrane potential. The uptake of K$^+$ would then lead to a conversion of $\Delta\psi$ into ΔpH.

However, what is true for protons is also true for cations such as Cd(II), Zn(II), and Cu(I). Bacteria have evolved to maintain very low concentrations of free metal cations in the cytoplasm (Changela et al., 2003). With intracellular concentrations in the nanometer-to-micrometer range, the gradient across the cytoplasmic membrane can be huge (at an external concentration of 1 mM copper or zinc and assuming 10^{-15} M free zinc or 10^{-21} M for free copper). When the membrane potential is oriented normally, as in a neutrophilic bacterium (i.e., negative inside), divalent cations are drawn into the cytoplasm at rates proportional to the magnitude of $\Delta\varphi$ and the passive permeability of the membrane. An inside-positive membrane potential would provide resistance to cation accumulation and decrease the energy needed to maintain a large gradient significantly, as the cations would be extruded against their concentration gradient but along their electrogenic gradient.

RESISTANCE TO ARSENIC IN ACIDOPHILES

As stated earlier, early life could have arisen under acidic, metal-rich conditions. This has led to the proposal that acidophiles were dominant in this early period. However, it is wrong to think that present-day acidophiles are remnants of that period because this would not take 3.5 to 3.8 billion years of evolution into account. It is perhaps easier, as has been attempted, to reconstruct development of metabolic pathways or metal and metalloid resistance mechanism. Arsenicals, and in particular arsenite, must have been a companion of life from the very beginning. Volcanic activities contributed to a constant supply of arsenic that was mainly in the form of arsenite under the prevailing reducing conditions. The advent of oxygen then shifted the redox balance toward arsenate. Cyanobacteria are thought to be responsible for the beginning of dioxygen production about 3 × 10^9 years ago. The appearance of a significant O$_2$ concentration in the atmosphere required another 200 to 300 million years because the produced oxygen was initially consumed by the oxidation of ferrous iron in oceans. This event irreversibly changed life on Earth and forced organisms to increasingly deal with arsenate, an analog of phosphate. It was therefore speculated that during this period of increased oxygen content in the atmosphere, genes encoding the intracellular arsenate reductase *arsC* were added to the ancestral *arsRB* operon (Rosen, 1999). ArsB is an arsenite and antimonite/proton antiporter that pumps these oxyanions out of cells using the PMF. In acidophiles, this protein would not have a problem pumping arsenite out because it solely uses the proton gradient and not the membrane potential. ArsR is a regulatory protein repressing the transcription of the *ars* operon in the absence of arsenite (or antimonite). Binding of arsenite leads to a conformational change of ArsR and decreased affinity to the *ars* promoter, so transcription of the *ars* operon is no longer repressed. In the acidophilic archaeon *Ferroplasma acidarmus*, the only known arsenic resistance gene present on the genome is *arsRB* (Gihring et al., 2003). In contrast, *A. ferrooxidans* contains four *ars* genes, *arsRCBH*, with an unusual operon arrangement (Butcher et al., 2000). ArsH has been recently shown in *Sinorhizobium meliloti* to confer resistance to arsenite but not arsenate (Yang et al., 2005), but the mechanism is not known (Fig. 1).

Acidophiles are not intrinsically arsenic or metal resistant but often achieve high levels of resistance through plasmids and/or transposons. Resistance against arsenic in *Leptospirillum ferriphilum* is clearly correlated with the number of *ars* operons on the chromosome. In a recently isolated strain of *L. ferriphilum*, high level of arsenic resistance could be linked to two copies of a transposon: Tn*LfArs*, with a size of 8,772 bp (Tuffin et al., 2006). Transposon Tn*LfArs* is related to a transposon, Tn*AtcArs*, identified in *Acidithiobacillus caldus* (Tuffin et al., 2005). Both acidophiles were isolated from an arsenopyrite biooxidation tank. Both transposons contain an arsenic resistance determinate composed of *arsRCDA* followed by an open reading frame (ORF) encoding a CBS domain-like protein and *arsB*. However, Tn*AtcArs* contains two copies of *arsD* and *arsA* as well as an ORF encoding a NADH oxidase-like protein. This is interesting because in *Halobacterium* sp. NRC-1 only *arsRCDA* but no gene encoding a homolog of ArsB were detected on the megaplasmid conferring arsenic resistance (Wang et al., 2004). It has been previously suggested that *arsRC* could have arisen separately (Rosen, 1999). In an As(V)-rich environment, intracellular reduction of

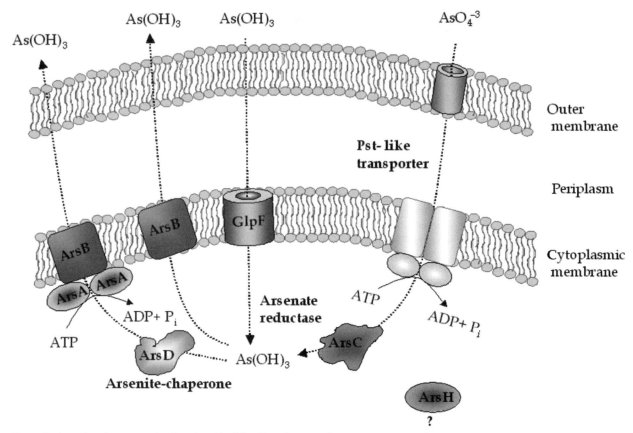

Figure 1. Arsenic resistance mechanisms in acidophiles. *F. acidarmus* only contains *arsR* encoding an As(III)-responsive repressor ArsR and *arsB*, encoding the putative As(III) efflux pump ArsB. The *ars* operon of *A. ferrooxidans* contains two additional genes, *arsC* and *arsH*, encoding the arsenate reductase and possibly an arsenite oxidase, respectively. The *ars* operon from *A. caldus* also encodes the ArsA, As(III)-, and Sb(III)-ATPase and the putative As(III)- and Sb(III)-chaperone ArsD.

As(V) to As(III) by ArsC would create a gradient that would favor extrusion. This efflux would not need an energy-dependent transport, but a mere channel would be sufficient, such as an aquaglyceroporin (Yang et al., 2005). ArsD always accompanies the As(III) and Sb(III) ATPase ArsA and appears to be a novel chaperone. ATP-dependent transport has only been shown by the complex ArsAB, but not by ArsA when forming a complex with another transporter such as Acr3 (B. P. Rosen, personal communication).

In *Acidiphilium*, plasmids were shown to be responsible for the extremely high resistance to cadmium, nickel, zinc, and arsenic (Ghosh et al., 1997; Mahapatra et al., 2002). A transfer of fragments of these plasmids was able to confer metal resistance in *Escherichia coli*. However, subsequent sequence analysis failed to show any conclusive determinant responsible for this slight resistance. *Acidiphilium multivorum* AIU301 was also shown to contain *arsADBCR* on one of its plasmids (Suzuki et al., 1997). This *ars* operon is closely related to similar operons in eubacteria, indicating lateral gene transfer.

COPPER RESISTANCE

Other metals such as copper and iron have also changed speciation and availability over the course of Earth's history because of oxygenation of the atmosphere. Biological systems did not utilize copper before the advent of atmospheric oxygen. In the prevailing reducing conditions before this event, copper was in the water-insoluble Cu(I) state, in the form of highly insoluble sulfides, and was not available for life. Although enzymes involved in anaerobic metabolism needed to work in the lower portion of the redox spectrum, the presence of dioxygen created the need for a redox active metal with $E_0 M^{n+1}/M^n$ from 0 to 0.8 V. The redox potential of copper [Cu(II)/Cu(I)] in enzymes is usually higher than that of iron [Fe(III)/Fe(II)] containing enzymes. Most copper enzymes work between 0.25 and 0.75 V. This high potential can be utilized for direct oxidation of easily oxidized substrates such as superoxide, ascorbate, catechol, or phenolates. This is one of the reasons why copper is extremely toxic, and all organisms exposed to copper

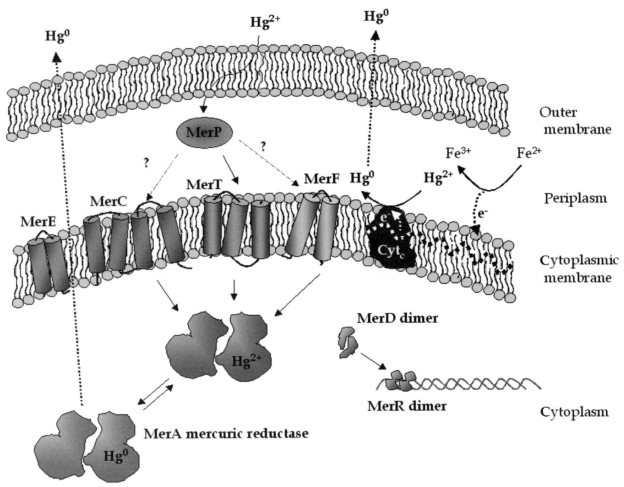

Figure 2. Mercury resistance mechanisms in acidophiles. Mercury resistance in acidophiles was first reported in *A. ferrooxidans* containing *merRECA*. MerR is a mercury-dependent activator, MerC a transporter involved in Hg(II) uptake, and MerA the mercuric reductase. MerE could be an alternative Hg(II) uptake system. Other strains contain *mer* determinants with *merPT* encoding a putative Hg(II) uptake system. MerP is a periplasmic Hg(II)-binding protein, and MerT is thought to accept Hg(II) from MerP and deliver it to MerA. In some strains, iron-dependent reduction of mercury was shown to be dependent on mercury-resistant cytochrome *c* oxidase.

have acquired or developed resistance mechanisms. *Acidithiobacillus* is highly copper resistant and contains a putative Cu(I)-translocating P-type ATPase (Barreto et al., 2003). However, homologs of this Cu ATPase are also found in almost every organism and are necessary for copper homoeostasis and resistance; therefore, *A. ferrooxidans* must contain additional genetic determinants to account for this high resistance. One possibility is that inorganic poly phosphates bind copper and this complex is subsequently trans ported out of cells. Alvarez and Jerez (2004) showed that copper stimulates polyphosphate degradation and phosphate efflux. Similar results were also reported for the thermoacidophilic archaeon *Sulfolobus metallicus* (Remonsellez et al., 2006). However, whether this mechanism confers any additional copper resistance remains to be determined.

MERCURY RESISTANCE

Mercury resistance in acidophiles has been studied extensively in *A. ferrooxidans*. It was shown to contain an operon *merRECA* (Shiratori et al., 1989; Inoue et al., 1991). MerR is an Hg(II)-dependent activator (Lund et al., 1986), and MerA is the mercuric reductase and the key enzyme of this resistance mechanism (Barkay et al., 2003). Interestingly, this *mer* determinant contains MerC as the transporter responsible for Hg(II) uptake and delivery to MerA and not MerT (Fig. 2). In other *mer* operons such as in Tn21, MerT appears to be the transporter responsible for the uptake of Hg(II) and the function of MerC was not clear (Osborn et al., 1997). Preliminary results indicate that MerT and MerF are better Hg(II) uptake systems under neutrophilic conditions. Both transporters contain a periplasmic Cys–Cys motif thought to be responsible for conferring metal specificity (DeSilva et al., 2002). In contrast, MerC contains a CXXC motif and is not a good Hg(II) transporter under neutrophilic conditions. Perhaps MerC is a much better transporter under acidic condition. However, the *mer* operon in *A. ferrooxidans* also encodes another putative mercury transporter, MerE, designated Urf-1 (#BAA14137). In addition, some strains of *A. ferrooxidans* contain a *mer* operon on transposon Tn*5037* encoding MerP and MerT for Hg(II) uptake (Kalyaeva et al., 2001).

In various strains of *A. ferrooxidans*, an additional mercury volatilization system that is dependent on iron oxidation can be found (Iwahori et al., 2000; Sugio et al., 2003). A mercury-resistant cytochrome *c* oxidase was shown to catalyze the iron-dependent reduction of Hg(II). Interestingly, successive cultivation of strain *A. ferrooxidans* SUG-2 led to an adapted highly mercury-resistant strain *A. ferrooxidans* MON-1 (Takeuchi et al., 2005). This strain showed an enhanced rate of iron-dependent mercury volatilization activity but no increase in NADPH-dependent MerA activity (Sugio et al., 2003). This is a clear example of adaptation under mercury stress.

Another example of a mercury-resistant acidophilic eubacterium is *Alicyclobacillus vulcanalis*, which was isolated from Coso Hot Springs in California (Simbahan et al., 2004). It was found that *A. vulcanalis* contains a mercury reductase, MerA, which is closely related to MerA from bacilli and streptococci. However, there are significant differences in the sequence of MerA from *A. vulcanalis*, which might be important for the thermostability of the protein (Simbahan et al., 2005). Which other proteins are involved in mercury resistance of *A. vulcanalis* is not yet known.

Also isolated from the mercury-rich Coso Hot Springs in California is the hyperthermophilic archaeon *S. solfataricus* (Simbahan et al., 2005). Like in eubacteria, resistance to mercury depends on the presence of the mercury reductase MerA. Analysis of mercury resistance revealed a gene transcribed in opposite direction of *merA*, encoding the transcription regulator MerR (Schelert et al., 2004). The same genetic organization can be found in *Sulfolobus acidocaldarius* (Chen et al., 2005). No genes encoding the known eubacterial mercury uptake systems can be found close to *merR* and *merA*. So it remains to be discovered which other proteins might be involved in archaeal mercury resistance.

CONCLUSION

Acidophiles had to develop highly efficient mechanisms to protect them from protons and metals and metalloids. At this stage, there is no unifying feature explaining the observed high resistance to metals/metalloids. The mechanisms characterized to date are very similar to those found in organisms that grow at neutral pH, and their genes are often found to be located on either plasmids or transposons that would facilitate their spread by interspecies gene transfer. However, with rapid advancements in different subfields of microbiology such as environmental genomics, microbial biochemistry, and novel applications spectroscopic methods such as Fourier transform infrared spectroscopy, significant progress in elucidating novel mechanisms of resistance can be expected in the coming years.

REFERENCES

Alvarez, S., and C. A. Jerez. 2004. Copper ions stimulate polyphosphate degradation and phosphate efflux in *Acidithiobacillus ferrooxidans*. *Appl. Environ. Microbiol.* 70:5177–5182.

Amaral-Zettler, L. A., F. Gomez, E. Zettler, B. G. Keenan, R. Amils, and M. L. Sogin. 2002. Microbiology: eukaryotic diversity in Spain's River of Fire. *Nature* 417:137.

Amaral-Zettler, L. A., M. A. Messerli, A. D. Laatsch, P. J. Smith, and M. L. Sogin. 2003. From genes to genomes: beyond biodiversity in Spain's Rio Tinto. *Biol. Bull.* 204:205–209.

Bada, J. L., and A. Lazcano. 2002. Origin of life: some like it hot, but not the first biomolecules. *Science* 296:1982–1983.

Baker, B. J., M. A. Lutz, S. C. Dawson, P. L. Bond, and J. F. Banfield. 2004. Metabolically active eukaryotic communities in extremely acidic mine drainage. *Appl. Environ. Microbiol.* 70:6264–6271.

Barkay, T., S. M. Miller, and A. O. Summers. 2003. Bacterial mercury resistance from atoms to ecosystems. *FEMS. Microbiol. Rev.* 27:355–384.

Barreto, M., R. Quatrini, S. Bueno, C. Arriagada, J. Valdes, S. Silver, E. Jedlicki, and D. S. Holmes. 2003. Aspects of the predicted physiology of *Acidithiobacillus ferrooxidans* deduced from an analysis of its partial genome sequence. *Hydrometallurgy* 71:97–105.

Butcher, B. G., S. M. Deane, and D. E. Rawlings. 2000. The chromosomal arsenic resistance genes of *Thiobacillus ferrooxidans* have an unusual arrangement and confer increased arsenic and antimony resistance to *Escherichia coli*. *Appl. Environ. Microbiol.* 66:1826–1833.

Butcher, B. G., and D. E. Rawlings. 2002. The divergent chromosomal *ars* operon of *Acidithiobacillus ferrooxidans* is regulated by an atypical ArsR protein. *Microbiology* 148:3983–3992.

Changela, A., K. Chen, Y. Xue, J. Holschen, C. E. Outten, T. V. O'Halloran, and A. Mondragon. 2003. Molecular basis of metal-ion selectivity and zeptomolar sensitivity by CueR. *Science* 301:1383–1387.

Chen, L., K. Brugger, M. Skovgaard, P. Redder, Q. She, E. Torarinsson, B. Greve, M. Awayez, A. Zibat, H. P. Klenk, and R. A. Garrett. 2005. The genome of *Sulfolobus acidocaldarius*, a model organism of the Crenarchaeota. *J. Bacteriol.* 187:4992–4999.

Das, A., J. M. Modak, and K. A. Natarajan. 1997. Studies on multi-metal ion tolerance of *Thiobacillus ferrooxidans*. *Miner. Eng.* 10:743–749.

DeSilva, T. M., G. Veglia, F. Porcelli, A. M. Prantner, and S. J. Opella. 2002. Selectivity in heavy metal-binding to peptides and proteins. *Biopolymers* 64:189–197.

Di Giulio, M. 2005. Structuring of the genetic code took place at acidic pH. *J. Theor. Biol.* 237:219–226.

Dopson, M., C. Baker-Austin, P. R. Koppineedi, and P. L. Bond. 2003. Growth in sulfidic mineral environments: metal resistance mechanisms in acidophilic microorganisms. *Microbiology* 149:1959–1970.

Futterer, O., A. Angelov, H. Liesegang, G. Gottschalk, C. Schleper, B. Schepers, C. Dock, G. Antranikian, and W. Liebl. 2004. Genome sequence of *Picrophilus torridus* and its implications for life around pH 0. *Proc. Natl. Acad. Sci. USA* 101:9091–9096.

Galtier, N., N. Tourasse, and M. Gouy. 1999. A nonhyperthermophilic common ancestor to extant life forms. *Science* 283:220–221.

Ghosh, S., N. R. Mahapatra, and P. C. Banerjee. 1997. Metal resistance in *Acidocella* strains and plasmid-mediated transfer of this characteristic to *Acidiphilium multivorum* and *Escherichia coli*. *Appl. Environ. Microbiol.* 63:4523–4527.

Gihring, T. M., P. L. Bond, S. C. Peters, and J. F. Banfield. 2003. Arsenic resistance in the archaeon "*Ferroplasma acidarmanus*": new insights into the structure and evolution of the *ars* genes. *Extremophiles* 7:123–130.

Golyshina, O. V., and K. N. Timmis. 2005. *Ferroplasma* and relatives, recently discovered cell wall-lacking archaea making a living in extremely acid, heavy metal-rich environments. *Environ. Microbiol.* 7:1277–1288.

Hopmans, E. C., S. Schouten, R. D. Pancost, M. T. van der Meer, and J. S. Sinninghe-Damste. 2000. Analysis of intact tetraether lipids in archaeal cell material and sediments by high performance liquid chromatography/atmospheric pressure chemical ionization mass spectrometry. *Rapid Commun. Mass Spectrom.* 14:585–589.

Inoue, C., K. Sugawara, and T. Kusano. 1991. The *merR* regulatory gene in *Thiobacillus ferrooxidans* is spaced apart from the *mer* structural genes. *Mol. Microbiol.* 5:2707–2718.

Iwahori, K., F. Takeuchi, K. Kamimura, and T. Sugio. 2000. Ferrous iron-dependent volatilization of mercury by the plasma membrane of *Thiobacillus ferrooxidans*. *Appl. Environ. Microbiol.* 66:3823–3827.

Johnson, D. B., and K. B. Hallberg. 2003. The microbiology of acidic mine waters. *Res. Microbiol.* 154:466–473.

Kalyaeva, E. S., G. Y. Kholodii, I. A. Bass, Z. M. Gorlenko, O. V. Yurieva, and V. G. Nikiforov. 2001. Tn5037, a Tn21-like mercury resistance transposon from *Thiobacillus ferrooxidans*. *Russ. J. Genet.* 37:972–975.

Kassen, R., and P. B. Rainey. 2004. The ecology and genetics of microbial diversity. *Annu. Rev. Microbiol.* 58:207–231.

Leduc, L. G., J. T. Ferroni, and J. T. Trevors. 1997. Resistance to heavy metals in different strains of *Thiobacillus ferrooxidans*. *World J. Microbiol. Biotechnol.* 13:453–455.

Lund, P. A., S. J. Ford, and N. L. Brown. 1986. Transcriptional regulation of the mercury-resistance genes of transposon Tn501. *J. Gen. Microbiol.* 132:465–480.

Macalady, J. L., M. M. Vestling, D. Baumler, N. Boekelheide, C. W. Kaspar, and J. F. Banfield. 2004. Tetraether-linked membrane monolayers in *Ferroplasma* spp: a key to survival in acid. *Extremophiles* 8:411–419.

Mahapatra, N. R., S. Ghosh, C. Deb, and P. C. Banerjee. 2002. Resistance to cadmium and zinc in *Acidiphilium symbioticum* KM2 is plasmid mediated. *Curr. Microbiol.* 45:180–186.

Mergeay, M., S. Monchy, T. Vallaeys, V. Auquier, A. Benotmane, P. Bertin, S. Taghavi, J. Dunn, D. van der Lelie, and R. Wattiez. 2003. *Ralstonia metallidurans*, a bacterium specifically adapted to toxic metals: towards a catalogue of metal-responsive genes. *FEMS. Microbiol. Rev.* 27:385–410.

Mukhopadhyay, R., B. P. Rosen, L. T. Phung, and S. Silver. 2002. Microbial arsenic: from geocycles to genes and enzymes. *FEMS. Microbiol. Rev.* 26:311–325.

Nisbet, E. G., and N. H. Sleep. 2001. The habitat and nature of early life. *Nature* 409:1083–1091.

Osborn, A. M., K. D. Bruce, P. Strike, and D. A. Ritchie. 1997. Distribution, diversity and evolution of the bacterial mercury resistance (*mer*) operon. *FEMS. Microbiol. Rev.* 19:239–262.

Remonsellez, F., A. Orell, and C. A. Jerez. 2006. Copper tolerance of the thermoacidophilic archaeon *Sulfolobus metallicus*: possible role of polyphosphate metabolism. *Microbiology* 152:59–66.

Rensing, C., and G. Grass. 2003. *Escherichia coli* mechanisms of copper homeostasis in a changing environment. *FEMS. Microbiol. Rev.* 27:197–213.

Rosen, B. P. 1999. Families of arsenic transporters. *Trends Microbiol.* 7:207–212.

Russell, M. J., and A. J. Hall. 1997. The emergence of life from iron monosulphide bubbles at a submarine hydrothermal redox and pH front. *J. Geol. Soc. Lond.* 154:377–402.

Schelert, J., V. Dixit, V. Hoang, J. Simbahan, M. Drozda, and P. Blum. 2004. Occurrence and characterization of mercury resistance in the hyperthermophilic archaeon *Sulfolobus solfataricus* by use of gene disruption. *J. Bacteriol.* 186:427–437.

Shiratori, T., C. Inoue, K. Sugawara, T. Kusano, and Y. Kitagawa. 1989. Cloning and expression of *Thiobacillus ferrooxidans* mercury ion resistance genes in *Escherichia coli*. *J. Bacteriol.* 171:3458–3464.

Simbahan, J., R. Drijber, and P. Blum. 2004. *Alicyclobacillus vulcanalis* sp. nov., a thermophilic, acidophilic bacterium isolated from Coso Hot Springs, California, USA. *Int. J. Syst. Evol. Microbiol.* 54:1703–1707.

Simbahan, J., E. Kurth, J. Schelert, A. Dillman, E. Moriyama, S. Jovanovich, and P. Blum. 2005. Community analysis of a mercury hot spring supports occurrence of domain-specific forms of mercuric reductase. *Appl. Environ. Microbiol.* 71:8836–8845.

Starkley, R. L., and S. A. Waksman. 1943. Fungi tolerant to extreme acidity and high concentrations of copper sulfate. *J. Bacteriol.* 45:509–519.

Sugio, T., M. Fujii, F. Takeuchi, A. Negishi, T. Maeda, and K. Kamimura. 2003. Volatilization of mercury by an iron oxidation enzyme system in a highly mercury-resistant *Acidithiobacillus ferrooxidans* strain MON-1. *Biosci. Biotechnol. Biochem.* 67:1537–1544.

Suzuki, K., N. Wakao, Y. Sakurai, T. Kimura, K. Sakka, and K. Ohmiya. 1997. Transformation of *Escherichia coli* with a large plasmid of *Acidiphilium multivorum* AIU 301 encoding arsenic resistance. *Appl. Environ. Microbiol.* 63:2089–2091.

Takeuchi, F., A. Negishi, S. Nakamura, T. Kanao, K. Kamimura, and T. Sugio. 2005. Existence of an iron-oxidizing bacterium *Acidithiobacillus ferrooxidans* resistant to organomercurial compounds. *J. Biosci. Bioeng.* 99:586–591.

Travisano, M., and P. B. Rainey. 2000. Studies of adaptive radiation using model microbial systems. *Am. Nat.* 156:35–44.

Tuffin, I. M., P. de Groot, S. M. Deane, and D. E. Rawlings. 2005. An unusual Tn21-like transposon containing an *ars* operon is present in highly arsenic-resistant strains of the biomining bacterium *Acidithiobacillus caldus*. *Microbiology* 151:3027–3039.

Tuffin, I. M., S. B. Hector, S. M. Dean, and D. E. Rawlings. 2006. Resistance determinants of a highly arsenic-resistant strain of *Leptospirillum ferriphilum* isolated from a commercial biooxidation tank. *Appl. Environ. Microbiol.* 72:2247–2253.

Wang, G., S. P. Kennedy, S. Fasiludeen, C. Rensing, and S. DasSarma. 2004. Arsenic resistance in *Halobacterium* sp. strain NRC-1 examined by using an improved gene knockout system. *J. Bacteriol.* 186:3187–3194.

Woese, C. R. 1987. Bacterial evolution. *Microbiol. Rev.* 51:221–271.

Yang, H. C., J. Cheng, T. M. Finan, B. P. Rosen, and H. Bhattacharjee. 2005. Novel pathway for arsenic detoxification in the legume symbiont *Sinorhizobium meliloti. J. Bacteriol.* 187:6991–6997.

Chapter 22

Genomics of Acidophiles

A. ANGELOV AND W. LIEBL

INTRODUCTION

Acidophily is a trait of organisms referring to the ability to survive and preferentially multiply at low pH (<3) (see Johnson, 1998). Extremely acidic environments are found in natural geothermal areas (solfataric fields, acidic vents, and pools) as well as in man-made habitats such as mining wastes (acid mine drainage [AMD] fluids, which are not only acidic but also rich in metal ion content). Acidophiles that thrive in these habitats are found among the archaeal, bacterial, and eukaryotic microorganisms. Primary production in extremely acidic environments takes place by chemolithoautotrophic prokaryotes (iron and sulfur oxidizers, the best-known being the bacterial species *Acidithiobacillus ferrooxidans*, previously known as *Thiobacillus ferrooxidans*, and *Leptospirillum* sp.) and at mesophilic temperatures sometimes phototrophic organisms (e.g., acidophilic algae) (Gonzalez-Toril et al., 2003; Johnson, 1998; Lopez-Archilla et al., 2001). Heterotrophic microorganisms in acidic habitats are believed to live as scavengers which utilize decomposing organic matter. Under moderately thermophilic conditions, archaeal and bacterial acidophiles can co-exist, sometimes forming interesting microbial communities. Among these, mesophilic and moderately thermophilic acidophiles which are present in AMD microbial communities and are involved in bioleaching applications have recently attracted interest. To this end, it was found that the recently described cell-wall-lacking archaea of the *Ferroplasma* kinship appear to be major players in biogeochemical sulfur and metal cycling in highly acidic environments and in the biotechnological metal solubilization and leaching from ores (Bond et al., 2000; Edwards et al., 1999; 2000; Golyshina and Timmis, 2005; Johnson and Hallberg, 2003; Rawlings, 2002). At the most extreme combinations of low pH and high temperature that are known to be tolerable for life, only archaeal species are found. These so-called thermoacidophiles are represented in both major phylogenetic lineages within the domain *Archaea*, i.e., in the *Crenarchaeota* and the *Euryarchaeota*, where they are grouped into the orders *Sulfolobales* and *Thermoplasmatales*, respectively. Among these, *Sulfolobus* species can tolerate the highest temperatures while *Picrophilus* species are the record holders in terms of the lowest pH (optimum growth at pH 0.7 at temperatures up to 65°C).

The significant occurrence of archaeal prokaryotes in highly acidic environments is in line with the concept about these organisms that was generally accepted at the time of the proposal of the domain *Archaea* as an above-kingdom-level taxon (Woese et al., 1990), i.e., it was thought that this group was confined to specialized or extreme habitats. Today, a different picture is emerging from various cultivation-independent and cultivation-based environmental studies which demonstrate the dissemination of *Archaea* also in more "standard" and less extreme biotopes (DeLong, 1992; Rudolph et al., 2001; Henneberger et al., 2006). Still, those archaea adapted to extremes of pH, temperature and salt concentrations remain fascinating research objects. Studies about their specific adaptation mechanisms at the levels of macromolecular structure/function-relationships, metabolism, regulation, and genome content and integrity help us to understand how cells have evolved to withstand and function in breathtakingly hostile surroundings. A powerful method to gather new information about the features and (putative) metabolic properties of organisms is to determine and interpret their genomic information content. Genome sequence analysis of prokaryotic microorganisms is proceeding at an enormous pace and in the past few years has been applied to several acid-adapted microbes. Since among acidophiles most

---

A. Angelov and W. Liebl • Institute of Microbiology and Genetics, University of Göttingen, Grisebachstrasse 8, D–37077 Göttingen, Germany.

genomic information is available on representatives of the thermoacidophilic archaeal lineages *Sulfolobales* and *Thermoplasmatales*, this chapter will focus mainly on these organisms.

FEATURES OF THE GENOME SEQUENCES OF ACIDOPHILIC *ARCHAEA*

Two approaches have been used in whole genome sequencing—the "shotgun sequencing" and the directed approach. In the first strategy, a high-coverage library of short (2 to 4 kb) genomic fragments is generated, sequenced, and the fragments are assembled in silico. In the case of acidophilic archaea, the usually small genomes allow to quickly acquire 90 to 98% of the sequence solely by shotgun sequencing, the remaining gaps are usually closed by polymerase chain reaction (PCR)-based methods (Fleischmann et al., 1995). A spectacular implementation of the random shotgun sequencing approach is the mentioned almost complete reconstruction of several genomes directly from an AMD environmental sample (Tyson et al., 2004). In the conventional shotgun sequencing all genomic fragments arise from the same genome (clone), while assembling fragments derived from a polyclonal population, as in an environmental sample, is complicated by the presence of several species as well as by intraspecies variation—gross genome rearrangements within a population of a species would make the assembly impossible. The directed approach requires an ordered genome library consisting of large overlapping clones (a genetic map), which in turn are sequenced by the shotgun method and primer walking (Himmelreich et al., 1996). Currently, the availability of high-throughput sequencing technologies has led to the use of the whole genome shotgun approach as the standard strategy in microbial genome sequencing.

From the 25 publicly available archaeal genomes (as of January 2006), 6 belong to acidophilic organisms: the genome sequences from three species of the crenarchaeal genus *Sulfolobus*, *S. solfataricus* (She et al., 2001), *S. acidocaldarius* (Chen et al., 2005), and *S. tokodaii* (Kawarabayasi et al., 2001), and two from the euryarchaeal genus *Thermoplasma*, *T. acidophilum* (Ruepp et al., 2000), and *T. volcanium* (Kawashima et al., 2000) have been completed, as well as the genome sequence from the record holder in acidophily, *Picrophilus torridus* (Fütterer et al., 2004), also an *Euryarchaeon*. The genome sequencing of another two acidophiles is in progress—that of the *Euryarchaeon Ferroplasma acidarmanus* (*Thermoplasmatales*) and of the *Crenarchaeon Acidianus brierleyi* (*Sulfolobales*). In addition to genome sequence data from cultured organisms, two near-complete genomes (*Leptospirillum* group II and *Ferroplasma* type II) have been reconstructed wholly by random shotgun sequencing of a DNA sample derived from a natural acidophilic biofilm, growing at pH of 0.83 (Tyson et al., 2004). Apart from being able to grow at extremely low pH values, all of these organisms are also, to a varying extent, thermophilic.

With the exception of the representatives of the *Sulfolobales*, acidophilic archaea have small genomes (Fig. 1). Small genome size is often associated with either a parasitic lifestyle or extreme habitats. Indeed, the genomes of thermophilic methanogenic archaea and hyperthermophilic bacteria are not much larger (about 1.6 to 1.8 Mb). One of the many effects of high temperature on the organisms living in such environments is an increased error rate in their nucleic acids due to cytosine deamination. It can be assumed that this has led to a selective pressure towards a reduced genome size. Most of the acidophilic archaea are also, to a different extent, thermophiles, thus they have to cope with two extreme conditions. For example, *P. torridus* is a moderate thermophile (T_{opt} 60°C), which is exposed to pH values around 0 in the medium and has the lowest-known intracellular pH (4.6). Another feature which can also be attributed to strong selective pressure from the hostile environment is the high coding density of the genomes of acidophiles. *P. torridus* (pH_{opt} 0.7), *T. volcanium* (pH_{opt} 2.0), *T. acidophilum* (pH_{opt} 2.0), and *S. tokodaii* (pH_{opt} 2.5) have very compact genomes with coding densities from 85% for *S. tokodaii* to 91.7% for *P. torridus* (see Fig. 1). Thus it is plausible to assume that the concurrent presence of two harsh conditions, pH and temperature, has exerted pressure in favor of small genomes with high coding density.

Other genome features that can be estimated computationally from whole genome sequence data are the amino acid composition, the protein length distribution, and the isoelectric point (p*I*) distribution of the predicted proteome. A characteristic trimodal p*I* distribution with two main peaks at pH 5.5 and 9.5 and a smaller one at pH 7.8 has been observed in a study of a large sample of 115 proteomes belonging to representatives of *Archaea*, *Eubacteria*, and *Eukaryota* (Weiller et al., 2004). Also, it has been noted for several organisms that inhabit extreme environments that these parameters differ significantly from the ones of the closely related taxa which are living in more "normal" habitats (Kennedy et al., 2001). Three noteworthy examples of unusual amino acid composition and p*I* distribution are the genome-derived proteomes of the halophilic and alkaliphilic *Halobacterium* sp., of the "conditional" acidophile *Helicobacter pylori* and of the extreme thermophile *Aeropyrum pernix* (Fig. 2). In the case of *Halobacterium* sp., the amino

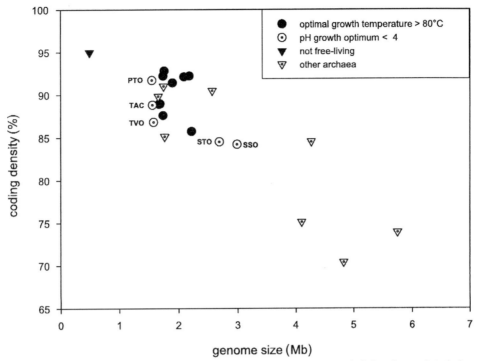

Figure 1. Size and coding density distribution of selected archaeal genomes. The genomes included in the graph include representatives of all archaeal lineages. Organismal designations are PTO, *Picrophilus torridus*; TAC, *Thermoplasma acidophilum*; TVO, *Thermoplasma volcanium*; STO, *Sulfolobus tokodaii*; SSO, *Sulfolobus solfataricus*.

acid distribution is strongly biased (e.g., in favor of aliphatic amino acids) and the isoelectric point distribution is down-shifted with an acid peak around pH 4.2 and a missing basic peak (see Fig. 2c). Conversely, in *H. pylori* more than 70% of the proteins have an isoelectric point greater than 7 (see Fig. 2e), and it has been speculated that this distribution reflects adaptation to high acidity (Tomb et al., 1997). The p*I* distribution of *Aeropyrum pernix* shows an additional fourth peak at pH 12 (see Fig. 2a); however, it is not clear if this can be linked to thermostability of its proteins as such a peak is not present in other hyperthermophiles (Weiller et al., 2004). The amino acid and p*I* distributions of the proteins from acidophilic archaea show no significant deviation from the average when these parameters were compared (see Fig. 2b, 2d, 2f). Therefore, we assume that the deviation from the consensus p*I* distribution in the case of *H. pylori* may reflect adaptation to low pH in the environment, but cannot be considered a general feature of acidophiles.

A more comprehensive understanding of the mechanisms that underlie protein stability at harsh conditions may be accessible by a comparative analysis of protein structures. In contrast to proteins from thermophilic organisms, very few structures of proteins which are actually exposed to acidic environment have been solved (Fushinobu et al., 1998; Matzke et al., 1997; Schäfer et al., 2004). This may partially be due to the fact that most of these proteins are secreted or periplasmatic and/or difficult to crystallize. A recent study of the structure of the maltose-maltodextrin-binding protein (MBP) from *Alicyclobacillus acidocaldarius* gives some insights into the molecular basis for protein acidostability (Schäfer et al., 2004). Among the observed features which could be linked specifically to the MBP's acidostability are reduced charge density on its surface with predominating positively charged residues and fewer buried salt bridges, while compactness, a high proline content in secondary structures and an increased number of polar but uncharged surface residues seem to be common also for thermophilic proteins. Amino acid distribution analysis of the proteome of *P. torridus* revealed a slight increase in the overall isoleucine content (Fig. 2f; Fütterer et al., 2004). In this respect, it is tempting to analyze the surface amino acid distribution of those *P. torridus* proteins that have close homologs with known three-dimensional (3-D) structure (known as 3-D neighbours).

ARE THERE PARALLELS BETWEEN TRUE ACIDOPHILY AND THE ACID RESISTANCE OF PATHOGENS?

In most cases the acidophiles manage to maintain their cytoplasmic pH in the normal range despite the

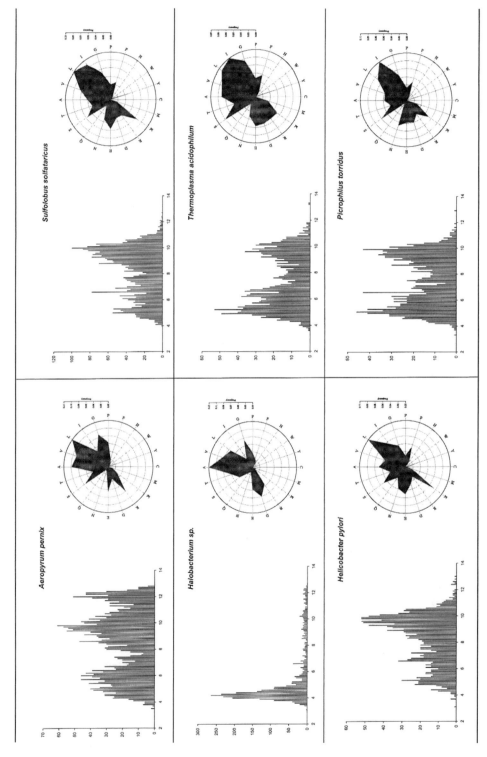

Figure 2. Proteome isoelectric point distribution and amino acid usage based on genome data. The isoelectric point distribution graphs represent plots of the number of proteins with a certain pI (y-axis) in intervals of 0.1 pH units (x-axis). The amino acid usage data is derived from the genome atlas database (https://www.cbs.dtu.dk/services/GenomeAtlas/index.php) (Hallin and Ussery, 2004).

presence of a huge proton gradient across the membrane. Remarkably, the most extreme among them, the thermoacidophilic archaea, manage to do this without the protection of a rigid cell wall. This is achieved by the concerted action of several mechanisms, which can be classified as passive and active ones. Among the passive mechanisms thought to be important for acidophiles are the low proton permeability of the cell membrane (Zychlinsky and Matin, 1983; van de Vossenberg et al., 1998), the presence of an S-layer and/or extracellular polysaccharides (EPS), which lower the actual proton activity, and special features of their extracellular and intracellular macromolecules. The active mechanisms include the operation of special metabolic pathways for consumption of protons inside the cell (acid resistance systems in *E. coli* and *H. pylori*, for a review see Foster, 2004), inversion of the membrane potential to positive inside, ATP-mediated K^+ import, and the utilization of transporter proteins for pH regulation. An example of a highly specialized, active acid protection mechanism is the urease system of *H. pylori*, which plays an important role for the long-term persistence of this pathogen in the human stomach (Marshall et al., 1990). Acid tolerance is thought to be mediated by the enzymatic cleavage of urea in the cell (Scott et al., 1998; Weeks et al., 2000), accompanied by an electrogenic export of the resulting ammonium ions from the cytoplasm (Stingl et al., 2002). It has to be stressed, however, that the functional investigation of these systems in extremophilic archaea is often hindered by the lack of genetic systems and the difficulties in cultivation typical for these organisms.

A very well characterized acid resistance system is that of the gastrointestinal pathogen *E. coli*. Therefore, it is of interest to ask if parallels exist between the acid resistance systems of *E. coli* and the situation deduced from the genomes of true (thermo)acidophiles. During its life cycle, *E. coli* is faced with the challenge of surviving the gastric acidic barrier, which imposes an acid stress of pH 1.5 to 2.5. Although *E. coli* is not actually an acidophile, the strategies which are involved in acid resistance may be operative also in thermoacidophilic archaea. Three acid resistance systems (AR or XAR) have been described in *E. coli*. The first one (AR1, also designated glucose-repressed AR) is dependent on the stationary phase alternative sigma factor (σ^s) and the cAMP receptor protein (CRP)(Castanie-Cornet et al., 1999) and the actual mechanisms it involves are currently not clear. The other two AR systems are each composed of an amino acid decarboxylase (glutamate for AR2 and arginine for AR3) and a corresponding antiporter (Lin et al., 1995). The decarboxylation of glutamate and arginine produces CO_2 and γ-aminobutiric acid (GABA) and agmatine, respectively, whereby a proton from the cytoplasm is consumed. The corresponding antiporters, GadC and AdiC, exchange the decarboxylation products for new amino acid substrates. In addition to these systems, Iyer et al. (2002) suggested that the ubiquitous prokaryotic CIC-type chloride transporters are also required for the survival of *E. coli* after acid challenge, and subsequently it was shown that these proteins actually function as H^+/Cl^- antiporters (Accardi and Miller, 2004). The currently accepted model for AR in enterobacteria presumes that at exposure to low pH, protons enter the cell either through protein channels or directly through the cell membrane (most probably as undissociated acids) and are consumed by decarboxylation. This process is, however, electrogenic—the protons are removed, but the positive charge remains in the decarboxylation products (Agm^{2+} vs. Arg^+, $GABA^0$ vs. Glu^-). This can only partially be compensated by the associated antiporters and leads to a transient inversion of the membrane potential ($\Delta\psi$) to positive inside, a situation found in obligate acidophiles (Richard and Foster, 2004). The proposed role of the CIC H^+/Cl^- antiporters is to assist in restoring of the membrane potential to negative inside by exchanging H^+ for Cl^- and in addition to further decrease the cytoplasmatic H^+ concentration. The range of native substrates of the CIC transporter homologs in other microorganisms is not yet clear; it makes sense to assume Cl^- for enterobacteria which survive the HCl-rich stomach, but other specificities should also be considered for prokaryotes that inhabit acid environments with different chemical setup.

Interestingly, the genomes of acidophilic archaea encode a surplus of both high-affinity amino acid transporters and of the CIC family of chloride channels. Specifically, homologs of the GadA/GadC proteins of *E. coli* with significant level of amino acid similarity are encoded in the genomes of diverse groups of acidophilic archaea (Table 1). Although the potential physiological relevance of these homologs in archaea is not yet investigated, an interesting observation comes from a work on the bioenergetics in the extreme acidophile *Picrophilus oshimae* (van de Vossenberg et al., 1998). The authors have measured the rate of amino acid uptake by energized cells of *P. oshimae* and the highest reported level of uptake is that of glutamate. Further, L-glutamate uptake was found to be driven by the ΔpH, which at the experimental pH of 1.0 constitutes the major contributor to the proton motive force (PMF). Whether the preferential glutamate uptake is a part of a mechanism reminiscent to that found in enterobacteria is not yet clear.

It has to be emphasized, however, that the challenges that are faced by acid-tolerant bacteria and obligate acidophiles, which have a pH growth optimum at

Table 1. Occurrence of homologs of the E. coli acid resistance genes among thermoacidophilic archaea[a]

Acid resistance gene in E. coli	Picrophilus torridus		Ferroplasma acidarmanus*		Thermoplasma acidophilum		Sulfolobus solfataricus		Sulfolobus acidocaldarius	
	E-value	% identity	E-value	% identity	E-value	% identity	E-value	% identity	E-value	% identity
Glutamate decarboxylase gadA	1×10^{-19}	25	6×10^{-75}	42	—	—	—	—	4×10^{-15}	25
GABA APC transporter gadC	9×10^{-6}	22	2×10^{-8}	24	2×10^{-6}	21	5×10^{-9}	23	1×10^{-9}	28
Arginine decarboxylase adiA	—	—	—	—	—	—	—	—	—	—
Arginine: agmatine antiporter adiC	2×10^{-17}	24	4×10^{-26}	28	4×10^{-14}	24	2×10^{-15}	22	2×10^{-14}	24
ClC chloride transporter clcB	1×10^{-10}	24	1×10^{-9}	31	2×10^{-6}	28	2×10^{-12}	24	3×10^{-13}	27

[a] An E-value of 1×10^{-5} has been used as a threshold for considering the presence of a homolog with significant level of amino acid sequence similarity.

less than 3, can be quite different. While an acidoresistant species like E. coli has to be able to survive acid stress only rarely and for a short period of time, an extremely acidophilic archaeon has to be able to constantly maintain its metabolism at a pH of around 0 in the medium, often at moderately high temperatures. Hence, different physiological mechanisms for acid resistance can be expected.

GENOME-BASED RECONSTRUCTION OF THE MAJOR METABOLIC PATHWAYS OF ACIDOPHILIC *ARCHAEA*

In order to maintain an intracellular pH at an acceptable value in the face of a huge proton gradient (ΔpH can reach more than 5 pH units), acidophiles need to generate a substantial amount of metabolic energy. The majority of acidophilic archaea, especially the extreme acidophiles of the *Thermoplasmatales*, have a scavenging life style—they rely on the decomposition of organic matter for their nutrition and usually require yeast, meat, or bacterial extracts to grow in culture (Smith et al., 1975; Schleper et al., 1995). The genomes of thermoacidophilic archaea encode a complete set of enzymes necessary for the utilization of extracellular peptides—secreted proteases, oligopeptide transport systems, and intracellular protease complexes (tricorn protease and cofactors). It is noteworthy that all of these proteins form a distinct cluster in acidophilic archaea that belong to distantly related taxa, i.e., the closest homologs of the protein utilization Open reading frames (ORFs), of the *Thermoplasma* group are found in acidophilic representatives of the distant branch of the *Crenarchaea*—*S. solfataricus* and *S. tokodaii*. This is also true for many other proteins, in most cases related to transport and metabolism. Alpha-linked EPSs are considered to partially hydrolyze at low pH and high temperature, which explains the absence of genes encoding secreted amylases in the genome of *S. solfataricus* (She et al., 2001) although starch can support growth of this organism (Grogan et al., 1990). In the genomes of *Thermoplasma acidophilum* and *Picrophilus torridus*, however, several genes could be identified that code for membrane-anchored amylases/pullulanases. Glucoamylases have been purified from culture fluids of *T. acidophilum*, *P. torridus*, and *P. oshimae* (Serour and Antranikian, 2002). Enzymatic activity was detected even at pH 1 and 100°C, and the enzymes displayed remarkable resistance against thermoinactivation, with half-lives of 20 to 24 h at 90°C. Utilization of the more-stable β-linked glucan polymers by thermoacidophiles seems to be scarce but putative β-glucanase genes are found in *S. solfataricus*.

The currently available genome and biochemical data for acidophilic archaea suppose that glucose is catabolized via a non-phosphorylated variant of the Entner-Doudoroff (ED) pathway, the phosphorylation taking place at the level of glycerate. This route is found in *S. solfataricus* (De Rosa et al., 1984), *S. acidocaldarius* (Selig et al., 1997) as well as in *T. acidophilum* (Budgen, 1986) and *P. torridus* (Fütterer et al., 2004) and differs from the semi-phosphorylated and the classical ED pathways only at the level at which phosphorylation takes place, i.e., at 2-keto 3-deoxygluconate for the semi-phosphorylated and glucose for the classical ED. Interestingly, at least in *S. solfataricus*, the non-phosphorylated ED pathway seems to be used for the catabolism of both glucose and galactose (Lamble et al., 2004) and this may be the case also in *P. torridus* (Angelov et al., 2005). This mode of metabolism may be an indication of an ancient origin, but may also suggest a selective pressure in favor of using a broader substrate range with a limited set of genes.

With several exceptions, the genes coding for the enzymes of the Embden-Meyerhoff-Parnas (EMP) pathway as well as for a complete citrate (in some cases also methylcitrate) cycle are present in the genomes of thermoacidophilic archaea.

An important metabolic feature for acidophiles, related to their life style, is the ability to efficiently metabolize weak organic acids. This is so because at the low pH of the environment these acids exist in their protonated (undissociated) forms and can easily diffuse into the cell where they readily dissociate at the higher intracellular pH, which can lead to acidification of the cytoplasm and dissipation of the PMF. For example, it has been reported that respiration in the extreme acidophile *P. oshimae* was inhibited by the addition of the acids formate, lactate, propionate, and acetate to the medium (van de Vossenberg et al., 1998). The genome of the other member of the genus, *P. torridus*, contains genes encoding enzymes for the degradation of acetate, lactate, and propionate, which are found also in the *Thermoplasma* and *Ferroplasma* genomes (Fütterer et al., 2004). Interestingly, these proteins have their closest homologs in such distant bacterial genera as *Clostridium* (formyltetrahydrofolate ligase) and *Mycobacterium* (lactate monooxygenase) but not in other archaea, which is an indication for lateral gene transfer events. It seems that these bacteria-like enzymes for organic acid degradation are restricted only to the genomes of the *Thermoplasmatales* members among archaea and it can be speculated that they serve mainly as protection against the uncoupling effect of the acids.

The huge proton gradient that extreme acidophiles are facing requires active respiration. Most of the thermoacidophilic archaea are aerobes, the available genome data reveals differing complexity of their respiratory chains which reflects their varying oxygen requirements. The members of the genus *Thermoplasma* are facultative aerobes and in the absence of oxygen they can use sulfur for respiration (Darland et al., 1970). Interestingly, the genes that are thought to mediate sulfur respiration in *Thermoplasma acidophilum* are most similar to bacterial and not to archaeal ones (Ruepp et al., 2000). The obligate aerobe *P. torridus* uses a more complex and multi-component respiratory chain, as is deduced from the genome data (Fütterer et al., 2004) (Fig. 3). This genome contains all genes for a type I NADH oxydoreductase except the ones for the input module, *nuoEFG* (*E. coli* nomenclature). The other members of this group, *T. acidophilum* and *F. acidarmanus*, also lack the input module which mirrors the situation found in most of the sequenced archaeal genomes (Ruepp et al., 2000). It is still unclear how electrons are fed into the respiratory chain in organisms that lack the NADH input module (Schäfer et al., 1999). The representatives of the *Sulfolobales* lack, in addition, some of the genes for the integral and electron transfer subunits (She et al., 2001). Finally, when the genes encoding the subunits of succinate dehydrogenase (SDH) are compared among the members of the thermoacidophilic archaea, an interesting observation can be made. While *T. acidophilum* has only classical type SDH genes, *P. torridus* and *F. acidarmanus* contain the non-classical type SDH typical for the crenarchaea which indicates that these genes were internalized in the *P. torridus* and

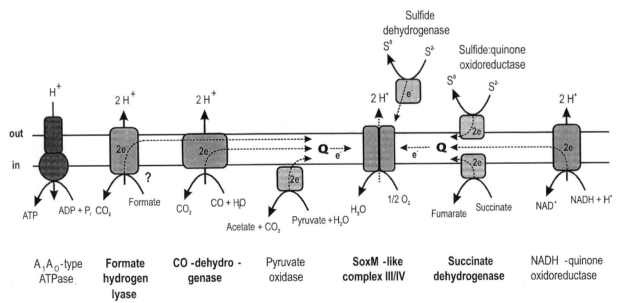

Figure 3. Genome-based reconstruction of the predicted respiratory chain of the euryarchaeon *P. torridus*. The components marked in bold are proposed to be either of bacterial or of crenarchaeal origin, based on protein similarity.

F. acidarmanus genomes by lateral gene transfer. Overall, as can be seen in Figure 3, a large proportion of the respiratory chain components of *P. torridus* show significant similarity to bacterial or crenarchaeal proteins and lack detectable homologs in the closely related *Euryarchaeota*.

An important asset of aerobic acidophiles is their ability to cope with oxygen stress. Not surprisingly, the genome of *P. torridus* encodes several key enzymes that are important in protecting the cell against oxidative damage. These include a superoxide dismutase, three putative peroxiredoxin-like proteins and an alkyl hydroperoxide reductase. In addition, a β-carotene biosynthetic operon could be identified that showed similarity to marine ε-proteobacteria and corynebacteria. Similar operons are not known to be present in other members of the *Thermoplasmatales*.

TRANSPORT

The lifestyle of acidophilic archaea imposes an important role of transport for their survival. For the uptake of organic compounds, such as sugars and peptides, these archaea rely both on gradient-driven secondary transporters (permeases) and high-affinity ABC-type (ATP-Binding Casette) transport systems. The essential role of transport proteins in thermoacidophiles becomes clear when the number of transport proteins encoded per megabase genome is compared among representatives of the *Archaea* (Fig. 4). For the 17 organisms listed in Figure 4, the average number of transporters encoded per megabase genome is 46.1 while for the *Thermoplasma* group (including *T. acidophilum*, *T. volcanium*, and *P. torridus*) this number is 70.6. This high transport coding density is further emphasized if the small genome size of these organisms is considered, for example for *P. torridus* this number means that 12% of the genome are devoted to genes for transport proteins. Also, it is evident that the thermoacidophiles have a substantially higher fraction of secondary (gradient-driven) transporters than the other archaeal groups, the most widely represented transport families in thermoacidophiles being MFS (Major Facilitator Superfamily) and APC (Amino Acid-Polyamine-Organocation Superfamily).

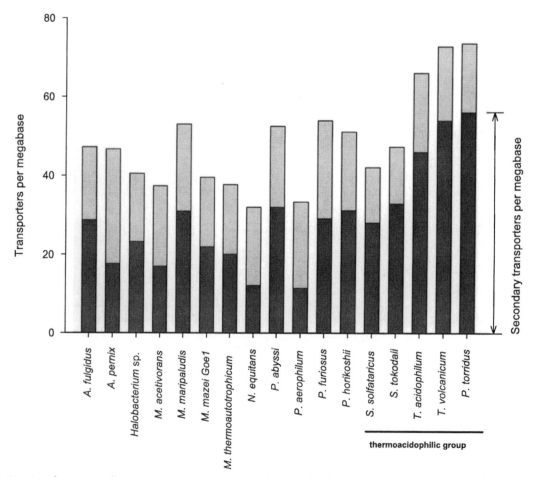

Figure 4. Density of genes encoding transporter proteins in the publicly available archaeal genomes. Data were obtained from http://membranetransport.org/ (Ren and Paulsen, 2005).

The secondary transporters of thermoacidophiles are believed to utilize protons and not Na$^+$ as a motive force. Considering that a permanent huge ΔpH exists across the membrane of acidophiles, this extensive use of the transmembrane proton gradient for transport is not surprising. In contrast, it is known that in a different group of extremophiles, the hyperthermophilic bacteria and archaea, i.e., organisms with growth optima of at least 80°C, primary uptake systems are preferred (Albers et al., 2001).

Another transport system that can be related to the thermoacidophilic lifestyle is the K$^+$-transporting ATPase. This ABC-type transporter contains three subunits and actively transports potassium ions into the cell. While it is broadly distributed in the bacterial kingdom, among *Archaea* it can be found only in the acidophilic group of *Thermoplasmatales*. As acidophiles typically have an unusual, positive inside charge distribution across the membrane, the physiological role of this transporter is probably to maintain this charge and in this way to regulate the PMF and cope with the high ΔpH.

GENES FOR DNA SUPERSTRUCTURE FORMATION, REPAIR, AND RESTRICTION IN ACIDOPHILIC *ARCHAEA*

It has been shown previously that the presence in the archaeal genomes of some genes related to DNA superstructure formation is related to the optimal growth temperature of the organism (Kawashima et al., 2000). Thus, all genomes of hyperthermophilic archaea carry genes for reverse gyrase and topoisomerase VI, responsible for creating and relaxing positive DNA superhelices. On the contrary, the genomes of acidophilic archaea that belong to the *Thermoplasma* group have genes for DNA gyrase and topoisomerase I for producing and relaxing negative DNA superhelicity (Table 2). The members of the *Sulfolobales* possess subunits from both classes of enzymes, which corresponds to their assignment as hyperthermophilic acidophiles. Overall, from the available genome data it seems likely that positive DNA superspiralization is restricted to hyperthermophiles only and is not found in other acidophilic archaea.

The distribution of archaeal histones and eubacterial histone-like HU proteins among chosen representatives of the sequenced archaeal genomes also shows a pattern that is more related to the T_{opt} and pH_{opt} of growth than to phylogenetic affiliation (see Table 2). Thus, while the eukarya-like archaeal histone is distributed in diverse groups of archaea, the histone-like HU is restricted only to the acidophilic group of *Euryarchaea*, suggesting that this typically bacterial protein has been laterally transferred to archaeal acidophiles. In *E. coli*, HU has been shown to be involved in both DNA compaction (Dame and Goosen, 2002) and transcriptional regulation (Aki and Adhya, 1997), but it is not clear yet whether its physiological role in the branch of acidophilic archaea is similar to the one in enterobacteria.

It was noted recently that the sequenced genomes of *Thermoplasmatales* reveal a couple of peculiarities when the encoded DNA integrity and repair proteins are analyzed in more detail (Ciaramella et al., 2005). In particular, both the *T. acidophilum* and *P. torridus* genomes neither contain homologs for the essential NER (nucleotide excision repair) nuclease XPF nor for UvrABC proteins. In addition, the absence of homologs for archaeal histone proteins in these genomes raises interesting questions about the mechanisms involved in maintaining genome structure and integrity under conditions that are expected to be highly mutagenic.

CHAPERONE GENES IN THE GENOMES OF ACIDOPHILES

It might be expected that molecular chaperones, having in mind the essential role they play in cell physiology, would be highly conserved in all three domains of life. However, in contrast to the *Bacteria* and *Eukaryota*, the chaperone families universally distributed in all *Archaea* are very limited. In particular, only small HSPs (heat shock proteins), chaperonin, AAA ATPases, prefoldin, NAC and folding catalysts (peptidyl-prolyl isomerase and disulfide isomerase) are found in all archaeal species whose whole genome sequences have become available. A very puzzling observation is the absence of the Hsp70 system (including Hsp40/DnaJ and GrpE cofactors) in some archaeal genomes. This absence becomes even more intriguing when the optimal growth temperature of the respective organism is taken into account: it seems that only hyperthermophiles do not posses homologs of the Hsp70 chaperone family (see Table 2). In contrast, the DnaK-DnaJ-GrpE system has been found in all bacteria, including hyperthermophilic ones. The genomes of the acidophilic archaea of the *Thermoplasma* group all encode homologs of this chaperone system and Hsp70-based phylogenetic trees place their protein together with *Thermotogales* and green nonsulfur bacteria, while the majority of archaeal Hsp70 proteins have their closest homologs in gram-positive bacteria (Gribaldo et al., 1999). The patched distribution of several chaperone families among the archaeal lineage suggests that these proteins are, to a reasonable degree, functionally redundant, with members of other families (or other yet unknown proteins) taking over their

Table 2. Distribution of ORFs coding for chaperones and DNA superstructure formation proteins in archaeal genomes[a]

Species	T_{opt}	pH_{opt}	Category	Chaperone proteins							DNA superstructure formation proteins				
				DnaJ	DnaK	GrpE	FtsJ	HtpX	Chaperonin	HU protein	archaeal histone	TOPO VI Subunit A	TOPO VI Subunit B	DNA reverse gyrase	TOPO I	DNA gyrase Subunit A	DNA gyrase Subunit B
Mth	65	7.4	Hyperthermophiles	X	X	X	X	X	X	–	X	X	X	–	X	–	–
Ape	95	7.0		–	–	–	–	X	X	–	–	X	X	X	X	–	–
Afu	83	7.0		–	–	–	X	X	X	–	X	X	X	X	X	X	X
Mja	85	6.0		–	–	–	X	X	X	–	X	X	X	X	X	–	–
Mka	98	6.5		–	–	–	X	X	X	–	X	X	X	X	X	–	–
Pae	100	7.0		–	–	–	–	X	X	–	–	X	X	X	X	–	–
Pho	98	7.0		–	–	–	–	X	X	–	X	X	X	X	X	–	–
Pfu	100	7.0		–	–	–	–	X	X	–	X	X	X	X	X	–	–
Pab	96	7.0		–	–	–	–	X	X	–	X	X	X	X	X	–	–
Sso	85	3.0		–	X	X	–	X	X	–	–	X	X	X	X	X	X
Sto	80	2.5		–	–	–	–	X	X	–	–	X	X	X	X	X	X
Tac	59	2.0	Acidophiles	X	X	X	X	X	X	X	–	–	–	–	X	X	X
Tvo	60	2.0		X	X	X	X	X	X	X	–	–	–	–	X	X	X
Fac	40	1.6		X	X	X	X	X	X	X	–	–	–	–	–	X	X
Pto	60	0.7		X	X	X	X	X	X	X	–	–	–	–	X	X	X

[a] Only hyperthermophilic and acidophilic archaea are included. The presence of an ORF is marked with an X; protein designations follow the *E. coli* nomenclature. *Afu*, *Archaeoglobus fulgidus*; *Ape*, *Aeropyrum pernix*; *Fac*, *Ferroplasma acidarmanus*; *Mja*, *Methanococcus jannaschii*; *Mka*, *Methanopyrus kandleri*; *Mth*, *Methanobacterium thermautotrophicus*; *Pab*, *Pyrococcus abyssi*; *Pae*, *Pyrobaculum aerophilum*; *Pfu*, *Pyrococcus furiosus*; *Pho*, *Pyrococcus horikoshii*; *Pto*, *Picrophilus torridus*; *Sso*, *Sulfolobus solfataricus*; *Sto*, *Sulfolobus tokodaii*; *Tac*, *Thermoplasma acidophilum*; *Tvo*, *Thermoplasma volcanium*.

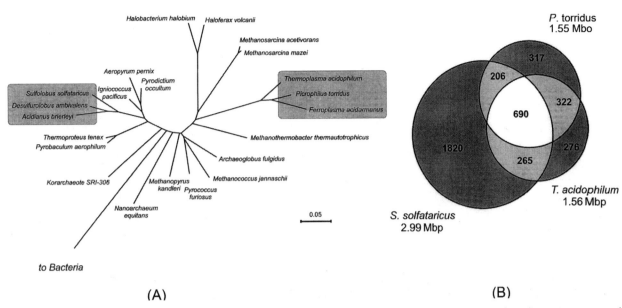

Figure 5. (A) A simplified 16S rRNA phylogenetic tree of Archaea. The two thermoacidophilic groups are boxed. (B) Occurrence of homologs in *P. torridus*, *T. acidophilum*, and *S. solfataricus*, the criterion for homology is a minimum of 30% amino acid sequence similarity. Orthologous and paralogous sequences were counted only once. The size of the circles is proportional to the genome size. Reproduced from Fütterer et al. (2004) with permission from the National Academy of Sciences.

essential role. Also, the presence of two distinct bacterial types of the Hsp70 protein in distant archaeal species raises interesting questions about the origin and evolution of these genes in the archaeal domain.

Apart from the Hsp70 system, the chaperone makeup of acidophilic archaea resembles more that of eukaryotes, and archaeal chaperones have played an important role in elucidating the structure and function of their more-complex eukaryal homologs (for a review see Macario et al., 1999; Hartl and Hayer-Hartl, 2002; Young et al., 2004).

THE EVOLUTION OF THERMOACIDOPHILES AND THE CONCEPT OF "LIFESTYLE GENES"

Thermophilic and acidophilic archaea form two distinct phylogenetic groups, one belonging to the euryarchaeal and the second one to the crenarchaeal lineage (Fig. 5a). The majority of these microorganisms are aerobic or microaerophilic heterotrophs which often share the same habitat (Gonzalez-Toril et al., 2003; Johnson et al., 2003; Okibe et al., 2003). It is becoming increasingly acknowledged that microorganisms that live together exchange genes at a higher frequency (DeLong, 2000; Papke et al., 2004). When the first genome of an acidophilic archaeon, that of *T. acidophilum*, was sequenced in 2000 it was realized that there has been extensive lateral gene transfer (LGT) between *Thermoplasma*, and the distantly related crenarchaeon *S. solfataricus*, which inhabits the same environment (Ruepp et al., 2000). Thus, the genes of *T. acidophilum* could be assigned to two classes: "housekeeping genes", which are consistent with the phylogenetic origin of the organism, and "lifestyle genes", which are generally related to metabolism and have their closest homologs in *S. solfataricus*. The latter class of genes includes for example the whole set of proteins necessary for the utilization of exogenous peptides—extracellular proteases, oligopeptide transport systems, tricorn protease and its cofactors. With the availability of several more genome sequences of acidophilic archaea, the large extent of LGT between crenarchaeal and euryarchaeal acidophiles was further substantiated (Fütterer et al., 2004). Figure 5b illustrates the number of shared genes among three representatives of these two groups: *P. torridus*, *T. acidophilum* (*Euryarchaeota*), and *S. solfataricus* (*Crenarchaeota*). When 30% amino acid sequence identity is used as a threshold, *P. torridus* shares nearly the same number of homologous genes with the ecologically similar and phylogenetically closely related *T. acidophilum* and also ecologically similar but phylogenetically distant *S. solfataricus* (66% and 58% respectively). In contrast, but significantly less homologs are shared with the phylogenetically and ecologically distinct *P. furiosus* (35%) which is a hyperthermophilic archaeon isolated from abyssal vents. A further 13.5% of the *P. torridus* ORFs have their homologues in the *S. solfataricus* but not in the *T. acidophilum* genome, suggesting that these genes were transferred relatively recently (after the *Picrophilus* and *Thermoplasma* lineages have separated). Importantly, these putative LGT events must have taken place in a hot and acidic environment which presents an obstacle for direct DNA transfer between cells. On the other hand, the large selection pressure exercised by these harsh conditions may be the reason for the observed high frequency of shared genes, as most of these genes can be related to the life-style of the extreme acidophiles.

Another convincing line of evidence for frequent genetic exchange in extremely acidic environments comes from the direct sequencing to near completion of two genomes of organisms forming an AMD biofilm (Tyson et al., 2004). Individual genomes of the *Ferroplasma* type II population, one of the species that dominated the biofilm, displayed a "mosaic" structure, i.e., these genomes contained various combinations of three main genotypes (in the form of nucleotide polymorphisms) found in the population. Further, the authors could estimate that the mosaic character of the individual genomes must have arisen by at least 400 recombination events, based on the frequency of transitions between the three nucleotide polymorphism patterns. Thus, at any one time, the *Ferroplasma* population consists of a multitude of slightly differing combinatorial genomic variants, generated most probably through homologous recombination, a situation more reminiscent of a sexually reproducing organism. Although in the above example the observed sequence variation is on the level of a population (intraspecies), it can be expected that the frequency at which microorganisms are integrating divergent DNA will increase with increasing beneficial effects of the genes that are transferred (Gogarten and Townsend, 2005). Importantly, these results indicate that genetic exchange, involving even large genomic regions, does occur in the hot and acid environments typical for acidophilic archaea. The actual mechanisms of gene transfer among extreme acidophiles are still unclear; however, conjugation and transduction can be thought of as more probable than transformation by uptake of naked DNA, as DNA is expected to have a short half-life in acid biotopes.

REFERENCES

Accardi, A., and C. Miller. 2004. Secondary active transport mediated by a prokaryotic homologue of ClC Cl-channels. *Nature* **427**:803–807.

Aki, T., and S. Adhya. 1997. Repressor induced site-specific binding of HU for transcriptional regulation. *Embo. J.* **16**:3666–3674.

Albers, S. V., J. L. van de Vossenberg, A. J. Driessen, and W. N. Konings. 2001. Bioenergetics and solute uptake under extreme conditions. *Extremophiles* **5**:285–294.

Angelov, A., O. Fütterer, O. Valerius, G. H. Braus, and W. Liebl. 2005. Properties of the recombinant glucose/galactose dehydrogenase from the extreme thermoacidophile, *Picrophilus torridus*. *Febs. J.* **272**:1054–1062.

Bond, P. L., G. K. Druschel, and J. F. Banfield. 2000. Comparison of acid mine drainage microbial communities in physically and geochemically distinct ecosystems. *Appl. Environ. Microbiol.* **66**:4962–4971.

Budgen, N. 1986. Metabolism of glucose via a modified Entner-Doudoroff pathway in the thermoacidophilic archaebacterium *Thermoplasma acidophilum*. *FEBS* **196**:207–210.

Castanie-Cornet, M. P., T. A. Penfound, D. Smith, J. F. Elliott, and J. W. Foster. 1999. Control of acid resistance in *Escherichia coli*. *J. Bacteriol.* **181**:3525–3535.

Chen, L., K. Brugger, M. Skovgaard, P. Redder, Q. She, E. Torarinsson, B. Greve, M. Awayez, A. Zibat, H. P. Klenk, and R. A. Garrett. 2005. The genome of *Sulfolobus acidocaldarius*, a model organism of the Crenarchaeota. *J. Bacteriol.* **187**:4992–4999.

Ciaramella, M., A. Napoli, and M. Rossi. 2005. Another extreme genome: how to live at pH 0. *Trends Microbiol.* **13**:49–51.

Dame, R. T., and N. Goosen. 2002. HU: promoting or counteracting DNA compaction? *FEBS Lett.* **529**:151–156.

Darland, G., T. D. Brock, W. Samsonoff, and S. F. Conti. 1970. A thermophilic, acidophilic mycoplasma isolated from a coal refuse pile. *Science* **170**:1416–1418.

De Rosa, M., A. Gambacorta, B. Nicolaus, P. Giardina, E. Poerio, and V. Buonocore. 1984. Glucose metabolism in the extreme thermoacidophilic archaebacterium *Sulfolobus solfataricus*. *Biochem. J.* **224**:407–414.

DeLong, E. F. 1992. Archaea in coastal marine environments. *Proc. Natl. Acad. Sci. USA* **89**:5685–5689.

DeLong, E. F. 2000. Extreme genomes. *Genome Biol.* **1**:1029.

Edwards, K. J., T. M. Gihring, and J. F. Banfield. 1999. Seasonal variations in microbial populations and environmental conditions in an extreme acid mine drainage environment. *Appl. Environ. Microbiol.* **65**:3627–3632.

Edwards, K. J., P. L. Bond, T. M. Gihring, and J. F. Banfield. 2000. An archaeal iron-oxidizing extreme acidophile important in acid mine drainage. *Science* **287**:1796–1799.

Fleischmann, R. D., M. D. Adams, O. White, R. A. Clayton, E. F. Kirkness, A. R. Kerlavage, C. J. Bult, J. F. Tomb, B. A. Dougherty, J. M. Merrick, and et al. 1995. Whole-genome random sequencing and assembly of *Haemophilus influenzae* Rd. *Science* **269**:496–512.

Foster, J. W. 2004. *Escherichia coli* acid resistance: tales of an amateur acidophile. *Nat. Rev. Microbiol.* **2**:898–907.

Fushinobu, S., K. Ito, M. Konno, T. Wakagi, and H. Matsuzawa. 1998. Crystallographic and mutational analyses of an extremely acidophilic and acid-stable xylanase: biased distribution of acidic residues and importance of Asp37 for catalysis at low pH. *Protein Eng.* **11**:1121–1128.

Fütterer, O., A. Angelov, H. Liesegang, G. Gottschalk, C. Schleper, B. Schepers, C. Dock, G. Antranikian, and W. Liebl. 2004. Genome sequence of *Picrophilus torridus* and its implications for life around pH 0. *Proc. Natl. Acad. Sci. USA* **101**:9091–9096.

Gogarten, J. P., and J. P. Townsend. 2005. Horizontal gene transfer, genome innovation and evolution. *Nat. Rev. Microbiol.* **3**:679–687.

Golyshina, O. V., and K. N. Timmis. 2005. *Ferroplasma* and relatives, recently discovered cell wall-lacking archaea making a living in extremely acid, heavy metal-rich environments. *Environ. Microbiol.* **7**:1277–1288.

Gonzalez-Toril, E., E. Llobet-Brossa, E. O. Casamayor, R. Amann, and R. Amils. 2003. Microbial ecology of an extreme acidic environment, the Tinto River. *Appl. Environ. Microbiol.* **69**:4853–4865.

Gribaldo, S., V. Lumia, R. Creti, E. C. de Macario, A. Sanangelantoni, and P. Cammarano. 1999. Discontinuous occurrence of the hsp70 (dnaK) gene among Archaea and sequence features of

HSP70 suggest a novel outlook on phylogenies inferred from this protein. *J. Bacteriol.* 181:434–443.

Grogan, D., P. Palm, and W. Zillig. 1990. Isolate B12, which harbours a virus-like element, represents a new species of the archaebacterial genus *Sulfolobus*, *Sulfolobus shibatae*, sp. nov. *Arch. Microbiol.* 154:594–599.

Hallin, P. F., and D. W. Ussery. 2004. CBS Genome Atlas Database: a dynamic storage for bioinformatic results and sequence data. *Bioinformatics* 20:3682–3686.

Hartl, F. U., and M. Hayer-Hartl. 2002. Molecular chaperones in the cytosol: from nascent chain to folded protein. *Science* 295:1852–1858.

Henneberger, R., C. Moissl, T. Amann, C. Rudolph, and R. Huber. 2006. New insights into the lifestyle of the cold-loving SM1 euryarchaeon: natural growth as a monospecies biofilm in the subsurface. *Appl. Environ. Microbiol.* 72:192–199.

Himmelreich, R., H. Hilbert, H. Plagens, E. Pirkl, B. C. Li, and R. Herrmann. 1996. Complete sequence analysis of the genome of the bacterium *Mycoplasma pneumoniae*. *Nucleic Acids Res.* 24:4420–4449.

Iyer, R., T. M. Iverson, A. Accardi, and C. Miller. 2002. A biological role for prokaryotic ClC chloride channels. *Nature* 419:715–718.

Johnson, D. B. 1998. Biodiversity and ecology of acidophilic microorganisms. *FEMS Microbiol. Ecol.* 27:307–317.

Johnson, D. B., and K. B. Hallberg. 2003. The microbiology of acidic mine waters. *Res. Microbiol.* 154:466–473.

Johnson, D. B., N. Okibe, and F. F. Roberto. 2003. Novel thermoacidophilic bacteria isolated from geothermal sites in Yellowstone National Park: physiological and phylogenetic characteristics. *Arch. Microbiol.* 180:60–68.

Kawarabayasi, Y., Y. Hino, H. Horikawa, K. Jin-no, M. Takahashi, M. Sekine, S. Baba, A. Ankai, H. Kosugi, A. Hosoyama, S. Fukui, Y. Nagai, K. Nishijima, R. Otsuka, H. Nakazawa, M. Takamiya, Y. Kato, T. Yoshizawa, T. Tanaka, Y. Kudoh, J. Yamazaki, N. Kushida, A. Oguchi, K. Aoki, S. Masuda, M. Yanagii, M. Nishimura, A. Yamagishi, T. Oshima, and H. Kikuchi. 2001. Complete genome sequence of an aerobic thermoacidophilic crenarchaeon, *Sulfolobus tokodaii* strain7. *DNA Res.* 8:123–140.

Kawashima, T., N. Amano, H. Koike, S. Makino, S. Higuchi, Y. Kawashima-Ohya, K. Watanabe, M. Yamazaki, K. Kanehori, T. Kawamoto, T. Nunoshiba, Y. Yamamoto, H. Aramaki, K. Makino, and M. Suzuki. 2000. Archaeal adaptation to higher temperatures revealed by genomic sequence of *Thermoplasma volcanium*. *Proc. Natl. Acad. Sci. USA* 97:14257–14262.

Kennedy, S. P., W. V. Ng, S. L. Salzberg, L. Hood, and S. DasSarma. 2001. Understanding the adaptation of *Halobacterium* species NRC-1 to its extreme environment through computational analysis of its genome sequence. *Genome Res.* 11:1641–1650.

Lamble, H. J., C. C. Milburn, G. L. Taylor, D. W. Hough, and M. J. Danson. 2004. Gluconate dehydratase from the promiscuous Entner-Doudoroff pathway in *Sulfolobus solfataricus*. *FEBS Lett.* 576:133–136.

Lin, J., I. S. Lee, J. Frey, J. L. Slonczewski, and J. W. Foster. 1995. Comparative analysis of extreme acid survival in *Salmonella typhimurium*, *Shigella flexneri*, and *Escherichia coli*. *J. Bacteriol.* 177:4097–4104.

Lopez-Archilla, A. I., I. Marin, and R. Amils. 2001. Microbial community composition and ecology of an acidic aquatic environment: the Tinto River, Spain. *Microb. Ecol.* 41:20–35.

Macario, A. J., M. Lange, B. K. Ahring, and E. C. De Macario. 1999. Stress genes and proteins in the archaea. *Microbiol. Mol. Biol. Rev.* 63:923–967.

Marshall, B. J., L. J. Barrett, C. Prakash, R. W. McCallum, and R. L. Guerrant. 1990. Urea protects *Helicobacter* (*Campylobacter*) *pylori* from the bactericidal effect of acid. *Gastroenterology* 99:697–702.

Matzke, J., B. Schwermann, and E. P. Bakker. 1997. Acidostable and acidophilic proteins: the example of the alpha-amylase from *Alicyclobacillus acidocaldarius*. *Comp. Biochem. Physiol. A Physiol.* 118:475–479.

Okibe, N., M. Gericke, K. B. Hallberg, and D. B. Johnson. 2003. Enumeration and characterization of acidophilic microorganisms isolated from a pilot plant stirred-tank bioleaching operation. *Appl. Environ. Microbiol.* 69:1936–1943.

Papke, R. T., J. E. Koenig, F. Rodriguez-Valera, and W. F. Doolittle. 2004. Frequent recombination in a saltern population of *Halorubrum*. *Science* 306:1928–1929.

Rawlings, D. E. 2002. Heavy metal mining using microbes. *Annu. Rev. Microbiol.* 56:65–91.

Ren, Q., and I. T. Paulsen. 2005. Comparative analyses of fundamental differences in membrane transport capabilities in Prokaryotes and Eukaryotes. *PLoS Comput. Biol.* 1:e27.

Richard, H., and J. W. Foster. 2004. *Escherichia coli* glutamate- and arginine-dependent acid resistance systems increase internal pH and reverse transmembrane potential. *J. Bacteriol.* 186:6032–6041.

Rudolph, C., G. Wanner, and R. Huber. 2001. Natural communities of novel archaea and bacteria growing in cold sulfurous springs with a string-of-pearls-like morphology. *Appl. Environ. Microbiol.* 67:2336–2344.

Ruepp, A., W. Graml, M. L. Santos-Martinez, K. K. Koretke, C. Volker, H. W. Mewes, D. Frishman, S. Stocker, A. N. Lupas, and W. Baumeister. 2000. The genome sequence of the thermoacidophilic scavenger *Thermoplasma acidophilum*. *Nature* 407:508–513.

Schäfer, G., M. Engelhard, and V. Muller. 1999. Bioenergetics of the Archaea. *Microbiol. Mol. Biol. Rev.* 63:570–620.

Schäfer, K., U. Magnusson, F. Scheffel, A. Schiefner, M. O. Sandgren, K. Diederichs, W. Welte, A. Hulsmann, E. Schneider, and S. L. Mowbray. 2004. X-ray structures of the maltose-maltodextrin-binding protein of the thermoacidophilic bacterium *Alicyclobacillus acidocaldarius* provide insight into acid stability of proteins. *J. Mol. Biol.* 335:261–274.

Schleper, C., G. Puehler, I. Holz, A. Gambacorta, D. Janekovic, U. Santarius, H. P. Klenk, and W. Zillig. 1995. *Picrophilus* gen. nov., fam. nov.: a novel aerobic, heterotrophic, thermoacidophilic genus and family comprising archaea capable of growth around pH 0. *J. Bacteriol.* 177:7050–7059.

Scott, D. R., D. Weeks, C. Hong, S. Postius, K. Melchers, and G. Sachs. 1998. The role of internal urease in acid resistance of *Helicobacter pylori*. *Gastroenterology* 114:58–70.

Selig, M., K. B. Xavier, H. Santos, and P. Schonheit 1997. Comparative analysis of Embden-Meyerhof and Entner-Doudoroff glycolytic pathways in hyperthermophilic archaea and the bacterium *Thermotoga*. *Arch. Microbiol.* 167:217–232.

Serour, E., and G. Antranikian. 2002. Novel thermoactive glucoamylases from the thermoacidophilic Archaea *Thermoplasma acidophilum*, *Picrophilus torridus* and *Picrophilus oshimae*. *Antonie Van Leeuwenhoek* 81:73–83.

She, Q., R. K. Singh, F. Confalonieri, Y. Zivanovic, G. Allard, M. J. Awayez, C. C. Chan-Weiher, I. G. Clausen, B. A. Curtis, A. De Moors, G. Erauso, C. Fletcher, P. M. Gordon, I. Heikamp-de Jong, A. C. Jeffries, C. J. Kozera, N. Medina, X. Peng, H. P. Thi-Ngoc, P. Redder, M. E. Schenk, C. Theriault, N. Tolstrup, R. L. Charlebois, W. F. Doolittle, M. Duguet, T. Gaasterland, R. A. Garrett, M. A. Ragan, C. W. Sensen, and J. Van der Oost. 2001. The complete genome of the crenarchaeon *Sulfolobus solfataricus* P2. *Proc. Natl. Acad. Sci. USA* 98:7835–7840.

Smith, P. F., T. A. Langworthy, and M. R. Smith. 1975. Polypeptide nature of growth requirement in yeast extract for *Thermoplasma acidophilum*. *J. Bacteriol.* 124:884–892.

Stingl, K., E. M. Uhlemann, R. Schmid, K. Altendorf, and E. P. Bakker. 2002. Energetics of *Helicobacter pylori* and its implications for the mechanism of urease-dependent acid tolerance at pH 1. *J. Bacteriol.* 184:3053–3060.

Tomb, J. F., O. White, A. R. Kerlavage, R. A. Clayton, G. G. Sutton, R. D. Fleischmann, K. A. Ketchum, H. P. Klenk, S. Gill, B. A. Dougherty, K. Nelson, J. Quackenbush, L. Zhou, E. F. Kirkness, S. Peterson, B. Loftus, D. Richardson, R. Dodson, H. G. Khalak, A. Glodek, K. McKenney, L. M. Fitzegerald, N. Lee, M. D. Adams, E. K. Hickey, D. E. Berg, J. D. Gocayne, T. R. Utterback, J. D. Peterson, J. M. Kelley, M. D. Cotton, J. M. Weidman, C. Fujii, C. Bowman, L. Watthey, E. Wallin, W. S. Hayes, M. Borodovsky, P. D. Karp, H. O. Smith, C. M. Fraser, and J. C. Venter. 1997. The complete genome sequence of the gastric pathogen *Helicobacter pylori*. *Nature* 388:539–547.

Tyson, G. W., J. Chapman, P. Hugenholtz, E. E. Allen, R. J. Ram, P. M. Richardson, V. V. Solovyev, E. M. Rubin, D. S. Rokhsar, and J. F. Banfield. 2004. Community structure and metabolism through reconstruction of microbial genomes from the environment. *Nature* 428:37–43.

van de Vossenberg, J. L., A. J. Driessen, W. Zillig, and W. N. Konings. 1998. Bioenergetics and cytoplasmic membrane stability of the extremely acidophilic, thermophilic archaeon *Picrophilus oshimae*. *Extremophiles* 2:67–74.

Weeks, D. L., S. Eskandari, D. R. Scott, and G. Sachs. 2000. A H+-gated urea channel: the link between *Helicobacter pylori* urease and gastric colonization. *Science* 287:482–485.

Weiller, G. F., G. Caraux, and N. Sylvester. 2004. The modal distribution of protein isoelectric points reflects amino acid properties rather than sequence evolution. *Proteomics* 4:943–949.

Woese, C. R., O. Kandler, and M. L. Wheelis. 1990. Towards a natural system of organisms: proposal for the domains Archaea, Bacteria, and Eucarya. *Proc. Natl. Acad. Sci. USA* 87:4576–4579.

Young, J. C., V. R. Agashe, K. Siegers, and F. U. Hartl. 2004. Pathways of chaperone-mediated protein folding in the cytosol. *Nat. Rev. Mol. Cell Biol.* 5:781–791.

Zychlinsky, E., and A. Matin. 1983. Cytoplasmic pH homeostasis in an acidophilic bacterium, *Thiobacillus acidophilus*. *J. Bacteriol.* 156:1352–1355.

VI. ALKALIPHILES

Chapter 23

Environmental and Taxonomic Biodiversities of Gram-Positive Alkaliphiles

ISAO YUMOTO

INTRODUCTION

Microorganisms are widely distributed in nature, and some of them live in extreme environments such as high and low temperatures, high osmotic conditions, and high hydrothermal pressure (Horikoshi and Grant, 1998). These microorganisms are called extremophiles, and most of them are adapted to specific environments and cannot grow in conventional laboratory conditions used for growth of microorganisms such as *Escherichia coli* (i.e., atmospheric pressure of 37°C, moderate ion strength, and neutral pH). Generally, such microorganisms have various special and specific physiological mechanisms for adaptation to such extreme environments. Compared with ordinary extremophiles, alkaliphiles are unique because typical alkaliphiles—the alkaliphilic *Bacillus* spp.—usually exist in conventional environments such as gardens and agricultural soil and manure (Guffanti et al., 1986; Horikoshi, 1991; Nielsen et al., 1995; Yumoto et al., 1997). Alkaliphilic microorganisms are also present in high pH environments such as naturally occurring alkaline environments, including soda lakes (Borsodi et al., 2005; Duckworth et al., 1996; Jones et al., 1998), underground alkaline water (Takami et al., 1997), relatively small alkaline niches such as intestines of insects (Thongaram et al., 2003), artificial alkaline environments such as liquid of indigo fermentation (Nakajima et al., 2005; Takahara and Tanabe, 1960; Yumoto et al., 2004b), and alkaline wastes as by-products of food-processing industries (Collins et al., 1983; Ntougias and Russel, 2000, 2001, Yumoto et al., 2002, 2004a).

There are two categories of bacteria that are able to grow at high pH, e.g., pH 9. The first group of bacteria is categorized as alkali-tolerant bacteria. They can grow at alkaline pH (pH 8 to 9), but their growth rate is highest around neutral pH. Many strains of nonalkaliphilic *Bacillus* spp. are able to grow at pH 9 but not at pH 10, and their optimum growth pH is around pH 7 to 8 (Priest et al., 1998). The second group of bacteria is categorized as alkaliphilic bacteria. Alkaliphilic bacteria can be defined as microorganisms that can grow at pH above 10 and/or grow equally or better in terms of growth intensity or velocity at pH higher than 9 compared with those at pH of 7 to 8. Alkaliphilic microorganisms can be further divided into facultative alkaliphiles, which can grow well at neutral pH, and obligate alkaliphiles, which cannot grow well at pH lower than 8. These facultative and obligate characteristics are related to the strategies of alkali-adaptation mechanisms, and these characteristics are reflected in the phylogenetic position constructed on the basis of 16S rRNA gene sequences. However, there are a few species that include both facultative and obligate alkaliphilic strains (Nielsen et al., 1995).

Since Vedder (1934) isolated the obligate alkaliphile *Bacillus alcalophilus*, many strains of obligate and facultative alkaliphiles have been isolated for industrial applications of their enzymes (Horikoshi, 1991, 1999), for ecological and taxonomic studies (Jones et al., 1998; Takami et al., 1997; Thongaram et al., 2003) and for physiological studies on environmental adaptation mechanisms (Goto et al., 2005; Krulwich and Guffanti, 1989; Krulwich et al., 1998, 2001; Yumoto, 2002, 2003). These alkaliphiles were isolated from ordinary soils, feces, manure of animals (Guffanti et al., 1986; Horikoshi, 1991; Nielsen et al., 1995), and indigo-dye ball (Ohta et al., 1975), and most of them were identified as *Bacillus* spp. During the past decade, many novel alkaliphilic microorganisms have been isolated not only from soils, feces, and manure of animals but also from soda lakes, the sea, artificial environments, and animals in sea waters.

I. Yumoto • National Institute of Advanced Industrial Science and Technology (AIST), Research Institute of Genome-Based Biofactory, Sapporo 062-8517, Japan.

The present understanding of the taxonomic distribution of alkaliphiles based on microorganisms so far isolated is as follows: *Cyanobacteria*, high G+C gram-positive bacteria, low G+C gram-positive bacteria, *Proteobacteria* (including alpha, gamma, and delta subdivisions), *Sphingobacteria*, *Cytophaga-Flexibacter-Bacteroides*, *Spirochaetales* (*Spirochaetaceae*), *Thermotogales* (*Thermopallium*), *Archaea*, and yeast (Horikoshi, 1991; Jones et al., 1998; Tiago et al., 2004; Zhilina et al., 2004). In addition to those isolated strains, bacteria belonging to beta subdivisions of *Proteobacteria*, *Verrucomicrobiales*, and *Sphingobacteria* were identified on the basis of results of analysis of DNA samples obtained from soda lakes (Humayoun et al., 2003). Although we cannot directly determine the taxonomic distribution of alkaliphiles on the basis of the analysis of a DNA clone library from DNA obtained from soda lakes, the results of previous studies indicate the possibility of an additional taxonomic diversities of alkaliphiles. Reviews on microbial diversity of soda lakes (Jones et al., 1998), chemolithotrophic haloalkaliphiles from soda lakes (Sorokin and Kuenen, 2005), and alkalithermophiles (Wiegel and Kevbrin, 2004) have been published. This chapter focuses on the environmental and taxonomic distributions of gram-positive alkaliphiles.

ENVIRONMENTAL DISTRIBUTIONS AND DIVERSITIES

Soil Samples

Alkaliphilic microorganisms can be isolated from ordinary garden soil by a conventional medium containing about 1% Na_2CO_3 as described in Horikoshi (1991). Alkaliphiles have been isolated not only from alkaline soils (e.g., pH 8 to 10) but also from acidic soils (e.g., pH 4). However, alkaliphiles have been isolated more frequently from soil samples with higher pH (Horikoshi, 1991). The reason why alkaliphiles exist in soils other than alkaline soils is not clear, though there are several possibilities. In the soil microbial process, ammonification of microorganisms might lead to localized alkalization in the soil, especially in soil to which nutrients containing nitrogen sources have been added. It is also possible that the intestines of insects living in the soil provide the microorganisms with a localized alkaline environment (Thongaram et al., 2003), and these microorganisms are scattered in the environment through feces of the insects. Furthermore, some alkaliphiles produce extracellular alkaline substances and make favorable the pH of their ambient environment (Horikoshi, 1991). Possible reasons for the distribution of alkaliphiles in conventional environments are summarized as follows: localized alkaline environments might be distributed in soil samples or endospore-forming alkaliphilic bacteria might be scattered in soil samples by, for example, dust storms from other places in which there are condensed niches of alkaliphiles.

Isolation of alkaliphilic microorganisms has been performed using soil samples collected from 7 sites in Hokkaido, the northernwest island of Japan. Twenty strains of *Bacillus* spp. and three strains of *Dietzia maris* were isolated. Isolated *Bacillus* spp. covered the DNA G+C mol% range from 34.4 to 45.3. Sixty percent of *Bacillus* strains exhibit DNA G+C mol% higher than 43.8, and they grow at 50°C and pH 7 (unpublished results). This relatively high G+C group contains isolates from every sampling station. Most of the isolates are facultative alkaliphiles. Only one strain, strain K241, is an obligate alkaliphile, suggesting that there are few obligate alkaliphiles in ordinary soil. However, some of the facultative alkaliphiles exhibit obviously better growth at pH 8 to 10 than at pH 7. The results described above indicated that the major group of alkaliphiles in soil isolated by using a conventional medium supplemented with 1% Na_2CO_3 is *Bacillus* spp. As shown in Table 1, most of the alkaliphilic *Bacillus* strains so far identified have been isolated from soil. In addition to *Bacillus* strains, *Paenibacillus campinasensis* (Yoon et al., 1998), *Paenibacillus daejeonensis* (Lee et al., 2002) and several strains in a high G+C gram-positive group, *Nocardiopsis kribbensis* (Yoon et al., 2005), *Nocardiopsis alkaliphila* (Hozzein et al., 2004) and *Dietzia maris* (Rainey et al., 1995), have been isolated from soil samples (Tables 3, 4).

Garbeva et al. (2003) developed a polymerase chain reaction (PCR) system for studying the diversity of the species of *Bacillus* and related taxa using DNA directly obtained from soil. This system is based on the specific amplification of 16S rRNA for *Bacillus* spp. By this PCR system, DNA clones of *Bacillus* spp. have been amplified using slightly acidic soil samples (pH 5.5 to 6.5). Among the amplified DNA clones, a clone identified as *Bacillus halodurans* has been obtained. It is thought that only major groups of *Bacillus* spp. were obtained as DNA clones by amplification of DNA obtained from the soil sample. Detection of *Bacillus halodurans* by this procedure indicated that although the soil samples were slightly acidic, *Bacillus halodurans* might be one of the major *Bacillus* species in the soil samples used in that study.

Many halophilic bacterial strains that can grow in a defined medium containing 20% NaCl were isolated from ordinary garden soil, yards, and roadways in an area surrounding Tokyo (Echigo et al., 2005). Although the purpose of the study was the isolation

Table 1. Characteristics of alkaliphilic *Bacillus* spp.

Species	Isolated location	pH range (optimum) (pH)	Temperature range (optimum) (°C)	Aerobe or anaerobe	Other characteristics	G+C content of DNA (%)	References
Bacillus cohnii	Soda lake, garden soil, indigo ball, horse feces meadow, feces	7–10 (9)	10–47	Aerobe	Grows at 7% NaCl	33.8–35.0	Nielsen et al., 1995; Ohta et al., 1975; Spanka and Fritze, 1993; Yumoto et al., 2000
Bacillus akibai	Soil	8–10 (9–10)	20–45 (37)	Aerobe	Grows at 7% NaCl	34.4	Nogi et al., 2005
Bacillus alcalophilus	Soil, faeces	8–10 (9–10)	10–40 (ND^a)	Aerobe	Grows at 8% NaCl	36.2–38.4	Nielsen et al., 1995; Nogi et al., 2005; Vedder, 1934
Bacillus agaradhaerens	Soil	8–11 (10 or above)	10–45 (ND)	Aerobe	Grows at 16% NaCl	36.5–36.8	Nielsen et al., 1995; Nogi et al., 2005
Bacillus hemicellulosilyticus	Soil	8–11 (10)	10–40 (37)	Aerobe	Grows at 12% NaCl	36.8	Nogi et al., 2005
Bacillus mannanilyticus	Soil	8–10 (9)	20–45 (37)	Aerobe	Grows at 3% NaCl	37.4	Nogi et al., 2005
Bacillus bongoriensis	Soda lake	8–11 (10)	10–40 (37)	Aerobe	Nonmotile	37.5	Vargas et al., 2005
Bacillus wakoensis	Soil	8–11 (9–10)	10–40 (37)	Aerobe	Grows at 10% NaCl	38.1	Nogi et al., 2005
Bacillus pseudoalcalophilus	Soil	8–10 (~10)	10–40 (ND)	Aerobe	Grows at 10% NaCl	38.2–39.0	Nielsen et al., 1995; Nogi et al., 2005
Bacillus vedderi	Bauxite-processing red mud tailing pond	Optimum temperature about 10	Maximum temperature for growth 40–50	Aerobe	Grows at 7.5% NaCl	38.3	Agnew␣et␣a.l, 1995
Bacillus pseudofirmus	Soil, animal manure	8–10 (~9)	10–45 (ND)	Aerobe	Grows at 16% NaCl	38.7–40.9	Nielsen et al., 1995
Bacillus halmapalus	Soil	7–10 (~8)	10–40 (ND)	Aerobe	No growth at 5% NaCl	38.6	Nielsen et al., 1995; Nogi et al., 2005
Bacillus oshimensis	Soil	7–10 (ND)	13–41 (28–32)	Aerobe	Grows at 0–20% NaCl, nonmotile	39.1–40.8	Nogi et al., 2005; Yumoto et al., 2005a
Bacillus cellulosilyticus	Soil	8–10 (9–10)	20–40 (37)	Aerobe	Grows at 12% NaCl	39.6	Nogi et al., 2005
Bacillus patagoniensis	Rhizosphere of shrub	7–10 (8)	5–40 (ND)	Aerobe	Grows at 15% NaCl	39.7	Olivera et al., 2005
Bacillus horti	Soil	7–10 (~10)	15–40 (ND)	Aerobe	Grows at 10% NaCl	40.2–40.9	Yumoto et al., 1998
Bacillus arsenicoselenatis	Bottom sediment of soda lake	7.5–10 (ND)	ND	Obligately anaerobe	Nonmotile, respiratory growth with Se(VI), As(V), Fe(III),nitrate, and fumarate	40.0	Switzer Blum et al., 1998
Bacillus krulwichiae	Soil	8–10 (ND)	20–45 (ND)	Facultatively anaerobe	Grows at 0–14% NaCl	40.6–41.5	Yumoto et al., 2003
Bacillus gibsonii	Soil	8–10 (ND)	10–37 (ND)	Aerobe	Grows at 9% NaCl	40.6–41.7	Nielsen et al., 1995; Nogi et al., 2005
Bacillus okuhidaensis	Hot spa	6.0–11.0 (ND)	30–60 (ND)	Aerobe	Grows at 10% NaCl	41.0–41.1	Li et al., 2002; Nogi et al., 2005
Bacillus horikoshii	Soil	7–10 (ND)	10–40 (ND)	Aerobe	Grows at 8–9% NaCl	41.1–42.0	Nielsen et al., 1995; Nogi et al., 2005
Bacillus clausii	Soil	7–10 (~8)	15–50 (ND)	Aerobe	Grows at 10% NaCl	41.7–43.5	Nielsen et al., 1995; Nogi et al., 2005
Bacillus halodurans	Soil	7–10 (ND)	15–55 (ND)	Aerobe	Grows at 12% NaCl	42.1–43.9	Nielsen et al., 1995; Nogi et al., 2005
Bacillus clarkii	Soil	8–11 (ND)	15–45 (ND)	Aerobe	Grows at 16% NaCl	42.4–43.0	Nielsen et al., 1995; Nogi et al., 2005
Bacillus saliphilus	Algal mat from mineral pool	7–10 (9.0)	4–50 (37)	Aerobe	Grows at 1–25% NaCl	48.4	Romano et al., 2005a
Bacillus selenitreducens	Bottom sediment of soda lake	8–11 (ND)	ND	Facultatively anaerobe	Nonmotile, reducing selenite	49.0	Switzer Blum et al., 1998

^{a} ND, no data.

of halophiles from ordinary soils, several alkaliphilic bacteria were isolated. Several isolates exhibited more than 99.5% similarity with *Alkalibacillus haloalkaliphilus* (Jeon et al., 2005) (formerly *Bacillus haloalkaliphilus* (Fritze et al., 1996). Although *Alkalibacillus haloalkaliphilus* was originally isolated from highly saline mud, the species was also isolated from ordinary soil. This bacterium is a spore-forming obligate alkaliphile and is strictly aerobe. In conclusion, most alkaliphiles in soil are gram-positive, spore-forming, and aerobic bacteria such as alkaliphilic *Bacillus* spp.

Gut of Higher Termites

It is known that termites harbor abundant microorganisms in their gut. Generally, the first proctodaeal segment of the gut in higher termites, especially in the subfamily *Termitinae*, has high pH (i.e., pH 10 to 12) and K^+ richness. Thongaram et al. (2003) analyzed the microflora of alkaliphilic bacteria in the first proctodaeal segment of five species of higher termites collected in Thailand by isolation of bacteria and a culture-independent method. The higher termites used in that study were *Pericapritermes latignathus* (*Termitidae*, soil-feeding), *Amitermes longignathus* (*Termitinae*, soil-feeding), *Termes comis* (*Termitinae*, soil/wood-interface), *Microcerotermes crassus* (*Termitinae*, wood-feeding), and *Speculitermes* sp. (*Apicotermitinae*, feeding unknown). Twenty-one alkaliphilic bacteria were isolated using a medium containing yeast extract, starch, several kinds of mineral, and K_2CO_3. All isolates except one strain were affiliated with the genus *Bacillus* in phylogenetic analysis based on the 16S rRNA gene sequences. Isolates belonging to the genus *Bacillus* were classified into six groups (Nielsen et al., 1994) and each group clustered with reported 16S rRNA gene sequences of alkaliphilic *Bacillus* as follows: eight strains with *Bacillus* sp. DSM 8717, six strains with *Bacillus gibsonii* or *Bacillus horikoshii*, two strains with *Bacillus clausii*, two strains with *Bacillus horti*, one strain with *Bacillus halmapalus*, and one strain with *Bacillus* sp. N-1. One strain not belonging to the genus *Bacillus* was closest to *Paenibacillus* sp. SM-XY60 (GenBank/EMBL/DDBJ accession number for 16S rRNA gene sequence: AB046581). Although bacterial flora in the termite gut has been studied extensively, very few bacilli clones have been detected in culture-independent analyses (Hongoh et al., 2003, Ohkuma and Kudo, 1996). The primers of PCR were made for specific amplification of DNA of alkaliphilic *Bacillus* spp., *Bacillus clausii*, *Bacillus gibsonii*, *Bacillus horikoshii*, *Bacillus patagoniensis* (formerly *Bacillus* sp. DSM 8714) and two strains of alkaliphilic isolates. By this method, many DNA clones related to *Bacillus* sp. DSM 8717, *Bacillus gibsonii* or *Bacillus horikoshii* and *Paenibacillus* sp. SM-XY60 were detected. Although those isolates exhibit characteristics similar to those of strains previously isolated from soil, some of the isolates from termites show a distinctive sensitivity to NaCl compared with corresponding soil isolates of alkaliphilic *Bacillus* spp. Both isolates and DNA clones obtained are related to alkaliphilic *Bacillus* spp. and related taxa isolated from soil. The results using termites suggest that several alkaliphilic *Bacillus* spp. in soil have been sharing ecological niches with those in the gut of termites in taxonomic viewpoint. Most of the alkaliphiles isolated from the gut of termites are facultative alkaliphiles.

Soda Lakes

Microbial diversities of soda lakes in Africa, Europe, and North America have been detected on the basis of the analysis of DNA clone libraries produced by amplification of obtained DNA as well as from the isolation of microorganisms from the environments (Duckworth et al., 1996; Humayoun et al., 2003, Jan-Roblero et al., 2004; Jones et al., 1998; López-García et al., 2005). Results of those studies indicate that the microbial diversities of soda lakes cover *Cyanobacteria*, high G+C gram-positive bacteria, low G+C gram-positive bacteria, *Verrucomicrobiales*, *Sphingobacteria*, *Proteobacteria* (including alpha, beta, gamma, and delta subdivisions), *Cytophaga-Flexibacter-Bacteroides*, *Spirochaetales* (*Spirochaetaceae*), *Thermotogales* (*Thermopallium*) and *Archaea*. In this review, ecological studies in East African soda lakes are focused upon.

Soda lakes are natural, high pH environments (pH 9.5 to 10.5) with Na^+, Cl^-, and HCO_3^-/CO_3^{2-} as the major ions in solution. Evaporation rates of soda lakes are high because of the scarcity of Mg^{2+} and Ca^{2+}. It is also known that soda lakes are among the most productive aquatic environments in the world. This is due to the almost unlimited supply of CO_2 combined with high ambient temperature and strong daily light. The primary productions in soda lakes are mainly supported by a dense population of cyanobacteria. The bacterial numbers are remarkably constant at about 10^5 cfu ml^{-1}. Analyses of aerobic bacteria populations in soda lakes have been performed by an isolation method using a conventional medium supplemented with Na_2CO_3. Most of the isolates were found to be obligate alkaliphiles. The major gram-negative isolates are members of the gamma subdivision of *Proteobacteria*. The isolates are related to the strain belonging to the genus *Halomonas* but are not identical to known species. In other gram-negative bacteria, two strains belong to the genus *Pseudomonas*, while the remaining six strains formed distinct groups.

Gram-positive isolates are divided into high G+C and low G+C groups. Two strains are related to the genus *Dietzia*, and they have been proposed as *Dietzia natronolimnaea* as the bacterium belongs to the high G+C group (Duckworth et al., 1998). Other strains of high G+C gram-positive types are related to known species belonging to the genera *Terrabacter*, *Micrococcus*, and *Arthrobacter*. However, these isolates are not very close to the corresponding known strains. Most of the low G+C group of isolates were associated with the genus *Bacillus*. Three strains are close but not identical to obligately alkaliphilic *Bacillus pseudofirmus* that was isolated from soil and manure (Fig. 1). Two strains in a distinct group (16S rRNA group 7) are related to facultatively alkaliphilic *Bacillus saliphilus*. However, the other six strains in group 7 are less related to known species. Isolates in this group are diversified also in phenotypic characteristics. One strain not belonging to the genus *Bacillus* exhibits highest similarity to a strain belonging to the genus *Carnobacterium*. This strain is related to the group of lactic acid bacteria. Strains related to *Bacillus pseudofirmus* seem to predominate in shoreline mud and dry foreshore soda soil that are subjected to fluctuating conditions in alkalinity and salinity with seasonal changes in water level. The strains in group 7 exhibiting high Na$^+$ requirement for growth appear to be more prevalent in lake waters and sediments where conditions were less changeable. Presently, isolates of alkaliphilic *Bacillus* strains from soda lakes are obviously different from known soil isolates in phylogenetic analysis based on 16S rRNA gene sequences (see Fig. 1).

From the analysis of anaerobic environments of soda lakes, six strains related to *Clostridium* "cluster XI" have been isolated. They formed a distinct group separated from previously described obligately anaerobic alkaliphiles, *Clostridium thermoalcaliphilum* and *Clostridium paradoxum*. Another isolate that is related to the genus *Eubacterium* might belong to a new genus. The remaining three haloalkaliphilic strains are closely related *Clostridium* "cluster VI". They represent a new genus of obligately anaerobic haloalkaliphiles.

The results described above suggested that although some of the alkaliphiles from soda lakes are closely related to known alkaliphiles, most of the isolates seem to be soda lake–specific taxa and they might have evolved separately within an alkaline environment.

The Sea and Related Samples

Six strains of alkaliphiles have been isolated from mud samples obtained from the Mariana Trench at a depth of 10,897 m. These strains grow in a pH range of 7.6 to 10.5 (four strains) or 7.6 to 9.5 (two strains). Four strains are able to grow at 30 MPa at 25°C and three strains grow slightly even at 100 MPa of hydrostatic pressure (Takami et al., 1997). Five of the six alkaliphiles are high G+C gram-positive. Two strains of isolates that are high G+C gram-positive are not closely related to reported sequences obtained by phylogenetic analysis based on 16S rRNA gene sequences. One strain is related to *Dietzia maris* (99% similarity), two strains are related to *Aureobacterium testaceum* (one of the two strains exhibiting 98% similarity) as high G+C gram-positive bacteria, and one strain is almost identical to proteobacterium, *Brevundimonas diminuta*. A halotolerant and facultative alkaliphile, *Oceanobacillus iheyensis*, has been isolated from deep-sea sediment collected at a depth of 1050 m on the Iheya Ridge of Nansei Islands (27°44.18′N, 126°54.15′E). This strain is a low G+C gram-positive and it is a strictly aerobic and spore-forming bacterium.

Marine lactic acid bacteria, *Marinilactibacillus psychrotolerans* and *Halolactibacillus halophilus*, have been isolated from living and decomposing marine organisms (Ishikawa et al., 2003, 2005). They are low G+C gram-positive bacteria. The former strain is a nonspore forming facultative alkaliphile that grows in the pH range of 6.0 to 10.5 and is a facultative anaerobe. It is similar to *Alkalibacterium* spp. (Nakajima et al., 2005, Ntougias and Russell, 2001, Yumoto et al., 2004b), an obligate alkaliphiles according to phylogenetic analysis based on 16S rRNA gene sequences. The latter strain is a nonspore forming facultative alkaliphile that grows in the pH range of 6 to 10 and is a facultative anaerobe. It lacks catalases, quinones, and cytochrome oxidase and is similar to *Amphibacillus* spp. (Niimura et al., 1990; Zhilina et al., 2001).

A typical gram-positive alkaliphilic *Bacillus* sp. has been isolated in soil samples as well as in samples of marine origin. Ivanova et al. (1999) isolated 20 aerobic endospore-forming bacilli from marine invertebrates and seawater in different areas of the Pacific Ocean. A group of four alkaliphiles were tentatively identified as facultative alkaliphilic *Bacillus horti*. On the other hand, a gram-negative facultatively alkaliphile, *Pseudomonas alcaliphila*, has been isolated from a seawater sample obtained from the coast of Japan Sea in Hokkaido (Yumoto et al., 2001). This strain belongs to the gamma subdivision *Proteobacteria*.

In conclusion, alkaliphiles belonging to high G+C gram-positive and low G+C gram-positive bacterial groups and *Proteobacteria* (including alpha and gamma subdivisions) are distributed in seawater, mud, or sediment and in living or decomposing marine organisms. Most of the alkaliphiles that are

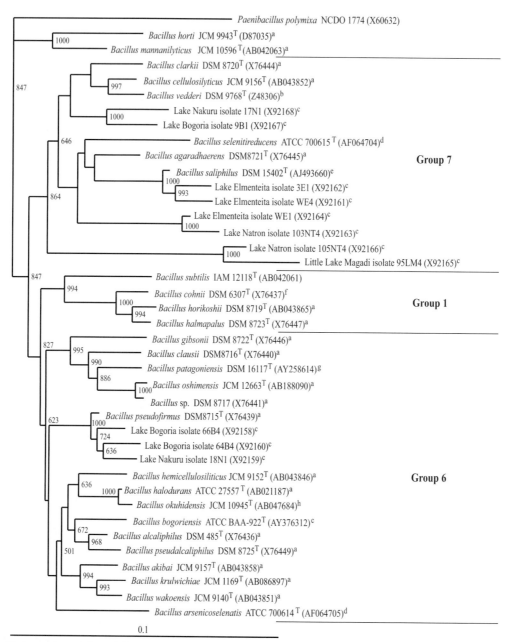

Figure 1. Phylogenetic tree derived from 16S rRNA gene sequences of alkaliphilic *Bacillus* spp., constructed using the neighbor-joining method. *Bacillus subtilis* IAM 12118[T] and *Paenibacillus polymixa* NCDO 1774 are used as a representative strain of group I species and outgroup of the phylogenetic tree, respectively. The two strains are not alkaliphiles. Origin of alkaliphiles: a, soil; b, bauxite-processing red mud tailing pond; c, soda lake; d, bottom sediment of soda lake; e, algal mat from mineral pool; f, soil and horse meadow; g, rhizosphere of shrub; h, hot spa. GenBank/EMBL/DDBJ accession numbers are given in parentheses. Numbers indicate bootstrap values greater than 500. Bar, 0.1 K_{nuc}.

distributed in the sea and its related samples are facultative alkaliphiles.

Indigo Fermentation

Traditionally, indigo blue dye has been produced from a Japanese indigo plant (*Polygonum tinctorium* Lour.) in Japan. The harvested indigo leaves are first air-dried and then appropriately wetted for proceeding to oxidation by aerobic activity of microorganisms. In this process the temperature reaches about 55°C but it is very difficult to control the appropriate water concentration for the microbial oxidation of indigo leaves. The product thus obtained is further processed by microbial reduction under alkaline conditions (pH values above 10), at which the original insoluble oxidized indigo is converted to the soluble form by the microorganisms. This process can be substituted by chemical reduction using dithionite. A procedure similar to that described above was also performed during the Middle Ages in Europe. The fermentation procedure for indigo reduction declined around 1960 in Japan because of difficulties in the production and maintenance processes; since then, the dye has been produced by a chemical process. Recently, traditional methods have been re-evaluated because they yield better products in terms of color than do chemical reduction.

Indigo-reducing bacteria have been isolated by Takahara and Tanabe (1960) and identified as *Bacillus* sp. they have been named *Bacillus alcalophilus*. The strain exhibits optimum growth at pH of 10.0 to 11.5 and requires, for its propagation, a peptide made of seven amino acid residues. The bacterium reduces indigo within 48 h fermentation in adjusted conditions, whereas its control without inoculation of the bacterium takes 168 h for the reduction. Since sodium dithionite used for reducing indigo is a burden to the environment, Padden et al. (1999) also isolated a thermophilic indigo-reducing bacterium, *Clostridium isatidis*, from a woad vat as a natural alternative. Although the bacterium can reduce indigo, it is not an alkaliphile according to the above definition.

Psychrotolerant, obligately alkaliphilic bacteria, *Alkalibacterium psychrotolerans*, and *Alkalibacterium iburiense*, have been isolated from indigo fermentation fluid of the same origin (Nakajima et al., 2005; Yumoto et al., 2004b). They are low G+C gram-positive facultative anaerobes, reduce indigo in laboratory medium, and produce DL-lactic acid. Although they share some characteristics, they exhibit different growth rates. It is thought that several kinds of indigo-reducing bacteria support the reduction of indigo in the actual indigo fermentation process. Normally, no microorganism is inoculated in the vat for fermentation since a microflora is spontaneously constructed in this artificial environment. Indigo-reducing *Alkalibacterium* spp. are obligate alkaliphiles and are well adapted to an alkaline environment. The origin of indigo-reducing *Alkalibacterium* spp. in the natural environment is not known.

Alkaliphilic bacteria other than indigo-reducing microorganisms, e.g., *Bacillus* spp., exist in the vat for indigo fermentation. The microflora involved in the indigo fermentation process, and the relationship between indigo-reducing bacteria and other microorganisms are currently unknown.

Other Environments

Alkaliphiles may be distributed not only in natural environments but also in alkaline wastes such as byproducts of agricultural or fish-processing plants. For example, *Exiguobacterium aurantiacum* has been isolated from potato-processing plant effluents (Collins et al., 1983). Facultatively alkaliphilic *Bacillus* sp. WW3-SN6 has been isolated from alkaline washwaters derived from the preparation of edible olives (Ntougias and Russell, 2000). Alkaline solution containing NaOH was used to precipitate a bitter compound of the fruit. The isolate is related to alkaliphilic *Bacillus* sp. DSM 8714 originated from soil in the analysis of 16S rRNA gene sequences. *Bacillus* sp. DSM 8714 was later identified as an alkaliphilic bacterium, *Bacillus patagoniensis*, that was isolated from the rhizosphere of the perennial shrub *Atiplex lampa* (Olivera et al., 2005). This is the only species that can grow at 5°C among the currently known alkaliphilic *Bacillus* spp. The facts described above suggest that niches of *Bacillus patagoniensis* are in soil and in rhizosphere of certain plants. In addition to alkaliphilic *Bacillus* sp. WW3-SN6, *Alkalibacterium olivoapovliticus* has also been isolated from washwater of edible-olive production (Ntougias and Russell, 2001). Another known *Alkalibacterium* spp. has been isolated from the indigo fermentation process (Nakajima et al., 2005; Yumoto et al., 2004b), and the most likely origin of these strains is the plant from which indigo dye is produced. These facts suggest that the genus *Alkalibacterium* originally is associated to indigo plant or is ubiquitously present in the process of the indigo blue dye production.

On the other hand, alkaliphilic bacteria, *Dietzia psychralcaliphila* (Yumoto et al., 2002) and *Exiguobacterium oxidotolerans* (Yumoto et al., 2004a), have been isolated from the drain of a fish-processing plant. The former one is a high G+C gram-positive bacterium and the latter one is a low G+C gram-positive bacterium. The relationship between these isolates and fish treated in the plant is not known. An alkaliphile species related to *Dietzia psychralcaliphila*, *Dietzia maris*, has been isolated from carp in freshwater as well as from halibut in the sea (Harrison, 1929; Nesterenko et al., 1982; Rainey et al., 1995), and *Oceanobacillus oncorhynchi* has been isolated from the skin of rainbow trout (*Oncorhynchus mykiss*) (Yumoto et al., 2005b), suggesting that certain alkaliphiles exist in the skin or intestines of fishes. However, it is not known why alkaliphiles are attached to the skin of certain fishes.

In addition to the samples described above, alkaliphiles have been isolated from animal manure. For example, obligate alkaliphiles belonging to *Bacillus pseudofirmus* have been isolated from deer and ostrich manure, and strains of the facultative alkaliphile *Bacillus halodurans* have been isolated from chicken, tiger, pigeon, and elephant manure (Nielsen et al., 1995). On the other hand, the facultative alkaliphile *Amphibacillus xylanus* has been isolated from compost of grass and rice straw (Niimura et al., 1990). Low G+C gram-positive alkaliphilic bacteria therefore appear to be distributed in manure from animal as well as plant origins. Several microorganisms, called hyperammonia-producing (HAP) bacteria, have been isolated from swine manure storage pits (Whitehead and Cotta, 2004) as well as from the ruminal ecosystem (Eschenlauer et al., 2002, Wallace et al., 2004). The distribution of alkaliphiles may be related to the distribution of such HAP bacteria.

The above results suggest that alkaliphiles are distributed in rhizosphere plants, skin of fishes, and manure from plants as well as animals sources. Other habitats of alkaliphiles such as a hot spa (Li et al. 2002) and feces (Spanka and Fritze, 1993) have been identified. It has been considered that most of the isolates belong to the genus *Bacillus*. The wide distribution of alkaliphilic *Bacillus* spp. might be related with their spore-forming characteristic.

TAXONOMIC DIVERSITIES

Gram-Positive Low G+C Bacteria

Bacillus spp.

Over 20 alkaliphilic *Bacillus* species have become approved species. The characteristics of these species are summarized in Table 1. A phylogenetic tree of alkaliphilic *Bacillus* spp., including known species and isolates from soda lakes, built on the basis of 16S rRNA gene sequences is shown in Fig. 1. Most of the known species have the following characteristics: isolated from soil samples, maximum pH for growth between 10 and 11, mesophilic growth characteristics, aerobes and ability to grow at more than 5% NaCl. G+C contents of DNA in these species range from 33.8 to 43.0% except for *Bacillus saliphilus* (Romano et al., 2005a) and *Bacillus selenitireducens* (Switzer Blum et al., 1998) (ca. 49.0%). Among these *Bacillus* spp., *Bacillus arsenicoselenatis* exhibits particularly distinct characteristics (Switzer Blum et al., 1998) (ca. 49%). Although catalase and oxidase reactions are positive, it is a strict anaerobe that demonstrates respiratory growth with Se(VI), As(V), Fe(III), nitrate, and fumarate as electron acceptors. Most of the reported alkaliphilic *Bacillus* spp. belong to a similar phylogenetic position derived from the 16S rRNA gene sequence of group 1, group 6, or group 7 including *Bacillus agaradhaerens* and *Bacillus clarkii* in the genus *Bacillus* (according to the grouping in refs Ash et al., 1991 and Nielsen et al., 1994) (Fig. 1). Obligate alkaliphilic *Bacillus* spp. were found to have a similar phylogenetic position (e.g., *Bacillus alcalophilus* and *Bacillus pseudoalcaliphilus* in group 6 and alkaliphiles in group 7).

Obligate anaerobes

Since the isolation of the obligately anaerobic alkaliphile *Clostridium paradoxum* (Li et al., 1993) from various sewage plants, about 10 species of obligately anaerobic alkaliphiles have been isolated to date. The approved species are listed in Table 2. All of the approved species except *Alkalibacter saccharofermentans* (Garnova et al., 2004) (belonging to cluster XV) belong to *Clostridium* cluster XI. Most of the approved species have been isolated from soda lakes. This fact indicates that although most of the strains belonging to *Clostridiaceae* are facultative alkaliphiles, they are distributed only in high pH environments. Some of the strains exhibit particular characteristics. For example, *Alkaliphilus transvaalensis* (Takai et al., 2001) grows at pH 7.5 to 12.5, the highest growth pH in alkaliphiles.

Lactic acid bacteria

Since the isolation of the alkaliphilic lactic acid bacterium *Alkalibacterium olivoapovliticus* (Ntougias and Russell, 2001), six approved species of alkaliphilic lactic acid bacteria have been isolated to date. The approved species are listed in Table 3. They have been isolated from artificial environments such as washwater of edible olive production as well as natural environments such as living and decomposing marine organisms. This means that this group of bacteria is widely distributed. This group contains both obligate and facultative alkaliphiles, and all strains are facultative anaerobes. The range of G+C contents of DNA is 36.2 to 43.2%. *Alkalibacterium psychrotolerans* and *Alkalibacterium iburiensis* have been isolated from a vat for indigo fermentation as indigo-reducing bacteria (Nakajima et al., 2005; Yumoto et al., 2004b). They are able to grow at very high pH, i.e., pH 12. This fact might be useful for the selection of indigo-reducing bacteria.

Other gram-positive low G+C bacteria

This group is closely related to the group of the genus *Bacillus*. The approved species are listed in Table 3. Some of the strains in this group were formally classified as *Bacillus* (Claus and Berkeley, 1986). Sources of their isolation are soil or similar environments. Their characteristics are also

Table 2. Characteristics of alkaliphilic obligately anaerobic and lactic acid bacterial groups

Species	Isolated location	pH range (optimum) (pH)	Aerobe or anerobe	Other characteristics	G+C content of DNA (%)	References
Obligate anaerobe						
Clostridium paradoxum	Various sewage plants	7.0–11.1 (around 10.1)	Obligate anerobe	Growth temperature range 30–63°C	30	Li et al., 1993
Clostridium thermoalcaliphilum	Sewage sludge	7.0–11.0 (9.6–10.1)	Obligate anerobe	Growth temperature range 27–57.5°C	32	Li et al., 1994
Natronoincola histidinovorans	Soda deposits in soda lake	8.0–10.5 (9.4)	Obligate anerobe	Acetogenic bacterium, moderately haloalkaliphile	32.3	Zhilina et al., 2001
Alkaliphilus metalliredigens	Alkaline leachate pond	7.5–11.0 (9.5)	Obligate anerobe	Metal-reducing bacterium	ND	Ye et al., 2004
Alkaliphilus transvaalensis	Alkaline water pool at depth of 3.2 kmbls in gold mine	8.5–12.5 (10)	Obligate anerobe	Growth temperature range 20–50°C optimum temperature 40°C	36.4	Takai et al., 2001
Alkalibacter saccharofermetans	Soda lake	7.2–10.2 (9.0)	Obligate anerobe	Saccharolytic	42.1	Garnova et al., 2004
Tindallia magadiensis	Soda deposites in soda lake	8.5–10.5 (7.5)	Obligate anerobe	Produce acetate, propionate, and ammonia	37.6	Kevbrin et al., 1998
Tindallia californiensis	Soda lake	8.0–10.5 (9.5)	Obligate anerobe	Mesophilic	44.4	Pikuta et al., 2003
Anoxynatronum sibiricum	Mixture of mud and surface cyanobacterial mud	7.1–10.1 (ND)[a]	Obligate anerobe	Reduces Fe^{3+} to Fe^{2+}	48.4	Garnova et al., 2003
Lactic acid bacteria						
Marinilactobacillus psychrotolerans	Decaying marine algae, living sponge	6.0–10.5 (8.5–9.0)	Facultative anerobe	DL-lactic acid producing	36.2	Ishikawa et al., 2003
Alkalibacterium olivoapoliticus	Wash-waters of edible olives	8.5–10.8 (9.0–10.2)	Facultative anerobe	Optimum growth at 3–5% NaCl	39.7±1.0	Ntougias and Russell, 2001
Halolactobacillus miurensis	Living and decaying marine organisms	6.0–10.0 (8.0–9.5)	Facultative anerobe	L-lactic acid is a major end-product from glucose in anaerobic cultivation	38.5–40.0	Ishikawa et al., 2003
Halolactobacillus halophilus	Living and decaying marine organisms	6.0–10.0 (8.0–9.5)	Facultative anerobe	L-lactic acid is a major end-product from glucose in anaerobic cultivation	39.6–40.7	Ishikawa et al., 2003
Alkalibacterium psychrotolerans	Fermented polygonum indigo	9–12 (ND)	Facultative anerobe	DL-lactic acid producing	40.6	Yumoto et al., 2004b
Alkalibacterium iburiense	Fermented polygonum indigo	9–12 (ND)	Facultative anerobe	DL-lactic acid producing	42.6–43.2	Nakajima et al., 2005

[a]ND, no data.

Table 3. Characteristics of alkaliphilic aerobic mesophiles other than *Bacillus* spp. and thermophiles

Species	Isolated location	pH range (optimum) (pH)	Aerobe or anaerobe	Other characteristics	G+C content of DNA (%)	References
Aerobic mesophile						
Cerasibacillus quisquilarum	Decomposing system of kitchen refuse	7.5–10 (8–9)	Obligate aerobe	Grows at <10% NaCl	35.5	Nakamura et al., 2004
Oceanobacillus oncorhnchi	Skin of a rainbow trout	9–10 (ND[a])	Facultative anaerobe	Grows at 0–22% NaCl	38.5	Yumoto et al., 2005b
Oceanobacillus iheyensis	Mud from Iheya Ridge	6.5–10 (7–9.5)	Obligate aerobe	Grows at 0–20% NaCl	35.8	Lu et al., 2001
Salibacillus salexigens	Hypersaline soils	6.0–11.0 (7.5)	Obligate aerobe	Grows at 7–20% NaCl	36.3–39.5	Garabito et al., 1997; Waino et al., 1999
Gracilibacillus halotolerans	Salt lake (Great Salt Lake, USA)	5–10 (7.5)	Obligate aerobe	Grows at 0–20% NaCl	38	Waino et al., 1999
Amphibacillus xylanus	Alkaline manure with grass and rice straw	8–10	Facultative anaerobe	Grows at <6% NaCl	36–38	Niimura et al., 1990
Amphibacillus tropicus	Sediment, soda lake	8.5–11.5 (9.5–9.7)	Facultative anaerobe	Grows at 1.0–20.9% NaCl	39.2	Zhilina et al., 2001
Amphibacillus fermentum	Sediment, soda lake	7.0–10.5 (8.5–9)	Facultative anaerobe	Grows at 1.0–19.7% NaCl	41.5	Zhilina et al., 2001
Alkalibacillus haloalkaliphilus	Highly saline mud	no growth at pH 7, good growth occurs at pH 9.7	Obligate aerobe	Grows at 20% NaCl	38.0	Fritze et al., 1996; Jeon et al., 2005
Alkalibacillus filiformis	Mineral pool	7.0–10.0 (9.0)	Obligate aerobe	Nonmotile, grows at 18% NaCl	39.5	Romano et al., 2005b
Sporosarcina pasteurii	Soil, water, sewage and incrustation and incrustations on urinals	opt. pH: 9	Facultative anaerobe	Maximum temperature 33–42°C	42	Claus and Berkeley 1986
Exiguobacterium oxidotolerans	Drain of a fish processing plant	7–10 (ND)	Facultative anaerobe	Growth temperature range 4–40°C	46.7	Yumoto et al., 2005a
Exiguobacterium marinum	Tidal flat of Yellow Sea	5–10.5 (6.5–8.5)	Facultative anaerobe	Growth temperature range 10–47°C	48.0	Kim et al., 2005
Exiguobacterium aestuarii	Tidal flat of Yellow Sea	5–10.5 (6.5–8.5)	Facultative anaerobe	Growth temperature range 10–47°C	48.4–48.6	Kim et al., 2005
Exiguobacterium aurantiacum	Potato-processing effluent	6.5–11.5 (8.5–9.5)	Facultative anaerobe	Growth temperature range 7–43°C	53.2–55.8	Collins et al., 1983
Paenibacillus campinasensis	Soil	7.5–10.5 (10)	Aerobe	Gram-variable	50.9	Yoon et al., 1998
Paenibacillus daejeonensis	Soil	7.0–13.0 (8.0)	Facultative anaerobe	Gram-variable	53	Lee et al., 2002
Thermophile						
Anaerobranca horikoshii	Water and soil samples (Yellowstone National Park)	6.9–10.3 (8.5)	Obligately anaerobe	Growth temperature range 30–66°C, Optimum temperature 57°C	34	Engle et al., 1995
Anoxybacillus pushchinensis	Manure	8.0–10.5 (9.5–9.7)	Obligately anaerobe	Optimum temperature 62°C	42.2±0.2	Pikuta et al., 2000
Anoxybacillus kestanbolensis	Hot spa	6.0–10.5 (7.5–8.5)	Facultatively anaerobe	Growth temperature range 40–70°C, Optimum temperature 50–55°C	50	Dulger et al., 2004
Anoxybacillus ayderensis	Hot spa	6.0–11.0 (7.5–8.5)	Facultatively anaerobe	Growth temperature range 30–70°C, Optimum temperature 50°C	54	Dulger et al., 2004
Anoxybacillus gonensis	Hot spa	6.0–10.0 (7.5–8.0)	Facultative anaerobe	Growth temperature range 40–70°C, Optimum temperature 55–60°C	57	Belduz et al., 2003

[a] ND, no data.

CHAPTER 23 • ENVIRONMENTAL AND TAXONOMIC BIODIVERSITIES OF GRAM-POSITIVE ALKALIPHILES 305

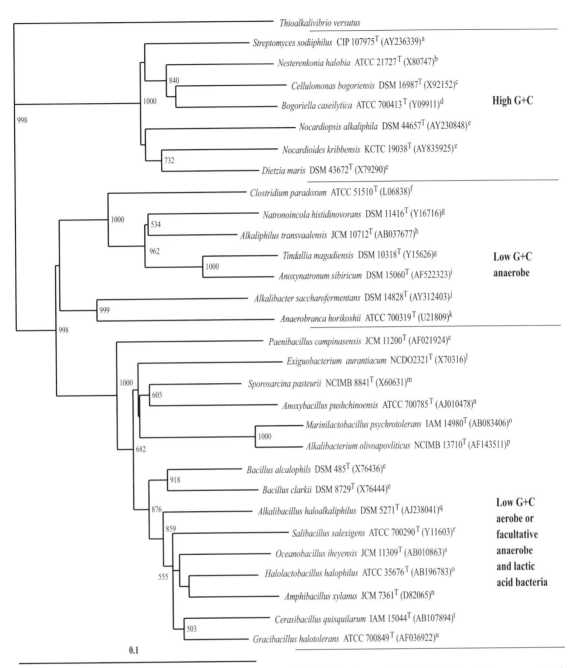

Figure 2. Phylogenetic tree derived from 16S rRNA gene sequences of alkaliphiles that are high G+C and low G+C gram-positive bacteria. *Thioalkalivibrio versutus* DSM 13738[T] is used as an outgroup of the phylogenetic tree, and the strain is not a gram-positive alkaliphile. Origin of alkaliphiles: a, muddy sample in salt lake; b, salt for soy source; c, sediment of littoral zone of soda lake; d, soda soil in soda lake; e, soil; f, sewage; g, soda deposits in soda lake; h, alkaline water pool in gold mine; i, mixture of mud and surface cyanobacterial mud; j, soda lake; k, water and soil sample; l, potato-processing effluent; m, unknown; n, manure; o, marine organisms; p, washwaters of edible olive; q, highly saline mud; r, hypersaline soil; s, mud from deep sea; t kitchen refuse; u, salt lake. GenBank/EMBL/DDBJ accession numbers are given in parentheses. Numbers indicate bootstrap values greater than 500. Bar, 0.1 K_{nuc}.

similar to those for the group of the genus *Bacillus*. Several strains can adapt to a wide pH range. For example, *Exiguobacterium aestuarii* grows at pH 5 to 10.5 (Kim et al., 2005). The number of facultative aerobes in this group is larger than that of alkaliphilic *Bacillus* spp. Several strains can grow even at 20% NaCl. The range of G+C contents of DNA is 35.5 to 55.8%, larger than the range for alkaliphilic *Bacillus* spp. These different characteristics from those of *Bacillus* spp. suggest differences in mechanisms of adaptation to alkaline environments.

Table 4. Characteristics of alkaliphilic gram-positive high G+C bacteria

Species	Isolated location	pH range (optimum) (pH)	Aerobe or anerobe	Other characteristics	G+C content of DNA (%)	References
Dietzia natronolimnaea	Soda lake	6–10 (9)	Aerobe	Grows at 20–40°C	66.1±0.4	Duckworth et al., 1998
Dietzia psychralcaliphila	Drain of a fish processing plant	7–10 (ND[a])	Aerobe	Grows at 5–30°C, grows at 0–10% NaCl	69.6	Yumoto et al., 2001
Dietzia maris	Halibut, skin and intestinal tract of carp, soil, deep sea sediments	7–10 (ND)	Aerobe	Mesophilic	73	Harrison, 1929; Nesterenko et al., 1982; Rainey et al., 1995
Cellulomonas bogoriensis	Sediment of the littorral zone of soda lake	6.0–10.5 (9–10)	Facultative anaerobe	Optimum temperature 30–37°C	71.5	Jones et al., 2005
Nocardiopsis alkaliphila	Soil sample collected from eastern desert of Egypt	7.0–12.0 (9.5–10.0)	Aerobe	Grows at 10–45°C, grows at 0–10.0% NaCl	65.5	Hozzein et al., 2004
Nacardioides kribbensis	Alkaline soil in Korea	6.0–11.0 (9.0)	Aerobe	Grows at 4 and 35°C	73–74	Yoon et al., 2005
Begoriella caseilytica	Soda soil in soda lake	Optimum pH 9–10	Aerobic or microaerophile	NaCl-tolerant	70	Groth et al., 1997
Streptomyces sodiiphilus	Muddy sample in salt lake	Optimum pH 9.0–10.0	Aerobe	Optimum temperature 28°C	70.5	Li et al., 2005b
Nesterenkonia halobia	Salt for brew soy sauce	6–10 (7.0)	Aerobe	Grows at 1–2 M NaCl	ND	Onishi and Kamekura, 1972; Stackebrandt et al., 1995
Nesterenkonia lutea	Saline soil	6.5–10.0 (8.0–9.0)	Aerobe	Grows at 1–15% NaCl	64	Li et al., 2005a

[a] ND, no data.

Thermophiles

Several alkaliphilic thermophiles have been isolated. Most of them have been isolated from hot spas. Their optimal pH range for growth is 7.5 to 8.5 and their optimal temperature range for growth is 50 to 60°C. Among this group, *Anaerobranca horikoshi* alone belongs to the position of the low G+C anaerobic group in a phylogenetic tree based on 16S rRNA gene sequences (Fig. 2). The species exhibiting pronounced characteristics in this group is *Anoxybacillus pushchinensis*. It has been isolated from manure, and its optimum pH range and optimum temperature for growth are 9.5 to 9.7 and 62°C, respectively (Pikuta et al., 2000).

Gram-positive high G+C bacteria

Since the isolation of *Dietzia maris* (Rainey et al., 1995), formally *Rhodococcus maris* (Nesterenko et al., 1982), several strains in the gram-positive high G+C group have been isolated. The approved species are listed in Table 4. The phylogenetic position of a representative strain based on 16S rRNA gene sequences is shown in Figure 2. Sources of these strains include sediment, soil, and soda lakes. Most of these bacteria are facultative alkaliphiles. All strains except *Cellulomonas bogoriensis* are aerobic or microaerophilic. The range of G+C contents of DNA is 65.5 to 74%. A wide distribution of *Dietzia maris* has been demonstrated by isolation of the bacteria from various natural environments. This species has been isolated from fishes from both seawater (i.e., halibut) and freshwater (i.e., carp), soil and deep sea sediments as described above. On the other hand, *Dietzia natronolimnaea* is distributed not only in African and Hungarian soda lakes but also in an alkaline groundwater environment generated by active serpentinization, which results in a $Ca(OH)_2^-$ enriched, extremely diluted groundwater with pH of 11.4 (Borsodi et al., 2005). The species is distributed widely but in quite specific environments.

CONCLUSIONS

In the past decade, many gram-negative as well as gram-positive alkaliphilic bacteria have been isolated and their taxonomic positions have been determined. A phylogenetic tree including every category of gram-positive bacteria is shown in Figure 2.

We now have a better understanding of the environmental and taxonomic diversities of alkaliphiles than a decade ago. In the natural environment, although there are several exceptions, obligate alkaliphiles exhibit better distribution in vast and obligately alkaline environments such as soda lakes than in conventional environments such as garden soil. Furthermore, several species of obligate alkaliphiles occupy similar phylogenetic positions based on 16S rRNA gene sequences. On the other hand, the original distribution in natural environment of obligate alkaliphiles isolated from artificial environments such as a vat for indigo fermentation and waste from plant manure is not yet known. It can be assumed that they ubiquitously exist in artificial or natural environments and that they have tolerance to unfavorable conditions such as dryness, inadaptable pHs, and nutritionally poor conditions. We also do not understand why facultative alkaliphiles are distributed in conventional environments such as soil. It can be assumed that there is a mosaic distribution of alkaline environments of very limited sizes. For example, certain larvae of insects such as the beetle, *Trypoxylus dichotoma*, live in soil and their gut exhibits high pH, i.e., pH 10. Such insects may scatter alkaliphiles through their feces.

In the next decade, our understanding of the distribution in the environment and of the taxonomic diversities of alkaliphiles will proceed further not only by isolation of novel species of alkaliphiles but also from results of analyses of DNA directly obtained from various environments.

REFERENCES

Agnew, M. D., S. F. Koval, and K. F. Jarrell. 1995. Isolation and characterization of novel alkaliphiles from bauxite-processing waste and description of *Bacillus vedderi* sp. nov., a new obligate alkaliphile. *Syst. Appl. Microbiol.* 18:221–230.

Ash, C., J. A. E. Farrow, S. Wallbanks, and M. D. Collins. 1991. Phylogenetic heterogeneity of the genus *Bacillus* revealed by comparative analysis of small subunit ribosomal RNA sequences. *Lett. Appl. Microbiol.* 13:202–206.

Belduz, A. O., S. Dulger, and Z. Demirbag. 2003. *Anoxybacillus gonensis* sp. nov., a moderately thermophilic, xylose-utilizing, edospore-forming. *Int. J. Syst. Evol. Microbiol.* 53:1315–1320.

Borsodi, A. K., A. Micsinai, A. Rusznyák, P. Kovács, E. M. Tóth, and K. Márialigeti. 2005. Diversity of alkaliphilic and alkalitolerant bacteria cultivated from decomposing reed rhizomes in a Hungarian soda lake. *Microb. Ecol.* 50:9–18.

Claus D., R. and C. W. Berkeley. 1986. Genus *Bacillus*, Cohn 1872, p. 1105–1139. *In* P. H. A. Sneath, N. S. Mair, M. E. Sharpe, and J. G. Holt. (ed.) *Bergey's Manual of Systematic Bacteriology*. Vol. 2. Baltimore: The Williams and Wilkins Co.

Collins, M. D., B. M. Lund, J. A. E. Farrow, and K. H. Schleifer. 1983. Chemotaxonomic study of an alkalophilic bacterium, *Exiguobacterium aurantiacum* gen. nov., sp. nov. *J. Gen. Microbiol.* 129:2037–2042.

Duckworth, A. W., W. D. Grant, B. E. Jones, and R. van Steenbergen. 1996. Phylogenetic diversity of soda lake alkaliphiles. *FEMS Microbiol. Ecol.* 117:61–65.

Duckworth, A. W., S. Grant, W. D. Grant, B. E. Jones, and D. Meijer. 1998. *Dietzia natronolimnaios* sp. nov., a new member of the genus *Dietzia* isolated from an East African soda lake. *Extremophiles* 2:359–366.

Dulger, S., Z. Demirbag, and A. O. Belduz. 2004. *Anoxybacillus ayderensis* sp. nov. and *Anoxybacillus kestanbolensis* sp. nov. *Int. J. Syst. Evol. Microbiol.* **54**:1499–1503.

Echigo, A., M. Hino, T. Fukushima, T. Mizuki, M. Kamekura, and R. Usami. 2005. Endospores of halophilic bacteria of the family Bacillaceae isolated from nonsaline Japanese soil may be transported by Kosa event (Asian dust storm). *Saline Systems* **1**:8.

Engle, M., Y. Li, C. Woese, and J. Wiegel. 1995. Isolation and characterization of a novel alkalitolerant thermophile, *Anaerobranca horikoshii* gen. nov., sp. nov. *Int. J. Syst. Bacteriol.* **45**:454–461.

Eschenlauer, S. C. P., N. McKain, N. D. Walker, N. R. McEwan, C. J. Newbold, and R. J. Wallace. 2002. Ammonia production by ruminal microorganisms and enumeration, isolation, and characterization of bacteria capable of growth on peptide and amino acids from the sheep rumen. *Appl. Environ. Microbiol.* **68**:4925–4931.

Fritze, D. 1996. *Bacillus haloalkaliphilus* sp. nov. *Int. J. Syst. Bacteriol.* **46**:98–101.

Garabito, M. J., D. R. Arahal, E. Mellado, M. C. Marquez, and A. Ventosa. 1997. *Bacillus salexigens* sp. nov., a new moderately halophilic *Bacillus* species. *Int. J. Syst. Bacteriol.* **47**:735–741.

Garbeva, P., J. A. van Veen, J. D. Van Elsas. 2003. Predominant *Bacillus* spp. In agricultural soil under different management regimes detected via PCR-DGGE. *Microb. Ecol.* **45**:302–316.

Garnova, E. S., T. N. Zhilina, T. P. Tourova, and A. M. Lysenko. 2003. *Anoxynatronum sibiricum* gen. nov., sp. nov. alkaliphilic saccharolytic anaerobe from cellulolytic community of Nizhnee Beloe (Transbaikal region). *Extremophiles* **7**: 213–220.

Garnova, E. S., T. N. Zhilina, T. P. Tourova, N. A. Kostrikina, and G. A. Zavarzin. 2004. Anaerobic, alkaliphilic, saccharolytic bacterium *Alkalibacter saccharofermentans* gen. nov., sp. nov. from a soda lake in the Transbaikal region of Russia. *Extremophiles* **8**:309–316.

Goto, T., T. Matsuno, M. Hishinuma-Narisawa, K. Yamazaki, H. Matsuyama, N. Inoue, and I. Yumoto. 2005. Cytochrome *c* and bioenergetic hypothetical model for alkaliphilic *Bacillus* spp. *J. Biosci. Bioeng.* **100**:365–379.

Groth, I., P. Schumann, F. A. Rainey, K. Martin, B. Schuetze, and K. Augusten. 1997. *Bogoriella caseilytica* gen. nov., sp. nov., a new alkaliphilic actinomycete from a soda lake in Africa. *Int. J. Syst. Bacteriol.* **47**:788–794.

Guffanti, A. A., O. Finkelthal, D. B. Hicks, L. Falk, A. Sidhu, A. Garro, and T. A. Krulwich. 1986. Isolation and characterization of new facultatively alkalophilic strains of *Bacillus* species. *J. Bacteriol.* **167**:766–773.

Harrison, F. C. 1929. The discoloration of halibut. *Can. J. Res.* **1**:214–239.

Hongoh, Y., M. Ohkuma, and T. Kudo. 2003. Molecular analysis of bacterial microbiota in the gut of the termite *Reticulitermes speratus* (*Isoptera; Rhinotermitidae*). *FEMS Microbiol. Ecol.* **44**:231–242.

Horikoshi, K. 1991. Microorganisms in Alkaline Environments. Kodansha, Tokyo, Japan/VCH, Weinheim, Germany.

Horikoshi, K., and W. D. Grant. 1998. *Extremophiles, Microbial Life in Extreme Environments.* Wiley-Liss, A John Wiley & Sons, New York, NY.

Horikoshi, K. 1999. Alkaliphiles: some applications of their products for biotechnology. *Microbiol. Mol. Biol. Rev.* **63**:735–750.

Hozzein, W. N., W.-J. Li, M. I. A. Ali, O. Hammouda, A. S. Mousa, L.-H. Xu, and C.-L. Jiang. 2004. *Nocardiopsis alkaliphila* sp. nov., a novel alkaliphilic actinomycete isolated from desert soil in Egypt. *Int. J. Syst. Evol. Microbiol.* **54**:247–252.

Humayoun, S. B., N. Bano, and J. T. Hollibaugh. 2003. Depth distribution of microbial diversity in Mono Lake, a meromictic soda lake in California. *Appl. Environ. Microbiol.* **69**:1030–1042.

Ishikawa, M., K. Nakajima, M. Yanagi, Y. Yamamoto, and K. Yamasato. 2003. *Marinilactobacillus psychrotolerans* gen. nov., ap. nov., a halophilic and alkaliphilic marine lactic acid bacterium isolated from marine organisms in temperate and subtropical areas of Japan. *Int. J. Syst. Evol. Microbiol.* **53**:711–720.

Ishikawa, M., K. Nakajima, Y. Itamiya, S. Furukawa, Y. Yamamoto, and K. Yamasato. 2005. *Halolactobacillus halophilus* gen. nov., sp. nov. and *Halolactibacillus miurensis* sp. nov., halophilic and alkaliphilic marine lactic acid bacteria constituting a phylogenetic lineage in *Bacillus* rRNA group 1. *Int. J. Syst. Evol. Microbiol.* **55**:2427–2439.

Ivanova, E. P., M. V. Vysotskii, V. I. Svetashev, O. I. Nedashkovskaya, N. M. Gorshkova, N. M., V. V. Mikhailov, N. Yumoto, Y. Shigeri, T. Taguchi, and S. Yoshikawa. 1999. Characterization of *Bacillus* strains of marine origin. *Int. Microbiol.* **2**:267–271.

Jan-Roblero, J., X. Magos, L. Fernández, and S. Le Borgne. 2004. Phylogenetic analysis of bacterial populations in waters of former Texcoco Lake, Mexico. *Can. J. Microbiol.* **50**:1049–1059.

Jeon, C. O., J.-M. Lim, J.-M. Lee, L.-H. Xu, C.-L. Jiang, and C.-J. Kim. 2005. Reclassification of *Bacillus haloalkaliphilus* Frite 1996 as *Alkalibacillus haloalkaliphilus* gen. nov., comb. nov. and the description of *Alkalibacillus salilacus* sp. nov., a novel halophilic bacterium isolated from a salt lake in China. *Int. J. Syst. Evol. Microbiol.* **55**:1891–1896.

Jones, B. E., W. D. Grant, A. W. Duckworth, and G. G. Owenson. 1998. Microbial diversity of soda lakes. *Extremophiles* **2**:191–200.

Jones, B. E., W. D. Grant, A. W. Duckworth, P. Schuman, N. Weiss, and E. Stackbrandt. 2005. *Cellulomonas bogoriensis* sp. nov., an alkaliphilic cellulomonad. *Int. J. Syst. Evol. Microbiol.* **55**:1711–1714.

Kevbrin, V. V., T. N. Zhilina, F. A. Rainey, and G. A. Zavarzin. 1998. *Tindallia magadii* gen. nov., sp. nov.: an alkaliphilic anaerobic ammonifier from soda lake deposits. *Curr. Microbiol.* **37**:94–100.

Kim, I.-G., M.-H. Lee, S.-Y. Jung, J. J. Song, T.-K. Oh, and J.-H. Yoon. 2005. *Exiguobacterium aestuarii* sp. nov. and *Exiguobacterium marinum* sp. nov., isolated from a tidal flat of the Yellow Sea in Korea. *Int. J. Syst. Evol. Microbiol.* **55**: 885–889.

Krulwich, T. A., and A. A. Guffanti. 1989. Alkalophilic bacteria. *Annu. Rev. Microbiol.* **43**: 435–463.

Krulwich, T. A., M. Ito, R. Gilmour, D. B. Hicks, and A. A. Guffanti. 1998. Energetics of alkaliphilic *Bacillus* species: physiology and molecules. *Adv. Microb. Physiol.* **40**:401–438.

Krulwich, T. A., M. Ito, and A. A. Guffanti. 2001. The Na^+-dependence of alkaliphily in *Bacillus*. *Biochim. Biophys. Acta.* **1505**:158–168.

Lee, J.-S., K. C. Lee, Y.-H. Chang, S. G. Hong, H. W. Oh, Y.-R. Pyun, and K. S. Bae. 2002. *Paenibacillus daejeonensis* sp. nov., a novel alkaliphilic bacterium from soil. *Int. J. Syst. Evol. Microbiol.* **57**:2107–2111.

Li, W.-J. H.-H. Chen, C.-J. Kim, Y.-Q. Zhang, D.-J. Park, J.-C. Lee, L.-H. Xu, and C.-L. Jiang. 2005a. *Nesterenkonia sandarakina* sp. nov., and *Nesterenkonia lutea* sp. nov., novel actinobacteria, and emended description of the genus *Nesterenkonia*. *Int. J. Syst. Evol. Microbiol.* **55**:463–466.

Li, W.-J., Y.-G. Zhang, Y.-Q. Zhang, S.-K. Tang, P. Xu, L.-H. Xu, and C.-L. Jiang. 2005b. *Streptomyces sodiiphilus* sp. nov., a novel alkaliphilic actinomycete. *Int. J. Syst. Evol. Microbiol.* **55**:1329–1333.

Li, Y., L. Mandelco, and J. Wiegel. 1993. Isolation and characterization of moderately thermophilic anaerobic alkaliphile, *Clostridium paradoxum* sp. nov. *Int. J. Syst. Bacteriol.* **43**:450–460.

Li, Y., M. Engle, N. Weiss, L. Mandelco, and J. Wiegel. 1994. *Clostridium thermoalcaliphilum* sp. nov., an anaerobic and thermotolerant facultative alkaliphile. *Int. J. Syst. Bacteriol.* **44**:111–118.

Li, Z. Y. Kawamura, O. Shida, S. Yamagata, T. Deguchii, and T. Ezaki. 2002. *Bacillus okuhidaensis* sp. nov., isolated from the Okuhida spa area of Japan. *Int. J. Syst. Evol. Microbiol.* 52:1205–1209.

López-García, P., J. Kazimierczak, K. Benzerara, S. Kempe, F. Guyot, and D. Moreira. 2005. Bacterial diversity and carbonate precipitation in the giant microbialites from the highly alkaline Lake Van, Turkey. *Extremophiles* 9:263–274.

Lu, J., Y. Nogi and H. Takami. 2001. *Oceanobacillus iheyensis* gen. nov., sp. nov., a deep-sea extremely halotolerant and alkaliphilic species isolated from a depth of 1050 m on the Iheya Ridge. *FEMS Microbiol. Lett.* 205:291–297.

Nakajima, K., K. Hirota, Y. Nodasaka, and I. Yumoto. 2005. *Alkalibacterium iburiense* sp. nov., an obligate alkaliphile that reduces an indigo dye. *Int. J. Syst. Evol. Microbiol.* 55:1525–1530.

Nakamura, K., S. Haruta, S. Ueno, M. Ishii, A. Yokota, and Y. Igarashi. 2004. *Cerasibacillus quisquiliarum* gen. nov., sp. nov., isolated from a semi-continuous decomposing system of kitchen refuse. *Int. J. Syst. Evol. Microbiol.* 54:1063–1069.

Nesterenko, O. A., T. M. Nogina, S. A. Kasumova, E. I. Kvasnikov, and S. G. Batrakov. 1982. *Rhodococcus luteus* nom. nov. and *Rhodococcus maris* nom. nov. *Int. J. Syst. Bacteriol.* 32:1–14.

Nielsen, P., F. A. Rainey, H. Outtrup, F. G. Priest, and D. Fritze. 1994. Comparative 16S rDNA sequence analysis of some alkaliphilic bacilli and the establishment of a sixth rRNA group within the genus *Bacillus*. *FEMS Microbiol. Lett.* 177:61–66.

Nielsen, P., D. Fritze, and F. G. Priest. 1995. Phenetic diversity of alkaliphilic *Bacillus* strains: proposal for nine new species. *Microbiology* 141:1745–1761.

Niimura, Y., E. Koh, F. Yanagida, K. Suzuki, K. Komagata, and M. Kozaki. 1990. *Amphibacillus xylanus* gen. nov., sp. nov., a facultatively anaerobic spore-forming xylan-digesting bacterium which lacks cytochrome, quinine, and catalase. *Int. J. Syst. Bacterial.* 40:297–301.

Nogi, Y., H. Takami, and K. Horikoshi. 2005. Characterization of alkaliphilic *Bacillus* strains used in industry: proposal of five novel species. *Int. J. Syst. Evol. Microbiol.* 55:2309–2315.

Ntougias, S., and N. J. Russell. 2000. *Bacillus* sp. WW3-SN6, a novel facultatively alkaliphilic bacterium isolated from the washwaters of edible olive. *Extremophiles* 4:201–208.

Ntougias, S., and N. J. Russell. 2001. *Alkalibacterium olivoapovliticus* gen. nov., sp. nov., a new obligately alkaliphilic bacterium isolated from edible-olive washwaters. *Int. J. Syst. Evol. Microbiol.* 51:1161–1170.

Ohkuma, M., and T. Kudo. 1996. Phylogenetic diversity of the intestinal bacterial community in the termite *Reticulitermes speratus*. *Appl. Environ. Microbiol.* 62:461–468.

Ohta, K., A. Kiyomiya, N. Koyama, and Y. Nosoh. 1975. The basis of the alkalophilic property of a species of *Bacillus*. *J. Gen. Microbiol.* 86:259–266.

Olivera, N., F. Siñeriz, and J. D. Breccia. 2005. *Bacillus patagoniensis* sp. nov., a novel alkalitolerant bacterium from *Atriplex lampa* rhizosphere, Patagonia, Argentina. *Int. J. Syst. Evol. Microbiol.* 55:443–447.

Onishi, H., and M. Kamekura. 1972. *Micrococcus halobius* sp. nov. *Int. J. Syst. Bacteriol.* 22:233–236.

Padden, A. N., V. M. Dillon, J. Edmonds, M. D. Collins, N. Alvarez, and P. John. 1999. An indigo-reducing moderate thermophile from a woad vat, *Clostridium isatidis* sp. nov. *Int. J. Syst. Bacteriol.* 49:1025–1031.

Pikuta, E., A. Lysenko, N. Chuvilskaya, U. Mendorock, H. Hippe, N. Suzina, D. Nikitin, G. Osipov, and K. Laurinavichius. 2000. *Anoxybacillus pushchinensis* gen. nov., sp. nov., a novel anaerobic alkaliphilic, moderately thermophilic bacterium from manure, and description of *Anoxybacillus flavithermus* comb. nov. *Int. J. Syst. Evol. Microbiol.* 50:2109–2117.

Pikuta, E. V., R. B. Hoover, A. K. Bej, D. Marsic, E. N. Detkova, W. B. Whitman, P. Krader. 2003. *Tindallia californiensis* sp. nov., a new anaerobic, haloalkaliphilic, spore-forming acetogen isolated from Mono Lake in California. *Extremophiles* 7:327–334.

Priest, F. G., M. Goodfellow, and C. Todd. 1998. A numerical classification of the genus *Bacillus*. *J. Gen. Micorbiol.* 134:1847–1882.

Rainey, F. A., S. Klatte, R. M. Kroppenstedt, and E. Stackebrandt. 1995. *Dietzia*, a new genus including *Dietzia maris* comb. nov., formerly *Rhodococcus maris*. *Int. J. Syst. Bacteriol.* 45:32–36.

Romano, I., L. Lama, B. Nicolaus, A. Gambacorta, and A. Garmbacorta, and A. Giordano. 2005a. *Bacillus saliphilus* sp. nov., isolated from a mineral pool in Campania, Italy. *Int. J. Syst. Evol. Microbiol.* 55:159–163.

Romano, I., L. Lama, B. Nicolaus, A. Gambacorta, and A. Giordano. 2005b. *Alkalibacillus filiformis* sp. nov., isolated from a mineral pool in Campania, Italy. *Int. J. Syst. Evol. Microbiol.* 55:2395–2399.

Sorokin, D. Y., and J. G. Kuenen. 2005. Chemolithotrophic haloalkaliphiles from soda lakes. *FEMS Microbiol. Ecol.* 52:287–295.

Spanka, R., and D. Fritze. 1993. *Bacillus cohnii* sp. nov., new, obligately alkaliphilic, oval-spore-forming *Bacillus* species with ornithine and aspartic acid instead of diaminopimeric acid in the cell wall. *Int. J. Syst. Bacteriol.* 43:150–156.

Stackebrandt, E., C. Koch, O. Gvozdiak, and P. Schumann. 1995. Taxonomic dissection of the genus *Micrococcus*: *Kocuria* gen. nov., *Nesterenkonia* gen. nov., *Kytococcus* gen. nov., *Dermacoccus* gen. nov., and *Micrococcus* Cohn 1872 gen. emend. *Int. J. Syst. Bacteriol.* 45:682–692.

Switzer Blum, J., A. Burns Bindi, J. Buzzelli, J. F. Stolz, and R. S. Oremland. 1998. *Bacillus arsenicoselenatis*, sp. nov., and *Bacillus selenitireducence*, sp. nov.: two haloalkaliphiles from Mono Lake, California that respire oxyanions of selenium and arsenic. *Arch. Microbiol.* 171:19–30.

Takahara, Y., and O. Tanabe. 1960. Studies on the reduction of indigo in industrial fermentation vat (VII). *J. Ferment. Technol.* 38:329–331.

Takai, K., D. P. Moser, T. C. Onstott, N. Spoelstra, S. M. Pfiffner, A. Dohnalkova, and J. K. Fredcikson. 2001. *Alkaliphilus transvaalensis* gen. nov., sp. nov., an extremely alkaliphilic bacterium isolated from a deep South Africa gold mine. *Int. J. Syst. Evol. Microbiol.* 51:1245–1256.

Takami, H., Inoue, A., Fujii, F., and K. Horikoshi. 1997. Microbial flora in the deepest sea mud of the Mariana Trench. *FEMS Mcrobiol. Lett.* 152:279–285.

Thongaram, T., S. Kosono, M. Ohkuma, Y. Hongoh, M. Kitada, T. Yoshinaka, S. Trakulnaleamsai, N. Noparatnaraporn, and T. Kudo. 2003. Gut of higher termites as a niche for alkaliphiles as shown by culture-based and culture-independent studies. *Microbes Environ.* 18:152–159.

Tiago, I., A. P. Chung, and A. Veríssimo. 2004. Bacterial diversity in a nonsaline alkaline environment: heterotrophic aerobic populations. *Appl. Environ. Microbiol.* 70:7378–7387.

Vargas, V. A., O. D. Delgado, R. Hatti-Kaul, and B. Mattiasson. 2005. *Bacillus bogoriensis* sp. nov., a new alkaliphilic halotolerant member of the genus *Bacillus* isolated from a Kenyan soda lake. *Int. J. Syst. Evol. Microbiol.* 55:899–902.

Vedder, A. 1934. *Bacillus alcalophilus* n. sp.; benevens enkele ervaringen met sterk alcalischevoedingbodems. *Antonie Leeuwenhoek J. Microbiol. Serol.* 1:143–147.

Wainø, M., B. J. Tindall, P. Schumann, and K. Ingvorsen. 1999. *Gracilibacillus* gen. nov., with description of *Gracilibacillus halotolerans* gen. nov. sp. nov.; transfer of Bacillus dipsosauri to *Gracilibacillus dipsosauri* comb. nov., and *Bacillus salexigens* to the genus *Salibacillus* gen. nov., as *Salibacillus salexigens* comb. nov. *Int. J. Syst. Bacteriol.* 49:821–831.

Wallace, R. J., L. C. Chaudhary, E. Miyagawa, N. Mckain, and N. D. Walker. 2004. Metabolic properties of *Eubacterium pyruvativorans*, a ruminal "hyper-ammonia-producing" anaerobe with metabolic properties analogous to those of *Clostridium kluyveri*. *Microbiology* 150:2921–2930.

Whitehead, T. R., and M. A. Cotta. 2004. Isolation and identification of hyper-ammonia producing bacteria from swine manure storage pits. *Curr. Microbiol.* 48:20–26.

Wiegel, J., and V. V. Kevbrin. 2004. Alkalithermophiles. *Biochem. Soc. Trans.* 32:193–198.

Ye, Q., Y. Roh, S. L. Carrol, B. Blair, J. Zhou, C. L. Zhang, and M. W. Fields. 2004. Alkaline anaerobic respiration: isolation and characterization of a novel alkaliphilic and metal-reducing bacterium. *Appl. Environ. Microbiol.* 70:5595–5602.

Yoon, J.-H., D. K. Yim, J.-S. Lee, K.-S. Shin, H. H. Sato, S. T. Lee, Y. K. Park, and Y.-H. Park. 1998. *Paenibacillus campinasensis* sp. nov., a cyclodextrin-producing bacterium isolated in Brazil. *Int. J. Syst. Bacteriol.* 48: 833–837.

Yoon, J.-H., I.-G. Kim, M.-H. Lee, and T.-K. Oh. 2005. *Nocardioides kribbensis* sp. nov., isolated from an alkaline soil. *Int. J. Syst. Evol. Microbiol.* 55:1611–1614.

Yumoto, I., K. Nakajima, and K. Ikeda. 1997. Comparative study on cytochrome content of alkaliphilic *Bacillus* strains. *J. Ferment. Bioeng.* 83:466–469.

Yumoto, I., K. Yamazaki, T. Sawabe, K. Nakano, K. Kawasaki, Y. Ezura, and H. Shinano. 1998. *Bacillus horti* sp. nov., a new Gram-negative alkaliphilic bacillus. *Int. J. Syst. Bacteriol.* 48: 565–571.

Yumoto, I., K. Yamazaki, M. Hishinuma, Y. Nodasaka, N. Inoue, and K. Kawasaki 2000. Identification of facultatively alkaliphilic *Bacillus* sp. strain YN-2000 and its fatty acid composition and cell-surface aspects depending on culture pH. *Extremophiles* 4:285–290.

Yumoto, I., K. Yamazaki, M. Hishinuma, Y. Nodasaka, A. Suemori, K. Nakajima, N. Inoue, and K. Kawasaki. 2001. *Pseudomonas alcaliphila* sp. nov., a novel facultatively psychrophilic alkaliphile isolated from seawater. *Int. J. Syst. Evol. Microbiol.* 51:349–355.

Yumoto, I. 2002. Bioenergetics of alkaliphilic *Bacillus* spp. *J. Biosci. Bioeng.* 93:342–353.

Yumoto, I., A. Nakamura, H. Iwata, K. Kojima, K. Kusumoto, Y. Nodasaka, and H. Matsuyama. 2002. *Dietzia psychralcaliphila* sp. nov., a novel facultatively psychrophilic alkaliphile that grows on hydrocarbons. *Int. J. Syst. Evol. Microbiol.* 52:85–90.

Yumoto, I. 2003. Electron transport system in alkaliphilic *Bacillus* spp. *Recent Res. Dev. Bacteriol.* 1:131–149.

Yumoto, I., S. Yamaga, Y. Sogabe, Y. Nodasaka, H. Matsuyama, K. Nakajima, and A. Suemori. 2003. *Bacillus krulwichae* sp. nov., a halotolerant obligate alkaliphile that utilizes benzoate and *m*-hydroxybenzoate. *Int. J. Syst. Evol. Microbiol.* 53:1531–1536.

Yumoto, I., M. Hishinuma-Narisawa, K. Hirota, T. Shingyo, F. Takebe, Y. Nodasaka, H. Matsuyama, and I. Hara. 2004a. *Exiguobacterium oxidotolerans* sp. nov., a novel alkaliphile exhibiting high catalase activity. *Int. J. Syst. Evol. Microbiol.* 54:2013–2017.

Yumoto, I., K. Hirota, Y. Nodasaka, Y. Yokota, T. Hoshino and K. Nakajima. 2004b. *Alkalibacterium psychrotolerans* sp. nov., a psychrotolerant obligate alkaliphile that reduces an indigo dye. *Int. J. Syst. Evol. Microbiol.* 54:2379–2383.

Yumoto, I., K. Hirota, T. Goto, Y. Nodasaka, and K. Nakajima. 2005a. *Bacillus oshimensis* sp. nov., a moderately halophilic, nonmotile alkaliphile. *Int. J. Syst. Evol. Microbiol.* 55:935–939.

Yumoto, I., K. Hirota, Y. Nodasaka, and K. Nakajima. 2005b. *Oceanobacillus oncorhynchi* sp. nov., a halotolerant obligate alkaliphile isolated from the skin of a rainbow trout (*Oncorhynchus mykiss*), and emended description of the genus *Oceanobacillus*. *Int. J. Syst. Evol. Microbiol.* 55:1521–1524.

Zhilina, T. N., E. N. Detkova, F. A. Rainey, G. A. Osipov, A. M. Lysenko, N. A. Kostrikina, G. A. Zavarzin. 1998. *Natronoincola histidinovorans* gen. nov., sp. nov., a new alkaliphilic acetogenic anaerobe. *Curr. Microbiol.* 37:177–185.

Zhilina, T. N., E. S. Garnova, T. P. Tourova, N. A. Kostrikina, and G. A. Zavarzin. 2001. *Amphibacillus fermentum* sp. nov. and *Amphibacillus tropicus* sp. nov., new alkaliphilic, facultatively anaerobic, saccharolytic bacilli from Lake Magadi. *Mikrobiologiia* 70:825–827.

Zhilina, T. N., R. Appel, C. Probian, E. L. Brossa, J. Harder, F. Widdel, and G. A. Zavarzin. 2004. *Alkaliflexus imshenetskii* gen. nov., a new alkaliphilic gliding carbohydrate-fermenting bacterium with propionate formation from a soda lake. *Arch. Microbiol.* 182:244–253.

Chapter 24

Bioenergetic Adaptations That Support Alkaliphily

TERRY ANN KRULWICH, DAVID B. HICKS, TALIA SWARTZ, AND MASAHIRO ITO

INTRODUCTION

Prokaryotes that thrive in highly alkaline environments have a remarkable ability to maintain their cytoplasmic pH lower than the external pH. For the aerobic alkaliphilic *Bacillus* species, the internal pH is kept 2 units lower than the external pH of 10 or higher. Here we will focus on the centrality and diverse mechanisms of cytoplasmic pH homeostasis to extreme alkaliphily. Additionally, the mechanisms of bioenergetic processes such as solute transport, motility, and ATP synthesis will be discussed, especially in relation to the energetic problem that results from pH homeostasis at high pH. Two themes that will run through the chapter are the whole-cell, systems biology aspects of alkaliphile bioenergetics and the diverse ion transporters, pumps, and channels that participate in this system, many of which were first discovered in alkaliphiles and many of which have alkaliphile-specific roles or adaptations.

Almost 3 decades ago, P. B. Garland predicted that studies of alkaliphiles could be "most rewarding" if alkaliphiles kept their cytoplasmic pH much lower than the external pH (Garland, 1977). If so they would have a problem carrying out respiration-dependent ATP synthesis. This problem results from an apparent insufficiency in the chemiosmotic energy available to energize ATP synthesis. Mitchell's chemiosmotic model posits that ATP synthesis during respiration is coupled to the bulk proton motive force (PMF or Δp, alkaline and negative inside a bacterial cell or mitochondrion) (Mitchell, 1961). The PMF or Δp is an electrochemical gradient of protons across the membrane that is established by the outward pumping of protons from the cytoplasm during electron transport in respiring bacteria. The PMF consists of a chemical component, the ΔpH (alkaline in), and an electrical component, the $\Delta \Psi$ (negative in). When an alkaliphile maintains a much lower cytoplasmic pH than the external pH (a ΔpH, acid in), this ΔpH is reversed from the chemiosmotically productive PMF component and reduces the total bulk PMF.

All alkaliphiles examined to date, including both anaerobes and aerobes, do indeed maintain a cytoplasmic pH much lower than the external pH. The molecular mechanisms whereby extreme aerobic alkaliphiles accomplish this provide a paradigm for alkali adaptation for other microbial cells (Krulwich, 1995; Krulwich et al., 1999; Padan et al., 2005). Garland foresaw that for aerobic alkaliphiles, the resulting energetic dilemma might be resolved by the production of a large enough $\Delta \Psi$ to offset the adverse ΔpH. Otherwise, these organisms would need alternative mechanisms to overcome the apparent energetic deficiency that results from H^+-coupled ATP synthesis, for which the PMF was the only known driving force (Garland, 1977).

Studies of this problem have so far been conducted only on alkaliphilic *Bacillus* species that do use a PMF-dependent ATP synthetic mechanism at high external pH. Although $\Delta \Psi$ is substantial, the total Δp is very small (Dimroth and Cook, 2004; Krulwich, 1995) (Fig. 1). The resolution of the oxidative phosphorylation (OXPHOS) conundrum, instead, depends upon specific adaptations of the proton-coupled ATP synthase itself that allow it to participate in proton-coupled synthesis that is partially sequestered from the alkaline bulk medium (Wang et al., 2004). As with pH homeostasis, these observations on alkaliphile OXPHOS have implications for the larger biological context in which this central energy-transducing process takes place. The growing amount of comparative genomic data between alkaliphiles and neutrophiles has made it much easier to identify putative alkaliphile-specific deviations in conserved and functionally important residues or motifs in proteins of

---

T. A. Krulwich, D. B. Hicks, and T. Swartz • Department of Pharmacology and Systems Therapeutics, Mount Sinai School of Medicine, New York, NY 10029. M. Ito • Graduate School of Life Sciences, Toyo University, Oura-gun, Gunma 374-0193, Japan.

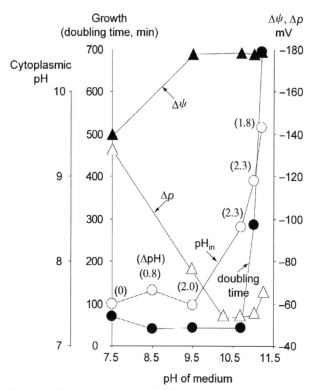

Figure 1. The cytoplasmic pH, doubling time, and proton motive force of alkaliphilic *B. pseudofirmus* OF4 growing at pH values from 7.5 to 11.2 in pH-controlled continuous cultures. Cells were grown in continuous cultures on malate-containing semidefined medium at the indicated, rigorously maintained pH values. Assays of the proton motive force parameters were conducted as described (Sturr et al., 1994). The closed and open circles indicate the doubling times (in minutes) in relation to the cytoplasmic pH at different growth pH; the numbers in parentheses are the values of the ΔpH, acid inside relative to the medium. The ΔΨ (closed triangles) and Δp (PMF in millivolt, open triangles) are also shown. (Modified from Krulwich, 1995, with permission from Blackwell Publishing Ltd.)

bioenergetic interest. Such data have been enormously valuable to the work on alkaliphile OXPHOS.

The critical factor that still limits broader progress on bioenergetic studies of alkaliphiles is that very few alkaliphiles have been rendered genetically tractable. Therefore, the effect of introducing targeted gene deletions or site-directed changes in alkaliphile-specific motifs of important proteins cannot be tested in the vast majority of alkaliphiles. The properties of interesting alkaliphile membrane proteins can still be characterized in vitro or in a functional heterologous system (e.g., as expressed in *Escherichia coli*), which has been elegantly applied to the study of light-driven ion pumps of archaeal alkaliphiles (Kamo et al., 2001; Klare et al., 2004; Muller and Oren, 2003; Spudich and Luecke, 2002). However, the role of novel proteins or novel features of conserved proteins in alkaliphile physiology must be studied in the extremophile itself where studies of whole cells or membrane vesicles can be conducted at high pH values of bioenergetic interest. Because these types of studies can only be carried out in a few strains, much of this discussion will focus on extremely alkaliphilic *Bacillus* species, especially facultatively alkaliphilic *Bacillus pseudofirmus* OF4, which is genetically tractable and grows nonfermentatively as well as fermentatively in a range of external pH values from 7.5 to >11.2 (Ito et al., 1997b; Sturr et al., 1994). The development of a larger number and array of genetically tractable alkaliphile strains represents an investment that would allow us to capture the diversity of adaptations that can support alkaliphily. There has been an increase in recognized alkaliphile genera and species during the past decade as teams have explored more of Earth's extremely alkaline niches (Yumoto, this volume, Chapter 23). It will be of great interest to explore how these diverse alkaliphiles solve the bioenergetic challenges discussed. We will therefore note alternative strategies that might meet these challenges in other settings when describing the strategies that have been found in *Bacillus*.

ALKALIPHILES ARE EXPERTS AT ALKALINE pH HOMEOSTASIS AND ALSO TOLERATE HIGHER CYTOPLASMIC pH THAN NEUTROPHILES

Alkaliphilic *B. pseudofirmus* OF4 maintains a cytoplasmic pH of 7.5 when grown on malate-containing medium in a rigorously pH-controlled continuous culture at external pH values from 7.5 to 9.5; the pH difference between the cytoplasm and the bulk medium thus goes from 0 at an external pH of 7.5 to 2 units at an external pH of 9.5 (Sturr et al., 1994) (Fig. 1). Most bacteria that are not extremophiles grow optimally when the cytoplasmic pH is ~7.5 and maintain cytoplasmic pH within a range of 7.4–7.8 (Harold and Van Brunt, 1977; Padan et al., 1981, 2005). At external pH values >9.5, *B. pseudofirmus* OF4 and other extremely alkaliphilic *Bacillus* species do not continue to increase the pH gradient (acid in) sufficiently to maintain the cytoplasmic pH at 7.5. The ΔpH (acid in) peaks at 2.3 units, i.e., the cytoplasmic pH of *B. pseudofirmus* OF4 is 8.2 in a continuous culture at pH 10.5, and as the external pH is raised further, the cytoplasmic pH rises (Guffanti and Hicks, 1991; Sturr et al., 1994) (Fig. 1).

Several features of the pH homeostasis profile of alkaliphilic *Bacillus* species are notable. First, the capacity of *B. pseudofirmus* OF4 and other extremely alkaliphilic aerobes to maintain a pH gradient of 2 to 2.3 units, acid in, reflects an extraordinary ability to accumulate cytoplasmic protons for respiring cells that are pumping protons out during respiration

(Krulwich, 1995; Padan et al., 2005; Yumoto, 2003). As discussed below, acidification of the cytoplasm relative to the outside is more straightforward for acid-producing fermentative alkaliphiles, but they do not regulate their cytoplasmic pH as well as respiring alkaliphiles, and they remain dependent upon multiple strategies, not only metabolic acid production (Cook et al., 1996) (Table 1). Second, the growth rate of extreme alkaliphiles ultimately slows (the doubling time increases) in parallel with the increase in cytoplasmic pH at the alkaline edge of the pH range (seen in Fig. 1 at pH >10.5). This indicates that pH homeostasis is a central challenge of alkaliphily, with the growth rate of the organism closely paralleling the success of pH homeostasis mechanisms up to external pH values >11. Third, the growth rate of *B. pseudofirmus* OF4 does not fall as soon as the capacity for pH homeostasis is exceeded and the cytoplasmic pH rises >8.0. The optimal growth rate (a doubling time of 38 min) is observed at external pH values from 8.5, at which the cytoplasmic pH is 7.5, to pH 10.6, at which cytoplasmic pH is 8.3; the growth rate at pH 7.5 is slower (doubling time of 54 min) (Fig. 1). Facultatively alkaliphilic *Bacillus cohnii* YN-2000 also grows at optimal rates when its cytoplasmic pH is >8 (Yumoto, 2003). By contrast, the growth of *E. coli* arrests when its cytoplasmic pH ≥8.0 (Padan et al., 2005; Zilberstein et al., 1984). Functions of alkaliphiles that depend upon enzymes on the outside cell surface obviously require special adaptation to high pH, e.g., exoenzymes (van der Laan et al., 1991), and the final cell-wall biosynthesis steps that are posited as one of the limiting factors in alkali stress in neutrophiles (Cao et al., 2002). Because alkaliphiles tolerate cytoplasmic pH values ≥8.0 that are not tolerated by neutrophiles, there must be key cytoplasmic processes in alkaliphiles that retain optimal function at such cytoplasmic pH values but not in neutrophiles. One such process may be cell division, because *B. pseudofirmus* OF4 cells form chains as the cytoplasmic pH moves into suboptimally high ranges (Sturr et al., 1994). Interestingly, the phosphoserine aminotransferase from alkaliphilic *Bacillus alcalophilus* exhibits distinct structural features that are proposed to be adaptive to alkaline pH (Dubnovitsky et al., 2005), consistent with strategies for multiple cytoplasmic processes to tolerate higher pH than found in neutrophiles.

The extraordinary capacity of extremely alkaliphilic *Bacillus* species for pH homeostasis relative to *E. coli* and other nonalkaliphiles is even more evident during sudden pH transitions than in the relationship between growth rate and external pH. An acute test of alkaline pH homeostasis is the response to a suddenly imposed alkaline shift in the external pH.

Table 1. Alkaliphilic *B. pseudofirmus* OF4 pH homeostasis after a sudden alkaline shift in the external pH

Strain	Energy	Cytoplasmic pH, 10 min after shift in pH_{out} from 8.5 to 10.5	
		NaCl (50 mM)	KCl (50 mM)
Wild type	Glucose	8.4	9.2
Wild type	Malate[a]	8.2	10.5
ΔslpA[b]	Malate	9.1	10.5
		NaCl (100 mM)	NaCl (2.5 mM)
Wild type	None	8.5	9.0
ΔncbA[b]	None	8.7	9.4
ΔmotPS[b]	None	8.5	9.0

Data in the top and bottom parts are, respectively, from experiments reported by Gilmour et al. (2000) and Ito et al. (2004b). Cells equilibrated at pH 8.5 were subjected to a sudden alkaline shift in external pH in the presence of the indicated energy source or buffer alone (energy source, none), and the cytoplasmic pH was determined after 10 min, as described by Ito et al. (1997b).
[a] When wild type was equilibrated at pH 7.5 and shifted to an external pH of 8.5 (malate energizing), the cytoplasmic pH was 8.4 to 8.5 inside after 10 min in the presence of 50 mM choline or KCl and pH 7.5 in the presence of 50 mM NaCl.
[b] *slpA* encodes the S layer; *ncbA* encodes the NavBP channel; and *motPS* encodes the Na^+-dependent stator-force generator for motility.

When alkaliphilic *B. pseudofirmus* OF4 is equilibrated at pH 8.5 and then subjected to a sudden alkaline shift in the external pH to 10.5, measurements of the cytoplasmic pH immediately after the shift or 10 min after the shift detect no change in cytoplasmic pH from the preshift value as long as adequate Na^+ is present (Krulwich et al., 1985, 1999; Padan et al., 2005; Wang et al., 2004) (Table 1). Perhaps there is postshift cytoplasmic alkalinization right after a large alkaline shift in pH_{out}, but it is sufficiently small and/or transient that the methods thus far applied cannot capture it. By contrast, when *E. coli* cells are equilibrated at pH 7.2 and then subjected to a shift in the medium to pH 8.3, similar measurements show that the cytoplasmic pH rises to 8.3 during the first postshift minutes and is still close to pH 8 when assayed after 10 min, with subsequent reestablishment of a cytoplasmic pH near 7.5 (Padan et al., 2005; Zilberstein et al., 1984). What mechanisms support this centrally important and remarkable capacity of alkaliphiles for alkaline pH homeostasis?

DIVERSE MECHANISMS CONTRIBUTE TO CYTOPLASMIC pH HOMEOSTASIS BY ALKALIPHILES AT pH ≥10

Metabolic Generation of Acid

Production of acid from fermentation of carbohydrates and deaminated amino acids is an established alkali-adaptive strategy of neutrophilic bacteria

(Blankenhorn et al., 1999; Bordi et al., 2003; Maurer et al., 2005; Stancik et al., 2002). The benefit of acid production is probably its direct effect on cytoplasmic pH rather than effective acidification of large external volumes (Padan et al., 2005). Alkaliphiles use metabolic acids too, but growth on fermentable carbon sources does not completely eliminate the dependence of extreme alkaliphiles upon transporter-based, Na^+-requiring mechanisms of pH homeostasis. For example, *B. pseudofirmus* OF4 exhibits a Na^+-independent component to pH homeostasis when energized by glucose as opposed to malate during a pH-shift experiment, but pH homeostasis remains significantly Na^+ dependent (Gilmour and Krulwich, 1997) (Table 1). Alkaline-tolerant *Oceanobacillus iheyensis* appears to grow almost exclusively on fermentative carbon sources (Lu et al., 2001), but its use of a Na^+ cycle for pH homeostasis has been inferred from genomic data (Takami et al., 2002).

Proton Trapping in the "Periplasm" and at the External Membrane Surface and Hypothesized Roles of Membrane Lipids

Compelling genomic and biochemical evidence attest to the fact that extreme alkaliphiles experience a low PMF at high pH, as discussed below in connection with the problem of OXPHOS. However, there are passive mechanisms for trapping protons and/or the Na^+ required for pH homeostasis between the cell-wall layer and the membrane (Takami et al., 2002) near the external cytoplasmic membrane surface, without which the low-PMF problem would be even more adverse (Koch, 1986). Alkaliphilic *Bacillus halodurans* C-125 produces negatively charged teichuronic acid and teichuronopeptides that are associated with the peptidoglycan as secondary cell-wall polymers (SCWPs). Aono and colleagues (Aono and Ohtani, 1990; Aono et al., 1999) showed that mutations in genes encoding both of these polymers result in a growth defect accompanied by elevated cytoplasmic pH. This growth defect is dwarfed, however, by the nonalkaliphilic phenotype that results from single-point mutations in the major Na^+/H^+ antiporter involved in active pH homeostasis (Hamamoto et al., 1994; Seto et al., 1995). Comparable SCWPs to those of *B. halodurans* C-125 have not been found in *B. pseudofirmus* OF4, a more extreme alkaliphile (Ito, 2002), but this strain has genes encoding an acidic S-layer (*slpA*) and an acidic polyglutamate capsule (*cap*) (Gilmour et al., 2000; Ito et al., 1997a). Conditions of capsule expression are not yet known, but the S-layer is abundantly produced both at pH 7.5 and at pH 10.5. S-layer production during growth at pH 7.5 is a hardwired adaptation of *B. pseudofirmus* OF4 that protects the extremophile in the event of a sudden upward shift in pH even though the "cost" of S-layer production is obviously disadvantageous at pH 7.5 (Gilmour et al., 2000). An *slpA* deletion mutant grows better than wild type at pH 7.5, both at optimal and at suboptimal Na^+ concentrations (Fig. 2). The presence of the S-layer lowers the lag time but does not impact the growth rate itself at pH 10.5, having an impact on both lag and exponential growth rate only once the

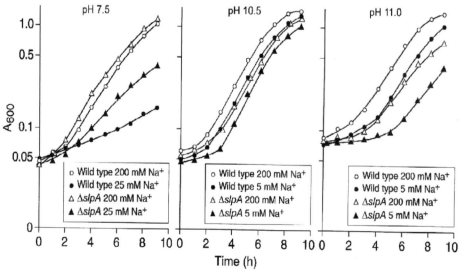

Figure 2. Growth curves of wild-type alkaliphilic *B. pseudofirmus* OF4 and a mutant disrupted in the S-layer-encoding gene *slpA*. The two strains were compared at two concentrations, an optimal Na^+ concentration, and a suboptimally low Na^+ concentration at three different pH values on semidefined malate-containing medium at 30°C. Although not shown, the wild-type strain grew negligibly at pH 7.5 when the added Na^+ was at 5 mM, a concentration that supported growth well at pH 10.5. (Modified from Gilmour et al., 2000, with permission from the publisher.)

pH is raised to 11. Notably, defects in the *slpA* mutant at pH 10.5 to 11 are exacerbated at suboptimal concentrations of Na$^+$, raising the possibility that increased Na$^+$ near the surface is part of the S-layer function (Fig. 2) (Gilmour et al., 2000). Absence of the S-layer significantly decreases the Na$^+$-dependent pH homeostatic capacity of the alkaliphile upon a sudden alkaline shift in external pH (Gilmour et al., 2000) (Table 1).

The lipid composition of alkaliphile membranes differs from neutrophiles in its high cardiolipin and squalene content as well as in a C_{40} isoprenoid fraction (Clejan et al., 1986). Cardiolipin may have roles in proton movements near the membrane and/or in fostering the formation of complexes and interaction among protein complexes in the membrane (Haines and Dencher, 2002; Pfeiffer et al., 2003; Zhang et al., 2002, 2005). Haines (2001) has suggested that squalene could limit proton and Na$^+$ leaks across the membrane. It will be of particular interest to determine whether archaeal alkaliphiles have membrane lipid adaptations that provide substantial protection against alkali and obviate the need for some of the strategies of alkali tolerance that are important in eubacterial alkaliphiles. Genomic data are beginning to emerge (Falb et al., 2005) but have not yet been extended to mutational and transcriptome analyses.

Transport-Based Mechanisms Are Crucial, Especially the Role of Na$^+$/H$^+$ Antiporters

Alkaliphily in bacteria depends upon one or more Na$^+$/H$^+$ antiporters that catalyze proton uptake in exchange for cytoplasmic Na$^+$ (Krulwich, 1995, 1998, 1999; Padan et al., 2005; West and Mitchell, 1974). Whereas a neutrophilic bacterium that is cultured within a narrow pH range around 7.5 can do without such antiporters (Harold and Van Brunt, 1977; Speelmans et al., 1993), antiporters are essential for optimally meeting a major alkaline challenge (Booth, 1985; Macnab and Castle, 1987). Most bacteria have multiple Na$^+$/H$^+$ antiporters (Padan et al., 2005), as shown for alkaliphilic *Bacillus* in Fig. 3. In neutrophilic bacteria such as *Bacillus subtilis* and *E. coli*, as well as in marine *Vibrio* species, both K$^+$/H$^+$ and Na$^+$/H$^+$ antiporters are used for pH

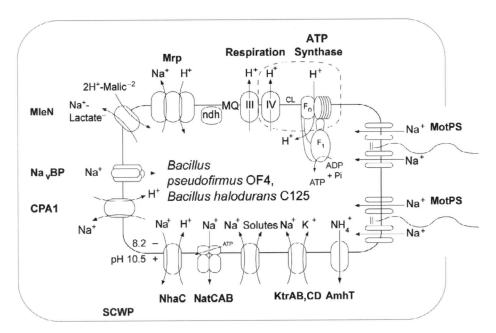

Figure 3. A diagrammatic representation of the Na$^+$ and proton cycles that support alkaline pH homeostasis, solute transport, and motility in extreme facultative alkaliphiles such as *B. halodurans* C-125 and *B. pseudofirmus* OF4. Oxidative phosphorylation by *B. pseudofirmus* OF4 is one of two processes that depend upon inward proton movements, as it utilizes a H$^+$-proton-pumping respiratory chain and a H$^+$-coupled ATP synthase; these complexes have alkaliphilic-specific features and function in a membrane that has a high content of cardiolipin (CL); the dashed lines around OXPHOS elements indicate a hypothesized use of kinetically sequestered proton transfers during OXPHOS at very high pH. The $\Delta\Psi$ generated by respiration energizes a complement of Na$^+$/H$^+$ antiporters: Mrp is shown as a multisubunit antiporter and has a dominant role in pH homeostasis (Swartz et al., 2005a), to which NhaC has also been shown to contribute (Ito et al., 1997b). Genomic evidence suggests that CPA1-type (Na$^+$/H$^+$) and MleN-type (2H$^+$ malate/Na$^+$ lactate) antiporters also contribute to alkaline pH homeostasis (Padan et al., 2005; Takami et al., 2000). Na$^+$-coupled solute uptake and motility, as well as the voltage-gated channel Na$_V$BP play roles in Na$^+$ reentry in support of pH homeostasis (Padan et al., 2005). Na$_V$BP also has a role in motility and chemotaxis (Ito et al., 2004b); the ABC transporter, NatCAB (Wei et al., 1999), presumably plays a role under excessively high Na$^+$ conditions; and AmhT has a role during growth on high-amine media but the transport mechanism is not yet established (Wei et al., 2003).

homeostasis (Fujisawa et al., 2005; Krulwich et al., 1999; Lewinson et al., 2004; Nakamura et al., 1992). By contrast, alkaliphiles use Na$^+$/H$^+$ antiport. As might be anticipated for active pH homeostasis at alkaline pH, alkaliphilic *B. pseudofirmus* OF4 also exhibits much higher aggregate levels of membrane antiport activity than neutrophilic *B. subtilis* (Krulwich et al., 1999; Liu et al., 2005). We hypothesize that the specificity for Na$^+$ protects the alkaliphile from loss of too much cytoplasmic K$^+$ during such active antiport (Krulwich, 1995). A high concentration of cytoplasmic K$^+$ is physiologically important (Epstein, 2003) and also ameliorates the increased cytotoxicity of Na$^+$ at high pH (Padan and Krulwich, 2000). Alkaliphile Na$^+$/H$^+$ antiporters are likely to be electrogenic, catalyzing the net inward movement of positive charges during a turnover because of their intrinsic coupling stoichiometry, i.e., with the H$^+$ moving inward/Na$^+$ moving outward >1. This property makes it possible for the transmembrane potential ($\Delta\Psi$, negative inside the cell relative to outside) to energize net proton accumulation by the antiport reaction (Macnab and Castle, 1987; Padan et al., 2005).

Almost all Na$^+$/H$^+$ antiporters are single-gene products, with a notable exception being the Mrp antiporter system, which has a dominant role in pH homeostasis in alkaliphilic *Bacillus* species as well as a role in Na$^+$ resistance (Hamamoto et al., 1994; Hiramatsu et al., 1998; Kitada et al., 2000; Swartz et al., 2005a). Mrp systems are widespread among bacteria and archaea, where they are also called Sha, Mnh, and Pha depending upon the species (Swartz et al., 2005a). Among the two major conserved groups that have thus far been studied, antiport activity is encoded by a polycistronic operon containing seven genes (group 1 Mrp systems) or six genes (group 2 Mrp systems in which MrpA and MrpB are fused) (Mathiesen and Hagerhall, 2003; Swartz et al., 2005a). The alkaliphile Mrp systems studied to date are all group 1 products, and they cluster together in sequence-similarity analyses, although some alkaliphilic species have second Mrp systems that fall outside that cluster (Fig. 4).

Two point mutants, one in *mrpA* and one in *mrpC* of *B. halodurans* C-125, result in a viable but nonalkaliphilic phenotype that led to the discovery of this antiporter locus (Hamamoto et al., 1994; Seto et al., 1995). Similarly, both *mrp*-null mutants and nonpolar mutants with deletions in single *mrp* genes in *B. subtilis* are also viable (Ito et al., 1999, 2000; Kosono et al., 2000). However, attempts to make single *mrp* gene deletions in *B. pseudofirmus* OF4 have been unsuccessful, suggesting that Mrp function is required at pH 7.5 as well as at higher pH (Swartz et al., 2005a). Among the intriguing features of Mrp systems are the following: (i) the apparent requirement of each of the six to seven hydrophobic gene products for full function in alkali and/or Na$^+$ resistance (Hiramatsu et al., 1998; Ito et al., 2000), even though MrpA and MrpD are considered the likely candidates for the antiport catalyst(s) (Hamamoto et al., 1994; Kosono et al., 2005; Mathiesen and Hagerhall, 2002); (ii) the strong sequence similarity between several Mrp proteins and membrane-associated subunits of the proton-pumping NADH dehydrogenase complex of mitochondria and bacteria (Hamamoto et al., 1994; Mathiesen and Hagerhall, 2002, 2003) that may relate to a shared capacity for translocating protons (Drose et al., 2005; Friedrich et al., 2005); and (iii) evidence for an anion export activity associated with Mrp of *B. subtilis* (Ito et al., 1999, 2000). These observations suggest that Mrp functions optimally as a multisubunit complex (Hiramatsu et al., 1998) and that such a complex may be a consortium of transporters that, together, provide value-added over other single-gene-product antiporters (Swartz et al., 2005a). The value-added could also involve an electron transport function that has been hypothesized for Mrp systems in several connections (Kashyap et al., 2006; Swartz et al., 2005b), but there is no direct evidence that supports this idea (Swartz et al., 2005a, 2005b). Another unanswered question about the Mrp Na$^+$/H$^+$ antiporters of extreme alkaliphiles is how they gather protons rapidly enough to support net cytoplasmic proton accumulation at low external proton concentrations. Perhaps the large protein surface constituted by a multiprotein Mrp complex at the outer cytoplasmic membrane surface plays a role in effective proton gathering, as has been described for surface features of other membrane proteins or complexes (Hosler, 2004; Marantz et al., 1998; Riesle et al., 1996).

The specific properties of the antiporters of alkaliphilic *Bacillus* that support its functions are not yet clear, but antiporter properties of interest in relation to alkaliphily have emerged for a different alkaliphile. The first study of a Na$^+$/H$^+$ antiporter from a soda lake alkaliphile is on the activity profile of an NhaD-type antiporter from the gram-negative bacterium *Alkalimonas amylolytica* N10. The antiporter, designated Aa-NhaD, was cloned by complementation of an antiporter-deficient *E. coli* strain and then studied in vesicles from such strains, expressing Aa-*nhaD* (Liu et al., 2005). The profile is highly adaptive to the natural environment and reflects the salt tolerance (7% NaCl) and alkali tolerance (up to pH 11) of the organism (Liu et al., 2005; Ma et al., 2004). Aa-NhaD exhibits low activity, is only active at pH ≥8.5, and functions optimally in the presence of 600 mM NaCl. All three of these properties are altered by a single

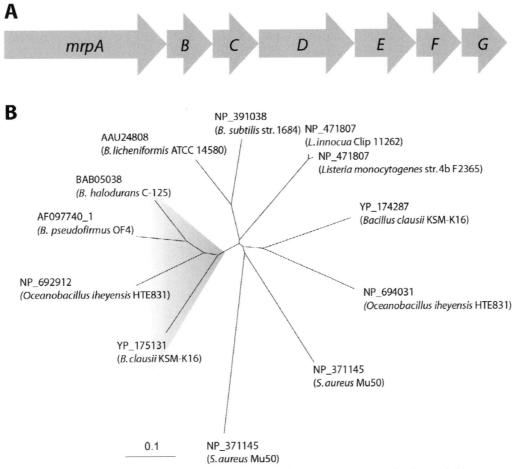

Figure 4. The Mrp Na$^+$/H$^+$ antiporter system. (A) The operon structure of a group 1 Mrp system that is encoded by a seven-gene operon (Swartz et al., 2005a). (B) Unrooted tree (TreeView) of ClustalW (DSGene) analysis indicating that alkaliphiles contain a group 1 Mrp system that shows closer sequence similarity to each other than to Mrp systems from other gram-positive bacteria and to second Mrp systems found, in some cases, in the alkaliphiles themselves.

mutation that arises during propagation in a triple antiporter mutant of *E. coli*, with a resulting activity profile that is closer to the native antiporters of *E. coli* (Liu et al., 2005).

Two additional transporter-based contributions to pH homeostasis have been documented. First, the H$^+$ cycle that supports OXPHOS is especially adapted to support the pH homeostasis needs of the alkaliphile, and these adaptations are not found in the fermentative *O. iheyensis* (discussed below in connection with OXPHOS). In several other aerobic bacteria in which the F$_1$F$_0$-ATPase acts primarily as an ATP synthase, an increased expression of the *atp* operon that encodes the synthase has been noted as a response to alkali challenge or mutational loss of a major Na$^+$/H$^+$ antiporter (Kosono et al., 2004; Maurer et al., 2005), whereas expression of the synthase decreases upon acid adaptation of *Helicobacter pylori* (Wen et al., 2003). Biochemical studies have shown a correlation between ATP synthesis capacity and pH homeostasis during a sudden alkaline shift. The results are consistent with the transcriptome studies of neutrophiles, showing a strong correlation between the capacity for ATP synthesis at pH 10.5 and the capacity to acidify the cytoplasm relative to the alkaline pH of the buffer (Wang et al., 2004).

The second transporter-based alkali adaptation distinct from Na$^+$/H$^+$ antiporters is one that plays a critical role when *B. pseudofirmus* OF4 is grown on amine-rich media. AmhT (for ammonium homeostasis transporter) facilitates ammonium efflux from the cells (Wei et al., 2003). At highly alkaline pH, this prevents trapping of ammonium in the cytoplasm by the high cytoplasmic proton concentration achieved by the antiporters and/or metabolic acids. This in turn prevents ammonium accumulation from subverting the homeostatic mechanism. Instead, upon efflux of ammonium into the high-pH medium, NH$_3$ is produced and largely volatilizes, leaving the proton in the medium. AmhT may function as a channel or as an antiporter, both of which have been proposed for other members of the cation:proton antiporter-2 family, of which it is a member (Busch and Saier, 2002). It will be of interest to study whether the numerous homologs of AmhT in other bacteria have similar catalytic activities and roles (Booth et al., 2005). We expect there to be additional transporter-based contributions to alkaline pH homeostasis in alkaliphiles, perhaps involving anions such as chloride and bicarbonate that are centrally involved in pH homeostasis in eukaryotic cells (Alper, 2002; Romero and Boron, 1999).

CYCLES THAT SUPPORT ALKALIPHILE pH HOMEOSTASIS ALSO SUPPORT ION-COUPLED SOLUTE UPTAKE, MOTILITY, AND CHEMOTAXIS

In high-pH, low-Na^+ environments, the availability of sufficient cytoplasmic Na^+ can become limiting for the high Na^+/H^+ antiport activity that supports pH homeostasis, thus making the adequacy of Na^+ reentry routes important (Krulwich et al., 1985; McLaggan et al., 1984; Padan et al., 2005). The Na^+ that serves as the coupling ion for all ion-coupled solute transport in alkaliphiles is a major source of cytoplasmic Na^+ for pH homeostasis. Use of Na^+ instead of protons is an apparent adaptation to the low PMF (Horikoshi, 1991; Krulwich, 1995). At alkaline pH, the sodium motive force (SMF) is higher than the PMF, as it is at temperatures that are high enough to increase membrane proton permeability (Lolkema et al., 1994; Tolner et al., 1997).Strong stimulation of pH homeostasis is observed during an alkaline shift when the buffer contains a nonmetabolizable aminoacid whose uptake is coupled to Na^+ uptake (Krulwich et al., 1985). This is evidence that the Na^+ entering with substrates supports pH homeostasis. Suboptimal but significant pH homeostasis is observed at low Na^+ concentrations even when such solutes are absent (Table 1, bottom); under those conditions, Na^+ channels that have important roles in chemotaxis and/or motility have roles in Na^+ reentry in support of pH homeostasis (Table 1, Fig. 3).

Na+/Solute Symporters

The Na^+/solute symporters (cotransporters) of alkaliphiles do not use H^+ as an alternate coupling ion, as is the case for many Na^+/solute symporters of neutrophiles (Horikoshi, 1999; Jung, 2001; Pourcher et al., 1995; Skulachev, 1989). Moreover, an alkaline pH optimum for the alkaliphile symporters suggests that protons are inhibitory to the ion-coupled transporters of alkaliphiles (Guffanti and Krulwich, 1992; Horikoshi, 1999). This would explain the otherwise paradoxical observation that facultatively alkaliphilic *Bacillus* strains require higher concentrations of Na^+ for optimal growth at pH 7.5 than at pH 10.5, although the requirement for Na^+/H^+ antiport in service of pH homeostasis is much lower. (This requirement is illustrated by growth rates at pH 7.5 with 25 mM added Na^+ versus 10.5 with 5 mM added Na^+ in Fig. 2.) We hypothesize that higher Na^+ is needed to overcome competitive inhibition of the Na^+/symporters by protons at low pH (Gilmour et al., 2000).

MotPS and NavBP, the Two Channels Involved in Motility Chemotaxis

Flagellar rotation in bacteria is another bioenergetic work function powered by an electrochemical ion gradient across the cytoplasmic membrane (Larsen et al., 1974). Mot assemblies consist of four MotA-like proteins, each of which has four transmembrane segments (TMS) and two single-TMS MotB-like proteins; these proteins form a Mot complex. This complex serves both the ion channel function and stator function for the flagellar rotor (Berg, 2003; Berry, 2000; Kojima and Blair, 2004). In most neutrophilic bacteria, motility is H^+ coupled and *motAB* is the sole gene locus encoding a Mot channel, but marine *Vibrio* species exhibit dual flagellar assemblies that are powered by distinct Mot-type systems, one of which, constituted with LafAB, utilizes H^+ (PMF) to power lateral flagella, whereas the other, constituted with PomAB,MotXY, uses Na^+ (SMF) to power a polar flagellum (McCarter, 2001, 2004; Yorimitsu and Homma, 2001). Extremely alkaliphilic *Bacillus* species use only the SMF to energize motility (Hirota et al., 1981; Hirota and Imae, 1983; Sugiyama et al., 1985). The sodium Mot for alkaliphilic *Bacillus* is encoded by a pair of *motAB*-like genes designated *motPS* (for pH and salt) (Ito et al., 2004a). They are downstream of the catabolite regulator gene *ccpA* in *B. halodurans* C-125, *B. pseudofirmus* OF4, and the modestly alkaliphilic *O. iheyensis* but not in moderately alkaliphilic *Bacillus clausii*, although they are found in the neutrophilic *B. subtilis* and other neutrophilic *Bacillus* (Fig. 5). MotPS is required for the motility of *B. pseudofirmus* OF4 and is sufficient to constitute the Na^+-translocating Mot of alkaliphilic *Bacillus*; Na^+ fluxes are mediated by purified MotPS reconstituted into proteoliposomes (Ito et al., 2004a). The sequences of the Mot systems of alkaliphiles are most closely related to each other but show sequence similarity to neutrophilic MotPS, to the Na^+-driven PomAB systems,

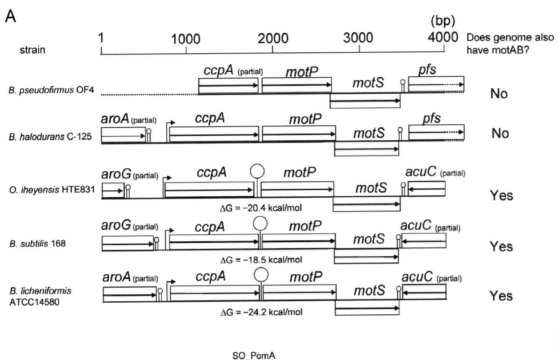
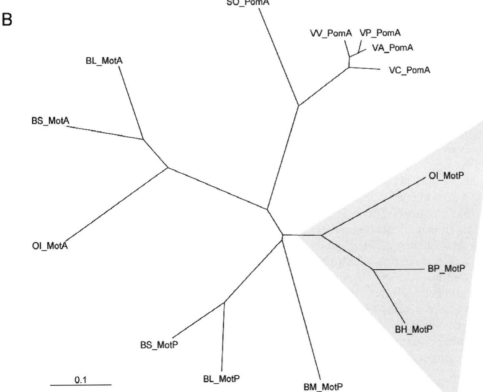

Figure 5. The Na$^+$-coupled MotPS of alkaliphilic bacilli. (A) The *ccpA–motPS* operon of the two extreme alkaliphiles, *B. halodurans* C 125 and *B. pseudofirmus* OF4. No stem loop is found between *ccpA* and *motPS* in these two alkaliphiles, for which MotPS is the sole Mot (based, respectively, on genomic and genetic evidence; Ito et al., 2004a; Takami et al., 2000). The stem loop is found in less alkaliphilic *O. iheyensis* and nonalkaliphilic *Bacillus* strains that also possess MotAB; the stem loop reduces transcription read-through between *ccpA* and *motPS* (Terahara et al., submitted for publication). (B) Unrooted tree (TreeView) of ClustalW (DSGene) analysis of MotP from alkaliphiles is shown to cluster relative to homologous PomA from a Na$^+$-coupled Mot of *Vibrio* and H$^+$-coupled MotA from moderately alkaliphilic *O. iheyensis* and other bacilli.

and to MotAB proteins, in that order of relatedness (Ito et al., 2004a) (Fig. 5B). *B. subtilis* MotPS is also a functional Na$^+$-driven Mot in an organism in which H$^+$-coupled MotAB is the dominant Mot, but the presence of MotPS increases motility under conditions of elevated Na$^+$, pH, and viscosity (Ito et al., 2004a, 2004b, 2005; Terahara et al., 2006). In the neutrophilic *Bacillus* species that possess *motPS* genes, there is a stem-loop structure in the *ccpA–motPS* operon that results in incomplete read-through from *ccpA* through *motPS*, an impediment that is largely removed in an up-motile MotPS locus that has a single mutation in the stem loop (Terahara et al., 2006). That stem loop is absent from the *ccpA–motPS* operons of extremely alkaliphilic *Bacillus* species (Fig. 5A), suggesting that there is selection against this transcriptional punctuation when MotPS serves as the sole Mot system. Motility in alkaliphilic *Bacillus* is restricted to high pH (Aono et al., 1992; Ito et al., 2004a; Sturr et al., 1994). The basis for this regulation is not yet known, but a role is indicated for a second Na$^+$ channel, the alkaliphilic channel of the Na$_V$Bac (prokaryotic voltage-gated Na$^+$ channel) superfamily (Ito et al., 2004b; Koishi et al., 2004; Ren et al., 2001; Yue et al., 2002).

The first member of the Na$_V$Bac superfamily was recognized in the genome of alkaliphilic *B. halodurans* C-125 as a putative calcium channel (Durell and Guy, 2001) and was then named NaChBac, after being shown to be a voltage-gated Na$^+$ channel (Ren et al., 2001). Electrophysiological studies conducted on NaChBac and its homolog from *B. pseudofirmus* OF4, Na$_V$BP, demonstrate its voltage-sensing properties and channel potentiation by alkaline pH (Chahine et al., 2004; Ito et al., 2004b; Kuzmenkin et al., 2004). Deletion of the *ncbA* gene that encodes Na$_V$BP in *B. pseudofirmus* OF4 results in impaired pH homeostasis when Na$^+$ is limiting and solutes are not present (Table 1, bottom); a small contribution of MotPS is indicated only when the Na$_V$Bac-type channel of *B. pseudofirmus* OF4 is also deleted (Ito et al., 2004a). A severe motility defect is also observed in the *ncbA* mutant. This defect is overcome upon selection of up-motile variants of the mutant strain, suggesting that some partially redundant channel in this alkaliphile compensates partially for loss of Na$_V$BP function. However, the up-motile *ncbA* mutant exhibits maladaptive inversed chemotaxis to a variety of different effectors, including movement away from the attractants proline, glucose, aspartate, and pH 10.5 in the presence of malate; and movement toward the repellent condition of pH 10.5 in the absence of nutrients (Ito et al., 2004b). We hypothesize that Na$_V$BP activation by high pH supports normal motility, chemotaxis, and pH homeostasis—functions of particular importance in alkaliphiles at very high pH—through a combination of direct effects (e.g., on the chemotaxis machinery) and indirect effects (e.g., providing a reentry route in the Na$^+$ cycle) (Ito et al., 2004b).

THE OXPHOS CONUNDRUM OF ALKALIPHILIC *BACILLUS*: ROBUST H$^+$-COUPLED ATP SYNTHESIS AT LOW PMF

Proton-Pumping Respiratory Chain: Alkaliphile Features

During respiration, alkaliphilic *Bacillus* species pump protons outward, concomitant with electron transport (Hicks and Krulwich, 1995; Lewis et al., 1983; Muntyan et al., 1993; Yumoto, 2003) (see Fig. 3). There are two notable features of the respiratory components of these bacteria: (i) they are present in very high abundance in the cytoplasmic membranes relative to other bacteria and (ii) the midpoint redox potentials are much lower than those of homologs from other organisms (Hicks and Krulwich, 1995; Lewis et al., 1980, 1981; Muntyan et al., 2005; Yumoto, 2002, 2003). Yumoto et al. (1991) proposed that the low midpoint redox potentials may be necessary to facilitate inward electron transport from cytochrome *c* on the outer membrane surface toward the binuclear center of the proton-pumping terminal oxidase under conditions of relatively high transmembrane potentials ($\Delta\Psi$, negative inside) generated by alkaliphilic *Bacillus* at high pH (Fig. 1); they suggested that cytochrome *c* may play a special role in electron flow through the alkaliphile respiratory chain.

Use of an H$^+$- Rather than Na$^+$-Coupled ATP Synthase: Low PMF Problem

OXPHOS differs from other bioenergetic work in that F$_1$F$_0$-ATP synthase of alkaliphilic *Bacillus* species is H$^+$ coupled (Cook et al., 2003; Hicks and Krulwich, 1990; Hoffmann and Dimroth, 1991b). The ATP synthase catalyzes PMF-driven ($\Delta\Psi$-dependent) movement of protons inward through the membrane-embedded F$_0$ subunits that set off rotational events, via the barrel of *c* subunits in the F$_0$, which result in conformational changes in the catalytic sites of the F$_1$ sector that result in ATP synthesis (Mitchell, 1961; Stock et al., 2000) (see Fig. 6A). Extreme alkaliphiles that grow nonfermentatively and carry out OXPHOS reap the benefit of maximal energy capture from metabolic substrates and add the ATP synthase as a contributor to pH homeostasis. However, the alkaliphile must prevent leakage or outward pumping of protons through the ATP synthase when cells are subjected to a particularly high-alkaline, low-PMF challenge and must also solve the energetic problem of synthesizing ATP at the low-bulk PMF that results from successful pH homeostasis. When *B. pseudofirmus* OF4 is growing at pH 7.5, the cytoplasmic pH is also 7.5, so the total PMF (expressed in millivolt as Δp in Fig. 1) is equal to the $\Delta\Psi$ of −140 mV. By contrast, at pH 10.5, the ΔpH (acid in) is a reversed, chemiosmotically adverse force equivalent to +136 mV, which when added to the $\Delta\Psi$ of −180 mV yields a low net Δp of −4 mV. A core conundrum of alkaliphile OXPHOS is that the phosphorylation potential (ΔG_p) that reflects the [ATP]/[ADP][P_i] achieved by OXPHOS is −530 mV at pH 10.5 and lower, −435 mV, at pH 7.5, even though the PMF is approximately three times higher at pH 7.5 than at pH 10.5 (Guffanti and Hicks, 1991; Sturr et al., 1994). This robust ATP synthesis at pH 10.5 is exhibited both by whole cells and by ADP + P_i-loaded, ascorbate-energized right-side-out membrane vesicles of *B. pseudofirmus* OF4. These vesicles are devoid of cell-wall components that would trap protons. Moreover, synthesis is not inhibited by addition of high concentrations of mobile buffer and/or by elevation of the ionic strength, both of which would be expected to equilibrate protons trapped near the membrane by surface charges (Guffanti and Krulwich, 1994). The discordance between the magnitude of the bulk PMF and the capacity for ATP synthesis via OXPHOS runs counter to the strict chemiosmotic expectation of complete coupling of OXPHOS to the bulk PMF (Cramer and Knaff, 1990; Dimroth and Cook, 2004; Mitchell, 1961). The same discordant pattern is observed in *B. alcalophilus* and in the moderately alkaliphilic *Bacillus* sp. strain TA2.A1 (Guffanti and Hicks, 1991; Hoffmann and Dimroth, 1991a; Olsson et al., 2003).

Why is not a Na^+-coupled cycle used for alkaliphile OXPHOS to bypass the problem of the low PMF as it is for solute transport and motility? We hypothesize that extreme alkaliphiles are unlikely to utilize a primary Na^+-pumping respiratory chain such as the NQR-type NADH dehydrogenase to support a Na^+-based OXPHOS cycle because of the primacy of coupling Na^+ efflux to net proton uptake in alkaliphiles, whereas marine bacteria that contain NQR have a more modest pH homeostasis burden that can be met by both Na^+/H^+ and K^+/H^+ antiporters (Nakamura et al., 1984; Tokuda and Unemoto, 1985). Another factor that may prevent the use of a Na^+-coupled OXPHOS cycle in alkaliphiles is that the synthase would have to be different from the known Na^+-coupled F_1F_0-ATP synthases because these enzymes also couple H^+-fluxes to ATP synthesis (Laubinger and Dimroth, 1989). The presence of the large ΔpH, acid in, that alkaliphiles maintain at high pH is expected to inhibit ATP synthesis by such an enzyme.

Is it possible that the pH near the membrane surface is not unusually high? This would make the PMF near the OXPHOS machinery right outside the membrane in the conventional range of neutrophilic bacteria and mitochondria. A calculation of kinetic trapping of protons near the charged membrane surface by an interfacial barrier at the membrane-external water interface has been modeled. The model predicts that the surface pH is significantly lower than the bulk pH as a general rule, tempering the magnitude of the low-PMF problem of alkaliphiles. This prompted the speculation that the pH near the protein complexes participating in alkaliphile OXPHOS could be low enough so that no special adaptations of the OXPHOS machinery would be needed (Cherepanov et al., 2003; Mulkidjanian et al., 2005). That prediction is negated by abundant genomic, genetic, and physiological data: (i) Even in alkaliphiles that grow in low-Na^+ environments, flagellar motility using membrane-embedded MotPS and ion-coupled solute uptake are Na^+ coupled, creating challenges of Na^+ acquisition that would be maladaptive if there was no low PMF problem; (ii) in addition to the unusual midpoint potentials of respiratory chain complexes, there are alkaliphile-specific adaptations of the F_0 subunits of the ATP synthase in regions within as well as just outside the membrane that are essential for ATP synthesis and for the retention of cytoplasmic protons at pH 10.5 (see below); and (iii) diverse membrane protein segments just outside the membrane have greatly reduced abundance of basic amino acids and increased acidic residues, especially in regions that must be charged in order to function (three examples are shown in Fig. 7) (Ito et al., 1997a; Krulwich, 2003). If the surface pH is not significantly higher in the alkaliphile than in neutrophiles in the region of the OXPHOS machinery, these pronounced deviations among homologous protein segments would be absent, and indeed they are less pronounced among homologs from moderate alkaliphiles (as shown for ATP synthase features in Fig. 6).

Clues to Mechanistic Strategies Are Supported by the OXPHOS Phenotypes of Mutants in Alkali-Specific ATP Synthase Features

Clues to the nature of the additional strategies underpinning alkaliphile OXPHOS come from the lack of equivalence between ATP synthesis by cells or ADP + P_i-loaded vesicles energized by respiration and by an imposed potential of the same magnitude. ATP synthesis is energized less well by an imposed

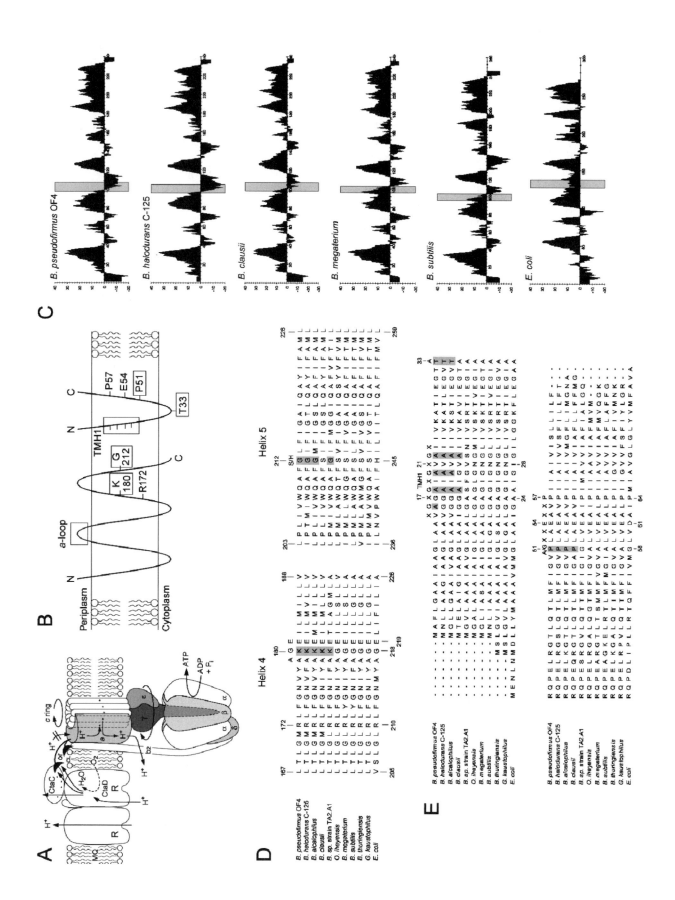

potential than by respiration, and with titration-like behavior, an imposed potential completely fails to energize ATP synthesis at pH ≥9.2 (shown in whole-cell experiments in Fig. 8A); controls show that the imposed potential is efficacious for other bioenergetic work and that the cells are still capable of responding to a respiration-generated potential if malate is added at pH 10.5 as long as respiratory chain inhibitors are omitted from the protocol (Guffanti and Krulwich, 1992, 1994). These observations show that respiratory chain pumping per se is required for OXPHOS above pH 9.2 and indicate the presence of a barrier to proton entry from the bulk in response to an imposed potential above this pH. We hypothesized that the apparent barrier to proton entry from the bulk results from a pH-dependent gating feature of the proton entry pathway into the ATP synthase; closure of this putative gate at very alkaline pH would protect the cell from loss of protons by outward proton flux when cells are exposed to particularly high pH.

If the entry of protons from the bulk phase is blocked at pH ≥9.2, the alkaliphile must have a path of proton flow from the respiratory chain to the ATP synthase that kinetically sequesters the protons from full equilibration with the bulk and bypasses this block at very alkaline pH. The existence of an energizing component for OXPHOS above pH 9.2 that is not present at lower pH is supported by the pH-dependent effect of lowering the $\Delta\Psi$ in ADP + P_i-loaded vesicles carrying out OXPHOS, i.e., ascorbate-energized ATP synthesis. At pH 9.5, ATP is synthesized much faster than at pH 7.8 when the $\Delta\Psi$ is −150 mV; as the $\Delta\Psi$ is titrated downward, ATP synthesis at the lower pH is more sensitive to initial reductions, although the $\Delta\Psi$ generated by respiration is a major and necessary energy component at both pH values (Fig. 8B). The extra energizing component available to alkaliphiles at high pH supports predictions by Williams of a sequestered path of protons in OXPHOS that would be fostered by complementary adaptations of the proton-pumping complexes and proton-utilizing complexes (Williams, 1977, 1978). In line with this prediction, alkaliphilic-specific features in regions of the F_0 a and c subunits are evident in extremely alkaliphilic Bacillus species and are depicted in Fig. 6 (Ivey and Krulwich, 1991, 1992; Wang et al., 2004).

The bioenergetic profiles of B. pseudofirmus OF4 mutants in which the alkaliphile-specific features are changed to the Bacillus consensus sequence in nonalkaliphiles demonstrate that three of them, c-subunit proline^{51}, a-subunit lysine^{180}, and the a-loop motif, are required for growth on malate and OXPHOS at pH 10.5 but not at pH 7.5 and had little or no effect on glucose growth at either pH (Wang et al., 2004) (Table 2). Mutation of the c-subunit threonine^{33} to the consensus results in a unique profile that suggests a role for threonine^{33} in optimizing interactions between the c barrel and the F_1 subunits; this profile probably does not relate directly to the path of the protons but to the elevated cytoplasmic pH at which OXPHOS occurs at an external pH of 10.5. By contrast, the a-subunit lysine^{180} as well as the putative interacting glycine^{212} that are in the proton-uptake pathway (Fillingame et al., 2003; Hartzog and Cain, 1994) are directly implicated in pH-dependent

Figure 6. Alkaliphile-specific features of the ATP synthase a and c subunits. (A) A model indicating how features of the ATP synthase could contribute to kinetically sequestered movements of protons from the respiratory chain to the ATP synthase of an extremely alkaliphilic Bacillus at pH ≥9.2. Protons in the bulk external phase are hypothesized to be blocked (hatched arrow) from entering the ATP synthase at pH ≥9.2 by gating that depends upon a-subunit lysine^{180} and glycine^{212}. This gating prevents proton loss through the ATP synthase when extremely high-pH, low-PMF conditions prevail and accounts for failure of an imposed potential to energize ATP synthesis above pH 9.2. Protons are pumped into the bulk by both Complex III and Cta (the caa_3-type cytochrome oxidase), but some protons, as shown for cytochrome oxidase, are hypothesized to reach the ATP synthase without equilibration with the bulk, assisted by some combination of: (i) the alkaliphile-specific ATP synthase features (which are detailed in B–E and one of which is shown here as a cup-like proton-gathering element); (ii) special features of the cytochrome oxidase (indicated by the negatively charged region of CtaC, see Fig. 8); (iii) and the negatively charged membrane lipid environment (CL, cardiolipin). (B) Topological models illustrate positions of both the a- and the c-subunit features that are boxed. (Modified from Wang et al., 2004, with permission from the publisher.). The features are: a-loop, aK180, aG212, cTMH1, cT33, and cP51. (C) Kyte–Doolittle hydropathy plots (Gene Runner) of different a subunits. The boxed region is the hypothesized periplasmic loop between aTMH2 and aTMH3 (designated as "a-loop") corresponding to E. coli residues 128–137 (Valiyaveetil and Fillingame, 1998), according to the ClustalW alignment (DS Gene). (D) An alignment showing the single, but probably interacting, amino acid features in aTMH4 and aTMH5 of the a subunit, aK180, and aG212. (E) An alignment illustrating the features in the c subunit displayed only by extreme alkaliphiles. Shown are a cTMH1 feature, in which the glycines of the conserved XGXGXGXGX region are largely or completely replaced by alanines, the cT33 residue instead of a conserved alanine, and cP51 in place of a glycine or alanine in other bacteria. Numbering refers to B. pseudofirmus OF4 at the top and to E. coli at the bottom. The accession numbers for the data shown are: AF330160, B. pseudofirmus OF4; NP_244627, B. halodurans C-125 a subunit; NP_244626, B. halodurans C-125 c subunit; M84712, B. alcalophilus; NC_006582, B. clausii a subunit; NC_006582, B. clausii c subunit; AF533147, B. sp. strain TA2.A1; NP_693903, O. iheyensis a subunit; NP_693902, O. iheyensis c subunit; M20255, Bacillus megaterium; Z28592, B. subtilis; NC_005957, Bacillus thuringiensis a subunit; NC_005957, B. thuringiensis c subunit; NC_006510, Geobacillus kaustophilus a subunit; NC_006510, G. kaustophilus c subunit; NP_290377, E. coli a subunit; NP_290376, E. coli c subunit. (Modified from Wang et al., 2004.)

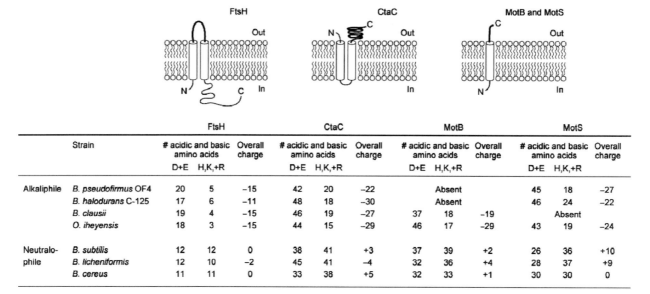

Figure 7. Segments of alkaliphile membrane proteins just outside the membrane surface, including a functionally important segment of cytochrome oxidase, show evident sequence adaptations relative to homologous regions from nonalkaliphiles. The content of acidic and basic residues and overall charge is displayed for the external segments of: the stress protein FtsH; the CtaC subunit of the caa_3-type cytochrome oxidase; and the smaller subunit, MotB (nonalkaliphiles except for *B. clausii*) or MotP (alkaliphiles) of the membrane-embedded flagellar stator-force generator.

gating of proton fluxes, as shown by the 10- to 20-fold increase in ATP synthesis in response to an imposed potential at pH 9.3 (and a significant increase at pH 8.3). Consistently, when cells of these two mutants are equilibrated at pH 8.5 and shifted to pH 10.5, the cytoplasmic pH rises immediately after the shift, whereas the cytoplasmic pH of wild-type *B. pseudofirmus* OF4 or the *a*-loop mutant remains steady at pH 8.5. The glycine[212] mutant also exhibits a much higher hydrolytic ATPase activity than wild type. Low hydrolytic activity is a general feature of alkaliphile ATP synthases (Dimroth and Cook,

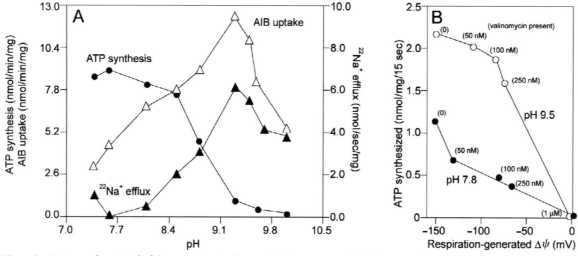

Figure 8. An imposed potential of the same magnitude as respiration-generated $\Delta\Psi$ values energizes ATP synthesis by *B. pseudofirmus* OF4 below but not above pH 9.2, whereas respiration-energized ATP synthesis is more robust and resistant to small drops in the $\Delta\Psi$ >pH 9.2 than below. (A) The efficacy of an imposed potential (a valinomycin-mediated K^+ diffusion potential of −160 mV) to energize ATP synthesis, AIB uptake, and $^{22}Na^+$ efflux (an assessment of aggregate Na^+/H^+ antiport) in energy-depleted whole cells that are reenergized by addition of malate. (Modified from Guffanti and Krulwich, 1992, with permission from the publisher.) (B) The effect of downward titration of the $\Delta\Psi$ on respiration-dependent ATP synthesis by ADP + P_i-loaded membrane vesicles from *B. pseudofirmus* OF4 at pH 7.8 or 9.5. Ascorbate-phenazine methosulfate is the electron donor, potassium-containing buffers are present both inside and outside the vesicles, and $\Delta\Psi$ is reduced by addition of the concentrations of valinomycin shown in parentheses. (Modified from Guffanti and Krulwich, 1994.)

Table 2. Alkaliphile-specific sequence features of the *a* and *c* subunits of ATP synthase are required for robust OXPHOS at high pH and prevention of proton loss to the outside[d]

Strain	Growth yield (%) wild type on malate[a]		Respiration-driven ATP synthesis (nmol ATP/mg/10 sec)[b]		Peak diffusion potential-dependent ATP synthesis (nmol/mg)[c]	
	pH 7.5	pH 10.5	pH 7.5	pH 10.5	pH 8.3	pH 9.3
Wild type	100	100	1.9	0.9	2.1	0.1
a-Loop mutant	82	20	1.4	0.1	2.5	0.2
*a*G212S	86	100	2.3	0.7	3.9	0.9
*a*K180G	86	16	2.0	0.2	3.9	1.1
*a*K180G/*a*G212S	70	14	1.4	0.3	4.0	1.8
*c*P51A	88	0	1.4	0.2	Not done	Not done

[a] Conducted in semidefined malate-containing medium in batch culture.
[b] Assayed in ADP + P_i-loaded membrane vesicles energized by ascorbate-phenazine methosulfate.
[c] Assayed in ADP + P_i-loaded membrane vesicles energized by imposition of a valinomycin-mediated K^+ diffusion potential.
[d] Data are from a study of mutants in the motifs and sequence features depicted in Fig. 4 (Wang et al., 2004). The effect of changing the alkaliphile residue or region on the consensus sequence for *Bacillus* ATP synthases was examined. The specific sequence of *Bacillus megaterium* was used because it was the closest homolog that was not an extremophile of any kind. The assembly of all of the mutant synthases for which data are shown was ≥98% of wild type; only the mutant synthase with an altered *c*-subunit motif in TMH-1, a change to XGXGXGXGX, failed to assemble.

2004). Together, studies of alkaliphile ATP synthases support the hypothesis that this synthase has a pH-dependent gating feature of the *a* subunit that protects the alkaliphile from proton loss upon a large, sudden alkalinization of the outside pH, that alkaliphile-specific motifs of both the *a* and *c* subunits are specifically required for OXPHOS at high pH, and that there may be adaptations needed in the *c* subunit to optimize interactions with the subunits of the F_1 sector with which it directly interacts.

The specific roles of the critical ATP synthase motifs and the actual path of protons into the synthase at pH ≥9.2 still need to be clarified. The proton transfer might involve direct protein–protein interactions with a respiratory chain complex, as suggested by Qiu et al. (1992) for mitochondria, and/or involve the abundant cardiolipin of the alkaliphile membrane (Fig. 6A). The positive charge of the *a*-subunit lysine[180] is likely to have a critical role in proton capture. Features of the *c* subunit are also likely to be critical. Nuclear magnetic resonance (NMR) studies of the purified alkaliphile *c* subunit found a higher pK_a of 7.7 than similar studies of nonalkaliphilic *c* subunits (Rivera-Torres et al., 2004). This higher pK_a may be an adaptation that prevents proton loss from the *c* rotor. Also of special interest is the possibility that alkaliphile ATP synthases have a *c* rotor with a very high number of *c* subunits per rotor oligomer. A high stoichiometry of protons translocated inward/ATP synthesized is predicted to correlate with a high phosphorylation potential relative to the bulk PMF ($\Delta G_p/\Delta p$ ratio) (Dimroth and Cook, 2004; Stock et al., 2000). The $\Delta G_p/\Delta p$ ratios in malate-grown *B. pseudofirmus* OF4 at pH 7.5 and 10.5 are, respectively, ~3 and 13 (Krulwich, 1995; Sturr et al., 1994). Notably, the $\Delta G_p/\Delta p$ ratios are higher during synthesis at pH 10.5 than at pH 7.5, even when cells are grown at pH 8.5 and then shifted to pH 7.5 or 10.5 (Wang et al., 2004) or when ADP + P_i-loaded vesicles are prepared from and assayed for ATP synthesis at pH 7.5 and 10.5 (Guffanti and Krulwich, 1994). Therefore, it is most unlikely that a large difference in *c*-subunit/holoenzyme stoichiometry is responsible for the lower ratio at pH 7.5 than at pH 10.5. The alkaliphile may well be found to have a high *c*-subunit stoichiometry, perhaps fostered by special packing properties involving replacement of glycines in the XGXGXGXGX motif of the N-terminal *c*-subunit helix of neutrophiles by alanines in alkaliphiles (Fig. 6). The crucial *c*-subunit proline[51] in extreme alkaliphiles may function in concert with this altered TMH1 motif. If this does result in a high *c*-subunit stoichiometry, the hardwired extremophile is likely to have that stoichiometry at all pH values, prepared to function optimally at high pH but perhaps functioning suboptimally at near-neutral pH. This would be analogous to negative effects of S-layer synthesis and use of an ATP synthase with lysine[180]–glycine[212], respectively, on growth and ATP synthesis at near-neutral pH, although they make a critical contribution at high pH.

CONCLUDING REMARKS

So far the focus of work on alkaliphile pH homeostasis has been on the active Na^+ cycles that support this homeostasis and on cell-surface properties that have supporting roles. The issue of the greater tolerance of alkaliphiles than neutrophiles to elevated cytoplasmic pH merits greater investigative interest to assess whether just a few key adaptations or a large number of adaptations underpin this tolerance and to explore the possibility that adapted alkaliphile enzymes will support greater tolerance of neutrophile strains in which particularly vulnerable targets of alkaline stress have been identified. As a more diverse group of genetically tractable alkaliphiles are developed, it will be possible to explore such questions as: whether any of them has a Na$^+$-coupled ATP synthase that is responsible for OXPHOS and, perhaps, has lost the capacity for H$^+$-coupled synthesis; whether some alkaliphiles grow optimally with cytoplasmic pH values >8.5; whether archaeal or other unusual alkaliphiles generate sufficiently high $\Delta\Psi$ values to energetically offset their ΔpH, acid in; and to further define the roles of NH$_4$+ in the bioenergetics of both aerobic and anaerobic alkaliphiles.

Acknowledgments. Work conducted in T.A.K.'s laboratory was supported by NIH grant GM28454, and work in M.I.'s laboratory was supported by a Grant-in-Aid for Scientific Research (C) 2005-17613004 for the 21st Century Center of Excellence program and high-technology centers organized by the Ministry of Education, Culture, Sports, Sciences and Technology.

REFERENCES

Alper, S. L. 2002. Genetic diseases of acid-base transporters. *Annu. Rev. Physiol.* **64:**899–923.

Aono, R., M. Ito, and T. Machida. 1999. Contribution of the cell wall component teichuronopeptide to pH homeostasis and alkaliphily in the alkaliphile *Bacillus lentus* C-125. *J. Bacteriol.* **181:**6600–6606.

Aono, R., H. Ogino, and K. Horikoshi. 1992. pH-dependent flagella formation by facultative alkaliphilic *Bacillus* sp. C-125. *Biosci. Biotechnol. Biochem.* **56:**48–53.

Aono, R., and M. Ohtani. 1990. Loss of alkalophily in cell-wall-component-defective mutants derived from alkalophilic *Bacillus* C-125. Isolation and partial characterization of the mutants. *Biochem. J.* **266:**933–936.

Berg, H. C. 2003. The rotary motor of bacterial flagella. *Annu. Rev. Biochem.* **72:**19–54.

Berry, R. M. 2000. Theories of rotary motors. *Philos. Trans. R. Soc. Lond. B* **355:**503–509.

Blankenhorn, D., J. Phillips, and J. L. Slonczewski. 1999. Acid- and base-induced proteins during aerobic and anaerobic growth of *Escherichia coli* revealed by two-dimensional gel electrophoresis. *J. Bacteriol.* **181:**2209–2216.

Booth, I. R. 1985. Regulation of cytoplasmic pH in bacteria. *Microbiol. Rev.* **49:**359–378.

Booth, I. R., M. D. Edwards, E. Murray, and S. Miller. 2005. The role of bacterial channels in cell physiology, p. 291–312. *In* A. Kubalski and B. Marinac (ed.), *Bacterial Ion Channels and Their Eukaryotic Homologs*. ASM Press, Washington, DC.

Bordi, C., L. Theraulaz, V. Mejean, and C. Jourlin-Castelli. 2003. Anticipating an alkaline stress through the Tor phosphorelay system in *Escherichia coli*. *Mol. Microbiol.* **48:**211–223.

Busch, W., and M. H. Saier, Jr. 2002. The transporter classification (TC) system, 2002. *Crit. Rev. Biochem. Mol. Biol.* **37:**287–337.

Cao, M., T. Wang, R. Ye, and J. D. Helmann. 2002. Antibiotics that inhibit cell wall biosynthesis induce expression of the *Bacillus subtilis* sigma(W) and sigma(M) regulons. *Mol. Microbiol.* **45:**1267–1276.

Chahine, M., S. Pilote, V. Pouliot, H. Takami, and C. Sato. 2004. Role of arginine residues on the S4 segment of the *Bacillus halodurans* Na$^+$ channel in voltage-sensing. *J. Membr. Biol.* **201:**9–24.

Cherepanov, D. A., B. A. Feniouk, W. Junge, and A. Y. Mulkidjanian. 2003. Low dielectric permittivity of water at the membrane interface: effect on the energy coupling mechanism in biological membranes. *Biophys. J.* **85:**1307–1316.

Clejan, S., T. A. Krulwich, K. R. Mondrus, and D. Seto-Young. 1986. Membrane lipid composition of obligately and facultatively alkalophilic strains of *Bacillus* spp. *J. Bacteriol.* **168:**334–340.

Cook, G. M., S. Keis, H. W. Morgan, C. von Ballmoos, U. Matthey, G. Kaim, and P. Dimroth. 2003. Purification and biochemical characterization of the F$_1$F$_0$-ATP synthase from thermoalkaliphilic *Bacillus* sp. strain TA2.A1. *J. Bacteriol.* **185:**4442–4449.

Cook, G. M., J. B. Russell, A. Reichert, and J. Wiegel. 1996. The intracellular pH of *Clostridium paradoxum*, an anaerobic, alkaliphilic, and thermophilic bacterium. *Appl. Environ. Microbiol.* **62:**4576–4579.

Cramer, W. A., and D. B. Knaff. 1990. Membrane structure and storage of free energy, p. 124–130, *Energy Transduction in Biological Membranes*. Springer-Verlag, New York, NY.

Dimroth, P., and G. M. Cook. 2004. Bacterial Na$^+$- or H$^+$-coupled ATP synthases operating at low electrochemical potential. *Adv. Microb. Physiol.* **49:**175–218.

Drose, S., A. Galkin, and U. Brandt. 2005. Proton pumping by complex I (NADH:ubiquinone oxidoreductase) from *Yarrowia lipolytica* reconstituted into proteoliposomes. *Biochim. Biophys. Acta* **1710:**87–95.

Dubnovitsky, A. P., E. G. Kapetaniou, and A. C. Papageorgiou. 2005. Enzyme adaptation to alkaline pH: atomic resolution (1.08 A) structure of phosphoserine aminotransferase from *Bacillus alcalophilus*. *Protein Sci.* **14:**97–110.

Durell, S. R., and H. R. Guy. 2001. A putative prokaryote voltage-gated Ca^{2+} channel with only one 6TM motif per subunit. *Biochem. Biophys. Res. Commun.* **281:**741–746.

Epstein, W. 2003. The roles and regulation of potassium in bacteria. *Prog. Nucleic Acids Res. Mol. Biol.* **75:**293–320.

Falb, M., F. Pfeiffer, P. Palm, K. Rodewald, V. Hickmann, J. Tittor, and D. Oesterhelt. 2005. Living with two extremes: conclusions from the genome sequence of *Natronomonas pharaonis*. *Genome Res.* **15:**1336–1343.

Fillingame, R. H., C. M. Angevine, and O. Y. Dmitriev. 2003. Mechanics of coupling proton movements to c-ring rotation in ATP synthase. *FEBS Lett.* **555:**29–34.

Friedrich, T., S. Stolpe, D. Schneider, B. Barquera, and P. Hellwig. 2005. Ion translocation by the *Escherichia coli* NADH:ubiquinone oxidoreductase (complex I). *Biochem. Soc. Trans.* **33:**836–839.

Fujisawa, M., A. Kusomoto, Y. Wada, T. Tsuchiya, and M. Ito. 2005. NhaK, a novel monovalent cation/H$^+$ antiporter of *Bacillus subtilis*. *Arch. Microbiol.* **183:**411–420.

Garland, P. B. 1977. Energy transduction and transmission in microbial systems, p. 1–21. *In* B. A. Haddock and W. A. Hamilton (ed.), *Microbial Energetics: Twenty-Seventh Symposium of the Society for General Microbiology*. Cambridge University Press, Cambridge, MA.

Gilmour, R., and T. A. Krulwich. 1997. Construction and characterization of a mutant of alkaliphilic *Bacillus firmus* OF4 with a disrupted *cta* operon and purification of a novel cytochrome *bd*. *J. Bacteriol.* **179:**863–870.

Gilmour, R., P. Messner, A. A. Guffanti, R. Kent, A. Scheberl, N. Kendrick, and T. A. Krulwich. 2000. Two-dimensional gel electrophoresis analyses of pH-dependent protein expression in facultatively alkaliphilic *Bacillus pseudofirmus* OF4 lead to characterization of an S-layer protein with a role in alkaliphily. *J. Bacteriol.* **182:**5969–5981.

Guffanti, A. A., and D. B. Hicks. 1991. Molar growth yields and bioenergetic parameters of extremely alkaliphilic *Bacillus* species in batch cultures, and growth in a chemostat at pH 10.5. *J. Gen. Microbiol.* **137:**2375–2379.

Guffanti, A. A., and T. A. Krulwich. 1992. Features of apparent nonchemiosmotic energization of oxidative phosphorylation by alkaliphilic *Bacillus firmus* OF4. *J. Biol. Chem.* **267:**9580–9588.

Guffanti, A. A., and T. A. Krulwich. 1994. Oxidative phosphorylation by ADP + P_i-loaded membrane vesicles of alkaliphilic *Bacillus firmus* OF4. *J. Biol. Chem.* **269:**21576–21582.

Haines, T. H. 2001. Do sterols reduce proton and sodium leaks through lipid bilayers? *Prog. Lipid Res.* **40:**299–324.

Haines, T. H., and N. A. Dencher. 2002. Cardiolipin: a proton trap for oxidative phosphorylation. *FEBS Lett.* **528:**35–39.

Hamamoto, T., M. Hashimoto, M. Hino, M. Kitada, Y. Seto, T. Kudo, and K. Horikoshi. 1994. Characterization of a gene responsible for the Na^+/H^+ antiporter system of alkaliphilic *Bacillus* species strain C-125. *Mol. Microbiol.* **14:**939–946.

Harold, F. M., and J. Van Brunt. 1977. Circulation of H^+ and K^+ across the plasma membrane is not obligatory for bacterial growth. *Science* **197:**372–373.

Hartzog, P. E., and B. D. Cain. 1994. Second-site suppressor mutations at glycine 218 and histidine 245 in the α-subunit of F_1F_0 ATP synthase in *Escherichia coli*. *J. Biol. Chem.* **269:**32313–32317.

Hicks, D. B., and T. A. Krulwich. 1990. Purification and reconstitution of the F_1F_0-ATP synthase from alkaliphilic *Bacillus firmus* OF4. Evidence that the enzyme translocates H^+ but not Na^+. *J. Biol. Chem.* **265:**20547–20554.

Hicks, D. B., and T. A. Krulwich. 1995. The respiratory chain of alkaliphilic bacteria. *Biochim. Biophys. Acta* **1229:**303–314.

Hiramatsu, T., K. Kodama, T. Kuroda, T. Mizushima, and T. Tsuchiya. 1998. A putative multisubunit Na^+/H^+ antiporter from *Staphylococcus aureus*. *J. Bacteriol.* **180:**6642–6648.

Hirota, N., and Y. Imae. 1983. Na^+-driven flagellar motors of an alkalophilic *Bacillus* strain YN-1. *J. Biol. Chem.* **258:**10577–10581.

Hirota, N., M. Kitada, and Y. Imae. 1981. Flagellar motors of alkalophilic *Bacillus* are powered by an electrochemical potential gradient of Na^+. *FEBS Lett.* **132:**278–280.

Hoffmann, A., and P. Dimroth. 1991a. The electrochemical proton potential of *Bacillus alcalophilus*. *Eur. J. Biochem.* **201:**467–473.

Hoffmann, A., and P. Dimroth. 1991b. The ATPase of *Bacillus alcalophilus*. Reconstitution of energy-transducing functions. *Eur. J. Biochem.* **196:**493–497.

Horikoshi, K. 1991. Microorganisms in alkaline environments. VCH Publishers Inc., New York, NY.

Horikoshi, K. 1999. Alkaliphiles: some applications of their products for biotechnology. *Microbiol. Mol. Biol. Rev.* **63:**735–750.

Hosler, J. P. 2004. The influence of subunit III of cytochrome *c* oxidase on the D pathway, the proton exit pathway and mechanism-based inactivation in subunit I. *Biochim. Biophys. Acta* **1655:**332–339.

Ito, M. 2002. Aerobic alkaliphiles, p. 133–140. *In* G. Bitton (ed.), *Encyclopedia of Environmental Microbiology*, vol. 1. Wiley, New York, NY.

Ito, M., B. Cooperberg, and T. A. Krulwich. 1997a. Diverse genes of alkaliphilic *Bacillus firmus* OF4 that complement K^+-uptake-deficient *Escherichia coli* include an *ftsH* homologue. *Extremophiles* **1:**22–28.

Ito, M., A. A. Guffanti, B. Oudega, and T. A. Krulwich. 1999. *mrp*, a multigene, multifunctional locus in *Bacillus subtilis* with roles in resistance to cholate and to Na^+ and in pH homeostasis. *J. Bacteriol.* **181:**2394–2402.

Ito, M., A. A. Guffanti, W. Wang, and T. A. Krulwich. 2000. Effects of nonpolar mutations in each of the seven *Bacillus subtilis mrp* genes suggest complex interactions among the gene products in support of Na^+ and alkali but not cholate resistance. *J. Bacteriol.* **182:**5663–5670.

Ito, M., A. A. Guffanti, J. Zemsky, D. M. Ivey, and T. A. Krulwich. 1997b. Role of the *nhaC*-encoded Na^+/H^+ antiporter of alkaliphilic *Bacillus firmus* OF4. *J. Bacteriol.* **179:**3851–3857.

Ito, M., D. B. Hicks, T. M. Henkin, A. A. Guffanti, B. Powers, L. Zvi, K. Uematsu, and T. A. Krulwich. 2004a. MotPS is the stator-force generator for motility of alkaliphilic *Bacillus* and its homologue is a second functional Mot in *Bacillus subtilis*. *Mol. Microbiol.* **53:**1035–1049.

Ito, M., N. Terahara, S. Fujinami, and T. A. Krulwich. 2005. Properties of motility in *Bacillus subtilis* powered by the H^+-coupled MotAB flagellar stator, Na^+-coupled MotPS or hybrid stators MotAS or MotPB. *J. Mol. Biol.* **352:**396–408.

Ito, M., H. Xu, A. A. Guffanti, Y. Wei, L. Zvi, D. E. Clapham, and T. A. Krulwich. 2004b. The voltage-gated Na^+ channel NavBP has a role in motility, chemotaxis, and pH homeostasis of an alkaliphilic *Bacillus*. *Proc. Natl. Acad. Sci. USA* **101:**10566–10571.

Ivey, D. M., and T. A. Krulwich. 1991. Organization and nucleotide sequence of the *atp* genes encoding the ATP synthase from alkaliphilic *Bacillus firmus* OF4. *Mol. Gen. Genet.* **229:**292–300.

Ivey, D. M., and T. A. Krulwich. 1992. Two unrelated alkaliphilic *Bacillus* species possess identical deviations in sequence from those of other prokaryotes in regions of F_0 proposed to be involved in proton translocation through the ATP synthase. *Res. Microbiol.* **143:**467–470.

Jung, H. 2001. Towards the molecular mechanism of Na^+/solute symport in prokaryotes. *Biochim. Biophys. Acta* **1505:**131–143.

Kamo, N., K. Shimono, M. Iwamoto, and Y. Sudo. 2001. Photochemistry and photoinduced proton-transfer by pharaonis phoborhodopsin. *Biochemistry (Mosc)* **66:**1277–1282.

Kashyap, D., L. M. Botero, C. Lehr, D. J. Hassett, and T. R. McDermott. 2006. A Na^+:H^+ antiporter and a molybdate transporter are essential for arsenite oxidation in *Agrobacterium tumefaciens*. *J. Bacteriol.* **188:**1577–1584.

Kitada, M., S. Kosono, and T. Kudo. 2000. The Na^+/H^+ antiporter of alkaliphilic *Bacillus* sp. *Extremophiles* **4:**253–258.

Klare, J. P., V. I. Gordeliy, J. Labahn, G. Buldt, H. J. Steinhoff, and M. Engelhard. 2004. The archaeal sensory rhodopsin II/transducer complex: a model for transmembrane signal transfer. *FEBS Lett.* **564:**219–224.

Koch, A. L. 1986. The pH in the neighborhood of membranes generating a protonmotive force. *J. Theor. Biol.* **120:**73–84.

Koishi, R., H. Xu, D. Ren, B. Navarro, B. W. Spiller, Q. Shi, and D. E. Clapham. 2004. A superfamily of voltage-gated sodium channels in bacteria. *J. Biol. Chem.* **279:**9532–9538.

Kojima, S., and D. F. Blair. 2004. The bacterial flagellar motor: structure and function of a complex molecular machine. *Int. Rev. Cytol.* **233:**93–134.

Kosono, S., K. Asai, Y. Sadaie, and T. Kudo. 2004. Altered gene expression in the transition phase by disruption of a Na^+/H^+ antiporter gene (*shaA*) in *Bacillus subtilis*. *FEMS Microbiol. Lett.* **232:**93–99.

Kosono, S., K. Haga, R. Tomizawa, Y. Kajiyama, K. Hatano, S. Takeda, Y. Wakai, M. Hino, and T. Kudo. 2005. Characterization of a multigene-encoded sodium/hydrogen antiporter (Sha) from *Pseudomonas aeruginosa*: its involvement in pathogenesis. *J. Bacteriol.* **187:**5242–5248.

Kosono, S., Y. Ohashi, F. Kawamura, M. Kitada, and T. Kudo. 2000. Function of a principal Na^+/H^+ antiporter, ShaA, is required for initiation of sporulation in *Bacillus subtilis*. *J. Bacteriol.* **182**:898–904.

Krulwich, T. A. 1995. Alkaliphiles: 'basic' molecular problems of pH tolerance and bioenergetics. *Mol. Microbiol.* **15**:403–410.

Krulwich, T. A. 2003. Alkaliphily. *In* C. Gerday (ed.), *Extremophiles (Life Under Extreme Environmental Conditions)*. Eolss Publishers, Oxford, United Kingdom.

Krulwich, T. A., J. G. Federbush, and A. A. Guffanti. 1985. Presence of a nonmetabolizable solute that is translocated with Na^+ enhances Na^+-dependent pH homeostasis in an alkalophilic *Bacillus*. *J. Biol. Chem.* **260**:4055–4058.

Krulwich, T. A., A. A. Guffanti, and M. Ito. 1999. pH tolerance in *Bacillus*: alkaliphile vs. non-alkaliphile, p. 167–182, *Mechanisms by which bacterial cells respond to pH*. Novartis Found. Symp. 221, Wiley, Chichester.

Krulwich, T. A., M. Ito, R. Gilmour, D. B. Hicks, and A. A. Guffanti. 1998. Energetics of alkaliphilic *Bacillus* species: physiology and molecules. *Adv. Microb. Physiol.* **40**:401–438.

Kuzmenkin, A., F. Bezanilla, and A. M. Correa. 2004. Gating of the bacterial sodium channel, NaChBac: voltage-dependent charge movement and gating currents. *J. Gen. Physiol.* **124**:349–356.

Larsen, S. H., J. Adler, J. J. Gargus, and R. W. Hogg. 1974. Chemomechanical coupling without ATP: the source of energy for motility and chemotaxis in bacteria. *Proc. Natl. Acad. Sci. USA* **71**:1239–1243.

Laubinger, W., and P. Dimroth. 1989. The sodium ion translocating adenosinetriphosphatase of *Propionigenium modestum* pumps protons at low sodium ion concentrations. *Biochemistry* **28**:7194–7198.

Lewinson, O., E. Padan, and E. Bibi. 2004. Alkalitolerance: a biological function for a multidrug transporter in pH homeostasis. *Proc. Natl. Acad. Sci. USA* **101**:14073–14078.

Lewis, R. J., S. Belkina, and T. A. Krulwich. 1980. Alkalophiles have much higher cytochrome contents than conventional bacteria and than their own non-alkalophilic mutant derivatives. *Biochem. Biophys. Res. Commun.* **95**:857–863.

Lewis, R. J., T. A. Krulwich, B. Reynafarje, and A. L. Lehninger. 1983. Respiration-dependent proton translocation in alkalophilic *Bacillus firmus* RAB and its non-alkalophilic mutant derivative. *J. Biol. Chem.* **258**:2109–2111.

Lewis, R. J., R. C. Prince, P. L. Dutton, D. B. Knaff, and T. A. Krulwich. 1981. The respiratory chain of *Bacillus alcalophilus* and its nonalkalophilic mutant derivative. *J. Biol. Chem.* **256**:10543–10549.

Liu, J., Y. Xue, Q. Wang, Y. Wei, T. H. Swartz, D. B. Hicks, M. Ito, Y. Ma, and T. A. Krulwich. 2005. The activity profile of the NhaD-type $Na^+(Li^+)/H^+$ antiporter from the soda lake haloalkaliphile *Alkalimonas amylolytica* is adaptive for the extreme environment. *J. Bacteriol.* **187**:7589-7595.

Lolkema, J. S., G. Speelmans, and W. N. Konings. 1994. Na^+-coupled versus H^+-coupled energy transduction in bacteria. *Biochim. Biophys. Acta* **1187**:211–215.

Lu, J., Y. Nogi, and H. Takami. 2001. *Oceanobacillus iheyensis* gen. nov., sp. nov., a deep-sea extremely halotolerant and alkaliphilic species isolated from a depth of 1050 m on the Iheya Ridge. *FEMS Microbiol. Lett.* **205**:291–297.

Ma, Y., Y. Xue, W. D. Grant, N. C. Collins, A. W. Duckworth, R. P. Van Steenbergen, and B. E. Jones. 2004. *Alkalimonas amylolytica* gen. nov., sp. nov., and *Alkalimonas delamerensis* gen. nov., sp. nov., novel alkaliphilic bacteria from soda lakes in China and East Africa. *Extremophiles* **8**:193–200.

Macnab, R. M., and A. M. Castle. 1987. A variable stoichiometry model for pH homeostasis in bacteria. *Biophys. J.* **52**:637–647.

Marantz, Y., E. Nachliel, A. Aagaard, P. Brzezinski, and M. Gutman. 1998. The proton collecting function of the inner surface of cytochrome *c* oxidase from *Rhodobacter sphaeroides*. *Proc. Natl. Acad. Sci. USA* **95**:8590–8595.

Mathiesen, C., and C. Hagerhall. 2002. Transmembrane topology of the NuoL, M and N subunits of NADH:quinone oxidoreductase and their homologues among membrane-bound hydrogenases and bona fide antiporters. *Biochim. Biophys. Acta* **1556**:121–132.

Mathiesen, C., and C. Hagerhall. 2003. The 'antiporter module' of respiratory chain Complex I includes the MrpC/NuoK subunit—a revision of the modular evolution scheme. *FEBS Lett.* **5459**:7–13.

Maurer, L. M., E. Yohannes, S. S. Bondurant, M. Radmacher, and J. L. Slonczewski. 2005. pH regulates genes for flagellar motility, catabolism, and oxidative stress in *Escherichia coli* K-12. *J. Bacteriol.* **187**:304–319.

McCarter, L. L. 2001. Polar flagellar motility of the Vibrionaceae. *Microbiol. Mol. Biol. Rev.* **65**:445–462.

McCarter, L. L. 2004. Dual flagellar systems enable motility under different circumstances. *J. Mol. Microbiol. Biotechnol.* **7**:18–29.

McLaggan, D., M. H. Selwyn, and A. P. Sawson. 1984. Dependence of Na^+ of control of cytoplasmic pH in a facultative alkalophile. *FEBS Lett.* **165**:254–258.

Mitchell, P. 1961. Coupling of phosphorylation to electron and hydrogen transfer by a chemiosmotic type of mechanism. *Nature* **191**:144–148.

Mulkidjanian, A. Y., D. A. Cherepanov, J. Heberle, and W. Junge. 2005. Proton transfer dynamics at membrane/water interface and mechanism of biological energy conversion. *Biochemistry (Mosc)* **70**:251–256.

Muller, V., and A. Oren. 2003. Metabolism of chloride in halophilic prokaryotes. *Extremophiles* **7**:261–266.

Muntyan, M. S., D. A. Bloch, V. S. Ustiyan, and L. A. Drachev. 1993. Kinetics of CO binding to H^+-motive oxidases of the caa_3-type from *Bacillus* FTU and of the o-type from *Escherichia coli*. *FEBS Lett.* **327**:351–354.

Muntyan, M. S., I. V. Popova, D. A. Bloch, E. V. Skripnikova, and V. S. Ustiyan. 2005. Energetics of alkalophilic representatives of the genus *Bacillus*. *Biochemistry (Mosc)* **70**:137–142.

Nakamura, T., S. Kawasaki, and T. Unemoto. 1992. Roles of K^+ and Na^+ in pH homeostasis and growth of the marine bacterium *Vibrio alginolyticus*. *J. Gen. Microbiol.* **138**:1271–1276.

Nakamura, T., H. Tokuda, and T. Unemoto. 1984. K^+/H^+ antiporter functions as a regulator of cytoplasmic pH in a marine bacterium, *Vibrio alginolyticus*. *Biochim. Biophys. Acta* **776**:330–336.

Olsson, K., S. Keis, H. W. Morgan, P. Dimroth, and G. M. Cook. 2003. Bioenergetic properties of the thermoalkaliphilic *Bacillus* sp. strain TA2.A1. *J. Bacteriol.* **185**:461–465.

Padan, E., E. Bibi, M. Ito, and T. A. Krulwich. 2005. Alkaline pH homeostasis in bacteria: new insights. *Biochim. Biophys. Acta* **1717**:67–88.

Padan, E., and T. A. Krulwich. 2000. Sodium stress, p. 117–130. *In* G. Storz and R. Hengge-Aronis (ed.), *Bacterial Stress Response*. ASM Press, Washington, DC.

Padan, E., D. Zilberstein, and S. Schuldiner. 1981. pH homeostasis in bacteria. *Biochim. Biophys. Acta* **650**:151–166.

Pfeiffer, K., V. Gohil, R. A. Stuart, C. Hunte, U. Brandt, M. L. Greenberg, and H. Schagger. 2003. Cardiolipin stabilizes respiratory chain supercomplexes. *J. Biol. Chem.* **278**:52873–52880.

Pourcher, T., S. Leclercq, G. Brandolin, and G. Leblanc. 1995. Melibiose permease of *Escherichia coli*: large scale purification and evidence that H^+, Na^+, and Li^+ sugar symport is catalyzed by a single polypeptide. *Biochemistry* **34**:4412–4420.

Qiu, Z. H., L. Yu, and C. A. Yu. 1992. Spin-label electron paramagnetic resonance and differential scanning calorimetry studies of the interaction between mitochondrial cytochrome *c* oxidase and adenosine triphosphate synthase complex. *Biochemistry* 31:3297–3302.

Ren, D., B. Navarro, H. Xu, L. Yue, Q. Shi, and D. E. Clapham. 2001. A prokaryotic voltage-gated sodium channel. *Science* 294:2372–2375.

Riesle, J., D. Oesterhelt, N. A. Dencher, and J. Heberle. 1996. D38 is an essential part of the proton translocation pathway in bacteriorhodopsin. *Biochemistry* 35:6635–6643.

Rivera-Torres, I. O., R. D. Krueger-Koplin, D. B. Hicks, S. M. Cahill, T. A. Krulwich, and M. E. Girvin. 2004. pK_a of the essential Glu54 and backbone conformation for subunit c from the H^+-coupled F_1F_0 ATP synthase from an alkaliphilic *Bacillus*. *FEBS Lett.* 575:131–135.

Romero, M. F., and W. F. Boron. 1999. Electrogenic Na^+/HCO_3^- cotransporters: cloning and physiology. *Annu. Rev. Physiol.* 61:699–723.

Seto, Y., M. Hashimoto, R. Usami, T. Hamamoto, T. Kudo, and K. Horikoshi. 1995. Characterization of a mutation responsible for an alkali-sensitive mutant, 18224, of alkaliphilic *Bacillus* sp. strain C-125. *Biosci. Biotechnol. Biochem.* 59:1364–1366.

Skulachev, V. P. 1989. The sodium cycle: a novel type of bacterial energetics. *J. Bioenerg. Biomembr.* 21:635–647.

Speelmans, G., B. Poolman, T. Abee, and W. N. Konings. 1993. Energy transduction in the thermophilic anaerobic bacterium *Clostridium fervidus* is exclusively coupled to sodium ions. *Proc. Natl. Acad. Sci. USA* 90:7975–7979.

Spudich, J. L., and H. Luecke. 2002. Sensory rhodopsin II: functional insights from structure. *Curr. Opin. Struct. Biol.* 12:540–546.

Stancik, L. M., D. M. Stancik, B. Schmidt, D. M. Barnhart, Y. N. Yoncheva, and J. L. Slonczewski. 2002. pH-dependent expression of periplasmic proteins and amino acid catabolism in *Escherichia coli*. *J. Bacteriol.* 184:4246–4258.

Stock, D., C. Gibbons, I. Arechaga, A. G. Leslie, and J. E. Walker. 2000. The rotary mechanism of ATP synthase. *Curr. Opin. Struct. Biol.* 10:672–679.

Sturr, M. G., A. A. Guffanti, and T. A. Krulwich. 1994. Growth and bioenergetics of alkaliphilic *Bacillus firmus* OF4 in continuous culture at high pH. *J. Bacteriol.* 176:3111–3116.

Sugiyama, S., H. Matsukura, and Y. Imae. 1985. Relationship between Na^+-dependent cytoplasmic pH homeostasis and Na^+-dependent flagellar rotation and amino acid transport in alkalophilic *Bacillus*. *FEBS Lett.* 182:265–268.

Swartz, T. H., S. Ikewada, O. Ishikawa, M. Ito, and T. A. Krulwich. 2005a. The Mrp system: a giant among monovalent cation/proton antiporters? *Extremophiles* 9:345–354.

Swartz, T. H., M. Ito, D. B. Hicks, M. Nuqui, A. A. Guffanti, and T. A. Krulwich. 2005b. The Mrp Na^+/H^+ antiporter increases the activity of the malate:quinone oxidoreductase of an *Escherichia coli* respiratory mutant. *J. Bacteriol.* 187:388–391.

Takami, H., K. Nakasone, Y. Takaki, G. Maeno, R. Sasaki, N. Masui, F. Fuji, C. Hirama, Y. Nakamura, N. Ogasawara, S. Kuhara, and K. Horikoshi. 2000. Complete genome sequence of the alkaliphilic bacterium *Bacillus halodurans* and genomic sequence comparison with *Bacillus subtilis*. *Nucleic Acids Res.* 28:4317–4331.

Takami, H., Y. Takaki, and I. Uchiyama. 2002. Genome sequence of *Oceanobacillus iheyensis* isolated from the Iheya Ridge and its unexpected adaptive capabilities to extreme environments. *Nucleic Acids Res.* 30:3927–3935.

Terahara, N., M. Fujisawa, B. D. Powers, T. M. Henkin, T. A. Krulwich, and M. Ito. 2006. An intergenic stem-loop mutation in the *Bacillus subtilis* cccp-motPS operon increases motPS transcription and the MotPS contribution to motility. *J. Bacteriol.* 188:2701–2705.

Tokuda, H., and T. Unemoto. 1985. The Na^+-motive respiratory chain of marine bacteria. *Microbiol. Sci.* 2:65–66.

Tolner, B., B. Poolman, and W. N. Konings. 1997. Adaptation of microorganisms and their transport systems to high temperatures. *Comp. Biochem. Physiol. A. Physiol.* 118:423–428.

Valiyaveetil, F. I., and R. H. Fillingame. 1998. Transmembrane topography of subunit a in the *Escherichia coli* F_1F_0 ATP synthase. *J. Biol. Chem.* 273:16241–16247.

van der Laan, J. C., G. Gerritse, L. J. Mulleners, R. A. van der Hoek, and W. J. Quax. 1991. Cloning, characterization, and multiple chromosomal integration of a *Bacillus* alkaline protease gene. *Appl. Environ. Microbiol.* 57:901–909.

Wang, Z., D. B. Hicks, A. A. Guffanti, K. Baldwin, and T. A. Krulwich. 2004. Replacement of amino acid sequence features of a- and c-subunits of ATP synthases of alkaliphilic *Bacillus* with the *Bacillus* consensus sequence results in defective oxidative phosphorylation and non-fermentative growth at pH 10.5. *J. Biol. Chem.* 279:26546–26554.

Wei, Y., A. A. Guffanti, and T. A. Krulwich. 1999. Sequence analysis and functional studies of a chromosomal region of alkaliphilic *Bacillus firmus* OF4 encoding an ABC-type transporter with similarity of sequence and Na^+ exclusion capacity to the *Bacillus subtilis* NatAB transporter. *Extremophiles* 3:113–120.

Wei, Y., T. W. Southworth, H. Kloster, M. Ito, A. A. Guffanti, A. Moir, and T. A. Krulwich. 2003. Mutational loss of a K^+ and NH_4^+ transporter affects the growth and endospore formation of alkaliphilic *Bacillus pseudofirmus* OF4. *J. Bacteriol.* 185:5133–5147.

West, I. C., and P. Mitchell. 1974. Proton/sodium ion antiport in *Escherichia coli*. *Biochem. J.* 144:87–90.

Williams, R. J. 1977. Fundamental features of proton-coupled transport. *Biochem. Soc. Trans.* 5:29–32.

Williams, R. J. 1978. The multifarious couplings of energy transduction. *Biochim. Biophys. Acta* 505:1–44.

Yorimitsu, T., and M. Homma. 2001. Na^+-driven flagellar motor of *Vibrio*. *Biochim. Biophys. Acta* 1505:82–93.

Yue, L., B. Navarro, D. Ren, A. Ramos, and D. E. Clapham. 2002. The cation selectivity filter of the bacterial sodium channel, NaChBac. *J. Gen. Physiol.* 120:845–853.

Yumoto, I. 2002. Bioenergetics of alkaliphilic *Bacillus* spp. *J. Biosci. Bioeng.* 93:342–353.

Yumoto, I. 2003. Electron transport system in alkaliphilic *Bacillus* spp. *Recent Res. Devel. Bacteriol.* 1:131–149.

Yumoto, I., Y. Fukumori, and T. Yamanaka. 1991. Purification and characterization of two membrane-bound c-type cytochromes from a facultative alkalophilic Bacillus. *J Biochem. (Tokyo)* 110:267–273.

Zhang, M., E. Mileykovskaya, and W. Dowhan. 2002. Gluing the respiratory chain together. *J. Biol. Chem.* 277:43553–43556.

Zhang, M., E. Mileykovskaya, and W. Dowhan. 2005. Cardiolipin is essential for organization of complexes III and IV into a supercomplex in intact yeast mitochondria. *J. Biol. Chem.* 280:29403–29408.

Zilberstein, D., V. Agmon, S. Schuldiner, and E. Padan. 1984. *Escherichia coli* intracellular pH, membrane potential, and cell growth. *J. Bacteriol.* 158:246–252.

VII. PIEZOPHILES

Chapter 25

Microbial Adaptation to High Pressure

DOUGLAS H. BARTLETT, FEDERICO M. LAURO, AND EMILEY A. ELOE

INTRODUCTION

Piezomicrobiology is one of the lesser studied areas in extremophilic microbiology, although it constitutes a significant field of research, considering that piezophilic microorganisms reside in the largest habitat on Earth—the deep sea. The rigor of sampling and the requirement for highly specialized laboratory equipment are the main limitations to greater participation in this field (Regnard, 1884; Yayanos, 1995). The term piezophile (formerly barophile) describes a microorganism that displays optimal growth rates at pressures above atmospheric pressure (1 atm = ~0.1 MPa) (Yayanos, 2001). Pressure is a fundamental and unique physical parameter that exerts its effects on chemical phases, equilibria, and rates as a result of system volume changes. It is one of the key parameters, along with temperature, water activity, electromagnetic radiation, oxidative stress, redox balance, and nutrient characteristics, that has governed the evolution and distribution of life both on and in Earth.

As a general rule, pressure increases by ~0.1 MPa (1 atm) with every 10-m increase in depth. This is a useful generality for freshwater and marine environments (Regnard, 1884; Yayanos, 1995) and may also be applied to aqueous interstitial spaces within and between rocks of Earth's upper crust (DeLong, 1992). Pressure in the marine environment extends to ~110 MPa (~1,100 atm) in the deepest ocean trench location, the Challenger Deep of the Mariana Trench (10,898 m). Deep-sea microbes capable of growing at pressures as high as 130 MPa have been reported (Yayanos et al., 1981). High-pressure microbial habitats include the abyssal and hadal deep-sea environments, which are typified by low temperatures, darkness, sporadic nutrient inputs, and high diversity (low biomass) of invertebrate and vertebrate life. The abyssal plain is commonly thought of as a barren desert, punctuated by the presence of reducing environments such as hydrothermal vents, cold seeps, and whale falls. Additional high-pressure environments include oceanic and terrestrial deep-subsurface habitats, where a high abundance of prokaryotic life exists (Whitman et al., 1998). Examples of deep-subsurface environments include Antarctic subglacial lakes such as Lake Vostok (Christner et al., 2001), the Great Artesian Basin of Australia (Kanso et al., 2002), the Taylorsville Triassic Basin in Virginia (Boone et al., 1995), deep-subsurface petroleum reservoirs (Slobodkin et al., 1999), and several South African mines that extend >3 km below land surface (Moser et al., 2005).

Interested readers are advised to consult additional reviews on high-pressure microbiology (Bartlett, 1992, 2002, 2005; Kato and Bartlett, 1997; Yayanos, 2001; Yayanos, 1995) and to take note that an American Society for Microbiology book encompassing high-pressure microbiology is forthcoming.

CULTURE-INDEPENDENT STUDIES

One of the limitations of deep-sea microbiology to date has been the lack of extensive information on microbial diversity. Culture-independent analyses of microbial diversity in low-temperature deep-ocean habitats have indicated the presence of particular groups of *Eukarya*, *Archaea*, and *Bacteria* (Edgcomb et al., 2002; Fuhrman and Davis, 1997; Karner et al., 2001; Kato et al., 1997; Li et al., 1999a, 1999b; López-García et al., 2001a, 2001b; Moon-van der Staay et al., 2001; Yanagibayashi et al., 1999). A particular cluster of the *Euryarchaea*, which branches off at the base of haloarchaea, is present in moderate-depth deep-seawater environments (López-García et al., 2001c). At the Hawaii Ocean Time-series station A Long-term Oligotrophic Habitat Assessment

---

D. H. Bartlett, F. M. Lauro, and E. A. Eloe • Marine Biology Research Division, Center for Marine Biotechnology and Biomedicine, Scripps Institution of Oceanography, University of California, San Diego, La Jolla, CA 92093-0202.

(ALOHA), the fraction of group I *Crenarchaea* in the total picoplankton increases with depth to the deepest site examined (4,750 m) (Karner et al., 2001). Both DNA sequencing and culture-based studies of shallow-water relatives suggest that group I *Crenarchaea* are autotrophic and gain energy via ammonia oxidation (Francis et al., 2005; Konneke et al., 2005; Treusch et al., 2005). Isotope studies suggest that in contrast to archaeal metabolism at surface waters, archaeal metabolism at depth is primarily autotrophic (Ingalls et al., 2006).

More recently, environmental genomic approaches have been applied to picoplankton diversity at the Hawaii Ocean Time-series station ALOHA down several kilometers in depth (DeLong et al., 2006; Worden et al., 2006). Both 16S rRNA libraries and fosmid libraries were prepared, the latter used for 5,000 end-sequencing reactions at seven depths extending to ~4 km. Depth-associated subgroups within both the *Crenarchaea* and *Euryarchaea* were identified, along with a greater abundance of *Chloroflexus*-like and *Planctomycetales*-like bacterial sequences at depth. This study additionally correlated gene functional groups with increased depth, finding an increase in certain types of transporters, transposases, type II secretion systems, a decrease in motility and chemotaxis, and an absence of light-activated system genes.

CULTURE-BASED STUDIES

Despite the fact that the origins of deep-sea microbiology can be traced back to the dawn of marine microbiology and the expeditions of the Travailleur and Talisman in 1882 to 1883 (Marquis, 1982; Prieur, 1992), the very existence of piezophiles was a matter of active debate until 1978.

The principal reason for this confusion is that many microbes in the deep sea are not autochthonous residents. Microbes from surface waters can descend to the deep sea as components of phytodetrital aggregates (Lochte and Turley, 1988). Because of these and other conduits to the deep sea and because of the preservation of such microbes in abyssal and hadal locations, water samples recovered from deep-ocean environments often have greater activity at atmospheric pressure than when measured at the in situ pressure (Jannasch and Taylor, 1984; Jannasch and Wirsen, 1982). A number of nonpiezophilic mesophiles and thermophiles have been isolated from sediment samples obtained from great water depths (Morita, 1976; Takai et al., 1999; Takami et al., 1997, 2004; Tamegai et al., 1997).

Isolates of deep-sea piezophilic *Archaea* (from deep-sea hydrothermal vents) or *Bacteria* and *Eukarya* (from cold deep-sea habitats) have been obtained (Fig. 1 and Table 1). Although only a handful of piezophilic *Archaea* have been cultured, these isolates span a broad collection of both the *Euryarchaea* and the *Crenarchaea* kingdoms (Bernhardt et al., 1988; Marteinsson et al., 1997, 1999a, 1999b; Miller et al., 1988). In contrast, culture-based studies of *Bacteria* (mostly from amphipods, fish, and mud samples) have thus far resulted in the isolation of a narrow phylogenetic assemblage of γ-proteobacteria within the orders *Alteromonadales* and *Vibrionales*, including *Colwellia*, *Moritella*, *Photobacterium*, *Psychromonas*, and *Shewanella* species (DeLong et al., 1997; Kato et al., 1995, 1996a, 1998; Li et al., 1998; Nogi and Kato, 1999; Nogi et al., 1998b, 1998c, 2002, 2004; Xu et al., 2003a, 2003b). Exceptions to the genera listed above include two reports of the isolation of a moderately piezophilic sulfate-reducing species, *Desulfovibrio profundus*, obtained from a deep sediment sample in the Japan Sea and from a hydrothermal vent chimney in the East Pacific Rise (Alazard et al., 2003; Barnes et al., 1998), and a thermophilic member of the *Thermotogales* isolated from a vent (Alain et al., 2002). Additionally, a piezophilic gram-positive species from the Aleutian Trench (isolated by Yayanos, 1980) has yet to be described (Lauro et al., unpublished data). The gram-positive *Carnobacterium* species is related to a piezotolerant relative isolated from the deep subseafloor sediment of the Nankai Trough (Toffin et al., 2005). One of the intriguing aspects of piezophilic microorganisms is their close relation to mesophilic microbes, indicating that the underlying genetic changes required to bring about this type of extremophily are relatively modest.

Little work has been done thus far on pressure adaptation in deep-sea eukaryotic microbes. An obligately piezophilic bactiverous flagellate of the genus *Bodo* was isolated from a 4.5-km abyssal depth in the northeastern Atlantic Ocean (Turley et al., 1988). Deep-sea foraminifera have been maintained at high pressure (Turley et al., 1993). Piezotolerant fungi have been isolated from various deep-sea environments, but no piezophilic isolates have yet been obtained (Damare et al., 2006; Miura et al., 2001; Raghukumar and Raghukumar, 1998).

The cultivation of microorganisms at high pressure requires pumps, pressure gauges, and pressure-retaining culture tubes. Detailed information is available for devices needed for psychrophilic/psychrotolerant piezophilic *Bacteria* (Yayanos, 2001), and a useful description of how the source of the pressure and the characteristics of the pressure vessels can influence microbial growth is included for the hyperthermophilic archaeon *Methanocaldococcus jannaschii* (Boonyaratanakornkit and Clark, in press).

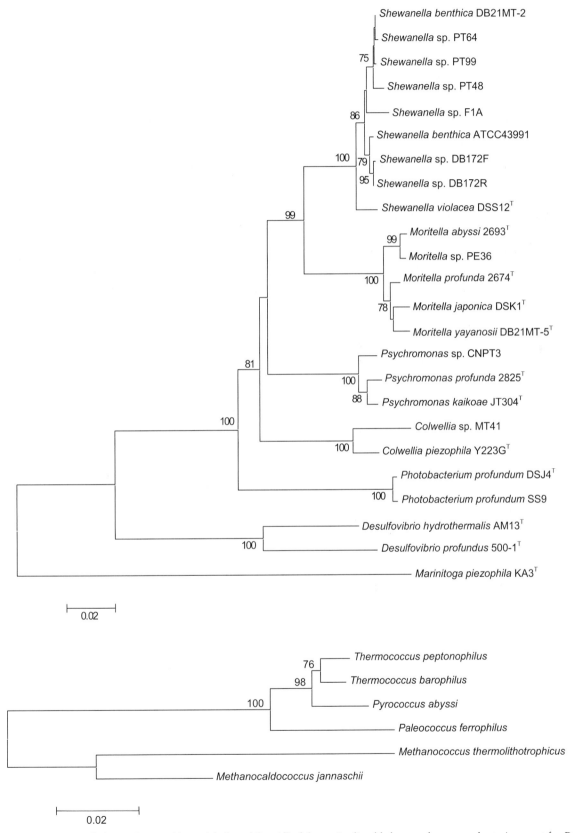

Figure 1. (Upper Tree) Phylogenetic tree of bacterial piezophiles. All of the strains listed belong to the γ-proteobacteria except for *Desulfovibrio*, which belong to the δ-proteobacteria, and *Marinitoga*, which is a member of the *Thermotogales*. (Lower Tree) Phylogenetic tree of archaeal piezophiles. All of the species listed are within the *Crenarchaea* except for *M. jannaschii*, which is within the *Euryarchaea*. Both trees were reconstructed using the neighbor-joining algorithm.

Table 1. Piezophilic *Bacteria* and *Archaea*

Strain	Source	P_{opt} (MPa)	T_{opt} (°C)	Reference	16S rRNA GenBank accession no.
Bacteria					
Colwellia piezophila Y223G^T	Japan Trench	60	10	Nogi et al., 2004	AB094412
Colwellia sp. MT41	Mariana Trench	103	8	Yayanos, 1986; Yayanos et al., 1981	DQ027051
Desulfovibrio hydrothermalis AM13T	East Pacific Rise	26	35	Alazard et al., 2003	AF458778
D. profundus 500-1T	Japan Sea	15	25	Bale et al., 1997	AF418172
Marinitoga piezophila KA3^T	East Pacific Rise	40	65	Alain et al., 2002	AF326121
Moritella abyssi 2693^T	Eastern Tropical Atlantic	30	10	Xu et al., 2003a	AJ252022
Moritella japonica DSK1^T	Japan Trench	50	15	Nogi et al., 1998a	D21224
M. profunda 2674^T	Eastern Tropical Atlantic	25	6	Xu et al., 2003a	AJ252023
Moritella sp. PE36	North Pacific Ocean	41	15	DeLong et al., 1997; Yayanos, 1986	DQ027053
Moritella yayanosii DB21MT-5^T	Mariana Trench	80	10	Kato et al., 1998	AB008797
P. profundum DSJ4^T	Ryukyu Trench	10	10	Nogi et al., 1998b	D21226
P. profundum SS9	Sulu Sea	28	15	Vezzi et al., 2005	AB003191
Psychromonas kaikoae JT7304^T	Japan Trench	50	10	Nogi et al., 2002	AB052160
Psychromonas profunda 2825T	Eastern Tropical Atlantic	25	10	Xu et al., 2003b	AJ416756
Psychromonas sp. CNPT3	Central North Pacific Ocean	52	8	DeLong et al., 1997	DQ027056
Shewanella benthica ATCC43991	Puerto Rico Trench	41	8	DeLong et al., 1997	U91594
S. benthica DB21MT-2	Mariana Trench	70	10	Kato et al., 1998	AB008796
S. benthica F1A	North Atlantic	41	8	DeLong et al., 1997	U91592
Shewanella sp. DB172F	Izu-Bonin Trench	70	10	Kato et al., 1996a	D63488
Shewanella sp. DB172R	Izu-Bonin Trench	60	10	Kato et al., 1996a	D63824
Shewanella sp. PT48	Philippine Trench	62	8	DeLong et al., 1997	DQ027059
Shewanella sp. PT64	Philippine Trench	90	9	Yayanos, 1986	DQ027060
Shewanella sp. PT99	Philippine Trench	62	8	DeLong et al., 1997; Li et al., 1998	AB003189
S. violacea DSS12	Ryukyu Trench	30	10	Nogi et al., 1998c	D21225
Archaea					
M. jannaschii	East Pacific Rise	75	86	Miller et al., 1988	L77117
M. thermolithotrophicus	Coastal Beach of Italy	50	65	Bernhardt et al., 1988; Huber et al., 1982	M59128
Paleococcus ferrophilus	Ogasawara Trough	30	83	Takai et al., 2000	AB019239
Pyrococcus abyssi	Southwestern Pacific	20	95	Marteinsson et al., 1997	L19921
Thermococcus barophilus	Mid-Atlantic Ridge	40	85	Marteinsson et al., 1999a	U82237
Thermococcus peptonophilus	Western Pacific Ocean	45	95	Canganella et al., 1997	AJ298871

Several factors influence microbial growth at any given pressure, including the medium, the source of elevated pressure, and the container used for cell cultures. Factors such as carbon source and the presence of divalent cations also have a pronounced influence on the upper pressure limits of microbial growth (Marquis, 1982). When elevated pressure is applied using gases, different effects on cell growth may be noted because of the effects the gases have on substrate solubility, free radical reactions, or gaseous anesthetic effects (Boonyaratanakornkit and Clark, in press; Finch and Kiesow, 1979; Kaminoh et al., 1998; Thom and Marquis, 1987). It has even been noted that different types of plastic containers used inside pressure vessels influence growth characteristics as a function of pressure (F. Wang, personal communication). Another factor is the rate of decompression used to obtain cell material for following growth. In one study, rapid decompression was found to result in large amounts of cell lysis (Park and Clark, 2002). Thus, many factors must be considered when evaluating the extent of pressure adaptation in a given microbial strain.

It should also be emphasized that the extent of piezophily does not reflect resistance to killing by ultra high pressure used in pressure pasteurization of foods today (Hauben et al., 1997). This may be because at ultra high pressures mesophiles have no metabolism, whereas piezophiles may exhibit unbalanced metabolism, which leads to cell inactivation via a process akin to thymineless death (Nakayama, 2005). The most ultra-high-pressure-resistant microbes are those that produce spores (Lauro et al., 2004).

Below are described some of the key processes and structures implicated in piezophily. Some of these have been identified exclusively on the basis of their

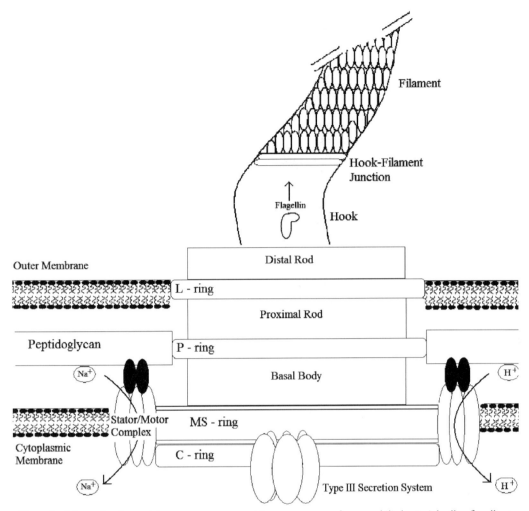

Figure 2. Schematic of one of the most pressure-sensitive components of a mesophilic bacterial cell: a flagellum.

disproportionate pressure sensitivity in mesophilic bacteria such as *Escherichia coli*, and in other cases supporting evidence is also available from investigations of piezophiles. These include motility, cell division, DNA, ribosome, and membrane structure and function.

MOTILITY

Figure 2 presents a schematic of a typical gram-negative bacterial flagellum. High-pressure effects on bacterial motility were first investigated by Claude ZoBell. He examined the presence or absence of flagella on bacterial isolates after various pressure treatments (ZoBell, 1970). He found that *E. coli*, two *Vibrio* spp., and four *Pseudomonas* spp., which are normally flagellated, after incubation at 40 MPa did not possess flagella. ZoBell additionally questioned whether pressure-adapted deep-sea bacteria were similarly flagellated as their non–pressure-adapted counterparts.

Meganathan and Marquis (1973) further explored the loss of motility in the mesophile *E. coli* during high-pressure treatment, citing pressure as an inhibitor of both the formation of new flagella and the functioning of previously assembled filaments. New flagella synthesis in *E. coli* was shown to be abolished at 20 MPa, indicating this as the most pressure-sensitive cellular process. Interestingly, in vitro studies of *Salmonella enterica* serovar Typhimurium flagellar filaments showed an increase in partial molar volume (340 cm^3/mol flagellin monomer) upon polymerization, indicating a preference for the monomeric state under high-pressure conditions (Tamura et al., 1997). Additionally, the filaments were shown to irreversibly depolymerize in vitro under high pressure. These in vitro findings are in contrast to *E. coli* in vivo results, indicating that pressure does not cause disaggregation of already assembled flagella even at pressures up to 60 MPa (Meganathan and Marquis, 1973).

To date, investigation of bacterial motility at high hydrostatic pressure has been limited to mesophilic species.

CELL DIVISION

One of the classic responses of mesophilic microbial cells to growth-permissive elevated pressure is the impairment of cell division (ZoBell, 1970). At pressures still permitting biomass accumulation (20 to 50 MPa for many mesophiles), cell enlargement, typically cell filamentation, results (Fig. 3). At supra- or superoptimal pressure, many piezophiles likewise tend to form larger cells (Jannasch, 1987; Yayanos and DeLong, 1987).

Insight into the mechanism of this phenomenon has been obtained from an elegant series of genetic experiments performed in the laboratory of Chris Michiels (Aertsen and Michiels, 2005a, 2005b). Their results indicate that at high pressure, *E. coli* induces a novel "SOS" DNA damage response, which then leads to cell filamentation. The SOS regulon includes genes whose products repair DNA damage as well as prevent cell division (Friedberg et al., 1995; Kuzminov, 1999). The source of the SOS induction in at least one *E. coli* strain appears to be double-stranded DNA damage arising from the induction of the Mrr cryptic restriction endonuclease (Aertsen and Michiels, 2005a). The connection between the SOS response and cell division proceeds via the DNA damage-induced autoproteolytic inactivation of the LexA transcriptional repressor, which leads to the derepression of the *sulA* gene, whose product functions as an inhibitor of the cell division protein FtsZ. Further evidence linking SulA to this phenomenon is that mutants deficient in the lon protease, which posttranslationally controls SulA abundance, become hyperfilamentous following high-pressure treatment in a SulA-dependent manner (Aertsen and Michiels, 2005b). Thus, at least one pathway for cell filamentation at high pressure is connected to a DNA damage response.

A further connection between DNA damage and cell division at high pressure is that *recD* mutants of the piezophile *Photobacterium profundum* strain SS9 are particularly impaired in cell division at high pressure (Bidle and Bartlett, 1999). These *recD* mutants exhibit increased rates of DNA recombination. Replacement of the *E. coli recD* gene with that of *P. profundum* strain SS9 enabled *E. coli* to grow at elevated pressure without filamentation.

However, DNA damage may not be *required* for all high-pressure-induced filamentation. It has been reported that RecA (required for SOS induction) and SulA are not required for the filamentation response (Aertsen et al., 2004; Ishii et al., 2004; Kawarai et al., 2004). The basis of filamentation in this case is not well understood. This SulA-independent pathway could result from direct pressure effects on FtsZ ring formation, as has been suggested for both *E. coli* and lactic acid bacteria (Kawarai et al., 2004; Molina-Hoppner et al., 2003). Indeed, high pressure does inhibit FtsZ ring formation in vivo and dissociates FtsZ polymers in vitro (Ishii et al., 2004). But in the absence of genetic evidence linking *ftsZ* directly to pressure-induced filamentation, it is difficult to ascertain whether the FtsZ results simply correlate with in vivo cell division defects or reflect the source of the problem. For example, the SulA-independent pressure-induced filamentation could stem from pressure effects on other systems connected to cell division, such as chromosome segregation (Bartosik and Jagura-Burdzy, 2005; Lockhart and Kendrick-Jones, 1998).

The FtsZ pressure experiments are also noteworthy for another reason. These studies are the first pressure experiments to have been performed with a bacterial cytoskeletal protein. The pressure sensitivity of the cytoskeletal proteins of the *Eukarya* often reflects organismal pressure adaptation (Morita, 2003).

ADDITIONAL PRESSURE STRESS RESPONSES

When the abundance or rate of synthesis of DNA, RNA, and protein was examined in mesophilic bacteria as a function of pressure, results have indicated that either DNA or protein synthesis is the most pressure-sensitive process, depending on the microbe studied and the assay conditions employed. ZoBell (1970) reported that at elevated pressures still permitting growth, many piezosensitive prokaryotes had a reduction in nuclear content, even when other intracellular structures, such as ribosomes, appeared to be unaffected. However, he also described a *Corynebacterium* strain that exhibited large amounts of nuclear material and reduced ribosome numbers. Under similar conditions, *E. coli* was found to produce substantially less DNA, increased RNA, and similar amounts

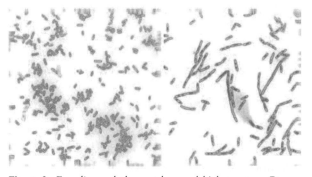

Figure 3. *E. coli* morphology at low and high pressure. Deconvolved fluorescence image of *E. coli* cells incubated at atmospheric pressure (left) and elevated pressure (right). The membrane has been stained with the fluorescent stain FM4-64 and the DNA with 4′,6′-diamidino-2-phenylindole (DAPI).

of protein, per cell (ZoBell and Cobet, 1963). When viewed at high pressure in a high-pressure microscope, *E. coli* chromosomal DNA is poorly segregated but quickly segregates following decompression (Ishii et al., 2004).

Based on the ability of an *E. coli* strain to accumulate radiolabeled substrates over time periods >100 min, Yayanos and Pollard (1969) concluded that DNA synthesis, protein synthesis, and RNA synthesis have upper limits of 50, 58, and 78 MPa, respectively. These values are very specific to the strain and assay conditions employed. ^{35}S-Methionine and cysteine pulse-labeling studies with a different *E. coli* strain, in a different medium, under anaerobic conditions, indicated that protein synthesis could continue at pressures up to 80 MPa (Welch et al., 1993). Despite the substantial differences present among different bacteria assayed under different conditions, the above results indicate that DNA synthesis and protein synthesis are major pressure points in mesophilic bacteria.

In order to gain further insight into the nature of elevated pressure as a stress, the response of *E. coli* to pressure has been examined. Moderate and extreme elevated pressure induces an oxidative stress (Aertsen et al., 2005), and loss of genes that confer protection from oxidative stress results in greater pressure sensitivity (Aertsen et al., 2005; Robey et al., 2001). At elevated pressures still permitting protein synthesis, *E. coli* surprisingly mounts both a heat-shock and a cold-shock response (Welch et al., 1993), and the transcript abundance of many cold-shock protein (CSP) genes remains high during high-pressure growth (Ishii et al., 2005). Heat-shock protein (HSP) induction could be a reflection of the well-known role of many HSPs in protecting or refolding proteins (Guisbert et al., 2004), whereas CSP production has been interpreted as arising from stress on nucleoid structure, transcription, or translation (Phadtare, 2004; Phadtare et al., 2006). Piezophiles produce HSPs in response to low-pressure stress (Marteinsson et al., 1999b; Vezzi et al., 2005).

The relationship between CSP induction and pressure effects on nucleic acids is noteworthy. Pressure effects on DNA, for example, can arise from influences on DNA, DNA-binding proteins, protein ligands, and water. Elevated pressure up to 200 MPa stabilizes nucleic acid base pairing and affects its spatial configuration (Tang et al., 1998). Many DNA-binding proteins display pressure-sensitive binding properties, and in many instances, this is due to hydration effects (Lynch and Sligar, 2002; Silva et al., 2001). Readers may also take some interest in knowing that it has been proposed that another function of DNA is mechanical, protecting cell shape at high turgor pressure (Norris et al., 1999).

Additional characterization of a handful of the pressure-inducible CSPs has been obtained. H-NS is a small cold-shock protein that plays a major role in the compaction of the chromosome and in the modulation of the expression of a number of genes. *E. coli* H-NS mutants exhibit increased pressure-sensitive growth (Ishii et al., 2005). The basis of this phenotype could stem from defects in H-NS oligomerization in vivo, because conditions such as low temperature and increased osmolarity have this effect (Stella et al., 2006). If true, this in turn could partially explain the influence of pressure on chromosome condensation and segregation and on the expression of certain genes. DNA gyrase is another CSP whose activity is modified by pressure. At low enzyme and magnesium concentrations, gyrase-mediated supercoiling activity is highly pressure sensitive (Chilukuri et al., 1995).

Although the details of pressure effects on H-NS and DNA gyrase are unknown, more information is available for a single-stranded DNA-binding protein (SSB). SSB is a small homotetrameric protein required for DNA replication, recombination, and repair. The stability of SSB from marine *Shewanella* species differing in the extent of their pressure adaptation has been examined (Chilukuri and Bartlett, 1997; Chilukuri et al., 2002). The more piezophilic the microbial source of SSB, the smaller the SSB volume change of association. Reduced helix-destabilizing glycine and helix-breaking proline residues correlate with greater tetramer stability at high pressure.

TRANSLATION

Figure 4 presents a schematic of a ribosome. Translation is another pressure-sensitive cellular process involving nucleic acid–protein interactions. The pioneering work of Landau in 1967 indicated that protein synthesis is pressure sensitive (Landau, 1967). Numerous studies have attempted to identify the rate-limiting step in this process. Initially, in vitro studies of ribosomes from *E. coli* demonstrated that ribosomal subunits dissociate under high pressure (Gross and Jaenicke, 1990; Schulz et al., 1976). More recently, the shift toward dissociation in the equilibrium of the ribosomal complex and its free components has been observed in vivo through differential scanning calorimetry (DSC) (Alpas et al., 2003; Niven et al., 1999) and correlated with pressure effects on cell viability. These observations lead to two conclusions: (i) ribosome dissociation is associated with a decrease in volume and therefore favored at high pressures; and (ii) the extent of ribosome stability correlates closely with cell viability at high pressure.

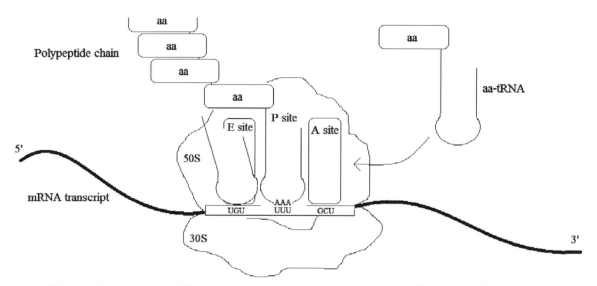

Figure 4. Schematic of one of the most pressure-sensitive components of a mesophilic bacterial cell: a ribosome.

The ribosome undergoes multiple conformational changes during the stepwise addition of each amino acid in the elongation cycle. Briefly, at the beginning of the cycle, the peptidyl site (P) is occupied by a tRNA carrying the growing polypeptide chain, while the aminoacyl site (A) is empty. The elongation factor EF-Tu loads a new aminoacyl tRNA into the A site, while hydrolyzing GTP. The binding of an aminoacyl tRNA into the A site initiates the transpeptidation reaction. Once the new peptide bond is formed, the last phase of the cycle (translocation) is fueled by the hydrolysis of an additional molecule of GTP and requires the presence of the elongation factor EF-G. The net result is that the empty tRNA is released from the ribosome, the new peptidyl tRNA moved to the P-site, and the mRNA moved one codon downstream (Savelsbergh et al., 2003). Among the ribosome structures present throughout this cycle the most pressure-sensitive one appears to be the posttranslocational complex (Gross et al., 1993). Its dissociation midpoint is at 70 MPa, close to the protein synthesis threshold values obtained by others.

However, ribosome dissociation appears to be only part of the story. *E. coli* overproduces the ribosomal proteins L7/L12 and L6 when exposed to a sudden pressure upshift (Welch et al., 1993). Similarly, microarray analysis of high-pressure shocked *Lactobacillus sanfranciscensis* cells (45 MPa for 30 min) (Pavlovic et al., 2005) indicated overexpression of genes coding for ribosomal proteins involved in the binding between 30S and aa-tRNA (S2, L6, and L11).

Interestingly, the levels of L7/L12, S6, and EF-G are negatively regulated by the amount of ppGpp in the cell during the stringent response (Jones et al., 1992). At the same time, a decrease in ppGpp levels has been previously associated with modulating levels of cold-shock proteins (Jones et al., 1992). It is therefore tempting to speculate that some of the physiological effects observed in *E. coli* cells at high pressure are caused by the drop in ppGpp. Moreover, proteins L7 and L12 are required for correctly binding the elongation factor EF-G during the translocation step of the ribosome (Savelsbergh et al., 2005), which, as mentioned, is the most pressure-sensitive process in the elongation cycle.

SIGNALING CHANGES IN GENE EXPRESSION AS A FUNCTION OF PRESSURE

The response of a piezophile to pressure changes was first explored in the methanogen *Methanococcus thermolithotrophicus*, where it was found that a shift from atmospheric pressure to 50 MPa resulted in a change of protein abundance (Jaenicke et al., 1988). This led to investigations into the pressure responsiveness of other piezophiles, eventually leading to the identification of ToxR/S transcriptional regulator in *P. profundum* and s^{54} regulated gene expression in *Shewanella* DSS12. ToxRS is a multimeric cytoplasmic membrane-associated protein complex that is present throughout the *Vibrionaceae* (Welch and Bartlett, 1998). The piezophile *P. profundum* strain SS9 has evolved modifications that enable it to regulate gene expression in response to pressure changes. This was first discovered in studies of the inverse pressure regulation of outer membrane protein gene expression (Bartlett et al., 1989; Welch and Bartlett, 1998; Welch and Bartlett, 1996) and then extended to additional genes (Bidle and Bartlett, 2001). Comparative genome analyses between *P. profundum* strain SS9 and *Vibrio cholerae*, a microbe whose ToxR regulon has been thoroughly characterized (Bina et al., 2003), suggest that there are dozens of pressure- and ToxR-regulated genes in *P. profundum*. Many of these genes appear to be important in starvation adaptation, which is consistent with the general observation that the deep sea is a nutrient-poor environment (Aluwihare et al., 2005). "Additionally, many of the genes encode proteins that localize to the membrane, possibly reflecting the relatively high compressibility of this compartment (see *Membranes and Transport* below)".

During the course of investigating the pressure regulation of gene expression in piezophilic *Shewanella* species, it was revealed that many of these genes appear to be under the control of the alternative RNA polymerase factor σ^{54} (Nakasone et al., 1998). In *Shewanella violacea*, σ^{54} levels do not change with pressure, but the abundance of the σ^{54} transcriptional activator NtrC does in a manner comparable to the genes under its control (Nakasone et al., 2002).

MEMBRANES AND TRANSPORT

Life at low temperature and high pressure requires major changes in phospholipid composition. The effective temperature for a phospholipid bilayer of a deep-sea microbe living near 2°C at 10-km depth is approximately −11 to −19°C. The principal effect of elevated pressure is to promote the lateral compaction of the fatty acids, increasing order and restricting motion (Braganza and Worcester, 1986). This can have adverse consequences for membrane transport, intracellular signaling and gene regulation, membrane protein dispersion, and electron transport (Bartlett, 1992, 1999, 2002; Bartlett and Bidle, 1999). Microbes adapt to temperature drops (or in piezophiles, to pressure increases) by increasing fatty acid unsaturation, decreasing fatty acid length, increasing methyl branching, *cis/trans* isomerization of fatty acid double bonds, or the ratio of *anteiso* to *iso* chain branching, acyl chain shuffling between the *sn*-1 and *sn*-2 positions, or changing phospholipid head groups. The ultimate basis of these changes has been a matter of considerable debate but has included homeoviscous adaptation, homeophasic adaptation, maintenance of an optimal proton permeability, and reduction of elastic stress within the membrane (Attard et al., 2000; Bartlett, 2002; Bartlett and Bidle, 1999).

Unequivocal evidence exists for the critical role of unsaturated fatty acids (UFAs) in the growth of piezophiles at high pressure. Many piezophiles increase their ratio of UFAs to saturated fatty acids as a function of growth pressure (Allen et al., 1999; DeLong and Yayanos, 1985, 1986; Kamimura et al., 1993; Yano et al., 1997). *P. profundum* mutants that produce reduced amounts of monounsaturated fatty acids are pressure sensitive in the absence of exogenous UFAs (Allen et al., 1999). The β-ketoacyl acyl carrier protein synthase II enzyme is required for piezophily (Allen and Bartlett, 2000). Membrane fatty acid composition and membrane fluidity are also a major determinant of the susceptibility of microbes to inactivation at high pressure (Casadei et al., 2002).

In addition to producing monounsaturated fatty acids, many piezophiles also produce ω3-polyunsaturated fatty acids (PUFAs) (Nichols et al., 1993) via a polyketide biosynthetic system (Allen and Bartlett, 2002; Metz et al., 2001). An excellent review on bacterial production of these fatty acids is available (Valentine and Valentine, 2004). *P. profundum* PUFA mutants do not display pressure-sensitive growth characteristics (Allen et al., 1999), and thus, the role of PUFAs in piezophiles is still a matter of conjecture.

Far fewer studies have been performed on the role of the lipid constituents of the membranes of piezophilic archaea in growth at high pressure. The hydrothermal vent piezophile *M. jannaschii* responds to pressure increases by increasing the proportion of macrocyclic archaeol lipid and reducing the amount of aracheol and caldarchaeol lipids (Kaneshiro and Clark, 1995). The significance of this change is unknown and is difficult to assess because of the lack of well-defined phase transitions in these and other archaeal membranes.

In addition to effects on membrane lipids, pressure also exerts a strong influence on membrane proteins. Indeed, so many membrane proteins are sensitive to elevated pressure that pressure treatments have been proposed as a general tool for their extraction (Orr et al., 1990). However, some categories of protein–membrane association appear to be particularly pressure sensitive. This includes the CydCD proteins that are needed for cytochrome *bd* assembly. *E. coli cydD* mutants are sensitive to various stresses, such as recovery from starvation and high temperature (Delaney et al., 1992; Siegele and Kolter, 1993) and also display increased pressure sensitivity (Kato et al., 1996b). *S. violacea* dramatically alters both *cydCD* expression and cytochrome levels as a function of pressure (Kato and Qureshi, 1999; Kato et al., 1996b). Within the piezophile *P. profundum*, *rseC* mutants are pressure sensitive (Chi and Bartlett, 1995), and these mutants, like *cydD* mutants, appear to also be defective in the assembly of certain types of membrane proteins.

Additional membrane proteins from piezosensitive microbes that have been found to be particularly sensitive to pressure include ATPases from *E. coli* and *Saccharomyces* (Abe and Horikoshi, 1998; Marquis and Bender, 1987), mechanosensitive ion channels from *E. coli* (MacDonald and Martinac, 2005), and tryptophan permease from *Saccharomyces cerevisiae* (Abe and Horikoshi, 2000). Expression of high levels of the Tat2 tryptophan permease in *S. cerevisiae* confers enhanced growth ability at high pressure (Abe and Horikoshi, 2000).

PROTEINS/EXTRINSIC FACTORS

The field of high-pressure protein biophysics is well developed. However, despite the fact that numerous investigations have utilized changes in pressure to explore the thermodynamics of protein folding, subunit interactions, ligand recognition, and formation of activated enzyme substrate complexes (Balny, 2006), these approaches have not yet made significant headway into studies of proteins derived from piezophiles. The few exceptions include the aforementioned studies of SSB proteins from piezophilic *Shewanella* (Chilukuri et al., 2002) and, more recently, a high-pressure nuclear magnetic resonance (NMR) study of dihydrofolate reductase (DHFR) from a deep-sea bacterium *Moritella profunda* (Hata et al., 2004). Adaptations to high pressure have also been reported for the RNA polymerase core enzyme from the piezophile *S. violacea*. RNA polymerase derived from *S. violacea*, particularly in the presence of sigma factor, is considerably more active and stable at high pressure than that of *E. coli*, but the basis of this tolerance is unknown (Kawano et al., 2004). The primary structure of malate dehydrogenase has also been examined from piezophilic *Photobacterium*, *Moritella*, and *Shewanella* species (Saito et al., 2006; Saito and Nakayam, 2004;

Welch and Bartlett, 1996). Homology modeling suggests that in *Moritella* and *Shewanella* enzymes the amino acid substitutions present could be related to protein hydration (Saito et al., 2006).

High-pressure biology is an area of extremophile research where considerable insight has been provided into protein function at high pressure from studies of the *Eukarya*. Analyses of α-actin from the skeletal muscle of deep-sea fish have identified residues that prevent ATP and Ca^{2+} dissociation (and thereby enable polymerization to occur) at high pressure (Morita, 2003). Studies of deep-sea metazoans also suggest a major role for extrinsic factors in protein function at high pressure. Many deep-sea animals have high levels of the compatible solute trimethylamine oxide (TMAO) in their tissues (Gillett et al., 1997; Kelly and Yancey, 1999), a compound that favors a more compact protein structure. At elevated pressure, TMAO enhances the activity and stability of a number of proteins (Yancey et al., 2001). The in vivo accumulation of organic solutes may also be a property of many piezophilic prokaryotes. With increasing hydrostatic and osmotic pressure, *P. profundum* accumulates oligomers of β-hydroxy butyrate (Martin et al., 2002).

PIEZOPHILE GENOMICS AND FUNCTIONAL GENOMICS

Two piezophile genome sequences are currently available: the hydrothermal vent methanogen *M. jannaschii* (Bult et al., 1996) and the psychrotolerant piezophile *P. profundum* strain SS9 (Vezzi et al., 2005) genomes. Others are in process. The genome sequence of *P. profundum* strain SS9 indicates the presence of many gene duplications, which could

Table 2. Genes influencing high-pressure (HP) resistance or high-pressure growth in *E. coli* or in *P. profundum*

Mutant	Gene function	Pressure-sensitive phenotype	Reference
E. coli strains			
rpoS	Alternative sigma factor	Sensitive to HP killing	Robey et al., 2001
katE, *katF*, *oxyR*, *sodAB*, *soxS*	Oxidative stress resistance	Sensitive to HP killing	Aertsen et al., 2005
lon, *sulA*	Proteolytic control of certain cytoplasmic regulatory proteins	Sensitive to HP killing	Aertsen and Michiels, 2005a, 2005b
mrr	Restriction endonuclease	Increased resistance to HP killing	Aertsen and Michiels, 2005a
cydD	Cytochrome *bd* assembly	Increased pressure-sensitive growth	Kato et al., 1996b
hns	Nucleoid structure	Increased pressure-sensitive growth	Ishii et al., 2005
P. profundum strains			
rseC	Membrane protein assembly (?)	Pressure-sensitive growth	Chi and Bartlett, 1995
recD	DNA replication, recombination, and repair	Pressure-sensitive growth	Bidle and Bartlett, 1999
fabF	Monounsaturated fatty acid biosynthesis	Pressure-sensitive growth	Allen and Bartlett, 2000

explain, in part, the broad pressure range for this microbe. Duplicated genes include functions known to be pressure sensitive in other bacteria, including motility, electron transport, and ATP biosynthesis. Transcriptome analyses indicate that SS9 turns on many genes for complex carbohydrate degradation at high pressure, a novel form of amino acid fermentation at high pressure (the Stickland reaction), and a number of stress-response genes at atmospheric pressure (Campanaro et al., 2005).

FUTURE FRONTIERS

The field of piezomicrobiology is expanding, with more researchers studying various aspects of these extremophilic microorganisms. However, broad access to these microbes and the equipment needed for their study still greatly limit the number of participants in this field. Future cultivation efforts from deep-sea and deep-subsurface samples will greatly benefit from enrichments for a physiologically diverse array of heterotrophs and autotrophs. Additional genome and environmental genome (metagenome) sequences will certainly aid in identifying functions particularly important for life at depth and will also provide an important platform for interrogating relevant genes and regulatory pathways. The description of genes required for high-pressure growth is now remarkably small (Table 2) but is likely to be greatly expanded as a result of ongoing genetic and genomic studies. Molecular investigation into piezophile macromolecular function at high pressure is still in its infancy, as are considerations of piezophile ecosystems biology and biotechnological application.

Acknowledgment. The authors gratefully acknowledge support from the National Science Foundation (MCB02-37059 and MCB05-44524).

REFERENCES

Abe, F., and K. Horikoshi. 1998. Analysis of intracellular pH in the yeast *Saccharomyces cerevisiae* under elevated hydrostatic pressure: a study in baro- (piezo-)physiology. *Extremophiles* 2:223–228.

Abe, F., and K. Horikoshi. 2000. Tryptophan permease gene TAT2 confers high-pressure growth in *Saccharomyces cerevisiae*. *Mol. Cell. Biol.* 20:8098–8102.

Aertsen, A., P. De Spiegeleer, K. Vanoirbeek, M. Lavilla, and C. W. Michiels. 2005. Induction of oxidative stress by high hydrostatic pressure in *Escherichia coli*. *Appl. Environ. Microbiol.* 71:2226–2231.

Aertsen, A., and C. W. Michiels. 2005a. Mrr instigates the SOS response after high pressure stress in *Escherichia coli*. *Mol. Microbiol.* 58:1381–1391.

Aertsen, A., and C. W. Michiels. 2005b. SulA-dependent hypersensitivity to high pressure and hyperfilamentation after high-pressure treatment of *Escherichia coli lon* mutants. *Res. Microbiol.* 156:233–237.

Aertsen, A., R. Van Houdt, K. Vanoirbeek, and C. W. Michiels. 2004. An SOS response induced by high pressure in *Escherichia coli*. *J. Bacteriol.* 186:6133–6141.

Alain, K., V. G. Marteinsson, M. L. Miroshnichenko, E. A. Bonch-Osmolovskaya, D. Prieur, and J.-L. Birrien. 2002. *Marinitoga piezophila* sp. nov., a rod-shaped, thermo-piezophilic bacterium isolated under high hydrostatic pressure from a deep-sea hydrothermal vent. *Int. J. Syst. Evol. Microbiol.* 52:1331–1339.

Alazard, D., S. Dukan, A. Urios, F. Verhe, N. Bouabida, F. Morel, P. Thomas, J. L. Garcia, and B. Ollivier. 2003. *Desulfovibrio hydrothermalis* sp. nov., a novel sulfate-reducing bacterium isolated from hydrothermal vents. *Int. J. Syst. Evol. Microbiol.* 53:173–178.

Allen, E. E., and D. H. Bartlett. 2000. FabF is required for piezoregulation of *cis*-vaccenic acid levels and piezophilic growth of the deep-sea bacterium *Photobacterium profundum* strain SS9. *J. Bacteriol.* 182:1264–1271.

Allen, E. E., and D. H. Bartlett. 2002. Structure and regulation of the omega-3 polyunsaturated fatty acid synthase from the deep-sea bacterium *Photobacterium profundum* strain SS9. *Microbiology* 148:1903–1913.

Allen, E. E., D. Facciotti, and D. H. Bartlett. 1999. Monounsaturated but not polyunsaturated fatty acids are required for growth at high pressure and low temperature in the deep-sea bacterium *Photobacterium profundum* strain SS9. *Appl. Environ. Microbiol.* 65:1710–1720.

Alpas, H., J. Lee, F. Bozoglu, and G. Kaletunç. 2003. Evaluation of high hydrostatic pressure sensitivity of *Staphylococcus aureus* and *Escherichia coli* O157:H7 by differential scanning calorimetry. *Int. J. Food Microbiol.* 87:229–237.

Aluwihare, L. I., S. P. Pantoja, C. G. Johnson, and D. J. Repeta. 2005. Two chemically distinct pools of organic nitrogen accumulate in the ocean. *Science* 308:1007–1010.

Attard, G. S., R. H. Templer, W. S. Smith, A. N. Hunt, and S. Jackowski. 2000. Modulation of CTP:phosphocholine cytidylyltransferase by membrane curvature elastic stress. *Proc. Natl. Acad. Sci. USA.* 97:9032–9036.

Bale, S. J., K. Goodman, P. A. Rochelle, J. R. Marchesi, J. C. Fry, A. J. Weightman, and R. J. Parkes. 1997. *Desulfovibrio profundus* sp nov, a novel barophilic sulfate-reducing bacterium from deep sediment layers in the Japan Sea. *Int. J. Syst. Bacteriol.* 47:515–521.

Balny, C. 2006. What lies in the future of high-pressure bioscience? *Biochim. Biophys. Acta* 1764:632–639.

Barnes, S. P., S. D. Bradbrook, B. A. Cragg, J. R. Marchesi, A. J. Weightman, J. C. Fry, and R. J. Parkes. 1998. Isolation of sulfate-reducing bacteria from deep sediment layers of the Pacific Ocean. *Geomicrobiology* 15:67–83.

Bartlett, D. H. 1992. Microbial life at high pressures. *Sci. Prog.* 76:479–496.

Bartlett, D. H. 1999. Microbial adaptations to the psychrosphere/piezosphere. *J. Mol. Microbiol. Biotechnol.* 1:93–100.

Bartlett, D. H. 2002. Pressure effects on in vivo microbial processes. *Biochim. Biophys. Acta* 1595:367–381.

Bartlett, D. H. 2005. Extremophilic Vibrionaceae, p. 156–171. *In* F. Thompson, B. Austin, and J. Swings (ed.), *The Biology of Vibrios*. ASM Press, Washington, D.C.

Bartlett, D. H., and K. A. Bidle. 1999. Membrane-based adaptations of deep-sea piezophiles., p. 501–512. *In* J. Seckbach (ed.), *Enigmatic Microorganisms and Life in Extreme Environments*. Kluwer Publishing, Dordrecht, The Netherlands.

Bartlett, D., M. Wright, A. Yayanos, and M. Silverman. 1989. Isolation of a gene regulated by hydrostatic pressure. *Nature* 342:572–574.

Bartosik, A. A., and G. Jagura-Burdzy. 2005. Bacterial chromosome segregation. *Acta Biochim Pol.* **52**:1–34.

Bernhardt, G., R. Jaenicke, H.-D. Ludemann, H. Koning, and K. O. Stetter. 1988. High pressure enhances the growth rate of the thermophilic archaebacterium *Methanococcus thermolithotrophicus* without extending its temperature range. *Appl. Environ. Microbiol.* **54**:1258–1261.

Bidle, K. A., and D. H. Bartlett. 1999. RecD function is required for high pressure growth in a deep-sea bacterium. *J. Bacteriol.* **181**:2330–2337.

Bidle, K. A., and D. H. Bartlett. 2001. RNA arbitrarily primed PCR survey of genes regulated by ToxR and ToxS in the deep-sea bacterium *Photobacterium profundum* strain SS9. *J. Bacteriol.* **183**:1688–1693.

Bina, J., J. Zhu, M. Dziejman, S. Faruque, S. Calderwood, and J. J. Mekalanos. 2003. ToxR regulon of *Vibrio cholerae* and its expression in vibrios shed by cholera patients. *Proc. Natl. Acad. Sci. USA* **100**:2801–2806.

Boone, D. R., Y. Liu, Z. J. Zhao, D. L. Balkwill, G. R. Drake, T. O. Stevens, and H. C. Aldrich. 1995. *Bacillus infernus* sp. nov., an Fe(III)- and Mn(IV)-reducing anaerobe from the deep terrestrial subsurface. *Int. J. Syst. Bacteriol.* **45**:441–448.

Boonyaratanakornkit, B. B., and D. S. Clark. Physiology and biochemistry of *Methanocaldococcus jannaschii* at elevated pressures. *In* A. Aertsen, D. H. Bartlett, C. W. Michiels, and A. A. Yayanos (ed.), *High Pressure Microbiology*, in press. ASM Press, Washington, D.C.

Braganza, L. F., and D. L. Worcester. 1986. Structural changes in lipid bilayers and biological membranes caused by hydrostatic pressure. *Biochemistry* **25**:7484–7488.

Bult, C. J., O. White, G. J. Olsen, L. Zhou, R. D. Fleischmann, G. G. Sutton, J. A. Blake, L. M. FitzGerald, R. A. Clayton, J. D. Gocayn, A. R. Kerlavage, B. A. Dougherty, J.-F. Tomb, M. D. Adams, C. I. Reich, R. Overbeek, E. F. Kirkness, K. G. Weinstock, J. M. Merrick, A. Glodek, J. L. Scott, N. S. M. Geoghagen, J. F. Weidman, J. L. Fuhrmann, D. Nguyen, T. R. Utterback, J. M. Kelley, J. D. Peterson, P. W. Sadow, M. C. Hanna, M. D. Cotton, K. M. Roberts, M. A. Hurst, B. P. Kaine, M. Borodovsky, H.-P. Klenk, C. M. Fraser, H. O. Smith, C. R. Woese, and J. C. Venter. 1996. Complete genome sequence of the methanogenic archaeon, *Methanococcus jannaschii*. *Science* **273**:1058–1073.

Campanaro, S., A. Vezzi, N. Vitulo, F. M. Lauro, M. D'Angeo, F. Simonato, A. Cestaro, G. Malacrida, G. Bertoloni, G. Valle, and D. H. Bartlett. 2005. Laterally transferred elements and high pressure adaptation in *Photobacterium profundum* strains. *BMC Genomics* **6**:122.

Canganella, F., J. M. Gonzalez, M. Yanagibayashi, C. Kato, and H. K. Horikoshi. 1997. Pressure and temperature effects on growth and viability of the hyperthermophilic archaeon *Thermococcus peptonophilus*. *Arch. Microbiol.* **168**:1–7.

Casadei, M. A., P. Mañas, G. Niven, E. Needs, and B. M. Mackey. 2002. Role of membrane fluidity in pressure resistance of *Escherichia coli* NCTC 8164. *Appl. Environ. Microbiol.* **68**:5965–5972.

Chi, E., and D. H. Bartlett. 1995. An *rpoE*-like locus controls outer membrane protein synthesis and growth at cold temperatures and high pressures in the deep-sea bacterium *Photobacterium* SS9. *Mol. Microbiol.* **17**:713–726.

Chilukuri, L. N., and D. H. Bartlett. 1997. Isolation and characterization of the gene encoding single-stranded-DNA-binding protein (SSB) from four marine *Shewanella* strains that differ in their temperature and pressure optima for growth. *Microbiology* **143**:1163–1174.

Chilukuri, L. N., D. H. Bartlett, and P. A. G. Fortes. 2002. Comparison of high pressure-induced dissociation of single-stranded DNA-binding protein (SSB) from high pressure-sensitive and high pressure-adapted marine *Shewanella* species. *Extremophiles* **6**:377–383.

Chilukuri, L. N., P. A. G. Fortes, and D. H. Bartlett. 1995. High pressure modulation of DNA gyrase activity. *Biochim. Biophys. Res. Comm.* **239**:552–556.

Christner, B. C., E. Mosley-Thompson, L. G. Thompson, and J. N. Reeve. 2001. Isolation of bacteria and 16S rDNAs from Lake Vostok accretion ice. *Environ. Microbiol.* **3**:570–577.

Damare, S., C. Raghukumar, and S. Raghukumar. 2006. Fungi in deep-sea sediments of the Central Indian Basin. *Deep-Sea Res. I* **53**:14–27.

Delaney, J. M., D. Ang, and C. Georgopoulos. 1992. Isolation and characterization of the *Escherichia coli htrD* gene, whose product is required for growth at high temperatures. *J. Bacteriol.* **174**:1240–1247.

DeLong, E. F. 1992. High pressure habitats, p. 405–417. *In* J. Lederberg (ed.), *Encyclopedia of Microbiology, Volume 2*. Academic Press, San Diego, CA.

DeLong, E. F., and A. A. Yayanos. 1985. Adaptation of the membrane lipids of a deep-sea bacterium to changes in hydrostatic pressure. *Science* **228**:1101–1103.

DeLong, E. F., and A. A. Yayanos. 1986. Biochemical function and ecological significance of novel bacterial lipids in deep-sea prokaryotes. *Appl. Environ. Microbiol.* **51**:730–737.

DeLong, E. F., D. G. Franks, and A. A. Yayanos. 1997. Evolutionary relationships of cultivated psychrophilic and barophilic deep-sea bacteria. *Appl. Environ. Microbiol.* **63**:2105–2108.

DeLong, E. F., C. M. Preston, T. Mincer, V. Rich, S. J. Hallam, N. U. Frigaard, A. Martinez, M. B. Sullivan, R. Edwards, B. R. Brito, S. W. Chisholm, and D. M. Karl. 2006. Community genomics among stratified microbial assemblages in the ocean's interior. *Science* **311**:496–503.

Edgcomb, V. P., D. T. Kysela, A. Teske, A. de Vera Gomez, and M. L. Sogin. 2002. Benthic eukaryotic diversity in the Guaymas Basin hydrothermal vent environment. *Proc. Natl. Acad. Sci. USA* **99**:7658–7662.

Finch, E. D., and L. A. Kiesow. 1979. Pressure, anesthetics, and membrane structure: a spin-probe study. *Undersea Biomed. Res.* **6**:41–53.

Francis, C. A., K. J. Roberts, J. M. Berman, and A. E. Santoro. 2005. Ubiquity and diversity of ammonia-oxidizing archaea in water columns and sediments of the ocean. *Proc. Natl. Acad. Sci. USA* **102**:14683–14688.

Friedberg, E. C., G. C. Walker, and W. Siede. 1995. *DNA Repair and Mutagenesis*. ASM Press, Washington, D.C.

Fuhrman, J. A., and A. A. Davis. 1997. Widespread Archaea and novel Bacteria from the deep sea as shown by 16S rRNA gene sequences. *Mar. Ecol. Prog. Ser.* **150**:275–285.

Gillett, M. B., J. R. Suko, F. O. Santoso, and P. H. Yancey. 1997. Elevated levels of trimethylamine oxide in muscles of deep-sea gadiform teleosts: a high-pressure adaptation? *J. Exp. Zool.* **279**:386–391.

Gross, M., and R. Jaenicke. 1990. Pressure-induced dissociation of tight couple ribosomes. *FEBS Lett.* **267**:239–241.

Gross, M., K. Lehle, R. Jaenicke, and K. H. Nierhaus. 1993. Pressure-induced dissociation of ribosomes and elongation cycle intermediates. Stabilizing conditions and identification of the most sensitive functional state. *Eur. J. Biochem.* **218**:463–468.

Guisbert, E., C. Herman, C. Z. Lu, and C. A. Gross. 2004. A chaperone network controls the heat shock response in *E. coli*. *Genes Dev.* **18**:2812–2821.

Hata, K., R. Kono, M. Fujisawa, R. Kitahara, Y. O. Kamatari, K. Akasaka, and Y. Xu. 2004. High pressure NMR study of dihydrofolate reductase from a deep-sea bacterium *Moritella profunda*. *Cell. Mol. Biol.* **50**:311–316.

Hauben, K. J. A., D. H. Bartlett, C. C. F. Soontjens, K. Cornelis, E. Y. Wuytack, and C. W. Michiels. 1997. *Escherichia coli* mutants resistant to inactivation by high hydrostatic pressure. *Appl. Environ. Microbiol.* 63:945–950.

Huber, H., M. Thomm, G. König, G. Thies, and K. O. Stetter. 1982. *Methanococcus thermolithotrophicus*, a novel thermophilic lithotrophic methanogen. *Arch. Microbiol.* 132:47–50.

Ingalls, A. E., S. R. Shah, R. L. Hansman, L. I. Aluwihare, G. M. Santos, E. R. Druffel, and A. Pearson. 2006. Quantifying archaeal community autotrophy in the mesopelagic ocean using natural radiocarbon. *Proc. Natl. Acad. Sci. USA* 103:6442.

Ishii, A., T. Oshima, T. Sato, K. Nakasone, H. Mori, and C. Kato. 2005. Analysis of hydrostatic pressure effects on transcription in *Escherichia coli* by DNA microarray procedure. *Extremophiles* 9:65–73.

Ishii, A., T. Sato, M. Wachi, K. Nagai, and C. Kato. 2004. Effects of high hydrostatic pressure on bacterial cytoskeleton FtsZ polymers *in vivo* and *in vitro*. *Microbiology* 150:1965–1972.

Jaenicke, R., G. Bernhardt, H.-D. Ludemann, and K. O. Stetter. 1988. Pressure-induced alterations in the protein pattern of the thermophilic archaebacterium *Methanococcus thermolithotrophicus*. *Appl. Environ. Microbiol.* 54:2375–2380.

Jannasch, H. W. 1987. Effects of hydrostatic pressure on growth of marine bacteria, p. 1–15. In H. W. Jannasch, R. E. Marquis, and A. M. Zimmerman (ed.), *Current Perspectives in High Pressure Biology*. Academic Press, Toronto, Ontario.

Jannasch, H. W., and C. D. Taylor. 1984. Deep-sea microbiology. *Annu. Rev. Microbiol.* 38:487–514.

Jannasch, H. W., and C. O. Wirsen. 1982. Microbial activities in undecompressed and decompressed deep sea water samples. *Appl. Environ. Microbiol.* 43:1116–1124.

Jones, P. G., M. Cashel, G. Glaser, and F. C. Neidhardt. 1992. Function of a relaxed-like state following temperature downshifts in *Escherichia coli*. *J. Bacteriol.* 174:3903–3914.

Kamimura, K., H. Fuse, O. Takimura, and Y. Yamaoka. 1993. Effects of growth pressure and temperature on fatty acid composition of a barotolerant deep-sea bacterium. *Appl. Environ. Microbiol.* 59:924–926.

Kaminoh, Y., H. Kamaya, C. Tashiro, and I. Ueda. 1998. The effects of temperature and pressure on the thermodynamic activity of anesthetics. *Toxicol. Lett.* 100-101:353–357.

Kaneshiro, S. M., and D. S. Clark. 1995. Pressure effects on the composition and thermal behavior of lipids from the deep-sea thermophile *Methanococcus jannaschii*. *Appl. Environ. Microbiol.* 177:3668–3772.

Kanso, S., A. C. Greene, and B. K. Patel. 2002. *Bacillus subterraneus* sp. nov., an iron- and manganese-reducing bacterium from a deep subsurface Australian thermal aquifer. *Int. J. Syst. Evol. Microbiol.* 52:869–874.

Karner, M. B., E. F. DeLong, and D. M. Karl. 2001. Archaeal dominance in the mesopelagic zone of the Pacific Ocean. *Nature* 409:507–510.

Kato, C., and D. H. Bartlett. 1997. The molecular biology of barophilic bacteria. *Extremophiles* 1:111–116.

Kato, C., L. Li, Y. Nogi, Y. Nakamura, J. Tamaoka, and K. Horikoshi. 1998. Extremely barophilic bacteria isolated from the Mariana Trench, Challenger Deep, at a depth of 11,000 meters. *Appl. Environ. Microbiol.* 64:1510–1513.

Kato, C., L. Li, J. Tamaoka, and K. Horikoshi. 1997. Molecular analyses sediment of the 11000-m deep Mariana Trench. *Extremophiles* 1:117–123.

Kato, C., N. Masui, and K. Horikoshi. 1996a. Properties of obligately barophilic bacteria isolated from a sample of deep-sea sediment from the Izu-Bonin Trench. *J. Mar. Biotech.* 4:96–99.

Kato, C., and M. H. Qureshi. 1999. Pressure response in deep-sea piezophilic bacteria. *J. Mol. Microbiol. Biotechnol.* 1:87–89.

Kato, C., T. Sato, and K. Horikoshi. 1995. Isolation and properties of barophilic and barotolerant bacteria from deep-sea mud samples. *Biodivers. Conserv.* 4:1–9.

Kato, C., H. Tamegai, A. Ikegami, R. Usami, and K. Horikoshi. 1996b. Open reading frame 3 of the barotolerant bacterium strain DSS12 is complementary with *cydD* in *Escherichia coli* — CydD functions are required for cell stability at high pressure. *J. Biochem.* 120:301–305.

Kawano, H., K. Nakasone, M. Natsumoto, Y. Yoshida, R. Usami, C. Kato, and F. Abe. 2004. Differential pressure resistance in the activity of RNA polymerase isolated from *Shewanella violacea* and *Escherichia coli*. *Extremophiles* 8:367–375.

Kawarai, T., M. Wachi, H. Ogino, S. Furukawa, K. Suzuki, H. Ogihara, and M. Yamasaki. 2004. SulA-independent filamentation of *Escherichia coli* during growth after release from high hydrostatic pressure treatment. *Appl. Mirobiol. Biotechnol.* 64:255–262.

Kelly, R. H., and P. H. Yancey. 1999. High contents of trimethylamine oxide correlating with depth in deep-sea teleost fishes, skates, and decapod crustaceans. *Biol. Bull.* 196:18–25.

Konneke, M., A. E. Bernhard, J. R. de la Torre, C. B. Walker, J. B. Waterbury, and D. A. Stahl. 2005. Isolation of an autotrophic ammonia-oxidizing marine archaeon. *Nature* 437:543–546.

Kuzminov, A. 1999. Recombinational repair of DNA damage in *Escherichia coli* and bacteriophage lambda. *Microbiol. Mol. Biol. Rev.* 63:751–813.

Landau, J. V. 1967. Induction, transcription, and translation in *Escherichia coli*: a hydrostatic pressure study. *Biochim. Biophys. Acta* 149:506–512.

Lauro, F. M., G. Bertoloni, A. Obraztsova, C. Kato, B. M. Tebo, and D. H. Bartlett. 2004. Pressure effects on *Clostridium* strains isolated from a cold deep-sea environment. *Extremophiles* 8:169–173.

Li, L., J. Guezennec, P. Nichols, P. Henry, M. Yanagibayashi, and C. Kato. 1999a. Microbial diversity in Nankai Trough sediments at a depth of 3,843 m. *J. Oceanogr.* 55:635–642.

Li, L. N., C. Kato, and K. Horikoshi. 1999b. Bacterial diversity in deep-sea sediments from different depths. *Biodivers. Conserv.* 8:659–677.

Li, L., C. Kato, Y. Nogi, and K. Horikoshi. 1998. Distribution of the pressure-regulated operons in deep-sea bacteria. *FEMS Microbiol. Lett.* 159:159–166.

Lochte, K., and C. M. Turley. 1988. Bacteria and cyanobacteria associated with phytodetritus in the deep sea. *Nature* 333:67–70.

Lockhart, A., and J. Kendrick-Jones. 1998. Interaction of the N-terminal domain of MukB with the bacterial tubulin homologue FtsZ. *FEBS Lett.* 430:278–282.

López-García, P., A. Lopez-Lopez, D. Moreira, and F. Rodriguez-Valera. 2001a. Diversity of free-living prokaryotes from a deep-sea site at the Antarctic Polar Front. *FEMS Microbiol. Ecol.* 36:193–202.

López-García, P., D. Moreira, A. López-López, and F. F. Rodríguez-Valera. 2001c. A novel haloarchaeal-related lineage is widely distributed in deep oceanic regions. *Environ. Microbiol.* 3:72–78.

López-García, P., F. Rodriguez-Valera, C. Pedros-Alio, and D. Moreira. 2001b. Unexpected diversity of small eukaryotes in deep-sea Antarctic plankton. *Nature* 409:603–607.

Lynch, T. W., and S. G. Sligar. 2002. Experimental and theoretical high pressure strategies for investigating protein–nucleic acid assemblies. *Biochim. Biophys. Acta* 1595:277–282.

MacDonald, A. G., and B. Martinac. 2005. Effect of high hydrostatic pressure on the bacterial mechanosensitive channel MscS. *Eur. Biophys. J.* 34:434–441.

Marquis, R. E. 1982. Microbial barobiology. *BioScience* 32:267–271.

Marquis, R. E., and G. R. Bender. 1987. Barophysiology of prokaryotes and proton-translocating ATPases, p. 65–73. *In* R. E. Marquis and A. M. Zimmerman (ed.), *Current Perspectives in High Pressure Biology*. Academic Press, London, United Kingdom.

Marteinsson, V. T., J. L. Birrien, A. L. Reysenbach, M. Vernet, D. Marie, A. Gambacorta, P. Messner, U. Sleytr, and D. Prieur. 1999a. *Thermococcus barophilus* sp. nov., a new barophilic and hyperthermophilic archaeon isolated under high hydrostatic pressure from a deep-sea hydrothermal vent. *Int. J. Syst. Bacteriol.* **49:**351–359.

Marteinsson, V. T., P. Moulin, J.-L. Birrien, A. Gambacorta, M. Vernet, and D. Prieur. 1997. Physiological responses to stress conditions and barophilic behavior of the hyperthermophilic vent archaeon *Pyrococcus abyssi*. *Appl. Environ. Microbiol.* **63:**1230–1236.

Marteinsson, V. T., A.-L. Reysenbach, J.-L. Birrien, and D. Prieur. 1999b. A stress protein is induced in the deep-sea barophilic hyperthermophile *Thermococcus barophilus* when grown under atmospheric pressure. *Extremophiles* **3:**277–282.

Martin, D. D., D. H. Bartlett, and M. E. Roberts. 2002. Solute accumulation in the deep-sea bacterium *Photobacterium profundum*. *Extremophiles* **6:**507–514.

Meganathan, R., and R. E. Marquis. 1973. Loss of bacterial motility under pressure. *Nature* **246:**526–527.

Metz, J. G., P. Roessler, D. Facciotti, C. Levering, F. Dittrich, M. Lassner, R. Valentine, K. Lardizabal, F. Domergue, A. Yamada, K. Yazawa, V. Knauf, and J. Browse. 2001. Production of polyunsaturated fatty acids by polyketide synthases in both prokaryotes and eukaryotes. *Science* **293:**290–293.

Miller, J. F., N. N. Shah, C. M. Nelson, J. M. Lulow, and D. S. Clark. 1988. Pressure and temperature effects on growth and methane production of the extreme thermophile *Methanococcus jannaschii*. *Appl. Environ. Microbiol.* **54:**3039–3042.

Miura, T., F. Abe, A. Inoue, R. Usami, and K. Horikoshi. 2001. Isolation of a highly copper-tolerant yeast, *Cryptococcus* sp., from the Japan Trench and the induction of superoxide dismutase activity by Cu^{2+}. *Biotechnol. Lett.* **23:**2027–2034.

Molina-Hoppner, A., T. Sato, C. Kato, M. G. Ganzle, and R. F. Vogel. 2003. Effects of pressure on cell morphology and cell division of lactic acid bacteria. *Extremophiles* **7:**511–516.

Moon-van der Staay, S. Y., R. De Wachter, and D. Vaulot. 2001. Oceanic 18S rDNA sequences from picoplankton reveal unsuspected eukaryotic diversity. *Nature* **409:**607–610.

Morita, R. Y. 1976. Survival of bacteria in cold and moderate hydrostatic pressure environments with special reference to psychrophilic and barophilic bacteria, p. 279–298. *In* R. G. Gray and J. R. Postgate (ed.), *The Survival of Vegetative Microbes*. Cambridge University Press, Cambridge, MA.

Morita, T. 2003. Structure-based analysis of high pressure adaptation of alpha-actin. *J. Biol. Chem.* **278:**28060–28066.

Moser, D. P., T. M. Gihring, F. J. Brockman, J. K. Fredrickson, D. L. Balkwill, M. E. Dollhopf, B. S. Lollar, L. M. Pratt, E. Boice, G. Southam, G. Wanger, B. J. Baker, S. M. Pfiffner, L. H. Lin, and T. C. Onstott. 2005. *Desulfotomaculum* and *Methanobacterium* spp. dominate a 4- to 5-kilometer-deep fault. *Appl. Environ. Microbiol.* **71:**8773–8783.

Nakasone, K., A. Ikegami, C. Kato, R. Usami, and K. Horikoshi. 1998. Mechanisms of gene expression controlled by pressure in deep-sea microorganisms. *Extremophiles* **2:**149–154.

Nakasone, K., A. Ikegami, H. Kawano, C. Kato, R. Usami, and K. Horikoshi. 2002. Transcriptional regulation under pressure conditions by RNA polymerase sigma54 factor with a two-component regulatory system in *Shewanella violacea*. *Extremophiles* **6:**89–95.

Nakayama, H. 2005. *Escherichia coli* RecQ helicase: a player in thymineless death. *Mutat. Res.* **577:**228–236.

Nichols, D. S., P. D. Nichols, and T. A. McMeekin. 1993. Polyunsaturated fatty acids in Antarctic bacteria. *Antarct. Sci.* **5:**149–160.

Niven, G. W., C. A. Miles, and B. M. Mackey. 1999. The effects of hydrostatic pressure on ribosome conformation in *Escherichia coli*: an in vivo study using differential scanning calorimetry. *Microbiology* **145:**419–425.

Nogi, Y., S. Hosoya, C. Kato, and K. Horikoshi. 2004. *Colwellia piezophila* sp. nov., a novel piezophilic species from deep-sea sediments of the Japan Trench. *Int. J. Syst. Evol. Microbiol.* **54:**1627–1631.

Nogi, Y., and C. Kato. 1999. Taxonomic studies of extremely barophilic bacteria isolated from the Mariana Trench and description of *Moritella yayanosii* sp. nov., a new barophilic bacterial isolate. *Extremophiles* **3:**71–77.

Nogi, Y., C. Kato, and K. Horikoshi. 1998a. *Moritella japonica* sp. nov., a novel barophilic bacterium isolated from a Japan trench sediment. *J. Gen. Appl. Microbiol.* **44:**289–295.

Nogi, Y., C. Kato, and K. Horikoshi. 2002. *Psychromonas kaikoae* sp. nov., a novel piezophilic bacterium from the deepest cold-seep sediments in the Japan Trench. *Int. J. Syst. Evol. Microbiol.* **52:**1527–1532.

Nogi, Y., N. Masui, and C. Kato. 1998b. *Photobacterium profundum* sp. nov., a new moderately barophilic bacterial species isolated from a deep-sea sediment. *Extremophiles* **2:**1–7.

Nogi, Y., N. Masui, and C. Kato. 1998c. Taxonomic studies of deep-sea barophilic *Shewanella* species, and *Shewanella violacea* sp. nov., a new moderately barophilic bacterial species. *Arch. Microbiol.* **170:**331–338.

Norris, V., T. Onoda, H. Pollaert, and G. Grehan. 1999. The mechanical advantages of DNA. *Biosystems* **49:**71–78.

Orr, N., E. Yavin, M. Shinitzky, and D. S. Lester. 1990. Application of high-pressure to subfractionate membrane protein–lipid complexes: a case study of protein kinase C. *Anal. Biochem.* **191:**80–85.

Park, C. B., and D. S. Clark. 2002. Rupture of the cell envelope by decompression of the deep-sea methanogen *Methanococcus jannaschii*. *Appl. Environ. Microbiol.* **68:**1458–1463.

Pavlovic, M., S. Hormann, R. F. Vogel, and M. A. Ehrmann. 2005. Transcriptional response reveals translation machinery as target for high pressure in *Lactobacillus sanfranciscensis*. *Arch. Microbiol.* **184:**11–17.

Phadtare, S. 2004. Recent developments in bacterial cold-shock response. *Curr. Issues Mol. Biol.* **6:**125–136.

Phadtare, S., V. Tadigotla, W. H. Shin, A. Sengupta, and K. Severinov. 2006. Analysis of *Escherichia coli* global gene expression profiles in response to overexpression and deletion of CspC and CspE. *J. Bacteriol.* **188:**2521–2527.

Prieur, D. 1992. Physiology and biotechnological potential of deep-sea bacteria, p. 163–202. *In* R. A. Herbert and R. J. Sharp (ed.), *Molecular Biology and Biotechnology of Extremophiles*. Chapman Hall, New York, NY.

Raghukumar, C., and S. Raghukumar. 1998. Barotolerance of fungi isolated from deep-sea sediments of the Indian Ocean. *Aquat. Microb. Ecol.* **15:**153–163.

Regnard, P. 1884. Note sur les conditions de la vir dans les profondeurs de la mer. *Compt. Rend. Soc. Biol.* **36:**164–168.

Robey, M., A. Benito, R. H. Hutson, C. Pascual, S. F. Park, and B. M. Mackey. 2001. Variation in resistance to high pressure and *rpoS* heterogeneity in natural isolates of *Escherichia coli* O157:H7. *Appl. Environ. Microbiol.* **67:**4901–4907.

Saito, R., C. Kato, and A. Nakayama. 2006. Amino acid substitutions in malate dehydrogenases of piezophilic bacteria isolated from intestinal contents of deep-sea fishes retrieved from the abyssal zone. *J. Gen. Appl. Microbiol.* 52:9–19.

Saito, R., and A. Nakayam. 2004. Differences in malate dehydrogenases from obligately piezophilic deep-sea bacterium *Moritella* sp. strain 2d2 and the psychrophilic bacterium *Moritella* sp. strain 5710. *FEMS Microbiol. Lett.* 233:165–172.

Savelsbergh, A., V. I. Katunin, D. Mohr, F. Peske, M. V. Rodnina, and W. Wintermeyer. 2003. An elongation factor G-induced ribosome rearrangement precedes tRNA–mRNA translocation. *Mol. Cell* 11:1517–1523.

Savelsbergh, A., D. Mohr, U. Kothe, W. Wintermeyer, and M. V. Rodnina. 2005. Control of phosphate release from elongation factor G by ribosomal protein L7/L12. *EMBO J.* 24:4316–4323.

Schulz, E., H.-D. Ludemann, and R. Jaenicke. 1976. High pressure equilibrium studies on the dissociation–association of *E. coli* ribosomes. *FEBS Lett.* 64:40–43.

Siegele, D. D., and R. Kolter. 1993. Isolation and characterization of an *Escherichia coli* mutant defective in resuming growth after starvation. *Genes Dev.* 7:2629–2640.

Silva, J. L., D. Foguel, and C. A. Royer. 2001. Pressure provides new insights into protein folding, dynamics and structure. *Trends Biochem. Sci.* 26:612–618.

Slobodkin, A. I., C. Jeanthon, S. L'Haridon, T. Nazina, M. Miroshnichenko, and E. Bonch-Osmolovskaya. 1999. Dissimilatory reduction of Fe(III) by thermophilic bacteria and archaea in deep subsurface petroleum reservoirs of western Siberia. *Curr. Microbiol.* 39:99–102.

Stella, S., M. Falconi, M. Lammi, C. O. Gualerzi, and C. L. Pon. 2006. Environmental control of the in vivo oligomerization of nucleoid protein H–NS. *J. Mol. Biol.* 355:169–174.

Takai, K., A. Inoue, and K. Horikoshi. 1999. *Thermaerobacter marianensis* gen. nov., sp. nov., an aerobic extremely thermophilic marine bacterium from the 11,000 m deep Mariana Trench. *Int. J. Syst. Bacteriol.* 49:619–628.

Takai, K., A. Sugai, T. Itoh, and K. Horikoshi. 2000. *Palaeococcus ferrophilus* gen. nov., sp nov., a barophilic, hyperthermophilic archaeon from a deep-sea hydrothermal vent chimney. *Int. J. Syst. Evol. Microbiol.* 50:489–500.

Takami, H., A. Inoue, F. Fuji, and K. Horikoshi. 1997. Microbial flora in the deepest sea mud of the Mariana Trench. *FEMS Microbiol. Lett.* 152:279–285.

Takami, H., S. Nishi, J. Lu, S. Shimamura, and Y. Takaki. 2004. Genomic characterization of thermophilic *Geobacillus* species isolated from the deepest sea mud of the Mariana Trench. *Extremophiles* 8:351–356.

Tamegai, H., L. Li, N. Masui, and C. Kato. 1997. A denitrifying bacterium from the deep sea at 11,000-m depth. *Extremophiles* 1:207–211.

Tamura, Y., K. Gekko, K. Yoshioka, F. Vonderviszt, and K. Namba. 1997. Adiabatic compressibility of flagellin and flagellar filament of *Salmonella typhimurium*. *Biochim. Biophys. Acta* 1335:120–126.

Tang, G.-Q., N. Tanaka, and S. Kunugi. 1998. In vitro increases in plasmid DNA supercoiling by hydrostatic pressure. *Biochim. Biophys. Acta* 1443:364–368.

Thom, S. R., and R. E. Marquis. 1987. Free radical reactions and the inhibitory and lethal actions of high-pressure gases. *Undersea Biomed. Res.* 14:485–501.

Toffin, L., K. Zink, C. Kato, P. Pignet, A. Bidault, N. Bienvenu, J. L. Birrien, and D. Prieur. 2005. *Marinilactibacillus piezotolerans* sp. nov., a novel marine lactic acid bacterium isolated from deep sub-seafloor sediment of the Nankai Trough. *Int. J. Syst. Evol. Microbiol.* 55:345–351.

Treusch, A. H., S. Leininger, A. Kletzin, S. C. Schuster, H. P. Klenk, and C. Schleper. 2005. Novel genes for nitrite reductase and Amo-related proteins indicate a role of uncultivated mesophilic Crenarchaeota in nitrogen cycling. *Environ. Microbiol.* 7:1985–1995.

Turley, C. M., A. J. Gooday, and J. C. Green. 1993. Maintenance of abyssal benthic foraminifera under high pressure and low temperature: Some preliminary results. I. *Deep-Sea Res. I* 40:643–652.

Turley, C. M., K. Lochte, and D. J. Patterson. 1988. A barophilic flagellate isolated from 4500 meters in the mid-North Atlantic. *Deep-Sea Res. A* 35:1079–1092.

Valentine, R. C., and D. L. Valentine. 2004. Omega-3 fatty acids in cellular membranes: a unified concept. *Prog. Lipid Res.* 43:383–402.

Vezzi, A., S. Campanaro, M. D'Angelo, F. Simonato, N. Vitulo, F. M. Lauro, A. Cestaro, G. Malacrida, B. Simionati, N. Cannata, C. Romualdi, D. H. Bartlett, and G. Valle. 2005. Life at depth: *Photobacterium profundum* genome sequence and expression analysis. *Science* 307:1459–1461.

Welch, T. J., and D. H. Bartlett. 1996. Isolation and characterization of the structural gene for OmpL, a pressure-regulated porin-like protein from the deep-sea bacterium *Photobacterium* species strain SS9. *J. Bacteriol.* 178:5027–5031.

Welch, T. J., and D. H. Bartlett. 1998. Identification of a regulatory protein required for pressure-responsive gene expression in the deep-sea bacterium *Photobacterium* species strain SS9. *Mol. Microbiol.* 27:977–985.

Welch, T. J., A. Farewell, F. C. Neidhardt, and D. H. Bartlett. 1993. Stress response in *Escherichia coli* induced by elevated hydrostatic pressure. *J. Bacteriol.* 175:7170–7177.

Whitman, W. B., D. C. Coleman, and W. J. Wiebe. 1998. Prokaryotes: the unseen majority. *Proc. Natl. Acad. Sci. USA* 95:6578–6583.

Worden, A. Z., M. L. Cuvelier, and D. H. Bartlett. 2006. In-depth analyses of marine microbial community genomics. *Trends Microbiol.* 14:331–336.

Xu, Y., Y. Nogi, C. Kato, Z. Liang, H.-J. Rueger, D. De Kegel, and N. Glansdorff. 2003a. *Moritella profunda* sp. nov. and *Moritella abyssi* sp. nov., two psychropiezophilic organisms isolated from deep Atlantic sediments. *Int. J. Syst. Evol. Microbiol.* 53:533–538.

Xu, Y., Y. Nogi, C. Kato, Z. Liang, H.-J. Rueger, D. De Kegel, and N. Glansdorff. 2003b. *Psychromonas profunda* sp. nov., a psychropiezophilic bacterium from deep Atlantic sediments. *Int. J. Syst. Evol. Microbiol.* 53:527–532.

Yanagibayashi, M., Y. Nogi, L. Li, and C. Kato. 1999. Changes in the microbial community in Japan Trench sediment from a depth of 6292 m during cultivation without decompression. *FEMS Microbiol. Lett.* 170:271–279.

Yancey, P. H., A. L. Fyfe-Johnson, R. H. Kelly, V. P. Walker, and M. T. Auñon. 2001. Trimethylamine oxide counteracts effects of hydrostatic pressure on proteins of deep-sea teleosts. *J. Exp. Zool.* 289:172–176.

Yano, Y., A. Nakayama, and K. Yoshida. 1997. Distribution of polyunsaturated fatty acids in bacteria present in intestines of deep-sea fish and shallow-sea poikilothermic animals. *Appl. Environ. Microbiol.* 63:2572–2577.

Yayanos, A. A. 1986. Evolutional and ecological implications of the properties of deep-sea barophilic bacteria. *Proc. Natl. Acad. Sci. USA.* 83:9542–9546.

Yayanos, A. A. 1995. Microbiology to 10,500 meters in the deep sea. *Annu. Rev. Microbiol.* 49:777–805.

Yayanos, A. A. 2001. Deep-sea piezophilic bacteria. *Methods Microbiol.* 30:615–637.

Yayanos, A. A., and E. F. DeLong. 1987. Deep-sea bacterial fitness to environmental temperatures and pressures, p. 17–32. *In* H. W. Jannasch, R. E. Marquis, and A. M. Zimmerman (ed.),

Current Perspectives in High Pressure Biology. Academic Press, Toronto, Ontario.

Yayanos, A. A., A. S. Dietz, and R. Van Boxtel. 1981. Obligately barophilic bacterium from the Mariana trench. *Proc. Natl. Acad. Sci. USA* 78:5212–5215.

Yayanos, A. A., and E. C. Pollard. 1969. A study of the effects of hydrostatic pressure on macromolecular synthesis in *Escherichia coli*. *Biophys. J.* 9:1464–1482.

ZoBell, C. E. 1970. Pressure effects on morphology and life processes of bacteria, p. 85–130. *In* A. Zimmerman (ed.), *High Pressure Effects on Cellular Processes*. Academic Press, London, United Kingdom/New York, NY.

ZoBell, C. E., and A. B. Cobet. 1963. Filament formation by *Escherichia coli* at increased hydrostatic pressures. *J. Bacteriol.* 87:710–719.

VIII. EXOBIOLOGY

Chapter 26

Astrobiology and the Search for Life in the Universe

GILES M. MARION AND DIRK SCHULZE-MAKUCH

INTRODUCTION

Astrobiology is the search for the origin, evolution, distribution, and future of life in the universe. The term astrobiology has largely superseded the term exobiology, which literally means life beyond Earth. A good starting point in the search for life beyond Earth is to focus on what limits life on Earth. Before beginning a discussion of life in the universe, we need to define exactly what we mean by "life." In this chapter, we will limit our discussion to life as we know it, based on the one Terran example of life we are familiar with. For a discussion of potentially more exotic forms of life based on silicon (rather than carbon) chemistries, solvents other than water, and different types of potentially exploitable energy sources such as osmotic gradients and magnetic energy, the reader is referred to Bains (2004), Benner et al. (2004), and Schulze-Makuch and Irwin (2004). Active life as we know it is based on metabolism, growth, and reproduction. This is to be distinguished from dormant life in protected forms, such as spores that apparently can survive for millions of years under adverse conditions (Cano and Borucki, 1995; Fischman, 1995; Vreeland et al., 2000). There are many different definitions of life. In this chapter, we will accept the definition given by Schulze-Makuch and Irwin (2004) that life is (i) composed of bounded microenvironments in thermodynamic disequilibrium with their external environment, (ii) capable of transforming energy and the environment to maintain a low-entropy (high-order) state, and (iii) capable of information encoding and transmission. This definition is designed to facilitate the search for and detection of life beyond Earth.

In this chapter, we will first examine the requirements for life as we know it. With these constraints as guidelines, we will then examine potential habitats for life in our solar system and, finally, elaborate on a strategy to search for life in the universe.

LIFE REQUIREMENTS

A good starting point in the search for life beyond Earth is to first examine what controls life on Earth. Most of the factors that ultimately limit life are summarized in Table 1. Above all else, Terran life requires energy, water, and nutrients. Energy fuels the metabolic processes that allow life to exist in disequilibrium with their environments. A life form in thermodynamic equilibrium with its environment is dead. The preponderance of life on Earth is based, directly or indirectly, on solar energy captured by autotrophic organisms through photosynthesis. The other autotrophic way of capturing energy is through chemosynthesis, which relies on transformations of inorganic chemicals. Most life forms on Earth (including most bacteria, all fungi, and all animals) are heterotrophs that live off energy captured by autotrophic organisms.

Table 1. Environmental requirements and limitations for active life on Earth

Limiting factor	Requirements/limitations[a]
Energy	Solar, geochemical, geothermal
Water	Presence, composition (see below)
Nutrients	C, O, H, N, K, Ca, P, Mg, S, Fe, Cl, Cu, Mn, Zn, Mo, B
Temperature	−20 to 121°C
Salinity	a_w = 0.6 to 1.0
Desiccation	RH = 60 to 100%
Acidity	pH ≈ 0 to >12
Toxic elements	Pb, Hg, Cd, Zn, Cu, As, Al, B, Ni, Se, Ag, Mg, Mn, Mo
Radiation	Microbes up to 4,000 × more tolerant than humans
Pressure	≈ 0.1 to >1100 bar

[a] Energy, water, and nutrients are requirements (resources) for life. The remaining properties are environmental factors that can potentially limit life; ranges for the latter are known tolerances of active Earth life forms. For details, see Marion et al. (2003).

G. M. Marion • Division of Earth and Ecosystem Sciences, Desert Research Institute, Reno, NV 89512. D. Schulze-Makuch • Department of Geology, Washington State University, Pullman, WA 99164.

Today in planetary sciences, the first step in the search for life is often the search for liquid water. The current mantra of astrobiology is "follow the water" (Kargel, 2004). Water is an ideal solvent for life as we know it (i) because it creates an environment that allows for the stability of some chemical bonds to maintain macromolecular structure, while (ii) promoting the dissolution of other chemical bonds with sufficient ease to enable frequent chemical interchange and energy transformations from one molecular state to another; (iii) because of its ability to dissolve many solutes, while enabling some macromolecules to resist dissolution, thereby providing boundaries, surfaces, interfaces, and stereochemical stability; (iv) because of a density sufficient to maintain critical concentrations of reactants and constrain their dispersal; (v) because of a medium that provides both an upper and a lower limit to the temperatures and pressures at which biochemical reactions operate, thereby funneling the evolution of metabolic pathways into a narrower range optimized for multiple interactions; and (vi) because of a buffer against environmental fluctuations (Schulze-Makuch and Irwin, 2004). Because its physical properties in the liquid state are matched optimally to the prevailing temperatures and pressures on Earth, water is such an efficient solvent for Earth environmental conditions, thus supplying any organism with energy and nutrients, but also with excess salts, acidity, and toxic metals that can limit life (Table 1).

Sixteen elements (nutrients) are considered essential for life (Table 1); the first nine elements (C to S) are macronutrients because they are needed in large amounts (in a percentage range), whereas the last seven elements are micronutrients (Fe to B) that are needed in trace amounts (in a parts per million range) (Raven et al., 1986). Potentially, any one of these nutrients can limit life. Nevertheless, the macronutrients are the more probable limiting nutrients in any environment because they are needed in larger amounts than micronutrients. For example, N, P, and K are the most common commercial fertilizer nutrients.

There are three environments on Earth where microbes have been identified with temperature tolerances in the range from 100°C to 121°C: (i) submarine hydrothermal vents, (ii) subterranean deep biosphere, and (iii) terrestrial hot springs (Marion et al., 2003). The highest temperature tolerances (110°C to 121°C) are found in microbes from marine hydrothermal vents and the subterranean deep biosphere; high pressures prevent these waters from boiling at 100°C, the normal boiling point of water at 1.01 bar (1 atm) pressure. From terrestrial hot springs, microbes have been isolated, which can tolerate temperatures up to 103°C (Stetter, 1999). Hyperthermophiles are invariably either bacteria or archaea. Eukaryotes have an upper temperature range of ~50°C to 60°C (Madigan and Marrs, 1997; Nealson, 1997; Nealson and Conrad, 1999; Rothschild and Mancinelli, 2001).

There are four types of environments on Earth where low temperatures are prevalent: (i) ice, (ii) cold terrestrial environments, (iii) the deep sea, and (iv) the troposphere. Ice includes snow, glaciers, frozen lakes, sea ice, and permafrost. Examples of cold terrestrial environments include the Dry Valleys of Antarctica and Arctic polar deserts. Temperatures in the oceanic abysses hover around 2°C, at a maximum hydrostatic pressure of 1,100 bar (10,660 m) in the Mariana Trench (Yayanos, 1995). Temperatures in the troposphere can drop to −50°C, but life in cloud droplets, claimed to independently grow and reproduce (Sattler et al., 2001), may only extend to slightly <0°C. There are a number of reports in recent years that have demonstrated that some microbes can metabolize, albeit slowly, at temperatures in the range from −17°C to −20°C (Rivkina et al., 2000; Junge et al., 2001, 2004; Gilichinsky, 2002; Junge, 2002). These organisms include bacteria, lichens (a symbiotic association of algae and fungi), and fungi (yeasts). Price and Sowers (2004) have, however, argued that there is no evidence of a minimum temperature for metabolism (growth, maintenance, or survival).

Salinity affects biological activity, in part, because it controls water availability. The higher the salinity, the more energy an organism must expend to maintain a favorable osmotic balance. Because salinity has been studied by scientists representing many different disciplines, measures of salinity (including molality, salt % [by weight or volume], and the thermodynamic activity of water) vary widely among disciplines. The activity of water (a_w) is probably the best measure of salinity as it relates directly to the osmotic gradient controlling flows of salts and water into and out of organisms. Table 1 outlines the approximate salinity limits for biological activity of bacteria/archaea and fungi. Salinity is one of the few limiting factors for life where a eukaryotic organism (fungi) has a higher tolerance ($a_w = 0.60$) than prokaryotic organisms (bacteria/archaea) ($a_w = 0.70$) (Marion et al., 2003). To place these limiting a_w values in perspective, the a_w for seawater is 0.98. Most prokaryotes and fungi can tolerate much higher salinities than seawater.

The desiccating power of the atmosphere is generally measured by relative humidity (RH), which is related to the activity of water (a_w) by:

$$a_W = \frac{\text{RH}}{100} \qquad (1)$$

Just as $a_w = 0.6$ is considered the lower limit for biological activity in saline solutions (Table 1), RH of

60% is considered the lower limit for biological activity under dry atmospheric conditions (Kushner, 1981; Dose et al., 2001). A clear distinction must be made between biological activity and survival under desiccating conditions. Some organisms can survive 99% loss of water with $a_w \sim 0$ (Mazur, 1980). *Bacillus sphaericus* spores were reported to have survived 25 million years of desiccation in amber through a process called anhydrobiosis (Fischman, 1995). Bacteria, fungi, plants, and insects have also been shown to survive extensive periods of dehydration (Rothschild and Mancinelli, 2001).

Acidity is typically quantified using the pH scale:

$$pH = -\log_{10}(a_{H+}) \quad (2)$$

where a_{H+} is the hydrogen ion activity. There are many studies demonstrating that a wide range of organisms can tolerate pH values <1.0. For example, bacteria, archaea, fungi, and algae have all been demonstrated to tolerate pH values ≤1.0 (Bachofen, 1986; Schleper et al., 1995; Johnson, 1998; Huber and Stetter, 1998; Schrenk et al., 1998; Edwards et al., 1999; Robbins et al., 2000). The current record holders are *Picrophilus oshimae* and *Picrophilus torridus* (archaea) that can grow at a pH of −0.06 (Schleper et al., 1995). There are fewer studies of high alkalinities (pH >10) than of high acidities (pH <1.0) probably because high alkalinities are more rare in nature on geological spatial scales. Nevertheless, there are reports of organisms tolerating pH values >11 and maybe even as high as 12.5 to 13 (Bachofen, 1986; Duckworth et al., 1996).

There are many toxic elements that can limit life, including mercury (Hg), lead (Pb), and arsenic (As) (Table 1). On Earth, high toxic element concentrations are often associated with high acidities because strong acids are very effective in dissolving primary minerals and releasing heavy metals into the environment (Robbins et al., 2000; Lopez-Archilla et al., 2001; Fernandez-Remolar et al., 2003). Therefore, organisms that tolerate strong acidity also generally tolerate high levels of heavy metals. Note that several of the toxic elements are also essential nutrients (Table 1). Often, there is a delicate balance between adequate nutrients and excess toxins. Also, some of the elements that are generally toxic, such as arsenic and selenium, are used by some bacteria in their metabolic pathways as a source of energy (Stolz and Oremland, 1999). This is a good example of how adaptable life forms are to a broad range of environments.

The types of radiation that can limit life are generally separated into UV and ionizing radiation (Raven et al., 1986; Horneck and Baumstark-Khan, 2002). In reality, a part of the UV spectra is ionizing (Raven et al., 1986). However, to maintain the normal biological separation of these radiation sources, we will handle the two types separately. UV radiation is a significant component of sunlight. Sources of ionizing radiation include cosmic rays, X-rays, and radioactive decay. Resistance to one form of radiation does not necessarily convey protection from other forms. Table 2 depicts resistance to UV and ionizing radiation for several organisms. *Deinococcus radiodurans* is well known to have high resistance to ionizing radiation. This resistance to radiation is thought to have evolved initially as a resistance to desiccation (Mattimore and Battista, 1996). The mechanism for conveying this resistance is believed to be due to the ability to quickly repair DNA damage (Kushner, 1981; Smith, 1982; Bachofen, 1986; Jawad et al., 1998; Rothschild and Mancinelli, 2001). Almost all organisms are prone to UV damage because macromolecules that propagate genetic information (DNA) absorb UV radiation. Mechanisms to protect organisms from UV radiation include the development of iron-enriched silica crusts (Phoenix et al., 2001), specialized organic pigments (Wynn-Williams et al., 2002), self-shading (Smith, 1982), and shielding by organic compounds derived from dead cells (Marchant et al., 1991). Also, both water and ice are effective in absorbing UV radiation (Baumstark-Khan and Facius, 2002).

High pressures can occur in both deep-earth and deep-sea environments, but there are some fundamental differences between these two systems. In the deep sea, hydrostatic pressures on organisms are easily calculated by the depth (m). For example, 1 atm = 1.01325 bar = 0.101325 MPa = 9.816 m (Yayanos, 1995). Deep in the terrestrial subsurface, the confining pressure could be atmospheric—with organisms growing in air pockets—or, in contrast, very high, as in brine pockets—where organisms may be subjected to both hydrostatic and lithostatic pressures. Another significant difference between deep-earth and

Table 2. Radiation dose giving ≈37% survival for UV and ionizing radiation[a]

Organism	UV radiation (J/m^2)	Ionizing radiation (Gy)
T_1-phage (virus)		2,600
E. coli (bacterium)	50	20–30
Bacillus subtilis (bacterium)		33
D. radiodurans (bacterium)	600	1,500–6,000
S. cerevisiae (yeast)	80	30–150
Chlamydomonas (alga)		24
Bodo marina (eukaryote: heterotrophic flagellate)	5000	
Human (eukaryote)		1.4

[a] Adapted from Kushner, 1981; Baumstark-Khan and Facius, 2002.

deep-sea environments is that deep terrestrial subsurface environments increase in temperature with increasing depth, but deep-sea environments decrease in temperature with increasing depth. Microorganisms have been isolated from the Mariana Trench in the Pacific (10,660 m depth) where pressures reach 1,100 bar (Yayanos, 1995; Kato et al., 1998; Abe et al., 1999). Two bacteria similar to *Moritella* and *Shewanella* are apparently obligately barophilic, with optimum pressures for growth occurring at 700 bar and no growth <500 bar (Kato et al., 1998). These Mariana Trench organisms grow at a temperature of 2°C. There are some archaea associated with deep-sea hydrothermal vents that can survive at pressures as high as 890 bar (Pledger et al., 1994). The high pressure of hydrothermal vents has a compensatory effect that allows stabilization of molecules, which allows growth at elevated temperatures up to at least 121°C (Table 1). Recently, it was demonstrated in a diamond anvil cell that *Shewanella oneidensis* and *Escherichia coli* strains remain physiologically and metabolically active at pressures of 680 to 16,800 bar for up to 30 h (Sharma et al., 2002). At pressures of 12,000 to 16,000 bar, living bacteria resided in fluid inclusions in ice VI crystals and continued to be viable when pressure returned to 1 bar. However, only 1% remained alive, and whether this constitutes viability or survival under pressure is contentious (Couzin, 2002). Nevertheless, such studies and the earlier cited deep-ocean studies (1,100 bar of pressure) demonstrate that pressure may not be much of an impediment for some life forms.

Organisms may be more sensitive to low pressures, however, since low pressure leads to rapid desiccation. It is difficult to envision how an organism would hold on to its environmental substrate on a planet with no or only a thin atmosphere. It has been shown that for Terran organisms under Martian surface conditions, the low pressure is at least as much of an environmental obstacle as UV radiation (Schulze-Makuch et al., 2004a; Diaz and Schulze-Makuch, 2005).

HABITATS FOR LIFE

In the past few decades, our concept of a potentially habitable world has broadened from the narrow range between Venus and Mars, traditionally referred to as the habitable zone, to cover most of the solar system. There is even evidence that Pluto's moon, Charon, may have a subsurface liquid water ocean that could provide a suitable habitat for life (Vogel, 1999). Two primary factors have contributed to broadening our concept of a habitable world.

First, there has been an explosion of research on the ability of microorganisms, especially bacteria and archaea, to grow in extreme environments with respect to temperature, salinity, acidity, radiation, and pressure (Table 1). Environments that would have been considered inhospitable for life a few decades ago have been found thriving with life (e.g., hydrothermal vents). The second factor is a better understanding of environments within which liquid water can exist. The two solar system bodies beyond Earth that have elicited the most interest as potential habitats for life are Mars and Europa because both have clearly been subjected to aqueous processes. We will examine these two solar-system bodies in more detail below.

By Earth environmental standards, the habitats for life on Mars, Europa, and other solar-system and most extrasolar-system bodies are likely to be extreme environments. On the other hand, extraterrestrial life, if it exists, may be well adapted to its environments and find Earth environments extreme. Here, we will argue conservatively from an Earth-centric perspective. Thus, we will judge environments by terrestrial life standards ("life as we know it") and not consider in detail possible exotic habitats such as Titan's surface or the Venusian atmosphere. (For details on the latter, the reader is referred to Schulze-Makuch and Grinspoon, 2005; McKay and Smith, 2005; and Schulze-Makuch et al., 2004b.) The extremes for active life outlined in Table 1 place relative, not absolute, constraints on environments for life beyond Earth. For example, planetary bodies that lack liquid water are unlikely to be habitats for Terran-type life.

In principle, there are three potential habitats for active life. They are the surface of planetary bodies, its subsurface, and its atmosphere. The surface has two advantages: the use of visible light as an energy resource and space to expand. The surface has two major disadvantages: vulnerability to UV and cosmic radiation and meteoritic impacts. Advantages for subsurface life include stable temperatures and vapor pressures and protection from damaging radiation and meteoritic impacts. The disadvantages are lack of sunlight as an energy source and limitations on organismic size. Because subsurface environments are more stable and secure abodes for life, subsurface life is more likely to be the rule on other worlds. Nevertheless, it is hard to visualize how complex life could have evolved on Earth had it not have been for the habitable surface, which provided the opportunity to utilize tremendous amounts of solar energy captured by photosynthesis and space to allow macroscopic organismic size. Life in the atmosphere is more problematic, although Schulze-Makuch and Irwin (2002b) and Schulze-Makuch et al. (2004b) made the case for possible life in the Venusian atmosphere. However, even in their hypothesis, the postulate is that life originated initially in a primordial liquid water ocean and then adapted gradually to live in a thick and nutrient-rich atmosphere because of deteriorating surface conditions and evolutionary pressures. Life in space can only occur temporarily in a dormant state and thus can be dismissed as a possible habitat for active life. A complete discussion of principal habitats is provided in Schulze-Makuch and Irwin (2004).

The cold, dry surface of Mars today is an unsuitable habitat for life because of a variety of stresses including UV radiation, oxidative stress, lack of liquid water, cold temperatures, and low-pressure conditions. Nevertheless, a putative warmer, wetter early Mars would have been suitable for life, which has stimulated considerable speculation about the prospects for life on Mars (McKay and Stoker, 1989; Klein et al., 1992; McKay et al., 1992, 1996; Gibson et al., 1997; Shock, 1997; Jakosky, 1998; Jakosky and Shock, 1998; Fisk and Giovannoni, 1999, Max and Clifford, 2000; Cabrol et al., 2001; Kargel, 2004).

Europa is a cold, ice-covered moon of Jupiter that might at first seem inhospitable for life. However, there is abundant evidence for the presence of a subsurface briny ocean (Pappalardo et al., 1999; Kargel et al., 2000; Stevenson, 2000). The putative ocean of Europa has focused considerable attention on the possible habitats for life on Europa (Reynolds et al., 1983; Jakosky and Shock, 1998; Gaidos et al., 1999; McCollom, 1999; Chyba, 2000; Kargel et al., 2000; Chyba and Hand, 2001; Chyba and Phillips, 2001; Navarro-Gonzalez et al., 2002; Pierazzo and Chyba, 2002; Schulze-Makuch and Irwin, 2002a; Marion et al., 2003).

Many factors need to be integrated to estimate the prospects for life beyond Earth. On Mars and Europa, energy is likely the most limiting factor for life. Estimates of geochemical and geothermal energy are sufficient for life to exist on either Mars or Europa but not sufficient to support large amounts of biomass (Jakosky and Shock, 1998; Gaidos et al., 1999; Chyba, 2000; Irwin and Schulze-Makuch, 2003; Zolotov and Shock, 2003). High radiation impacts the surface of both Mars and Europa (Baumstark-Khan and Facius, 2002). If life exists on either Mars or Europa, it is almost certainly subsurface life. Schulze-Makuch et al. (2005), however, reported recently on the possibility that life on Mars could have adapted to a life style between cycles of dormant and active life, with life becoming active during the warmer and wetter periods of Mars that would occur during periods of increased volcanic activity or a tilt of its axis. To date, there is no definitive evidence that proves or negates the existence of life on Mars or Europa. As extensive as the investigations for life on these planetary bodies have been, it is unlikely that definitive evidence for life, or its absence, will be obtained before samples are returned from these planetary bodies (sample-return missions). Nevertheless, the search for geosignatures and biosignatures of life, which we will examine next, can at least isolate niches with higher probabilities for life in the universe.

A SEARCH STRATEGY FOR LIFE IN THE UNIVERSE

Table 1 outlined environmental limitations for active life. Energy, water, and nutrients are resources that all life forms require. The other seven factors in Table 1 are environmental constraints for active life as we know it. Some of these factors are more easily sensed remotely than others. For example, detecting the presence or absence of surface liquid water is in principle easier than estimating whether or not all the essential nutrients are present in appropriate amounts.

In the search for extraterrestrial signatures of life, we need to distinguish among solar-system sites, extrasolar-system sites, and the search for intelligent life.

It is likely that within the twenty-first century most potential life-bearing bodies in our solar system will be explored by lander or sample-return missions. Such missions provide high-technology assessments of whether life is present or at least refine the potential for life on the planetary body. Exploration for life beyond the solar system will have to rely entirely upon remote sensing for the foreseeable future. At present, the search for intelligent life in the universe is largely restricted to radiowave monitoring by Search for Extraterrestrial Intelligence (SETI). The latter program is likely the best approach in the search for intelligent life in the universe, no matter how rare such life might be (Ward and Brownlee, 2000).

There are two biosignatures for life on Earth that could be especially important in the search for life beyond Earth: (i) the presence of complex organic chemicals and (ii) disequilibrium concentrations of O_2 (or O_3) and methane in the atmosphere. Among the complex organic chemicals that are biosignatures are chlorophyll, proteins, polypeptides, and phospholipids. The presence of high concentrations of O_2 and O_3 in the atmosphere is a powerful indicator of disequilibrium because these oxygen molecules are highly reactive (Leger et al., 1993; Akasofu, 1999; Schidlowski, 2002; Frey and Lummerzheim, 2002). On Earth, atmospheric oxygen levels are maintained at a high level (21%) because O_2 production through photosynthesis is in balance with O_2 reduction reactions.

Table 3. Astrobiology plausibility categories

Category	Definition	Examples
I	Demonstrable presence of liquid water, readily available energy, and organic compounds	Earth
II	Evidence for the past or present existence of liquid water, availability of energy, and inference of organic compounds	Mars, Europa
III	Physically extreme conditions, but with evidence of energy sources and complex chemistry possibly suitable for life forms unknown on Earth	Venus,[a] Titan,[a] Triton, Enceladus
IV	Persistence of life very different from on Earth, conceivable in isolated habitats or reasonable inference of past conditions suitable for the origin of life prior to the development of conditions so harsh as to make its perseverance at present unlikely, but conceivable in isolated habitats	Mercury, Jupiter
V	Conditions so unfavorable for life by any reasonable definition that its origin or persistence cannot be rated a realistic probability	Sun, Moon

[a] Venus and Titan have been upgraded to Category II in Schulze-Makuch and Irwin (2004) because of recent insights into their dynamic activity, planetary/lunar history, and presence of organic compounds. Adapted from Irwin and Schulze-Makuch, 2001.

The atmosphere per se is a geosignature of a favorable environment for life. Atmospheres help moderate climate, protect the surface from radiation and meteoritic strikes, and prevent liquids and gases from dissipating into space (Schulze-Makuch et al., 2002).

A plausibility scale for life beyond Earth was developed by Irwin and Schulze-Makuch (2001) (Table 3). Earth-like planetary bodies with liquid water, readily available energy, and organic compounds were listed as Plausibility of Life (POL) Category I (high). At the other extreme are solar-system bodies such as Sun and Moon, with conditions so harsh that life is unrealistic, which is Category V (remote). The previously discussed planetary bodies, Mars and Europa, are Category II (favorable), which is why so much attention has been placed on these two solar-system bodies.

In the search for life in our solar system, all of the above important resources for life, constraints for life, biosignatures, and geosignatures are appropriate criteria. In the search for life beyond our solar system, attention should focus on (i) energy (especially solar) sources, (ii) liquid water, (iii) complex organic chemistry, (iv) the presence of an atmosphere, and (v) O_2 (O_3) disequilibrium. The first three of these properties are those of Category I (Table 3). The other two properties are strong indicators of life for Terran-type life as it presently occurs on Earth. Here, we propose that planets or moons that possess these five attributes should be assigned a SupraEarth designation, Category Ia (very high, Earth-like). Other planetary bodies on which the plausibility of life is high, but which are not Earth-like, could be assigned a POL Category of Ib. Category Ia will become especially relevant when we are able to detect Earth-type planets in other solar systems. Our extrasolar-system search for life should focus on Earth-like planets where there are strong signatures of life. Such planets would also supplement the SETI search for intelligent life. Intelligent life may be rare in our universe, as suggested by Ward and Brownlee (2000), but our immediate search, especially the search in other solar systems, must continue and focus primarily on properties that are easily measured and are clear-cut necessities and strong indicators of life. Fortunately, there are several planned missions that will contribute to the search for extrasolar-system life, including Gaia, Convection, Rotation, & Planetary Transits (COROT), Eddington, Kepler, and Darwin (Foing, 2002). The first four missions are primarily designed to detect planets around stars. The Darwin mission will determine the habitability of planets around nearby stars, including spectral signatures of gases such as CH_4 and O_3 that are potential biosignatures of life (Foing, 2002). Coupled with ongoing and future solar-system missions (e.g., Cassini, Mars Exploration Rover [MER], Mars Express, Europa Orbiter), we may be on the verge of major breakthroughs in our search for life beyond Earth.

REFERENCES

Abe, F., C. Kato, and K. Horikoshi. 1999. Pressure-regulated metabolism in microorganisms. *Trends Microbiol.* 7:447–453.

Akasofu, S. 1999. Auroral spectra as a tool for detecting extraterrestrial life. *EOS* 35:397.

Bachofen, R. 1986. Microorganisms in extreme environments: introduction. *Experientia* 42:1179–1182.

Bains, W. 2004. Many chemistries could be used to build living systems. *Astrobiology* 4:137–167.

Baumstark-Khan, C., and R. Facius. 2002. Life under conditions of ionizing radiation, p. 261–284. *In* G. Horneck and C. Baumstark-Khan (ed.), *Astrobiology: The Quest for the Conditions of Life.* Springer, Berlin, Germany.

Benner, S. A., A. Ricardo, and M. A. Carrigan. 2004. Is there a common chemical model for life in the universe? *Curr. Opin. Cell Biol.* 8:672–689.

Cabrol, N. A., D. D. Wynn-William, D. A. Crawford, and E. A. Grin. 2001. Recent aqueous environments in Martian impact craters: an astrobiological perspective. *Icarus* 154:98–112.

Cano, R. J., and M. K. Borucki. 1995. Revival and identification of bacterial spores in 25- to 40-million-year-old Dominican amber. *Science* **268:**1060–1064.

Chyba, C. F. 2000. Energy for microbial life on Europa. *Nature* **403:**381–382.

Chyba, C. F., and K. P. Hand. 2001. Life without photosynthesis. *Science* **292:**2026–2027.

Chyba, C. F., and C. B. Phillips. 2001. Possible ecosystems and the search for life on Europa. *Proc. Natl. Acad. Sci. USA* **98:**801–804.

Couzin, J. 2002. Weight of the world on microbes' shoulders. *Science* **295:**1444–1445.

Diaz, B., and D. Schulze-Makuch. 2005. Microbial survival rates of *Escherichia coli* and *Deinococcus radiodurans* under low temperature, low pressure, and UV-irradiation conditions, and their relevance to possible Martian life. *Astrobiology* **6:**332–347.

Dose, K., A. Bieger-Dose, B. Ernst, U. Feister, B. Gomez-Silva, A. Klein, S. Risi, and C. Stridde. 2001. Survival of microorganisms under the extreme conditions of the Atacama Desert. *Orig. Life Evol. Biosph.* **31:**287–303.

Duckworth, A. W., W. D. Grant, B. E. Jones, and R. van Steenbergen. 1996. Phylogenetic diversity of soda lake alkaliphiles. *FEMS Microbiol. Ecol.* **19:**181–191.

Edwards, K. J., T. M. Gihring, and J. F. Banfield. 1999. Seasonal variations in microbial populations and environmental conditions in an extreme acid mine drainage environment. *Appl. Environ. Microbiol.* **65:**3627–3632.

Fernandez-Remolar D. C., N. Rodriquez, and F. Gomez. 2003. Geological record of an acidic environment driven by iron hydrochemistry: the Tinto River system. *J. Geophys. Res.* **108:**5080. doi:10.1029/2002JE001918.

Fischman, J. 1995. Have 25-million-year-old bacteria returned to life? *Science* **268:**977.

Fisk, M. R., and S. J. Giovannoni. 1999. Sources of nutrients and energy for a deep biosphere on Mars. *J. Geophys. Res.* **104:**11805–11811, 11815.

Foing, B. H. 2002. Space activities in exo-astrobiology, p. 389–398. *In* G. Horneck and C. Baumstark-Khan (ed.), *Astrobiology: The Quest for the Conditions of Life*. Springer, Berlin, Germany.

Frey, H. U., and D. Lummerzheim. 2002. Can conditions for life be inferred from optical emissions of extra-solar-system planets, p. 381–388. *In* M. Mendillo, A. Nagy, and J. H. Waite (ed.), *Atmospheres in the Solar System: Comparative Aeronomy*. Geophysical Monograph 130. American Geophysical Union, Washington, DC.

Gaidos, E. J., K. H. Nealson, and J. L. Kirschvink. 1999. Life in ice-covered oceans. *Science* **284:**1631–1633.

Gibson, Jr., E. K., D. S. McKay, K. Thomas-Keprta, and C. S. Romanek. 1997. The case for relic life on Mars. *Sci. Am.* **277:**58–65.

Gilichinsky, D. A. 2002. Permafrost model of extraterrestrial habitat, p. 125–142. *In* G. Horneck and C. Baumstark-Khan (ed.), *Astrobiology: The Quest for the Conditions of Life*. Springer, Berlin, Germany.

Horneck, G., and C. Baumstark-Khan. 2002. *Astrobiology: The Quest for the Conditions of Life*. Springer, Berlin, Germany.

Huber, H., and K. O. Stetter. 1998. Hyperthermophiles and their possible potential in biotechnology. *J. Biotechnol.* **64:**39–52.

Irwin, L. N., and D. Schulze-Makuch. 2001. Assessing the plausibility of life on other worlds. *Astrobiology* **1:**143–160.

Irwin, L. N., and D. Schulze-Makuch. 2003. Strategy for modeling putative multilevel ecosystems on Europa. *Astrobiology* **3:**813–821.

Jakosky, B. M. 1998. *The Search for Life on other Planets*. Cambridge University Press, Cambridge, MA.

Jakosky, B. M., and E. L. Shock. 1998. The biological potential of Mars, the early Earth, and Europa. *J. Geophys. Res.* **103:**19359–19364.

Jawad, A., A. M. Snelling, J. Heritage, and P. M. Hawkey. 1998. Exceptional desiccation tolerance of *Acinetobacter radioresistens*. *J. Hosp. Infect.* **39:**235–240.

Johnson, D. B. 1998. Biodiversity and ecology of acidophilic microorganisms. *FEMS Microbiol. Ecol.* **27:**307–317.

Junge, K. 2002. Bacterial abundance, activity, and diversity at extremely cold temperatures in Arctic sea ice. Ph.D. dissertation. University of Washington, Washington, DC.

Junge, K., J. W. Deming, and H. Eicken. 2004. Bacterial activity at −2 to −20°C in Arctic wintertime sea ice. *Appl. Environ. Microbiol.* **70:**550–557.

Junge, K., C. Krembs, J. W. Deming, A. Stierle, and H. Eicken. 2001. A microscopic approach to investigate bacteria under in-situ conditions in sea-ice samples. *Ann. Glaciol.* **33:**304–310.

Kargel, J. S. 2004. *Mars—A Warmer, Wetter Planet*. Springer/Praxis, London, United Kingdom.

Kargel, J. S., J. Kaye, J. W. Head III, G. M. Marion, R. Sassen, J. Crowley, O. Prieto, S. A. Grant, and D. Hogenboom. 2000. Europa's salty crust and ocean: origin, composition, and the prospects for life. *Icarus* **148:**226–265.

Kato, C., L. Li, Y. Nogi, Y. Nakamura, J. Tamaoka, and K. Horikoshi. 1998. Extremely barophilic bacteria isolated from the Mariana Trench, Challenger Deep, at a depth of 11,000 meters. *Appl. Environ. Microbiol.* **64:**1510–1513.

Klein, H. P., N. H. Horowitz, and K. Biemann. 1992. The search for extant life on Mars, p. 1221–1233. *In* H. H. Kieffer, C. W. Snyder, and M. S. Mathews (ed.), *Mars*. The University of Arizona Press, Tucson, AZ.

Kushner, D. 1981. Extreme environments: are there any limits to life? p. 241–248. *In* C. Ponnamperuma (ed.), *Comets and the Origin of Life*. D. Reidel Publishing Company, Dordrecht, Netherlands.

Leger, A., M. Pirre, and F. J. Marceau. 1993. Search for primitive life on a distant planet: relevance of O_2 and O_3 detections. *Astron. Astrophys.* **277:**309–313.

Lopez-Archilla, A. I., I. Marin, and R. Amils. 2001. Microbial community composition and ecology of an acidic aquatic environment: the Tinto River, Spain. *Microbiol. Ecol.* **41:**20–35.

Madigan, M. T., and B. L. Marrs. 1997. Extremophiles. *Sci. Am.* **276:**82–87.

Marchant, J., A. T. Da Vison, and G. J. Kelly. 1991. UV-B protecting compounds in the marine alga *Phaeocystis pouchetii* from Antarctica. *Mar. Biol.* **109:**391–395.

Marion, G. M., C. H. Fritsen, H. Eicken, and M. C. Payne. 2003. The search for life on Europa: limiting environmental factors, potential habitats, and Earth analogues. *Astrobiology* **3:**785–811.

Mattimore, V., and J. R. Battista. 1996. Radioresistance of *Deinococcus radiodurans*: functions necessary to survive prolonged desiccation. *J. Bacteriol.* **178:**633–637.

Max, M. D., and S. M. Clifford. 2000. The state, potential distribution, and biological implications of methane in the Martian crust. *J. Geophys. Res.* **105:**4165–4171.

Mazur, P. 1980. Limits to life at low temperatures and at reduced water contents and water activities. *Orig. Life* **10:**137–159.

McCollom, T. M. 1999. Methanogenesis as a potential source of chemical energy for primary biomass production by autotrophic organisms in hydrothermal systems on Europa. *J. Geophys. Res.* **104:**30729–30730, 30742.

McKay, C. P., R. L. Mancinelli, C. R. Stoker, R. A. Wharton, Jr. 1992. The possibility of life on Mars during a water-rich past, p. 1234–1245. *In* H. H. Kieffer, C. W. Snyder, and M. S. Matthews (ed.), *Mars*. The University of Arizona Press, Tucson, AZ.

McKay, C. P., and H. D. Smith. 2005. Possibilities for methanogenic life in liquid methane on the surface of Titan. *Icarus* 178: 274–276.

McKay, C. P., and C. R. Stoker. 1989. The early environment and its evolution on Mars: implications for life. *Rev. Geophys.* 27:189–214.

McKay, D. S., E. K. Gibson, Jr., K. L. Thomas-Keprta, H. Vali, C. S. Romanek, S. J. Clemett, X. D. F. Chillier, C. R. Maechling, and R.N. Zare. 1996. Search for past life on Mars: possible relic biogenic activity in Martian meteorite ALH84001. *Science* 273:924–930.

Navarro-Gonzalez, R., L. Montoya, W. Davis, and C. McKay. 2002. Laboratory support for a methanogenesis driven biosphere in Europa, p. 35. *In* R. Greeley (ed.), *Europa Focus Group Workshop 3*. Flagstaff, Arizona State University, Tempe, AZ. Available at http://astrobiology.asu.edu/focus/europa/intro.html.

Nealson, K. H. 1997. The limits of life on Earth and searching for life on Mars. *J. Geophys. Res.* 102:23675–23686.

Nealson, K. H., and P. G. Conrad. 1999. Life: past, present, and future. *Phil. Trans. R. Soc. Lond. B.* 354:1923–1939.

Pappalardo, R. T., M. J. S. Belton, H. H. Breneman, M. H. Carr, C. R. Chapman, G. C. Collins, T. Denk, S. Fagents, P. E. Geissler, B. Giese, R. Greeley, R. Greenberg, J. W. Head, P. Helfenstein, G. Hoppa, S. D. Kadel, K. P. Klaasen, J. E. Klemaszewski, K. Magee, A. S. McEwen, J. M. Moore, W. B. Moore, G. Neukum, C. B. Phillips, L. M. Prockter, G. Shubert, D. A. Senske, R. J. Sullivan, B. R. Tufts, E. P. Turtle, R. Wagner, and K. K. Williams. 1999. Does Europa have a subsurface ocean? Evaluation of the geological evidence. *J. Geophys. Res.* 104:24015–24055.

Phoenix, V. R., K. O. Konhauser, D. G. Adams, and S. H. Bottrell. 2001. Role of biomineralization as an ultraviolet shield: implications for Archean life. *Geology* 29:823–826.

Pierazzo, E., and C. F. Chyba. 2002. Cometary delivery of biogenic elements to Europa. *Icarus* 157:120–127.

Pledger, R. J., B. C. Crump, and J. A. Baross. 1994. A barophilic response by two hyperthermophilic, hydrothermal vent Archaea: an upward shift in the optimal temperature and acceleration of growth rate at supra-optimal temperatures by elevated pressure. *FEMS Microbiol. Ecol.* 14:233–242.

Price, P. B., and T. Sowers. 2004. Temperature dependence of metabolic rates for microbial growth, maintenance, and survival. *Proc. Natl. Acad. Sci. USA* 101:4631–4636.

Raven, P. H., R. F. Evert, and S. E. Eichhorn. 1986. *Biology of Plants*, 4th ed. Worth Publishers, New York, NY.

Reynolds, R. T., S. W. Squyres, D. S. Colburn, and C. P. McKay. 1983. On the habitability of Europa. *Icarus* 56:246–254.

Rivkina, E. M., E. I. Friedmann, C. P. McKay, and D. A. Gilichinsky. 2000. Metabolic activity of permafrost bacteria below the freezing point. *Appl. Environ. Microbiol.* 66: 3230–3233.

Robbins, E. I., T. M. Rodgers, C. N. Alpers, and D. K. Nordstrom. 2000. Ecogeochemistry of the subsurface food web at pH 0–2.5 in Iron Mountain, California, U.S.A. *Hydrobiologia* 433:15–23.

Rothschild, L. J., and R. L. Mancinelli. 2001. Life in extreme environments. *Nature* 409:1092–1101.

Sattler, B., H. Puxbaum, and R. Psenner. 2001. Bacterial growth in supercooled cloud droplets. *Geophys. Res. Lett.* 28:239–242.

Schidlowski, M. 2002. Search for morphological and biogeochemical vestiges of fossil life in extraterrestrial settings: utility of terrestrial evidence, p. 373–386. *In* G. Horneck and C. Baumstark-Khan (ed.), *Astrobiology: The Quest for the Conditions of Life*. Springer, Berlin, Germany.

Schleper, C., G. Pühler, B. Kühlmorgen, and W. Zillig. 1995. Life at extremely low pH. *Nature* 375:741–742.

Schrenk, M. O., K. J. Edwards, R. M. Goodman, R. J. Hamers, and J. F. Banfield. 1998. Distribution of *Thiobacillus ferrooxidans* and *Leptospirillum ferrooxidans*: implications for generation of acid mine drainage. *Science* 279:1519–1522.

Schulze-Makuch, D., B. Diaz, and L. N. Irwin. 2004a. Environmental history and stresses on Mars and their effect on microbial populations. Paper presented at AGU Fall Meeting, San Francisco, CA, December 2004.

Schulze-Makuch, D., D. H. Grinspoon, O. Abbas, L. N. Irwin, and M. Bullock. 2004b. A sulfur-based UV adaptation strategy for putative phototrophic life in the Venusian atmosphere. *Astrobiology* 4:11–18.

Schulze-Makuch, D., and D. H. Grinspoon. 2005. Biologically energy and carbon cycling on Titan? *Astrobiology* 5:560–567.

Schulze-Makuch, D., and L. N. Irwin. 2002a. Energy cycling and hypothetical organisms in Europa's ocean. *Astrobiology* 2: 105–121.

Schulze-Makuch, D., and L. N. Irwin. 2002b. Reassessing the possibility of life on Venus: proposal for an astrobiology mission. *Astrobiology* 2:197–202.

Schulze-Makuch, D., and L. N. Irwin. 2004. *Life in the Universe*. Springer, Berlin, Germany.

Schulze-Makuch, D., L. N. Irwin, and H. Guan. 2002. Search parameters for the remote detection of extraterrestrial life. *Planet. Space Sci.* 50:675–683.

Schulze-Makuch, D., L. N. Irwin, J. H. Lipps, D. LeMone, J. M. Dohm, and A. G. Fairén. 2005. Scenarios for the evolution of life on Mars. *J. Geophys. Res.* 110:E12S23.

Sharma, A., J. H. Scott, G. D. Cody, M. L. Fogel, R. M. Hazen, R. J. Hemley, and W. T. Huntress. 2002. Microbial activity at gigapascal pressures. *Science* 295:1514–1516.

Shock, E. L. 1997. High-temperature life without photosynthesis as a model for Mars. *J. Geophys. Res.* 102:23687–23694.

Smith, D. W. 1982. Extreme natural environments, p. 555–574. *In* R. G. Burns and J. H. Slater (ed.), *Experimental Microbial Ecology*. Blackwell Scientific Publications, Oxford, United Kingdom.

Stetter, K. O. 1999. Extremophiles and their adaptation to hot environments. *FEBS Lett.* 452:22–25.

Stevenson, D. 2000. Europa's ocean—the case strengthens. *Science* 289:1305–1307.

Stolz, J. F., and R. S. Oremland. 1999. Bacterial respiration of arsenic and selenium. *FEMS Microbiol. Rev.* 23:615–627.

Vogel, G. 1999. Expanding the habitable zone. *Science* 286:70–71.

Vreeland, R. H., W. D. Rosenzweig, and D. W. Powers. 2000. Isolation of a 250 million-year-old halotolerant bacterium from a primary salt crystal. *Nature* 407:897–900.

Ward, P. D., and D. Brownlee. 2000. *Rare Earth: Why Complex Life is Uncommon in the Universe*. Copernicus, New York, NY.

Wynn-Williams, D. D., H. G. M. Edwards, E. M. Newton, and J. M. Holder. 2002. Pigmentation as a survival strategy for ancient and modern photosynthetic microbes under high ultraviolet stress on planetary surfaces. *Int. J. Astrobiol.* 179: 174–183.

Yayanos, A. A. 1995. Microbiology to 10,500 meters in the deep sea. *Annu. Rev. Microbiol.* 49:777–805.

Zolotov, M. Y., and E. L. Shock. 2003. Energy for biologic sulfate reduction in a hydrothermally formed ocean on Europa. *J. Geophys. Res. Planets* 108:5022.

IX. BIOTECHNOLOGY

Chapter 27

Extremophiles, a Unique Resource of Biocatalysts for Industrial Biotechnology

GARABED ANTRANIKIAN AND KSENIA EGOROVA

SUMMARY

There is an interest in the production of robust enzymes because of their potential application in various industrial fields, particularly in the chemical, pharmaceutical, food, feed, beverage, and textile industries. Furthermore, efficient enzyme systems are needed for the breakdown of biomass (renewable resources), which contains polymeric substrates such as starch, cellulose, and hemicellulose. Microorganisms living in extreme habitats are a good source for such enzymes (extremozymes), which allow to perform biotransformation reactions under non-conventional conditions under which many proteins are completely denatured. These enzymes are in general superior to the traditional catalysts, because they provide proteins with unique properties that are active at extremes of temperature 0 and 120°C, pH 0 and 12, and in the presence of organic solvents (up to 99%). In this chapter, enzymes and molecules that are produced by thermophilic, thermoacidophilic, thermoalkaliphilic, and halophilic microorganisms including archaea and bacteria will be introduced and their industrial relevance will be discussed. Attempt will be made to identify the factors that enhance or limit the exploitation of extremophiles for industrial biotechnology (white biotechnology).

INTRODUCTION

Extremophiles are defined as organisms that can thrive optimally in habitats that are hostile for human life such as elevated temperatures (up to 110°C), high pressures in the deep sea (up to 1000 bar) and extremes of pH (0 or 12). Most of the extremophiles identified to date belong to *Archaea* and *Bacteria*. Since many representatives of this group are able to grow optimally under extreme conditions, they are an interesting source of stable enzymes (extremozymes). Based on the unique stability of their enzymes at high temperature, extremes of pH, high pressure, salt, organic solvent, detergents, and metal tolerance, they are expected to be a powerful tool in industrial biotransformation processes that run at harsh conditions. In order to ensure that energy and raw material for various industries will be supplied in the future, a new strategy based on biomass has to be developed. New technologies should allow the efficient conversion of renewable resources which contain polymeric substrates, e.g., starch, cellulose, hemicellulose, and oils to high value products. The application of robust enzymes and microorganisms for the sustainable production of chemicals, biopolymers, materials, and fuels from renewable resources—also defined as industrial (white) biotechnology—will offer great opportunities for the chemical and pharmaceutical industries. White biotechnology aims at the reduction of waste, energy input and raw material and the development of highly efficient and environmentally friendly processes.

The majority of the industrial enzymes known to date have been derived from bacteria and fungi. The global annual enzyme market has been estimated to be around 5 billion Euros. In the case of extremophiles only few enzymes have found their way to the market. Tailor-made enzymes are needed for various industrial applications such as food, feed, textile, paper, pharmaceutical industries and fine chemistry. The rapid development in the fields of genomics, metagenomics, proteomics, metabolomics, systems biology, and directed evolution will provide biocatalysts with the desired properties. In order to meet the future challenges, innovative technologies for the production of

---

G. Antranikian and K. Egorova • Institute of Technical Microbiology, Hamburg University of Technology, Kasernenstr. 12, 21073 Hamburg, Germany.

new generation of enzymes and bioprocesses are needed. In this chapter, we will focus on extremophilic archaea and bacteria and their relevance for industrial biotechnology.

EXTREMOPHILIC ARCHAEA AND BACTERIA

Extreme environments, such as geothermal sites (80 to 121°C), polar regions (−20 to 20°C), acidic (pH < 4), and alkaline (pH > 8) springs, saline lakes (2 to 5 M NaCl), and the cold-pressurized depths of the oceans (<5°C, up to 1100 bar), are promising sources of unique microorganisms and enzymes. The majority of extremophiles identified to date belong to the archaeal domain (nearly 300 species), which consists of four kingdoms: *Crenarchaeota*, *Euryarchaeota*, *Korarchaeota*, and *Nanoarchaeota*. The recently discovered nanosized parasitic hyperthermophilic archaeon *Nanoarchaeum equitans* grows in coculture with the crenarchaeon *Ignicoccus* sp. Another symbiotic psychrophilic crenarchaeon *Cenarchaeum symbiosum* inhabits a marine sponge. The majority of the cold-adapted microorganisms belong to bacteria. Polar regions are of interest since they provide diverse marine and terrestrial habitats (Groudieva et al., 2004). The majority of the strains were isolated at 4°C but optimal growth is between 15 and 25°C. The bacteria, isolated from the Arctic region, fall in different phylogenetic groups such as *Proteobacteria*, *Bacillus-Clostridium* group, the order *Actinomycetales* and the *Cytophaga-Flexibacter-Bacteroides* phylum (Groudieva et al., 2004). Thermophiles (which grow at 50 to 60°C), extreme thermophiles (60 to 80°C), and hyperthermophiles (80 to 113°C) are widely distributed in the archaeal domain. A large number of bacterial species, such as the anaerobic bacteria belonging to the genera *Clostridium*, *Thermoanaerobacter*, *Thermobacteroides*, *Anaerobranca*, *Thermotoga* and *Fervidobacterium*, are able to grow at up to 75°C (Bertoldo and Antranikian, 2002; Antranikian et al., 2005). Other thermophilic or extreme thermophilic bacteria live under aerobic conditions; the major representatives belong to the genus *Thermus* and *Alicyclobacillus* (da Costa et al., 1998). Only few bacterial species are able to grow at up to 100°C. Interestingly, methanogenic archaea are able to grow at up to 110°C (*Methanopyrus kandleri*) or below 0°C (*Methanogenium frigidum*). Some members of archaea can also survive under both extreme conditions, namely high temperature and low pH (50 to 90°C, pH 0 to 4) (e.g., *Acidianus*, *Ferroplasma*, *Picrophilus*, *Sulfolobus*, and *Thermoplasma*). Only few bacteria are able to survive at extreme acidic conditions at elevated temperatures, e.g., *Alicyclobacillus acidocaldarius* and *Thiobacillus caldus* (Bertoldo et al., 2004). The microbial diversity at alkaline conditions and high temperature is very limited (Horikoshi, 1999). These extremophiles provide interesting enzyme systems, cell components, and other organic compounds that are active at elevated temperatures and extremes of pH. Various extremophiles can also tolerate other extreme conditions such as high pressure (*Paleococcus ferrophilus*, *Thermococcus barophilus*), high levels of radiation or toxic compounds (*Pyrococcus furiosus*) or low water and nutrient supply (*Halobacterium* sp.).

BIOCATALYSTS FROM EXTREMOPHILES

Starch-Processing Enzymes: Biochemistry at the Boiling Point of Water

Many microbial enzymes involved in carbohydrate metabolism, particularly those of the glucosyl hydrolase family, are of industrial interest. The starch-processing industry, which converts starch into more valuable products such as dextrins, glucose, fructose, maltose, and trehalose, can profit from thermostable enzymes. In all starch-converting processes, high temperatures are required to liquefy starch and make it accessible to enzymatic hydrolysis. The synergetic action of thermostable amylolytic enzymes brings an advantage to those processes, lowering the cost of sugar syrup production. Furthermore, the use of thermostable enzymes can lead to other valuable products, which include innovative starch-based materials with gelatin-like characteristics and defined linear dextrins that can be used as fat substitutes, texturizers, aroma stabilizers, and prebiotics (Gupta et al., 2003). Extremophilic archaea and bacteria have been shown to be a good source for the production of a number of starch modifying enzymes (Tables 1 and 2).

α-Amylases

α-Amylase (α-1,4-glucan-4-glucanohydrolase; EC 3.2.1.1) hydrolyzes linkages in the starch polymer, which leads to the formation of linear and branched oligosaccharides. The sugar-reducing groups are liberated in the α-anomeric configuration. Most of starch-hydrolyzing enzymes belong to the α-amylase family that contains a characteristic catalytic $(\beta/\alpha)_8$-barrel domain. Throughout the α-amylase family, only eight amino acid residues are invariant, seven at the active site and a glycine in a short turn. A variety of amylolytic enzymes have been detected in thermophilic and halophilic archaea (Table 1). Extremely thermostable α-amylases have been characterized from a number of hyperthermophilic archaea belonging to the genera

Table 1. Starch-processing enzymes produced by extremophilic archaea

Enzymes	Strain[a]	MW (kDa)	T_{opt} (°C)	pH_{opt}	Thermostability (half-life)	Possible applications	References
α-Amylase (EC 3.2.1.1)	Desulfurococcus mucosus	43.3	100	5.5		Bread and baking industry, starch liquefaction and saccharification, production of glucose, textile desizing, paper industry, synthesis of oligosaccharides, detergent application, gelling and thickening in food industry	Antranikian et al., 2005
	Haloarcula hispanica		50	6.5	0.5 h at 60°C		Hutcheon et al., 2005
	Haloarcula sp. S-1	70	50	7	0.5 h at 70°C		Fukushima et al., 2005
	Halobacterium salinarum						Margesinand Schinner, 2001
	Haloferax mediterranei	58	50–60	7–8	10 h at 50°C		Perez-Pomares et al., 2003
	Halothermothrix orenii	62	65	7.5	>1 h at 70°C		Mijts and Patel, 2002
	Methanocaldococcus jannaschii		120	5–8	50 h at 100°C		Kim et al., 2001
	Natronococcus sp. Ah-36	74	55	9	0.5 h at 55°C		Antranikian et al., 2005
	Pyrococcus furiosus (intracellular)	76	92	7			Antranikian et al., 2005
	Pyrococcus furiosus (extracellular)	100	100	5.5–6	13 h at 98°C		Antranikian et al., 2005
	Pyrococcus woesei (struct.)	68	100	5.5	11 h at 90°C		Antranikian et al., 2005
	Pyrodictium abyssi		100	5			Antranikian et al., 2005
	Staphylothermus marinus		100	5			Antranikian et al., 2005
	Sulfolobus solfataricus	240					Antranikian et al., 2005
	Thermococcus aggregans		95	6.5			Antranikian et al., 2005
	Thermococcus celer		90	5.5			Antranikian et al., 2005
	Thermococcus hydrothermalis	49	75–85	5–5.5	4 h at 90°C		Antranikian et al., 2005
	Thermococcus profundus (amyS)	43	80	5–6	3 h at 80°C		Antranikian et al., 2005
	Thermococcus kodakaraensis	49.5	90	6.5	24 h at 70°C		Antranikian et al., 2005
Glucoamylase (EC 3.2.1.3)	Methanocaldococcus jannaschii	140	80	6.5	20 h at 90°C		Uotsu-Tomita et al., 2001
	Picrophilus oshimae (extracellular)	133	90	2	24 h at 90°C		Serour and Antranikian, 2002
	Picrophilus torridus (extracellular)	312	90	2	4 h at 55°C		Serour and Antranikian, 2002
	Picrophilus torridus (intracellular)	141	50	5	24 h at 90°C		Antranikian, unpublished
	Thermoplasma acidophilum (extracellular)	140	90	2	40 h at 60°C		Serour and Antranikian, 2002
	Thermoplasma acidophilum (intracellular)	250	75	5			Antranikian, unpublished
	Sulfolobus solfataricus		90	5.5–6			Kim et al., 2004
α-Glucosidase (EC 3.2.1.20)	Ferroplasma acidiphilum	57	55–60	2.4–3.5			Ferrer et al., 2005
	Pyrococcus furiosus	125	115	5.5			Eichler, 2001
	Pyrococcus woesei	90	110	5–5.5			Antranikian et al., 2005b
	Sulfolobus shibatae	313	85	5.5	6 h at 80°C		Leveque et al., 2000b
	Sulfolobus solfataricus	80	105	4.5	39 h at 85°C		Rolfsmeier et al., 1998
	Thermococcus hydrothermalis	57	120	5.5			Antranikian et al., 2005
	Thermococcus sp. AN1	63	120	7			Eichler, 2001
	Thermococcus zilligii		75	7			Leveque et al., 2000a

Continued on following page

Table 1. *Continued*

Enzymes	Strain[a]	MW (kDa)	T_{opt} (°C)	pH_{opt}	Thermostability (half-life)	Possible applications	References
Pullulanase (EC 3.2.1.41)	*Desulfurococcus mucosus*	74	85	5	0.8 h at 85°C		Antranikian et al., 2005
	Pyrococcus furiosus (pullulanase)	90	105	6			Antranikian et al., 2005
	Pyrococcus furiosus (pullulan-hydrolase)	77	90	5	2 h at 95°C		Yang et al., 2004
	Pyrococcus woesei (struct.)	90	100	6			Antranikian et al., 2005
	Pyrodictium abyssi		100	9			Niehaus et al., 1999
	Thermococcus aggregans	80	95	6.5	2.5 h at 100°C		Niehaus et al., 2000
	Thermococcus celer		90	5.5			Niehaus et al., 1999
	Thermococcus hydrothermalis	128	95	5.5			Antranikian et al., 2005
	Thermococcus litoralis	119	98	5.5			Antranikian et al., 2005
	Thermococcus profundus (amyL)	43	90	5.5	6.7 h at 90°C		Antranikian et al., 2005
	Thermococcus sp. ST489		80–95	8			Leveque et al., 2000b
Cyclodextrin glucosyltransferase (EC 2.4.1.19)	*Thermococcus kodakaraensis*	77	80	5.5–6	0.33 h at 100°C		Rashid et al., 2002
	Thermococcus sp.	83	90–110	5–5.5	0.66 h at 110°C		Tachibana et al., 1999
Amylomaltase (EC 2.4.1.25)	*Pyrobaculum aerophilum*		95	6.7			Kaper et al., 2005
	Thermococcus aggregans	80	100	6.8	20 h at 90°C		Antranikian, unpublished
	Thermococcus litoralis (struct.)	87	90	6			Imamura et al., 2001

[a] (struct.), the protein has been crystallized and the three-dimensional structure determined.

Methanocaldococcus, Pyrococcus, and *Thermococcus* (Egorova and Antranikian, 2005). The thermostability of the enzymes often enhances by the presence of divalent metal ions. The optimal temperatures for the activity of these enzymes range between 80 and 100°C. The high thermostability of the pyrococcal extracellular α-amylase (thermal activity even at 130°C and after autoclaving for 4 h at 120°C) and α-amylase from *Methanocaldococcus jannaschii* (temperature optimum 120°C, half-life of 50 h at 100°C) makes these enzymes interesting candidates for industrial application (Kim et al., 2001). Further investigations have shown that an extreme marine hyperthermophilic archaeon *Pyrodictium abyssi* can grow on various polysaccharides and also secretes a heat stable α-amylase (Andrade et al., 2001). α-Amylases with lower thermostability have been isolated from the archaea *Thermococcus profundus, Thermococcus kodakaraensis* and the extreme thermophilic bacteria *Dictyoglomus thermophilum, Thermotoga maritima, Thermus filiformis,* and *Rhodothermus marinus* (Table 2). In general the bacterial enzymes are less thermoactive than the archaeal enzymes. Similar to the amylase from *Bacillus licheniformis*, which is commonly used in liquefaction of starch in the industry, most of the enzymes from extreme thermophilic bacteria require calcium for activity. The use of α-amylases in detergents for medium-temperature laundering demands enzymes with high stability and activity at alkaline conditions. The enzyme from the thermoalkaliphile *Anaerobranca gottschalkii* is optimally active at pH 8.0 and has high transglycosylation activity on maltooligosaccharides. Interestingly, the enzyme also exhibits significant β-cyclodextrin glycosyltransferase (CGTase, EC 2.4.1.19) activity (Ballschmiter et al., 2005). On the other hand, an acid-stable amylase was purified and characterized from the thermoacidophilic bacterium *Alyciclobacillus acidocaldarius* (Matzke et al., 1997). This enzyme with a molecular mass of 160 kDa exhibits highest activity at pH 3.0 and 75°C.

Halophilic archaea of the genera *Haloarcula, Haloferax, Halothermothrix,* and *Natronococcus* sp. Ah-36 are also a good source of amylolytic enzymes (see Table 1). Amylases from halophiles could be used in processes containing high salt concentrations and hydrophobic organic solvents. The α-amylase from *Haloarcula* sp. S-1 exhibits its maximal activity at 4.3 M NaCl and is stable in benzene, toluene, and chloroform (Fukushima et al., 2005). The α-amylase from *Haloferax mediterranei* is highly stable, having half-life of 240 days at pH 10 (Perez-Pomares et al., 2003). A maltotriose-forming amylase from *Natronococcus* sp. Ah-36 with maximal activity at pH 8.7, 55°C and 2.5 M NaCl was expressed in the halophilic archaeon *Haloferax volcanii* (Kobayashi et al., 1994; Kobayashi et al., 1992).

β-Amylases

β-Amylase (1,4-alpha-D-glucan maltohydrolase; EC 3.2.1.2) hydrolyzes 1,4-alpha-D-glucosidic linkages in polysaccharides removing successive maltose units from the non-reducing ends of the chains. For the efficient production of maltose syrups an additional debranching enzyme is needed. To date only few thermoactive bacterial β-amylases are known (see Table 2). The enzyme from *Thermotoga maritima* is active in the absence of calcium at low pH (pH 4.3 to 5.5) and high temperature (95°C) (Vieille and Zeikus, 2001). The enzyme from *Thermoanaerobacter thermosulfurigenes* retains 70% of its activity at pH 4.0 and 70°C (Vieille and Zeikus, 2001).

Glucoamylases

Unlike α-amylase, the production of glucoamylase seems to be rare in archaea and bacteria (see Tables 1 and 2). Glucoamylases (EC 3.2.1.3) hydrolyze terminal α-1,4-linked-D-glucose residues successively from non-reducing ends of the chains, releasing β-D-glucose. An ideal catalyst for starch liquefaction should be optimally active at 100°C and pH 4.0 to 5.0 without the requirement of calcium ions for the stabilization of the enzyme. Recently, it has been shown also that the thermoacidophilic archaea *Thermoplasma acidophilum, Picrophilus torridus,* and *Picrophilus oshimae* produce heat- and acid-stable extracellular glucoamylases. The purified archaeal glucoamylases are optimally active at pH 2 and 90°C. Catalytic activity is still detectable at pH 0.5 and 100°C (Serour and Antranikian, 2002). These enzymes are more thermostable than already described glucoamylases from bacteria, yeasts, and fungi. They are of interest for application in the beverage industry. However, the lack of suitable genetic methods for thermoacidophiles have precluded structural studies aimed to discover their adaptation at very low pH. Recently, the gene encoding a putative glucoamylase from *Sulfolobus solfataricus* was cloned and expressed in *E. coli,* and the properties of the recombinant protein were examined in relation to the glucose production process (Kim et al., 2004). This recombinant enzyme is extremely thermostable, with an optimal temperature at 90°C; however, it is most active in a slight acidic pH range of 5.5 to 6.0. The tetrameric enzyme liberates β-D-glucose from maltotriose, and the substrate preference for maltotriose distinguishes this enzyme

Table 2. Starch-processing enzymes produced by extremophilic bacteria

Enzymes	Strain[a]	MW (kDa)	T_{opt} (°C)	pH_{opt}	Thermostability (half-life)	Possible applications	References
α-Amylase (EC 3.2.1.1)	Alicyclobacillus acidocaldarius	136	75	3	48 h at 70°C	Bread and baking industry, starch liquefaction and saccharification, production of glucose, textile desizing, paper industry, synthesis of oligosaccharides, detergent application, and gelling and thickening in food industry	Matzke et al., 1997
	Anaerobranca gottschalkii (AmyA)	58	70	8	0.2 h at 70°C		Ballschmiter et al., 2005
	Anaerobranca gottschalkii (AmyB)	108	55	6–6.5			Ballschmiter et al., 2005
	Dictyoglomus thermophilum	75	90	5.5			Antranikian et al., 2005
	Rhodothermus marinus		85	6.5			Gomes et al., 2003
	Thermotoga maritima (AmyA)	61	85	7	4 h at 80°C		Liebl et al., 1997
	Thermotoga maritima (AmyC) (struct.)		90	8.5			Ballschmiter et al., 2006
	Thermus filiformis	60	95	6	8 h at 85°C		Egas et al., 1998
β-Amylase (EC 3.2.1.2)	Thermoanaerobacter thermosulfurigenes	180	70	6	2 h at 80°C		Vieille and Zeikus, 2001
	Thermotoga maritima		95	4–5.5	0.5 h at 90°C		Vieille and Zeikus, 2001
Glucoamylase (EC 3.2.1.3)	Bacillus sp.		70	5	3.8 h at 70°C		Gill and Kaur, 2004
	Clostridium thermohydrosulfuricum		75	4–6	1 h at 85°C		Hyun et al., 1985
	Clostridium thermosaccharolyticum	75	70	5	>6 h at 70°C		Specka et al., 1991
	Thermoactinomyces vulgaris	75	60	6.8	0.5 h at 60°C		Ichikawa et al., 2004
	Thermoanaerobacter thermosaccharolyticum	75	65	4–5.5	8 h at 65°C		Ganghofner et al., 1998
α-Glucosidase (EC 3.2.1.20)	Thermoanaerobacter ethanolicus		75	5–5.5			Bertoldo and Antranikian, 2002
	Thermoanaerobacter thermosaccharolyticum	60	65	5.5			Ganghofner et al., 1998
	Thermotoga maritima	110	90	7.5	48 h at 50°C		Raasch et al., 2000
Pullulanase (EC 3.2.1.41)	Anaerobranca gottschalkii	98	70	8	22 h at 70°C		Bertoldo et al., 2004
	Caldicellulosiruptor saccharolyticus	96	85	6			Albertson et al., 1997
	Clostridium thermohydrosulfuricum		85	5.6–6	1 h at 90°C		Hyun et al., 1985
	Fervidobacterium pennivorans	190	80	7	2 h at 80°C		Bertoldo et al., 1999
	Rhodothermus marinus		80				Gomes et al., 2003
	Thermoanaerobacter ethanolicus	109	90	5–5.5			Lin and Leu, 2002
	Thermoanaerobacter thermosaccharolyticum	150	65	6	0.5 h at 65°C		Ganghofner et al., 1998

Enzyme	Organism				Reference
	Thermoanaerobacter thermosulfurigenes	100	75	6	2 h at 80°C	Ramesh et al., 2001
	Thermotoga maritima	93	90	6.4	3.5 h at 90°C	Bertoldo and Antranikian, 2002
	Thermus aquaticus	83	80–95	5.5	4.5 h at 95°C	Bertoldo and Antranikian, 2002
	Thermus caldophilus	65	75		72 h at 78°C	Bertoldo and Antranikian, 2002
	Thermus thermophilus	80	70			Tomiyasu et al., 2001
Cyclodextrin glucosyltransferase (EC 2.4.1.19)	Anaerobranca gottschalkii	78	65–70	6–9		Thiemann et al., 2004
	Thermoanaerobacter sp.		90–95	6		Alcalde et al., 2001
	Thermoanaerobacterium thermosulfurigenes	68	80–85	4.5–7	0.5 h at 100°C	Thiemann et al., 2004
Cyclomaltodextrinase (EC 3.2.1.54)	Alicyclobacillus acidocaldarius	66		5.5		Matzke et al., 2000
	Clostridium thermohydrosulfuricum	66	65	6		Podkovyrov and Zeikus, 1992
Amylomaltase (EC 2.4.1.25)	Dictyoglomus thermophilum	75	80	6		Nakajima et al., 2004
	Thermotoga maritima	53	55–80	6	2.5 h at 90°C	Liebl et al., 1992
	Thermus aquaticus	57	75	5.5–6	24 h at 80°C	Terada et al., 1999
Branching enzyme (EC 2.4.1.18)	Aquifex aeolicus	74	75	7.5–8	>0.5 h at 70°C	Thiemann et al., 2006
	Anaerobranca gottschalkii	72	50	7		Thiemann et al., 2006
	Geobacillus stearothermophilus	78	50	7.5		Takata et al., 1996
	Rhodothermus obamensis	72	65	6–6.5	16 h at 80°C	Shinohara et al., 2001

[a] (struct.), the protein has been crystallized and the three-dimensional structure determined.

from fungal glucoamylases. Genome analysis of other thermoacidophiles revealed further putative glucoamylases, which were cloned and expressed in *E. coli*. Thus, the recombinant intracellular glucoamylase from *Methanocaldococcus jannashii* is active at pH 6.5 and 80°C (Uotsu-Tomita et al., 2001). In our laboratory, the intracellular glucoamylases from the extreme thermoacidophiles *Picrophilus torridus* and *Thermoplasma acidophilum* have been recently cloned and expressed in *E. coli*; the recombinant enzymes are optimally active at 50 to 75°C and pH 5. Thermophilic anaerobic bacteria such as *Clostridium thermohydrosulfuricum*, *Clostridium thermosaccharolyticum*, and *Thermoanaerobacterium thermosaccharolyticum* produce glucoamylases, which have been purified and characterized (Specka et al., 1991; Ganghofner et al., 1998).

α-Glucosidases

These enzymes (EC 3.2.1.20) attack the α-1,4 linkages of oligosaccharides that are produced by the action of other amylolytic enzymes. Unlike glucoamylase, α-glucosidase prefers smaller oligosaccharides, e.g., maltose, maltotriose and liberates glucose with an α-anomeric configuration. An intracellular and an extracellular α-glucosidases have been purified and characterized from archaea, belonging to the genera *Pyrococcus*, *Sulfolobus*, and *Thermococcus* (see Table 1) (Antranikian et al., 2005). The enzymes exhibit optimal activity at pH 4.5 to 7.0 over a temperature range of 105 to 120°C. An α-glucosidase gene and flanking sequences from *Sulfolobus solfataricus* were cloned in *E. coli* and the product was characterized (Rolfsmeier et al., 1998). The purified recombinant enzyme with a calculated size of 80.5 kDa hydrolyzes *p*-nitrophenyl-D-glucopyranoside. At pH 4.5, it exhibits a pH optimum for maltose hydrolysis. Unlike maltose hydrolysis, glycogen was hydrolyzed efficiently at the intracellular pH of the organism (pH 5.5). The recombinant α-glucosidase exhibits greater thermostability than the native enzyme, with a half-life of 39 h at 85°C at a pH of 6.0. Recently, a novel α-glucosidase from the extreme acidophilic archaeon *Ferroplasma acidiphilum* was characterized (Ferrer et al., 2005). Maximal activity was detected at pH 2.4 to 3.5 (>70% activity at pH 1.5). Iron was found to be essential for enzymatic activity and His30 was shown to be responsible for iron binding. Less thermostable α-glucosidases were detected in the bacteria *Thermoanaerobacter ethanolicus* (Bertoldo and Antranikian, 2002), *Thermoanaerobacter thermosaccharolyticum* (Ganghofner et al., 1998) and *Thermotoga maritima* (Raasch et al., 2000).

Pullulanases

Enzymes capable of hydrolyzing α-1,6 glucosidic bonds in pullulan and branched oligosaccharides are defined as pullulanases. On the basis of substrate specificity and product formation, pullulanases have been classified into three groups: pullulanase type I, pullulanase type II and pullulan hydrolases (type I, II and III). Pullulanase type I (EC 3.2.1.41) specifically hydrolyzes the α-1,6-linkages in pullulan as well as in branched oligosaccharides (debranching enzyme), and its degradation products are maltotriose and linear oligosaccharides, respectively. Pullulanase type I is unable to attack α-1,4-linkages in α-glucans. Pullulanase type II (amylopullulanase) attacks α-1,6-glycosidic linkages in pullulan and branched polysaccharides. Unlike pullulanase type I, this enzyme also attacks α-1,4-linkages in branched and linear oligosaccharides and is able to fully convert polysaccharides (e.g., amylopectin) to small sugars (e.g., glucose, maltose, maltotriose) in the absence of amylases.

Thermostable and thermoactive pullulanases type II from the archaea *Desulfurococcus mucosus*, *Pyrococcus furiosus*, *Pyrococcus woesei*, and *Thermococcus hydrothermalis* have been reported to have temperature optima between 85 and 105°C (see Table 1), as well as remarkable thermostability even in the absence of substrate and calcium ions. In the presence of calcium ions, pullulanase activity was also detected at 130 to 140°C (Brown and Kelly, 1993). Site-directed mutagenesis performed on pullulanase from *Thermococcus hydrothermalis* reveals that the residues E291 and D394 are indeed critical for the pullulanolytic and amylolytic activities of the pullulanase (Zona et al., 2004). The crucial role of E291 as the catalytic nucleophile has also been confirmed for the pullulanase from *Pyrococcus furiosus* (Kang et al., 2004). The apparent catalytic efficiencies (K_{cat}/K_m) of mutants E291Q and D394N on pullulan were 123 and 24 times lower than that of the native enzyme. The hydrolytic patterns for pullulan and starch were the same, while the hydrolysis rates differed as reported. Therefore, these data strongly suggest that the bifunctionality of the pullulanase type II is determined by a single catalytic center. Due to the dual specificity of pullulanases type II to degrade both α-1,4- and α-1,6-glucosidic linkages they cannot be used as debranching enzymes in maltose and glucose syrup production. The archaeal enzymes are promising candidates to optimize starch liquefaction for the production of maltose, maltotriose, and maltotetraose syrups.

Interestingly, pullulanase type I has not been identified in archaea so far, whereas the enzyme has been characterized in several thermophilic bacteria belonging to the genera *Fervidobacterium, Thermoanaerobacter, Thermotoga* and *Thermus* (see Table 2). The aerobic thermophilic bacterium *Thermus caldophilus* GK-24 produces a thermostable pullulanase of type I when grown on starch (Bertoldo and Antranikian, 2002). The pullulanase is optimally active at 75°C and pH 5.5, is thermostable up to 90°C, and does not require calcium ions for either activity or stability. The first debranching enzyme (pullulanase type I) from anaerobic thermophilic bacteria was found in *Fervidobacterium pennivorans* (Bertoldo et al., 1999). This enzyme, cloned and expressed in *E. coli*, forms long chain linear polysaccharides from amylopectin. Similar enzyme was also characterized from *Thermotoga maritima* (Kriegshauser and Liebl, 2000). All known bacterial debranching enzymes are active in the acidic or neutral pH range. Until very recently, no reports have been presented on the ability of thermoalkaliphiles to produce heat and alkaline stable pullulanase type I. After sequencing the whole genome of the thermoalkaliphile *Anaerobranca gottschalkii*, a pullulanase-encoding gene was cloned and expressed in *E. coli*; this enzyme is optimally active at pH 8.0 and 70°C (Bertoldo et al., 2004).

The third class of pullulan-hydrolyzing enzymes includes pullulan hydrolases type I, II, and III. Pullulan hydrolases type I and II are active toward α-1,4 linkages of amylose, starch, pullulan, but are unable to hydrolyze α-1,6 linkages. An exception is pullulan hydrolase type III. This enzyme attacks α-1,4 as well as α-1,6 linkages of pullulan. The enzymes from *Pyrococcus furiosus* (AmyL), *Thermococcus aggregans* and *Thermococcus profundus* (see Table 1) exhibit maximal activity at 90°C and pH 5.5 to 6.5 and are stable for several hours at 95 to 100°C. In addition, the pullulan hydrolase from *Pyrococcus furiosus* degrades β–cyclodextrin.

CGTases

Cyclodextrin glucosyltransferase (CGTase; EC 2.4.1.19) converts α-glucans into cyclodextrins, which are composed of 6 (α), 7 (β) or 8 (γ) α-1,4 linked glucose molecules. The internal cavities of cyclodextrins are hydrophobic and they can encapsulate hydrophobic molecules. Thermostable CGTases are generally found in bacteria and was recently discovered in archaea. The archaeal enzyme found in *Thermococcus* sp. is optimally active at 90 to 110°C (see Table 1). Incubation of the enzyme with 30% corn-starch (wt/vol) for 24 h at 96°C and pH 4.5 resulted in the production of α-cyclodextrin (69%), β-cyclodextrin (20%), and γ-cyclodextrin (11%) (Tachibana et al., 1999). The major cyclodextrin formed by the action of the CGTase from *Thermococcus kodakaraensis* is β-cyclodextrin (Rashid et al., 2002). Bacterial CGTases, isolated mostly from the species of genus *Bacillus*, are already used in industry for the production of cyclodextrins (Horikoshi, 1999). The use of more thermoactive CGTases will allow the development of a single step process at temperatures above 90°C (Schiraldi et al., 2002). The CGTases active at the temperatures of 80 to 90°C are produced by some anaerobic thermophilic bacteria (see Table 2). After sequencing of the genome of the anaerobic bacterium *Anaerobranca gottschalkii* the CGTase gene was cloned and expressed (Thiemann et al., 2004).

Branching enzymes

Branching enzyme (EC 2.4.1.18) catalyzes the formation of α-1,6 branching points from linear oligosaccharides and polysaccharides, determining the final structures and properties of amylopectin and glycogen. It was shown that the branching enzymes increase the solubility and stability of starch solutions and shelf life and loaf volume of baked goods (Thiemann et al., 2006). The most thermoactive branching enzyme was isolated from the thermophilic bacterium *Aquifex aeolicus* (>90% activity after 10-min. treatment at 90°C). The enzyme is also able to produce large cyclic glucans using amylopectin as substrate (Takata et al., 2003). The branching activity of the enzyme from *Rhodothermus obamensis* is higher toward amylose than amylopectin (Shinohara et al., 2001). The branching enzyme from the thermoalkaliphilic bacterium *Anaerobranca gottschalkii* displays at 50°C high transglycosylation activity with extremely low hydrolytic activity (Thiemann et al., 2006).

Amylomaltases

Amylomaltases (EC 2.4.1.25, 4-α-glucanotransferase) catalyze the transfer of a segment of an α-1,4-D-glucan to a new 4-position of an acceptor, which may be glucose or another α-1,4-D-glucan. Acting upon starch, amylomaltases can produce products of commercial interest, such as cycloamylose, a thermoreversible starch gel, which can be used as a substitute for gelatin. In combination with α-amylase the amylomaltase produces syrups of isomalto-oligosaccharides with reduced sweetness and low viscosity (Kaper et al., 2004). The thermostable amylomaltase from the archaeon *Pyrobaculum aerophilum* produces a thermoreversible starch product with gelatin-like properties (Kaper et al., 2005). The enzyme from *Thermococcus litoralis* produces linear α-1,4-glucans and a cycloamylose (cyclic α-1,4-glucan) with a high degree (up to hundreds) of polymerization (Imamura et al., 1999; 2001, Jeon et al., 1997). Recently, a heat-stable amylomaltase was characterized from the archaeon *Thermococcus aggregans*. The recombinant enzyme is stable at 90°C for more than 22 h (Antranikian, unpublished results). The combined use of the amylomaltase from the bacterium *Thermotoga maritima* with a maltogenic amylase resulted in the production of isomalto-oligosaccharides from starch (Kaper et al., 2004). Cyclic glucans can be produced using the thermostable bacterial amylomaltase from *Thermus aquaticus* (Terada et al., 1999; Fujii et al., 2005).

The finding of novel thermostable starch-modifying enzymes will be a valuable contribution to the starch-processing industry. By using robust enzymes from thermophiles, innovative and environmentally friendly processes can be developed, aiming at the formation of products of high added value for the food and pharmaceutical industries. At elevated temperatures starch is more soluble (30 to 35% wt/vol) and the risk of contamination is reduced. This is of advantage when starch will be converted to high glucose and fructose syrups. The application of thermostable enzymes that are active and stable above 100°C and at acidic pH values can simplify the complicated multistage starch-conversion process. The use of the extremely thermostable amylolytic enzymes can lead to other valuable products, which include innovative starch-based materials with gelatin-like characteristics and defined linear dextrins that can be used as fat substitutes, texturizers, aroma stabilizers, and prebiotics. CGTases are used for the production of cyclodextrins that can be used as a gelling, thickening or stabilizing agent in jelly desserts, dressing, confectionery, and dairy and meat products. Due to the ability of cyclodextrins to form inclusion complexes with a variety of organic molecules, cyclodextrins improve the solubility of hydrophobic compounds in aqueous solution. This is of interest for the pharmaceutical and cosmetic industries. Cyclodextrin production is a multistage process in which starch is first liquefied by a heat-stable amylase and in the second step a less-thermostable CGTase from *Bacillus* sp. is used. The application of heat-stable CGTase from the *Thermococcus* species in jet cooking, where temperatures up to 105°C could be achieved, will allow liquefaction and cyclization to take place in one step (Biwer et al., 2002).

Another promising application of an archaeal enzyme is the production of a disaccharide trehalose, a stabilizer of enzymes, antibodies, vaccines and hormones. The use of thermoactive enzymes in the process would eliminate problems associated with viscosity and sterility. The process was developed to produce trehalose from dextrins using *Sulfolobus* enzymes at 75°C in a continuous bioreactor, with a final conversion of 90% (Schiraldi et al., 2002). Recently, the trehalose biosynthetic pathway was identified in *Sulfolobales* and the responsible enzymes were cloned and expressed in *E. coli* (de Pascale et al., 2002; Fang et al., 2004; Gueguen et al., 2001a).

Cellulose-Degrading Enzymes

Cellulose is the most abundant organic biopolymer in nature, since it is the structural polysaccharide of the cell wall in the plant kingdom. It consists of glucose units linked by β–1,4-glycosidic bonds with a polymerization grade of up to 15,000 glucose units in a linear mode. The minimal molecular weight of cellulose from different sources has been estimated to vary from about 50,000 to 2,500,000 in different species. Although cellulose has a high affinity to water, it is completely insoluble in it. Natural cellulose compounds are structurally heterogeneous and have both amorphous and highly ordered crystalline regions. The degree of crystallinity depends on the source of the cellulose and the higher crystalline regions are more resistant to enzymatic hydrolysis. Cellulose can be hydrolyzed into glucose by the synergistic action of different enzymes: endoglucanase (cellulase), exoglucanase (cellobiohydrolase), and β-glucosidase (cellobiase). Endoglucanase (E.C. 3.2.1.4) hydrolyzes cellulose in a random manner as endohydrolase, producing various oligosaccharides, cellobiose, and glucose. Exoglucanases (EC 3.2.1.91) hydrolyze β-1,4 D-glycosidic linkages in cellulose and cellotetraose, releasing cellobiose from the non-reducing end of the chain. β-glucosidases (EC 3.2.1.21) catalyze the hydrolysis of terminal, non-reducing β-D-glucose residues releasing β-D-glucose.

Several cellulose-degrading enzymes from various thermophilic organisms have been investigated (Tables 3 and 4). Thermostable endoglucanases, which degrade β-1,4 or β-1,3 linkages of β-glucans and cellulose, have been identified in archaea *Pyrococcus furiosus*, *Pyrococcus horikoshii*, and *Sulfolobus solfataricus* (Table 3) (Egorova and Antranikian, 2005). The purified recombinant endoglucanase from *Pyrococcus furiosus* is active at 100°C and hydrolyzes β-1,4 but not β-1,3 glycosidic linkages with the highest specific activity on cellopentaose and cellohexaose (Bauer et al., 1999). Another thermoactive glucanase (laminarinase) (T_{opt} 100°C) from this strain catalyzes the hydrolysis of mixed-linked oligosaccharides with both β-1,4 and β-1,3 specificities (Gueguen et al., 1997). The E170A mutant of the enzyme is additionally

Table 3. Cellulolytic, hemicellulolytic, chitinolytic, and proteolytic enzymes from archaea

Enzymes	Strain[a]	MW (kDa)	T_{opt} (°C)	pH_{opt}	Thermostability (half-life)	Possible applications	References
Cellulose-degrading enzymes							
β-Glucosidase (EC 3.2.1.21)	Pyrococcus furiosus	232	102	5	13 h at 110°C	Color brightening, color extraction of juice, saccharification of agricultural and industrial wastes, animal feed, biopolishing of cotton products, bioethanol, synthesis of sugars, and optically pure heterosaccharides	Lebbink et al., 2001
	Pyrococcus horikoshii		>100	6	15 h at 90°C		Matsui et al., 2002
	Sulfolobus acidocaldarius	224		7–8			Antranikian et al., 2005
	Sulfolobus shibatae						Antranikian et al., 2005
	Sulfolobus solfataricus (struct.)	240	95	6.5	15 h at 85°C		Antranikian et al., 2005
	Thermosphaera aggregans (struct.)				>130 h at 80°C		Chi et al., 1999
Endoglucanase (EC 3.2.1.4)	Pyrococcus furiosus (EglA)	36	100	6	40 h at 95°C		Bauer et al., 1999
	Pyrococcus furiosus (LamA)	31	100	6–6.5	19 h at 100°C		van Lieshout et al., 2004
	Pyrococcus horikoshii	43–52	97	5.6	>3 h at 97°C		Kashima et al., 2005
	Sulfolobus solfataricus MT4 (struct.)	40	65	6			Limauro et al., 2001
	Sulfolobus solfataricus P2	37	80	1.8	8 h at 80°C and pH 1.8		Huang et al., 2005
Xylan-degrading enzymes							
Endo-1,4-β-xylanase (EC 3.2.1.8)	Halorhabdus utahensis	45–67	55–70	6.5	0.3 h at 60°C	Paper bleaching	Waino and Ingvorsen, 2003
	Pyrodictium abyssi		110	5.5			Antranikian et al., 2005
	Sulfolobus solfataricus	57	100	7	0.8 h at 90°C		Cannio et al., 2004
	Thermococcus zilligii AN1	95		6	4 h at 95°C		Antranikian et al., 2005
1,4-β-xylosidase (EC 3.2.1.37)	Halorhabdus utahensis	45–67	65	7.6	0.3 h at 60°C		Waino and Ingvorsen, 2003
β-D-Mannosidase (EC 3.2.1.25)	Pyrococcus furiosus	240	105	7.4	60 h at 90°C		Bauer et al., 1996
Xylose dehydrogenase (EC 1.1.1.175)	Haloarcula marismortui	175	50	8.3			Johnsen and Schonheit, 2004
Chitin-degrading enzymes							
Endochitinase (EC 3.2.1.14)	Pyrococcus furiosus (ChiA)	40	90–95	6	1 h at 120°C	Utilization of biomass of marine environment	Gao et al., 2003
	Thermococcus chitonophagus (Chi70)	70	70	7			Andronopoulou and Vorgias, 2004b
	Thermococcus kodakaraensis (ChiA)	134	85	5			Tanaka et al., 1999
Exochitinase (EC 3.2.1.52)	Thermococcus chitonophagus (Chi50)	50	80	6			Andronopoulou and Vorgias, 2004b
	Pyrococcus furiosus (ChiB)	55	90–95	6			Gao et al., 2003

Continued on following page

Table 3. Continued

Enzymes	Strain[a]	MW (kDa)	T_{opt} (°C)	pH_{opt}	Thermostability (half-life)	Possible applications	References
Chitobiase (EC 3.2.1.30)	*Thermococcus chitonophagus* (Chi90)	90					Andronopoulou and Vorgias, 2004b
	Thermococcus kodakaraensis (GlmA)	193	80	6			Tanaka et al., 2003
Diacetylchitobiose Deacetylase	*Thermococcus kodakaraensis* (Tk-Dac)	160	75	8.5			Tanaka et al., 2004
Proteolytic enzymes (EC 3.4.21.x)							
Serine protease	*Aeropyrum pernix*	34	90	8–9	1 h at 100°C	Detergents, baking, brewing, amino acids production	Catara et al., 2003
	Desulfurococcus mucosus	43–54	95	7.5	4.3 h at 95°C		Ward et al., 2002
	Halobacterium halobium	66					Ryu et al., 1994
	Haloferax mediterranei		37				Antranikian et al., 2005
	Natrialba asiatica		37				Antranikian et al., 2005
	Natrialba magadii		37	8–10			Oren, 2002
	Natronococcus occultus		60				Horikoshi, 1999
	Natronomonas pharaonis		60	10			Stan-Lotter et al., 1999
	Pyrobaculum aerophilum	401	>100	7–9			Ward et al., 2002
	Pyrococcus abyssi	60	95	9			Ward et al., 2002
	Pyrococcus furiosus	150	115	6–9	0.33 h at 105°C		Antranikian et al., 2005
	Thermococcus aggregans		90	7			Eichler, 2001
	Thermococcus celer		95	7.5			Eichler, 2001
	Thermococcus kodakaraensis	44	80	9.5			Ward et al., 2002
	Thermococcus litoralis		95	9.5			Eichler, 2001
	Thermococcus stetteri	142	85	8.5–9	22 h at 90°C		Ward et al., 2002
	Thermoplasma acidophilum (struct.)	120	65	8.5			Ward et al., 2002
	Staphylothermus marinus	150	90	9			Antranikian et al., 2005
	Sulfolobus acidocaldarius	46–51	90	2			Ward et al., 2002
	Sulfolobus solfataricus (struct.)	118	>90	6.5–8			Ward et al., 2002
Thiol protease	*Thermococcus kodakaraensis*	45	110	7	1 h at 100°C		Morikawa and Imanaka, 2001
Acidic protease	*Sulfolobus acidocaldarius*	46–51	90	2			Ward et al., 2002
Metalloprotease	*Aeropyrum pernix*	52	100	6–8			Ward et al., 2002
	Pyrococcus furiosus	128	100	6.5			Ward et al., 2002
	Pyrococcus furiosus	79	75	7			Ward et al., 2002
	Pyrococcus horikoshii OT3	95	>95	7.5			Ishikawa et al., 2001
	Sulfolobus solfataricus	320	75	6.7			Ward et al., 2002
	Sulfolobus solfataricus	170	85	5.5–7			Ward et al., 2002
	Thermococcus sp. NA1	37	100	5	>1.5 h at 80°C		Lee et al., 2006

[a] (struct.), the protein has been crystallized and the three-dimensional structure determined.

active as a glycosynthase, catalyzing the condensation of α–laminaribiosyl fluoride to different acceptors at pH 6.5 and 50°C (van Lieshout et al., 2004). Depending on the acceptor, the synthase generates either β-1,4 or β-1,3 linkage. Recently, a recombinant endoglucanase from *Pyrococcus horikoshii* was characterized (Ando et al., 2002). This enzyme is active even toward crystalline cellulose. Its activity was recently improved by protein engineering (Kashima et al., 2005). This enzyme is expected to be useful for industrial hydrolysis of cellulose, particularly in biopolishing of cotton products. Very recently, a novel acid-stable endoglucanase from *Sulfolobus solfataricus* P2 was cloned and expressed in *E. coli* (Huang et al., 2005). The purified recombinant enzyme with optimal activity at 80°C and pH 1.8 hydrolyzes carboxymethylcellulose and cello-oligomers, with cellobiose and cellotriose as main products. The presence of a signal peptide indicates that this secreted protein enables *Sulfolobus solfataricus* to utilize cellulose as carbon source. The unique enzyme could be applicable for the large-scale hydrolysis of cellulose under acidic conditions.

Unlike endoglucanases, several β-glucosidases have been detected in archaea. These enzymes have been detected in strains of the genera *Sulfolobus*, *Pyrococcus*, and *Thermosphaera*. The β-glucosidase from *Pyrococcus furiosus* is very stable with optimal activity at 103°C and it also exhibits a β-mannosidase activity (Nagatomo et al., 2005). The β-glucosidase from *S. solfataricus* MT4 is very resistant to various denaturants with activity up to 85°C. The gene for this β-glucosidase has been cloned and expressed in *E. coli* (Pouwels et al., 2000) and *Saccharomyces cerevisiae* (Morana et al., 1995). Using a mixture of both β-glucosidases from *Pyrococcus furiosus* and *Sulfolobus solfataricus*, an ultra-high temperature process for the enzymatic production of novel oligosaccharides from lactose was developed (Schiraldi and De Rosa, 2002). For the production of glucose from cellobiose a bioreactor system with immobilized recombinant β-glucosidase from *Sulfolobus solfataricus* was developed (Schiraldi and De Rosa, 2002). The system runs at a high flow rate and has a high degree of conversion, productivity, and operational stability. The thermoactive β-glucosidase from *Pyrococcus horikoshii* is active in organic solvents and it synthesizes a heterosaccharide with high optical purity (Egorova and Antranikian, 2005).

Bacteria belonging to the genera *Thermotoga*, *Thermobifida*, *Rhodothermus*, and *Clostridium* were found to be good cellulose degraders (see Table 4). Thermostable endoglucanases from *Thermotoga maritima* and *T. neapolitana* are rather small with a molecular mass of 27 kDa and are optimally active at 95 to 106°C and between pH 6.0 and 7.0 (Liebl et al., 1996). Cellulase and hemicellulase genes have been found to be clustered together on the genome of the thermophilic anaerobic bacterium *Caldocellum saccharolyticum*, which grows on cellulose and hemicellulose as sole carbon sources (Te'o et al., 1995). The gene for one of the cellulases was isolated and was found to consist of 1,751 amino acids. This is the largest cellulase gene sequenced to date. Another large cellulolytic enzyme with the ability to hydrolyze microcrystalline cellulose was isolated from the extremely thermophilic bacterium *Anaerocellum thermophilum* (Zverlov et al., 1998). This enzyme has an apparent molecular mass of 230 kDa and exhibits significant activity toward Avicel and is most active toward soluble substrates such as carboxymethylcellulose and β-glucan. Maximal activity was observed at pH 5 to 6 and 85 to 100°C. Endo-1,4-β-D-glucanase from *Acetivibrio cellulolyticus*, was imported into chloroplasts, and an active enzyme was recovered (Jindou et al., 2006). The thermophilic bacterium *Rhodothermus marinus* produces a thermostable endoglucanase, with a temperature optimum of around 80°C (Halldorsdottir et al., 1998). A 100-kDa protein with endoglucanase activity was purified from Triton X-100 extract of cells of the thermoacidophilic gram-positive bacterium *Alicyclobacillus acidocaldarius* (Eckert and Schneider, 2003). The enzyme exhibits activity toward carboxy-methyl-cellulose and oat spelt xylan with pH and temperature optima of 4 and 80°C, respectively. Remarkable stability was observed at pH values between 2 and 6 and 60% of activity was retained after incubation at 80°C for 1 h. Another glucanase purified from the same microorganism is less acid stable, having maximum activity at pH 5.5 (Eckert et al., 2002).

There is a great demand for robust cellulolytic enzymes for various applications such as bioethanol production, improvement of juice yield, and effective color extraction of juices. Other suitable applications of cellulases include the pre-treatment of cellulose biomass and forage crops to improve nutritional quality and digestibility. Furthermore, cellulases are useful tools for the saccharification of agricultural and industrial wastes and production of fine chemicals.

Xylan-Degrading Enzymes

Xylan is a heterogeneous molecule that constitutes the main polymeric compound of hemicellulose, a fraction of the plant cell wall, which is a major reservoir of fixed carbon in nature. The main chain of the heteropolymer is composed of xylose residues linked by β-1,4-glycosidic bonds. Approximately half of the xylose residues have substitution at O-2 or O-3 positions with acetyl-, arabinosyl and

Table 4. Cellulolytic, hemicellulolytic, chitinolytic, pectinolytic and proteolytic enzymes from bacteria

Enzymes	Strain[a]	MW (kDa)	T_{opt} (°C)	pH_{opt}	Thermostability (half-life)	Possible applications	References
Cellulose-degrading enzymes							
β-Glucosidase (EC 3.2.1.21)	*Microbispora bispora*	52	60	6.2	>48 h at 60°C	Polymer degradation, color brightening, color extraction of juice, saccharification of agricultural and industrial wastes, animal feed, biopolishing of cotton products, synthesis of sugars, and optically pure heterosaccharides	Wright et al., 1992
	Thermotoga maritima						Yip et al., 2004
	Thermotoga neapolitana	56	95	5–7	3.6 h at 100°C		Park et al., 2005
	Thermus caldophilus		80	6			Choi et al., 2003
	Thermus nonproteolyticus	49	90	5.6	48 h at 85°C		Wang et al., 2003
	Thermus sp. Z-1		70	7			Akiyama et al., 2001
	Thermus thermophilus	49	60	7	0.2 h at 90°C		Fourage and Colas, 2001
Endoglucanase (EC 3.2.1.4)	*Acidothermus cellulolyticus*						Jin et al., 2003
	Alicyclobacillus acidocaldarius (CelA)	58	70	5.5	0.5 h at 75°C		Eckert et al., 2002
	Alicyclobacillus acidocaldarius (CelB)	100	80	4	1 h at 80°C		Eckert and Schneider, 2003
	Anaerocellum thermophilum	230	95–100	5–6.0	0.8 h at 100°C		Zverlov et al., 1998
	Aquifex aeolicus	39	80	7	2 h at 100°C		Kim et al., 2000a
	Caldocellum saccharolyticum						Te'o et al., 1995
	Clostridium cellulovorans	79	40–50	5–6			Han et al., 2005
	Clostridium thermocellum	56		6			Zverlov et al., 2005
	Rhodothermus marinus (struct.)	30	85	7			Halldorsdottir et al., 1998
	Thermobifida fusca		77	8.2	>24 h at 60°C		Posta et al., 2004
	Thermotoga maritima	27–29	95	6–7.5	>6 h at 80°C		Liebl, 2001
	Thermotoga neapolitana	29–30	95–106	6–6.6	>2 h at 106°C		Bok et al., 1998
Cellobiohydrolase (EC 3.2.1.91)	*Clostridium stercorarium*	102	75	5	1–2 h at 75°C		Bronnenmeier et al., 1997
	Clostridium thermocellum	75	65	6.6	>16 h at 55°C		Zverlov et al., 2002a
	Thermomonospora fusca	60	55	7–8	0.5 h at 95°C		Zhang et al., 1995
	Thermotoga maritima		95	6–7.5	1.2 h at 108°C		Vieille and Zeikus, 2001
	Thermotoga sp. FjSS3-B.1.	36	105	7–8			Ruttersmith and Daniel, 1991
Xylan-degrading enzymes							
Endo-1,4-β-xylanase (EC 3.2.1.8)	*Acidobacterium capsulatum*	41	65	5		Paper bleaching	Beg et al., 2001
	Caldicellulosiruptor sp.	36	70	7	>12 h at 70°C		Morris et al., 1999
	Clostridium cellulovorans	57	60	5			Kosugi et al., 2002
	Clostridium thermocellum	110	70	6.5	0.6 h at 70°C		Kim et al., 2000
	Dictyoglomus thermophilum		70–85	7.5			Morris et al., 1998
	Rhodothermus marinus	48	80	8.5	1.6 h at 80°C		Ramchuran et al., 2005
	Thermoactinomyces sacchari		50				Antranikian et al., 2005
	Thermoactinomyces thalophilus		65	8.5–9	2 h at 65°C		Kohli et al., 2001

Enzyme	Organism	(col3)	(col4)	pH	Stability	Notes	Reference
	Thermoanaerobacter saccharolyticum	130	70	5.5	1 h at 75°C		Beg et al., 2001
	Thermoanaerobacterium sp.	180	80	6.2			Antranikian et al., 2005
	Thermobifida fusca	36	70	7	>3 h at 70°C		Irwin et al., 2003
	Thermotoga maritima (XynA)	40,120	92–105	5–6	22 h at 90°C		Wassenberg et al., 1997
	Thermotoga maritima (XynB)		87	6.5	8 h at 90°C		Reeves et al., 2000
	Thermotoga neapolitana	119	102	5.5	2 h at 100°C		Antranikian et al., 2005
	Thermus thermophilus		100	6			Lyon et al., 2000
Acetyl xylan esterase (EC 3.1.1.6)	Clostridium cellulovorans	33	50	6			Kosugi et al., 2002
	Clostridium stercorarium		65	8	0.1 h at 75°C		Donaghy et al., 2000
	Thermoanaerobacterium sp. JW/SL (A)	195	80	7	1 h at 75°C		Shao and Wiegel, 1995
	Thermoanaerobacterium sp. JW/SL (B)	106	84	7.5	1 h at 100°C		Shao and Wiegel, 1995
	Thermomonospora fusca	80		5.7			Bachmann and McCarthy, 1991
1,4-β-xylosidase (EC 3.2.1.37)	Thermoanaerobacter ethanolicus	165	82	5–5.5	0.25 h at 85°C		Shao and Wiegel, 1992
	Thermotoga maritima		90	6.1	0.5 h at 95°C		Xue and Shao, 2004
β-D-Mannosidase (EC 3.2.1.25)	Thermobifida fusca	94	53	7.2	30 h at 40°C		Beki et al., 2003
β-Mannanase (EC 3.2.1.78)	Caldocellulosyruptor sp. Rt8B.4.	40	80	5	>16 h at 80°C		Gibbs et al., 1996
	Dictyoglomus thermophilum	113	85	5.4	>1 h at 90°C		Gibbs et al., 1999
	Rhodothermus marinus	120	65–75	5.8	6 h at 80°C		Politz et al., 2000
	Thermoanaerobacterium polysaccharolyticum						Cann et al., 1999b
	Thermomonospora fusca	38	80				Hilge et al., 1998
α-L-Arabinofuranosidase (EC 3.2.1.55)	Rhodothermus marinus		85	5.5–7	8.3 h at 85°C		Gomes et al., 2000
	Thermomicrobia sp.	350	70	6	>1 h at 70°C		Birgisson et al., 2004
	Thermotoga maritima	332	90	7	2.7 h at 100°C		Miyazaki, 2005b
Chitin-degrading enzymes							
Endochitinase (EC 3.2.1.14)	Clostridium thermocellum	53	60	5–6.5	10 h at 60°C	Utilization of biomass of marine environment	Zverlov et al., 2002
	Streptomyces thermoviolaceus		80	9			Kawase et al., 2004
Exochitinase (EC 3.2.1.52)	Microbispora sp. V2	35	60	3	24 h at 50°C		Nawani et al., 2002
	Rhodothermus marinus	42	70	4.5–5	3 h at 90°C		Hobel et al., 2005
Pectin-degrading enzymes (EC 3.1.1.11)	Clostridium stercorarium	135	65	7	0.2 h at 70°C		Si Si Hla et al., 2005
	Clostridium thermosulfurigenes		75	5.5	0.5 h at 70°C		Vieille and Zeikus, 2001
	Thermoanaerobacter italicus (Pel A)	135	80	9	>1 h at 80°C		Kozianowski et al., 1997

Continued on following page

Table 4. *Continued*

Enzymes	Strain[a]	MW (kDa)	T_{opt} (°C)	pH_{opt}	Thermostability (half-life)	Possible applications	References
	Thermoanaerobacter italicus (Pel B)	251	80	9	>1 h at 80°C		Kozianowski et al., 1997
	Thermomonospora fusca	56	60	10.5			Stutzenberger, 1987
	Thermotoga maritima (PelA)	151	90	9	2 h at 95°C		Kluskens et al., 2003
	Thermotoga maritima (PelB)		80				Kluskens et al., 2005
	Thermotoga maritima (exo-PG)	51	95	6	>5 h at 90°C		Parisot et al., 2002
Proteolytic enzymes (EC 3.4.21)	*Aquifex aeolicus*	54	80	8–8.5	>0.5 h at 110°C	Detergents, baking, brewing, amino acids production	Khan et al., 2000
	Aquifex pyrophilus	43	85	7–9	6 h at 105°C		Ward et al., 2002
	Alicyclobacillus sendaiensis	37		3.9			Tsuruoka et al., 2003
	Fervidobacterium pennivorans (struct.)	58	80	10			Kluskens et al., 2002
	Fervidobacterium islandicum		80	8			Godde et al., 2005
	Fervidobacterium islandicum AW-1	>200	100	9	1.5 h at 100°C		Nam et al., 2002
	Geobacillus caldoproteolyticus		70–80	8–9.0	1 h at 80°C		Chen et al., 2004b
	Thermoanaerobacter yonseiensis		92	9			Jang et al., 2002
	Thermoanaerobacter keratinophilus		70	6.8			Riessen and Antranikian, 2001
	Thermoactinomyces vulgaris	279	60–65	7.5–9			Ward et al., 2002
	Thermoactinomyces sp.	31	85	11			Gupta and Ramnani, 2006
	Thermomonospora fusca	21.7	80	8.5	0.3 h at 85°C		Kim and Lei, 2005
	Thermotoga maritima	>669	90–93	6–9.0			Ward et al., 2002
	Thermus aquaticus	281	80	10			Ward et al., 2002
	Thermus sp.	178	70	8			Ward et al., 2002

[a] (struct.), the protein has been crystallized and the three-dimensional structure determined.

glucuronosyl groups. The complete degradation of xylan requires the action of several enzymes. The endo-β-1,4-xylanase (E.C.3.2.1.8) hydrolyzes β-1,4-xylosydic linkages in xylans, while β-1,4-xylosidase (EC 3.2.1.37) hydrolyzes β-1,4-xylans and xylobiose by removing the successive xylose residues from the non-reducing termini (Jeffries, 1996).

To date, only few extreme thermophilic microorganisms are able to grow on xylan and secrete thermoactive xylanolytic enzymes (see Tables 3 and 4). Among the thermophilic archaea, a xylanase from *Pyrodictium abyssi* has been characterized with an optimum temperature of 110°C—one of the highest values reported for a xylanase (Andrade et al., 2001). The crenarchaeon *Thermosphaera aggregans* was shown to grow on heat-treated, but not native, xylan (Eichler, 2001). The xylanase from *Thermococcus zilligii* AN1 is active up to 100°C, can attack different xylans, and is not active toward cellulose (Uhl and Daniel, 1999). Recently, a thermoactive endoxylanase from *Sulfolobus solfataricus* was purified and characterized (Cannio et al., 2004). The products of xylan hydrolysis were xylooligosaccharides and xylobiose. The production of endo-β-1,4-xylanase and β-xylosidase from the extremely halophilic archaeon, *Halorhabdus utahensis* was also reported (Waino and Ingvorsen, 2003).

Thermophilic anaerobic bacteria such as *Clostridium*, *Dictyoglomus*, *Rhodothermus*, *Thermotoga*, and *Thermus* are also able to secrete heat-stable xylanases (Table 4). These enzymes are active between 80 and 105°C and are mainly cell associated and most probably localized within the toga, which covers the cells. Several genes encoding thermostable xylanases, e.g., *Thermotoga maritima*, have been already cloned and expressed in *E. coli*. Comparison of the recombinant xylanase and the commercially available enzyme, Pulpenzyme™ indicates that thermostable xylanases are of interest for application in pulp and paper industry (Antranikian et al., 2005). Other hemicellulases (glucoronidase, β-mannanase, β-mannosidase, galactosidase, acetyl xylan esterase, feruloyl esterase and α-arabinofuranosidase), isolated from extremophiles, are efficient enzymes for the complete saccharification of plant cell wall (see Table 4).

Robust xylanases are attractive candidates for various biotechnological applications. Enzymes from bacteria and fungi are already produced on industrial scale and are used as food additives in poultry, for increasing feed efficiency diets and in wheat flour for improving dough handling and the quality of baked products. In the last decade, the major interest in thermostable xylanases was in enzyme-aided bleaching of paper. The chlorinated lignin derivatives generated by this process constitute a major environmental problem. Recent investigations have demonstrated the feasibility of enzymatic treatments as alternatives to chlorine bleaching for the removal of residual lignin from pulp. A treatment of craft pulp with cellulase-free thermostable xylanases leads to a release of xylan and residual lignin without undue loss of other pulp components. Xylanase treatment at elevated temperatures opens up the cell wall structure, thereby facilitating lignin removal in subsequent bleaching stages and thus enhance the development of environmentally friendly processes (Vieille and Zeikus, 2001).

Chitin-Degrading Enzymes

Chitin is a linear β-1,4 homopolymer of N-acetyl-glucosamine residues and it is one of the most abundant natural biopolymer on earth. It has been estimated that the annual worldwide formation rate and steady state amount of chitin is in the order of 10^{10} to 10^{11} tons per year (Antranikian et al., 2005; Fukamizo, 2000). Particularly in the marine environment, chitin is produced in enormous amounts and its turnover is due to the action of chitinolytic enzymes. Chitin is the major structural component of most fungi and some invertebrates (crustacia and insects), while for soil or marine bacteria chitin serves as a nutrient. Chitin degradation is known to proceed with the endo-acting chitin hydrolase (EC 3.2.1.14), the chitin oligomer degrading exo-acting hydrolases (EC 3.2.1.52) and the N-acetyl-D-glucosaminidase (chitobiase; EC 3.2.1.30).

Hyperthermophilic archaea, *Thermococcus chitonophagus* (Andronopoulou and Vorgias, 2004), *Thermococcus kodakaraensis* (Tanaka et al., 1999; 2004), and *Pyrococcus furiosus* (Gao et al., 2003) have been shown to grow on chitin and produce chitinolytic enzymes (see Table 3). The extreme thermophilic anaerobic archaeon *Thermococcus chitonophagus* possesses a multicomponent enzymatic system, consisting of an extracellular exochitinase (Chi50), a periplasmic chitobiase (Chi90), and a cell-membrane-anchored endochitinase (Chi70) (Andronopoulou and Vorgias, 2004b). The chitinolytic system is strongly induced by chitin, although a low-level constitutive production of the enzymes in the absence of any chitinous substrates was detected. The archaeal chitinase (Chi70) was purified and characterized. It is a monomeric enzyme with an apparent molecular weight of 70 kDa and appears to be associated with the outer surface of the cell membrane. The enzyme is optimally active at 70°C and pH 7.0 and is thermostable, maintaining 50% activity at 120°C even after 1 h. The enzyme was not inhibited by allosamidin, the natural inhibitor of chitinolytic activity, and was also resistant to denaturation by urea and SDS.

The chitinase has a broad substrate specificity for several chitinous substrates and derivatives and has been classified as an endochitinase due to its ability to release chitobiose from colloidal chitin (Andronopoulou and Vorgias, 2003). The purified recombinant chitinase from the hyperthermophile *Thermococcus kodakaraensis* is optimally active at 85°C and pH 5.0 and produces chitobiose as the major end product (Tanaka et al., 1999). The thermostable chitinase from *T. kodakaraensis* is active in the presence of detergents and organic solvents and can be applied, e.g., for the production of N-acetyl-chitooligosaccharides with biological activity (Tanaka et al., 1999). This unique multidomain protein consists of two active sites with different cleavage specificities and three substrate-binding domains, which are related to two families of cellulose-binding domains (Tanaka et al., 2001). A chitin-degrading pathway involves unique enzymes diacetylchitobiose deacetylase and exo-β-d-glucosaminidase. After the hydrolysis of chitin by chitinase, diacetylchitobiose will be deacetylated and then successively hydrolyzed to glucosoamine (Tanaka et al., 2004). *Pyrococcus furiosus* was also found to grow on chitin, adding this polysaccharide to the inventory of carbohydrates utilized by this hyperthermophilic archaeon. Two open reading frames (*ChiA* and *ChiB*) were identified in the genome of *P. furiosus*, which encode chitinases with sequence similarity to proteins from the glycosyl hydrolase family 18 in less-thermophilic organisms (Gao et al., 2003). The two chitinases share little sequence homology to each other, except in the catalytic region, where both have the catalytic glutamic acid residue that is conserved in all family 18 bacterial chitinases. The pH optimum of both recombinant chitinases is pH 6.0 with a temperature optimum between 90 and 95°C. The chitinase A (ChiA) melts at 101°C, whereas chitinase B (ChiB) has a melting temperature of 114°C. ChiA exhibits no detectable activity toward chitooligomers smaller than chitotetraose, indicating that the enzyme is an endochitinase whereas ChiB is a chitobiosidase, processively cleaving off chitobiose from the nonreducing end of chitin or other chitooligomers. The synergetic action of both thermoactive chitinases on colloidal chitin allows *P. furiosus* to grow on chitin as sole carbon source.

Although a large number of bacterial chitin-hydrolyzing enzymes has been isolated and their corresponding genes have been cloned and characterized, only few thermostable chitin-hydrolyzing enzymes are known (Kawase et al., 2004). Those enzymes have been isolated from the thermophilic bacteria *Rhodothermus marinus*, *Microbispora* sp. and *Clostridium thermocellum* (see Table 4).

Pectin-Degrading Enzymes

Pectin is a branched heteropolysaccharide consisting of a main chain of α-1,4-D-polygalacturonate, which is partially methyl esterified. Along the chain, L-rhamnopyranose residues are present that are the binding sites for side chains composed of neutral sugars. Pectin is an important plant material that is present in the middle lamellae as well as in the primary cell walls. Pectin is degraded by pectinolytic enzymes, which can be classified into two major groups. The first group comprises methylesterases, whose function is to remove the methoxy groups from pectin. The second group comprises the depolymerases (hydrolases and lyases), which attack both pectin and pectate (polygalacturonic acid). Unlike archaea, a few pectinolytic enzymes from thermophilic anaerobic bacteria have been reported (see Table 4). The enzymes usually act at alkaline pH and are calcium dependent. Thus, a spore-forming anaerobic microorganism *Thermoanaerobacter italicus* is able to grow at 70°C on citrus pectin and pectate. After growth on citrus pectin, two pectate lyases were induced, purified and biochemically characterized (Kozianowski et al., 1997). Both enzymes display similar catalytic properties and can function at temperatures up to 80°C. An increase in the enzymatic activity of both pectate lyases was observed after the addition of calcium ions. The ability of the hyperthermophilic bacterium *Thermotoga maritima* to grow on pectin as a sole carbon source coincides with the secretion of an extracellular pectate lyase. The corresponding gene was functionally expressed in *E. coli* as the first heterologously produced thermophilic pectinase (Kluskens et al., 2003). Highest activity was demonstrated on polygalacturonic acid, whereas pectins with an increasing degree of methylation were degraded at a decreasing rate. The tetrameric enzyme requires calcium ions for stability and activity. The enzyme is highly thermoactive and thermostable, operating optimally at 90°C and pH 9.0, with a half-life for thermal inactivation of almost 2 h at 95°C, and an apparent melting temperature of 102.5°C. Detailed characterization of the product formation with polygalacturonic acid indicated that PelA has a unique eliminative exo-cleavage pattern liberating unsaturated trigalacturonate as the major product, in contrast with unsaturated digalacturonate for other exopectate lyases known. *Thermotoga maritima* also produces exopolygalacturonase, which has been rarely described in bacteria (Kluskens et al., 2005). Pectin degrading enzymes from the bacteria of the genus *Clostridium* are not so thermostable (Vieille and Zeikus, 2001). Enzymatic pectin degradation is widely applied in food technology processes, as in fruit juice extraction and wine making, in order to increase the juice yield, to reduce its viscosity, improve color extraction from the fruit skin, and to macerate fruit and vegetable tissues.

Proteolytic Enzymes

Proteases are involved in the conversion of proteins into amino acids and peptides. They have been classified according to the nature of their catalytic site in the following groups: serine, cysteine, aspartic, or metalloproteases. Proteases and proteasomes play a key role in the cellular metabolism of archaea, and a variety of heat-stable proteases has been identified in hyperthermophilic archaea belonging to the genera *Aeropyrum*, *Desulfurococcus*, *Sulfolobus*, *Staphylothermus*, *Thermococcus*, *Pyrobaculum*, and *Pyrococcus* (see Table 3). It has been found that most proteases from extremophilic archaea and bacteria belong to the serine type and are stable at high temperatures even in the presence of high concentrations of detergents and denaturing agents (Antranikian et al., 2005). Those properties of extracellular serine proteases are reported in a number of *Thermococcus* species and could be well illustrated by the extracellular enzyme from *Thermococcus stetteri*, which is highly stable (half-life of 2.5 h at 100°C) and resistant to chemical denaturation such as 1% SDS (Niehaus et al., 1999). Heat-stable serine proteases were isolated from the cell-free supernatant of the hyperthermophilic archaea *Desulfurococcus* strain $Tok_{12}S_1$ and *Desulfurococcus* sp. SY (Antranikian et al., 2005). A globular serine protease from *Staphylothermus marinus* was found to be extremely thermostable and is heat-resistant up to 125°C in the stalk-bound form (Mayr et al., 1996). A novel intracellular serine protease (pernisine) from the aerobic hyperthermophilic archaeon *Aeropyrum pernix* K1 is active at 90°C. The enzyme has a broad pH profile with an optimum at pH 9.0 for peptide hydrolysis (Chavez Croocker et al., 1999; Sako et al., 1997). A gene encoding a serine protease, named aerolysin, has been cloned from *Pyrobaculum aerophilum* and the protein was modeled based on structures of subtilisin-type proteases (Ward et al., 2002). Multiple proteolytic activities have been observed in *Pyrococcus furiosus*. The cell-envelope associated serine protease of *P. furiosus* called pyrolysin was found to be highly stable with a half-life of 20 min at 105°C (Voorhorst et al., 1996). Proteases have also been characterized from the thermoacidophilic archaea *Sulfolobus solfataricus* (Burlini et al., 1992) and *Sulfolobus acidocaldarius* (Ward et al., 2002). Serine proteases were also characterized from halophilic archaea (see Table 3). An extracellular serine protease from *Halobacterium halobium* is highly salt dependent, active in dimethylformamide/water mixtures and is expected to be an excellent tool for the synthesis of glycine-containing peptides (Ryu et al., 1994). A serine protease from *Natrialba magadii* has a broad pH profile (pH 6 to 12) with an optimum at pH 8 to 10 and a high dependence on salt, which is required for enzymatic activity and stability (Oren, 2002). The salt-dependent proteolytic activity has also been detected in the halophile *Natronococcus occultus* (Horikoshi, 1999). A chymotrypsinogen B-like protease, which is optimally active at 60°C and pH 10, was isolated from the halophile *Natronomonas pharaonis*. In contrast to other haloarchaeal enzymes, which lose their catalytic activity at low salt concentrations, this protease can function at salt concentrations lower than 3 mM. This property makes the enzyme as a suitable detergent additive.

Thermostable serine proteases were also detected in a number of extreme thermophilic bacteria belonging to the genera *Aquifex*, *Thermotoga*, *Thermus*, and *Fervidobacterium* (see Table 4). The enzyme from *Fervidobacterium pennivorans* is able to hydrolyze feather keratin forming high value products like amino acids and peptides. The enzyme, which has been named fervidolysin, is optimally active at 80°C and pH 10.0 (Kluskens et al., 2002). The gene encoding fervidolysin was cloned and successfully expressed in *E. coli*. The gene encodes for a 73-kDa fervidolysin precursor, a 58-kDa mature fervidolysin, and a 14-kDa fervidolysin propeptide. Using site-directed mutagenesis, the active-site histidine residue at position 79 was replaced by an alanine residue. The resulting fervidolysin showed a single protein band corresponding in size to the 73-kDa fervidolysin precursor, indicating that its proteolytic cleavage resulted from an autoproteolytic process. From the thermoacidophile *Alicyclobacillus sendaiensis*, a novel thermostable collagenolytic member of the serine-carboxyl proteinase family was characterized (Tsuruoka et al., 2003). This enzyme, with a molecular mass of 37 kDa, can be applied for the production of peptides from collagen. Aminopeptidase resistant to organic solvents was described from the hyperthermophilic bacterium *Aquifex aeolicus* (Khan et al., 2000).

In addition to serine proteases other subclasses of proteases have been identified in archaea (see Table 3): aminopeptidases, metalloproteases, a thiol protease from *Thermococcus kodakaraensis* KOD1, an acidic protease from *Sulfolobus acidocaldarius*, and a propylpeptidase and a new type of protease from *Pyrococcus furiosus*. Indeed, the *Pyrococcus furiosus* strain contains at least 13 different proteins with proteolytic activity.

The amount of proteolytic enzymes produced worldwide on commercial scale is the largest. Heat-stable proteases are useful enzymes, especially for the detergent industry. Serine alkaline proteases from thermophiles could be used as additives for laundering, where they have to resist denaturation by detergents and alkaline conditions. Proteases are also applied for peptide synthesis using their reverse reaction, mainly because of their compatibility with organic solvents. A number of heat-stable proteases are now used in molecular biology and protein chemistry. The protease S from *Pyrococcus furiosus* is used to fragment proteins before peptide sequencing (TaKaRa Biomedicals). Carboxypeptidases and aminopeptidases from *Pyrococcus furiosus* and *Sulfolobus solfataricus* are used for protein N- or C-terminal sequencing (Cheng et al., 1999; Vieille and Zeikus, 2001).

DNA-Processing Enzymes

PCR

DNA polymerases (EC 2.7.7.7) are the key enzymes in the replication of cellular information present in all life forms. They catalyze, in the presence of Mg^{2+} ions, the addition of a deoxyribonucleoside 5'-triphosphate onto the growing 3'-OH end of a primer strand, forming complementary base pairs to a second strand. More than 100 DNA polymerase genes have been cloned and sequenced from various organisms, including thermophilic bacteria and archaea (Perler et al., 1996). Thermostable DNA polymerases play a major role in a variety of molecular biological applications, e.g., DNA amplification, sequencing, or labelling (Table 5). The first described PCR procedure utilized the Klenow fragment of *E.coli* DNA polymerase I, which was heat-labile and had to be added during each cycle following the denaturation and primer hybridization steps. Introduction of thermostable DNA polymerases in PCR facilitated the automation of the thermal cycling part of the procedure. The DNA polymerase I from the bacterium *Thermus aquaticus*, called *Taq* polymerase, was the first thermostable DNA polymerase characterized and applied in PCR. Due to the absence of a 3'-5'-exonuclease activity, this enzyme is unable to excise mismatches and as a result, the base insertion fidelity is low. The use of high fidelity DNA polymerases is essential for reducing the increase of amplification errors in PCR products that will be cloned, sequenced, and expressed. Several thermostable DNA polymerases with 3'-5'-exonuclease-dependent proofreading activity have been described and the error rates (number of misincorporated nucleotides per base synthesized) for these enzymes have been determined. Archaeal polymerases from *Pyrococcus* or *Thermococcus* species with stringent proofreading abilities are of widespread use. Archaeal proofreading polymerases such as *Pwo* pol from *Pyrococcus woesei*, *Pfu* pol from *Pyrococcus furiosus*, Deep Vent™ pol from *Pyrococcus* strain GB-D or Vent™ pol from *Thermococcus litoralis* have an error rate, which is up to 10-fold lower than that of *Taq* polymerase. The 9°N-7 DNA polymerase from *Thermococcus* sp. strain 9°N-7 has a 5-fold higher 3'-5'-exonuclease activity than *Thermococcus litoralis* DNA polymerase. However, *Taq* polymerase was not replaced by these DNA polymerases because of their low extension rates among other factors. DNA polymerases with higher fidelity are not necessarily suitable for amplification of long DNA fragments because of their potentially strong exonuclease activity. The recombinant KOD1 DNA polymerase from *Thermococcus kodakaraensis* KOD1 has been reported to show low error rates, high processivity, and highest-known extension rate, resulting in an accurate amplification of target DNA sequences up to 6 kb (Takagi et al., 1997; Hashimoto et al., 2001). Recently, the PCR technique has been improved to allow low error synthesis of long amplificates (20 to 40 kb) by adding small amounts of thermostable archaeal proofreading DNA polymerases, containing 3'-5'-exonuclease activity, to *Taq* or other non-proofreading DNA polymerases. In this long PCR, the reaction conditions are optimized for long extension by adding different components such as gelatine, Triton X-100 or bovine serum albumin to stabilize the enzymes and mineral oil to prevent evaporation of water in the reaction mixture. In order to enhance specificity, glycerol or formamide are added. The supplement of the PCR reaction mixtures with recombinant *P. woesei* dUTPase improves the efficiency of the reaction and allows amplification of longer targets (Dabrowski and Kiaer Ahring, 2003). Low fidelity mutants of *P. furiosus* polymerase were also created for the performance in error-prone PCR (Biles and Connolly, 2004). The first DNA polymerase from non-thermophilic uncultured crenarchaeote *Cenarchaeum symbiosum* is remarkably less thermostable (half-life 10 min. at 46°C) than closely related homologs from thermophilic and hyperthermophilic crenarchaeotes (Schleper et al., 1997).

The ssDNA-binding proteins are known to be involved in eliminating DNA secondary structure, and are key components in DNA replication, recombination and repair. The archaeal ssDNA-binding proteins derived from *Methanocaldococcus jannashii*, *Methanothermobacter thermautotrophicum*, and *Archaeoglobus fulgidus* are therefore useful reagents

Table 5. Application of DNA-modifying enzymes from extremophiles[a]

Enzymes	Strain[a]	MW (kDa)	T_{opt} (°C)	pH_{opt}	Thermostability (half-life)	References
DNA Polymerase (EC 2.7.7.7)	*Archaea:*					
	Aeropyrum pernix (pol I)	108			0.5 h at 85°C	Cann et al., 1999
	Aeropyrum pernix (pol II)	88			0.5 h at 100°C	Cann et al., 1999
	Cenarchaeum symbiosum	96	38		0.17 h at 46°C	Schleper et al., 1997
	Pyrobaculum islandicum	90	70–80	7.5	>5 h at 90°C	Kahler and Antranikian, 2000
Pfu pol	*Pyrococcus abyssi*	90	70–80	7.3	5 h at 100°C	Dietrich et al., 2002
Deep Vent pol	*Pyrococcus furiosus*	90	72–78	9	4 h at 95°C	Antranikian et al., 2005
Pwo pol	*Pyrococcus* sp. GB-D	90.6	70–80	8–9	8 h at 100°C	Antranikian et al., 2005
	Pyrococcus woesei	90				Antranikian et al., 2005
	Sulfolobus acidocaldarius	100	65–75		0.25 h at 87°C	Nastopoulos et al., 1998
	Sulfolobus solfataricus	101	75	7.5	0.1 h at 90°C	Bohlke et al., 2000
KOD1	*Thermococcus aggregans*	90	70–80	6.8	1.2 h at 90°C	Antranikian et al., 2005
Vent pol	*Thermococcus kodakaraensis* (struct.)	90	75	7.5	12 h at 95°C	Antranikian et al., 2005
9°N-7 pol	*Thermococcus litoralis*	98.9	70–80	6.5	2 h at 100°C	Antranikian et al., 2005
	Thermococcus sp. 9°N-7	90	70–80	8.8	6.7 h at 95°C	Antranikian et al., 2005
	Bacteria:					
	Carboxydothermus hydrogenoformans	104.8	60–70			Roche Molecular Biochemicals
	Rhodothermus marinus	97	55		2 min at 90°C	Blondal et al., 2001
	Thermoanaerobacter yonseiensis					Kim et al., 2002
Tay	*Thermus aquaticus*		75	9		Roche Molecular Biochemicals
Taq	*Thermus caldophilus*		75	8.7		Vieille and Zeikus, 2001
Tca	*Thermus filiformis*					Roche Molecular Biochemicals
Tfi	*Thermus thermophilus*	87	70–75	8.8		Roche Molecular Biochemicals
Tth	*Methanopyrus kandleri*					Margesin and Schinner, 2001
DNA topoisomerase type I-group B					Modeling of novel drugs	
DNA Ligase (EC 6.5.1.1)	*Archaea:*					
	Pyrococcus furiosus		45–80	6–7.0	>1 h at 95°C	Stratagene
	Sulfolobus shibatae	62	50–70	7	0.15 h at 90°C	Lai et al., 2002
	Thermococcus fumicolans		65	8		Rolland 2004
	Thermococcus kodakaraensis	52	100			Nakatani et al., 2002
	Bacteria:					
	Aquifex pyrophilus	82	65	8–8.6	>1 h at 95°C	Lim et al., 2001
	Rhodothermus marinus	80	>55		0.1 h at 90°C	Thorbjarnardottir et al., 1995
	Thermus scodoductus		65		0.5 h at 90°C	Thorbjarnardottir et al., 1995

Continued on following page

Table 5. Continued

Enzymes	Strain[a]	MW (kDa)	T_{opt} (°C)	pH_{opt}	Thermostability (half-life)	References
Alkaline phosphatase (EC 3.1.3.1)	**Archaea:**					
	Haloarcula marismortui	160	25	8.5		Marhuenda-Egea et al., 2002
	Halobacterium salinarum					Marhuenda-Egea et al., 2002
	Pyrococcus abyssi	108	70	11	18 h at 100°C	Zappa et al., 2001
	Bacteria:					
	Thermotoga maritima		65	8	5 h at 90°C	Wojciechowski et al., 2002
	Thermotoga neapolitana		85	9.9	4 h at 90°C	Vieille and Zeikus, 2001
	Thermus sp. 3041					Ji et al., 2001
	Thermus yunnanensis	104	70–80	8–10.0		Gong et al., 2005
Inorganic pyrophosphatase (EC 3.6.1.1)	**Archaea:**					
	Sulfolobus acidocaldarius	80	56	6.5		Wakagi et al., 1992
	Thermoplasma acidophilum (struct.)					Vander Horn et al., 1997
	Bacteria:					
	Aquifex pyrophilus	105	80–95	7.5–8	at 80–95°C	Hoe et al., 2002
ssDNA-binding proteins	**Archaea:**					
	Archaeoglobus fulgidus					Kowalczykowski et al., 2005
	Methanocaldococcus jannashii					Kowalczykowski et al., 2005
	Methanothermobacter thermautotrophicum					Kowalczykowski et al., 2005

[a] (struct.), the protein has been crystallized and the three-dimensional structure determined.

for genetic engineering and other procedures involving DNA recombination, such as PCR (Kowalczykowski et al., 2005).

DNA sequencing

DNA sequencing by the Sanger method has undergone countless refinements in the last twenty years (Sanger et al., 1992). A major step forward was the introduction of thermostable DNA polymerases leading in the cycle sequencing procedure. This method uses repeated cycles of temperature denaturation, annealing, and extension with dideoxy-termination to increase the amount of sequencing product by recycling the template DNA. Due to this "PCR-like" amplification of the sequencing products, several problems could have been overcome. Caused by the cycle denaturation, only fmoles of template DNA are required, no separate primer annealing step is needed and unwanted secondary structures within the template are resolved at high temperature elongation. The first enzyme used for cycle sequencing was the thermostable DNA polymerase I from *Thermus thermophilus* or *Thermus flavus* (Bechtereva et al., 1989; Rao and Saunders, 1992). The enzyme displays 5'-3'-exonuclease activity that is undesirable because of the degradation of sequencing fragments. A combination of thermostable enzymes has been developed that produces higher quality cycle sequences. Thermo Sequenase DNA polymerase is a thermostable enzyme engineered to catalyze the incorporation of ddNTPs with an efficiency of several thousandfold better than other thermostable DNA polymerases. Since the enzyme also catalyzes pyrophosphorolysis at dideoxy termini, a thermostable inorganic pyrophosphatase is needed to remove the pyrophosphate produced during sequencing reactions. *Thermoplasma acidophilum* inorganic pyrophosphatase (TAP) is thermostable and effective for converting pyrophosphate to orthophosphate. The combination of Thermo Sequenase polymerase and TAP for cycle sequencing yields sequence data with uniform band intensities and allow the determination of longer, more accurate sequence reads. Uniform band intensities also facilitate interpretation of sequence anomalies and the presence of mixed templates. Sequencing PCR products of DNA amplified from heterozygous diploid individuals results in signals of equal intensity from each allele (Vander Horn et al., 1997). Another extremely stable inorganic pyrophosphatase was purified from *Sulfolobus acidocaldarius*. The complete activity of the enzyme remained after incubation at 100°C for 10 min. (Wakagi et al., 1992). Highly thermostable alkaline phosphatases, which dephosphorylate linear DNA fragments, were also identified in archaea (see Table 5). The alkaline phosphatase from *Pyrococcus abyssi* dephosphorylates linear DNA fragments with efficiencies of 94 and 84% regarding to cohesive and blunt ends, respectively (Zappa et al., 2001).

Ligase chain reaction

A variety of analytical methods is based on the use of thermostable ligases. Of considerable potential is the construction of sequencing primers by high temperature ligation of hexameric primers, the detection of trinucleotide repeats through repeat expansion detection or DNA detection by circularization of oligonucleotides (Landegren et al., 2004). Up to now several archaeal DNA ligases, displaying nick joining and blunt-end ligation activities using either ATP or NAD+ as a cofactor, have been identified and characterized in detail (see Table 5). Over the years, several additional thermostable DNA ligases were discovered. The ligase from the archaeon *Acidianus ambivalens* is NAD+-independent but ATP-dependent similar to the enzymes from bacteriophages, eukaryotes, and viruses (Kletzin, 1992). The DNA ligase from a hyperthermophilic archaeon *Thermococcus kodakaraensis* is also ATP-dependent (Nakatani et al., 2000). Sequence comparison with previously reported DNA ligases indicated that the ligase is closely related to the ATP-dependent DNA ligase from *Methanobacterium thermoautotrophicum* H, a moderate thermophilic archaeon, along with putative DNA ligases from *Euryarchaeota* and *Crenarchaeota*. The optimum pH of the recombinant monomeric enzyme is 8.0, and the optimum concentration of Mg^{2+} is 14 to 18 mM and of K^+ is 10 to 30 mM. The protein does not display single-stranded DNA ligase activity. At enzyme concentrations of 200 nM, a significant DNA ligase activity is observed even at 100°C. Surprisingly, the protein displays a DNA ligase activity also when NAD+ is added as the cofactor (Nakatani et al., 2002). The ability for DNA ligases, to use either ATP or NAD+, as a cofactor, appears to be specific of DNA ligases from *Thermococcales*. Also a DNA ligase from *Thermococcus fumicolans* displays nick joining and blunt-end ligation activity using either ATP or NAD+, as a cofactor (Rolland et al., 2004). The optima of temperature and pH of the ligase are 65°C and 7.0, respectively. The presence of $MgCl_2$ (optimally at 2 mM) is required for the enzymatic activity. In contrast to that the recombinant ATP-dependent ligase from the thermoacidophilic crenarchaeon, *Sulfolobus shibatae* is more active in the presence of Mn^{+2} ions than in the presence of other divalent cations such as Mg^{+2} or Ca^{+2} (Lai et al., 2002). A splicing ligase activity was characterized in cell extracts of the halophile *Haloferax volcanii* (Oren, 2002). Bacterial enzymes were derived and cloned from *Aquifex pyrophilus* (Lim et al., 2001), *Thermus scotoductus,* and *Rhodothermus marinus* (Thorbjarnardottir et al., 1995).

The archaeal strains *Pyrococcus furiosus, Thermococcus marinus, T. radiotolerans,* and the bacterium *Deinococcus radiodurans* are resistant to high levels of ionizing and ultraviolet radiation and therefore may have a unique method of removing damaged DNA (Jolivet et al., 2004). A thermostable flap endonuclease from *P. furiosus* is described, which cleaves the replication fork-like structure endo/exonucleolytically (Matsui et al., 1999). The 06-methylguanine-DNA methyltransferase is the most common form of cellular defense against the biological effects of O6-methylguanine in DNA. The thermostable recombinant 06-methylguanine-DNA methyltransferase from *Thermococcus kodakaraensis* is functional in vivo and complements the mutant phenotype, making the cells resistant to the cytotoxic properties of the alkylating agent N-methyl-N'-nitro-N-nitrosoguanidine (Leclere et al., 1998).

A thermostable type I-group B DNA topoisomerase has been isolated and purified from the hyperthermophilic methanogen *Methanopyrus kandleri* (Slesarev et al., 2002). The enzyme is active over a wide range of temperatures and salt concentrations and does not require magnesium or ATP for its activity, which makes manipulations on DNA more convenient and more efficient. Exploitation of the common features and the differences of topoisomerases will be important for modeling of novel drugs and understanding of the action of cancer chemotherapeutic agents.

Lipases and Esterases

Lipases, triacylglycerol hydrolases, are an important group of biotechnologically relevant enzymes and they find immense applications in food, dairy, detergent, and pharmaceutical industries. Lipases are produced by microbes, and specifically bacterial lipases play a vital role in commercial ventures. Lipases are generally produced on carbon sources, such as oils, fatty acids, glycerol, or tweens in the presence of an organic nitrogen source. Bacterial lipases are mostly extracellular and are produced by submerged fermentation. Most lipases can act in a wide range of pH and temperature, though alkaline bacterial lipases are more common. Bacterial lipases generally have temperature optima in the range 30 to 60°C. Lipases are serine hydrolases and have high stability in organic solvents. In addition, some lipases exhibit chemo-, regio- and enantio-selectivity. Very recently, more than five anaerobic thermophilic bacteria were found to produce extremely heat-stable lipases. They are active at a broad temperature (50 to 95°C) and pH (3 to 11) range (Antranikian, unpublished results). The latest trend in lipase research is the development of novel and improved lipases through molecular approaches such as directed evolution and exploring natural communities by the metagenomic approach.

In the field of industrial biotechnology, esterases too are gaining increasing attention because of their application in organic biosynthesis. In aqueous solution, esterases catalyze the hydrolytic cleavage of esters to form the constituent acid and alcohol, whereas in organic solutions, transesterification reaction is promoted. Both the reactants and the products of transesterification are usually highly soluble in the organic phase and the reactants may even form the organic phase itself. Several archaeal and bacterial esterases were successfully cloned and expressed in mesophilic hosts (Table 6). Esterases from archaea *Aeropyrum pernix, Pyrobaculum calidifontis,* and *Sulfolobus tokodaii* exhibit high thermoactivity and thermostability and are active also in a mixture of a buffer and water-miscible organic solvents, such as acetonitrile and dimethyl sulfoxide (Egorova and Antranikian, 2005). The optimal activity for ester cleavage of the esterase from *Sulfolobus tokodaii* strain 7 is at 70°C and pH 7.5 to 8.0. From the kinetic analysis, *p*-nitrophenyl butyrate is the better substrate than caproate and caprylate (Suzuki et al., 2004). The *Pyrococcus furiosus* esterase is the most thermostable (a half-life of 50 min. at 126°C) and thermoactive (temperature optimum 100°C) esterase known to date (Ikeda and Clark, 1998). A carboxylesterase from *Pyrobaculum calidifontis,* stable against heating and organic solvents, is active toward tertiary alcohol esters, a very rare feature among previously reported lipolytic enzymes (Hotta et al., 2002). A novel thermostable esterase from *Aeropyrum pernix* K1 with an optimal temperature at 90°C exhibits additionally a phospholipase activity (Wang et al., 2004). In our laboratory two thermoactive esterases from the thermoacidophilic archaeon *Picrophilus torridus* have been recently characterized after successful expression in *E. coli* (Antranikian, unpublished results). Both esterases are active at 50 to 60°C and neutral pH. A gene coding the esterase from *Archaeoglobus fulgidus* was subjected to error-prone PCR in an effort to increase the low enantioselectivity toward the racemic mixture of *p*-nitrophenyl-2-chloropropionate to produce the *S*-2-chloropropionic acid. This compound is an important intermediate in the synthesis of some optically pure compounds, such as a herbicide mecoprop (Manco et al., 2002). A double mutant, Leu101Ile/Asp117Gly was obtained with increased preference in the opposite direction. The esterase from *Sulfolobus solfataricus* P1 has been studied in detail for the chiral resolution of

Table 6. Other thermoactive enzymes of industrial relevance

Enzymes	Strain[a]	MW (kDa)	T_{opt} (°C)	pH_{opt}	Thermostability (half-life)	Possible applications	References
Alcohol dehydrogenase (EC 1.1.1.1)						Stereoselective transformation of ketones to pure chiral alcohols	Guy et al., 2003
	Archaea:						
	Aeropyrum pernix (struct.)		90				Radianingtyas and Wright, 2003
	Methanoculleus thermophilicus		70				van der Oost et al., 2001
	Pyrococcus furiosus	55	90	7.5	150 h at 80°C		Radianingtyas and Wright, 2003
	Sulfolobus solfataricus (struct.)	71	95	7.5			Antoine et al., 1999
	Thermococcus hydrothermalis	80.5	80	7.5	0.25 h at 80°C		Ma and Adams, 2001
	Thermococcus litoralis	192	80	8.8	2 h at 85°C		Li and Stevenson, 2001
	Thermococcus sp. AN1	200	85	7			Ma and Adams, 2001
	Thermococcus sp. ES-1						Radianingtyas and Wright, 2003
	Thermococcus zilligii	184	85	7	0.25 h at 80°C		
	Bacteria:						
	Thermoanaerobacter brockii (struct.)	160					Radianingtyas and Wright, 2003
	Thermoanaerobacter ethanolicus	160	90		1.7 h at 90°C		Radianingtyas and Wright, 2003
	Thermomicrobium roseum	86	70	10			Yoon et al., 2002
Aldolase (EC 4.1.2.13, EC 4.1.2.14)						Synthesis of chiral carbohydrates	
	Archaea:						
	Haloarcula vallismortis	280					Krishnan and Altekar, 1991
	Methanocaldococcus jannaschii	271	80	7–8.5	24 h at 80°C		Soderberg and Alver, 2004
	Pyrococcus furiosus	272	50				Lorentzen et al., 2004
	Sulfolobus solfataricus (struct.)	133	100		2.5 h at 100°C		Buchanan et al., 1999
	Thermoproteus tenax	245	50				Lorentzen et al., 2004
	Bacteria:						
	Thermus aquaticus (struct.)	165	80		>1 h at 97°C		Izard and Sygush, 2004
Amidase (EC 3.5.1.4)						Synthesis of fine chemicals	
	Archaea:						
	Sulfolobus solfataricus (struct.)	56	95	7.5	25 h at 80°C		Antranikian et al., 2005
	Bacteria:						
	Pseudonocardia thermophila	108	70	4–8	1.2 h at 70°C		Egorova et al., 2004
Aminoacylase (EC 3.5.1.14)						Pharmaceutical industry (production of stereoisomers)	
	Archaea:						
	Pyrococcus furiosus	170	100	6.5			Story et al., 2001
	Pyrococcus horikoshii OT3	95	95	7.5	>48 h at 90°C		Ishikawa et al., 2001
	Thermococcus litoralis	172	85	8	1.7 h at 85°C		Taylor et al., 2004

Continued on following page

Table 6. Continued

Enzymes	Strain[a]	MW (kDa)	T_{opt} (°C)	pH_{opt}	Thermostability (half-life)	Possible applications	References
Arabinose isomerase (EC 5.3.1.3)	**Bacteria:**					Sweeteners in food industry	
	Alicyclobacillus acidocaldarius	224	65	6–6.5	10 h at pH 5.0		Lee et al., 2005
	Geobacillus stearothermophilus	225	80	7.5			Rhimi and Bejar, 2005
	Thermoanaerobacter mathranii	212	65	8			Jorgensen et al., 2004
	Thermotoga maritima	232	90	7.5	4 h at 80°C		Lee et al., 2004
	Thermotoga neapolitana	230	85	7.5			Kim et al., 2002
Catalase (EC 1.11.1.6)	**Bacteria:**					Industrial bleaching	
	Thermus brockianus	178	90	8	3 h at 90°C		Thompson et al., 2003
Cysteine synthase (EC 4.2.1.22)	**Archaea:**					Synthesis of sulfur-organic compounds	
	Aeropyrum pernix	65	>60	7.5–8	>6 h at 100°C		Ishikawa and Mino, 2004
Esterase (EC 3.1.1.1)	**Archaea:**					Biotransformation in organic solvents	
	Aeropyrum pernix (struct.)	18	90		1 h at 100°C		Wang et al., 2004
	Archaeoglobus fulgidus (struct.)	35.5	70	7			Manco et al., 2000
	Methanocaldococcus jannaschii (struct.)		70	9.5			Chen et al., 2004
	Picrophilus torridus	21	70–80	6.5			Antranikian, unpublished
	Pyrobaculum calidifontis	35	90	7	2 h at 100°C		Hotta et al., 2002
	Pyrococcus furiosus		100		2 h at 120°C		Sehgal and Kelly, 2003
	Sulfolobus acidocaldarius	128	90				Antranikian et al., 2005
	Sulfolobus shibatae		90	6	0.33 h at 120°C		Antranikian et al., 2005
	Sulfolobus solfataricus P1	33	100	5–6			Sehgal and Kelly, 2003
	Sulfolobus solfataricus P2	34	80	7.4	0.66 h at 80°C		Kim and Lee, 2004
	Sulfolobus tokodaii		70	7.5–8			Suzuki et al., 2004
	Bacteria:						
	Thermoanaerobacter tengcongensis	43	70	9	>10 h at 50°C		Zhang et al., 2003
α-Galactosidase (EC 3.2.1.22)	**Bacteria:**					Sugar processing	
	Thermotoga maritima		90–95	5–5.5	1.2 h at 90°C		Vieille and Zeikus, 2001
	Thermus sp.	54	75		1 h at 70°C		Ishiguro et al., 2001
β-Galactosidase (EC 3.2.1.23)	**Archaea:**					Synthesis of oligosaccharides, production of dietary milk products	
	Haloferax alicantei	156	22	7.2	3.5 h at 100°C		Holmes and Dyall-Smith, 2000
	Pyrococcus woesei	61	90	4			Dabrowski et al., 2000

Enzyme (EC)	Source	MW (kDa)	T opt (°C)	pH opt	Stability	Application	References
	Bacteria:						
	Thermotoga maritima		80	5.3			Vieille and Zeikus, 2001
	Thermus sp. A4	75	70	6.5	>2 h at 85°C		Ohtsu et al., 1998
β-Glucuronidase (EC 3.2.1.31)	**Bacteria:**					Synthesis of oligosaccharides	
	Thermotoga maritima		85	6.5	3 h at 85°C		Salleh et al., 2006
Glucose isomerase (EC 5.3.1.5)	**Bacteria:**					Sweeteners in food industry	
	Thermoanaerobacter ethanolicus	200	70				Erbeznik et al., 2004
	Clostridium thermosulfurigenes	200	65	7	0.6 h at 85°C		Vieille and Zeikus, 2001
	Thermoanaerobacterium saccharolyticum		80	7–7.5			Vieille and Zeikus, 2001
	Thermoanaerobacterium sp.	200	80	6.8	1 h at 80°C		Liu et al., 1996
	Thermotoga maritima		>100	7	0.2 h at 120°C		Vieille and Zeikus, 2001
	Thermotoga neapolitana	200	97	7.1	2 h at 90°C		Vieille et al., 1995
	Thermus aquaticus		70	5.5	240 h at 70°C		Vieille and Zeikus, 2001
	Thermus caldophilus (struct.)		90				Chang et al., 1999
	Thermus flavus	185	90	7	2 h at 95°C		Park et al., 1997
	Thermus thermophilus	200	90	7			Dekker et al., 1991
β-Fructosidase (EC 3.2.1.26)	**Bacteria:**					Confectionery industry	
	Thermotoga maritima		90–95	5.5			Vieille and Zeikus, 2001
Laccase (EC 1.10.3.2)	**Bacteria:**					Polymer synthesis, biosensors	
	Thermus thermophilus HB27	53	92	4.5–5.5	>14 h at 80°C		Miyazaki, 2005
Lipase (EC 3.1.1.3)	**Bacteria:**					Biotransformation	
	Caldanaerobacter subterraneus	28	75	7	2 h at 80°C		Antranikian, unpublished
	Thermoanaerobacter thermohydrosulfuricus	28	75	8	2 h at 85°C		Antranikian, unpublished
N-Methyltransferase (EC 2.1.1.17)	**Archaea:**					Synthesis of phosphadylcholine for medicine and food	
	Pyrococcus horikoshii	23	90–100	8.5	>2 h at 100°C		Matsui et al., 2002
Maltooligosyl trehalose synthase (EC 5.4.99.15)	**Archaea:**						
	Sulfolobus acidocaldarius	84	75	5	72 h at 80°C		Gueguen et al., 2001
	Sulfolobus solfataricus	87	75	5	2 h at 85°C		Fang et al., 2004
Nitrilase (EC 3.5.5.1)	**Archaea:**					Fine chemicals (mononitriles)	
	Pyrococcus abyssi	60	60–90	6–8	6 h at 90°C		Mueller et al., 2006

[a] (struct.), the protein has been crystallized and the three-dimensional structure determined.

2-arylpropionic esters (Sehgal and Kelly, 2003). Thus, the application of the esterase toward R,S-naproxen methyl ester yields highly optically pure S-naproxen (ee(p) > 90%) (Sehgal and Kelly, 2002; 2003). The enzyme is activated by DMSO to various extents, due to small changes in the enzyme structure resulting in an increase in its conformational flexibility. Thus, the addition of cosolvents, which is useful for solubilization of hydrophobic substrates in water, also serves as activators in applications involving thermostable biocatalysts at suboptimal temperatures (Sehgal et al., 2002). Interestingly, experimental data on kinetic resolution of α-arylpropionic acid revealed that a carboxylesterase from *Sulfolobus solfataricus* P2 hydrolyzes the R-ester of racemic ketoprofen methylester with enantiomeric excess of 80% (Kim and Lee, 2004). A gene encoding a thermostable esterase was cloned from the bacterium *Thermoanaerobacter tengcongensis* and over-expressed in *E. coli*. The recombinant esterase, with a molecular mass of 43 kDa hydrolyzes tributyrin but not olive oil. The esterase is optimally active at 70°C (over 15 min) and at pH 9. It is highly thermostable, with a residual activity greater than 80% after incubation at 50°C for more than 10 h (Zhang et al., 2003).

Alcohol Dehydrogenases

Dehydrogenases are enzymes belonging to the class of oxidoreductases. Within this class, alcohol dehydrogenases (ADHs) (EC 1.1.1.1, also named ketoreductases) represent an important group of biocatalysts due to their ability to stereospecifically reduce prochiral carbonyl compounds. ADHs can be used efficiently in the synthesis of optically active alcohols, which are key building blocks in the synthesis of chirally pure pharmaceutical agent. From a practical point of view, ADHs that use NADH as cofactor are of particular importance, because they represent an established method to regenerate NADH efficiently. By contrast, for NADP-dependent enzymes the cofactor-recycling systems that are available are much less efficient (Radianingtyas and Wright, 2003). The secondary specific ADH, which catalyzes the oxidation of secondary alcohols and, less readily, the reverse reaction (the reduction of ketones) has a promising future in biotechnology. Although ADHs are widely distributed among microorganisms, only few examples derived from hyperthermophilic microorganisms are currently known (see Table 6). The ADH from the archaeon *Sulfolobus solfataricus* requires NAD as cofactor and contains Zn ions (Radianingtyas and Wright, 2003). In contrast to the enzyme from *Thermococcus litoralis*, it lacks metal ions and catalyzes preferentially the oxidation of primary alcohols, using NADP as cofactor. The enzyme is thermostable, having half-lives of 15 min. at 98°C and 2 h at 85°C (Ma and Adams, 2001). The pyrococcal ADH is the most thermostable short-chain ADH (half-life of 150 h at 80°C) known to date (van der Oost et al., 2001). The NADP-dependent ADH from *T. hydrothermalis* oxidizes a series of primary aliphatic and aromatic alcohols preferentially from C_2 to C_8 but is also active toward methanol and glycerol and is stereospecific for monoterpenes (Antoine et al., 1999). The enzyme structure is pH-dependent, being a tetramer (45 kDa per subunit) at pH 10.5 (pH optimum for alcohol oxidation), and a dimer at pH 7.5 (pH optimum for aldehyde reduction). Among the extreme thermophilic bacteria, *Thermoanaerobacter ethanolicus* 39E and *T. brockii* were shown to produce an ADH, whose gene was cloned and expressed in *E. coli*. Interestingly, a mutant has been found to posses an advantage over the wild-type enzyme by using the more stable cofactor NAD instead of NADP (Radianingtyas and Wright, 2003). An ADH was purified from an extremely thermophilic bacterium, *Thermomicrobium roseum*. The pI of the homodimeric enzyme (43 kDa/subunit) was determined to be 6.2, while its optimum pH and temperature are 10.0 and 70°C, respectively (Yoon et al., 2002). The enzyme oxidizes mainly primary aliphatic alcohols.

Glucose and Arabinose Isomerases

Glucose isomerase or xylose isomerase (D-xylose ketol-isomerase; EC 5.3.1.5) catalyzes the reversible isomerization of D-glucose and D-xylose to D-fructose and D-xylulose, respectively. The enzyme has the largest market in the food industry because of its application in the production of high-fructose corn syrup (HFCS). Glucose isomerases are widely distributed in mesophilic microorganisms and intensive research efforts were directed toward improving their suitability for industrial application. In order to reach fructose concentration of 55% the reaction must approach 110°C. Improved thermostable glucoses isomerases have been engineered from mesophilic enzymes (Crabb and Bolin, 1999). Mostly thermophilic bacteria were found to produce glucose isomerases (see Table 6). The gene encoding a xylose isomerase of *Thermus flavus* AT62 was cloned and the DNA sequence was determined. The enzyme has an optimum temperature at 90°C and pH 7.0; divalent cations are required for enzyme activity (Park et al., 1997). *Thermoanaerobacterium* strain JW/SL-YS 489 forms a xylose isomerase, which is optimally active at pH 6.4 and 60°C or pH 6.8 and 80°C. Like other xylose isomerases, this enzyme requires divalent cations for thermal stability (stable for 1 h at

82°C in the absence of substrate). The gene encoding the xylose isomerase of *Thermoanaerobacterium* strain JW/SL-YS 489 was cloned and expressed in *E. coli* (Vieille and Zeikus, 2001). Comparison of the deduced amino acid sequence with sequences of other xylose isomerases showed that the enzyme has 98% homology with a xylose isomerase from a closely related bacterium *Thermoanaerobacterium saccharolyticum* B6A-RI. A thermostable glucose isomerase was purified and characterized from *Thermotoga maritima*. This enzyme is stable up to 100°C, with a half-life of 10 min. at 115°C (Vieille and Zeikus, 2001). Interestingly, the glucose isomerase from *Thermotoga neapolitana* displays a catalytic efficiency at 90°C, which is 2 to 14 times higher than any other thermoactive glucose isomerases at temperatures between 60 and 90°C (Vieille et al., 1995).

Arabinose isomerase (EC 5.3.1.4) catalyzes the reversible isomerization of arabinose to ribulose. Thermoactive enzymes have been reported to convert D-galactose to D-tagatose, a novel and natural sweetener (Lee et al., 2005). Such enzymes have been described from the thermophilic bacteria *Alicyclobacillus acidocaldarius*, *Geobacillus stearothermophilus*, *Thermoanaerobacter mathranii*, *Thermotoga maritima* and *T. neapolitana* (see Table 6).

C–C Bond-Forming Enzymes

Synthetic building blocks bearing hydroxylated chiral centers are important targets for biocatalysis. C–C bond-forming enzymes, such as aldolases and transketolases, have been investigated for new applications, and various strategies for the synthesis of sugars and related oxygenated compounds have been developed (Fessner and Helaine, 2001). The use of aldolases in stereoselective C–C bond-forming reactions is applicable for asymmetric synthesis of carbohydrates, leading to the development of new therapeutics and diagnostics. However, many aldolases display a narrow specificity, often prefer phosphorylated substrates, which can limit the product range of chiral aldols. In contrast, an extremely thermostable aldolase (half-life 2.5 h at 100°C) from *Sulfolobus solfataricus*, actively expressed in *E. coli*, possesses a broad specificity for non-phosphorylated substrates and has a great potential for use in asymmetric aldol reactions (see Table 6) (Demirjian et al., 2001). This aldolase represents a rare example of an enzyme that exhibits no diastereocontrol for the aldol condensation of its natural substrates pyruvate and glyceraldehyde. Recently, it was demonstrated that the stereoselectivity of the enzyme has been induced by employing the substrate engineering procedure (Lamble et al., 2005). In another application, thermostable pentose phosphate enzymes, e.g., the transaldolases from *Methanocaldococcus jannaschii*, could greatly increase the efficiency of an enzymatic hydrogen production that employs a novel archaeal hydrogenase (Soderberg and Alver, 2004). The aerobic bacterium *Thermus aquaticus* produces fructose aldolase, which is stable after heating at 90°C for 2 hours (Demirjian et al., 2001).

Nitrile-Degrading Enzymes

Nitrile-degrading enzymes are of considerable importance in industrial biotransformations, and to date several processes have been developed for chemical and pharmaceutical industries for the production of optically pure compounds, drugs, acrylic, and hydroxamic acids (Egorova et al., 2004). Nitrile-degrading enzymes also play a significant role in protecting the environment by their capability to eliminate highly toxic nitriles. Thermostable amidases and nitrilases are gaining more attention, especially in enzymatic processes in mixtures of organic solvent or in the formation of highly pure products with a concomitant reduction of wastes (Alcantara et al., 2000). A number of bacterial amidases and nitrilases have been purified and characterized. Very little, however, is known on the enzymes that are active at high temperatures. Amidases (EC 3.5.1.4) catalyze the conversion of amides to the corresponding carboxylic acids and ammonia. Amidases are highly S-enantioselective, usually forming the optically pure acids with an enantiomeric excess above 99%. The only amidase derived from archaea is the amidase from the thermoacidophile *Sulfolobus solfataricus* (see Table 6). This enzyme is S-stereoselective with a broad substrate spectrum and is optimally active at 95°C (d'Abusco et al., 2001). Very recently, the first thermoactive and thermostable amidase from the thermophilic actinomycete *Pseudonocardia thermophila* has been purified and characterized (Egorova et al., 2004). The amidase is active at a broad pH (4 to 9) and temperature range (40 to 80°C) and has a half-life of 1.2 h at 70°C. The amidase has a broad substrate spectrum, including aliphatic, aromatic, and amino acid amides. The amidase is highly S-stereoselective for 2-phenylpropionamide with an enantiomeric excess of >95% at 50% conversion of the substrate. Nitrilases (EC 3.5.5.1) are thiol enzymes that convert nitriles directly to the corresponding carboxylic acids with release of ammonia. Recently, the first archaeal nitrilase from the hyperthermophile *Pyrococcus abyssi*, regiospecific toward aliphatic dinitriles, was cloned and characterized in our laboratory (Mueller et al., 2006). The enzyme is highly thermostable, having a half-life at 90°C for 6 h. Thermoactive nitrilases described so far were isolated from the bacteria *Acidovorax facilis* 72W and *Bacillus pallidus* Dac521 (Mueller et al., 2006).

Other Extremozymes of Industrial Relevance

Due to their chiral specificity in the synthesis of acylated amino acids, aminoacylases (EC 3.5.1.14) are attractive candidates for application in fine chemistry (Story et al., 2001). The L-aminoacylase from *Thermococcus litoralis* accepts a wide range of amino acid side chains and N-protecting groups and was recently commercialized (Toogood et al., 2002). The application of the thermostable enzyme reduces the process time, simplifies filtration procedure, improves substrate solubility, and increases the enantiomeric excess to 99%. In contrast to the chemical process, the reaction completes overnight at 70°C, which avoids boiling in 20 equivalent volumes of 6 M HCl for 2 days (Taylor et al., 2004). Two thermostable zinc-containing aminoacylases were also characterized from *Pyrococcus* species (Ishikawa et al., 2001; Story et al., 2001) (see Table 6).

Thermostable β-galactosidase is potentially useful for whey utilization and for the preparation of low-lactose milk and other dairy products or it can be used as a catalyst in the synthesis of galactooligosaccharides, using lactose as substrate and a nucleophile. A β-galactosidase from the halophilic archaeon *Haloferax lucentensis* was cloned and expressed in *Haloferax volcanii*, a widely used strain that lacks detectable β-galactosidase activity (Holmes and Dyall-Smith, 2000). The extremely halotolerant β-galactosidase is optimally active at 4 M NaCl, cleaves several different β-galactoside substrates, and does not exhibit β-glucosidase, β-arabinosidase or β–xylosidase activities. Also a β-galactosidase from *Pyrococcus woesei* was cloned and characterized (Dabrowski et al., 2000).

In recent years carotenoids have gained importance in nutraceutical field. These pigments have been shown to possess physiological function in the prevention of cancer and heart diseases, enhancing in vitro antibody production and as precursors for vitamins. The majority of carotenoids are synthesized from lycopene. β-carotene, the precursor of vitamin A, is biosynthesized directly from lycopene by β-cyclization at both termini, and the reaction is catalyzed by lycopene β-cyclase. Recently, lycopene β-cyclase was predicted in the carotenogenic gene cluster in the genome of the thermoacidophilic archaeon *Sulfolobus solfataricus* (Hemmi et al., 2003). The recombinant expression of the gene in *E. coli* resulted in the accumulation of lycopene β-carotene in the cells. Due to its great antioxidant activity canthaxanthin has been used as food and feed additive, in cosmetics and pharmaceuticals. Recently, the production of this pigment by the archaeon *Haloferax alexandrinus* was investigated. The highest production of the pigment (2.19 µg/l) was obtained during batch fermentation of the strain on a medium with 25% salinity. Other carotenoids such as 3-hydroxy echinenone and trans-astaxanthin are also produced by archaea (Asker and Ohta, 2002).

Sulfur-containing organic compounds have been synthesized mainly chemically. Due to the side reactions, the chemical synthesis of those molecules results in the unavoidable production of impurities in the product and environmental pollution by the formation of by-products such as sulfur oxides. In order to overcome these problems, a method for the synthesis of sulfur-containing organic compounds using O-acetylserine sulfhydrylase has been proposed (Ishikawa and Mino, 2004). A recombinant cysteine synthase from *Aeropyrum pernix* is highly stable within pH 6 to 10 and resistant to organic solvents. Due to its high heat resistance, the enzyme can act on highly concentrated substrate solutions compared to mesophilic thermolabile cysteine synthases.

Phosphatidylethanolamine N-methyltransferase plays a key role in the synthesis of phosphatidylcholine, a main component of liposomal membrane, which is present in various foods as digestible surfactant. It plays an important role in medicine as a component of microcapsule for drugs. A phosphatidylethanolamine N-methyltransferase from *Pyrococcus horikoshii* was cloned and expressed in *E. coli*. The enzyme is thermostable and is active in organic solvents. This opens the possibility to develop a new process for the synthesis of polar lipids with high optical purity (Matsui et al., 2002b).

WHOLE CELL BIOCATALYSIS

Biomining

The development of industrial mineral processing has been established in several countries, such as South Africa, Brazil, and Australia. Iron- and sulfur-oxidizing microorganisms are used to release occluded gold from mineral sulfides. Most industrial plants for biooxidation of gold-bearing concentrates have been operated at 40°C with mixed cultures of mesophilic bacteria of the genera *Thiobacillus* or *Leptospirillum*. In subsequent studies a dissimilatory iron-reducing archaea *Pyrobaculum islandicum* and *Pyrococcus furiosus* were shown to reduce gold chloride to insoluble gold (Lloyd et al., 2003). The potential of thermophilic sulfide-oxidizing archaea in copper extraction has attracted interest due to the efficient extraction of metals from sulfide ores that are recalcitrant to dissolution (Schiraldi et al., 2002). The acidophilic archaea *Sulfolobus metallicus* and *Metallosphaera sedula* tolerate up to 4% of copper and have been exploited for mineral biomining (Norris et al., 2000). The efficiency of copper extraction from chalcopyrite by thermoacidophilic archaea was influenced by the characteristics of mineral concentrates. Between 40 and 60% copper extraction was achieved in primary reactors and more than 90% extraction in secondary reactors with overall residence times of about 6 days (Norris et al., 2000).

The handling and recycling of spent tyres are a significant and worldwide problem. The reuse of rubber material is preferable from an economic and environmental point of view. The anaerobic sulfate-reducing thermophilic archaeon *Pyrococcus furiosus* was investigated for its capacity to desulfurize rubber. The tyre rubber treated with *P. furiosus* for 10 days was subsequently vulcanized with virgin rubber material (15%, wt/wt). This results in the desulfurization of ground rubber and leads to a product with good mechanical properties (Bredberg et al., 2001). The thermoacidophilic archaeon *Sulfolobus acidocaldarius* has also been tested for desulfurization of rubber material (Bredberg et al., 2001).

Decontamination and Hydrogen Production

Microorganisms, growing in the presence of toxic chemicals, heavy metals, halogenated solvents and radionuclides, can be used to detoxify those compounds during treatment of wastes. The biological treatment of a synthetic saline wastewater, containing diluted molasses and ures, was investigated. High removal efficiencies were obtained at salt concentrations above 4% using immobilized cells of *Halobacterium halobium*. Extremely halophilic archaea, such as *Haloferax mediterranei*, were found to utilize crude oil even at high salinities and therefore could be used in bioremediation of polluted sites. The strain *Halobacterium* sp. degrades *n*-alkanes with a C_{10}–C_{30} composition in the presence of 30% NaCl. Halophilic archaea belonging to genera *Haloarcula*, *Halobacterium* and *Haloferax* degrade halogenated hydrocarbons, such as trichlorophenols or the insecticides lindane and DDT. It has been recently demonstrated that the hyperthermophilic anaerobic archaeon *Ferroglobus placidus* is capable of oxidizing aromatic compounds with the reduction of Fe^{3+}. Such microorganisms can be useful when heat treatment is employed to aid in the extraction of organic contaminants that are trapped in sediments. Textile wastewater is often of high salinity and therefore problematic for the conventional biological treatment. The use of halophilic microorganisms for the biodegradation of the segregated dye bath has been reported (Oren, 2002).

There is an increasing interest in the utilization of renewable sources to satisfy the exponentially growing energy needs. Research on biological hydrogen production became attractive due to the possible use of biohydrogen as a clean energy carrier and raw material. The production of hydrogen in photobiological or heterotrophic fermentation routes depends on the supply of organic substrates and could therefore be ideally suited for coupling energy production with treatment of organic wastes. A two-stage fermentation system was constructed for the production of biohydrogen from keratin-rich biowaste (Balint et al., 2005). First, the bacterial strain *Bacillus licheniformis* KK1 was employed to convert keratin-containing waste into a fermentation product that is rich in amino acids and peptides. In the next stage the thermophilic anaerobic archaeon *Thermococcus litoralis* was fermented on the hydrolysate and hydrogen was produced. Also archaeal hydrogenases have been the target of intensive research. A cytosolic NiFe-hydrogenase from the hyperthermophilic archaeon *Thermococcus kodakaraensis* is optimally active at 90°C for hydrogen production with methyl viologen as the electron carrier (Kanai et al., 2003). A membrane bound NiFe-hydrogenase, responsible for hydrogen production, was also identified in the anaerobic bacterium *Thermoanaerobacter tengcongensis* (Soboh et al., 2004). Other thermophilic bacteria of the order *Thermotogales* have demonstrated the ability to produce hydrogen (van Ooteghem, et al., 2002; 2004; Kadar et al., 2003).

MEMBRANE LIPIDS, PROTEINS AND POLYMERS

Lipids

Liposomes are artificial, spherical, closed vesicles consisting of one or more lipid bi-layers. Liposomes made from ether phospholipids have been studied extensively over the last thirty years as artificial membrane models with remarkable thermostability and tightness against solute leakage. Considerable interest has been generated for applications of liposomes in medicine, including their use as diagnostic agents, as carrier vehicles in vaccine formulations, or as delivery systems for drugs, genes or cancer-imaging agents (Sprott et al., 2003). In general, archaeosomes (liposomes from archaea) demonstrate higher stability to oxidative stress, high temperature, alkaline pH, to attack phospholipases, bile salts, and serum proteins. Some archaeosome formulations can be sterilized by autoclaving without problems of fusion or aggregation of the vesicles. The uptake of archaeosomes by phagocytic cells can be up to 50-fold higher than that of conventional liposomes (Patel and Sprott, 1999).

Ether-linked lipids from halophilic archaea have high chemical stability and resistance against esterases and thus a higher survival rate than liposomes based on fatty acid derivatives (Margesin and Schinner, 2001). Novel, patented ether lipids, prepared by pressure extrusion from the halophile *Halobacterium cutirubrum*, were resistant to attack by phospholipases and could be stored for more than 60 days. Cyclic and acyclic dibiphytanylglycerol tetraether lipids were identified in nonthermophilic crenarchaeotes (DeLong et al., 1998). The immune stimulating activity of the lipid vesicles, prepared from an archaeon *Haloferax volcanii*, was investigated and an increase of immune responses was observed. The unique ability of the archaeosomes, consisting of sulfoglycolipids, phosphoglycerols, and cardiolipins, to maintain antigen-specific T cell immunity may be attribute to a property of the archaeal 2,3-diphytanylglycerol lipid core (Sprott et al., 2003). Due to bipolar tetraether structure, archaeal lipids have also been proposed as monomers for bioelectronics (De Rosa et al., 1994).

Crystalline cell surface layers (S-layers) that are composed of protein and glycoprotein subunits are one of the most commonly observed cell envelope structures of bacteria and archaea. S-layers could be produced in large amounts by continuous cultivation of S-layer-carrying microorganisms and used as isoporous ultrafiltration membranes or as matrices for immobilization of biologically active macromolecules such as enzymes, ligands or monoclonal and polyclonal antibodies (Sleytr et al., 1997). S-layers have been shown to be excellent patterning structures in molecular nanotechnology because of their high molecular order, high binding capacity and ability to recrystallize with perfect uniformity on solid surfaces, at the water/air interface or on lipid films. The two-dimensionally organized S-layers of *Sulfolobus acidocaldarius* are suggested to be of practical use as biomimetic templates for material deposition and fabrication of advanced materials (Sleytr et al., 1997).

Proteins and Peptides

Production of antibiotic peptides and proteins is a near-universal feature of living organisms regardless of phylogenetic classification. Antimicrobial agents from bacteria and eucarya have been studied for more than fifty years. However, archaeal strains are just in the beginning of investigation for the production of peptide antibiotics. A variety of halocins have been detected in halophilic archaea. These antimicrobial agents are diverse in size, consisting of proteins as large as 35 kDa and peptide "microhalocins" as small as 3.6 kDa (O'Connor and Shand, 2002). Microhalocins with unclear mechanism of action are hydrophobic and robust, withstanding heat, desalting and exposure to neutral residues and are not cationic. The microhalocins S8 and R1 lack the biochemical and structural properties demonstrated by other antibiotics, suggesting that their mechanisms of action should be novel. The halocin H7 has been suggested for reducing injury during organ transplantation. Archaeocins are also produced by a thermoacidophilic *Sulfolobus* strain. The 20 kDa protein antibiotics are not excreted and are associated with small particles apparently derived from the cells' S-layer (O'Connor and Shand, 2002).

Chaperones and chaperonins of extremophilic microorganisms are useful agents in protein refolding, stabilization, and solubilization of recombinant proteins (Maruyama et al., 2004; Ideno et al., 2004). Peptidyl prolyl *cis-trans* isomerase (PPIase) is involved in many processes such as regeneration of denatured protein, stabilization of proteins, production of recombinant protein and development of novel immunosuppressant and physiologically active substances. A novel cyclophilin-type PPIase derived from *Halobacterium cutirubrum* was characterized (Margesin and Schinner, 2001).

Extremely halophilic archaea often contain membrane-bound retinal pigments, bacteriorhodopsin and halorhodopsin, which enable them to use light energy directly for bioenergetic processes by the generation of proton and chloride gradients. Bacteriorhodopsin is a protein with seven helical protein segments in its transmembrane domain. The excellent thermodynamic and photochemical stability of bacteriorhodopsin has led to many technical applications based on its protonmotive, photoelectric, and photochemical properties. The applications comprise holography, spatial light modulators, artificial retina, neutral network optical computing and volumetric and associative optical memories. The optical properties of bacteriorhodopsin could be exploited to manufacture electronic ink for laptop displays, which will be an important contribution to the problem of battery lifetime in portable computing. Another application of bacteriorhodopsin is the renewal of biochemical energy by conversion of ADP to ATP. Such a solar-driven recycling system could be of interest for biotechnological processes that require large amounts of expensive ATP. A patented ATP-synthesizing device, useful for bioelements, has been obtained by using bacteriorhodopsin and ATP synthase (Margesin and Schinner, 2001). Bacteriorhodopsin is commercially offered in the form of purple membrane patches, isolated from *Halobacterium salinarum* S9.

Biopolymers

Polyhydroxyalkanoates are microbial storage compounds with properties comparable to those of polyethylene and polypropylene. Such biodegradable plastics could replace oil-derived thermoplastics. The archaeon *Haloferax mediterranei* accumulates poly(β-hydroxy butyric acid) up to 60% of cell dry weight. The production could be enhanced to 6 g/l using phosphate limitation and starch as carbon source (Hezayen et al., 2000). The polymer can be easily recovered using cell lysis caused by an exposure of the cultures to low salt concentrations. The maximum production of poly(β-hydroxy butyric acid) (up to 53% of cell dry weight) by another halophilic archaeon was reached after 11 days of fermentation on *n*-butyric acid and sodium acetate as carbon sources (Hezayen et al., 2000). In contrast to *Haloferax mediterranei*, phosphate does not affect the polyester production by the strain. The accumulated polyester was recovered at 87% using chloroform extraction. The exopolymer poly(γ-D-glutamic acid) can be used as a biodegradable thickener, humectant sustained-release material or drug carrier in the food or pharmaceutical industries. The extreme halophilic archaeon *Natrialba aegyptiaca* starts to produce the polymer at 20% NaCl and the maximum production is reached at NaCl saturation. After 90 h of growth in a corrosion-resistant bioreactor, 470 mg/l of the polymer was produced (Hezayen et al., 2000; 2001).

Microbial exopolysaccharides are high-molecular mass polymers composed mainly of carbohydrates excreted by bacteria and archaea. Among the family *Halobacteriaceae* several species of the genus *Haloferax* have been described as producers of extracellular polysaccharides. The high viscosity at low concentrations, excellent rheological properties, and the remarkable tolerance to high pH, temperature, and salinity make these polymers suitable as emulsifying agents and mobility controller in microbial enhanced oil recovery. Furthermore, several biosurfactant producers such as *Methanothermobacter thermoautotrophicum* produce bioemulsifiers active over a wide range of pH (5 to 10) and at high salinities (up to 20%) (Oren, 2002).

COMPATIBLE SOLUTES

Accumulation of osmotically active substances, the so-called compatible solutes, by uptake or de novo synthesis, enables microorganisms to reduce the difference between osmotic potentials of the cell cytoplasm and the extracellular environment. Those compounds are highly water-soluble sugars or sugar alcohols, other alcohols, amino acids or their derivatives. They gained an increasing attention in biotechnology due to their action as stabilizers of biomolecules (enzymes, DNA, membranes, tissues) and stress-protecting agents (Table 7) (Borges et al., 2002). Additionally, compatible solutes support the high-yield periplasmic production of functional active recombinant proteins in different expression systems (Barth et al., 2000). Di-*myo*-inositol-1,1′-phosphate is the most widespread solute of hyperthermophilic archaea and was not detected in a mesophile. This thermoprotective compound was found in a variety of strains, belonging to the genera *Methanococcus, Pyrococcus, Pyrodictium, Pyrolobus,* and *Thermococcus* (Santos and da Costa, 2002). In most of these archaea, an increase of the solute concentrations is observed at growth temperatures above the optimum, reaching 20-fold in case of *Pyrococcus furiosus* grown at 101°C. In contrast, the concentration of mannosylglycerate, detected in the euryarchaeotes of the genera *Archaeoglobus, Pyrococcus, Thermococcus, Methanothermus fervidus* and in the crenarchaeote *Aeropyrum pernix*, increases concomitantly with the salinity of the medium and serves therefore as a compatible solute under salt stress. Mannosylglycerate has been observed to have a profound effect on thermoprotection and protection against desiccation of enzymes from mesophilic, thermophilic and hyperthermophilic microorganisms (Egorova and Antranikian, 2005). The biosynthetic routes for the synthesis of mannosylglycerate in the archaeon *Pyrococcus horikoshii* and di-*myo*-inositol-1,1′-phosphate in *Pyrococcus woesei* and *Methanococcus igneus* have been investigated (Empadinhas et al., 2001; Chen et al., 1998, Scholz et al., 1998). The hyperthermophilic archaeon *Archaeoglobus fulgidus* accumulates a very rare compound diglycerol phosphate under salt and temperature stress. This solute demonstrated a considerable stabilizing effect against heat inactivation of various dehydrogenases and a strong protective effect on bacterial rubredoxins (with a four-fold increase in the half-lives) (Lamosa et al., 2000). A compatible solute, cyclic 2,3-bisphosphoglycerate, has been detected only in methanogenic archaea such as *Methanopyrus kandleri*. The thermoprotective role of this solute was proven by in vitro studies showing that the solute protects selected enzymes from *Methanopyrus kandleri* against thermal denaturation (Santos and Da Costa, 2001). Another compatible solute, restricted to methanogenic archaea, is a derivative of β-amino acid (N^ε-acetyl-β-lysine), which is synthesized in response to increasing osmotic stress in both marine and nonmarine species of methanogenes. The first genes involved in the biosynthesis of the solute have

Table 7. Other applications of extremophilic microorganisms

Product	Strain	Application	References
Bacteriorhodopsin	*Halobacterium salinarum*	Holography, color-sensors, neural network optical computing, spatial light modulators	Margesin and Schinner, 2001
Carotenoids (canthaxanthin)	*Haloferax alexandrinus*	Food and feed additives, cosmetics	Asker and Ohta, 2002
Chaperons, chaperonins, maltodextrin-binding proteins, peptidyl-prolyl *cis-trans* isomerases	*Halobacterium cutirubrum* *Haloferax volcanii* *Methanothermococcus* sp. *Pyrococcus* spp. *Pyrococcus horikoshii* *Sulfolobus shibatae* *Thermococcus* spp. *Thermus* spp.	Stabilization and solubilization of recombinant proteins	Laksanalamai et al., 2003; Maruyama et al., 2004; Suzuki et al., 2003; Iida et al., 2000; Ideno et al., 2004; Lund et al., 2003; Fox et al., 2003; Yan et al., 1997
Compatible solutes (mannosylglycerate, mannosylglyceramide, diglycerol phosphate, di-*myo*-inositol-phosphate, N-acetyl-β-lysine, trehalose, 2-sulfotrehalose, cyclic-2, 3-bisphosphoglycerate)	*Aeropyrum pernix* *Archaeoglobus* spp. *Halomonas elongata* *Methanococcus igneus* *Methanopyrus kandleri* *Methanosarcina thermophila* *Methanothermus fervidus* *Natronobacterium* spp. *Natronococcus* sp. *Pyrobaculum aerophilum* *Pyrococcus* spp. *Pyrodictium occultum* *Pyrolobus fumarii* *Rhodothermus marinus* *Thermococcus* spp. *Thermus* spp.	Cosmetics, biomolecules and tissue stabilizers, molecular biology	Santos and da Costa, 2002; Borges et al., 2002; Pfluger et al., 2003; Barth et al., 2000; Desmarais et al., 1997
Cytochrome P450	*Sulfolobus solfataricus* *Thermus thermophilus*	Selective regiospecific and stereospecific hydroxylations in chemical synthesis	Vieille and Zeikus, 2001; Yano et al., 2003
Expression systems (vector/host)	*Haloarcula hispanica* *Halobacterium salinarum* *Haloferax volcanii* *Methanococcus* spp. *Sulfolobus solfataricus*	Production of recombinant proteins	Jonuscheit et al., 2003; Zhou et al., 2004; Aagaard et al., 1996; Holmes et al., 1994; Cannio et al., 2001; Kaczowka et al., 2005
Ni-Fe-hydrogenase	*Thermococcus kodakaraensis* *Caldicellulosiruptor saccharolyticus* *Carboxydothermus hydrogenoformans*	H_2 production	Kanai et al., 2003; Kadar et al., 2003; Soboh et al., 2002; Soboh et al., 2004; van Ooteghem et al., 2002; van Ooteghem et al., 2004

Peptides, proteins: Sulfolobicin, Halocins	*Fervidobacterium pennavorans* *Thermoanaerobacter tengcongensis* *Thermotoga elfii* *Thermotoga neapolitana* *Sulfolobus islandicus* *Halobacterium salinarum* *Haloferax mediterranei* *Haloferax gibbonsii* Strain S8a (archaea)	Antibiotics	O'Connor and Shand, 2002
Polymers: exopolysaccharides,	*Haloferax* spp., *Halobacterium* spp.	Emulsifiers	Margesin and Schinner, 2001
poly(γ-D-glutamic acid),	*Natrialba aegyptiaca*	Food and pharmaceuticals	Hezayen et al., 2001
poly(β-hydroxy butyric acid)	*Haloferax mediterranei* Strain 56 (archaea)	Bioplastics	Hezayen et al., 2000
S-Layer proteins, lipids, liposomes	*Halobacterium* spp. *Haloferax volcanii* *Haloarcula japonica* *Methanobrevibacter smithii* *Methanococcus* spp. *Methanothermus* spp. *Natronobacterium* spp. *Staphylothermus marinus* *Sulfolobus solfataricus*	Vaccine development, diagnostics, biomimetics, drugs, nanotechnology	Sprott et al., 2003; Eichler, 2003; DeLong et al, 1998; Patel and Sprott, 1999; Sleytr et al., 1997; Upreti et al, 2003
Whole cell biocatalysis	*Palaeococcus ferrophilus* *Thermococcus barophilus*	Formation of gels and starch granules	Abe and Horikoshi, 2001; Gomes and Steiner, 2004
	Methanogenic archaea *Deinococcus radiodurans*	Methane production, low temperature waste treatment	Schiraldi et al., 2002; Scherer et al, 2000 Norris et al., 2000; Jolivet et al., 2004; Jolivet et al., 2003
	Haloarcula spp. *Halobacterium* spp. *Haloferax* spp. *Thermococcus gammatolerans* *Thermococcus marinus* *Thermococcus radiotolerans* *Sulfolobus metallicus*	Detoxification of halogenated organic compounds and toxic chemicals, heavy metals, nuclear waste treatment	
	Pyrococcus furiosus	Rubber recycling	Bredberg et al., 2001

been recently identified (Pfluger et al., 2003). A novel compound, 2-sulfotrehalose, was found to be the major organic solute accumulated by halophilic archaea from the genera *Natronococcus* and *Natronobacterium*, which grow under high salt und high pH conditions (Desmarais et al., 1997).

ADVANTAGES AND LIMITATIONS OF USING EXTREMOPHILES

Running processes under extreme conditions such as elevated temperature has many advantages including increased solubility, decreasing viscosity, increased bioavailability of polymeric substrates and the decreased risk of contamination. Stable enzymes derived from extremophilic archaea are active in organic solvents and detergents, and they are more resistant to proteolytic attack. The efficient separation of recombinant extremozymes from mesophilic production hosts can be achieved by simple treatment such as heat denaturation, extraction with organic solvents or increased salinity. Consequently, the delivery of sufficient amount of enzymes for industrial trials can be obtained. The availability of extremozymes capable of catalysis at high pressures will also offer a novel biotechnological alternative to currently running processes, especially in food industry. During processing and sterilization of food materials, high pressure can be used to induce the formation of gels and starch granules, the denaturation/coagulation of proteins or the transition of lipid phases. The use of high pressure leads to better flavor and color preservation. On the other hand, a number of limitations did not allow so far the broad application of extremophiles. These limitations include difficulties associated with large-scale cultivation of extremophiles, nonefficient systems for the overexpression of archaeal genes and unknown factors that confer enzyme stability under extremes of temperature, pH, and pressure.

Molecular cloning of the archaeal genes and their expression in heterologous hosts circumvent the problem of its insufficient expression in natural hosts. A variety of archaeal enzymes have been cloned and successfully expressed in mesophilic hosts such as *E. coli, Bacillus subtilis* and yeasts. In *E. coli* the overexpression of proteins was optimized using cotransformation of the host cells with a plasmid encoding tRNA synthetases for low-frequency codons. Recombinant enzymes could also be obtained by cloning and expressing a synthetic gene with a codon usage optimized for the corresponding host. To overcome the limitations of different codon usage, the production of recombinant proteins can be increased using extremophilic microorganisms as hosts for autologous gene expression. Therefore, expression systems using extremophilic archaea as production hosts have to be developed, including hyperthermophiles, thermoacidophiles, and halophiles. Expression systems in *Haloferax, Methanococcus,* and *Sulfolobus* species have been recently constructed (Gardner and Whitman, 1999).

CONCLUSIONS

Owing to their properties such as activity over a wide temperature and pH range, substrate specificity, stability in organic solvents, diverse substrate range and enantioselectivity, extremophiles and their enzymes will represent the choice for future countless applications in industry. The growing demand for more robust biocatalysts has shifted the trend toward improving the properties of existing proteins for established industrial processes and producing new enzymes tailor-made for entirely new areas of application. The new technologies such as genomics, metanogenomics, gene shuffling, and mutagenesis provide valuable tools for improving or adapting enzyme properties to the desired requirements. However, the success of these techniques demands the production of recombinant enzymes at high level allowing experimental trials and application tests. Thus, the modern methods of genetic engineering combined with an increasing knowledge of structure and function and process engineering will allow further adaptation of biocatalysts to industrial needs, the exploration of novel applications and protection of the environment.

REFERENCES

Aagaard, C., I. Leviev, R. N. Aravalli, P. Forterre, D. Prieur, and R. A. Garrett. 1996. General vectors for archaeal hyperthermophiles: strategies based on a mobile intron and a plasmid. *FEMS Microbiol. Rev.* 18:93–104.

Abe, F., and K. Horikoshi. 2001. The biotechnological potential of piezophiles. *Trends Biotechnol.* 19:102–108.

Akiyama, K., M. Takase, K. Horikoshi, and S. Okonogi. 2001. Production of galactooligosaccharides from lactose using a beta-glucosidase from *Thermus* sp. Z-1. *Biosci. Biotechnol. Biochem.* 65:438–441.

Albertson, G. D., R. H. McHale, M. D. Gibbs, and P. L. Bergquist. 1997. Cloning and sequence of a type I pullulanase from an extremely thermophilic anaerobic bacterium, *Caldicellulosiruptor saccharolyticus. Biochim. Biophys. Acta.* 1354:35–39.

Alcalde, M., F. J. Plou, M. Teresa Martin, I. Valdes, E. Mendez, and A. Ballesteros. 2001. Succinylation of cyclodextrin glycosyltransferase from *Thermoanaerobacter* sp. 501 enhances its transferase activity using starch as donor. *J. Biotechnol.* 86:71–80.

Alcantara, A. R., J. M. Sanchez-Montero, and J. V. Sinisterra. 2000. Chemoenzymatic preparation of enantiomerically pure S(+)-2-arylpropionic acids with antiinflammatory activity.

In: R. Patel (ed.) *Stereoselective biocatalysis*. Marcel Dekker, New York and Basel. 659-702.

Ando, S., H. Ishida, Y. Kosugi, and K. Ishikawa. 2002. Hyperthermostable endoglucanase from *Pyrococcus horikoshii*. *Appl. Environ. Microbiol.* 68:430–433.

Andrade, C. M., W. B. Aguiar, and G. Antranikian. 2001. Physiological aspects involved in production of xylanolytic enzymes by deep-sea hyperthermophilic archaeon *Pyrodictium abyssi*. *Appl. Biochem. Biotechnol.* 91–93:655–669.

Andronopoulou, E., and C. E. Vorgias. 2004a. Isolation, cloning, and overexpression of a chitinase gene fragment from the hyperthermophilic archaeon *Thermococcus chitonophagus*: semi-denaturing purification of the recombinant peptide and investigation of its relation with other chitinases. *Protein Expr. Purif.* 35:264–271.

Andronopoulou, E., and C. E. Vorgias. 2004b. Multiple components and induction mechanism of the chitinolytic system of the hyperthermophilic archaeon *Thermococcus chitonophagus*. *Appl. Microbiol. Biotechnol.* 65:694–702.

Andronopoulou, E., and C. E. Vorgias. 2003. Purification and characterization of a new hyperthermostable, allosamidin-insensitive and denaturation-resistant chitinase from the hyperthermophilic archaeon *Thermococcus chitonophagus*. *Extremophiles* 7:43–53.

Antoine, E., J.-L. Rolland, J.-P. Raffin, and J. Dietrich. 1999. Cloning and over-expression in *Escherichia coli* of the gene encoding NADPH group III alcohol dehydrogenase from *Thermococcus hydrothermalis*. *Eur. J. Biochem.* 264: 880–889.

Antranikian, G., C. Vorgias, and C. Bertoldo. 2005. Extreme environments as a resource for microorganisms and novel biocatalysts. *Adv. Biochem. Eng. Biotechnol.* 96:In press.

Asker, D., and Y. Ohta. 2002. Production of canthaxanthin by *Haloferax alexandrinus* under non-aseptic conditions and a simple, rapid method for its extraction. *Appl. Microbiol. Biotechnol.* 58:743–750.

Bachmann, S. L., and A. J. McCarthy. 1991. Purification and cooperative activity of enzymes constituting the xylan-degrading system of *Thermomonospora fusca*. *Appl. Environ. Microbiol.* 57:2121–2130.

Balint, B., Z. Bagi, A. Toth, G. Rakhely, K. Perei, and K. L. Kovacs. 2005. Utilization of keratin-containing biowaste to produce biohydrogen. *Appl. Microbiol. Biotechnol.* 69:404–410

Ballschmiter, M., M. Armbrecht, K. Ivanova, G. Antranikian, and W. Liebl. 2005. AmyA, an alpha-amylase with beta-cyclodextrin-forming activity, and AmyB from the thermoalkaliphilic organism *Anaerobranca gottschalkii*: two alpha-amylases adapted to their different cellular localizations. *Appl. Environ. Microbiol.* 71:3709–3715.

Ballschmiter, M., O. Futterer, and W. Liebl. 2006. Identification and characterization of a novel intracellular alkaline alpha-Amylase from the hyperthermophilic bacterium *Thermotoga maritima* MSB8. *Appl. Environ. Microbiol.* 72:2206–2211.

Barth, S., M. huhn, B. Matthey, A. Klimka, E. Galinski, and A. Engert. 2000. Compatible-solute-supported periplasmic expression of functional recombinant proteins under stress conditions. *Appl. Environ. Microbiol.* 66:1572–1579.

Bauer, M. W., L. E. Driskill, W. Callen, M. A. Snead, E. J. Mathur, and R. M. Kelly. 1999. An endoglucanase, EglA, from the hyperthermophilic archaeon *Pyrococcus furiosus* hydrolyzes beta-1,4 bonds in mixed-linkage $(1 \rightarrow 3),(1 \rightarrow 4)$-beta-D-glucans and cellulose. *J. Bacteriol.* 181:284–290.

Bauer, M. W., E. J. Bylina, R. V. Swanson, and R. M. Kelly. 1996. Comparison of a beta-glucosidase and a beta-mannosidase from the hyperthermophilic archaeon *Pyrococcus furiosus*. Purification, characterization, gene cloning, and sequence analysis. *J. Biol. Chem.* 271,39:23749–23755.

Bechtereva, T. A., Y. I. Pavlov, V. I. Kramorov, B. Migunova, and O. I. Kiselev. 1989. DNA sequencing with thermostable Tet DNA polymerase from *Thermus thermophilus*. *Nucleic Acids Res.* 17:10507.

Beg, Q. K., M. Kapoor, L. Mahajan, and G. S. Hoondal. 2001. Microbial xylanases and their industrial applications: a review. *Appl. Microbiol. Biotechnol.* 56:326–338.

Beki, E., I. Nagy, J. Vanderleyden, S. Jager, L. Kiss, L. Fulop, L. Hornok, and J. Kukolya. 2003. Cloning and heterologous expression of a beta-D-mannosidase (EC 3.2.1.25)-encoding gene from *Thermobifida fusca* TM51. *Appl. Environ. Microbiol.* 69:1944–1952.

Bertoldo, C., F. Duffner, P. L. Jorgensen, and G. Antranikian. 1999. Pullulanase type I from *Fervidobacterium pennavorans* Ven5: cloning, sequencing, and expression of the gene and biochemical characterization of the recombinant enzyme. *Appl. Environ. Microbiol.* 65:2084–2091.

Bertoldo, C., M. Armbrecht, F. Becker, T. Schafer, G. Antranikian, and W. Liebl. 2004. Cloning, sequencing, and characterization of a heat- and alkali-stable type I pullulanase from *Anaerobranca gottschalkii*. *Appl. Environ. Microbiol.* 70:3407–3416.

Bertoldo, C., and G. Antranikian. 2002. Starch-hydrolyzing enzymes from thermophilic archaea and bacteria. *Curr. Opin. Chem. Biol.* 6:151–160.

Biles, B. D., and B. A. Connolly. 2004. Low-fidelity *Pyrococcus furiosus* DNA polymerase mutants useful in error-prone PCR. *Nucleic Acids Res.* 32:e176.

Birgisson, H., O. Fridjonsson, F. K. Bahrani-Mougeot, G. O. Hreggvidsson, J. K. Kristjansson, and B. Mattiasson. 2004. A new thermostable alpha-L-arabinofuranosidase from a novel thermophilic bacterium. *Biotechnol. Lett.* 26:1347–1351.

Biwer, A., G. Antranikian, and E. Heinzle. 2002. Enzymatic production of cyclodextrins. *Appl. Microbiol. Biotechnol.* 59: 609–617.

Blondal, T., S. H. Thorbjarnardottir, J. Kieleczawa, S. Hjorleifsdottir, J. K. Kristjansson, J. M. Einarsson, and G. Eggertsson. 2001. Cloning, sequence analysis and functional characterization of DNA polymerase I from the thermophilic eubacterium *Rhodothermus marinus*. *Biotechnol. Appl. Biochem.* 34:37–45.

Bohlke, K., F. M. Pisani, C. E. Vorgias, B. Frey, H. Sobek, M. Rossi, and G. Antranikian. 2000. PCR performance of the B-type DNA polymerase from the thermophilic euryarchaeon *Thermococcus aggregans* improved by mutations in the Y-GG/A motif. *Nucleic Acids Res.* 28:3910–3917.

Bok, J. D., D. A. Yernool, and D. E. Eveleigh. 1998. Purification, characterization, and molecular analysis of thermostable cellulases CelA and CelB from *Thermotoga neapolitana*. *Appl. Environ. Microbiol.* 64:4774–4781.

Borges, N., A. Ramos, N. D. Raven, R. J. Sharp, and H. Santos. 2002. Comparative study of the thermostabilizing properties of mannosylglycerate and other compatible solutes on model enzymes. *Extremophiles* 6:209–216.

Bredberg, K., J. Persson, M. Christiansson, B. Stenberg, and O. Holst. 2001. Anaerobic desulfurization of ground rubber with the thermophilic archaeon *Pyrococcus furiosus*—a new method for rubber recycling. *Appl. Microbiol. Biotechnol.* 55:43–48.

Bronnenmeier, K., K. Kundt, K. Riedel, W. H. Schwarz, and W. L. Staudenbauer. 1997. Structure of the *Clostridium stercorarium* gene celY encoding the exo-1,4-beta-glucanase Avicelase II. *Microbiology* 143:891–898.

Brown, S., and R. Kelly. 1993. Characterization of amylolytic enzymes, having both a-1,4 and a-1,6 hydrolytic activity, from the thermophilic archaea *Pyrococcus furiosus* and *Thermococcus litoralis*. *Appl. Environ. Microbiol.* 59:2614–2621.

Buchanan, C. L., H. Connaris, M. J. Danson, C. D. Reeve, and D. W. Hough. 1999. An extremely thermostable aldolase from

Sulfolobus solfataricus with specificity for non-phosphorylated substrates. *Biochem. J.* **343 Pt 3:**563–570.

Burlini, N., P. Magnani, A. Villa, F. Macchi, P. Tortora, and A. Guerritore. 1992. A heat-stable serine proteinase from the extreme thermophilic archaebacterium *Sulfolobus solfataricus*. *Biochim. Biophys. Acta.* **1122:**283–292.

Cann, I. K., S. Ishino, N. Nomura, Y. Sako, and Y. Ishino. 1999a. Two family B DNA polymerases from *Aeropyrum pernix*, an aerobic hyperthermophilic crenarchaeote. *J. Bacteriol.* **181:** 5984–5992.

Cann, I. K., S. Kocherginskaya, M. R. King, B. A. White, and R. I. Mackie. 1999b. Molecular cloning, sequencing, and expression of a novel multidomain mannanase gene from *Thermoanaerobacterium polysaccharolyticum*. *J. Bacteriol.* **181:**1643–1651.

Cannio, R., P. Contursi, M. Rossi, and S. Bartolucci. 2001. Thermoadaptation of a mesophilic hygromycin B phosphotransferase by directed evolution in hyperthermophilic Archaea: selection of a stable genetic marker for DNA transfer into *Sulfolobus solfataricus*. *Extremophiles* **5:**153–159.

Cannio, R., N. Di Prizito, M. Rossi, and A. Morana. 2004. A xylan-degrading strain of *Sulfolobus solfataricus*: isolation and characterization of the xylanase activity. *Extremophiles* **8:**117–124.

Catara, G., G. Ruggiero, F. La Cara, F. A. Digilio, A. Capasso, and M. Rossi. 2003. A novel extracellular subtilisin-like protease from the hyperthermophile *Aeropyrum pernix* K1: biochemical properties, cloning, and expression. *Extremophiles* **7:**391–399.

Chang, C., H. K. Song, B. C. Park, D. S. Lee, and S. W. Suh. 1999. A thermostable xylose isomerase from *Thermus caldophilus*: biochemical characterization, crystallization and preliminary X-ray analysis. *Acta Crystallogr. D. Biol. Crystallogr.* **55:**294–296.

Chavez Croocker, P., Y. Sako, and A. Uchida. 1999. Purification and characterization of an intracellular heat-stable proteinase (pernilase) from the marine hyperthermophilic archaeon *Aeropyrum pernix* K1. *Extremophiles* **3:**3–9.

Chen, L., E. T. Spiliotis, and M. F. Roberts. 1998. Biosynthesis of di-myo-inositol-1,1′-phosphate, a novel osmolyte in hyperthermophilic archaea. *J. Bacteriol.* **180:**3785–3792.

Chen, S., A. F. Yakunin, E. Kuznetsova, D. Busso, R. Pufan, M. Proudfoot, R. Kim, and S. H. Kim. 2004a. Structural and functional characterization of a novel phosphodiesterase from *Methanococcus jannaschii*. *J. Biol. Chem.* **279:**31854–31862.

Chen, X. G., O. Stabnikova, J. H. Tay, J. Y. Wang, and S. T. Tay. 2004b. Thermoactive extracellular proteases of *Geobacillus caldoproteolyticus*, sp. nov., from sewage sludge. *Extremophiles* **8:**489–498.

Cheng, T. C., V. Ramakrishnan, and S. I. Chan. 1999. Purification and characterization of a cobalt-activated carboxypeptidase from the hyperthermophilic archaeon *Pyrococcus furiosus*. *Protein Sci.* **8:**2474–2486.

Chi, Y. I., L. A. Martinez-Cruz, J. Jancarik, R. V. Swanson, D. E. Robertson, and S. H. Kim. 1999. Crystal structure of the beta-glycosidase from the hyperthermophile *Thermosphaera aggregans*: insights into its activity and thermostability. *FEBS Lett.* **445:**375–383.

Choi, J. J., E. J. Oh, Y. J. Lee, D. S. Suh, J. H. Lee, S. W. Lee, H. T. Shin, and S. T. Kwon. 2003. Enhanced expression of the gene for beta-glycosidase of *Thermus caldophilus* GK24 and synthesis of galacto-oligosaccharides by the enzyme. *Biotechnol. Appl. Biochem.* **38:**131–136.

Crabb, W. D., and J. Bolin. 1999. Protein technologies and commercial enzymes. *Curr. Opin. Biotechnol.* **10:**321–323.

d'Abusco, A. S., S. Ammendola, R. Scandurra, and L. Politi. 2001. Molecular and biochemical characterization of the recombinant amidase from hyperthermophilic archaeon *Sulfolobus solfataricus*. *Extremophiles* **5:**183–192.

da Costa, M. S., H. Santos, and E. A. Galinski. 1998. An overview of the role and diversity of compatible solutes in Bacteria and Archaea. *Adv. Biochem. Eng. Biotechnol.* **61:**117–153.

Dabrowski, S., and B. Kiaer Ahring. 2003. Cloning, expression, and purification of the His6-tagged hyper-thermostable dUTPase from *Pyrococcus woesei* in *Escherichia coli*: application in PCR. *Protein Expr. Purif.* **31:**72–78.

Dabrowski, S., G. Sobiewska, J. Maciunska, J. Synowiecki, and J. Kur. 2000. Cloning, expression, and purification of the His-tagged thermostable beta-Galactosidase from *Pyrococcus woesei* in *Escherichia coli* and some properties of the isolated enzyme. *Prot. Exp. Pur.* **19:**107–112.

de Pascale, D., I. Di Lernia, M. P. Sasso, A. Furia, M. De Rosa, and M. Rossi. 2002. A novel thermophilic fusion enzyme for trehalose production. *Extremophiles* **6:**463–468.

De Rosa, M., A. Morana, A. Riccio, A. Gambacorta, A. Trincone, and O. Incani. 1994. Lipids of the archaea: a new tool for bioelectronics. *Biosens. Bioelectron.* **9:**669–675.

DeLong, E. F., L. L. King, R. Massana, H. Cittone, A. Murray, C. Schleper, and S. G. Wakeham. 1998. Dibiphytanyl ether lipids in nonthermophilic crenarchaeotes. *Appl. Environ. Microbiol.* **64:**1133–1138.

Dekker, K., H. Yamagata, K. Sakaguchi, and S. Udaka. 1991. Xylose (glucose) isomerase gene from the thermophile *Thermus thermophilus*: cloning, sequencing, and comparison with other thermostable xylose isomerases. *J. Bacteriol.* **173:**3078–3083.

Demirjian, D. C., F. Moris-Varas, and C. S. Cassidy. 2001. Enzymes from extremophiles. *Curr. Opin. Chem. Biol.* **5:**144–151.

Desmarais, D., P. Jablonski, N. Fedarko, and M. Roberts. 1997. 2-Sulfotrehalose, a novel osmolyte in haloalkaliphilic archaea. *J. Bacteriol.* **179:**3146–3153.

Dietrich, J., P. Schmitt, M. Zieger, B. Preve, J. L. Rolland, H. Chaabihi, and Y. Gueguen. 2002. PCR performance of the highly thermostable proof-reading B-type DNA polymerase from *Pyrococcus abyssi*. *FEMS Microbiol. Lett.* **217:**89–94.

Donaghy, J. A., K. Bronnenmeier, P. F. Soto-Kelly, and A. M. McKay. 2000. Purification and characterization of an extracellular feruloyl esterase from the thermophilic anaerobe *Clostridium stercorarium*. *J. Appl. Microbiol.* **88:**458–466.

Eckert, K., and E. Schneider. 2003. A thermoacidophilic endoglucanase (CelB) from *Alicyclobacillus acidocaldarius* displays high sequence similarity to arabinofuranosidases belonging to family 51 of glycoside hydrolases. *Eur. J. Biochem.* **270:** 3593–3602.

Eckert, K., F. Zielinski, L. Lo Leggio, and E. Schneider. 2002. Gene cloning, sequencing, and characterization of a family 9 endoglucanase (CelA) with an unusual pattern of activity from the thermoacidophile *Alicyclobacillus acidocaldarius* ATCC27009. *Appl. Microbiol. Biotechnol.* **60:**428–436.

Eichler, J. 2001. Biotechnological uses of archaeal extremozymes. *Biotechnol. Adv.* **19:**261–278.

Eichler, J. 2003. Facing extremes: archaeal surface-layer (glyco)proteins. *Microbiology* **149:**3347–3351.

Egas, M. C., M. S. da Costa, D. A. Cowan, and E. M. Pires. 1998. Extracellular alpha-amylase from *Thermus filiformis* Ork A2: purification and biochemical characterization. *Extremophiles* **2:**23–32.

Egorova, K., and G. Antranikian. 2005. Industrial relevance of thermophilic Archaea. *Curr. Opin. Microbiol.* **8:**649–655.

Egorova, K., H. Trauthwein, S. Verseck, and G. Antranikian. 2004. Purification and properties of an enantioselective and thermoactive amidase from the thermophilic actinomycete *Pseudonocardia thermophila*. *Appl. Microbiol. Biotechnol.* **65:**38–45.

Empadinhas, N., J. D. Marugg, N. Borges, H. Santos, and M. S. da Costa. 2001. Pathway for the synthesis of mannosylglycerate in the hyperthermophilic archaeon *Pyrococcus horikoshii*.

Biochemical and genetic characterization of key enzymes. *J. Biol. Chem.* **276**:43580–43588.

Erbeznik, M., S. E. Hudson, A. B. Herrman, and H. J. Strobel. 2004. Molecular analysis of the xylFGH operon, coding for xylose ABC transport, in *Thermoanaerobacter ethanolicus*. *Curr. Microbiol.* **48**:295–299.

Fang, T. Y., X. G. Hung, T. Y. Shih, and W. C. Tseng. 2004. Characterization of the trehalosyl dextrin-forming enzyme from the thermophilic archaeon *Sulfolobus solfataricus* ATCC 35092. *Extremophiles* **8**:335–343.

Ferrer, M., O. V. Golyshina, F. J. Plou, K. N. Timmis, and P. N. Golyshin. 2005. A novel alpha-glucosidase from the acidophilic archaeon, *Ferroplasma acidiphilum* Y with high transglycosylation activity and an unusual catalytic nucleophile. *Biochem. J*:In press.

Fessner, W. D., and V. Helaine. 2001. Biocatalytic synthesis of hydroxylated natural products using aldolases and related enzymes. *Curr. Opin. Biotechnol.* **12**:574–586.

Fox, J. D., K. M. Routzahn, M. H. Bucher, and D. S. Waugh. 2003. Maltodextrin-binding proteins from diverse bacteria and archaea are potent solubility enhancers. *FEBS Lett.* **537**:53–57.

Fourage, L., and B. Colas 2001. Synthesis of beta-D-glucosyl- and beta-D-fucosyl-glucoses using beta-glycosidase from *Thermus thermophilus*. *Appl. Microbiol. Biotechnol.* **56**:406–410.

Fuji, K., H. Minagawa, Y. Terada, T. Takaha, T. Kuriki, J. Shimada, and H. Kaneko. 2005. Use of random and saturation mutagenesis to improve the properties of *Thermus aquaticus* amylomaltase for efficient production of cycloamyloses. *Appl. Environ. Microbiol.* **71**:5823-27.

Fukamizo, T. 2000. Chitinolytic enzymes: catalysis, substrate binding, and their application. *Curr. Protein Pept. Sci.* **1**:105–124.

Fukushima, T., T. Mizuki, A. Echigo, A. Inoue, and R. Usami. 2005. Organic solvent tolerance of halophilic alpha-amylase from a haloarchaeon, *Haloarcula* sp. strain S-1. *Extremophiles* **9**:85–89.

Ganghofner, D., J. Kellermann, W. L. Staudenbauer, and K. Bronnenmeier. 1998. Purification and properties of an amylopullulanase, a glucoamylase, and an alpha-glucosidase in the amylolytic enzyme system of *Thermoanaerobacterium thermosaccharolyticum*. *Biosci. Biotechnol. Biochem.* **62**:302–308.

Gao, J., M. W. Bauer, K. R. Shockley, M. A. Pysz, and R. M. Kelly. 2003. Growth of hyperthermophilic archaeon *Pyrococcus furiosus* on chitin involves two family 18 chitinases. *Appl. Environ. Microbiol.* **69**:3119–3128.

Gardner, W., and W. Whitman. 1999. Expression vestors for *Methanococcus maripalidus*. Overexpression of acetohydroxyacid synthase and ß-galactosidase. *Genetics* **152**:1439–1447.

Gibbs, M. D., A. U. Elinder, R. A. Reeves, and P. L. Bergquist. 1996. Sequencing, cloning and expression of a beta-1,4-mannanase gene, manA, from the extremely thermophilic anaerobic bacterium, *Caldicellulosiruptor* Rt8B.4. *FEMS Microbiol. Lett.* **141**:37–43.

Gibbs, M. D., R. A. Reeves, A. Sunna, and P. L. Bergquist. 1999. Sequencing and expression of a beta-mannanase gene from the extreme thermophile *Dictyoglomus thermophilum* Rt46B.1, and characteristics of the recombinant enzyme. *Curr. Microbiol.* **39**:351–357.

Gill, R. K., and J. Kaur. 2004. A thermostable glucoamylase from a thermophilic *Bacillus* sp.: characterization and thermostability. *J. Ind. Microbiol. Biotechnol.* **31**:540–543.

Godde, C., K. Sahm, S. J. Brouns, L. D. Kluskens, J. van der Oost, W. M. de Vos, and G. Antranikian. 2005. Cloning and expression of islandisin, a new thermostable subtilisin from *Fervidobacterium islandicum*, in *Escherichia coli*. *Appl. Environ. Microbiol.* **71**:3951–3958.

Gomes, J., I. I. Gomes, K. Terler, N. Gubala, G. Ditzelmuller, and W. Steiner. 2000. Optimisation of culture medium and conditions for alpha-L-arabinofuranosidase production by the extreme thermophilic eubacterium *Rhodothermus marinus*. *Enzyme Microb. Technol.* **27**:414–422.

Gomes, I., J. Gomes, and W. Steiner. 2003. Highly thermostable amylase and pullulanase of the extreme thermophilic eubacterium *Rhodothermus marinus*: production and partial characterization. *Bioresour. Technol.* **90**:207–214.

Gomes, J., and W. Steiner. 2004. The biocatalytic potential of extremophiles and extremozymes. *Food Technol. Biotechnol.* **42**:223–235.

Gong, N., C. Chen, L. Xie, H. Chen, X. Lin, and R. Zhang. 2005. Characterization of a thermostable alkaline phosphatase from a novel species *Thermus yunnanensis* sp. nov. and investigation of its cobalt activation at high temperature. *Biochim. Biophys. Acta.* **1750**:103–111.

Groudieva, T., M. Kambourova, H. Yusef, M. Royter, R. Grote, H. Trinks, and G. Antranikian. 2004. Diversity and cold-active hydrolytic enzymes of culturable bacteria associated with Arctic sea ice, Spitzbergen. *Extremophiles* **8**:475–488.

Gueguen, Y., J. L. Rolland, O. Lecompte, P. Azam, G. Le Romancer, D. Flament, J. P. Raffin, and J. Dietrich. 2001a. Characterization of two DNA polymerases from the hyperthermophilic euryarchaeon *Pyrococcus abyssi*. *Eur. J. Biochem.* **268**:5961–5969.

Gueguen, Y., J. L. Rolland, S. Schroeck, D. Flament, S. Defretin, M. H. Saniez, and J. Dietrich. 2001b. Characterization of the maltooligosyl trehalose synthase from the thermophilic archaeon *Sulfolobus acidocaldarius*. *FEMS Microbiol. Lett.* **194**:201–206.

Gueguen, Y., W. G. Voorhorst, J. van der Oost, and W. M. de Vos. 1997. Molecular and biochemical characterization of an endo-beta-1,3-glucanase of the hyperthermophilic archaeon *Pyrococcus furiosus*. *J. Biol. Chem.* **272**:31258–31264.

Gupta, R., and P. Ramnani. 2006. Microbial keratinases and their prospective applications: an overview. *Appl. Microbiol. Biotechnol.* **70**:21–33.

Gupta, R., P. Gigras, H. Mohapatra, V. Goswami, and B. Chauhan. 2003. Microbial alpha-amylases: a biotechnological perspective. *Process Biochemistry* **38**:1599–1616.

Guy, J. E., M. N. Isupov, and J. A. Littlechild. 2003. The structure of an alcohol dehydrogenase from the hyperthermophilic archaeon *Aeropyrum pernix*. *J. Mol. Biol.* **331**:1041–1051.

Halldorsdottir, S., E. T. Thorolfsdottir, R. Spilliaert, M. Johansson, S. H. Thorbjarnardottir, A. Palsdottir, G. O. Hreggvidsson, J. K. Kristjansson, O. Holst, and G. Eggertsson. 1998. Cloning, sequencing and overexpression of a *Rhodothermus marinus* gene encoding a thermostable cellulase of glycosyl hydrolase family 12. *Appl. Microbiol. Biotechnol.* **49**:277–284.

Han, S. O., H. Yukawa, M. Inui, and R. H. Doi. 2005. Molecular cloning and transcriptional and expression analysis of engO, encoding a new noncellulosomal family 9 enzyme, from *Clostridium cellulovorans*. *J. Bacteriol.* **187**:4884–4889.

Hashimoto, H., M. Nishioka, S. Fujiwara, M. Takagi, T. Imanaka, T. Inoue, and Y. Kai. 2001. Crystal structure of DNA polymerase from hyperthermophilic archaeon *Pyrococcus kodakaraensis* KOD1. *J. Mol. Biol.* **306**:469–477.

Hemmi, H., S. Ikejiri, T. Nakayama, and T. Nishino. 2003. Fusion-type lycopene beta-cyclase from a thermoacidophilic archaeon *Sulfolobus solfataricus*. *Biochem. Biophys. Res. Commun.* **305**: 586–591.

Hezayen, F. F., B. H. Rehm, R. Eberhardt, and A. Steinbuchel. 2000. Polymer production by two newly isolated extremely halophilic archaea: application of a novel corrosion-resistant bioreactor. *Appl. Microbiol. Biotechnol.* **54**:319–325.

Hezayen, F. F., B. H. Rehm, B. J. Tindall, and A. Steinbuchel. 2001. Transfer of *Natrialba asiatica* B1T to *Natrialba taiwanensis* sp. nov. and description of *Natrialba aegyptiaca* sp. nov., a novel extremely halophilic, aerobic, non-pigmented member of the

Archaea from Egypt that produces extracellular poly(glutamic acid). *Int. J. Syst. Evol. Microbiol.* 51:1133–1142.

Hilge, M., S. M. Gloor, W. Rypniewski, O. Sauer, T. D. Heightman, W. Zimmermann, K. Winterhalter, and K. Piontek 1998. High-resolution native and complex structures of thermostable beta-mannanase from *Thermomonospora fusca* – substrate specificity in glycosyl hydrolase family 5. *Structure* 6:1433–1444.

Hobel, C. F., G. O. Hreggvidsson, V. T. Marteinsson, F. Bahrani-Mougeot, J. M. Einarsson, and J. K. Kristjansson. 2005. Cloning, expression, and characterization of a highly thermostable family 18 chitinase from *Rhodothermus marinus*. *Extremophiles* 9:53–64.

Hoe, H. S., H. K. Kim, and S. T. Kwon. 2002. Cloning, analysis, and expression of the gene for inorganic pyrophosphatase of *Aquifex pyrophilus* and properties of the enzyme. *Mol. Cells* 13:296–301.

Holmes, M., F. Pfeifer, and M. Dyall-Smith. 1994. Improved shuttle vectors for Haloferax volcanii including a dual-resistance plasmid. *Gene* 146:117–121.

Holmes, M. L., and M. L. Dyall-Smith. 2000. Sequence and expression of a halobacterial beta-galactosidase gene. *Mol. Microbiol.* 36:114–122.

Horikoshi, K. 1999. Alkaliphiles: some applications of their products for biotechnology. *Microbiol. Mol. Biol. Rev.* 63:735–750.

Hotta, Y., S. Ezaki, H. Atomi, and T. Imanaka. 2002. Extremely stable and versatile carboxylesterase from a hyperthermophilic archaeon. *Appl. Environ. Microbiol.* 68:3925–3931.

Huang, Y., G. Krauss, S. Cottaz, H. Driguez, and G. Lipps. 2005. A highly acid-stable and thermostable endo-beta-glucanase from the thermoacidophilic archaeon *Sulfolobus solfataricus*. *Biochem. J.* 385:581–588.

Hutcheon, G. W., N. Vasisht, and A. Bolhuis. 2005. Characterisation of a highly stable alpha-amylase from the halophilic archaeon *Haloarcula hispanica*. *Extremophiles* 9:487–495.

Hyun, H. H., G. J. Shen, and J. G. Zeikus. 1985. Differential amylosaccharide metabolism of *Clostridium thermosulfurogenes* and *Clostridium thermohydrosulfuricum*. *J. Bacteriol.* 164:1153–1161.

Ichikawa, K., T. Tonozuka, H. Uotsu-Tomita, H. Akeboshi, A. Nishikawa, and Y. Sakano. 2004. Purification, characterization, and subsite affinities of *Thermoactinomyces vulgaris* R-47 maltooligosaccharide-metabolizing enzyme homologous to glucoamylases. *Biosci. Biotechnol. Biochem.* 68:413–420.

Ideno, A., M. Furutani, T. Iwabuchi, T. Iida, Y. Iba, Y. Kurosawa, H. Sakuraba, T. Ohshima, Y. Kawarabayashi, and T. Maruyama. 2004. Expression of foreign proteins in *Escherichia coli* by fusing with an archaeal FK506 binding protein. *Appl. Microbiol. Biotechnol.* 64:99–105.

Iida, T., T. Iwabuchi, A. Ideno, S. Suzuki, and T. Maruyama. 2000. FK506-binding protein-type peptidyl-prolyl cis-trans isomerase from a halophilic archaeum, *Halobacterium cutirubrum*. *Gene* 256:319–326.

Ikeda, M., and D. S. Clark. 1998. Molecular cloning of extremely thermostable esterase gene from hyperthermophilic archaeon *Pyrococcus furiosus* in *Escherichia coli*. *Biotechnol. Bioeng.* 57:624–629.

Imamura, H., S. Fushinobu, B. S. Jeon, T. Wakagi, and H. Matsuzawa. 2001. Identification of the catalytic residue of *Thermococcus litoralis* 4-alpha-glucanotransferase through mechanism-based labeling. *Biochemistry* 40:12400–12406.

Imamura, H., B. Jeon, T. Wakagi, and H. Matsuzawa. 1999. High level expression of *Thermococcus litoralis* 4-alpha-glucanotransferase in a soluble form in *Escherichia coli* with a novel expression system involving minor arginine tRNAs and GroELS. *FEBS Lett.* 457:393–396.

Irwin, D. C., M. Cheng, B. Xiang, J. K. Rose, and D. B. Wilson. 2003. Cloning, expression and characterization of a family-74 xyloglucanase from *Thermobifida fusca*. *Eur. J. Biochem.* 270:3083–3091.

Ishiguro, M., S. Kaneko, A. Kuno, Y. Koyama, S. Yoshida, G. G. Park, Y. Sakakibara, I. Kusakabe, and H. Kobayashi 2001. Purification and characterization of the recombinant *Thermus* sp. strain T2 alpha-galactosidase expressed in *Escherichia coli*. *Appl. Environ. Microbiol.* 67:1601–1606.

Ishikawa, K., H. Ishida, I. Matsui, Y. Kawarabayasi, and H. Kikuchi. 2001. Novel bifunctional hyperthermostable carboxypeptidase/aminoacylase from *Pyrococcus horikoshii* OT3. *Appl. Environ. Microbiol.* 67:673–679.

Ishikawa, K., and K. Mino. January 1, 2004. Heat resistant cysteine synthase. Japan patent 20040002075.

Izard, T., and J. Sygusch. 2004. Induced fit movements and metal cofactor selectivity of class II aldolases: structure of *Thermus aquaticus* fructose-1,6-bisphosphate aldolase. *J. Biol. Chem.* 279:11825–11833.

Jang, H. J., C. H. Lee, W. Lee, and Y. S. Kim. 2002. Two flexible loops in subtilisin-like thermophilic protease, thermicin, from *Thermoanaerobacter yonseiensis*. *J. Biochem. Mol. Biol.* 35:498–507.

Jeffries, T. W. 1996. Biochemistry and genetics of microbial xylanases. *Curr. Opin. Biotechnol.* 7:337–342.

Jeon, B. S., H. Taguchi, H. Sakai, T. Ohshima, T. Wakagi, and H. Matsuzawa. 1997. 4-Alpha-glucanotransferase from the hyperthermophilic archaeon *Thermococcus litoralis*—enzyme purification and characterization, and gene cloning, sequencing and expression in *Escherichia coli*. *Eur. J. Biochem.* 248:171–178.

Ji, C. N., T. Jiang, M. Q. Chen, X. Y. Sheng, and Y. M. Mao. 2001. Purification, crystallization and preliminary X-ray studies of thermostable alkaline phosphatase from *Thermus* sp. 3041. *Acta. Crystallogr. D. Biol. Crystallogr.* 57:614–615.

Jin, R., S. Richter, R. Zhong, and G. K. Lamppa. 2003. Expression and import of an active cellulase from a thermophilic bacterium into the chloroplast both in vitro and in vivo. *Plant Mol. Biol.* 2003 Mar;51:493–507.

Jindou, S., Q. Xu, R. Kenig, M. Shulman, Y. Shoham, E. A. Bayer, and R. Lamed. 2006. Novel architecture of family-9 glycoside hydrolases identified in cellulosomal enzymes of *Acetivibrio cellulolyticus* and *Clostridium thermocellum*. *FEMS Microb. Lett.* 254:308–316.

Johnsen, U., and P. Schonheit. 2004. Novel xylose dehydrogenase in the halophilic archaeon *Haloarcula marismortui*. *J. Bacteriol.* 186:6198–6207.

Jolivet, E., E. Corre, S. L'Haridon, P. Forterre, and D. Prieur. 2004. *Thermococcus marinus* sp. nov. and *Thermococcus radiotolerans* sp. nov., two hyperthermophilic archaea from deep-sea hydrothermal vents that resist ionizing radiation. *Extremophiles* 8:219–227.

Jolivet, E., S. L'Haridon, E. Corre, P. Forterre, and D. Prieur. 2003. *Thermococcus gammatolerans* sp. nov., a hyperthermophilic archaeon from a deep-sea hydrothermal vent that resists ionizing radiation. *Int. J. Syst. Evol. Microbiol.* 53:847–851.

Jorgensen, F., O. C. Hansen, and P. Stougaard. 2004. Enzymatic conversion of D-galactose to D-tagatose: heterologous expression and characterisation of a thermostable L-arabinose isomerase from *Thermoanaerobacter mathranii*. *Appl. Microbiol. Biotechnol.* 64:816–22.

Jonuscheit, M., E. Martusewitsch, K. M. Stedman, and C. Schleper. 2003. A reporter gene system for the hyperthermophilic archaeon *Sulfolobus solfataricus* based on a selectable and integrative shuttle vector. *Mol. Microbiol.* 48:1241–1252.

Kaczowka, S. J., C. J. Reuter, L. A. Talarico, and J. A. Maupin-Furlow. 2005. Recombinant production of *Zymomonas mobilis* pyruvate decarboxylase in the haloarchaeon *Haloferax volcanii*. *Archaea* **1**:327–334.

Kadar, Z., T. De Vrije, M. A. Budde, Z. Szengyel, K. Reczey, and P. A. Claassen. 2003. Hydrogen production from paper sludge hydrolysate. *Appl. Biochem. Biotechnol.* **105–108**:557–566.

Kahler, M., and G. Antranikian. 2000. Cloning and characterization of a family B DNA polymerase from the hyperthermophilic crenarchaeon *Pyrobaculum islandicum*. *J. Bacteriol.* **182**:655–663.

Kanai, T., S. Ito, and T. Imanaka. 2003. Characterization of a cytosolic NiFe-hydrogenase from the hyperthermophilic archaeon *Thermococcus kodakaraensis* KOD1. *J. Bacteriol.* **185**: 1705–1711.

Kang, S., C. Vieille, and J. G. Zeikus. 2004. Identification of *Pyrococcus furiosus* amylopullulanase catalytic residues. *Appl. Microbiol. Biotechnol.* **66**:408–413.

Kaper, T., B. Talik, T. J. Ettema, H. Bos, M. J. van der Maarel, and L. Dijkhuizen. 2005. Amylomaltase of *Pyrobaculum aerophilum* IM2 produces thermoreversible starch gels. *Appl. Environ. Microbiol.* **71**:5098–5106.

Kaper, T., M. J. van der Maarel, G. J. Euverink, and L. Dijkhuizen. 2004. Exploring and exploiting starch-modifying amylomaltases from thermophiles. *Biochem. Soc. Trans.* **32**:279–282.

Kashima, Y., K. Mori, H. Fukada, and K. Ishikawa. 2005. Analysis of the function of a hyperthermophilic endoglucanase from *Pyrococcus horikoshii* that hydrolyzes crystalline cellulose. *Extremophiles* **9**:37–43.

Kawase T., A. Saito, T. Sato, R. Kanai, T. Fujii, N. Nikaidou, K. Miyashita, and T. Watanabe 2004. Distribution and phylogenetic analysis of family 19 chitinases in Actinobacteria. *Appl. Environ. Microbiol.* **70**:1135–1144.

Kim, T., and X. G. Lei. 2005. Expression and characterization of a thermostable serine protease (TfpA) from *Thermomonospora fusca* YX in *Pichia pastoris*. *Appl. Microbiol. Biotechnol.* **68**:355–359.

Kim, J. W., L. O. Flowers, M. Whiteley, and T. L. Peeples. 2001. Biochemical confirmation and characterization of the family-57-like alpha-amylase of *Methanococcus jannaschii*. *Folia Microbiol (Praha)* **46**:467–473.

Kim, M. S., J. T. Park, Y. W. Kim, H. S. Lee, R. Nyawira, H. S. Shin, C. S. Park, S. H. Yoo, Y. R. Kim, T. W. Moon, and K. H. Park. 2004. Properties of a novel thermostable glucoamylase from the hyperthermophilic archaeon *Sulfolobus solfataricus* in relation to starch processing. *Appl. Environ. Microbiol.* **70**:3933–3940.

Kim, B. C., Y. H. Lee, H. S. Lee, D. W. Lee, E. A. Choe, and Y. R. Pyun. 2002a. Cloning, expression and characterization of L-arabinose isomerase from *Thermotoga neapolitana*: bioconversion of D-galactose to D-tagatose using the enzyme. *FEMS Microbiol. Lett.* **212**:121–126.

Kim, D. J., H. J. Jang, Y. R. Pyun, and Y. S. Kim. 2002b. Cloning, expression, and characterization of thermostable DNA polymerase from *Thermoanaerobacter yonseiensis*. *J. Biochem. Mol. Biol.* **35**:320–329.

Kim, S., and S. B. Lee. 2004. Thermostable esterase from a thermoacidophilic archaeon: purification and characterization for enzymatic resolution of a chiral compound. *Biosci. Biotechnol. Biochem.* **68**:2289–2298.

Kim, J. O., S. R. Park, W. J. Lim, S. K. Ryu, M. K. Kim, C. L. An, S. J. Cho, Y. W. Park, J. H. Kim, and H. D. Yun. 2000a. Cloning and characterization of thermostable endoglucanase (Cel8Y) from the hyperthermophilic *Aquifex aeolicus* VF5. *Biochem. Biophys. Res. Commun.* **279**:420–426.

Kim, H., K. H. Jung, and M. Y. Pack. 2000b. Molecular characterization of xynX, a gene encoding a multidomain xylanase with a thermostabilizing domain from *Clostridium thermocellum*. *Appl. Microbiol. Biotechnol.* **54**:521–527.

Khan, A. R., S. Nirasawa, S. Kaneko, T. Shimonishi, and K. Hayashi. 2000. Characterization of a solvent resistant and thermostable aminopeptidase from the hyperthermophillic bacterium, *Aquifex aeolicus*. *Enzyme Microb. Technol.* **27**: 83–88.

Kletzin, A. 1992. Molecular characterisation of a DNA ligase gene of the extremely thermophilic archaeon *Desulfurolobus ambivalens* shows close phylogenetic relationship to eukaryotic ligases. *Nucleic Acids Res.* **20**:5389–5396.

Kluskens, L. D., W. G. Voorhorst, R. J. Siezen, R. M. Schwerdtfeger, G. Antranikian, J. van der Oost, and W. M. de Vos. 2002. Molecular characterization of fervidolysin, a subtilisin-like serine protease from the thermophilic bacterium *Fervidobacterium pennivorans*. *Extremophiles* **6**:185–194.

Kluskens, L. D., G. J. van Alebeek, A. G. Voragen, W. M. de Vos, and J. van der Oost. 2003. Molecular and biochemical characterization of the thermoactive family 1 pectate lyase from the hyperthermophilic bacterium *Thermotoga maritima*. *Biochem. J.* **370**:651–659.

Kluskens, L. D., G. J. van Alebeek, J. Walther, A. G. Voragen, W. M. de Vos, and J. van der Oost 2005. Characterization and mode of action of an exopolygalacturonase from the hyperthermophilic bacterium *Thermotoga maritima*. *FEBS J.* **272**:5464–5473.

Kobayashi, T., H. Kanai, R. Aono, K. Horikoshi, and T. Kudo. 1994. Cloning, expression, and nucleotide sequence of the alpha-amylase gene from the haloalkaliphilic archaeon *Natronococcus* sp. strain Ah-36. *J. Bacteriol.* **176**:5131–5134.

Kobayashi, T., H. Kanai, T. Hayashi, T. Akiba, R. Akaboshi, and K. Horikoshi. 1992. Haloalkaliphilic maltotriose-forming alpha-amylase from the archaebacterium *Natronococcus* sp. strain Ah-36. *J. Bacteriol.* **174**:3439–3444.

Kohli, U., P. Nigam, D. Singh, and K. Chaudhary. 2001. Thermostable, alkalophilic and cellulase free xylanase production by *Thermoactinomyces thalophilus* subgroup C. *Enzyme Microb. Technol.* **28**:606–610.

Kosugi, A., K. Murashima, and R. H. Doi. 2002. Xylanase and acetyl xylan esterase activities of XynA, a key subunit of the *Clostridium cellulovorans* cellulosome for xylan degradation. *Appl. Environ. Microbiol.* **68**:6399–6402.

Kowalczykowski, S., F. Chedin, and E. Seitz. 8 February 2005. Single stranded DNA binding proteins from archaea. USA patent 6,852,832.

Kozianowski, G., F. Canganella, F. A. Rainey, H. Hippe, and G. Antranikian 1997. Purification and characterization of thermostable pectate-lyases from a newly isolated thermophilic bacterium, *Thermoanaerobacter italicus* sp. nov. *Extremophiles* **1**:171–182.

Kriegshauser, G., and W. Liebl. 2000. Pullulanase from the hyperthermophilic bacterium Thermotoga maritima: purification by β-cyclodextrin affinity chromatography. *J Chromat B: Biomedical Sciences and Applications* **737**:245–251.

Krishnan, G., and W. Altekar. 1991. An unusual class I (Schiff base) fructose-1,6-bisphosphate aldolase from the halophilic archaebacterium *Haloarcula vallismortis*. *Eur. J. Biochem.* **195**: 343–350.

Lai, X., H. Shao, F. Hao, and L. Huang. 2002. Biochemical characterization of an ATP-dependent DNA ligase from the hyperthermophilic crenarchaeon *Sulfolobus shibatae*. *Extremophiles* **6**:469–477.

Laksanalamai, P., A. Jiemjit, Z. Bu, D. L. Maeder, and F. T. Robb. 2003. Multi-subunit assembly of the *Pyrococcus furiosus* small heat shock protein is essential for cellular protection at high temperature. *Extremophiles* **7**:79–83.

Lamble, H. J., M. J. Danson, D. W. Hough, and S. D. Bull. 2005. Engineering stereocontrol into an aldolase-catalysed reaction. *Chem. Commun.* (Camb)**1**:124–126.

Lamosa, P., A. Burke, R. Peist, R. Huber, M. Y. Liu, G. Silva, C. Rodrigues-Pousada, J. LeGall, C. Maycock, and H. Santos. 2000. Thermostabilization of proteins by diglycerol phosphate, a new compatible solute from the hyperthermophile *Archaeoglobus fulgidus*. *Appl. Environ. Microbiol.* 66:1974–1979.

Landegren, U., E. Schallmeiner, M. Nilsson, S. Fredriksson, J. Baner, M. Gullberg, J. Jarvius, S. Gustafsdottir, F. Dahl, O. Soderberg, O. Ericsson, and J. Stenberg. 2004. Molecular tools for a molecular medicine: analyzing genes, transcripts and proteins using padlock and proximity probes. *J. Mol. Recognit.* 17:194–197.

Lebbink, J. H., T. Kaper, S. W. Kengen, J. van der Oost, and W. M. de Vos. 2001. Beta-glucosidase CelB from *Pyrococcus furiosus*: production by *Escherichia coli*, purification, and in vitro evolution. *Methods Enzymol.* 330:364–379.

Leclere, M. M., M. Nishioka, T. Yuasa, S. Fujiwara, M. Takagi, and T. Imanaka. 1998. The O6-methylguanine-DNA methyltransferase from the hyperthermophilic archaeon *Pyrococcus* sp. KOD1: a thermostable repair enzyme. *Mol. Gen. Genet.* 258:69–77.

Lee, D. W., H. J. Jang, E. A. Choe, B. C. Kim, S. J. Lee, S. B. Kim, Y. H. Hong, and Y. R. Pyun. 2004. Characterization of a thermostable L-arabinose (D-galactose) isomerase from the hyperthermophilic eubacterium *Thermotoga maritima*. *Appl. Environ. Microbiol.* 70:1397–1404.

Lee, H. S., Y. J. Kim, S. S. Bae, J. H. Jeon, J. K. Lim, B. C. Jeong, S. G. Kang, and J. H. Lee. 2006. Cloning, expression, and characterization of aminopeptidase P from the hyperthermophilic archaeon *Thermococcus* sp. strain NA1.*Appl. Environ. Microbiol.* 72:1886–1890.

Lee, S. J., D. W. Lee, E. A. Choe, Y. H. Hong, S. B. Kim, B. C. Kim, and Y. R. Pyun. 2005. Characterization of a thermoacidophilic L-arabinose isomerase from *Alicyclobacillus acidocaldarius*: role of Lys-269 in pH optimum. *Appl. Environ. Microbiol.* 71:7888–7896.

Leveque, E., B. Haye, and A. Belarbi. 2000a. Cloning and expression of an alpha-amylase encoding gene from the hyperthermophilic archaebacterium *Thermococcus hydrothermalis* and biochemical characterisation of the recombinant enzyme. *FEMS Microbiol. Lett.* 186:67–71.

Leveque, E., S. Janecek, B. Haye, and A. Belarbi. 2000b. Thermophilic archaeal amylolytic enzymes. *Enzyme Microb. Technol.* 26:3–14.

Li, D., and K. J. Stevenson. 2001. Alcohol dehydrogenase from *Thermococcus* strain AN1. *Methods Enzymol.* 331:201–207.

Liebl, W., I. Stemplinger, and P. Ruile.1997. Properties and gene structure of the *Thermotoga maritima* alpha-amylase AmyA, a putative lipoprotein of a hyperthermophilic bacterium. *J. Bacteriol.* 179:941–948.

Liebl, W., P. Ruile, K. Bronnenmeier, K. Riedel, F. Lottspeich, and I. Greif. 1996. Analysis of a *Thermotoga maritima* DNA fragment encoding two similar thermostable cellulases, CelA and CelB, and characterization of the recombinant enzymes. *Microbiology* 142:2533–2542.

Liebl, W., R. Feil, J. Gabelsberger, J. Kellermann, and K. H. Schleifer. 1992. Purification and characterization of a novel thermostable 4-alpha-glucanotransferase of *Thermotoga maritima* cloned in *Escherichia coli*. *Eur. J. Biochem.* 207: 81–88.

Liebl, W. 2001. Cellulolytic enzymes from *Thermotoga* species. *Methods Enzymol.* 330:290–300.

Lim, J. H., J. Choi, S. J. Han, S. H. Kim, H. Z. Hwang, D. K. Jin, B. Y. Ahn, and Y. S. Han. 2001. Molecular cloning and characterization of thermostable DNA ligase from *Aquifex pyrophilus*, a hyperthermophilic bacterium. *Extremophiles* 5:161–168.

Limauro, D., R. Cannio, G. Fiorentino, M. Rossi, and S. Bartolucci. 2001. Identification and molecular characterization of an endoglucanase gene, celS, from the extremely thermophilic archaeon *Sulfolobus solfataricus*. *Extremophiles* 5:213–219.

Lin, F. P., and K. L. Leu. 2002. Cloning, expression, and characterization of thermostable region of amylopullulanase gene from *Thermoanaerobacter ethanolicus* 39E. *Appl. Biochem. Biotechnol.* 97:33–44.

Liu, S. Y., J. Wiegel, and F. C. Gherardini. 1996. Purification and cloning of a thermostable xylose (glucose) isomerase with an acidic pH optimum from *Thermoanaerobacterium* strain JW/SL-YS 489. *J. Bacteriol.* 178:5938–5945.

Lloyd, J., D. R. Loveley, and L. E. Macaskie. 2003. Biotechnological application of metal-reducing microorganisms. *Adv. Appl. Microbiol.* 53:85–128.

Lorentzen, E., B. Siebers, R. Hensel, and E. Pohl. 2004. Structure, function and evolution of the Archaeal class I fructose-1,6-bisphosphate aldolase. *Biochem. Soc. Trans.* 32:259–263.

Lund, P. A., A. T. Large, and G. Kapatai. 2003. The chaperonins: perspectives from the Archaea. *Biochem. Soc. Trans.* 31: 681–685.

Lyon, P. F., T. Beffa, M. Blanc, G. Auling, and M. Aragno. 2000. Isolation and characterization of highly thermophilic xylanolytic *Thermus thermophilus* strains from hot composts. *Can. J. Microbiol.* 46:1029–1035.

Ma, K., and M. W. Adams. 2001. Alcohol dehydrogenases from *Thermococcus litoralis* and *Thermococcus* strain ES-1. *Methods Enzymol.* 331:195–201.

Manco, G., G. Carrea, E. Giosue, G. Ottolina, G. Adamo, and M. Rossi. 2002. Modification of the enantioselectivity of two homologous thermophilic carboxylesterases from *Alicyclobacillus acidocaldarius* and *Archaeoglobus fulgidus* by random mutagenesis and screening. *Extremophiles* 6:325–331.

Manco, G., E. Giosue, S. D'Auria, P. Herman, G. Carrea, and M. Rossi. 2000. Cloning, overexpression, and properties of a new thermophilic and thermostable esterase with sequence similarity to hormone-sensitive lipase subfamily from the archaeon *Archaeoglobus fulgidus*. *Arch. Biochem. Biophys.* 373: 182–192.

Margesin, R., and F. Schinner. 2001. Potential of halotolerant and halophilic microorganisms for biotechnology. *Extremophiles* 5:73–83.

Marhuenda-Egea, F. C., S. Piera-Velazquez, C. Cadenas, and E. Cadenas. 2002. Mechanism of adaptation of an atypical alkaline p-nitrophenyl phosphatase from the archaeon *Halobacterium salinarum* at low-water environments. *Biotechnol. Bioeng.* 78: 497–502.

Maruyama, T., R. Suzuki, and M. Furutani. 2004. Archaeal peptidyl prolyl cis-trans isomerases (PPIases) update 2004. *Front Biosci.* 9:1680–1720.

Matsui, E., S. Kawasaki, H. Ishida, K. Ishikawa, Y. Kosugi, H. Kikuchi, Y. Kawarabayashi, and I. Matsui. 1999. Thermostable flap endonuclease from the archaeon, *Pyrococcus horikoshii*, cleaves the replication fork-like structure endo/exonucleolytically. *J. Biol. Chem.* 274:18297–18309.

Matsui, I., K. Ishikawa, H. Ishida, and Y. Kosugi. 2002a. Methods for making and using a thermophilic enzyme as a beta-glycosidase. US patent 222866/1998.

Matsui, I., K. Ishikawa, H. Ishida, Y. Kosugi, and Y. Tahara. 2002b. Thermostable enzyme having phosphatidylethanolamine N-methyltransferase activity. Japan patent 6,391,604.

Matzke, J., B. Schwermann, and E. P. Bakker. 1997. Acidostable and acidophilic proteins: the example of the alpha-amylase from *Alicyclobacillus acidocaldarius*. *Comp. Biochem. Physiol. A Physiol.* 118:475–479.

Matzke, J., A. Herrmann, E. Schneider, and E. P. Bakker. 2000. Gene cloning, nucleotide sequence and biochemical properties of a cytoplasmic cyclomaltodextrinase (neopullulanase) from

Alicyclobacillus acidocaldarius, reclassification of a group of enzymes. *FEMS Microbiol. Lett.* **183**:55–61.

Mayr, J., A. Lupas, J. Kellermann, C. Eckerskorn, W. Baumeister, and J. Peters. 1996. A hyperthermostable protease of the subtilisin family bound to the surface layer of the archaeon *Staphylothermus marinus*. *Curr. Biol.* **6**:739–749.

Mijts, B. N., and B. K. Patel. 2002. Cloning, sequencing and expression of an alpha-amylase gene, amyA, from the thermophilic halophile *Halothermothrix orenii* and purification and biochemical characterization of the recombinant enzyme. *Microbiol.* **148**:2343–2349.

Miyazaki, K. 2005a. A hyperthermophilic laccase from *Thermus thermophilus* HB27. *Extremophiles* **9**:415–425.

Miyazaki, K. 2005b. Hyperthermophilic alpha-L: -arabinofuranosidase from *Thermotoga maritima* MSB8: molecular cloning, gene expression, and characterization of the recombinant protein. *Extremophiles* **9**:399–406.

Morana, A., M. Moracci, A. Ottombrino, M. Ciaramella, M. Rossi, and M. De Rosa. 1995. Industrial-scale production and rapid purification of an archaeal beta-glycosidase expressed in *Saccharomyces cerevisiae*. *Biotechnol. Appl. Biochem.* **22**:261–268.

Morikawa, M., and T. Imanaka. 2001. Thiol protease from *Thermococcus kodakaraensis* KOD1. *Methods Enzymol.* **330**:424–433.

Morris, D. D., M. D. Gibbs, M. Ford, J. Thomas, and P. L. Bergquist. 1999. Family 10 and 11 xylanase genes from *Caldicellulosiruptor* sp. strain Rt69B.1. *Extremophiles* **3**:103–111.

Morris, D. D., M. D. Gibbs, C. W. Chin, M. H. Koh, K. K. Wong, R. W. Allison, P. J. Nelson, and P. L. Bergquist. 1998. Cloning of the xynB gene from *Dictyoglomus thermophilum* Rt46B.1 and action of the gene product on kraft pulp. *Appl. Environ. Microbiol.* **64**:1759–1765.

Mueller, P., K. Egorova, C. E. Vorgias, E. Boutou, H. Trauthwein, S. Verseck, and G. Antranikian 2006. Cloning, overexpression, and characterization of a thermoactive nitrilase from the hyperthermophilic archaeon *Pyrococcus abyssi*. *Protein Expr. Purif.* In press.

Nakajima, M., H. Imamura, H. Shoun, S. Horinouchi, and T. Wakagi 2004. Transglycosylation activity of *Dictyoglomus thermophilum* amylase A. *Biosci. Biotechnol. Biochem.* **68**: 2369–2373.

Nagatomo, H., Y. Matsushita, K. Sugamoto, and T. Matsui. 2005. Preparation and properties of gelatin-immobilized beta-glucosidase from *Pyrococcus furiosus*. *Biosci. Biotechnol. Biochem.* **69**:128–136.

Nakatani, M., S. Ezaki, H. Atomi, and T. Imanaka. 2000. A DNA ligase from a hyperthermophilic archaeon with unique cofactor specificity. *J. Bacteriol.* **182**:6424–6433.

Nakatani, M., S. Ezaki, H. Atomi, and T. Imanaka. 2002. Substrate recognition and fidelity of strand joining by an archaeal DNA ligase. *Eur. J. Biochem.* **269**:650–656.

Nam, G. W., D. W. Lee, H. S. Lee, N. J. Lee, B. C. Kim, E. A. Choe, J. K. Hwang, M. T. Suhartono, and Y. R. Pyun. 2002. Native-feather degradation by *Fervidobacterium islandicum* AW-1, a newly isolated keratinase-producing thermophilic anaerobe. *Arch. Microbiol.* **178**:538–547.

Nastopoulos, V., F. M. Pisani, C. Savino, L. Federici, M. Rossi, and D. Tsernoglou. 1998. Crystallization and preliminary X-ray diffraction studies of DNA polymerase from the thermophilic archaeon *Sulfolobus solfataricus*. *Acta Cryst.* **54**:1002–1004.

Nawani, N. N., B. P. Kapadnis, A. D. Das, A. S. Rao, and S. K. Mahajan. 2002. Purification and characterization of a thermophilic and acidophilic chitinase from *Microbispora* sp. V2. *J. Appl. Microbiol.* **93**:965–975.

Niehaus, F., C. Bertoldo, M. Kahler, and G. Antranikian. 1999. Extremophiles as a source of novel enzymes for industrial application. *Appl. Microbiol. Biotechnol.* **51**:711–729.

Niehaus, F., A. Peters, T. Groudieva, and G. Antranikian. 2000. Cloning, expression and biochemical characterisation of a unique thermostable pullulan-hydrolysing enzyme from the hyperthermophilic archaeon *Thermococcus aggregans*. *FEMS Microbiol. Lett.* **190**:223–229.

Norris, P., N. Burton, and N. Foulis. 2000. Acidophiles in bioreactor mineral processing. *Extremophiles* **4**:71–76.

O'Connor, E., and R. Shand. 2002. Halocins and sulfolobicins: The emerging story of archaeal protein and peptide antibiotics. *J. Indust. Microbiol. Biotechnol.* **28**:23–31.

Ohtsu, N., H. Motoshima, K. Goto, F. Tsukasaki, and H. Matsuzawa. 1998. Thermostable beta-galactosidase from an extreme thermophile, *Thermus* sp. A4: enzyme purification and characterization, and gene cloning and sequencing. *Biosci. Biotechnol. Biochem.* **62**:1539–1545.

Oren, A. 2002. Diversity of halophilic microorganisms: environments, phylogeny, physiology, and applications. *J. Ind. Microbiol. Biotechnol.* **28**:56–63.

Parisot, J., A. Ghochikyan, V. Langlois, V. Sakanyan, and C. Rabiller. 2002. Exopolygalacturonate lyase from *Thermotoga maritima*: cloning, characterization and organic synthesis application. *Carbohydr. Res.* **337**:1427–1433.

Park, B. C., S. Koh, C. Chang, S. W. Suh, D. S. Lee, and S. M. Byun. 1997. Cloning and expression of the gene for xylose isomerase from *Thermus flavus* AT62 in *Escherichia coli*. *Appl. Biochem. Biotechnol.* **62**:15–27.

Park, T. H., K. W. Choi, C. S. Park, S. B. Lee, H. Y. Kang, K. J. Shon, J. S. Park, and J. Cha. 2005. Substrate specificity and transglycosylation catalyzed by a thermostable beta-glucosidase from marine hyperthermophile *Thermotoga neapolitana*. *Appl. Microbiol. Biotechnol.* **69**:411–422.

Patel, G. B., and G. D. Sprott. 1999. Archaeobacterial ether lipid liposomes (archaeosomes) as novel vaccine and drug delivery systems. *Crit. Rev. Biotechnol.* **19**:317–357.

Perez-Pomares, and F., V. Bautista, J. Ferrer, C. Pire, F. C. Marhuenda-Egea, and M. J. Bonete. 2003. Alpha-amylase activity from the halophilic archaeon *Haloferax mediterranei*. *Extremophiles* **7**:299–306.

Perler, F. B., S. Kumar, and H. Kong. 1996. Thermostable DNA polymerases. *Adv. Protein Chem.* **48**:377–435.

Pfluger, K., S. Baumann, G. Gottschalk, W. Lin, H. Santos, and V. Muller. 2003. Lysine-2,3-aminomutase and beta-lysine acetyltransferase genes of methanogenic archaea are salt induced and are essential for the biosynthesis of N-epsilon-acetyl-beta-lysine and growth at high salinity. *Appl. Environ. Microbiol.* **69**: 6047–6055.

Podkovyrov, S. M., and J. G. Zeikus. 1992. Structure of the gene encoding cyclomaltodextrinase from *Clostridium thermohydrosulfuricum* 39E and characterization of the enzyme purified from *Escherichia coli*. *J. Bacteriol.* **174**:5400–5405.

Politz, O., M. Krah, K. K. Thomsen, and R. Borriss. 2000. A highly thermostable endo-(1,4)-beta-mannanase from the marine bacterium *Rhodothermus marinus*. *Appl. Microbiol. Biotechnol.* **53**:715–721.

Pouwels, J., M. Moracci, B. Cobucci-Ponzano, G. Perugino, J. van der Oost, T. Kaper, J. H. G. Lebbink, W. M. de Vos, M. Ciaramella, and M. Rossi. 2000. Activity and stability of hyperthermophilic enzymes: a comparative study on two archaeal β-glycosidases. *Extremophiles* **4**:157–164.

Posta, K., E. Beki, D. B. Wilson, J. Kukolya, and L. Hornok. 2004. Cloning, characterization and phylogenetic relationships of cel5B, a new endoglucanase encoding gene from *Thermobifida fusca*. *J. Basic Microbiol.* **44**:383–399.

Raasch, C., W. Streit, J. Schanzer, M. Bibel, U. Gosslar, and W. Liebl. 2000. *Thermotoga maritima* AglA, an extremely thermostable NAD^+-, Mn^{2+}-, and thiol-dependent alpha-glucosidase. *Extremophiles* **4**:189–200.

Radianingtyas, H., and P. C. Wright. 2003. Alcohol dehydrogenases from thermophilic and hyperthermophilic archaea and bacteria. *FEMS Microbiol. Rev.* 27:593–616.

Ramchuran, S. O., B. Mateus, O. Holst, and E. N. Karlsson. 2005. The methylotrophic yeast *Pichia pastoris* as a host for the expression and production of thermostable xylanase from the bacterium *Rhodothermus marinus*. *FEMS Yeast Res.* 5:839–850.

Ramesh, B., P. R. Reddy, G. Seenayya, and G. Reddy. 2001. Effect of various flours on the production of thermostable beta-amylase and pullulanase by *Clostridium thermosulfurogenes* SV2. *Bioresour. Technol.* 76:169–171.

Rao, V. B., and N. B. Saunders. 1992. A rapid polymerase-chain-reaction-directed sequencing strategy using a thermostable DNA polymerase from *Thermus flavus*. *Gene* 113:17–23.

Rashid, N., J. Cornista, S. Ezaki, T. Fukui, H. Atomi, and T. Imanaka. 2002. Characterization of an archaeal cyclodextrin glucanotransferase with a novel C-terminal domain. *J. Bacteriol.* 184:777–784.

Reeves, R. A., M. D. Gibbs, D. D. Morris, K. R. Griffiths, D. J. Saul, and P. L. Bergquist. 2000. Sequencing and expression of additional xylanase genes from the hyperthermophile *Thermotoga maritima* FjSS3B.1. *Appl. Environ. Microbiol.* 66:1532–1537.

Riessen, S., and G. Antranikian. 2001. Isolation of *Thermoanaerobacter keratinophilus* sp. nov., a novel thermophilic, anaerobic bacterium with keratinolytic activity. *Extremophiles* 5:399–408.

Rhimi, M., and S. Bejar. 2006. Cloning, purification and biochemical characterization of metallic-ions independent and thermoactive l-arabinose isomerase from the *Bacillus stearothermophilus* US100 strain. *Biochim. Biophys. Acta.* 1760:191–199.

Rolfsmeier, M., C. Haseltine, E. Bini, A. Clark, and P. Blum. 1998. Molecular characterization of the alpha-glucosidase gene (malA) from the hyperthermophilic archaeon *Sulfolobus solfataricus*. *J. Bacteriol.* 180:1287–1295.

Rolland, J. L., Y. Gueguen, C. Persillon, J. M. Masson, and J. Dietrich. 2004. Characterization of a thermophilic DNA ligase from the archaeon *Thermococcus fumicolans*. *FEMS Microbiol. Lett.* 236:267–273.

Ruttersmith, L. D., and R. M. Daniel. 1991. Thermostable cellobiohydrolase from the thermophilic eubacterium *Thermotoga* sp. strain FjSS3-B.1. Purification and properties. *Biochem. J.* 277:887–890.

Ryu, K., J. Kim, and J. S. Dordick. 1994. Catalytic properties and potential of an extracellular protease from an extreme halophile. *Enzyme Microb. Technol.* 16:266–275.

Sako, Y., P. C. Croocker, and Y. Ishida. 1997. An extremely heat-stable extracellular proteinase (aeropyrolysin) from the hyperthermophilic archaeon *Aeropyrum pernix* K1. *FEBS Lett.* 415:329–334.

Salleh, H. M., J. Mullegger, S. P. Reid, W. Y. Chan, J. Hwang, R. A. Warren, and S. G. Withers. 2006. Cloning and characterization of *Thermotoga maritima* beta-glucuronidase. *Carbohydr. Res.* 341:49–59.

Sanger, F., S. Nicklen, and A. R. Coulson. 1992. DNA sequencing with chain-terminating inhibitors. 1977. *Biotechnology* 24:104–108.

Santos, H., and M. da Costa. 2001. Organic solutes from thermophiles and hyperthermophiles. *Methods Enzymol.* 334:302–315.

Santos, H., and M. S. da Costa. 2002. Compatible solutes of organisms that live in hot saline environments. *Environ. Microbiol.* 4:501–509.

Scherer, P. A., G. R. Vollmer, T. Fakhouri, and S. Martensen. 2000. Development of a methanogenic process to degrade exhaustively the organic fraction of municipal "grey waste" under thermophilic and hyperthermophilic conditions. *Water Sci. Technol.* 41:83–91.

Schiraldi, C., and M. De Rosa. 2002. The production of biocatalysts and biomolecules from extremophiles. *Trends Biotechnol.* 20:515–521.

Schiraldi, C., M. Giuliano, and M. De Rosa. 2002. Perspectives on biotechnological applications of archaea. *Archaea* 1:75–86.

Schleper, C., R. V. Swanson, E. J. Mathur, and E. F. DeLong. 1997. Characterization of a DNA polymerase from the uncultivated psychrophilic archaeon *Cenarchaeum symbiosum*. *J. Bacteriol.* 179:7803–7811.

Scholz, S., S. Wolff, and R. Hensel. 1998. The biosynthesis pathway of di-myo-inositol-1,1'-phosphate in *Pyrococcus woesei*. *FEMS Microbiol. Lett.* 168:37–42.

Sehgal, A. C., and R. M. Kelly. 2002. Enantiomeric resolution of 2-aryl propionic esters with hyperthermophilic and mesophilic esterases: contrasting thermodynamic mechanisms. *J. Am. Chem. Soc.* 124:8190–8191.

Sehgal, A. C., and R. M. Kelly. 2003. Strategic selection of hyperthermophilic esterases for resolution of 2-arylpropionic esters. *Biotechnol. Prog.* 19:1410–1416.

Sehgal, A. C., R. Tompson, J. Cavanagh, and R. M. Kelly. 2002. Structural and catalytic response to temperature and cosolvents of carboxylesterase EST1 from the extremely thermoacidophilic archaeon *Sulfolobus solfataricus* P1. *Biotechnol. Bioeng.* 80:784–793.

Serour, E., and G. Antranikian. 2002. Novel thermoactive glucoamylases from the thermoacidophilic archaea *Thermoplasma acidophilum*, *Picrophilus torridus* and *Picrophilus oshimae*. *Antonie Van Leeuwenhoek* 81:73–83.

Si Si Hla, J. Kurokawa, Suryani, T. Kimura, K. Ohmiya, and K. Sakka. 2005. A novel thermophilic pectate lyase containing two catalytic modules of *Clostridium stercorarium*. *Biosci. Biotechnol Biochem.* 69:2138–2145.

Shao, W., and J. Wiegel. 1992. Purification and characterization of a thermostable beta-xylosidase from *Thermoanaerobacter ethanolicus*. *J. Bacteriol.* 174:5848–5853.

Shao, W., and J. Wiegel. 1995. Purification and characterization of two thermostable acetyl xylan esterases from *Thermoanaerobacterium* sp. strain JW/SL-YS485. *Appl. Environ. Microbiol.* 61:729–733.

Shinohara, M. L., M. Ihara, M. Abo, M. Hashida, S. Takagi, and T. C. Beck. 2001. A novel thermostable branching enzyme from an extremely thermophilic bacterial species, *Rhodothermus obamensis*. *Appl. Microbiol. Biotechnol.* 57:653–659.

Slesarev, A. I., K. V. Mezhevaya, K. S. Makarova, N. N. Polushin, O. V. Shcherbinina, V. V. Shakhova, G. I. Belova, L. Aravind, D. A. Natale, I. B. Rogozin, R. L. Tatusov, Y. I. Wolf, K. O. Stetter, A. G. Malykh, E. V. Koonin, and S. A. Kozyavkin. 2002. The complete genome of hyperthermophile *Methanopyrus kandleri* AV19 and monophyly of archaeal methanogens. *Proc. Natl. Acad. Sci. USA* 99:4644–4649.

Sleytr, U. B., H. Bayley, M. Sara, A. Breitwieser, S. Kupcu, C. Mader, S. Weigert, F. M. Unger, P. Messner, B. Jahn-Schmid, B. Schuster, D. Pum, K. Douglas, N. A. Clark, J. T. Moore, T. A. Winningham, S. Levy, I. Frithsen, J. Pankovc, P. Beale, H. P. Gillis, D. A. Choutov, and K. P. Martin. 1997. Applications of S-layers. *FEMS Microbiol. Rev.* 20:151–175.

Soboh, B., D. Linder, and R. Hedderich. 2002. Purification and catalytic properties of a CO-oxidizing:H_2-evolving enzyme complex from *Carboxydothermus hydrogenoformans*. *Eur. J. Biochem.* 269:5712–5721.

Soboh, B., D. Linder, and R. Hedderich. 2004. A multisubunit membrane-bound [NiFe] hydrogenase and an NADH-dependent Fe-only hydrogenase in the fermenting bacterium *Thermoanaerobacter tengcongensis*. *Microbiology* 150:2451–2463.

Soderberg, T., and R. Alver. 2004. Transaldolase of *Methanocaldococcus jannaschii*. *Archaea* 1:255-262.

Specka, U., F. Mayer, and G. Antranikian. 1991. Purification and properties of a thermoactive glucoamylase from *Clostridium thermosaccharolyticum*. *Appl. Environ. Microbiol.* 57:2317–2323.

Sprott, G. D., G. B. Patel, and L. Krishnan. 2003. Archaeobacterial ether lipid liposomes as vaccine adjuvants. *Methods Enzymol.* 373:155–172.

Stan-Lotter, H., E. Doppler, M. Jarosch, C. Radax, C. Gruber, and K. I. Inatomi. 1999. Isolation of a chymotrypsinogen B-like enzyme from the archaeon *Natronomonas pharaonis* and other halobacteria. *Extremophiles* 3:153–161.

Story, S. V., A. M. Grunden, and M. W. W. Adams. 2001. Characterization of an aminoacylase from the hyperthermophilic archaeon *Pyrococcus furiosus*. *J. Bacteriol.* 183:4259–4268.

Stutzenberger, F. J. 1987. Inducible thermoalkalophilic polygalacturonate lyase from *Thermomonospora fusca*. *J. Bacteriol.* 169:2774–2780.

Suzuki, R., K. Nagata, F. Yumoto, M. Kawakami, N. Nemoto, M. Furutani, K. Adachi, T. Maruyama, and M. Tanokura. 2003. Three-dimensional solution structure of an archaeal FKBP with a dual function of peptidyl prolyl *cis-trans* isomerase and chaperone-like activities. *J. Mol. Biol.* 328:1149–1160.

Suzuki, Y., K. Miyamoto, and H. Ohta. 2004. A novel thermostable esterase from the thermoacidophilic archaeon *Sulfolobus tokodaii* strain 7. *FEMS Microbiol. Lett.* 236:97–102.

Tachibana, Y., A. Kuramura, N. Shirasaka, Y. Suzuki, T. Yamamoto, S. Fujiwara, M. Takagi, and T. Imanaka. 1999. Purification and characterization of an extremely thermostable cyclomaltodextrin glucanotransferase from a newly isolated hyperthermophilic archaeon, a *Thermococcus* sp. *Appl. Environ. Microbiol.* 65:1991–1997.

Takagi, M., M. Nishioka, H. Kakihara, M. Kitabayashi, H. Inoue, B. Kawakami, M. Oka, and T. Imanaka. 1997. Characterization of DNA polymerase from *Pyrococcus* sp. strain KOD1 and its application to PCR. *Appl. Environ. Microbiol.* 63:4504–4510.

Takata, H., T. Takaha, S. Okada, S. Hizukuri, M. Takagi, and T. Imanaka. 1996. Structure of the cyclic glucan produced from amylopectin by *Bacillus stearothermophilus* branching enzyme. *Carbohydr. Res.* 295:91–101.

Takata, H., K. Ohdan, T. Takaha, T. Kuriki, and S. Okada. 2003. Properties of branching enzyme from hyperthermophilic bacterium, *Aquifex aeolicus*, and its potential for production of highly-branched cyclic dextrin. *J. Appl. Glycosci.* 50:15–20.

Tanaka, T., S. Fujiwara, S. Nishikori, T. Fukui, M. Takagi, and T. Imanaka. 1999. A unique chitinase with dual active sites and triple substrate binding sites from the hyperthermophilic archaeon *Pyrococcus kodakaraensis* KOD1. *Appl. Environ. Microbiol.* 65:5338–5344.

Tanaka, T., T. Fukui, H. Atomi, and T. Imanaka. 2003. Characterization of an *exo*-beta-D-glucosaminidase involved in a novel chitinolytic pathway from the hyperthermophilic archaeon *Thermococcus kodakaraensis* KOD1. *J. Bacteriol.* 185:5175–5181.

Tanaka, T., T. Fukui, S. Fujiwara, H. Atomi, and T. Imanaka. 2004. Concerted action of diacetylchitobiose deacetylase and *exo*-beta-D-glucosaminidase in a novel chitinolytic pathway in the hyperthermophilic archaeon *Thermococcus kodakaraensis* KOD1. *J. Biol. Chem.* 279:30021–30027.

Tanaka, T., T. Fukui, and T. Imanaka. 2001. Different cleavage specificities of the dual catalytic domains in chitinase from the hyperthermophilic archaeon *Thermococcus kodakaraensis* KOD1. *J. Biol. Chem.* 276:35629–35635.

Taylor, I., R. Brown, M. Bycroft, G. King, J. A. Littlechild, M. Lloyd, C. Praguin, and S. Taylor. 2004. Application of thermophilic enzymes in commercial biotransformation processes. *Biochem. Soc. Trans.* 32:290–292.

Te'o, V. S., D. J. Saul, and P. L. Bergquist. 1995. celA, another gene coding for a multidomain cellulase from the extreme thermophile *Caldocellum saccharolyticum*. *Appl. Microbiol. Biotechnol.* 43:291–296.

Terada, Y., K. Fujii, T. Takaha, and S. Okada. 1999. *Thermus aquaticus* ATCC 33923 amylomaltase gene cloning and expression and enzyme characterization: production of cycloamylose. *Appl. Environ. Microbiol.* 65:910–915.

Thiemann, V., C. Donges, S. G. Prowe, R. Sterner, and G. Antranikian. 2004. Characterisation of a thermoalkali-stable cyclodextrin glycosyltransferase from the anaerobic thermoalkaliphilic bacterium *Anaerobranca gottschalkii*. *Arch. Microbiol.* 182:226–235.

Thiemann, V., B. Saake, A. Vollstedt, T. Schafer, J. Puls, C. Bertoldo, R. Freudl, and G. Antranikian. 2006. Heterologous expression and characterization of a novel branching enzyme from the thermoalkaliphilic anaerobic bacterium Anaerobranca gottschalkii. *Appl. Microbiol. Biotechnol.* 12:1–12.

Thompson, V. S., K. D. Schaller, and W. A. Apel. 2003. Purification and characterization of a novel thermo-alkali-stable catalase from *Thermus brockianus*. *Biotechnol. Prog.* 19:1292–1299.

Thorbjarnardottir, S. H., Z. O. Jonsson, O. S. Andresson, J. K. Kristjansson, G. Eggertsson, and A. Palsdottir 1995. Cloning and sequence analysis of the DNA ligase-encoding gene of *Rhodothermus marinus*, and overproduction, purification and characterization of two thermophilic DNA ligases. *Gene* 161:1–6.

Tomiyasu, K., K. Yato, M. Yasuda, T. Tonozuka, A. Ibuka, and H. Sakai 2001. Cloning and nucleotide sequence of the pullulanase gene of *Thermus thermophilus* HB8 and production of the enzyme in *Escherichia coli*. *Biosci. Biotechnol. Biochem.* 65:2090–2094.

Toogood, H. S., E. J. Hollingsworth, R. C. Brown, I. N. Taylor, S. J. Taylor, R. McCague, and J. A. Littlechild. 2002. A thermostable L-aminoacylase from *Thermococcus litoralis*: cloning, overexpression, characterization, and applications in biotransformations. *Extremophiles* 6:111–122.

Tsuruoka, N., T. Nakayama, M. Ashida, H. Hemmi, M. Nakao, H. Minakata, H. Oyama, K. Oda, and T. Nishino. 2003. Collagenolytic serine-carboxyl proteinase from *Alicyclobacillus sendaiensis* strain NTAP-1: purification, characterization, gene cloning, and heterologous expression. *Appl. Environ. Microbiol.* 69:162–169.

Uhl, A. M., and R. M. Daniel. 1999. The first description of an archaeal hemicellulase: the xylanase from *Thermococcus zilligii* strain AN1. *Extremophiles* 3:263–267.

Uotsu-Tomita, R., T. Tonozuka, H. Sakai, and Y. Sakano. 2001. Novel glucoamylase-type enzymes from *Thermoactinomyces vulgaris* and *Methanococcus jannaschii* whose genes are found in the flanking region of the alpha-amylase genes. *Appl. Microbiol. Biotechnol.* 56:465–473.

Upreti, R., M. Kumar, and V. Shankar. 2003. Bacterial glycoproteins: functions, biosynthesis and applications. *Proteomics* 3:363–379.

van der Oost, J., W. G. Voorhorst, S. W. Kengen, A. C. Geerling, V. Wittenhorst, Y. Gueguen, and W. M. de Vos. 2001. Genetic and biochemical characterization of a short-chain alcohol dehydrogenase from the hyperthermophilic archaeon *Pyrococcus furiosus*. *Eur. J. Biochem.* 268:3062–3068.

van Lieshout, J., M. Faijes, J. Nieto, J. van der Oost, and A. Planas. 2004. Hydrolase and glycosynthase activity of *endo*-1,3-beta-glucanase from the thermophile *Pyrococcus furiosus*. *Archaea* 1:285–292.

Van Ooteghem, S. A., A. Jones, D. Van Der Lelie, B. Dong, and D. Mahajan. 2004. H(2) production and carbon utilization by *Thermotoga neapolitana* under anaerobic and microaerobic growth conditions. *Biotechnol. Lett.* 26:1223–1232.

Van Ooteghem, S. A., S. K. Beer, and P. C. Yue. 2002. Hydrogen production by the thermophilic bacterium *Thermotoga neapolitana*. *Appl. Biochem. Biotechnol.* **98–100:**177–189.

Vander Horn, P. B., M. C. Davis, J. J. Cunniff, C. Ruan, B. F. McArdle, S. B. Samols, J. Szasz, G. Hu, K. M. Hujer, S. T. Domke, S. R. Brummet, R. B. Moffett, and C. W. Fuller. 1997. Thermo Sequenase DNA polymerase and *T. acidophilum* pyrophosphatase: new thermostable enzymes for DNA sequencing. *Biotechniques* **22:**758–762, 764–765.

Vieille, C., J. M. Hess, R. M. Kelly, and J. G. Zeikus. 1995. xylA cloning and sequencing and biochemical characterization of xylose isomerase from Thermotoga neapolitana. *Appl. Environ. Microbiol.* **61:**1867–1875.

Vieille, C., and G. J. Zeikus. 2001. Hyperthermophilic enzymes: sources, uses, and molecular mechanisms for thermostability. *Microbiol. Mol. Biol. Rev.* **65:**1–43.

Voorhorst, W. G., R. I. Eggen, A. C. Geerling, C. Platteeuw, R. J. Siezen, and W. M. Vos. 1996. Isolation and characterization of the hyperthermostable serine protease, pyrolysin, and its gene from the hyperthermophilic archaeon *Pyrococcus furiosus*. *J. Biol. Chem.* **271:**20426–20431.

Waino, M., and K. Ingvorsen. 2003. Production of beta-xylanase and beta-xylosidase by the extremely halophilic archaeon *Halorhabdus utahensis*. *Extremophiles* **7:**87–93.

Wakagi, T., C. H. Lee, and T. Oshima. 1992. An extremely stable inorganic pyrophosphatase purified from the cytosol of a thermoacidophilic archaebacterium, *Sulfolobus acidocaldarius* strain 7. *Biochim. Biophys. Acta* **1120:**289–296.

Wang, B., D. Lu, R. Gao, Z. Yang, S. Cao, and Y. Feng. 2004. A novel phospholipase A2/esterase from hyperthermophilic archaeon *Aeropyrum pernix* K1. *Protein Expr. Purif.* **35:**199–205.

Wang, X., X. He, S. Yang, X. An, W. Chang, and D. Liang. 2003. Structural basis for thermostability of beta-glycosidase from the thermophilic eubacterium *Thermus nonproteolyticus* HG102. *J. Bacteriol.* **185:**4248–4255.

Ward, D. E., K. R. Shockley, L. S. Chang, R. D. Levy, J. K. Michel, S. B. Conners, and R. M. Kelly. 2002. Proteolysis in hyperthermophilic microorganisms. *Archaea* **1:**63–74.

Wassenberg, D., H. Schurig, W. Liebl, and R. Jaenicke. 1997. Xylanase XynA from the hyperthermophilic bacterium *Thermotoga maritima*: structure and stability of the recombinant enzyme and its isolated cellulose-binding domain. *Protein Sci.* **6:**1718–1726.

Wright, R. M., M. D. Yablonsky, Z. P. Shalita, A. K. Goyal, and D. E. Eveleigh. 1992. Cloning, characterization, and nucleotide sequence of a gene encoding *Microbispora bispora* BglB, a thermostable beta-glucosidase expressed in *Escherichia coli*. *Appl. Environ. Microbiol.* **58:**3455–3465.

Wojciechowski, C. L., J. P. Cardia, and E. R. Kantrowitz. 2002. Alkaline phosphatase from the hyperthermophilic bacterium *Thermotoga maritima* requires cobalt for activity. *Protein Sci.* **11:**903–911.

Xue, Y., and W. Shao. 2004. Expression and characterization of a thermostable beta-xylosidase from the hyperthermophile, *Thermotoga maritima*. *Biotechnol. Lett.* **26:**1511–1515.

Yan, Z., S. Fujiwara, K. Kohda, M. Takagi, and T. Imanaka. 1997. In vitro stabilization and in vivo solubilization of foreign proteins by the beta subunit of a chaperonin from the hyperthermophilic archaeon *Pyrococcus* sp. strain KOD1. *Appl. Environ. Microbiol.* **63:**785–789.

Yang, S. J., H. S. Lee, C. S. Park, Y. R. Kim, T. W. Moon, and K. H. Park. 2004. Enzymatic analysis of an amylolytic enzyme from the hyperthermophilic archaeon *Pyrococcus furiosus* reveals its novel catalytic properties as both an alpha-amylase and a cyclodextrin-hydrolyzing enzyme. *Appl. Environ. Microbiol.* **70:**5988–5995.

Yano, J. K., F. Blasco, H. Li, R. D. Schmid, A. Henne, and T. L. Poulos. 2003. Preliminary characterization and crystal structure of a thermostable cytochrome P450 from *Thermus thermophilus*. *J. Biol. Chem.* **278:**608–616.

Yip, V. L., A. Varrot, G. J. Davies, S. S. Rajan, X. Yang, J. Thompson, W. F. Anderson, and S. G. Withers. 2004. An unusual mechanism of glycoside hydrolysis involving redox and elimination steps by a family 4 beta-glycosidase from *Thermotoga maritima*. *J. Am. Chem. Soc.* **126:**8354–8355.

Yoon, S. Y., H. S. Noh, E. H. Kim, and K. H. Kong. 2002. The highly stable alcohol dehydrogenase of *Thermomicrobium roseum*: purification and molecular characterization. *Comp. Biochem. Physiol. B. Biochem. Mol. Biol.* **132:**415–422.

Zappa, S., J. L. Rolland, D. Flament, Y. Gueguen, J. Boudrant, and J. Dietrich. 2001. Characterization of a highly thermostable alkaline phosphatase from the euryarchaeon *Pyrococcus abyssi*. *Appl. Environ. Microbiol.* **67:**4504–4511.

Zhang, J., J. Liu, J. Zhou, Y. Ren, X. Dai, and H. Xiang. 2003. Thermostable esterase from *Thermoanaerobacter tengcongensis*: high-level expression, purification and characterization. *Biotechnol. Lett.* **25:**1463–1467.

Zhang, S., G. Lao, and D. B. Wilson. 1995. Characterization of a *Thermomonospora fusca* exocellulase. *Biochemistry* **34:**3386–3395.

Zhou, M., H. Xiang, C. Sun, and H. Tan, 2004. Construction of a novel shuttle vector based on an RCR-plasmid from a haloalkaliphilic archaeon and transformation into other haloarchaea. *Biotechnol. Lett.* **26:**1107–1113.

Zona, R., F. Chang-Pi-Hin, M. J. O'Donohue, and S. Janecek. 2004. Bioinformatics of the glycoside hydrolase family 57 and identification of catalytic residues in amylopullulanase from *Thermococcus hydrothermalis*. *Eur. J. Biochem.* **271:**2863–2872.

Zverlov, V., S. Mahr, K. Riedel, and K. Bronnenmeier. 1998. Properties and gene structure of a bifunctional cellulolytic enzyme (CelA) from the extreme thermophile "*Anaerocellum thermophilum*" with separate glycosyl hydrolase family 9 and 48 catalytic domains. *Microbiology* **144:**457–465.

Zverlov, V. V., G. A. Velikodvorskaya, and W. H. Schwarz. 2002a. A newly described cellulosomal cellobiohydrolase, CelO, from *Clostridium thermocellum*: investigation of the exo-mode of hydrolysis, and binding capacity to crystalline cellulose. *Microbiology* **148:**247–255.

Zverlov, V. V., K. P. Fuchs, and W. H. Schwarz. 2002b. Chi18A, the endochitinase in the cellulosome of the thermophilic, cellulolytic bacterium *Clostridium thermocellum*. *Appl. Environ. Microbiol.* **68:**3176–3179.

Zverlov, V. V., J. Kellermann, and W. H. Schwarz. 2005. Functional subgenomics of *Clostridium thermocellum* cellulosomal genes: identification of the major catalytic components in the extracellular complex and detection of three new enzymes. *Proteomics* **5:**3646–3653.

X. LESSONS FROM EXTREMOPHILES

Chapter 28

Lessons from Extremophiles: Early Evolution and Border Conditions of Life

YING XU AND NICOLAS GLANSDORFF

The greatest challenge to any thinker is stating the problem in a way that will allow a solution.

—Bertrand Russell

INTRODUCTION

This chapter is not a summary nor a synthesis of the contributions gathered in this volume. Rather, the subject of the book itself was a starting point for considerations inspired by the very existence of extremophiles, their emergence with respect to the last universal common ancestor (LUCA), the origin of life, astrobiology, the natural limits of life, and, in a few closing remarks, the regrettable but presently unavoidable controversy with the advocates of so-called "intelligent design."

Before extremophiles ceased to be curiosities and became a fascinating new world, life used to be considered as fragile, easily disturbed, or destroyed by adverse physical and chemical conditions. For biologists, the realization that many forms of life are actually confined to environments that are severely hostile by human standards will remain one of the most significant scientific achievements of the second half of the twentieth century. Almost every topic in biochemistry and biophysics has broadened to an extent that would have been inconceivable only a few decades ago ever since studying the effects of extremes of pressure, temperature, salinity, water activity, acidity, or alkalinity on macromolecules and metabolism became biologically relevant in a general sense. Moreover, in retrospect, the conjunction between the study of some forms of extremophily—hyperthermophily in particular—and the discovery of a third domain of life—the *Archaea* (Woese and Fox, 1977)—appear a strikingly fruitful one since it sparked new ideas on the origin of life and its ramification into different forms of cellular organization. From the astrobiological point of view, the discovery on Mars and Europa of environments resembling the biotopes of certain terrestrial extremophiles and the possibility that highly resistant life forms may be able to survive interplanetary journeys in meteorites led to reappraise the question of the cradle of life in the solar system.

The questions raised by the molecular basis and the emergence of different forms of extremophily are not only deep and varied, but, not surprisingly, many remain controversial. Whether thermophily is relevant to the origin of life is certainly one of those.

THERMOPHILY, THE LUCA, AND THE ORIGIN OF LIFE

The Universal Tree of Life

The universal tree of life based on comparative analyses of the small subunit of rRNA is often presented in the rooted form with *Bacteria* branching out first, *Archaea* second, and *Eukarya* third. It is important to keep in mind that the rooting operation had nothing to do with the small subunit of rRNA itself but was an exercise based on paralogous protein genes (Gogarten et al., 1989; Iwabe et al., 1989). Whereas the separation into three domains is generally agreed upon, using different proteins may suggest different rootings, including some with the best fit in the eukaryotic branch (Forterre and Philippe, 1999; Lopez et al., 1999), as a phylogenetic analysis of RNA secondary structure also suggests (Caetano-Anolles, 2002). Nevertheless, the so-called "universal" tree is frequently but abusively presented as

Y. Xu • Marine Sciences Research Center, Stony Brook University, Stony Brook, NY 11794. N. Glansdorff • JM Wiame Research Institute for Microbiology, Vrije Universiteit Brussel, Brussels B-1070, Belgium.

established. Several authors have discussed or reviewed this topic in depth, but despite the doubts lingering on the canonical rooting, it is rarely discussed anymore and is mentioned in textbooks without further comments (see above and Glansdorff, 2000). Perhaps the most pernicious result of this state of affairs is that a tree where the two prokaryotic branches are the lower ones suggests to many that the first forms of life to emerge were prokaryotic (this belief being reinforced by the very ambiguity of the prefix "pro"), whereas nothing of the kind can actually be inferred from such a phylogeny. Indeed, the branching pattern of the tree, constructed with modern organisms, has no bearing on the cellular organization of the LUCA. Even if the rooting of the "universal" tree were to prove correct, with the prokaryotes branching out first, the LUCA could have been a form of life announcing *Eukarya* in many respects (a protoeukaryote) except of course the presence of organelles acquired by endosymbiosis. Such an organism could have undergone reductive evolution toward the cells we recognize today as *Bacteria* and *Archaea*, an idea expressed a long time ago in the wake of the discovery of introns (see Darnell and Doolittle, 1986) and rediscussed in a modern context by several authors (Forterre, 1995, 1996; Glansdorff, 2000; Penny and Poole, 1999; Poole et al., 1998).

Was Extreme Thermophily Primary or Secondary?

Modern prokaryotes are the only forms of life featuring organisms capable of growth above 62°C, and inside each domain, the first phylogenetic analyses singled out the most extreme of these thermophiles as the earliest lines of descent: Aquificales and Thermotogales among *Bacteria*, different Euryarchaeota and Crenarchaeota among *Archaea* (Stetter, 1996). These results supported the notion that the LUCA and perhaps the primeval cells themselves had been hyperthermophiles. However, resolving the order of the numerous and deeply branching divisions inside the bacterial domain is a difficult matter (Hugenholtz et al., 1998). If recent analyses appear to confirm that all presently known, early branching *Archaea* are indeed hyperthermophilic (Brochier et al., 2005), they produce conflicting patterns for the *Bacteria*. A revised 16S rRNA analysis focusing on sites without phylogenetic noise placed an essentially mesophilic group, the Planctomycetes, on the first line of divergence] (Brochier and Philippe, 2002). Several properties make the Planctomycetes particularly interesting: presence of a membrane-bound nucleoid similar to the eukaryotic nucleus (in particular, the double membrane-bound nuclear body of *Gemmata*), capacity to synthesize sterols, absence of peptidoglycan in the cell wall, division by budding, presence of large genomes, large variety of ecological types, and metabolic capabilities (see Fuerst, 2005). All this points to a possible relationship with the LUCA that would be closer than for any other living organism.

The method followed by Brochier and Philippe has been criticized by Di Giulio (2003b) who concluded on the basis of another 16S rRNA site selection that the former phylogeny remained correct. A bacterial phylogeny based on E-Tu factor emphasized the uniqueness of Planctomycetes but could not conclude regarding their position among early bacterial clades due to instability of reconstructed trees (Jenkins and Fuerst, 2001). Studies on one C_1-compound transfer pathway confirmed the antiquity of the Planctomycetes (Chistoserdova et al., 2004) but without resolving the branching pattern. Phylogenomic supertrees constructed either with 730 genes or with a core of 310 genes supporting similar species phylogenies from 45 organisms relegated the extreme thermophiles *Aquifex* and *Thermotoga* in a non-basal position while placing Spirochaetes and Chlamydiae on the lowest branch (Daubin et al., 2002). This work did not include Planctomycetes, but there is evidence indicating that they are related to Chlamydiae and Spirochaetes (Teeling et al., 2004). In the latter study, concatenated ribosomal proteins and DNA-dependent RNA polymerase (DdRp) sequences tended to shift the Planctomycete species studied (*Rhodopirellula baltica*) toward a deeper branching position when highly conserved sites were considered; however, whole-genome trees tended to separate *R. baltica* from the Chlamydiae while placing *Aquifex* and *Thermotoga* in a basal position, in contrast to Daubin et al. (2002). The exact position of *R. baltica* in these trees was, however, not well resolved. Interestingly, the DdRp tree disjoined *Aquifex* from *Thermotoga*; a study by Griffiths and Gupta (2004) using signature sequences in several conserved proteins including DdRp led to the same conclusion and pointed to a late divergence of the Aquificales. Lakshminarayan et al. (2004) emphasized that the domain architecture of DdRp subunits suggests phylogenetic clustering of *Bacteria* in two groups, again separating *Thermotoga* from *Aquifex*. The most parsimonious evolutionary explanation of the data roots the bacterial tree between the two groups, which suggests a mesophilic common ancestor for *Bacteria*. A recent study by Bern and Golberg (2005) used a new algorithm automatically picking up "representative" proteins, including both ubiquitous and non-ubiquitous but rather well-conserved proteins; here *Thermotoga* and *Aquifex* remained together in a basal position with a weak bootstrap support. In contrast to the study by Daubin et al. (2002), this analysis incorporated many fast-evolving and non-ubiquitous genes, perhaps more liable to long-branch attraction and convergent evolution. As the authors themselves pointed out, using "locally reliable genes" to resolve certain parts of the tree of life may therefore be in order; this emphasizes the importance of the DdRp data. It should be noted, however, that the "genomic core" approach of Daubin et al. has proved very useful to establish a coherent phylogeny of *Archaea* (Brochier et al., 2005). Another deeply branching lineage, moderately related to the Planctomycetes, was discovered recently, the sponge-associated Poribacteria. These organisms also contain a membrane-bound nucleoid (Fieseler et al., 2004).

The resolution of the controversy regarding the respective phylogenetic positions of the Planctomycetes/Chlamydiae/Spirochaetes superclade and of the most extreme thermophilic bacteria is clearly a matter of great evolutionary significance. In this respect, it is noteworthy that the nature of the membrane lipids found in thermophilic *Bacteria* suggests that these organisms evolved by convergence from a non-thermophilic ancestor toward thermophily (Xu and Glansdorff, 2002; Glansdorff and Xu, 2004). Whereas *Archaea* are characterized by *sn*-2,3 glycerol-ether (thus relatively thermostable) isoprenoid lipids, thermophilic *Bacteria* present a bewildering variety of membrane lipids, some of them with a glycerol-ether linkage but with the *sn*-1,2 stereochemistry characteristic of both *Bacteria* and *Eukarya*. Moreover, the maximal growth rate temperatures (so-called "optimal" temperatures) of the most extreme thermophilic bacteria remain well below archaeal performances. These observations were taken into account in the hypothesis (Glansdorff and Xu, 2004, and further, *Closing Comments*) presenting the emergence of *Archaea* and *Bacteria* as two separate events of reductive evolution from a non-thermophilic LUCA.

There are other reasons to surmise that thermophily, at least as it appears in modern organisms, may be an acquired trait: the molecular adaptations of tRNA to thermophily look rather evolved (Forterre, 2004; Grosjean, this volume). The structure of reverse gyrase (a thermophilic exclusivity) appears as an evolutionary novelty associating two domains coming from non-thermophilic ancestors (Forterre, 1996). A mesophilic (or at least non-extreme thermophilic) LUCA was inferred by Galtier et al. (1999) from correlations between rRNA G+C content and growth temperatures of microorganisms. The results of this analysis were confirmed by taking into account site-specific variation (covarion model, Galtier, 2001). Galtier's conclusions were, however, challenged by Di Giulio (2000, 2003a): using a thermophily index based on the number of codons attributed to each amino acid in the genetic code and the propensity of certain amino acids to be represented in thermophilic proteins (albeit with a so widely scattered distribution of individual values that quantitative correlations should be taken with caution even if linear regressions appear significant), this author concluded that the LUCA was (hyper)thermophilic, but still within a wide range of possible temperatures; a relatively moderate thermophile like *Geobacillus stearothermophilus* is at the lower limit of this range.

At this point, we would like to emphasize that even if the LUCA had been a thermophile, it does not follow that it grew in the very same range of temperatures as the modern extreme and hyperthermophilic organisms which may appear related to it in certain phylogenies. Modern thermophiles are the result of more than 3 billion years evolution, during which further adaptation has certainly occurred, and, as already mentioned above, molecular adaptations to thermophily look rather elaborated in the only living organisms we can investigate (see also below, *Adaptation to Thermophily*).

It remains that, in contrast to conflicting results of phylogenetic analyses and extrapolations based on protein composition, several biological considerations suggest that the LUCA may have been a mesophilic (or moderately thermophilic) proto-eukaryote rather than an extreme thermophilic prokaryote, perhaps an organism already endowed with a nucleus (as the Planctomycetes may suggest) and other eukaryotic characteristics such as integrin alpha-V, tubulin, actin, and dynamin, whose putative homologs were found among prokaryotes such as the Planctomycetes themselves (homolog of integrin in *Gemmata obscuriglobus*; Jenkins et al., 2002a), the somewhat related Verrucobacteria (eukaryotic tubulin homolog in *Prosthecobacter dejongeii*; Jenkins et al., 2002b), and many *Bacteria* and *Archaea* (FtsZ, ubiquitous homolog of tubulin; ParM and MreB, homologs of actin; MinD, homolog of dynamin; Amos et al., 2004). The presence of these cytoskeleton molecules in modern prokaryotes could indeed indicate their conservation from a protoeukaryotic ancestor rather than qualify them as ancestors of their eukaryotic counterparts. With a cytoskeleton, the LUCA may have been what C. de Duve (1991, 2005) calls the primitive phagocyte, capable of absorbing the bacteria that were to become its endosymbionts and perhaps also capable of evolving a nuclear membrane endogeneously. The presence of nucleoids not only in some *Bacteria* but possibly also in sponge-associated *Archaea*-like organisms suggests that the LUCA itself may already have been nucleated (Fuerst, 2005).

Adaptation to Thermophily

One of Wachtershauser's (1998) strong contentions is that evolution took place from thermophily toward mesophily and psychrophily because the converse, according to him, would be impossible. However, several phylogenetic analyses of bacterial phylogeny suggest a non-thermophilic origin of this domain, even if the question remains unresolved (see above). At any rate, it does not appear likely that all thermophiles belong to ancestral cell lines. Besides, evolution from mesophily toward thermophily could have occurred in steps and have been facilitated by at least two factors: the general increase in mutation rate that accompanies exposure to stress and the possibility of transient exposure in the course of adaptation. One interesting model in this respect is the annelid worm *Alvinella pompejana* which lives in the walls of hydrothermal vents where it is transiently exposed to temperatures as high as 80°C. Moreover, the acquisition in only a few steps of pleiotropic traits like reverse gyrase, more efficient chaperonins, and intracellular production of stabilizing solutes could have played a determinant role in evolution toward thermophily. Also, single genetic events can lead to dramatic changes in the G+C content of DNA (see Danchin, this volume). Of course, once thermophilic cells begin to emerge, horizontal gene transfer under strong selective pressure may become a factor in evolution toward thermophily.

Moreover, it should be taken into account that a given cell may contain more than one exemplar of the same gene or type of gene, which would have been a fundamental asset for adaptation to thermophily, as for any kind of evolutionary diversification. This is of course true of *Alvinella*, which displays genetic polymorphism (Hurtado et al., 2004), but it could also be the case with prokaryotes, which are abusively presented as "haploid," whereas the occurrence of merodiploids and even mechanisms of cell fusion may be more common than generally assumed by researchers who usually focus on stable recombinants in their genetic experiments (Gratia, 2003; Guillen et al., 1985; Lederberg, 1949). Moreover, the LUCA was probably a promiscuous community (Kandler, 1994; Woese, 1998) where cells would have been genetically redundant as can actually be inferred from the analysis of some paralogous genes (Labedan et al., 1999, 2004; Glansdorff, 2000). An ancestral population having evolved successfully to give rise to the diversity of metabolic functions observed in the three domains would indeed be expected to have had in store more than single copies of every essential gene. We suggest this redundancy could be the result of still imperfect cell division mechanisms allowing only redundant cells to survive on the average.

Regarding the question of progressive adaptation to thermophily, it is interesting that the genome of *Pyrococcus furiosus*, a hyperthermophilic archaeon, contains a sizeable proportion of genes that could have originated from mesophiles (Cohen et al., 2003). It should also be noted that the notion of "adaptation to high temperature" is perhaps more subtle than it appears at first sight: many microbiologists use the parameter "optimal temperature" (maximal growth rate temperature) to define the adaptation of an organism to a particular range of temperatures. Yet, this may be misleading because the temperature at which an organism grows the fastest is essentially a kinetic effect and more often than not may bear little relation to its physiological "well-being"; as a matter of fact fermenter studies have shown that *P. furiosus* is unhappy at its "optimal" growth temperature (Biller and Märkl, 2002; H. Märkl, personal communication), whereas, at the other end of the scale, adaptation to cold does not necessarily mean fast growth in the cold (see discussion in Glansdorff and Xu, 2002). The most physiological range of growth temperatures of *P. furiosus* is unknown but is probably lower than suggested by its "optimal temperature." In other words, even if the LUCA had been thermophilic, the range of temperatures where most of its evolution did occur may have been lower (perhaps by 15 or even 20°C) than suggested by correlations between protein composition (see Di Giulio, 2003a) and "optimal" temperatures of known thermophiles; the latter indicate an upper range which was probably important during transient exposures and perhaps for competition with other organisms, but not an optimal niche. The data on this are scarce concerning thermophiles, but the distinction between so-called "optimal" and more physiological (let us say "comfort") temperatures is documented in *Pseudomonas* sp. (Guillou and Guespin-Michel, 1996; Chablain et al., 1997). Besides, we already pointed out above that even a thermophilic LUCA may not have been as extreme as suggested by the behavior of modern hyperthermophiles.

Before the LUCA

Whether life itself originated at high temperature is an entirely different question. Wachtershauser (1988, 1992, 1998) has discussed in considerable detail how life could have found a hyperthermophilic, autotrophic origin in a volcanic environment such as the abyssal hydrothermal vents. This theory contrasts sharply and is irreducible with the view that organic matter, whether arising from prebiotic chemistry on Earth itself or delivered by meteorites, is at the origin of life (see de Duve, 1991, 2005, for a thorough discussion) and that life arose at low or moderate temperatures (Miller and Lazcano, 1998). Di Giulio's statistical studies on the origin of the genetic code are compatible with a thermophilic (and possibly piezophilic) origin (Di Giulio, 2000, 2005), but the data again allow for a considerable margin in the predicted temperatures and are based on correlations inferred from widely scattered values.

Other aspects militate against the notion of a thermophilic origin of life; one of them is the difficulty that RNA would have met to fold properly at high temperatures (Moulton et al., 2000). Another one is the high thermolability of countless metabolites in the aqueous phase (Glansdorff, 2000; Massant, this volume). Furthermore, the necessity for RNA to be sufficiently rich in G+C in order to resist thermodenaturation would have constituted an obstacle to the diversification of ribozymes into many specific catalysts (Forterre, 2004).

To speculate further on what really happened at the time the first cells evolved, a critical information would be when the primeval ocean (which formed probably as early as 4.3 or 4.4 billion years ago) (Mojszsiz et al., 2001; Wilde et al., 2001) would have cooled down to a temperature compatible with non-thermophilic life. Even if life originated in or on the walls of a hydrothermal vent (but not necessarily in the hyperthermophilic temperature range), the presence of a cooling ocean would have exerted a strong selective pressure to promote the emergence of a mesophilic or even a psychrophilic LUCA (Glansdorff and Xu, 2002). Some authors envisage the formation of a frigid ocean very early in Earth history (Bada et al., 1994; Vincent et al., 2004).

The LUCA, a Complex Community Evolving Under Selective Pressure

The notion of a complex LUCA was considerably enriched by the recent suggestion that viruses may have been active participants in the dynamics of this primeval population: the "three viruses–three domains" (3V3D) hypothesis (Forterre, 2005, 2006), which assumes that each cellular domain originated with an RNA–DNA transition engineered by a different DNA virus in a LUCA population with RNA genomes, may explain phylogenetic observations on DNA replication proteins that had remained rather puzzling until now. Moreover, the great diversity of RNA-cell lineages that could have emerged from this population could explain why the RNA-cell ancestor of *Eukarya* had "bacterial-like" lipids but "archaeal-like" ribosomes (Forterre, 2005, 2006). If we also consider that the LUCA was a protoeukaryotic cell, as argued above, reductive evolution toward the bacterial and archaeal cellular patterns could have begun before the viral infections ensuring the RNA–DNA transition; the emergence of *Archaea* and *Eukarya* would be separate events triggered by two related viruses, as predicted by the 3V3D hypothesis. However, the notion of an already nucleated DNA-LUCA (suggested by the presence of a nucleus in certain *Bacteria* and perhaps in certain *Archaea* as well, see Fuerst, 2005) is not in keeping with the 3V3D hypothesis as such, since the latter assumes that the emergence of each domain as a DNA lineage is related to a particular viral infection.

It has been argued that the very concept of a "community" is anti-Darwinian (P. Forterre, quoted in Whitfield, 2004). It would be so indeed if "community" also meant some kind of sustained "cooperation." This would not only sound strange at the level of evolution we are considering here, but it would be illogical simply because natural selection is unavoidable in whatever circumstances and whatever the extent of horizontal transfer assumed to have occurred in the community. "Darwin's dangerous idea" is indeed like a "universal acid" biting through everything (Dennett, 1995). Therefore, the concept of a "Darwininian threshold" (Woese, 2002, 2004) above which "vertically generated novelty can and does assume greater importance" (and species thus become recognizable entities) may sound ambiguous since "Darwinism" is often understood as synonymous with "natural selection," which must have been operating at the earliest stages of life emergence and even before.

We can view the LUCA as a community evolving under selective pressure, a population of promiscuous cells devoid of cell walls, constantly exchanging genes and perhaps already diversified from the metabolic point of view ("multiphenotypical") as proposed by Kandler (1994). The metabolically sophisticated LUCA that was inferred by Castresana (2001) from his phylogenetic analysis of energy metabolism may in fact reflect the existence of an evolving and diversified community. Such a community would certainly be unstable and very dynamic but is it not what we need to understand evolution toward progressive stabilization in diverse cell lineages? The latter may have been much more numerous than the three domains we recognize today as survivors of this early differentiation.

Closing Comments

Figure 1 summarizes two possible schemes among those that can be presently conjectured (see Forterre, 2005, 2006 for detailed implications of the possible role of viruses). With Wachtershauser (1992; see also

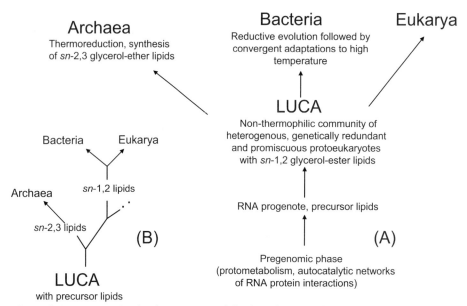

Figure 1. Alternative evolutionary scenarios for the emergence of the three domains of life from a protoeukaryotic LUCA. (A) Cells with sn-2,3 glycerol-ether lipids (Archaea) emerge by thermoreduction (Forterre, 1995) from a LUCA population with sn-1,2 glycerol-ester lipids; Bacteria emerge by reductive evolution but not thermoreduction; (B) simplified alternative scheme where both sn-1,2 and sn-2,3 glycerol lipids emerge at a different time from a LUCA with precursor lipids. The figure does not specify whether the LUCA had a DNA or an RNA genome (see text and Forterre, 2005, 2006). The "pregenomic" phase refers to concepts developed by de Duve (1991, 2005) and Kauffman (1993).

Koga et al., 1998), we consider the difference in glycerol lipid stereochemistry between *Bacteria* and *Eukarya* on the one hand and *Archaea* on the other, as a fundamental evolutionary divide between two types of modern cells; simpler molecules (perhaps sphingolipids) may have preceded the bifurcation toward sn-1,2 and sn-2,3 glycerol lipids. However, we presented sn-2,3 glycerol-ether isoprenoid lipids (more stable than ester lipids, with a higher hydrophobicity and microviscosity than alkyl chains and more proton impermeable) as an archaeal invention generated under strong selective pressure for adaptation to thermophily from a non-thermophilic LUCA (Glansdorff, 2000; Xu and Glansdorff, 2002; Glansdorff and Xu, 2004), whereas Wachtershauser (2003) argued that the sn-1,2 and sn-2,3 lipids were both already present at a very early stage, forming non-miscible membrane domains in pre-cells (according to Kandler's terminology, 1994). The recent observation that glycerol-1-phosphate dehydrogenase (G1DPH, which catalyzes a necessary step for the formation of sn-2,3 membrane lipids) appears to be an archaeal invention (Boucher et al., 2004) gives some support to the former view. G1DPH could have been recruited from an ancestral glycerol dehydrogenase (Boucher et al., 2004). Because of the stereochemical incompatibility between sn-1,2 and sn-2,3 membrane lipids (Sackmann, 1982 and other references in Wachtershauser, 2003), it is unlikely, however, that the LUCA could ever have been a homogeneous population of cells with heterochiral lipids in their membranes, as assumed by Pereto et al. (2004). We rather think that if the first archaeal cell line with a new type of membrane emerged from a cell with sn-1,2 lipids, it was a relatively sudden event, accelerated by the instability of a membrane that would have begun to contain the two types of lipids. When Wachtershauser (2003) argues against such a transition as being counterselective, it is with the assumption that it would have been gradual, whereas we postulate that the transition occurred under strong selection pressure for adaptation to high temperature from a non-thermophilic LUCA. It is, however, possible that both chiral lipid forms emerged at different times in a less sudden way from an organism with simpler lipids (such as sphingolipids) that would have become extinct later on (Fig. 1B). A more extreme form of this alternative would have consisted in the replacement of a primeval membrane made of proteins (Zillig et al., 1992) by the two types of lipids in two independent events.

Our alternative to a tree of life rooted in an extreme thermophilic bacterial cell line thus rests on the conjunction of ideas on reductive evolution with the notion of a non-extreme thermophilic, protoeukaryotic LUCA announcing the *Eukarya* by a variety of properties such as the stereochemistry of lipids, the presence of cytoskeletal proteins, and the filiation which we previously suggested between the composite, redundant, and flexible eukaryal gene promoters and their RNA world ancestors, poorly defined originally but endowed with a high combinatorial potential (Glansdorff, 2000). The presence of nucleus-like structures in Planctomycetes, the related Poribacteria, and perhaps certain *Archaea*-like organisms raises the intriguing possibility that the LUCA itself may already have been nucleated or at least on the way to compartimentalization of its genetic material (Fuerst, 2005).

The fact that an evolution from prokaryotes to eukaryotes assumes as a matter of course a progression from the simple to the complex is another reason to question this canonical scheme. There is indeed no obvious selective advantage for an alleged post-prokaryotic advent of mRNA splicing and spliceosomes, a complex machinery indeed (Poole et al., 1998), but there is impressive evidence suggesting that some introns date back to the LUCA (Roy and Gilbert, 2005); the emergence of the splicing mechanism may even have been a very ancient event, dating as far back as the RNA world (Poole et al., 1998; Jeffares et al., 2006). The intellectual bias for increasing complexity in evolution is traditional but could be based on a misunderstanding: what increases up the evolutionary scale is order, complexity coming along as a rather ill-defined and incidental aspect of this increase. For example, a biofilm can be more complex than the first metazoans to have emerged but remains considerably less ordered. Unicellular *Eukarya* are undoubtedly more complex than prokaryotes, but this is largely a result of endosymbiosis, and there is now evidence that even prokaryotes can have a nucleus. The notion of an already intricate and sophisticated LUCA is in keeping with the conditions enunciated by Kauffman for the emergence of pregenomic autoreproducible systems by catalytic closure from a dynamic mixture of peptides and nucleic acids coupled to an ambient protometabolism (Kauffman, 1993). Superimposing a coding device on this network of interactions would have produced an organism presenting from the start a higher degree of complexity than in most models discussing the appearance of the first cells.

Questioning the prejudice in favor of an evolutionary trend toward increasing complexity has a more subtle consequence regarding the cradle of life in the solar system. It is now frequently assumed that particularly resistant life forms (spore-forming organisms, perhaps the most thermophilic ones among them or even some other forms of tough extremophiles) may have been able to resist a journey in space, migrate from one planet to another in meteorites (Mars to Earth or vice-versa), and colonize it (Nicholson et al., 2000; Rettberg et al., 2004). In our present state of knowledge, these organisms would have to be prokaryotes, unless some particularly hardy fungal spores could be considered as candidates as well (de Vera et al., 2004). We consider that a colonization by space-traveling prokaryotes, even if it could have led to widespread colonization by their descendants, would not have produced anything fundamentally "more evolved," because prokaryotes are evolutionary dead ends, a 3 billion-year-old testimony of what reductive evolution and intensive specialization can generate.

LIMITS OF LIFE AND COMBINED EXTREMOPHILIES

Temperature is an all-pervasive factor with straightforward effects on the physical state of the universal life solvent, which has to remain in the liquid state to allow suitably adapted organisms to grow. The upper limit is probably somewhere below 150°C (under high hydrostatic pressure) as both natural observations and theoretical considerations suggest (Blochl et al., 1997; Kashefi and Lovley, 2003; White, 1984). The lower limit for active life (metabolic activity and growth) is more difficult to determine experimentally, mainly because of practical limitations to measurements of extremely low growth rates. There is, however, evidence that in Siberian permafrost, cells can remain viable and metabolically active for extremely long periods in an unfrozen state, around −12°C, surrounded by a film of brine (Gilichinsky et al., 2005; Rivkina et al., 2004; earlier data discussed in Glansdorff and Xu, 2002). Some of the organisms recovered from ancient permafrost are not psychrophilic, and it is therefore doubtful that they can actually grow under those circumstances. Others may be growing extremely slowly though this remains to be documented. At any rate, these observations are relevant to the possible occurrence of life in extraterrestrial environments such as the deep horizons of Martian permafrost.

To what extent is an organism adapted to a particular environment? More precisely, has the adaptation reached an evolutionary optimum or are there some fundamental limits we have to take into account? We would like to address this question first in relation to psychrophily and then consider the challenge raised by the conjunction of psychrophily and piezophily (a term to be preferred to "barophily" from the etymological point of view). This conjunction is ecologically dominant on Earth because more than 60% of the biosphere is oceanic, at no more than 5°C and at a hydrostatic pressure of at least 100 atm (10 MPa at a depth of 1,000 m) (Yayanos, 1995).

The Challenge of Adaptation to Low Temperatures

As the question is quantitative, it can be approached readily in simple biological systems such as unicellular organisms whose enzymes can be studied in terms of catalytic efficiency (k_{cat}/K_m). Enzyme kinetics predicts that the velocity of an enzyme-catalyzed reaction will depend on the k_{cat}/K_m ratio at low substrate concentrations ($v = k_{cat}/K_m[E][S]$). This is precisely the range of concentrations which is significant from the physiological point of view because in a metabolically active cell, most biosynthetic intermediates are present at very low concentrations, of the order of the K_m or even lower.

Kinetic optimization of an enzyme to make it efficient at low temperature is a challenge because improving k_{cat} by decreasing the activation enthalpy may have a cost in terms of affinity. Indeed, this operation implies a concomitant increase in conformational entropy (Fields and Somero, 1998; Lonhienne et al., 2001), thus an increase in the number of conformational states of the enzyme, some of which might bind the substrate only poorly. This is probably the reason why many psychrophilic intracellular enzymes display higher K_m's than their thermophilic or mesophilic homologs (see Glansdorff and Xu, 2002; Xu et al., 2003a); it also explains why in vivo selected mutations which substantially increase the k_{cat} of a thermophilic enzyme at a lower temperature simultaneously make the enzyme considerably more labile and increase the K_m (Xu et al., 2003a; Roovers et al., 2001). Therefore, are cold-active metabolic enzymes really optimized to function at low temperature or, rather, does the enthalpy–entropy balance preclude at least some of these enzymes to reach optimization?

The kinetic parameters and the catalytic efficiencies of two key metabolic enzymes, dihydrofolate reductase (DHFR) and ornithine carbamoyltransferase (OTCase), purified from two novel species of strictly psychrophilic bacteria (*Moritella abyssi* and *M. profunda*, with maximal growth rates close to 0°C and maximal growth temperatures of about 12°C) (Xu et al., 2003d) were compared with the values obtained for their counterparts from mesophilic and thermophilic species (Xu et al., 2003a, 2003b). For both enzymes, the K_m for at least one of the substrates was found to be much higher, and the k_{cat}/K_m ratios were considerably lower in psychrophiles. In addition, the sensitivity to non-substrate inhibitors (phosphonoacetylornithine for OTCase, methotrexate and trimethoprim for DHFR) was reduced in psychrophiles, suggesting that the actual affinity of the enzymes for their natural substrates and not only the K_m was actually altered.

The data therefore indicate that DHFR and OTCase adaptation to low temperature entailed a marked trade-off between affinity for the substrates of the reaction and catalytic velocity. The kinetic features of both enzymes in these strict psychrophiles thus suggest that enzyme adaptation to low temperature may be constrained by natural limits to optimization of catalytic efficiency. Incidentally, this analysis also showed that optimization was achieved more readily for a thermophilic enzyme than for its psychrophilic counterpart, a conclusion at variance with early predictions regarding life at high temperature (Brock, 1967).

Do such observations reflect a barrier to optimization that cannot be transgressed for thermodynamic reasons only, or also historical constraints, for example the fact that the ancestral protein was adapted to a higher temperature regime? Indeed, it might be difficult, even over eons, to perfect adaptation to cold from a thermophilic ancestor. As mentioned above, recent revisions of prokaryotic phylogeny suggest a non-thermophilic bacterial ancestor. It is therefore possible that the enthalpy–entropy balance is the main factor in this case.

It is unlikely that the results obtained with *Moritella* DHFR and OTCase are mere coincidences. Nevertheless, by comparing thermodynamic, kinetic, and structural features of series of homologous enzymes active in different ranges of temperatures, it may be possible to determine whether the above deficit in optimization is widespread among psychrophilic enzymes. At any rate, only a few critical enzymes displaying this optimization deficit in any particular organism would be enough to produce a global effect. It is already known that some enzymes, when made of more than one domain, can override the contradictory optimization of both k_{cat} and K_m by a subtle strategy: improving substrate binding by maintaining a stable and rigid domain while using a labile and flexible unit to favor the reaction rate (Bentahir et al., 2000).

The kinetic results on *Moritella* OTCase and DHFR were obtained in vitro with enzymes purified in only a few steps from recombinant *Escherichia coli* cells grown well below the "optimal" *E. coli* temperature range. Moreover, kinetic constants determined in cell-free extracts did not differ significantly from the values obtained with pure enzymes. Therefore, the proteins do not appear to have been damaged by purification. However, even if the analysis of these proteins does reflect their genuine properties, we may wonder if in vivo, the catalytic efficiency of the whole biosynthetic chain is affected to the same extent. It is possible that interactions assembling the enzymes of a pathway in a metabolon could at least in part compensate for reduced catalytic efficiency of individual enzymes. This problem could be studied by the biochemical approaches used to demonstrate channeling in the metabolism of thermolabile substrates in thermophilic prokaryotes (Massant et al., 2002; Massant and Glansdorff, 2005 and Massant, this volume). Metabolic channeling could therefore be as important in psychrophiles as it is in thermophiles but for very different reasons.

Combining Psychrophily and Piezophily

The genus *Moritella* is an interesting experimental system because it contains psychrophilic species adapted to different levels of the oceanic water column, from the surface to the deepest abysses (Xu et al., 2003d). It is thus excellent material to carry out comparative analyses of model enzymes to study molecular features of piezophily. Moreover, many of the cultivated piezophilic microorganisms belong to the family Vibrionaceae or are closely related to it, which facilitates comparative studies.

Trends in molecular adaptations toward piezophily are already apparent (Allen and Bartlett, 2004; Bartlett, 2002 and this volume; Glansdorff and Xu, 2002; Kato and Horikoshi, 2004; Vezzi et al., 2005). From the point of view of membrane fluidity, adaptation to both high pressure and low temperature (see Russell, this volume) could rely for a large part on the same basic mechanisms, a relative increase in the proportion of unsaturated fatty acids playing an important role. There are, however, some discordant observations which require further investigations; the increase in pressure tolerance observed when a moderately halophilic bacterium was exposed to increasing NaCl concentrations is at first sight puzzling since increasing osmolarity is expected to be associated with a decrease in the proportion of unsaturated fatty acids (Tanaka et al., 2001).

With proteins, however, adaptation to both high pressure and low temperature raises a challenge because they would be expected to be rigid enough to resist compression while retaining the flexibility characteristic of psychrophilic proteins. The difficulty of the compromise could explain, at least in part, why abyssal psychropiezophiles grow relatively slowly, the maximum growth rates appearing to decrease with increasing pressure adaptation (Jannasch and Wirsen, 1984; Yayanos, 1995).

Data are still relatively scarce regarding the detailed behavior of enzymes from piezophilic sources. The first nuclear magnetic resonance study of a protein from a deep-sea organism showed that the core structure of folate-bound *Moritella* DHFR resists pressure at least partially up to 295 MPa (Hata et al., 2004). When assessing the folded–unfolded equilibrium at different P and T from the intensities of the nuclear magnetic resonance signals coming from the folded part of the protein, the ΔG_O at atmospheric pressure turned out to be in agreement with the results of earlier urea denaturation studies (Xu et al., 2003b), and, most importantly, the stability expressed as ΔG was low and reached a maximum at 5°C. This is probably related to the need for molecular flexibility at low temperature. Also the value of ΔV (volume change upon unfolding) turned out to be low (–66 ml/mol), whereas for a protein of that size (18 kDa), one would expect about –100 ml/mol. This is probably advantageous for a high-pressure adapted protein so as to avoid the threat of pressure-induced denaturation. It is not yet known how *Moritella* DHFR activity is influenced by pressure, but DHFR from *Shewanella violacea*, a comparable, moderately piezophilic marine bacterium, was shown to increase its activity under increasing pressure by up to 30% at 100 MPa, before starting to decline gradually; at 250 MPa, the enzyme was still more active than at atmospheric pressure. In contrast, *E. coli* DHFR activity decreased monotonously with increasing pressure (Ohmae et al., 2004). Further studies of this kind, in particular with enzymes from obligate piezophiles, would be extremely valuable in order to assess the strategies implemented by proteins to meet the challenge of remaining active at both low temperature and high pressure.

M. profunda and *M. abyssi* are both moderate piezophiles. At 6°C, their maximal growth rates are reached at a pressure of approximately 20 MPa; at 10°C, their piezophily is markedly enhanced (Xu et al., 2003d). Enhancement of piezophily at higher temperature was also observed for *Psychromonas profunda* (Xu et al., 2003c) and deep-sea hyperthermophilic Archaea (Marteinsson et al., 1997). It may be due to partial compensation of the "gelling" effect of high pressure on membrane lipids by the increase in molecular mobility brought about by a rise in temperature. Alternatively, an increase in temperature could compensate for inhibition of enzyme activity due to pressure-induced compression. Some enzymes might be particularly sensitive to the opposite effects exerted by an increase in pressure and an increase in temperature.

The upper hydrostatic pressure limit for life is presently unknown. At the bottom of the putative 100-km deep brine ocean of the Jovian moon Europa, the pressure is probably no more than 1.5 to twice that in the deepest recesses of the Marianna Trench, considering Europa's density. A few surviving *E. coli* cells were retrieved from an exposure to approximately 10 times higher pressures, and some enzyme activity was detected under those conditions (Sharma et al., 2002). However, these experiments do not tell us much more about extreme piezophily than the recovery of a few mesophilic cells surviving a heat shock would tell us about extreme thermophily.

EXTREMOPHILES AND "INTELLIGENT DESIGN"

If there is a true director, there is no evidence of him

—Chuang-tzu, early Taoist philosopher

The discovery of various kinds of extremophiles offers a unique opportunity to question the pretentions of supernatural creationism by exploring the creative power of natural selection. This may appear a loss of time and energy to many of us, but we should not forget that we live in a world where the very nature of science is misunderstood by the majority. The rising importance of cultures which have not been dominated by biblical fundamentalism, as in China and India, may have a positive impact in this respect.

In apparent contrast with the strict, biblical version of creationism, "intelligent design" (ID) is presented by its partisans as a scientific alternative to the dominant theory of evolution, Darwinism. It does not deserve this status of course, because having recourse to supranatural explanations before having exhausted all natural possibilities (an unlimited quest) is by definition unscientific. In this respect, ID is but creationism reintroduced through the back door. Nevertheless, such is the power of logical analysis that it can put to test even unscientific proposals.

The ID concept is of course adverse to the notion of evolutionary descent. A simple example discussed by Penny et al. (2003) illustrates this: consider species of cactus and grasses adapted to hot and dry conditions. An "intelligent designer" would be expected to provide these plants with similar, "ideal" adaptations; this would be reflected in the genes of their photosynthetic apparatus, which would look more similar to each other than the genes of the xerophilic grass would to the cognate genes of moist-temperate grasses. Evolution as we understand would of course offer evidence of filiation among grasses, whether xerophilic or not. The world of extremophiles offers similar test cases; some of the most obvious ones concern extreme halophily and thermophily, conditions that impose adaptation to the whole proteome. In both cases, the selection pressure is particularly strong, and the relevant molecular adaptations begin to be understood. The study of many proteins shows that there is no fundamental difficulty in recognizing the filiation of extreme halophilic proteins from *Archaea* with their counterparts among non-halophilic Euryarchaeota (see for example Daubin et al., 2002; Brochier et al, 2005). Moreover, the recently discovered genus of extreme halophilic bacteria *Salinibacter* can be affiliated by whole-genome comparative analysis with other, non-halophilic bacteria as its closest relatives, even though some genes appear to have been imported from haloarchaea by horizontal gene transfer (Mongodin et al., 2005). Similarly, the filiation of genes belonging to thermophilic *Bacilli* with mesophilic ones rather than with thermophilic *Archaea* is clear (barring those instances where horizontal transfer or differential extinction of paralogs can be invoked; Daubin et al., 2002), even though thermophilic proteins can be identified by their amino acid compositions alone (Kreil and Ouzounis, 2001). Of course, as also noted by Penny et al. (2003), the pattern of filiation that we take as evidence of evolution may be the making of an "intelligent deceiver," perhaps the Devil himself.

This essay was written in the spirit of B. Russell's introductory quotation; we hope not having lost sight of it.

Acknowledgments. N.G. gratefully acknowledges the hospitality of Professor Gordon T. Taylor at the Marine Sciences Research Center of Stony Brook University (New York State) during the preparation of this chapter.

REFERENCES

Allen, E., and D. H. Bartlett. 2004. Microbial adaptations to the deep-sea environment. *In Extremophilies*, C. Gerday, and N. Glansdorff (ed.), in *Encyclopedia of Life Support Systems (EOLSS)*, Developed under the Auspices of the UNESCO, Eolss Publishers, Oxford, UK, http://www.eolss.net.

Amos, L. A., F. van den Ent, and J. Lowe. 2004. Structural/functional homology between the bacterial and eukaryotic cytoskeletons. *Curr. Opin. Cell Biol.* **16:**24–31.

Bada, J. L., C. Bigham, and S. L. Miller. 1994. Impact melting of frozen oceans on the early Earth: implications for the origin of life. *Proc. Natl. Acad. Sci. USA* **91:**1248–1250.

Bartlett, D. H. 2002. Pressure effects on *in vivo* microbial processes. *Biochim. Biophys. Acta* **1595:**367–381.

Bentahir, M., G. Feller., M. Aittaleb, J. Lamotte-Brasseur, T. Himori, J. P. Chessa, and C. Gerday. 2000. Structural, kinetic and calorimetric characterization of the cold-active phosphoglycerate kinase from the Antarctic *Pseudomonas* sp. TACII 18. *J. Biol. Chem.* **275:**11147–11153.

Bern, M., and D. Golberg. 2005. Automatic selection of representative proteins for bacterial phylogeny. *BMC Evol. Biol.* **5:**34.

Biller, K., and H. Märkl. 2002. A new mathematical model for the description of the growth of the hyperthermophilic archaeon *Pyrococcus furiosus*. Int. Congr. on Extremophiles, Naples 2002, Book of abstracts:LA1.

Blochl, E., R. Rachel, S. Burggraf, D. Hafenbradl, W. H. Jannasch, and K. O. Stetter. 1997. *Pyrolobus fumarii*, gen. and sp. nov., represents a novel group of archaea, extending the upper temperature limit for life to 113 degrees C. *Extremophiles* **1:**14–21.

Boucher, Y., M. Kamakura, and W. F. Doolittle. 2004. Origins and evolution of isoprenoid lipid biosynthesis in archaea. *Mol. Microbiol.* **52:**515–527.

Brochier, C., P. Forterre, and S. Gribaldo. 2005. An emerging phylogenetic core of Archaea: phylogenies of transcription and translation machineries converge following addition of new genome sequences. *BMC Evol. Biol.* **5:**36.

Brochier, C., and H. Philippe. 2002. Phylogeny: a non-hyperthermophilic ancestor for bacteria. *Nature* **417:**244.

Brock, T. D. 1967. Life at high temperatures. *Science* **158:**1012–1019.

Caetano-Anolles, G. 2002. Evolved RNA secondary structure and the rooting of the universal tree. *J. Mol. Evol.* **54:**333–345.

Castresana, J. 2001. Comparative genomics and bioenergetics. *Biochim. Biophys. Acta* **1506(3):**147–162.

Chablain, P. A., G. Philippe, A. Groboillot, N. Truffaut, and J. F. Guespin-Michel. 1997. Isolation of a soil psychrotrophic toluene-degrading *Pseudomonas* strain: influence of temperature on the growth chacteristics on different substrates. *Res. Microbiol.* **148:**153–161.

Chistoserdova, L., C. Jenkins, M. G. Kalyuznaya, C. J. Marx, A. Lapidus, J. A. Vorholt, J. T. Staley, and M. E. Lidstrom. 2004. The enigmatic planctomycetes may hold a key to the origins of methanogenesis and methylotrophy. *Mol. Biol. Evol.* **21:**1234–1241.

Cohen, G., V. Barbe, D. Flament, M. Galperin, R. Heilig, O. Lecompte, O. Poch, D. Prieur, J. Querellou, R. Ripp, J. C. Thierry, J. Van der Oost, J. Weissenbach, I. Zivanovic, and P. Forterre. 2003. An integrated analysis of the genome of the hyperthermophilic archaeon *Pyrococcus abyssi. Mol. Microbiol.* **47:**1495–1512.

Darnell, J. E., and W. F. Doolittle. 1986. Speculations on the early course of evolution. *Proc. Natl. Acad. Sci. USA* **83:**1271–1275.

Daubin, V., M. Gouy, and G. Perriere. 2002. A phylogenomic approach to bacterial phylogeny: evidence of a core of genes sharing a common history. *Genome Res.* **12:**1080–1090.

de Duve, C. 1991. *Blueprint for a Cell: The Nature and Origin of Life.* 275p. Patterson Publishers, Carolina Biological Supply Company, Burlington, NC.

de Duve, C. 2005. *Singularities: Landmarks on the Pathways of Life.* 256p. Cambridge University Press, Cambridge, UK.

Dennett, D. 1995. *Darwin's Dangerous Idea: Evolution and the Meanings of Life.* 586p. New York, Simon and Schuster.

de Vera J. P., G. Horneck, P. Rettberg, and S. Ott. 2004. The potential of the lichen symbiosis to cope with the extreme conditions of outer space. II. Germination capacity of lichen ascospores in response to simulated space conditions. *Adv. Space Res.* **33:**1236–1243.

Di Giulio, M. 2000. The late stage of genetic code structuring took place at a high temperature. *Gene* **261:**189–195.

Di Giulio, M. 2003a. The universal ancestor was a thermophile or a hyperthermophile: tests and further evidence. *J. Theor. Biol.* **221:**425–436.

Di Giulio, M. 2003b. The ancestor of the Bacteria domain was a hyperthermophile. *J. Theor. Biol.* **224:**277–283.

Di Giulio, M. 2005. A comparison of proteins from *Pyrococcus furiosus* and *Pyrococcus abyssi*: barophily in the physicochemical properties of amino acids and in the genetic code. *Gene* **346:**1–6.

Fields, P. A., and G. Somero. 1998. Hot spots in cold adaptation: localized increases in conformational flexibility in lactate dehydrogenase A(4) orthologs of Antarctic notothenioid fishes. *Proc. Natl. Acad. Sci. USA* **95:**11476–11481.

Fieseler, L., M. Horn, M. Wagner, and U. Hentschel. 2004. Discovery of the novel candidate phylum Poribacteria in marine sponges. *Appl. Environ. Microbiol.* **70:**3724–3732.

Forterre, P. 1995. Thermoreduction, a hypothesis for the origin of prokaryotes. *C. R. Acad. Sci III* **318:**415–422.

Forterre, P. 1996. A hot topic: the origin of hyperthermophiles. *Cell* **85:**789–792.

Forterre, P. 2005. The two ages of the RNA world, and the transition to the DNA world: a story of viruses and cells. *Biochimie* **87:**793–803.

Forterre, P. Three RNA cells for ribosomal lineages and three DNA viruses to replicate their genomes: a hypothesis for the origin of cellular domain. *Proc. Natl. Acad. Sci. USA* **103:**3669–3674.

Forterre, P. 2004. Strategies of extremophily in nucleic acids adaptation to high temperature. In *Extremophilies*, C. Gerday, and N. Glansdorff (ed.), in *Encyclopedia of Life Support Systems (EOLSS)*, Developed under the Auspices of the UNESCO, Eolss Publishers, Oxford, UK, http://www.eolss.net.

Forterre, P., and H. Philippe. 1999. Where is the root of the universal tree of life? *BioEssays* **21:**871–879.

Fuerst, J. A. 2005. Intracellular compartmentation in Planctomycetes. *Annu. Rev. Microbiol.* **59:**299–328.

Galtier, N. 2001. Maximum-likelihood phylogenetic analysis under a covarion-model. *Mol. Biol. Evol.* **18:**866–873.

Galtier, N., N. J. Tourasse, and M. Gouy. 1999. A non-hyperthermophilic ancestor to extant life forms. *Science* **283:**220–221.

Gilichinsky, D., E. Rivkina, C. Bakermans, V. Shcherbakova, L. Petrovskaya, S. Ozerskaya, N. Ivanushkina, G. Kochkina, K. Laurinavichius, S. Pecheritsina, R. Fattakhova, and J. M. Tiedje. 2005. Biodiversity of cryopegs in permafrost. *FEMS Microbiol. Ecol.* **53:**117–128.

Glansdorff, N. 2000. About the last common ancestor, the universal life-tree and lateral gene transfer: a reappraisal. *Mol. Microbiol.* **38:**177–185.

Glansdorff, N., and Y. Xu. 2002. Microbial life at low temperatures: mechanisms of adaptation and extreme biotopes. Implications for exobiology and the origin of life. *Recent Res. Dev. Microbiol.* **6:**1–21.

Glansdorff, N., and Y. Xu. 2004. Phylogeny of extremophiles. In *Extremophilies*, C. Gerday, and N. Glansdorff (ed.), in *Encyclopedia of Life Support Systems (EOLSS)*, Developed under the Auspices of the UNESCO, Eolss Publishers, Oxford, UK, http://www.eolss.net.

Gogarten, J. P., H. Kibak, P. Dittrich, L. Taiz, E. J. Bowman, B. J. Bowman, M. F. Manolson, R. J. Poole, T. Oshima, I. Konoshi, K. Denda, and M. Yoshida. 1989. Evolution of the vacuolar H^+ATPase: implications for the origin of eukaryotes. *Proc. Natl. Acad. Sci. USA* **86:**6661–6665.

Gratia, J. P. 2003. Spontaneous zygogenesis in *Escherichia coli*, a form of true sexuality in prokaryotes. *Microbiology* (UK) **149:**2571–2584.

Griffiths, E., and R. S. Gupta. 2004. Signature sequences in diverse proteins provide evidence for the late divergence of the order Aquificales. *Int. Microbiol.* **7:**41–52.

Guillen, N., M. Amar, and L. Hirschbein. 1985. Stabilized non-complementing diploids (Ncd) from fused protoplast products of *Bacillus subtilis. EMBO J.* **4:**1333–1338.

Guillou, C., and J. F. Guespin-Michel. 1996. Evidence for two domains of growth temperature for psychrotrophic bacterium *Pseudomonas fluorescens* MF0. *Appl. Environ. Microbiol.* **62:**3319–3324.

Hata, K., R. Kono, M. Fujisawa, R. Kitahara, Y. Katamari, K. Akasaka, and Y. Xu. 2004. High pressure NMR study of dihydrofolate reductase from a deep-sea bacterium *Moritella profunda. Cell. Mol. Biol.* **50:**311–316.

Hugenholtz, P., B. M. Goebell, and N. R. Pace. 1998. Impact of culture-independent studies on the emerging phylogenetic view of bacterial diversity. *J. Bacteriol.* **180:**4765–4774.

Hurtado, L. A., R. A. Lutz, and R. C. Vrijenhoek. 2004. Distinct patterns of genetic differentiation among annelids of eastern Pacific hydrothermal vents. *Mol. Ecol.* **13**:2603–2615.

Iwabe, N., K. Kuma, M. Hasegawa, S. Osawa, and T. Miyata. 1989. Evolutionary relationship of archaebacteria, eubacteria, and eukaryotes inferred from phylogenetic trees of duplicated genes. *Proc. Natl. Acad. Sci USA* **86**:9355–9359.

Jannasch, H. W, and C. O. Wirsen. 1984. Variability of pressure adaptation in deep sea bacteria. *Arch. Microbiol.* **139**:281–288.

Jeffares, D. C., T. Mourier, and D. Penny. 2006. The biology of intron gain and loss. *Trends Genet.* **22**:16–22.

Jenkins, C., and J. A. Fuerst. 2001. Phylogenetic analysis of evolutionary relationships of the Planctomycete division of the domain Bacteria based on amino acid sequences of elongation factor Tu. *J. Mol. Evol.* **52**:405–418.

Jenkins, C., V. Kedar, and J. A. Fuerst. 2002a. Gene discovery within the planctomycete division of the domain Bacteria using sequence tags from genomic DNA libraries. *Genome Biol.* **3**:RESEARCH0031.

Jenkins, C. R. Samudrala, I. Anderson, B. P. Hedlund, G. Petroni, N. Michailova, N. Pinel, R. Overbeek, G. Rosati, and J. T. Staley. 2002b. Genes for the cytoskeletal protein tubulin in the bacterial genus *Prosthecobacter*. *Proc. Natl. Acad. Sci. USA.* **99**:17049–17054.

Kandler, O. 1994. The early diversification of life, p. 152–160. *In* S. Bengston (ed.), *Nobel Symposium No 84. Early Life on Earth*. Columbia University Press, New York, NY.

Kashefi, K., and D. R. Lovley. 2003. Extending the upper temperature limit for life. *Science* **301**:934.

Kato, C., and K. Horikoshi. 2004. Characteristics of deep-sea environments and biodiversity of piezophilic organisms. *In Extremophiles*, C. Gerday, and N. Glansdorff (ed.), in *Encyclopedia of Life Support Systems (EOLSS)*, Developed under the Auspices of the UNESCO, Eolss Publishers, Oxford, UK, http://www.eolss.net.

Kauffman, S. A. 1993. *The Origins of Order: Self-Organization and Selection in Evolution*. 709 p. Oxford University Press, New York, NY.

Koga, Y., T. Kyuragi, M. Nishihara, and N. Sone. 1998. Did archaeal and bacterial cells arise independently from noncellular precursors? A hypothesis stating that the advent of membrane phospholipids with enantiomeric glycerophosphate backbones caused the separation of the two lines of descent. *J. Mol. Evol.* **46**:54–63.

Kreil, D. P., and C. A. Ouzounis. 2001. Identification of thermophilic species by the amino acid compositions deduced from their genomes. *Nucleic Acids Res.* **29**:1608–1615.

Labedan, B., A. Boyen, M. Baetens, D. Charlier, C. Pingguo, R. Cunin, V. Durbecq, N. Glansdorff, G. Herve, C. Legrain, Z. Liang, C. Purcarea, M. Roovers, R. Sanchez, T. L. Toong, M. Van de Casteele, F. Van Vliet, Y. Xu, and Y. F. Zhang. 1999. The evolutionary history of carbamoyltransferases: A complex set of paralogous genes was already present in the last universal common ancestor. *J. Mol. Evol.* **49**:461–473.

Labedan, B., Y. Xu, D. Naumoff, and N. Glansdorff. 2004. Using quaternary structures to assess the evolutionary history of proteins: the case of the aspartate carbamoyltransferase. *Mol. Biol. Evol.* **21**:364–372.

Lakshminarayan, M. Y., E. V. Koonin, and L. Aravind. 2004. Evolution of RNA polymerase: implications for large-scale bacterial phylogeny, domain accretion, and horizontal gene transfer. 2004. *Gene* **335**:73–88.

Lederberg, J. 1949. Aberrant heterozygotes in *Escherichia coli*. *Proc. Natl. Acad. Sci. USA* **35**:178–184.

Lonhienne, T., C. Gerday, and G. Feller. 2001. Psychrophilic enzymes: revisiting the thermodynamic parameters of activation may explain local flexibility. *Biochim. Biophys. Acta* **1543**:1–10.

Lopez, P., P. Forterre, and H. Philippe. 1999. The root of the tree of life in the covarion model. *J. Mol. Evol.* **49**:496–508.

Marteinsson, V. T., P. Moulin, J. Birrien, A. Gambacorta, M. Vernet, and D. Prieur. 1997. Physiological responses to stress condotions and barophilic behavior of the hyperthermophilic vent archaeon *Pyrococcus abyssi*. *Appl. Environ. Microbiol.* **63**:1230–1236.

Massant, J., and N. Glansdorff. 2005. New experimental approaches for investigating interactions between *Pyrococcus furiosus* carbamate kinase and carbamoyltransferases, enzymes involved in the channeling of thermolabile carbamoyl phosphate. *Archaea* **1**:365–373.

Massant, J., P. Verstreken, V. Durbecq, A. Kholti, C. Legrain, S. Beeckmans, P. Cornelis, and N. Glansdorff. 2002. Metabolic channeling of carbamoyl phosphate, a thermolabile intermediate: evidence for physical interaction between carbamate kinase-like carbamoyl-phosphate synthetase and ornithine carbamoyltransferase from the hyperthermophile Pyrococcus furiosus. *J. Biol. Chem.* **277**:18517–18522.

Miller, S. L., and A. Lazcano. 1998. Facing up to realities: life did not begin at the growth temperatures of hyperthermophiles, p. 127–133. *In* J. Wiegel, and M. W. W. Adams (ed.), *Thermophiles: The Keys to Molecular Evolution and the Origin of Life?* Taylor and Francis, London, United Kingdom.

Mojzsis, S. J., T. M. Harrison, and R. T. Pidgeon. 2001. Oxygen-isotope evidence from ancient zircons for liquid water at the Earth's surface 4,300 Myr ago. *Nature* **409**:178–181.

Mongodin, E. F., K. E. Nelson, S. Daugherty, R. T. DeBoy, J. Wister, H. Khouri, J. Weidman, D. A. Walsh, R. T. Papke, G. Sanchez Perez, A. K. Sharma, C. L. Nesbo, D. MacLeod, E. Bapteste, W. F. Doolittle, R. L. Charlebois, B. Legault, and F. Rodriguez-Valera. 2005. The genome of *Salinibacter ruber*: convergence and gene exchange among hyperhalophilic bacteria and archaea. *Proc. Natl. Acad. Sci. USA* **102**:18147–18152.

Moulton, V., P. P. Gardner, R. F. Pointon, L. K. Creamer, G. B. Jameson, and D. Penny. 2000. RNA folding argues against a hot-start origin of life. *J. Mol. Evol.* **51**:416–421.

Nicholson W. L., H. Munakata, G. Horneck, H. J. Melosh, and P. Setlow. 2000. Resistance of *Bacillus* endospores to extreme terrestrial and extraterrestrial environments. *Microbiol. Mol. Biol. Rev.* **64**:548–572.

Ohmae, E., K. Kubota, K. Nakasone, C. Kato, and K. Gekko. 2004. Pressure-dependent activity of dihydrofolate reductase from a deep-sea bacterium *Shewanella violacea* strain DSS12. *Chem. Lett.* **33**:798–799.

Penny, D, M. D. Hendy, and A. M. Poole. 2003. Testing fundamental evolutionary hypotheses. *J. Theor. Biol.* **223**:377–385.

Penny, D., and A. M. Poole. 1999. The nature of the last common ancestor. *Curr. Opin. Genet. Dev.* **9**:672–677.

Pereto, J., P. Lopez-Garcia, and D. Moreira. 2004. Ancestral lipid biosynthesis and early membrane evolution. *Trends Biochem Sci.* **29**:469–497.

Poole, A. M., D. C Jeffares, and M. Penny. 1998. The path from the RNA world. *J. Mol. Evol.* **46**:1–17.

Rettberg, P., E. Rabow, C. Panitz, and G. Horneck. 2004. Biological space experiments for the simulation of Martian conditions: UV radiation and Martian soil analogues. *Adv. Space Res.* **33**:1294–1301.

Rivkina, E., K. Laurinavichius, J. McGrath, J. Tiedje, V. Shcherbakova, and D. Gilichinsky. 2004. Microbial life in permafrost. *Adv. Space Res.* **33**:1215–1221.

Roovers, M., R. Sanchez, C. Legrain, and N. Glansdorff. 2001. Experimental evolution of enzyme temperature activity profile: selection in vivo and characterization of low-temperature-adapted mutants of *Pyrococcus furiosus* ornithine carbamoyltransferase. *J. Bacteriol.* **183**:1101–1105.

Roy, S. W., and W. Gilbert. 2005. Complex early genes. *Proc. Natl. Acad. Sci. USA* **102:**1986–1991.

Sackmann, E. 1982. Physikalische Grundlagen der molekularen Organisation und Dynamik von Membranen, p. 439–471. *In* W. Hoppe, W. Lohmann, H. Markl, and H. Ziegler (ed.), *Biophysik*, 2nd ed. Springer, Berlin, Germany.

Sharma, A., J. H. Scott, G. D. Cody, G. D. Fogel, R. M. Hazea, R. J. Hemley, and W. T. Huntress. 2002. Microbial activity at Gigapascal pressures. *Science* **295:**1514–1516.

Stetter, K. 1996. Hyperthermophilic prokaryotes. *FEMS Microbiol. Rev.* **18:**149–158.

Tanaka, T., J. G. Burgess, and P. C. Wright. 2001. High-pressure adaptation by salt stress in a moderately halophilic bacterium from open seawater. *Appl. Microbiol. Biotechnol.* **57:**200–204.

Teeling, H., T. Lombardot, M. Bauer, W. Ludwig, and F. O. Glockner. 2004. Evaluation of the phylogenetic position of the planctomycete *Rhodopirellula baltica* SH1 by means of concatenated ribosomal protein sequences, DNA-directed RNA polymerase subunit sequences and whole genome trees. *Int. J. Syst. Evol. Microbiol.* **54:**791–801.

Vezzi, A., S. Campanaro, M. D'Angelo, F. Simonato, N. Vitulo, FM Lauro, A. Cestaro, G. Malacrida, B. Simionati, N. Cannata, C. Romualdi, and D. H. Bartlett. 2005. Life at depth: *Photobacterium profundum* genome sequence and expression analysis. *Science* **307:**1459–1461.

Vincent, W. F., D. Mueller, P. Van Hove, and C. Howard-Williams. 2004. Glacial periods on early earth and implications for the evolution of life, p. 481–501. *In* J. Sekbach (ed.), *Origins: Genesis, Evolution and Diversity of Life*. Kluwer Academic Publishers, Dordrecht, The Netherlands.

Wachtershauser, G. 1988. Before enzymes and templates: theory of surface metabolism. *Microbiol. Rev.* **52:**452–484.

Wachtershauser, G. 1992. Groundworks for an evolutionary biochemistry—the iron-sulfur world. *Prog. Biophys. Mol. Biol.* **58:**85–201.

Wachtershauser, G. 1998. The case for a hyperthermophilic, chemolithotrophic origin of life in an iron-sulfur world, p. 47–57. *In* J. Wiegel, and M. W. W Adams (ed.), *Thermophiles: The Keys to Molecular Evolution and the Origin of Life?* Taylor and Francis, London, United Kingdom.

Wachtershauser, G. 2003. From pre-cells to eukarya—a tale of two lipids. *Mol. Microbiol.* **47:**13–22.

White, R. H. 1984. Hydrolytic stability of biomolecules at high temperatures and ist implications for life at 250°C. *Nature* **310:**430–432.

Whitfield, J. 2004. Origins of life: born in a watery commune. *Nature* **427:**674–676.

Wilde, S. A., J. W. Valley, W. H. Peck, and C. M. Graham. 2001. Evidence from detrital zircons for the existence of continental crust and oceans on the Earth 4.4 Gyr ago. *Nature* **409:**175–178.

Woese, C. R. 1998. The universal ancestor. *Proc. Natl. Acad. Sci. USA* **95:**6854–6859.

Woese, C. R. 2002. On the evolution of cells. *Proc. Natl. Acad. Sci. USA* **99:**8742–8747.

Woese, C. R. 2004. A new biology for a new century. *Microbiol. Mol. Biol. Rev.* **68:**173–186.

Woese, C. R., and G. E. Fox. 1977. Phylogenetic structure of the prokaryotic domain: the primary kingdoms. *Proc. Natl. Acad. Sci. USA* **74:**5088–5090.

Xu, Y., G. Feller, C. Gerday, and N. Glansdorff. 2003a. Metabolic enzymes from psychrophilic bacteria: challenge of adaptation to low temperatures in ornithine carbamoyltransferase from *Moritella abyssi*. *J. Bacteriol.* **185:**2161–2168.

Xu, Y., G. Feller, C. Gerday, and N. Glansdorff. 2003b. *Moritella* cold-active dihydrofolate reductase: are there natural limits to optimization of catalytic efficency at low temperature? *J. Bacteriol.* **185:**5519–5526.

Xu, Y., and N. Glansdorff. 2002. Was our ancestor a hyperthermophilic procaryote? *Comp. Biochem. Physiol. A* **133:**677–688.

Xu, Y., Y. Nogi, C. Kato, Z. Liang, H.-J. Ruger, D. De Kegel, and N. Glansdorff. 2003c. *Psychromonas profunda* sp. nov., a psychropiezophilic bacterium from deep Atlantic sediments. *Int. J. Syst. Evol. Microbiol.* **53:**527–532.

Xu, Y., Y. Nogi, C. Kato, Z. Liang, H.-J. Ruger, D. De Kegel, and N. Glansdorff. 2003d. *Moritella profunda* sp. nov. and *Moritella abyssi* sp. nov., two psychropiezophilic organisms isolated from deep Atlantic sediments. *Int. J. Syst. Evol. Microbiol.* **53:**533–538.

Yayanos, A. A. 1995. Microbiology to 10,500 meters in the deep sea. *Annu. Rev Microbiol.* **49:**1356–1361.

Zillig, W., P. Palm, and H. P. Klenk. 1992. The nature of the common ancestor of the three domains of life and the origin of the Eucarya, p. 181–193. *In* J. Tran Thanh Van, J. C. Mounolou, J. Schneider, and C. McKay (ed.), *Frontiers of Life*. Editions Frontiers, Gif-sur-Yvette.

INDEX

Accretion ice, 146
Acidimicrobium/Ferrimicrobium spp., 264–265
Acidiphilium spp., 265–266, 274
Acidithiobacillus ferrooxidans, 260, 262, 263–264, 271, 279
Acidithiobacillus spp., 262–263
Acidocella organivorus, 261, 266
Acidocella spp., 265–266
Acidophiles
    adaptations of living at low pH, 259–260
    arsenic resistance mechanisms, 273, 274
    copper resistance, 274–275
    energy transformations and carbon assimilation, 260–261
    extremely acidic environments, 257–259
    mercury resistance mechanisms, 275
    phylogenetic diversity and physiological characteristics, 259–261
    protection from metal toxicity, 272–273
Acidophiles, genomics of, 279–280
    acidophily and acid resistance of pathogens, 281–284
    chaperone genes, 287–289
    evolution of thermoacidophiles and concept of "lifestyle genes," 289–290
    genes for DNA superstructure formation, repair, and restriction, 287
    genome sequences of acidophilic archaea, 280–281
    genome-based reconstruction of major metabolic pathways, 284–286
    transport, 286–287
Acidophilic communities
    macroscopic growths and streamer communities, 267–268
    relationships and interactions, 266–267
Acidophilic prokaryotes
    dissimilatory oxidoreduction of iron, 263–265
    iron- and sulfur-oxidizing aerobes, 261–263
    iron-reducing heterotrophs, 265–266
    sulfur-oxidizing archaea, 263
    sulfur-reducing acidophiles, 266
Aeromonas hydrophila, 186
Alicyclobacillus spp. And other acidophilic *Firmicutes*, 264
Alkaliphiles, 295–296
    aerobic mesophiles other than *Bacillus* spp. and thermophiles, 304
    anaerobic and lactic acid bacterial groups, 303
    features in the ATP synthase, 318
    gram-positive high G+C bacteria, 307
    gram-positive low G+C bacteria, 302–306
    gut of higher termites, 298
    indigo fermentation, 300–301
    MotPS and NavBP in motility chemotaxis, 320–322
    Na^+/solute symporters, 319–320
    other environments, 301–302
    sea and related samples, 299–300
    segments of membrane proteins, 324
    soda lakes, 298–299
    soil samples, 296–298
    thermophiles, 307
Alkaliphilic *Bacillus* spp., 321–325
    B. pseudofirmus, 313, 320
    characteristics, 297
    phylogenetic tree derived from 16S rRNA gene sequences, 300
Alkaliphily, bioenergetic adaptations, 311–312
    alkaliphiles versus neutrophils, 312–313
    contribution to cytoplasmic pH homeostasis, 313–317
    cycles supporting alkaliphile pH homeostasis, 317–322
α-Amylases, 362–365
β-Amylases, 365
Amylomaltases, 369–370
Anthropogenic environments, 17
Archaea, 104, 409, 411. *See also* Hyperthermophiles
    archaeal versus bacterial membrane lipids, 107–110
    cellulolytic, hemicellulolytic, chitinolytic, and proteolytic enzymes, 371–372
    classification according to optimal growth temperatures, 14
    Crenarchaeota, 105
    ether linkages in archaeal lipids, 110
    Euryarchaeota, 105
    membrane adaptations to heat stress, 111
Archaea–Bacteria–Eukarya phylogenetic tree, 225
Archaeal genomes
    density of genes encoding transporter proteins, 286
    size and coding density distribution of selected, 281
Arctic winter permafrost, 134, 135
Arctic winter sea ice
    concentration of EPS scaled to brine volume of ice sections, 138
    microscopic images, 135
    viral and bacterial dynamics in section of, 140
Arthrobacter globiformis, 186

Astrobiology
    habitats for life, 354–355
    life requirements, 351–354
    perspectives in, 152
    plausibility categories, 356
    searching for life in the universe, 355–356
ATP synthase (alkaliphile-specific features), 318
Autonomous IS elements, 36

Bacteria, 409, 411
    archaeal versus bacterial membrane lipids, 107–110
    cellulolytic, hemicellulolytic, chitinolytic, pectinolytic and proteolytic enzymes, 374–376
    classification according to optimal growth temperatures, 14
    membrane adaptations to heat stress, 111
    membranes of, 107, 110
Bacterial membrane fluidity, low-temperature-induced changes, 194–195
    changes in fatty acid desaturation, 195
    changes in proportion of *cis* and *trans* fatty acids, 197–199
    perception of temperature and signal transduction, 200–201
    role of carotenoids, 199
    role of cold-inducible genes, 199–200
    sensing of temperature, 202–203
    sensors and transducers of low temperature, 201–202
    temperature-dependent changes, 195–197
Bacterial piezophiles, phylogenetic tree of, 335
Barophilicity, 113
Biocatalysts from extremophiles
    alcohol dehydrogenases, 388
    C–C bond-forming enzymes, 389
    cellulose-degrading enzymes, 370–373
    chitin-degrading enzymes, 377–378
    DNA-processing enzymes, 380–384
    extremozymes of industrial relevance, 390
    glucose and arabinose isomerases, 388–389
    lipases and esterases, 384–388
    nitrile-degrading enzymes, 389
    pectin-degrading enzymes, 378–379
    proteolytic enzymes, 379–380
    starch-processing enzymes, 362–370
    xylan-degrading enzymes, 373–377
Biodiversity, 13
Bioenergetics
    function of cytoplasmic membrane, 105–107
Biogeochemistry and thermal environments, 15–16
Biogeography and thermal environments, 15–16
Biomining, 390–391
Biopolymers, 393
Biotechnological applications (extremophiles)
    alcohol dehydrogenases, 388
    α-amylases, 362–365
    β-amylases, 365
    amylomaltases, 369–370
    branching enzymes, 369
    C–C bond-forming enzymes, 389
    cellulose-degrading enzymes, 370–373

CGTases, 369
    chitin-degrading enzymes, 377–378
    DNA-processing enzymes, 380–384
    extremophilic archaea and bacteria, 362
    glucoamylases, 365–368
    glucose and arabinose isomerases, 388–389
    α-glucosidases, 368
    β-glucosidases, 370, 373
    lipases and esterases, 384–388
    nitrile-degrading enzymes, 389
    pectin-degrading enzymes, 378–379
    proteolytic enzymes, 379–380
    pullulanases, 368–369
    starch-processing enzymes, 362
    xylan-degrading enzymes, 373–377
Biotechnological applications (psychrophiles)
    bioremediation and natural cycle processes, 125
    biotechnologically important properties, 125
    cleaning additives, 128
    food industry, 125–128
    molecular biology, 128
    textile industry, 128
Branching enzymes, 369

Caloramator fervidus, 112–113
Carbamoyl phosphate metabolism in hyperthermophiles, 65–69
Carbamoyl phosphate, 66
Cellulose-degrading enzymes, 370–373
Cenarchaeum symbiosum, 105
CGTases, 369
Chasmoendolithic organisms, 123
Chemical evolution, 6–7
Chemolithotrophy, 21–22, 136
Chitin-degrading enzymes, 377–378
Cold environments
    man-made psychrophilic environments, 124–125
    marine environments, 120–122
    soil environments, 122–124
Cold-adapted enzymes
    catalysis at low temperatures, 167–170
    cold denaturation, 174
    importance of flexibility, 174–176
    inactivation and conformational stability, 171–172
    mutagenesis studies, 173–174
    properties, 165–167
    structural modifications, 172–173
    thermodynamics of activity, 170–171
Cold-adapted extracellular protease activity, frequency of detection, 139
Cold-adapted microorganisms. See Psychrophiles
Cold-growing bacteria with complete genomes, 211
Cold-shock response
    adenylate metabolism, 186
    Aeromonas hydrophila, 186
    Arthrobacter globiformis, 186
    of cold-adapted bacteria, 184–185
    cold-adapted enzymes, 185
    cold-shock proteins of psychrotrophic bacteria, 186
    of CspA family of *E. coli*, 181–184

cyanobacteria, 187–188
E. coli versus psychrotrophs, 185
global transcript profiling, 188–189
membrane-associated changes, 185
of mesophilic bacteria, 180–181
protective effect of sugars during cold shock, 184
protein folding at low temperature, 184
Pseudomonas fragi, 186
thermophilic bacteria, 187
translation, 185–186
Compact tertiary structures, generation of, 48–49
Compatible solutes, 86, 88, 393–396
distribution in hyperthermophilic microorganisms, 88
hyperthermophiles preference, 96–97
mannosylglycerate biosynthesis, 91–95
molecular basis of protein stabilization, 97–101
occurrence in hyperthermophiles, 87–91
physiological relevance of compatible solute accumulation, 96
polyol-phosphodiesters biosynthesis, 95–96
Covalent modification of nucleosides, 44–48
Crenarchaeota, 105
Cryoprotection, 137–138
Cryptoendolithic organisms, 123, 147
Cytophaga-Flavobacterium-Bacteroides (CFB) group, 138
Cytoplasmic membrane
adaptations to heat stress, 111
in bioenergetics, 105–107
permeability under physiological conditions, 111–113

De novo purine nucleotide biosynthesis, 63–65
Decontamination and hydrogen production, 391
DNA repair, controlling DNA damages by efficient, 50
DnaJ cochaperone, 237, 238, 240
DnaK chaperone, 235, 236, 237
DNA-processing enzymes
DNA sequencing, 383
ligase chain reaction, 383–384
PCR, 380–383
DNAs, stabilizing strategies of, 51–52

E. coli
genes influencing high-pressure (HP) resistance or high-pressure growth in, 342
morphology at low and high pressure, 338
Earth
environmental requirements and limitations for active life on, 351
life on, 3
primitive, as an extreme environment, 4
Embden–Meyerhof (EM) glucose-degrading pathway, 58
Endoliths, 123–124
Entner–Doudoroff (ED) glucose-degrading pathway, 58
non-phosphorylative and semi-phosphorylative, 59
EPS
concentration scaled to brine volume of sections from Arctic winter sea ice, 138
Cytophaga-Flavobacterium-Bacteroides (CFB) group, 138
potential roles in winter sea-ice brine pores, 141

Eukarya, 44, 45, 50, 91, 104, 148, 200, 210, 216, 225, 240, 287, 333, 334, 338, 342, 409, 410, 411, 413, 414, 415
Euryarchaeota, 105
Exobiology. *See* Astrobiology
Extracellular enzyme activity, temperature-dependent, 139
Extreme environment
primitive Earth, 4
chemical evolution and, 6–7
Extremophiles
advantages and limitations of using, 396
applications, xii
biocatalysts from. *See* Biocatalysts from extremophiles
DNA-modifying enzymes from, 381–382
definition, xii
early history, xi
and "intelligent design," 418
limits of life and combined, 415–418
Extremophilic archaea and bacteria, 362
Extremophilic microorganisms, applications of, 394

Facultative anaerobic acidophiles catalyzing dissimilatory iron oxidoreduction
Acidimicrobium/Ferrimicrobium spp., 264–265
Acidithiobacillus ferrooxidans, 263–264
Alicyclobacillus spp. and other acidophilic *Firmicutes*, 264
Ferroplasma spp., 265
Sulfobacillus spp., 264
Fatty acid composition
glycerolipids from wild type and mutant cells of *Synechocystis* sp. PCC 6803 and *Synechococcus* sp. PCC 7002, 198
mesophilic versus psychrophile, 196
Fatty acyl chain length of lipids, 160
Fatty acyl unsaturation, 157–159, 160
Ferroplasma spp., 265
Flagellum, components of, 337
Functional genomics
adaptations required for thermophily and structural genomics, 31–32
definition, 30
future trends, 36
LGT, genome plasticity, and phylogeny, 34–36
transcriptional analysis of stress and responses and metabolic regulation, 32–34

G+C content in RNA, high, 41–42
Genomic DNA of hyperthermophiles, 41, 47, 48, 49, 50, 149
Glacial ice, 146
Glucoamylases, 365–368
Glucose catabolism, 57–61
α-glucosidases, 368
β-glucosidases, 370, 373

Haloarchaea, 232–233
adaptation to osmotic stress, 235–238
osmoregulation in, 233–235
Haloferax volcanii genome, transcriptional map of, 234

Halophiles
    diversity of hypersaline environments, 223–224
    metabolic diversity, 226–227
    microbial community structure in hypersaline environments, 228–229
    phylogenetic diversity, 225–226
    Salinibacter ruber, 227–228
Halophilic archaea. *See* Haloarchaea
Halophilic genome sequences, 240
    archaea, 241
    eubacteria, 241
Halophilic proteins
    from nonextreme halophile with available high-resolution structures, 245
    ions detected and net charge of (X-ray structure), 243
    with available high-resolution structures, 242
Homeo-proton permeability adaptation, 112, 162
Hydrogen production, 391
Hydrogenobaculum acidophilum, 263
Hypersaline environments, 223–224. *See also* Halophiles
Hypersolutes, 87, 96, 101
Hyperthermophiles, 104. *See also* Cytoplasmic membrane
    classification according to optimal growth temperatures, 14
    compatible solutes, 87–91
    compatible solutes restricted to, 90
    function of cytoplasmic membrane in bioenergetics, 105–107
    identifying heat-shock proteins, 33
    importance of cell-specific polyamines, 44
    membrane adaptations to heat stress, 111
    membrane lipids, 107–110
    membrane permeability under physiological conditions, 111–113
    physiochemical properties of membranes, 110–111
    preference for negatively charged compatible solutes, 96–97
    tree of life, 91
Hypolithic microbial communities, 123

Ice
    accretion, 146
    benefits of exopolymers in very cold ice, 138–140
    comparison of very cold glacial ice, permafrost, and sea ice, 135–137
    formations, 134
    freeze-concentration effect in very cold ice, 137–138
    glacial, 146
    life in Vostok ice core, 148–151
    potential for discovery, 141–142
    transition, 146
Insertion sequence (IS) elements, 34–36, 234
Iron- and sulfur-oxidizing aerobes
    Acidithiobacillus spp., 262–263
    Hydrogenobaculum acidophilum, 263
    Leptospirillum spp., 261–262
    Thiobacillus spp., 263
Iron oxidation and CO_2 fixation, 260
Iron-reducing heterotrophs
    Acidiphilium spp., 265

Acidocella organivorus, 266
Acidocella spp., 265–266

Last universal common ancestor (LUCA), 48, 409–410, 411
Leptospirillum spp., 261–262, 266, 267
    phenotypic characteristics of, 262
Life
    at high temperature, 30–31
    high temperatures and, 13–16
    in ice formations at very cold temperatures, 133–135
    in subglacial environments, 147–148
    in Vostok ice core, 148–151
    molecular cladistics and origin and early evolution, 4–6
    on Earth, 3
    timescale for origin and early evolution, 4
Lipids, 391–392
LUCA. *See* Last universal common ancestor

Man-made psychrophilic environments, 124–125
Mannosyl-3-phosphoglycerate synthase genes
    biochemical properties of recombinant, 93
    unrooted phylogenetic tree based on known or putative sequences, 94
Mannosylglycerate, 87
    biosynthesis, 91
    distribution, 91
    genomic organization, 95
    pathways for synthesis, 92
Mannosylglycerate biosynthesis, 91–94
    genes for, 94–95
    synthesis, 95
Marine environments, 16–17
    Antarctic marine environments, 121
    deep marine environments, 120–121
    glacial ice, 122
    sea ice, 121–122
Mesobiotic environments, 17–18
Mesophilic versus psychrophile
    enzymes, 170
    fatty acid composition, 196
Methanopyrus kandleri, 109–110
Methyl branched fatty acids, 159–160
Microbial adaptation to high pressure
    cell division, 338
    culture-based studies, 334–337
    culture-independent studies, 333–334
    membranes and transport, 341–342
    motility, 337
    pressure stress responses, 338–339
    protein/extrinsic factors, 342
    signaling changes in gene expression, 340–341
    translation, 339–340
Microecology, 23
Molecular adaptation
    by psychrophilic proteins, 80–82
    by thermophilic/hyperthermophilic proteins, 76–80
Molecular adaptation to high salt
    crystallographic studies of halophilic proteins, 242–245
    evolutionary mechanisms, 242

halophilic genome sequences, 240–242
protein–DNA interactions in a halophilic context, 245–246
salt and halophilic protein stability, 247
solvation, 246–247
Molecular cladistics and origin and early evolution of life, 4–6
Mrp Na^+/H^+ antiporter system, 316

Na^+ and proton cycles supporting alkaline pH homeostasis, 315
Nonautonomous IS elements, 36
Nucleic acids—effects on heat
in vitro studies, 39–40
stabilization of, by thermophiles, 41–50
strategies of thermostabilization, 42
thermoresistance (in vivo versus in vitro), 40–41

Oligotrophic environments, 17, 138, 150, 155

Pectin-degrading enzymes, 378–379
Phospholipids
core structures in bacteria and tetraether lipids, 108
freeze-fracture and freeze-etch, 109
mobility of acyl chains in fluid membrane phase, 107
Photobacterium profundum SS9, 211, 213, 214, 216, 217, 338
genes influencing high-pressure (HP) resistance or high-pressure growth in, 342
Piezophile genomics and functional genomics, 342–343
Piezophiles. *See* Microbial adaptation to high pressure
Piezophilic *Bacteria* and *Archaea*, 336
Planctomycetes, 123, 410, 411, 415
Polar glacial ice, 134
Polyol-phosphodiesters biosynthesis
DGP biosynthesis, 96
di-*myo*-inositol phosphate, 95–96
Primeval cell, xv, 410
Prokaryotes, modern, 410
Protein stabilization by solutes
probing effects of solutes on protein dynamics, 100–101
probing effects of solutes on unfolding pathway of proteins, 99–100
types of molecular interactions, 97–99
Proteins
molecular adaptation by psychrophilic, 80–82
molecular adaptation by thermophilic/hyperthermophilic, 76–80
and peptides, 392
Proteolytic enzymes, 379–380
Protoeukaryote, 410, 411
Protoeukaryotic LUCA, domains of life from (alternative evolutionary scenarios), 414
Proton motive force (PMF), 105–106, 111, 113
Pseudomonas fragi, 186
Psychrophiles. *See also* Cold environments
biotechnical applications, 125–128
cardinal temperatures, 119
genome sequencing projects, 126
genomics, 125
habitats, 119
Psychrophiles, membrane adaptations
future trends, 162–163
genomics and, 161–162
membrane stability and temperature, 156–157
regulation of adaptive changes in lipid composition, 160–161
solute transport across membranes, 162
thermal changes in lipid fatty acyl composition, 157–160
Psychrophilic habitats, microbial genera isolated from, 120
Psychrophilic prokaryotes genomes, 208–209
bacteria under cold conditions, 213
bacterial genome organization, 213–214
early studies, 210–211
genome programs, 211–213
psychrophily, 209–210
RNAs, 214–217
Psychrophilic versus mesophilic
enzymes, 170
fatty acid composition, 196
Psychrophily, 128, 168, 209, 213, 412, 415, 417
Psychrotrophs, 119, 128, 185, 186
Pullulanases, 368–369
Pyrite oxidation by consortia of acidophilic prokaryotes, 267
Pyrolobus fumarii, 14, 105

Radiation dose giving ~37% survival for UV and ionizing radiation, 353
Reverse gyrase, 8, 48, 49, 50, 287, 411, 412
Ribosome, components of, 340
RNA damages by RNA turnover, eliminating, 50
RNAs, 214–215
factors allowing tRNA molecules to function at high temperatures, 51
membrane fluidity, protein export, and secretion, 217
metabolic constraints, 216–217
phylogenetic distribution of modified nucleosides, 45
proteome, 215–216
reactive oxygen species, 216
schematic representation of tertiary interactions in tRNA structure, 46
stabilizing strategies of, 51–52
RNP particles with thermostable proteins, generation of, 49
16S rRNA gene sequences
phylogenetic and genetic diversity, 20
restriction enzyme phylotypes, 20–21
16S rRNA gene sequences, phylogenetic tree derived from
alkaliphilic *Bacillus* spp., 300
alkaliphiles that are high G+C and low G+C gram-positive bacteria, 305

Salt and halophilic protein stability
electrostatic contributions, 247
neutron-scattering studies of molecular dynamics in extreme halophiles, 250–251
salt bridges, ion binding, and stabilization of active dime and tetramer of *Hm* MalDH by salt, 247–248

salts and stabilization of halophilic proteins, 248–249
solubility, 249–250
Small ligand binding, 42–44
Sodium motive force (SMF), 106, 111, 112–113
Soil environments
    fellfield communities, 124
    lithic communities, 123–124
    montane environments, 123
    ornithogenic environments, 124
    polar soils, 122–123
Starch-processing enzymes, 362
    α-amylases, 362–365
    β-amylases, 365
    amylomaltases, 369–370
    branching enzymes, 369
    CGTases, 369
    extremophilic archaea, 363–364
    extremophilic bacteria, 366
    glucoamylases, 365–368
    α-glucosidases, 368
    β-glucosidases, 370, 373
    pullulanases, 368–369
Subglacial environments, 147–148
Subglacial lakes of Antarctica. *See also* Vostok, Lake
    life in subglacial environments, 147–148
    scientific objectives, 146–147
Subsurface environments, 17, 354
Sulfobacillus spp., 264
Sulfolobus acidocaldarius, 105, 110
Sulfolobus solfataricus, 109
Sulfur-oxidizing archaea, 263
Sulfur-reducing acidophiles, 266
Synthetic membrane spanning lipids, 110

Temperature profiles (generic), 134
Temperature, high, and life, 13–16
Temporary environments, 17
Terrestrial, nonanthropogenic environments, 16
Tetraether liposomes, 110–111, 113
Thermal adaptation and macromolecular organization of metabolism, 69–70
Thermal environments
    cultural diversity, 13–16, 18–20
    ecological diversity, 22–23
    high temperatures and life, 13–14
    metabolic diversity, 21–22
    phylogenetic and genetic diversity, 20–21
    types, 16–18
Thermoacidophilic archaea, *E. coli* acid resistance genes among, 284
Thermoactive enzymes of industrial relevance, 385–386
Thermolabile metabolites
    CP metabolism in hyperthermophiles, 65–69
    de novo purine nucleotide biosynthesis, 63–65
    glucose catabolism, 57–61
    thermal adaptation and macromolecular organization of metabolism, 69–70
    tryptophan biosynthesis, 61–63
Thermophile genomes, sequenced, 31
Thermophiles, stabilization of their nucleic acids
    controlling DNA damages by efficient DNA repair, 50
    eliminating RNA damages by RNA turnover, 50
    generation of compact tertiary structures, 48–49
    generation of RNP particles with thermostable proteins, 49
    high temperatures and life, 13–16
    stabilization of nucleic acid structures, 42–48
    use of high G+C content in RNA, 41–42
Thermophiles and thermolabile metabolites
    CP metabolism in hyperthermophiles, 65–69
    de novo purine nucleotide biosynthesis, 63–65
    glucose catabolism, 57–61
    thermal adaptation and macromolecular organization of metabolism, 69–70
    tryptophan biosynthesis, 61–63
Thermophilic archaea, 19, 20, 22
    analysis of genomic data, 77
    branched polyamines in, 44
    changes in lipid composition, 111
    and chitinolytic enzymes, 377
    Di-*myo*-inositol-1,1′-phosphate, 393
    DnaK system, 236
    genomes, 287
    glucose catabolism, 57
    heat-stable proteases, 379
    heat-stress response, 96
    recombinational DNA-repair system, 50
    reverse gyrase, 48
    salt concentration in cytosol, 245
    tetraether lipids, 113
    UAG codons and, 36
    and xylanolytic enzymes, 377
Thermophilic bacteria, 19
    amino acid positions, 236
    anaerobic, 22
    cellular machinery, 180
    chemolithoautotrophic, 22
    chitin-hydrolyzing enzymes, 378
    cold-shock response, 187
    composition, 107
    CspA homologs, 181
    genomes, 280
    glucose isomerases, 388
    growth temperature membranes, 112
    heat-stable lipases, 384
    life at high temperature, 30
    maximal growth rate temperatures, 411
    metabolic channeling, 65
    nature of membrane lipids, 411
    phylogenetic and genetic diversity, 20
    polyamines in, 43–44
    proton permeability of membranes, 112
    pullulanase type I, 369
    reverse gyrase, 48
    thermoactive enzymes, 389
    thermostable serine proteases, 379
Thermophilicity, 78, 104
Thermophily
    adaptation, 412
    before LUCA, 412–413

extreme, 410–412
LUCA, 413
Thiobacillus ferrooxidans See *Acidithiobacillus ferrooxidans*
Thiobacillus spp., 263
Transition ice, 146
Tryptophan biosynthesis, 61–63
Tryptophan biosynthetic pathway, 62

Universal tree of life, 409–410

Vostok, Lake, 145–146
    life in subglacial environments, 147–148
    scientific objectives, 146–147

Whole-cell biocatalysis
    biomining, 390–391
    decontamination and hydrogen production, 391

Xylan-degrading enzymes, 373–377